MOLECULAR AND COLLOIDAL
ELECTRO-OPTICS

SURFACTANT SCIENCE SERIES

MOLECULAR AND COLLOIDAL ELECTRO-OPTICS

Stoyl P. Stoylov
Bulgarian Academy of Sciences
Sofia, Bulgaria

Maria V. Stoimenova
Bulgarian Academy of Sciences
Sofia, Bulgaria

CRC Press
Taylor & Francis Group
Boca Raton London New York

CRC Press is an imprint of the
Taylor & Francis Group, an **informa** business

CRC Press
Taylor & Francis Group
6000 Broken Sound Parkway NW, Suite 300
Boca Raton, FL 33487-2742

© 2007 by Taylor & Francis Group, LLC
CRC Press is an imprint of Taylor & Francis Group, an Informa business

First issued in paperback 2019

No claim to original U.S. Government works

ISBN 13: 978-0-367-45348-0 (pbk)
ISBN 13: 978-0-8493-9811-7 (hbk)

Library of Congress Cataloging-in-Publication Data

Molecular and colloidal electro-optics / edited by Stoyl P. Stoylov and Mariyka V. Stoimenova.
   p. cm. -- (Surfactant science series)
  Includes bibliographical references and index.
  ISBN 0-8493-9811-8 (978-0-8493-9811-7)
  1. Colloids--Electric properties. 2. Colloids--Optical properties. 3. Molecules--Electric properties.
4. Molecules--Optical properties. I. Stoylov, S. P. II. Stoimenova, Mariyka V. III. Series.

QD549.M67 2006
541'.345--dc22
                                                           2006008363

Visit the Taylor & Francis Web site at
http://www.taylorandfrancis.com

and the CRC Press Web site at
http://www.crcpress.com

Preface

Electro-optics deals with the variations of optical properties of matter induced by the action of electric fields. Electro-optic investigations on molecular and colloidal systems started in 1875 with the discovery of electric field-induced birefringence by John Kerr. His simple experiments yielded essential information on the molecular characteristics of a variety of solids, simple liquids, and colloids. They gave birth to a new branch of science and stimulated the investigation and wide practical applications of electro-optics in both everyday life and advanced technologies—liquid crystal displays, optical location and optical communication systems, laser electronics, etc. Electro-optic methods are the basic source of information on the electric and optic parameters of colloidal particles, which are electric polarizability, permanent electric dipole moments, and optical anisotropy. In parallel, they present unique advantages in particle sizing analysis.

The fields of application of electro-optic methods in science and technology are numerous. Clays and oxides (widely spread in nature and widely used in modern technologies, such as paper, ceramics, and drilling liquids) are typical objects of electro-optics. Other fields where electro-optic methods have been applied for decades are in the (1) characterization of biocolloids, polymers and polyelectrolytes, synthetic polymers, and liquid crystals; (2) analysis of the structural properties of complexes between small ligands and macromolecules; and (3) studies of ion migration in the process of gel electrophoresis. Electro-optics finds wide applications in biological colloidal systems: cells, bacteria, phages, viruses, synthetic biosystems, etc.

The aim of this book is to present modern approaches to the electro-optic theory and experiment, to make it possible to compare different viewpoints, and to introduce the reader to new aspects in the development of electro-optic methods and especially to their new applications. It is addressed to researchers from academia and industry working in the field of physical chemistry of colloids, polymers and nanostructures and on their various applications.

The book combines fundamental and applied aspects. It is organized into two parts preceded by a historical review written by Professor Henri Benoit—one of the three pillars (Benoit, O'Konski, Tolstoy) of modern electro-optic research after the end of World War II. Special issues of *Colloids and Surfaces*, following the last two electro-optic conferences, were dedicated to the late Professor Nikita Tolstoy and to Professor Chester T. O'Konski.

Part I of this book explains the theory and the methodological aspects of colloid electro-optic science. Part II presents recent applications of the methods to colloids, polymers, composite and hybrid particles, liquid crystals, biological macromolecules, biological membranes, bacteria, etc. The contributors were selected from a group of internationally recognized researchers, many of whom presented their achievements in the symposium on "Colloidal and Molecular Electro-Optics," held as part of the ACS National Meeting in New Orleans in 2003. Additional contributions were included that summarize important results, which were not presented at the latest electro-optic meetings.

One of the basic problems in the electro-optic behavior of charged colloids is the origin of the large low-frequency electro-optic relaxation. Part I focuses on the latest achievements in electro-optic theory, concerning the low-frequency relaxation. It presents different viewpoints on the origin of the low-frequency effects, different theoretical constructions, and discussions on the comparative studies of theories and experimental data. Studies concerning simulation analyses, modeling of the electro-optic effects, and approaches permitting to avoid optical complications in the electro-optic spectroscopy measurements are included.

Most of the recent applications of electro-optic methods involve the analysis of the particle surface electric state (ion mobilities on the particle–medium interface, dipole interactions, colloid stability) and in the determination of particle size, shape, and polydispersity. Besides the determination of a particle's geometrical, optical and electrical parameters, different processes like deformation, structuring, and electroporation of a particle are also followed. Part IIA presents a number of chapters concerning similar applications. A new trend in colloid electro-optics is the investigation of adsorbed polyelectrolytes, amphiphiles, and nanoparticles (including multilayers) on the surface of colloidal particles. Part IIB includes chapters presenting the unique advantages of electro-optics for similar analyses. The final section, Part IIC, includes chapters demonstrating the sensitivity of the electro-optic methods to interparticle interactions. Different types of phase transitions in colloidal systems are detected and analyzed.

Editors

Stoyl P. Stoylov is a professor of physical chemistry at the Institute of Physical Chemistry of the Bulgarian Academy of Sciences, Sofia, Bulgaria, and also a professor of biophysics in the medical faculty of the Medical University, Sofia, Bulgaria. He is a member of the Bulgarian Academy of Sciences. He authored a book on colloid electro-optics in 1991, which was published by Academic Press, U.K. He has numerous professional publications in this field. He graduated from the University of Sofia and received his Ph.D. in 1965 and D.Sc. in 1973 from the Institute of Physical Chemistry of the Bulgarian Academy of Sciences.

Maria Stoimenova is an associate professor at the Institute of Physical Chemistry of the Bulgarian Academy of Sciences, Sofia, Bulgaria. She has published over 40 journal articles and book chapters on the light scattering theory of colloidal particles and the methodology of their electro-optic investigation. She received her Ph.D. in 1980 from the Institute of Physical Chemistry of the Bulgarian Academy of Sciences.

Contributors

S. Ahualli
Department of Applied Physics
University of Granada
Granada, Spain

Alexander Angersbach
Institute of Applied Microbiology
Obolensk, Russia

Jan M. Antosiewicz
Department of Biophysics
Warsaw University
Warsaw, Poland

Sergio Aragon
Department of Chemistry
   and Biochemistry
San Francisco State University
San Francisco, California

F.J. Arroyo
Department of Physics
University of Jaén
Jaén, Spain

Niloofar Asgharian
Department of Chemistry and
   Biochemistry
University of Texas at Arlington
Arlington, Texas

F.G. Díaz Baños
Department of Physical Chemistry
Murcia University
Murcia, Spain

H. Benoît
Institut Charles Sadron
Strasbourg, France

Yu. B. Borkovskaja
Institute of Biocolloid Chemistry
National Academy of
   Sciences of Ukraine
Kiev, Ukraine

S.N. Budankova
Institute of Biocolloid Chemistry
National Academy of Sciences of Ukraine
Kiev, Ukraine

M. Buleva
Rostislaw Kaischew Institute of Physical
   Chemistry
Bulgarian Academy of Sciences
Sofia, Bulgaria

Victor Bunin
Institute of Applied Microbiology
Obolensk, Russia
and
Biotronix GmbH
Berlin, Germany

F. Carrique
Department of Applied Physics
Faculty of Science
University of Málaga
Málaga, Spain

N. Mariano Correa
Department of Chemistry
National University of Rio Quarto
Cordoba, Argentina

J. García de la Torre
Department of Physical Chemistry
Murcia University
Murcia, Spain

A.V. Delgado
Department of Applied Physics
University of Granada
Granada, Spain

Don Eden
Department of Chemistry and
   Biochemistry
San Francisco State University
San Francisco, California

Reinosuke Hayakawa
School of Engineering
University of Tokyo
Tokyo, Japan

H. Hoffmann
University of Bayreuth
BZKG
Bayreuth, Germany

S. Holzheu
University of Bayreuth
BITÖK
Bayreuth, Germany

Oleg Ignatov
Institute of Applied
   Microbiology
Obolensk, Russia

M.L. Jiménez
Department of
   Applied Physics
University of Granada
Granada, Spain

Kazuo Kikuchi
Tokyo, Japan

Yasuyuki Kimura
Department of Physics
Kyushu University
Fukuoka, Japan

Igor' P. Kolomiets
Institute of Physics
St. Petersburg State University
St. Petersburg, Russia

David Lacey
Department of Chemistry
The University of Hull
Hull, U.K.

Peter N. Lavrenko
Institute of Macromolecular
   Compounds
Russian Academy of Sciences
St. Petersburg, Russia

Yongjun Lu
Department of Internal Medicine
University of Iowa
Iowa City, Iowa

Kerwin Ng
Department of Chemistry and
   Biochemistry
San Francisco State University
San Francisco, California

Tsuneo Okubo
Cooperative Research Center
Yamagata University
Yamagata, Japan

Viktor Peikov
American Pharmaceutical Partners
Melrose Park, Illinois

Martin Perez
Department of Chemistry and
   Biochemistry
San Francisco State University
San Francisco, California

Ivana B. Petkanchin
Rostislaw Kaischew Institute of Physical
   Chemistry
Bulgarian Academy of Sciences
Sofia, Bulgaria

Dietmar Porschke
Biomolecular Dynamics Department
Max Planck Institute of Biophysical
   Chemistry
Göttingen, Germany

Tsetska Radeva
Rostislaw Kaischew Institute of Physical
   Chemistry
Bulgarian Academy of Sciences
Sofia, Bulgaria

H.E. Pérez Sánchez
Department of Physical Chemistry
Murcia University
Murcia, Spain

Zoltan A. Schelly
Department of Chemistry and Biochemistry
The University of Texas at Arlington
Arlington, Texas

P. Schmiedel
Henkel KG
Düsseldorf, Germany

V.N. Shilov
Institute of Biocolloid Chemistry
National Academy of Sciences
    of Ukraine
Kiev, Ukraine

A.A. Spartakov
Faculty of Physics
St. Petersburg State University
St. Petersburg, Russia

Nancy C. Stellwagen
Department of Biochemistry
University of Iowa
Iowa City, Iowa

Maria Stoimenova
Rostislaw Kaischew Institute of Physical
    Chemistry
Bulgarian Academy of Sciences
Sofia, Bulgaria

Stoyl P. Stoylov
Rostislaw Kaischew Institute of Physical
    Chemistry
Bulgarian Academy of Sciences
Sofia, Bulgaria

A.A. Trusov
Faculty of Physics
St. Petersburg State University
St. Petersburg, Russia

Akira Tsuchida
Faculty of Engineering
Gifu University
Gifu, Japan

Darrell Velegol
Department of Chemical
    Engineering
Pennsylvania State University
University Park, Pennsylvania

A.V. Voitylov
Faculty of Physics
St. Petersburg State University
St. Petersburg, Russia

V.V. Vojtylov
Faculty of Physics
St. Petersburg State University
St. Petersburg, Russia

Hitoshi Washizu
Tribology Laboratory
Toyota Central R&D Labs.,
    Inc.
Aichi, Japan

Sixin Wu
Lab for Special Functional
    Materials
Henan University
Kaifeng, P.R, China

Natalia P. Yevlampieva
V.A. Fock Institute of Physics
St. Petersburg State University
St. Petersburg, Russia

Hongxia Zeng
Penwest Pharmaceutical
Patterson, New York

Hongguang Zhang
ST Microelectronics
Carrollton, Texas

Alexandar M. Zhivkov
Rostislaw Kaischew Institute of
    Physical Chemistry
Bulgarian Academy of Sciences
Sofia, Bulgaria

Table of Contents

Part II
New Applications

A. Rigid Particles, Polyelectrolytes, Biosystems

B. Adsorption on Particles

The First Years of Modern Electro-Optics—A Historical Review

H. Benoît

CONTENTS

0.1 INTRODUCTION

The theory, techniques, and applications of colloid electro-optics were described by Stoylov in 1991.[1] A historical account of the first electrical birefringence (EB) studies on biopolymers 50 years ago is also given. The EB of classical polymers solutions for polymers and monomer units were practically the same and very high voltage was required to achieve EB. This was the finding of my thesis.[2] This led me to look at rigid molecules like DNA and tobacco mosaic virus (TMV). Due to the high conductivity of the solutions, I decided to use electric pulses. The observed optical signal was asymmetric and deformed by the nonnegligible value of the rotary diffusion constant D. Hence, I developed the theory of this effect. It shows that, regardless of the mechanism of orientation, the decay of the signal is proportional to $\exp(-6Dt)$.

As I am writing about the work I did more than 50 years ago, the presentation will be more historical than scientific. The project I took up for my thesis project was devoted to EB of classical organic polymers solutions. Since it was impossible to work on flexible polymers, I decided to try to work on rigid polymers. The only samples that were available were tobacco mosaic virus and DNA. It was impossible to apply a dc field (the solution began to boil) due to the high conductivity of solutions, so I decided to use electric pulses and, to my surprise, the optical signal observed during the pulse was asymmetric and deformed by the high value of the rotary diffusion constant D.

This initially led me to build an apparatus enabling quantitative experiments to be performed. A theory for the interpretation of the data was able to show that, regardless of the mechanism of orientation, the decay of the signal is proportional to $\exp(-6Dt)$. Very high values of D were found for the tobacco mosaic virus (which was to be expected) and for DNA. This last result was difficult to explain since nothing was known on the structure of DNA at that time.

0.2 SOME HISTORICAL DATA

After studying in Paris during World War II, I arrived in Strasbourg at the beginning of October 1945, having obtained the position of assistant lecturer at the physics department of the Strasbourg University.

At the beginning of the War, Strasbourg was completely evacuated because of its location at the border between France and Germany. There was a serious risk of bombing because of its location. The city remained empty until the arrival of the Germans. So approximately two years even after the end of the war, the city was almost dead, and was under the supervision of the French army. The university buildings were closed and successively used by the Germans and later by Americans and French soldiers as a caserne and finally left empty with all the doors open. It was practically impossible to make any experimental work in those buildings, which were half destroyed. Anyway, full of enthusiasm I wanted to work on a thesis in order to obtain a Ph.D. degree but the problem of finding a domain of science to work with and a laboratory in which I could do some experimental work was difficult to solve. I had discussion with different professors with the hope of obtaining suggestions. The head of the institute told me that the magnetic properties of metals and alloys were a dead subject. I was not particularly interested in nuclear physics.

Eventually I met Professor C. Sadron who had spent one part of the War in a concentration camp in Germany and was full of projects. His idea was that the field of macromolecules was a new domain and that many interesting problems remained to be solved in this area. He was dreaming about a laboratory devoted to this subject, assuming that natural and synthetic polymers could be studied by similar methods, and wanted to obtain the approval for the construction of a new laboratory devoted to this field from the authorities.

So I decided to work with him and I did participate in the birth of the Centre de Recherches sur les Macromolecules. I became an assistant director in 1954 and was the director from 1967 to 1978 in this center. I will not describe here the history of the laboratory and I shall limit myself to my thesis project.

Professor Sadron's thesis was on the comparison of the magnetic properties of different metals and went as a post doc to the laboratory of Prof. Von Karman in Caltech (Pasadena, California) where he decided to study the behavior of the flow lines of a liquid along a surface by measuring its flow birefringence.

The experiences were difficult and, in order to increase the sensitivity, he did use macromolecular solutions. Back in Strasbourg he proposed the study of electrical birefringence of polymers to me, assuming that, since a flow field orients macromolecules, an electrical field should also orient them. This idea was a reasonable one and I began to study the literature on the subject and tried to make classical experiments to get acquainted with the subject.

It was not easy. When I complained that I had no material to make any experiment, my thesis advisor told me that there were batteries dating from before the War in the basement and a few polarizing quartz and that they were sufficient. I remember that, in order to have a slit, I slicked two razor blades on a piece of cork. It was perfect but unfortunately adjusting its width was impossible. With this material I began to conduct experiments on pure CS_2, which is known to present one of the largest effect. I did realize that the effect was small and difficult to observe (at least with the material I was using). After that I did try on classical polymer solutions in organic solvents. Usually the monomers (or the analogous molecules) with no double bond have a very small Kerr effect. My hope was to find that even in these conditions the polymers could have a measurable Kerr effect. So I spent a lot of time trying to measure different polymers in different solvents and concentrations without any success.

0.3   FIRST EXPERIMENTS: THE PULSE TECHNIQUE AND THE APPARATUS

The lack of success made me very unhappy and was amplified by the remark of a professor in rheology, who once when visiting my laboratory, told me that in his laboratory people were finding a large Kerr effect on the most classical homopolymer polystyrene (even in molten state). I tried many times to achieve that result by changing the sensitivity of my machine without any success. I was ashamed not to be able to reproduce the rheology professor's result. Meeting him later in a colloquium, I asked him the details of his experimental method and he told me that it was a mistake and that Kerr effect on polystyrene did not exist (morality: be very cautious about oral results and believe only what has been published). This absence of electrical birefringence on flexible polymers is now easy to understand. The Kerr effect is due to the orientation of the dipoles (permanent or induced) by the electrical field. These dipoles are essentially distributed on the side groups, which mean that if the chain is really flexible, dipoles are oriented without influencing the chain orientation, which is practically not modified. At this time Professor Sadron suggested to me to change my research project and work on classical problems in viscosity with which he was very familiar. However I did refuse and he told me I was taking a risk.

The situation was that, in order to observe a large effect, one has to deal with large anisodiametric molecules with strong electrical moments. The difficulty was that I did not find any macromolecules with a strong moment without being insoluble in water. But with the available samples, the conductivity of the solution of water-soluble polymer was usually high due to the presence of ions and counterions, and salt and, if you apply a high voltage, the current is high and, after a few seconds the liquid begins to boil, making measurements impossible. The only solution was to use a pulse technique and this forced me to become an expert in electronics. At that time, it was almost impossible to buy the equipment in a completely destroyed country and I had to spend a lot of time to find the equipment indispensable for my work. However I did manage to make square wave pulses with mechanical devices and to observe an optical signal from a photomutiplier cell. To my surprise the optical signal was completely deformed, and, instead of a rectangular signal one obtained a signal similar to what is described in Figure 0.1 and Figure 0.2.[2]

I was very surprised and the only conclusion was that the equilibrium state and its disappearance were not instantaneous. There was a relaxation phenomenon, which was worth studying. This led me to creating a theory of this effect, which will be described further. In fact I made the theory after the measurements in order to make my thesis look more serious but it is easier to expose it first, since it gives a tool for rapid interpretation of the results.

0.4   THEORY OF THE KERR EFFECT FOR THE PULSED ELECTRIC FIELD

This is a classical problem of diffusion and its theory has been understood for many years. If one calls D the rotary diffusion constant of the molecules assuming for simplicity to have an axis of revolution and $f(\theta,t)$ the probability of the orientation at time t, with θ as the polar angles defining the orientation of the molecules, and assuming that the ellipsoid has a

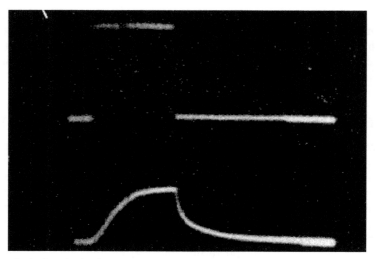

FIGURE 0.1 Typical signal obtained on tobacco mosaic virus in water. The figure represents the oscilloscope obtained with a double beam tube. The upper curve gives the intensity of the electric field, the lower curve gives the signal obtained between cross. The effect of the relaxation time of the motion of the molecules is clearly visible the speed on the x-axis is of the order of 10^{-2} s for the whole screen.

symmetry of revolution and the orientation is independent of ϕ, one obtains the general equation for the orientation function (normalized to unity). The problem is the evaluation of $f(\theta,t)$ as a function of time.

The laws of Brownian motion require that $f(\theta,t)$ satisfy the general diffusion equation:

$$D\,\Delta^2 f + \frac{1}{k_\mathrm{B}T}\,\mathrm{div}\,\vec{c} = \frac{\partial f}{\partial t}, \tag{0.1}$$

calling D the rotary diffusion constant of the molecules evaluated by Perrin[3] for ellipsoids and $\vec{c}$ the velocity of the particles in the absence of Brownian motion. This last quantity can be written as:

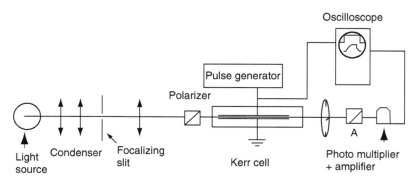

FIGURE 0.2 Schematic diagram of the equipment: The light source is a 12 V DC current bulb. After focalization, in order to have a parallel beam, the beam is polarized with a Nicol prism and passes between the two electrodes of the Kerr cell (two electrodes of 10 cm). Length is one or two tenths of a millimeter distance. The relaxation time is clearly identified.

$$\vec{c} = \frac{\vec{\gamma}}{\varsigma},$$ (0.2)

where $\vec{\gamma}$ is the orientation couple and ζ the angular velocity. The rotary diffusion constant D is also bound to ς by the relation:

$$D = \frac{kT}{\varsigma}$$ (0.3)

This allows to write the differential equation as:

$$\nabla^2 f + \frac{1}{kT} \operatorname{div} f \vec{\gamma} = \frac{1}{D} \frac{\partial f}{\partial t}$$ (0.4)

And if $\vec{\gamma}$ depends of a potential w $\gamma = grad\vec{w}$ this equation becomes:

$$\nabla^2 f + \frac{1}{kT} \operatorname{div}(f \operatorname{grad} \vec{w}) = \frac{1}{D} \frac{\partial f}{\partial t}$$ (0.5)

This equation allows the evaluation of the average position of the molecules and its birefringence as a function of the time t.

In order to solve the problem we have to first express w as a function of the mechanism of orientation of the particles (assumed to be rodlike and having a dipole making a fixed angle with the axis of the rodlike particle).. This is accomplished as follows.

If you have an electrically charged rigid molecule it is always surrounded by counter ions. When an electric field is applied to the charges, which are isotropically placed in the vicinity of the particle, they will move in the direction of the field, producing a couple, exactly in a permanent moment. Of course there could be nonlinear effects, but because of the low values of the electrical field, this can be neglected. This means that, as a first approximation, one can assume that the effect of the field on a rigid polyion is the same as on a simple molecule with permanent dipole moment.

Since an analogy with small molecules seems plausible, one can also consider another way of orienting the particles. If the polarizability of a particle is not the same in the direction of its axis, it is known that it tends to orient itself in the direction of the maximum polarizability axis in the perpendicular direction.

Moreover if the particles are not anisotropic, i.e., if the polarizabilities are the same in all directions, one can also have an orientation. This is due to a classical electrostatic problem of the effect of a uniform field on an ellipsoid. An ellipsoid placed in a constant electrostatic medium has a uniform field inside of it but not in the direction of the applied field. This produces a couple on the ellipsoid and therefore its orientation.

Without entering into the details of the calculations, I will only give the result for the energy w here. In the first case (no dipole) one obtains[4-6]

$$P_i = v \frac{\varepsilon_i - \varepsilon_0}{4\pi + \frac{\varepsilon_i - \varepsilon_0}{\varepsilon_0} L_i} E = v g_0^i E,$$ (0.6)

where v is the volume of the particle, ε_0 the dielectric constant of the solvent and ε_i (with $i = 1$ or 2) the dielectric constant of the particle in the direction of its axis and in the perpendicular direction.

It has been proved that the parameter L_i is a form factor given by:

$$L_1 = \frac{4\pi}{3}(1 - 2e)$$
$$L_2 = \frac{4\pi}{3}(1 + e) \tag{0.7}$$

with

$$e = \frac{1}{4(p^2 - 1)}\left[2p^2 + 4 - \frac{3p}{\sqrt{p^2 - 1}}\log\frac{p + \sqrt{p^2 - 1}}{p - \sqrt{p^2 - 1}}\right] \tag{0.8}$$

if $p > 1$, and

$$e = \frac{1}{2(1 - p^2)}\left[-p^2 - 2 + \frac{3p}{\sqrt{1 - p^2}}\text{Arctg}\frac{\sqrt{1 - p^2}}{p}\right] \tag{0.8}$$

when $p < 1$.

Assuming that the axis of the molecule makes an angle θ with the field, the energy in the case of a permanent dipole μ (in the direction of the axis of the molecule) is given by:

$$w = -\tfrac{1}{2}v[g_1^0 \cos^2\theta + g_2^0 \sin^2\theta]E^2 + \mu E\cos\theta \tag{0.9}$$

where g_i^0 represents the optical factors that are explained in the original paper.

If there is no induced dipole the Equation 1.9 reduces to its last term.

In the second case we assume that it is parallel to the axis of the molecule and we obtain:

$$w_\mu = E\mu\cos\theta \tag{0.10}$$

Assuming the coexistence of both effects one obtains:

$$w = -\frac{1}{2}v(g_1^0\cos^2\theta + g_2^0\sin^2\theta)E^2 + \mu E\cos\theta \tag{0.11}$$

Of course this is the simplest version of the orientation of a molecule in a electrical field but it is sufficient for trying to explain at least qualitatively the majority of results.

In fact we are not interested in the solution of the equation since what we want to know is the birefringence as a function of time: $\Delta n(t)$. This quantity depends of the distribution of the orientations $f(t)$ which will be renormalized as:

$$f(\theta, t) = \frac{1}{4\pi}\rho(\theta, t) \tag{0.12}$$

(in order to define it as the probability of finding the rod with the orientation θ), since we have simplified the problem having only one parameter θ for the orientation as a function of time t. The ρ function is equal to 1 in the absence of orientation. With these notations the equation becomes

$$\frac{1}{\mathrm{Sin}\,\theta}\frac{\partial}{\partial\theta}\left(\sin\theta\frac{\partial\rho}{\partial\theta}\right)+\frac{1}{kT}\left[\frac{\partial\rho}{\partial\theta}\frac{\partial w}{\partial\theta}+\frac{\rho}{\sin\theta}\frac{\partial}{\partial\theta}\left(\sin\theta\frac{\partial w}{\partial\theta}\right)\right]=\frac{1}{D}\frac{\partial\rho}{\partial t} \tag{0.13}$$

If one uses $u=\cos\theta$ as variable and define for simplifications:

$$p=\frac{E\mu}{kT}\ \text{and}\ q=v\frac{g_1^0-g_2^0}{kT}E^2 \tag{0.14}$$

one obtains for the differential equation:

$$\left(1-u^2\frac{\partial^2\rho}{\partial u^2}\right)-2u\frac{\partial\rho}{\partial u}-(1-u^2)\,(p+qu)+\rho2pu+q(3u^2-1)=\frac{1}{D}\frac{\partial\rho}{\partial t} \tag{0.15}$$

Our problem is first to solve this differential equation for the orientation factor and, from it to obtain the birefringence $\Delta n(t)$. The only factor we know is that $\rho(u,t)=1$ for $t=0$.

Since a rigorous solution is impossible we develop $\rho\,(u,t)$ as a series of Legendre polynomials writing

$$\rho(u,t)=\sum_n a_n(t)P_n(u), \tag{0.16}$$

where $P_n\,(u)$ represents the Legendre polynomial of order n and of the variable u and knowing that for $t=0$, $\rho(u,0)=1$.

After a simple but tedious calculation, one obtains a recurrence relation for the coefficient a_n.

In order to obtain the magnitude of the Kerr effect as a function of time, it is sufficient to integrate Equation 0.15 and one obtains, in the case where there is just an induced dipole:

$$\frac{\Delta n(t)}{\Delta n_0}=1-e^{-6Dt},\ \text{for}-1\leq u\leq 1 \tag{0.17}$$

For the establishment of the birefringence, one obtains the same expression; the curve is symmetrical.

Here we shall discuss only two cases: return to equilibrium and orientation when the electric field orientates molecules having a permanent dipole and also an anisotropic structure as well from an electrical and an optical point of view.

I do not reproduce here the calculation of the birefringence, which does not present any difficulty and I give only the result:

$$\Delta n=\Delta n_O e^{-6Dt} \tag{0.18}$$

This formula can be used for any kind of system.

I am credited with this this result, but the credit should be given to Prof. A. Peterlin,[5,6] from Ljubljana and later at the NBS, who had derived it first. But I think that it was not taken into account by scientists interested in fast motions since it was applied only to flow birefringence where it is impossible to stop the flow instantaneously (let us say in less than 10^{-3} s).

The calculation of the anisotropy is straightforward but it is rather lengthy. One has to introduce the optical anisotropy of the particle, which is usually different from the electrical anisotropy due to the variation of the index of refraction with frequency.

The details of the calculation are given in my thesis and I reproduce only the results here.

In a simplified form one can write in the general case where both mechanisms are taken into account:

$$\frac{\Delta n(t)}{\Delta n_0} = 1 - \frac{3\alpha}{2(1 + \alpha)}\, e^{-2Dt} + \frac{\alpha - 2}{2(\alpha + 1)} e^{-6Dt} \tag{0.19}$$

In this expression Δn_0 is the birefringence in the steady state ($t = \infty$), $\Delta n(t)$ the birefringence at time t. The quantity $\alpha = P/Q$ is the ratio of a term due to the dipole energy and the other due to anisotropy

$$P = \frac{\mu^2}{k^2 T^2} \quad \text{and} \quad Q = \frac{g_1 - g_2}{kT} \tag{0.20}$$

where P is the term due to the the permanent dipole and Q is the term due to the effect of induced dipoles, $\Delta n(t)/\Delta n_0$ is the ratio of the birefringence at time t to the steady state birefringence, μ is the permanent dipole moment, and $g_1 - g_2$ is the term characterizing the electric anisotropy (difference between the two main electric polarizabilities), which can be either positive or negative. For instance benzene has a larger polarizability in the direction perpendicular to its revolution axis and the quantity Q is negative. As an example, Figures 0.3 and 0.4 show the theoretical values of the establishment of birefringence as a function of the ratio α following Equation 0.19.

The factor α characterizes the ratio of the permanent dipole to the induced dipole. If there is no permanent dipole ($\alpha = 0$) one recovers the classical result:

$$\Delta n(t) = \Delta n_0 e^{-6DT} \tag{0.21}$$

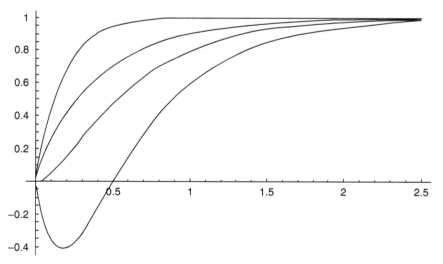

FIGURE 0.3 The quantity Δn as a function of time t for different values of α. Going from top to bottom corresponds to the following values of α: O ,1, 2, ∞, −2. The negative values of α correspond to situations in which both effects (dipole and induced dipole) lead to opposite orientations.

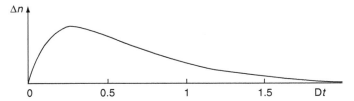

FIGURE 0.4 The case where the permanent and the induced dipoles have the same absolute value but opposite signs. Since the time evolutions of the two mechanisms are different, it is normal to obtain a curve starting from zero and going back to zero.

In this case the establishment and the disappearance are given by similar expressions. This result is quite general and has been demonstrated by A. Peterlin in the case of flow birefringence. Unfortunately the experiments are difficult as it is impossible to stop the flow in a short time. For positive α one has the simple superposition of two exponentials with a ratio of 3 for the decay times. If α is negative there is a competition between a positive and negative values of the birefringence, the curves become more complicated having possibly a maximum (i.e. for $\alpha = 1/2$). The requirements for such a curve to be obtained are difficult to satisfy since one needs P and Q to have the same value but different signs. It is to be remarked that when both mechanisms are present the initial curve is never a simple exponential.

For the sake of simplicity we have assumed that the anisotropy of the optical dipole had the same orientation as the electrical induced dipole. This could change the results but one can guess that qualitatively one would obtain the same type of results.

0.5 EXPERIMENTAL RESULTS ON TMV AND DNA

0.5.1 TOBACCO MOSAIC VIRUS

Going back to experiments I present a curve obtained on tobacco mosaic virus dissolved in distilled water. This virus is a rodlike particle having a length of the order of 3000 and a diameter of the order of 200 Å. It is a perfect model for rodlike particles.

The electrical field was of the order of 100 V/mm and applied for milliseconds. The cell had a length of 10 cm, and the distance between the electrodes was of the order of less than one 1 mm. Due to the imprecision of these data I did not try to fit this curve with an exponential since I did not know with enough precision the distribution of the length of the viruses in the sample. It appears that there is a simple exponential decay, and faster than the establishment as predicted by the theory, which is the probably due to the effect of polydispersity.

0.5.2 DNA

Another experiment[7] made on DNA is shown in Figure 0.6. All these calculations have been made assuming that the system is monodisperse. Of course if it is polydisperse, the interpretation of the data becomes more complex. One can work either on the rise of the birefringence or on its decay. This last procedure is easier since surely only one exponential by species of molecules can be obtained. I did not try to work on it as the lack of precision of the apparatus and the polydispersity of the samples did not allow such studies but I am sure that now better results could be obtained with a better precision. Due to the structure of the tobacco mosaic virus, it seems improbable to have a permanent dipole in the direction of the axis of the

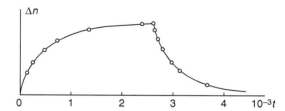

FIGURE 0.5 The signal is obtained for a solution in water of tobacco mosaic virus. The units for the value of the birefringence are arbitrary and the time units are milliseconds.[2] The electrical field is of the order of 100 V/cm.

particle. The origin of the effect is probably due to the migration of the counterions along the direction of the axis of the molecule but this has to be proved experimentally.

It is important to remark that placing the cell between two crossed Nicols gives a signal, which is proportional to the intensity, i.e., to the square of the birefringence. One could, using a quarter wave plate, transform the signal. In order to obtain Δn one has to work with Nicols prisms, which are not crossed; the parasitic light decreases the sensibility and the machine is not sensitive enough to allow this procedure.

If one tries to examine the value of the rotary diffusion constant of DNA, it is found that the value is very large and cannot be used if one assumes that the DNA is a coil. The only possibility is to assume that it is more similar to a rod than to a coil. This study was done in 1949, when the double helix of Crick and Watson was not yet discovered and the value of the rotary diffusion constant we were proposing was not sufficient to prove the existence of this double helix.

0.6   KERR EFFECT IN AN ALTERNATIVE ELECTRIC FIELD

At the end of my work, I had increased the sensibility of the apparatus and I thought it could be interesting to see what happens if an an alternative current is used instead of an electric field. The only difference is to use, instead of a pulse, a sinusoidal current of variable frequency.

The method is the following: attaching the two Nicol prisms simultaneously, with one polarizing the beam horizontally and the other vertically. The advantage of this technique is to use a system in which one can easily change the frequency. The results corresponding to the case of pure permanent and induced dipoles can be observed in Figure 0.7.

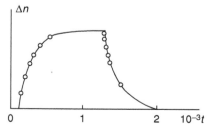

FIGURE 0.6 Results obtained on DNA with similar experimental conditions. There is much similarity between the establishment and the disappreance of the birefringence, indicating the absence of a permanent dipole.

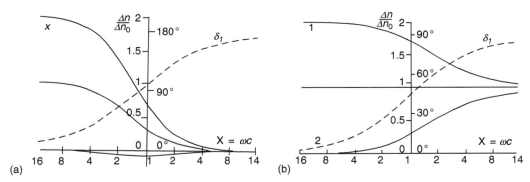

(a)
(b)

FIGURE 0.7 (a and b) The apparatus is the same as in Figure 0.2. The generator is replaced by an AC generator of variable frequency. The curves a and b present the theoretical result of the influence of frequency on the orientation of the particles. Figure (a) corresponds to a permanent dipole, while figure (b) to an induced dipole on the effect of anisotropy. The full lines represent the maximum, the minimum, and the average value of $\Delta n/\Delta n_0$ as a function of frequency. In the abscissa is frequency of the AC generator and in the ordinate is the intensity of the signal read one the oscillograph screen.

Another way of looking at the data is to use the Lissajous curves shown in Figure 0.8. Plotting Δn and E on the axes (x and y) one does not obtain the classical ellipses because the light transmitted through the crossed Nicol prisms is a quadratic function of the electrical field. This method gives directly the two components and does allow to determine the phase difference between the two components of the transmitted light but our results were not precise enough to allow a quantitative discussion of these data.

The only experiments made by us was to study tobacco mosaic virus drawing Lissajous curves. From these curves it is very easy to determine the amplitude of the two types of signal (amplitude on the oscilloscope screen and the maximum amplitude). These results were studied in the simple case of TMV, giving results, shown in Figure 0.9. This method gives rapidly the contribution of the permanent dipole and the induced dipole as well as the relaxation time. The apparatus was built by myself and was not precise. I am sure that, at the present time, it could be very easy to increase the quality of the results.

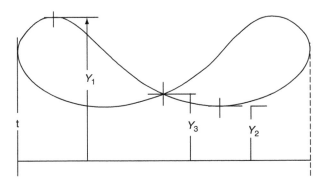

FIGURE 0.8 The types of curves obtained using the Lissajous representation, i.e., plotting Δn horizontal of the x-axis light and the vertically polarized light on the ordinate. From these curves it is very easy to determine the amplitude of the two types of signal (amplitude on the oscilloscope screen and the maximum amplitude).

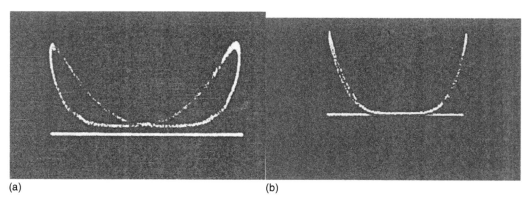

(a)   (b)

FIGURE 0.9 (a and b) Oscilloscope of the experiment. Experiments on TMV at two frequencies: (a) 50 cycles per second and (b) 2000 cycles per second.

This method is similar to the one proposed by Tinoco and Yamaoka[8] but it seems that the method using Lissajous curves could be able to give more precise data (see Figure 0.7).

As an example we give some preliminaries results obtained on tobacco mosaic virus. We used the Kerr cell and branched the x and y plates of the oscilloscope on the generator and y on the photomultiplier of the Kerr cell. The Kerr cell is filled with nitrobenzene (see Figure 0.9a and b).

The frequency is respectively 50 cycles per second in Figure 0.9a and 2000 cycles per second in Figure 0.9b. One sees clearly that the behavior is completely different; at low frequency the effect is in phase with the voltage but at high frequency the dipole cannot follow the current and reach an average value. The same type of experiment has done on DNA and the results are shown on Figure 0.10a and b in similar situation as in Figure 0.9.

The results are not very exciting but one sees that both mechanisms play a role in the results. At high frequency the rod is too slow to follow the voltage and takes an average value. At low frequency the effect is independent of the phase meaning that there is a permanent dipole.

I did try to obtain quantitative results but the precision was very limited. I am sure that now it would be possible to make these experiments with much more accuracy but my focus has shifted to other problems and I leave this method to younger people.

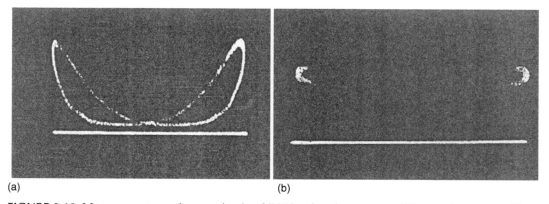

(a)   (b)

FIGURE 0.10 Measurements made on a simple of DNA using the same conditions as in the preceding figure. One sees clearly that the only effect is essentially due to an orientation similar to what is obtained with a permanent dipole.

Tinoco and Yamaoka[8,9] used a similar technique using a reversion of the field. If the pulses are very short, nothing should happen. These techniques are important since they should close the dispute about permanent dipole and induced dipole due to the displacement of the counterions. It could give a quantitative estimation of the mobility of the counterions, which it is difficult to measure precisely.

0.7 CONCLUSION

This short account of my work in the early 1950s is of course a really preliminary effort to use the possibilities of the Kerr effect to study colloidal particles. I did not stop this activity completely and due to the organization of the laboratory, I did not have time to develop more activity in this field and from this time I worked much more on light scattering and problems of the behavior of flexible chains solutions.

As a conclusion I would like to say that I am very proud of having been invited to write this chapter. It is very rare that work published more than 50 years ago is judged interesting enough to be considered again for presentation.

REFERENCES

1. Stoylov, S.P., *Colloid Electro-Optics: Theory, Techniques and Applications*, Academic Press, London, 1991, pp. 1–280.
2. Benoît, H., Thesis, *Sur un dispositif de mesure de l'effet Kerr par impulsions electriques isoles*, Universite Louis Pasteur, Strasbourg, 1949.
3. Perrin, J, *J. de Phys.* (*France*) 5, (1934) 497.
4. Maxwell, J.C., *A Treatise on Electricity and Magnetism*, Oxford University Press, Oxford, 1873.
5. Peterlin, A., *Zeit. für. Physik.*, 111 (1938) 232.
6. Peterlin , H. and Stuart, H.A., Doppelbrechung, inbesondere kunstliche Doppelbrechung, in Eucken, A. and Wolf, K.L., (eds.), *Hand- und Jahrbuch der Chemischen Physik*, , Akademische Verlagsgesellschaft, 1943, Bd. 8, Abt. IB, pp. 1–115.
7. Benoît, H., *J. Chim. Phys.*, 47 (1950) 718–721; 48 (1951) 612–617.
8. Tinoco, I., Circular dichroism and optical rotation in an electric field, in O'Konski, C.T. (ed.), *Molecular Electro-Optics,* Marcel Dekker , New York, 1976, Chapter 10.
9. Tinoco, I. and Yamaoka, K., The reverse pulse technique in electric birefringence, *J. Phys. Chem.*, 63 (1959) 423.
10. Benoît, H., Etude de temps de relaxation de l'effet Kerr en courant alternatif, *J. Chim. Phys.*, 19 (1952) 517.

Part I

Electro-Optic Methods—
Advances in Theory
and Methodology

1 Polar Nanoparticles

Stoyl P. Stoylov

CONTENTS

1.1 INTRODUCTION

The understanding of the behavior of nanoparticles in electric fields requires a knowledge of not only their average electric charge but also their dipole moments. An analysis of the present state of knowledge of the dipole moments of nanoparticles is made on the basis of the numerous studies and results on the dipole moments of colloid particles [1], the sizes of most of which are in the nanoscale range. The different methods of studying the dipole moments are considered and the special place of the electro-optic methods is analyzed.

The description of the electric dipole moments of nanoparticles, including macromolecules is much more complicated than that for simple (low molecular mass) molecules. It seems that the main source of complication is the particle–medium interface. An attempt of a simple classification of the electric dipole moments of nanoparticles is presented. Examples of experimental and theoretical investigations are discussed without going into detail. Details are not discussed for experiments because there are practically missing papers, in which an adequate description of the objects and the experimental conditions are made. Much detail is not available on the theoretical aspects because they are rarely experimentally verified, partly because of the problems in the experimental studies.

Special attention is paid to the two of the induced interfacial dipole moments (IDMs) (the Maxwell–Wagner and the charge dependent) and to the permanent IDMs and to their possible coupling. This is justified because the successful attempts are very rare, at present, for even a qualitative evaluation of the contribution of the different polarization mechanisms.

1.2   POLAR PARTICLES AND ELECTRO-OPTICS

Permanent or electrically induced polar nanoparticles are characterized by permanent and induced dipole moments. These dipole moments, at present, are best investigated by macromolecular and colloid electro-optic [1–14] and dielectric [15] measurements. Most of the macromolecules and colloid particles are of nanometer dimensions and thus they might be considered to be typical nanoparticles. The greatest contribution to the magnitude of the dipole moments, especially in aqueous medium, comes from their interfacial components known as the IDM. Probably IDMs are one of the central problems not only in the double electric layer (DEL) in colloid science but also in the electric phenomena in nanoscience. It seems that the higher electric field strengths in the electro-optical measurements compared to those in the dielectric measurements (mostly tens and hundreds of V/cm) and the electrically independent optical detection of the effect of the orientation of single particles leads in the lower frequency range (below 100 kHz) to more reproducible results [1,15]. All that is in spite of the strong charge, pH, ionic strength, particle size, and polydispersity dependencies of the electric-charge-dependent IDM (CHDIDM). These properties are one of the principal causes for the considerable differences in the values for the interfacial electric polarizability reported by different authors on samples of different origins (method of preparation) and at different experimental conditions [1].

The magnitude of the CHDIDM of the particles is often the prevailing component at lower frequencies. In the nonelectro-optic literature under interfacial polarization, quite often the researchers understand only the Maxwell–Wagner (MW) polarization mechanism. Thus the MWIDM represent only the highest frequency part of the IDM. In many cases this is a smaller part of the magnitude of the IDM. The latter essentially depends on the dielectric properties of the particles and the medium and not on the electric charge of the particle. It relaxes at higher frequencies (about and above 1 MHz) than the CHDIDM components and may essentially influence its formation. The charge-dependent component of the interfacial polarization (CHDIDM) very often may be several times bigger in magnitude than the MW component MWIDM.

1.3   INDUCED DIPOLE MOMENT AND ζ-POTENTIAL

Following the discovery of electrophoresis at the beginning of the 19th century [16] the characterization of the electric properties of colloid particles of various origins (inorganic, biological, etc.) was done experimentally by measurement of their electrophoretic mobility. The sign of the electric charge of the particles may be qualitatively determined from the direction of the electrophoretic motion of the particles v with respect to the direction of the applied electric field directly (Figure 1.1). In the figure the spherical nanoparticle moves toward the positive electrode, which means that its electric charge is negative.

Starting from the beginning of the 20th century the electric charge and respectively the ζ-potential of colloid particles were introduced as their basic electric characteristics and respectively as the most important quantitative electric parameters describing their behavior in an electric field. This followed the theoretical derivation of the relations among electrophoretic mobility, particle surface electric charge, or the particle ζ-potential [16].

About 100 years later, the situation is practically unchanged. At present the surface electric charge and the ζ-potential are widely accepted as the most relevant electric parameters of

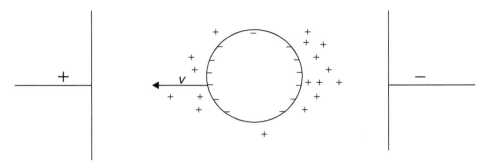

FIGURE 1.1 Electrophoretic movement of a nonconducting spherical nanoparticle with a negative surface charge and velocity v. The diffuse electric layer is deformed, i.e., there are more excess positive charges on the right than on the left of the particle.

colloid particles [17]. Probably, one of the causes for the wide acceptance is the often-observed good correlation of ζ-potential with the colloid stability. The latter is one of the basic problems of colloid science and is related to many important applications.

In the second half of the last (20th) century the first attempts aimed at the consideration of the inhomogeneity (transient, stationary, or permanent) of the surface electric charge (respectively of their ζ-potential) were made. As it might be expected they started with the consideration of the simplest dipolar distribution, often expressed by electric dipole moments [1,15,18,19]. Although the dipole moments present a number of advantages the predominance of the electrokinetic methods of characterization is evident in the beginning of the 21st century as well. At present there are no indications that the situation will be changed soon.

For spherical, isotropic particles their electric charge is a basic characteristic. Depending on the method of measurement, it might be an average characteristic that is very important for the representation of the behavior of electrostatically stabilized colloid systems in electric fields. However, for real particles, that very often deviate from the spherical shape and are anisotropic, a deeper, more detailed characterization is needed.

Following the general description of the electric characteristics of molecules by Debye [20], let us have on the particle surface some electric charges q_i with coordinates (ξ_i, η_i, ζ_i) in a rectangular coordinate system (x, y, z) (Figure 1.2). The origin of the coordinates has to be inside the particle. If φ is the potential of the external electric field in the origin of the coordinate system, the potential in point (ξ_i, η_i, ζ_i) φ_i may be presented as

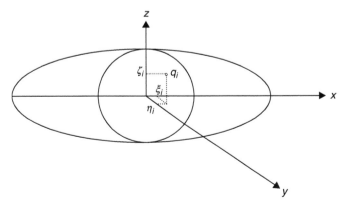

FIGURE 1.2 Coordinate system x, y, z for description of the electric potential φ_i in a point on the surface of an ellipsoidal particle with electric charge q_i with coordinates ξ_i, η_i, and ζ_i.

$$\varphi_i = \varphi + (\xi_i \partial\varphi/\partial x + \eta_i \partial\varphi/\partial y + \zeta_i \partial\varphi/\partial z) + (1/2)(\xi_i^2 \partial^2\varphi/\partial x^2 + \eta_i^2 \partial^2\varphi/\partial y^2 + \zeta_i^2 \partial^2\varphi/\partial z^2$$
$$+ 2\xi_i\eta_i \partial^2\varphi/\partial x\partial y + 2\eta_i\zeta_i \partial^2\varphi/\partial y\partial z + 2\zeta_i\xi_i \partial^2\varphi/\partial z\partial x) + \cdots$$

The potential energy of the particle due to the influence of the external electric field is

$$\Sigma q_i\varphi_i = \Sigma q_i\varphi + \Sigma q_i(\xi_i \partial\varphi/\partial x + \eta_i \partial\varphi/\partial y + \zeta_i \partial\varphi/\partial z) + (1/2)\Sigma q_i(\xi_i^2 \partial^2\varphi/\partial x^2 + \eta_i^2 \partial^2\varphi/\partial y^2$$
$$+ \zeta_i^2 \partial^2\varphi/\partial z^2 + 2\xi_i\eta_i \partial^2\varphi/\partial x\partial y + 2\eta_i\zeta_i \partial^2\varphi/\partial y\partial z + 2\zeta_i\xi_i \partial^2\varphi/\partial z\partial x) + \cdots$$

$$(1.1)$$

In Equation 1.1 the electric potential φ and its derivatives are constants (with respect to particles' electric parameters that characterize the properties of the potential of the external electric field). The behavior of the particle in the external electric field is characterized by its electric parameters that in Equation 1.1 represent sum of products of the charges on the particle times their coordinates. The first sum Σq_i is the total electric charge of the particle q. The second sum in Equation 1.1 represents practically the three sums $\Sigma q_i\xi_i$, $\Sigma q_i\eta_i$, $\Sigma q_i\zeta_i$ and might be considered as a sum of the components of electric dipole moment μ_x, μ_y, and μ_z of the particle. The first derivatives of the potential φ with negative sign are the components of the electric field strength E. Thus the second sum in Equation 1.1 can be represented as a scalar product of the vectors μ and E (μE). In the same way the third sum gives the particle electric quadrupolar moment and so on.

If the particles are electrically charged, the magnitude of the calculated electric dipole moment depends on the choice of the origin of the coordinate system. If particles are polar, the calculation of the magnitude of the electric quadrupolar moments depends on the choice of the origin of the coordinate system. Investigations are much simplified if model particles or experimental conditions are chosen, so that lower order particle electric parameters are zero or very small. Quadrupolar (third order) and higher order polar particle parameters may be studied in nonhomogeneous electric fields. Anderson [18] published a very interesting theoretical analysis of the effect of dipolar and quadrupolar moments on the translation and the rotation of spherical colloid particles with nonhomogeneous distribution of the electric charge on their surface (Figure 1.3). Such particles have been successfully designed for electronic paper [21] and other high-technology applications.

There are numerous experimental studies of nanoparticle electric charge (and ζ-potential) and of the dipole moments of polar molecules [17,20]. However, the experimental investigations of the electric dipole moments of colloid particles are relatively few [1–14] and practically there are no experimental studies on their quadrupolar moments. This situation to some extent is similar to that with the experimental studies of molecular and quadrupolar electric moments of molecules in the early 1920s, when the *Polar Molecules* of Debye was published [20].

1.4   ELECTRIC DIPOLE MOMENTS

The electric dipole may be presented in a simplified way as a system of two equal in magnitude electric charges but of opposite sign $+q$ and $-q$ separated by a distance d (Figure 1.4). The electric dipole moment μ is described by a vector directed from the negative to the positive charge and its magnitude is given as

$$\mu = qd. \tag{1.2}$$

The concept of the electric dipole has found considerable application in the study of simple (low molecular mass) molecular structures [20]. Later some more experimental studies were performed

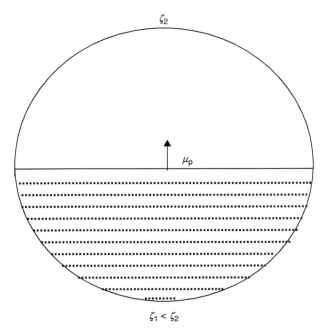

FIGURE 1.3 The simplest nonhomogeneous distribution of the surface electric charge of a spherical nanoparticle. The ζ-potential of the two halves (lower and upper hemispheres) is respectively ζ_1 and ζ_2, and μ_p is the permanent dipole moment of the particle.

on the dipole moments of macromolecules mostly of biological origin like proteins and nucleic acids. The introduction of the dipolar concept to colloid particles has followed the experience gained in the study of molecules of low molecular mass of polymers and biopolymers.

Like in the case of molecules of low molecular mass, the electric dipole moments of colloid particles may be subdivided into permanent and induced electric dipole moments. The magnitude of the permanent electric dipole moments does not depend on the electric field applied. They lead only to the orientation of the dipolar particles when a homogeneous electric field is applied or to orientation and migration when the applied electric field is nonhomogeneous. The magnitude of the induced dipole moment is zero before an electric field is applied. It appears and disappears some time after the electric field is switched on and off, respectively. The orientation of the particles by a permanent dipole represents an ordering of the particles with their specific parts pointing in one and the same direction, while in the orientation by an induced dipole moment the specific parts of the particles may point in different (opposite) directions (Figure 1.5).

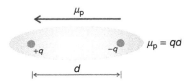

FIGURE 1.4 Electric dipole moment of an ellipsoidal nanoparticle. For simplicity it is represented as the permanent dipole moment μ_p with q the center of gravity of the electric charges of different signs and d is the distance between the two poles. This is equivalent to the presentation of the charges with the ζ-potentials. The latter may be preferable when particle orientation is considered.

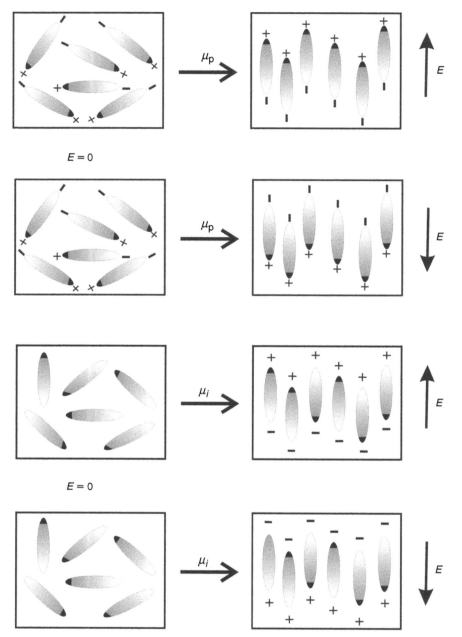

FIGURE 1.5 Difference between the behavior of ellipsoidal particles only with a permanent μ_p or an induced electric dipole moment μ_i on application of an electric field E. The permanent dipole moment μ_p orders the particles (black ends of the particles are oriented in one and the same direction), while μ_i only orients the particles (black ends are in different directions).

1.5 COLLOID PARTICLES WITH ZERO AVERAGE ELECTRIC CHARGE

The induced dipole moments of colloid particles with zero average surface charge may have various origins. In the simplest (and relatively rare) case of a nonconducting particle dispersed in a nonconducting medium, the induced dipole moment of the particle (and their orientation, respectively) is determined basically by the difference in the dielectric constants of the particle and the medium. In this case the dielectric (volume) properties of the particle are of essential importance for the particle's behavior in electric fields. Therefore often in the more general case of conducting media this mechanism of electric polarization (and orientation) is discussed as its (dielectric) volume component.

Now it is accepted that the three contributions to the noninterfacial (often called dielectric or volume) electric dipole moments are related to electron polarization (deformation of electron clouds of the atoms), the atomic polarization, related to the displacements of the atoms in the molecules and orientational polarization related to orientation of molecular dipoles under the action of the applied (induced) electric field. All these take place inside (in the volume) of the colloid particles. On the other hand, the interfacial electric-induced moments are exclusively related to the space near the particle interface and mainly to the ions on or around it. It has little to do directly with the dielectric properties of the colloid particles and the medium. Several components of the electric-induced dipole moment are most often considered. Further in this chapter every component will be shortly discussed.

1.5.1 Volume (Dielectric) Electric Polarizability

The induced dipole moment of a dielectric colloid particle in a dielectric medium with the shape of a rotational ellipsoid (Figure 1.6) may be written as [23]

$$\mu_i = (g_1 - g_2)VE = (g_1 - g_2)VE = (\gamma_1 - \gamma_2)E = \Delta\gamma E, \qquad (1.3)$$

where $g_1 - g_2$ is the anisotropy factor (in this case it may be both the optical and the electrical anisotropy factors, V is the volume of the particle, E is the strength of the applied electric field, and $\Delta\gamma$ is the electric polarizability anisotropy of the particle. The induced electric dipole moment μ_i is not characteristic of the particle. The electric characteristic of the particle that corresponds to its induced dipole moment is the anisotropy of the electric polarizability $\Delta\gamma$, similar to particle's electrophoretic mobility being an electric characteristic of the particle, whereas its electrophoretic velocity is not. The volume component of the particle's electric polarizability is its dielectric component, which may be presented as [1,6]

$$\Delta\gamma_v = f(v, \varepsilon_1, \varepsilon_2, \varepsilon_m, L_1, L_2), \qquad (1.4)$$

where ε_1, ε_2, and ε_m are respectively the dielectric permittivities of the particle along the symmetry axis, perpendicular to it and of the disperse medium (solvent), V is the volume of

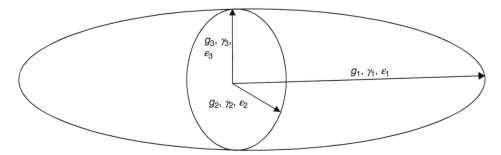

FIGURE 1.6 Optical and electrical parameters of a nanoparticle with a shape and properties that may be represented as a rotational ellipsoid. The electric optic parameters are respectively the polarizabilities per unit volume g_1 and g_2, γ_1 and γ_2, ε_1 and ε_2 are the respective permittivities.

the particle, and L_1, L_2 are known as the depolarization factors [24] or form factors. They take into account the depolarization electric field created by the bound charges on the particles' surface. Thus the electric field in the particles is smaller than the applied electric field, existing outside the particles [24]. The depolarization factors represent elliptical integrals whose sum is equal to 1.

The dispersion (relaxation) of this polarizability starts at frequencies above 10^9 Hz, where the relaxation of the orientational polarizability (e.g., of water) begins. At higher frequencies the atomic polarizability and the electronic polarizability relaxe.

1.5.2 Maxwell–Wagner–Sillars Interfacial Polarizability

MW [1,22] polarization mechanism is independent of the magnitude of the surface electric charge of the colloid particles. The magnitude of polarization and its relaxation time depend exclusively on the difference in the dielectric properties of the particle and the medium (more precisely their complex permittivities i.e., their dielectric constants and specific conductivities). It seems that this mechanism of polarization (or type of polarizability) might be prevailing for slightly anisometric micrometric particles. It is important to know that it might be the only independent polarization mechanism, which could be of opposite sign to all other interfacial electric dipole moments at certain particle and medium electric properties (Figure 1.7). In his original derivation Maxwell [25] used real dielectric permittivities. Wagner generalized the Maxwell theory by replacing the real with complex dielectric permittivities [26]. Later Sillars extended the theory to ellipsoids [27]. Thus the difference of the dielectric component $\Delta\gamma_d$ from the Maxwell–Wagner–Sillars (MWS) component of the electric polarizability $\Delta\gamma_{MWS}$ is the existence of free electric charges (ions) in the particles and the medium.

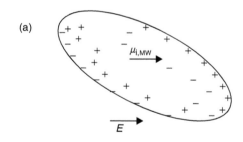

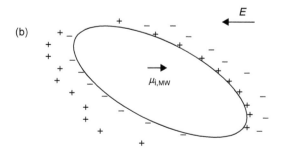

FIGURE 1.7 Maxwell–Wagner induced dipole moment $\mu_{i,MW}$ of an ellipsoidal conductive nanoparticle in a nonconductive medium (a) and a nonconductive particle in a conductive medium (b). In both cases the particles are without electric charge. The direction of the induced dipole moment $\mu_{i,MW}$ with respect to the applied electric field E changes from collinear in (a) to opposite in (b).

This is taken into account through the conductivities of the particles and of the medium k_1, k_2, and k_m:

$$\Delta\gamma_{MWS} = f(\varepsilon_1, \varepsilon_2, \varepsilon_m, k_1, k_2, k_m, V, L_1, L_2). \qquad (1.5)$$

The relaxation time of the MWS electric polarizability may be presented in the general case by the expression [1,23]:

$$\tau_{MWS} = f(\varepsilon_1, \varepsilon_2, \varepsilon_m, k_1, k_2, k_m, L_1, L_2, a_1, a_2), \qquad (1.6)$$

where a_1 and a_2 are the sizes of the particle along the symmetry axis and perpendicular to it. In most practically encountered cases the dispersion, corresponding to this relaxation time is around and above 1 MHz.

It seems that the size of the particle is important in the case of conducting particles [24]. Depending on the ratio of the dielectric permittivities and conductivities, a complex variety of directions of the induced dipole and respectively of orientation might be possible [28]. For example even for homogeneous dielectric particles and medium, depending on the sign of the difference in permittivities, the symmetry axis of the particles will orient along or perpendicular to the applied electric field. Thus conductive particles in dielectric medium will have an opposite direction of the induced dipole moment to that of dielectric particles in conductive medium, etc.

The mechanism of the MW polarization is induced by the applied electric field accumulation on one side and depletion on the opposite side through migration of free surface charge near (inside or outside the particles) the boundaries separating different permittivities and conductivities. The charge accumulation–depletion will be at the internal side for conductive particles and respectively on the external side for conducting medium (Figure 1.7a and Figure 1.7 b).

The spatial distribution of this space electric charge may be characterized like in the case of charged colloid particles by a thickness of the induced diffuse charge layer $1/\chi$ (respectively $1/\chi_0$ and $1/\chi_E$) in Figure 1.8b.

In the MW mechanism it will be the distance from the center of excess charge concentration (in the general case it will coincide with the maximum excess charge concentration and will be close to the boundary) at which the electrical potential has fallen to $1/e$ (i.e., $1/2.72 = 36.8\%$) [24]. In the case of charged particles (Figure 1.8) this induced space charge would be coupled to existing ("permanent," "reversible") space charge that might lead to some complications. The representation of two separate diffuse layers in Figure 1.8 is an oversimplification, because one and the same ion may participate in both polarization mechanisms.

This electric field induced charging of the particles' interface might have some overlooked implications even outside electro-optics.

1.5.3 ELECTRICALLY CHARGED COLLOID PARTICLES

Here we have in mind only the "permanent" surface electric charge that exists before the application of the electric field. It is usually accepted to be homogeneously distributed on the surface of the particles. In the case of electrically charged colloid particles when electrostatic stabilization makes possible the existence of stable dispersed systems the mechanisms and theories of dipole moments get much more complicated. Generally, in the experiment [1,15,29] greater dielectric or electro-optic effects are observed that relax at lower frequencies. Here electrode polarization is believed to play some more important unpleasant role, especially in the dielectric measurements [15].

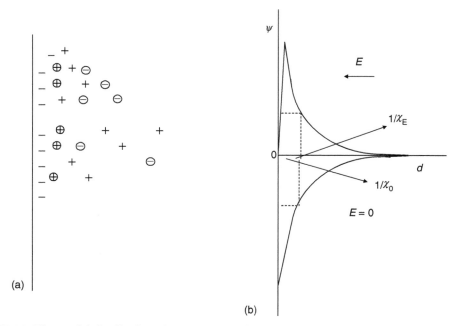

(a)

(b)

FIGURE 1.8 The spatial distribution of the space electric charge (diffuse layer) of a flat charged surface, separating a conductive and a nonconductive medium. The encircled electric charges "⊕" and "⊖" are those accumulated by the Maxwell–Wagner mechanism (a) and potential diagrams $\psi(d)$ for electric field E and no field (b). The distinction of the ions related to charge-dependent and Maxwell–Wagner induced IDMs is made for simplifying the illustration. The thicknesses of the "two" diffuse electric layers, with and without electric field, are $1/\chi_E$ and $1/\chi_0$, respectively.

1.5.4   CHARGE-DEPENDENT (IONIC OR COUNTERION) ELECTRICALLY INDUCED IDM (CHDIDM)

The mechanism of formation of the electric-charge-dependent component of the IDM may be simply visualized as a deformation of the DEL of the colloid particles (Figure 1.9, see also Figure 1.1). This qualitative model was suggested as early as 1935 on the basis of electric

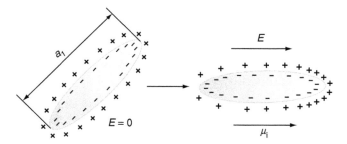

FIGURE 1.9 Charge-dependent induced IDM of an ellipsoid nanoparticle $\mu_{i,chd}$ and its orientation in a strong electric field E (energy of orientation is much greater than energy of thermal motion). The "−" charges are from the "permanent" electric surface charge of the particle and the "+" charges from the diffuse electric layer. The electrically induced double electric layer (Figure 1.8) here is not shown for simplicity.

birefringence experiments with colloids [30]. This is the main component of the interfacial electric-induced dipole moment for many submicron, strongly anisometric particles [1]. It is clearly distinguished by a big sub-MHz dispersion, strong dependence on size ($\mu_i \sim 1^{(1.5\div3)}$) and on the surface electric charge.

A considerable number of theoretical models have been suggested [1,22], which considered mechanisms from bound ions adsorbed on the particles [31] to free ions beyond the limits of the diffuse part of DEL [22]. At present there is no strict experimental verification of any of the theoretical models. This is related to problems in the used approximations related mainly to the particle shape and structure and thickness of DEL. The great diversity of theoretical models might in some way correspond to the diversity of the properties of the studied colloid particles.

Interfacial electric permanent and induced dipole moments are coupled because the same ions whose displacements lead to induced IDM influence the magnitude of the permanent IDM, e.g., through affecting the magnitude of the electrokinetic charge and potential of the two poles of the permanent dipole [22]. The IDMs determine exclusively the magnitude of the orientation of the colloid particles.

Disperse systems with predominant CHDIDM are often with highly charged particles in a low conducting medium. There are a considerable number of proposed mechanisms and theories, which differ mainly in the location of the counterions, whose displacement creates the induced IDM of the particle. The electric polarizability may be presented in a general way in order to satisfy most mechanisms (models) as

$$\Delta\gamma_{chd} = f(a_1q, \psi_d, \zeta, k_s, 1/\chi, \Delta\gamma_{MWS}), \tag{1.7}$$

where q is the electric charge of the particle, coinciding with the total charge of DEL and a_1 is the largest particle axis [31]. Depending on the model used by different authors instead, q in Equation 1.7 may, be substituted by ψ_d, the potential of DEL [22], or ζ, the potential at the slipping plane [18], or k_s, the surface conductivity [32]. $1/\chi$ is the thickness of DEL and $\Delta\gamma_{MWS}$ is the MWS component of the electric polarizability. The latter reflects the coupling of the "permanent" electric charge with the "induced" electric charge. Thus independent of the MWS component of the total particle polarizability, probably there will be an influence of the "induced" electric charge on the magnitude of the (CHDIDM). The inverse might be also true. That is in Equation 1.5 $\Delta\gamma_{chd}$ might be also added to the variables. Probably to a smaller extent, there might be some coupling among the other components of the electric polarizability as well. That means that the magnitude of the total particle electric polarizability in the general case is not an additive quantity of the magnitude of its components.

The relaxation time of the permanent (equilibrium) electric-charge-dependent (counterion) electric polarizability may be presented in the case of polarization of a strongly elongated ellipsoidal particle with longest dimensions along the symmetry axis a_1 by the expression [1,22]

$$\tau_{chd} = f(a_1^2, D_i, 1/\chi, q, \psi_d, \zeta, k_s, \tau_{MWS}), \tag{1.8}$$

where D_i is the counterion mobility and τ_{MWS} reflects the coupling (interaction) of the counterions compensating the permanent (equilibrium) charge of the particle with the electric field induced electric charge on the particle surface related to the MWS mechanism of polarization of ellipsoidal particles.

There are two more types of interfacial-induced electric dipole moments that deserve separate attention. These are the slow- [1,34] and the saturated induced IDMs [1]. Both are used in the interpretation of experiments that in some cases could be taken as evidence for the existence of a permanent dipole moment.

1.5.5 SLOW ELECTRIC POLARIZABILITIES

Slow electric polarizabilities in the kilohertz and hertz ranges were mainly introduced in the interpretation of the electro-optic experimental results that showed some evidence for the existence of a permanent electric dipole moment, while the structure of the particles was highly symmetric. At this time there was little experimental evidence that there might be other sources of permanent dipole moment than "volume" structure. The concept of permanent IDM moment was to a large extent still unknown.

The slow electric polarizability interpretations were mainly related to the experimental investigations of the dispersion dependencies of the electro-optic effects and of the reverse pulse measurements. In the first type of measurements, dispersions at frequencies lower than that corresponding to the particle rotation were observed. For reverse electric pulses changes in the electro-optic effect were detected that followed the reversal of the applied electric field [33]. The slow electric polarizability seems to be mainly a part of the CHDIDM.

A comprehensive review on the slow interfacial electric polarizabilities till the end of the 1980s including some experimental results and theories may be found in Ref. [1]. A recent example is the experimental study of purple membrane fragments [34]. Based on the increasing magnitude of the electric polarizability going from 100 kHz to d.c., the author concludes that there is a broad spectrum of polarization time constants, starting with that for the charge-dependent IDM of the fragments.

Another recent example may be found in the electro-optic studies of Radeva and coworkers on ellipsoidal β-FeOOH particles, covered with polyelectrolyte layers [35–37]. In this case the ion condensation on the highly charged polymer chains of the coat might be responsible for the observed slow polarization phenomenon.

1.5.6 SATURATED INTERFACIAL-INDUCED ELECTRIC DIPOLE MOMENTS

The saturated induced IDM appears at high electric field strengths, where it is no more a linear function of the applied electric field. At still higher electric fields the saturated interfacial-induced electric dipole moment may remain constant with the rise of the electric field strength, i.e., it saturates. These dipole moments were introduced mainly in the interpretation of the results from the investigation of the field strength dependence of the electro-optic effects of DNA solutions, where the interpretation with permanent dipole moments seemed much less plausible.

The development of the concept and theories for the saturated induced electric IDM is related with the investigations of Neumann and Katchalsky [38], Sokerov and Weill [39–41], Diekmann [42], Yoshioka [43], Morita and Watanabe [44], Atig et al. [45], Stellwagen [46], and others. Yoshioka [43] theoretically introduced a sharp transition in the electric field dependence of the induced IDM from linear at weak fields to saturated (independent on E) at higher fields [41] (Figure 1.10, curve 1). At low fields proposed by Yoshioka, the expression corresponds to that proposed earlier by Sokerov and Weill [39–41]. Later Atig et al. [45] considered a more realistic, gradual transition (Figure 1.10, curve 2) from linear to saturated electric field dependence of the induced IDM.

A more detailed review on this topic till the end of the 1980s may be found in Ref. [1]. A recent example may be found in the paper of Porschke [34]. The evidence for the existence of saturated IDM there at relatively weak electric field strengths (about 5 kV/m) comes from the impossibility to fit the experimental electric dichroism versus electric field strength curve to

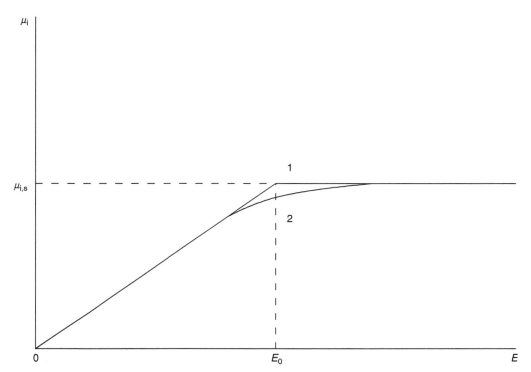

FIGURE 1.10 Variation of the induced IDM μ_i with the electric field strength E for saturating μ_i. The saturated induced IDM is $\mu_{i,s}$ and E_0 is the electric field at (about) which the saturation occurs. Curves 1 and 2 correspond respectively to theories for sharp and gradual transitions.

the Shah [47,48] disk model theoretical curve. When the author includes a charge-dependent IDM saturation component in the theory extended by him, the fit is much better. Uncertainties in the particle shape, flexibility, and other approximations might cast some doubts on the conclusions made.

The difference of the saturated induced electric IDM from the slow-induced electric IDM is that we are not really sure that the first was directly and experimentally observed, whereas for the latter there are a number of experimental investigations that are difficult to doubt. Another question is what the exact origin of the slow IDM is. At present it is not completely clear whether there is some general mechanism or only specific mechanisms for the slowing down of the counterion mobility for different objects. It is also not clear whether the MW component of the IDM may contribute to the saturated and slow IDMs.

1.5.7 PERMANENT ELECTRIC IDMs

We will consider interfacial electric dipole moments, which are due only to the permanent, inhomogeneous distribution of the electrical charges on the surface or around the colloid particles, i.e., in the interfacial region. The permanent dipole moments due to the inhomogeneous distribution of the electrical charges in the volume of the particles, mainly related to the structural asymmetry of the particles will not be considered here. As the main source of data for the electric dipole moments comes from the particle orientation, the natural timescale for determining dipole moment is according to its lifetime. A dipole moment may be treated as permanent when its lifetime is much longer than the times of disorientation (the so-called

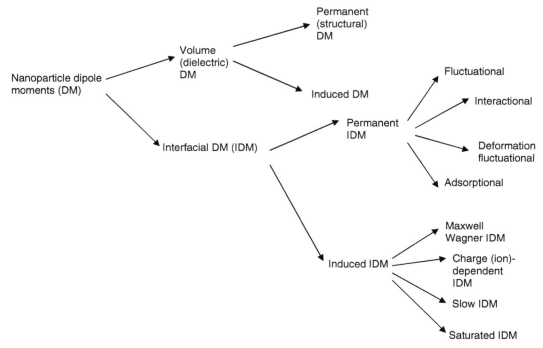

FIGURE 1.11 A simplified classification of nanoparticle electric dipole moments. For examples see the text.

decay times). It follows from this definition that with respect to the time, the slow-induced electric dipole moment may be regarded in some respects as permanent if its relaxation time is much longer than particles' rotational relaxation time. However this is not a real permanent electric dipole moment, because it might be proportional to the applied electric field, like all "normal" induced electric dipole moments. Generally the permanent IDM will be more variable than the volume (structural) permanent dipole moment.

All permanent electric dipole moments, both interfacial and volume, will not be completely independent from the applied electric field if they are accompanied by a considerable interfacial-induced dipole moments as often observed. Depending on the relative directions of the permanent and induced dipoles and their magnitudes, the applied electric field will more or less influence the shielding of the permanent dipoles. This may bring to some dependence of the permanent IDM on the electric field strength.

In Figure 1.11 a simplified classification of the greater part of the known at present electric dipole moments of nanoparticles are presented.

1.6   IDMs AND PARTICLES' ORIENTATION

The problem of the orientation of colloid particles in electric fields through their IDMs seems to be far from fully resolved. Thus for example Lyklema [49] in his capital standard reference five volume books stresses the importance of the size and charge-dependent induced IDM for the understanding of the nonequilibrium DEL as one of its basic characteristics and at the same time does not make its basic role clear in particle orientation. One does not have the same impression for the permanent IDM and its influence on the particle orientation. All existing electro-optic experimental data show that similar to the case of the interfacial permanent dipole moment the particle orientation is inevitable under the action of all types of IDMs. Thus both permanent and induced IDMs in all cases lead to particle orientation. It should be stressed that the orientation due to the induced IDM in many cases is predominant. There are no serious considerations for opposing particles with permanent, nonhomogeneous distribution of the ζ-potential (permanent IDM) to particles with a nonhomogeneous distribution of the ζ-potential induced by an applied electric field (induced IDM). This discrepancy comes probably from the theoretical and experimental investigations of Anderson [18] on big (micrometers) particles and the unawareness of the results and potential of colloid electro-optics. In Anderson's theoretical description of big colloid particles based on classical mechanics, the contribution of the induced IDM drops out in the linear approximation, i.e., when developing in series all terms with E on a power higher than 1 are neglected. For Brownian particles, in the probabilistic treatment, the term with E^1 is zero and the first contributing terms both for permanent and induced IDMs are with E^2, like in most electro-optic theoretical treatments [1]. It could be of interest to study separately the limits of application of both mechanic and probabilistic treatments, because it could be related to some older results of Tolstoy and collaborators. There are some doubts in the possibility of particle orientation in dielectric measurements at higher particle concentrations, where their rotation might be hindered [50]. Electro-optic measurements have shown that orientations persist even at higher concentrations [51,52]. This analysis of the orientation behavior of colloid particles may have some impact on the theory and experiment of electrophoresis of nonspherical colloid particles and probably on the better understanding of dielectric dispersion (DD) and dipolophoresis (dielectrophoresis). Electrophoresis of anisometric particles without permanent dipole moment have been investigated by taking into account only the distribution of orientation of the colloid particles existing before the application of an electric field, leading to particles' migration [53,54].

In 1968 Petkanchin and Stoylov [55] took into account the orientation of the particles and made an estimation of the degree of orientation of palygorskite particles based on the data from an electro-optic study of the same particles. The result was that the orientation in this case is negligible. However, there are other particles where the orientation might be not negligible. An example is the microelectrophoresis of spermatozoids, aggregates of purple membrane fragments, etc., where the orientation is easily observed [1].

1.7 POLARITY AND SYMMETRY

In the last 50 years there are numerous experimental dielectric [15] and electro-optic [1] data for the existence of considerable permanent electric dipole moment, which in many cases is difficult to explain with the well-known internal structure of the colloid particles or biopolymers. The particle sizes are often in the range of nanometers. Moreover there are some experimental indications for the dependence of the permanent IDM on ionic strength, pH, and valence of ions that confirms the concept of the interfacial nature of the interfacial permanent electric dipole moments of many types of colloid particles. In the last 15 years after the publication of *Colloid Electro-Optics* there seems to be no essential progress in this field except for the cases of purple membranes [56] and composite particles [57].

Various explanations have been provided regarding the mechanisms about the formation of the permanent electric dipole moments of colloid particles, mostly for those of biological origin (proteins, nucleic acids, viruses, membrane fragments, etc.). For the case of proteins, Kirkwood and Schumaker [58] have suggested the proton fluctuation mechanism. For disk-shaped membrane fragments (e.g., the so-called purple membranes) Wang and Porschke [56] have suggested the disk bending fluctuation mechanism. This mechanism is related to the flexoelectricity of biological membranes suggested earlier by Petrov [59,60]. Other rather speculative mechanisms include interactions with neighboring particle or surfaces, suggested by Stoylov [1], inhomogeneities in adsorption, in slipping plane distances due to adsorption of nonionogenic surfactants and polymers, in ionization, in adsorbed dipole layer in polar media [1], etc. In the case of composite nanoparticle, constructed from a big one (about 100 nm) and a small one (about 10 nm) of opposite sign of the electric charge, a permanent dipole moment, as expected, was observed [57]. This permanent IDM depends on the properties of the initial single particles and the structure of the composite particle. In most cases the permanent IDM is connected with a lower or decreased symmetry of the composite or hybrid particle and the properties of its interface.

We are aware of few attempts to estimate the magnitude of permanent electric dipole moment of colloid particles on the basis of their structure. A good example is the calculation of the magnitude of the permanent electric dipole moment of purple membrane fragments by Wang and Porschke [56], which obliged them to introduce a complimentary component—the fluctuating bending permanent dipole moment. The other two components, the asymmetric distribution of the charged lipids on the two sides of the membrane and the dipole moment of the only protein, the bacteriorhodopsin were unable alone to fit the experimental data. Another example is the estimation of the magnitude of the permanent electric IDM of simple composite particles mentioned above. Reasonable magnitudes near the experimental data are found. For the more detailed calculation of the permanent dipole moment of charged colloid particles there might be some problems with the choice of the origin of the coordinate system.

1.8   COMPARATIVE ANALYSIS OF METHODS OF MEASUREMENTS OF IDMs

The experimental separation of the different polarization mechanisms and the determination of their relative importance are done mostly through the investigation of the dispersion curves. Unfortunately only in a few studies, the dependence on the particle surface charge was investigated or attempts were made to study particle's suspensions or macromolecular solutions near the isoelectric point. Such studies may give the possibility, for example, to distinguish between MW polarization and charge-dependent polarization and slow polarizability.

A faster way of analysis is to study the electro-optic effects in reverse electric pulses, which have been introduced by Oosawa [33]. Further Yamaoka and collaborators introduced essential improvements in this method and applied it to various biological and inorganic systems [61,62].

The experimental determination of the permanent dipole moment meets much more difficulties because one has to work at low frequencies. The frequency should be below that of the rotational particle dispersion. Thus it is possible to measure the full contribution of the permanent dipole moment but unfortunately it appears in a sum containing the contributions of all electric dipole moments that have relaxed at higher frequencies.

1.9   DIELECTRIC, ELECTRO-OPTIC, AND "DIELECTROPHORETIC" SPECTROSCOPIES

The DD studies of colloid suspensions measure the macroscopic polarization of the suspension. The determination of the induced dipole moment and from it the particle polarizability might essentially depend on the mechanism of polarization of the suspension, the axis of preferential orientation, etc. Thus there might exist essential differences between dielectric spectroscopy of simple molecules and colloid particles. For the molecules no polarizability-dependent orientation is expected whereas for anisometric colloid particles there is always an electric polarizability anisotropy that might lead together with the permanent dipole moment to considerable orientational polarization of the suspension. On the other hand, the orientation of the permanent molecular dipoles inside the particles and in the medium are the main components of the dielectric (volume, noninterfacial) electric polarization of the particles and the medium. Further the direction of orientation of the interfacial permanent and induced dipoles of the colloid particles might be an important component of the suspension polarization.

Essential differences between the results from electro-optic and dielectric spectroscopies are found in the few complex experimental studies on the same type of colloid systems [50,63,64]. The ideal case is when the measurements are made on the same object by similar experimental conditions. Unfortunately such measurements are practically absent in the literature. Generally dielectric measurements are made at higher particle concentrations and in higher frequency ranges. When the frequency ranges overlap, there seems to be a fairly good correspondence in the form of the dispersion curve, i.e., in the two (dielectric and electro-optic) dispersions [1,65] observed.

However this correspondence is practically lost when the effect of ionic strength, particle electric charge, and particle concentration on the critical frequency and the magnitude of the dispersions are studied [63]. Thus the magnitude of the electro-optic dispersion (EOD) seems to be smaller than that of the DD, the relaxation times of EOD are smaller and depend on the ζ-potential, whereas of the DD are bigger and do not depend on the ζ-potential, the magnitude of the dielectric constant at high ζ-potentials rises with the increase in the ionic strength, while the Kerr constant is decreasing, etc. The explanation for the differences certainly requires more experiments, at the same experimental conditions and with other anisometric particles and different ways of changing the electric charge of the particle. It deserves mention that as it could be expected, Delgado [63] found considerable deformation of the DD curves, which he attributes to electrode polarization, whereas the EOD curves are practically not deformed. Here, as it was pointed out in the above text, probably the existence and the effect of particle orientation on the DD interpretation needs some further attention. It is also not clear whether the interpretation of the higher frequency DD with transversal particle polarizability is the only reasonable one. The MW polarization might be a good alternative.

In this connection very interesting are the recent time-domain DD studies of the electric dipole moments of tobacco mosaic virus (TMV) performed by Ermolina et al. [66]. The molecular structure and the properties of TMV are comparatively very well studied. The TMV particles are strongly elongated with an aspect ratio of 20 and a length of about 300 nm. They showed that in the frequency range 100 kHz to 70 MHz a well-expressed DD is observed. The authors correctly discuss the mechanism of the electric polarizabilities in terms of the CHDIDM (counterion polarization) and MWIDM polarizations. Whereas the experimentally observed relaxation time of the dispersion is much closer to that of the MWIDM polarization model (e.g., for the CHDIDM (counterion polarization) theoretical models of Mandel [31] and Oosawa [67] the relaxation times are about ten times lower), the experimentally observed magnitude of the dispersion is closer to the ECDIDM (counterion polarization) theoretical models.

The uncertainty in the interpretation of the dielectric spectroscopy experiment [66] might be successfully resolved if the numerous EOD data on TMV are considered.

Starting with the first electro-optic study of Lauffer [68], the main experimental EOD results were obtained in the last 50 years, using different electro-optic techniques: electric birefringence [69] and electric light scattering [70–72]. In all experimental studies, a considerable EOD was observed about 10–100 kHz. The experimentally observed relaxation times are close to that of the CHDIDM (counterion polarization) theoretical models. The electro-optic effect was strongly electric charge dependent [73]. The ratio of the magnitude of the experimentally observed electric polarizability [73] to what is left at the isoelectric point (that should be essentially MW polarizability) is about 1–2. Similar is the ratio of the magnitude of the experimentally observed EOD to the value obtained by relaxing at higher frequencies. There the principal contribution for TMV might be expected from the MWIDM polarization. So for TMV, the contribution of the MWIDM polarization might be expected to be similar in magnitude to that of the CHDIDM. From here follow two conclusions. First the magnitudes of the EOD and DD cannot be used to distinguish CHDIDM from MWIDM polarization for TMV. Second, the magnitude of the experimentally observed EOD might correspond well to the ECDIDM models. There is a good electro-optical experimental basis to suggest that the experimentally observed DD by Ermolina et al. [66] corresponds to the MWIDM polarization mechanism.

Unfortunately we do not know experimental results from EOD measurements above 1 MHz and DD measurements below 100 kHz for TMV. Thus direct comparisons of the two main IDM polarization mechanisms on the same experimental dispersion curve (EOD or DD) for TMV at present is difficult. To our knowledge the same is the situation with other similar objects, where EOD or DD has been studied experimentally [74]. This makes still more important the correct interpretation of the higher frequency DD reported by Osborn et al. [64].

At present the TMV seems to be a very appropriate object for studying of the two main IDM polarization mechanisms on the same experimental dispersion curve as the magnitudes of the two dispersions are similar. There are objects, like fractal aggregates of Al_2O_3, where at the point of zero charge, achieved by changing pH, the EOD reaches practically zero [75]. So the magnitude of the MWIDM dispersion might be very small and thus difficult to follow experimentally. Some compensation mechanism, e.g., with the volume (dielectric) polarizability, is not impossible in this case. TMV is appropriate for the study of IDM because volume (dielectric) polarization in TMV is expected to be very small (because of the large aspect ratio). However an important complication (linked to the same large aspect ratio) exists, which is the experimentally observed strong interparticle interactions at very low particle concentrations about 0.1% (that is close to the semidiluted state C*) by electro-optic methods. Thus it might be recommended to perform the experimental studies at particle concentrations below 0.1%. Unfortunately this might create some serious sensitivity problems with the experimental measurements of DD. For the EOD measurements no similar problems should be expected.

Another possible problem in the EOD studies of TMV become clear from the "dielectrophoretic" studies of Morgan and Green [76]. There the dependencies of the crossover (transition from accumulation to depletion of TMV particles at the point of the highest electric field strength) and of the threshold field strength (the field at which accumulation starts) on the medium conductivity were mainly studied. The crossover effect appeared in the frequency range 1.5–20 MHz for medium conductivities between 0.17 and 14 mS/m. For anisometric isotropic dielectric particles, this might lead to changes in the direction of orientation of the particles—from longitudinal to transversal orientation. There is no electro-optic evidence for such a transition. Probably the higher conductivity of the virus particle than the medium [76] and its anisotropy do not allow the transition in the direction of the orientation to happen. Thus once more there will be a situation where the dielectric properties of the particle are not the real basis for its translation in the inhomogeneous electric field but its IDM. Unfortunately the "dielectrophoresis" is not sensitive to particle anisotropy and average parameters may lead to good fits of the experiment to theory even for some unrealistic assumptions. On the other hand, it gives an independent approach to the understanding of particle behavior in electric fields and is much nearer to some important applications. Anyway the complex approach to the investigation of particles' IDM by DD, EOD, and dielectrophoresis will be the best way to the understanding of nano- and microparticle polarity and to its most successful applications.

There are some overlooked possibilities for the existence of mechanisms of polarization and the changes of the electrostatics of individual nano- and microstructures (NMS) that might be relevant to the effect of low frequency of modulation, weak electromagnetic fields on biological systems. One is the predominant magnitude of the components of the interfacial, electric-charge dependent, electric polarizability of NMS and the other is the high magnitude of the interfacial permanent dipole moment of NMS. Both are recently suggested and manifest their action in the low frequency, sub-kHz range. At present most of the researchers consider only the interfacial MW polarization mechanism, which is a higher frequency phenomenon. There is a great amount of experimental evidence, mainly electro-optical, on various colloidal and biological particles. In the case of interfacial permanent dipole moment, its great sensitivity to pH, ionic composition, and adsorption of very small quantities of ionic and charged substances may considerably change the electrostatics of individual NMS. This might be relevant to the hormetic responses observed at very small doses that may lead both to useful or harmful effects [77–80].

The deformation of the aqueous–air interface from strongly inhomogeneous electric field [81] to some extent is linked with the "dielectrophoresis." The amplitude of deformation was observed interferometrically and was of the order of micrometers and radius of curvature about 10 mm. For a sharpened electrode of the order of 70 μm and applied electric field up to 100 V, the deformation corresponds to positive "dielectrophoretic" effect and relaxes in the frequency range above 100 kHz. The relaxation effect and the independence of the deformation on the changes of the electric charge of the interface caused by the adsorption of NaLS were measured. It is probable that in this case the polarization responsible for the deformation is purely MW in nature. From the magnitude of the deformation, which has not relaxed at highest frequencies (about 2 MHz), it might be possible to evaluate the relative contribution of the MW polarization and the higher frequency polarization mechanisms, including the dielectric polarization.

The existence of a rising EOD curve when going to higher frequencies for bentonite [82] is interesting. Such rising dispersion curves are very rare at higher frequencies in the EOD studies and are principally impossible in the DD experiment. Similar rising EOD curves are met more often for lower frequencies—below 1 kHz. They are almost always attributed to relaxing transversal orientation due to permanent electric dipole moment and slow electric polarizability of the particles. The rising EOD curve at higher frequencies might be similarly explained with the relaxation of a component of the electric polarizability, leading to transversal (in the case of the platelike bentonite along the symmetry axis) of the particles.

1.10 CONCLUSION

The increased complexity of the electric description of polar nanoparticles in aqueous medium and especially of their IDM has led to considerable misunderstanding, starting from their terminology and their classification. An attempt is made to propose a simplified classification of the DM of nanoparticles. The importance of the IDM is underlined and especially of the charge-dependent component of the induced IDM that is often neglected. A possible coupling of the MW and the charge-dependent components is considered. The causes for the poor correspondence of the values for the IDM communicated by different authors are discussed and the need for a more detailed characterization and description of the samples is highly recommended.

REFERENCES

1. Stoylov, S.P., *Colloid Electro-Optics, Theory, Techniques and Applications*, Academic Press, London, 1991, pp. 1–280.
2. Fredericq, E. and Houssier, C., *Electric Dichroism and Electric Birefringence*, Oxford University Press, London, 1973, pp. 1–219.
3. O'Konski, C.T., Ed., *Molecular Electro-Optics, Part 1, Theory and Method*, Marcel Dekker, New York, 1976, pp. 1–528.
4. O'Konski, C.T., Ed., *Molecular Electro-Optics, Part 2, Applications to Biopolymers*, Marcel Dekker, New York, 1976, pp. 529–868.
5. Jennings, B.R., Ed., *Electro-Optics and Dielectrics of Macromolecules and Colloids*, Plenum Press, New York, 1979, pp. 1–408.
6. Krause, S., Ed., *Molecular Electro-Optics, Electro-Optic Properties of Macromolecules and Colloids in Solution*, Plenum Press, New York, 1981, pp. 1–520.
7. Watanabe, H., Ed., *Dynamic Behavior of Macromolecules, Colloids, Liquid Crystals and Biological Systems by Optical and Electro-Optical Methods*, Hirokawa Publishing Company, Tokyo, 1988, pp. 1–412.
8. Jennings, B.R. and Stoylov, S.P., Eds., *Colloid and Molecular Electro-Optics 1991*, Institute of Physics Publishing, Bristol, 1992, pp. 1–264.
9. Schwarz, G. and Neumann, E., Eds., Electro-Opto '94 (special issue), *Biophys. Chem.*, 58, 1–210, 1996.
10. Petkanchin, I.B., Ed., Special Issue in Honor of N.A. Tolstoy, Eighth International Symposium Colloid and Molecular Electro-Optics, St. Petersburg, June 30–July 4, *Colloids Surf.*, 148, 1–187, 1999.
12. Petkanchin, I.B., Ed., (Special Issue in Honor of C.T. O'Konski,) Ninth International Symposium Colloid and Molecular Electro-Optics, Pamporovo, Bulgaria, October 7–12, 2000, *Colloids Surf.*, 209, 105–338, 2002.
13. Schelly, Z., Ed., http://www.uta.edu/elopto-2003/proceedings.htm; *Tenth International Symposium of Colloidal and Molecular Electrooptics*, New Orleans, LA, March 24–27, 2003.
14. Trusov, A. and Vojtylov, V.V., *Electrooptics and Conductometry of Polydisperse Systems*, CRC Press, Boca Raton, 1993, pp. 1–144.
15. Takashima, S., *Electrical Properties of Biopolymers and Membranes*, Adam Hilger, Bristol, 1989, pp. 1–396.
16. Sheludko, A., *Colloid Chemistry*, Elsevier, Amsterdam, 1966, pp. 1–274.
17. Hunter, R., *Zeta Potential in Colloid Science, Principles and Applications*, Academic Press, London, 1988, pp. 1–386.
18. Anderson, J.L., Effect of nonuniform zeta potential on the particle movement in electric fields, *J.Colloid Interface Sci.*, 105, 45, 1985.
19. Jones, J.F. et al., Charge nonuniformity light scattering, *Colloids Surf. A*, 267, 79–85, 2005.
20. Debye, P., *Polar Molecules*, Dover, New York, 1929.
21. www.sciencecentral.com/articles
22. Stoylov, S.P. et al., S.S. Dukhin, Ed., *Kolloidnaya Electro-Optika Kolloidov*, Naukova Dumka, Kiev, 1977, pp. 1–200 (in Russian).
23. Peterlin, H. and Stuart, H.A., Doppelbrechung, inbesondere kunstliche Doppelbrechung, in *Hand- und Jahrbuch der Chemischen Physik*, Eucken, A. and Wolf, K.L., Eds, Akademische Verlagsgesellschaft, 1943, Bd. 8, Abt. IB, pp. 1–115.
24. Pethig, R., *Dielectric and Electronic Properties of Biological Materials*, John Wiley, New York, 1979, pp. 1–376, Chap. 5.
25. Maxwell, J.C., *A Treatise on Electricity and Magnetism*, Oxford University Press, Oxford, 1873.
26. Wagner, K.W., Erklarung der dielektischen Nachwirkungforgange auf grund Maxwellscher Forstellung, *Ann. Phys.*, 40, 817, 1914.
27. Sillars, R.W., The properties of dielectric containing semi-conducting particles of various shapes, *J. Inst. Electr. Eng.*, 80, 378, 1936.
28. Jones, T.B., *Electromechanics of Particles*, Cambridge University Press, Cambridge, 1995, pp. 1–265.
29. Chelidze, T.L., Derevyanko, A.I., and Kurilenko, O.D., *Electric Spectroscopy of Heterogeneous Systems*, Naukova Dumka, Kiev, 1977 (in Russian).

30. Errera, J., Overbeek, J.Th.G., and Sack, H., Dispersion de l'effer Kerr par certaines solutions colloidales, *J. Chim. Phys.*, 32, 681–704, 1935.
31. Mandel, M., The electric polarization of rod-like charged macromolecules, *Mol. Phys.*, 4, 489, 1961.
32. O'Konski, C.T., Electric properties of macromolecules. V. Theory of ionic polarization in polyelectrolytes, *J. Phys. Chem.*, 564, 605, 1960.
33. Oosawa, F., Counterion fluctuation and dielectric dispersion in linear polyelectrolytes, *Biopolymers*, 9, 677, 1960.
34. Porschke, D., Slow modes of polarization in purple membranes, *Phys. Chem. Chem. Phys.*, 6, 165, 2004.
35. Radeva, Ts., Electric light scattering of ferric oxide particles in sodium carboxymethylcellulose solutions, *J. Colloid Interface Sci.*, 174, 368, 1995.
36. Radeva, Ts. and Petkanchin, I., Electric properties of adsorbed polystyrenesulfonate, Dependence on the polyelectrolyte molecular weight, *J. Colloid Interface Sci.*, 220, 112, 1999.
37. Radeva, Ts., Overcharging of ellipsoidal particles by oppositely charged polyelectrolytes, *Colloids Surf. A*, 209, 219, 2002.
38. Neumann, E. and Katchalsky, A., Long-lived conformation changes induced by electric pulses in biopolymers, *Proc. Natl. Acad. Sci. USA*, 4, 993–997, 1972.
39. Sokerov, S. and Weill, G., Polarized fluorescence in an electric field: comparison with the other electro-optical effects for rod-like fragments of DNA and the problem of the saturation of the induced moment in polyelectrolytes, *Biophys. Chem.*, 10, 161–171, 1979.
40. Sokerov, S. and Weill, G., Polarized fluorescence in an electric field: steady-state and transient values for the fourth moment of the orientation function at arbitrary field, *Biophys. Chem.*, 10, 41–46, 1979.
41. Sokerov, S. and Weill, G., Polarized fluorescence in an electric field: theoretical calculation at arbitrary fields. Experimental comparison with other electro-optical effects. Saturation of the induced dipole moment in polyelectrolytes, in *Electro-Optics and Dielectrics of Macromolecules and Colloids*, Plenum Press, New York, 1979, pp. 109–119.
42. Diekmann, S. et al., Electric properties and structure of DNA restriction fragments from measurements of the electric dichroism, *Biophys. Chem.*, 15, 157–167, 1982.
43. Yoshioka, K., Orientation function of the electric birefringence and dichroism of rod-like polyelectrolytes on the basis of the saturating dipole mechanism, *J. Chem. Phys.*, 79, 3482–3486, 1983.
44. Morita, A. and Watanabe, H., Dynamics behaviour of the induced polarization arising from the motion of counterions bound on a rod-like polyion, *Macromolecules*, 17, 1545–1550, 1984.
45. Atig, J.A., Wesenberg, G.E., and Vaughan, W.E., Dielectric behaviour of polyelectrolytes. IV. Electric polarizability of rigid biopolymers in electric fields, *Biophys. Chem.*, 24, 221–234, 1986.
46. Stellwagen, N.C., Electric birefringence of restriction enzyme fragments of DNA: optical factor and electric polarizability as a function of molecular weight, *Biopolymers*, 20, 399–434, 1981.
47. Shah, M.J., Electric birefringence of bentonite. II. An extension of saturation birefringence theory, *J. Phys. Chem.*, 67, 2215–2219, 1963.
48. Shah, M.J. and Hart, C.M., Investigations of the electro-optical birefringence of polydisperse bentonite suspensions, *IBM J. Res. Dev.*, 7, 44–57, 1963.
49. Lyklema, J., *Fundamentals of Interface and Colloid Science*, Vol. II, Academic Press, New York, 1995, Chap. 2.
50. Foster, K., Osborn, A.J., and Wolfe, M.S., Electric birefringence of poly(tetrafluoroethylene) whiskers, *J. Phys. Chem.*, 96, 5483–5487, 1992.
51. Hoffmann, H., Kramer, U., and Thurn, H., Anomalous behaviour of micellar solutions in electric birefringence measurements, *J. Phys. Chem.*, 94, 2027–2033, 1990.
52. Peikov, V.Ts., Electro-Optic Studies of Semidiluted and Concentrated Disperse Systems, Ph.D. thesis, Academy of Sciences, Sofia, 1993.
53. Keizer, A.D., Van der Drift, W.P., and Overbeek, J.T.G., Electrophoresis of randomly oriented cylindrical particles, *Biophys. Chem.*, 3, 107–108, 1975.
54. Van Der Drift, W.P.J.T., De Keizer, A., and Overbeek, J.Th.G., Electrophoretic mobility of a cylinder with high surface charge density, *J. Colloid Interface Sci.*, 71, 67–78, 1979.
55. Petkanchin, I. and Stoylov, S., Microelectrophoresis of palygorskite colloidal particles, *God. Sof. Univ.*, 61, 233–241, 1966/1967.

56. Wang, G. and Porschke, D., Dipole reversal in bacteriorhodopsin and separation of dipole components, *J. Phys. Chem. B*, 107, 4632–4638, 2003.
57. Buleva, M.V., Electric Moments of Colloid Particles and the Structure of the Double Electric Layer, Ph.D. thesis, Academy of Sciences, Sofia, 1981.
58. Kirkwood, J.G. and Schumaker, J.B., The influence of dipole moment fluctuations on the dielectric increment of protein in solution, *Proc. Natl. Acad. Sci. USA*, 38, 855–862, 1952.
59. Petrov, A.G., *The Lyotropic State of Matter. Molecular Physics and Living Matter Physics*, Gordon and Breach, NY, 1999, Chaps. 5–7.
60. Petrov, A.G., Flexoelectric model for active transport, in: *Physical and Chemical Bases of Biological Information Transfer*, Plenum Press, New York, 1975, p. 111.
61. Yamaoka, K. and Matsuda, K., Reversing-pulse electric birefringence of poly(*p*-styrenesulfonate) in aqueous solutions: effect of molecular weight and concentration on anomalous signal patterns arising from last- and slow-induced ionic dipole moments, *J. Phys. Chem.*, 89, 2779–2785, 1985.
62. Yamaoka, K. and Ueda, K., Reversing-pulse electric birefringence study of helical poly(α-L-glutamic acid) in *N,N*-dimethylformamide with emphasis on a new data analysis for the polydisperse system, *J. Phys. Chem.*, 86, 406–413, 1982.
63. Delgado, A.V., Frequency dependence of the dielectric and electro-optic response in suspensions of charged rod-like colloidal particles, *Colloids Surf. A*, 140, 157–167, 1998.
64. Osborn, A., Foster, K., and Wolfe, M., Dielectric properties of polytetrafluoroethylene whiskers, *J. Phys. Chem.*, 95, 5915–5918, 1991.
65. Stoylov, S.P., Electric polarizability of polyelectrolyte and colloid media: dielectric vs. electro-optic approach, *Biophys. Chem.*, 58, 165–172, 1996.
66. Ermolina, I. et al., Dielectric spectroscopy of tobacco mosaic virus, *Biochim. Biophys. Acta*, 622, 57, 2003.
67. Oosawa, F., Counterion fluctuation and dielectric dispersion in linear polyelectrolytes, *Biopolymers*, 9, 677, 1960.
68. Lauffer, M.A., The electro-optic effect in certain viruses, *J. Am. Chem. Soc.*, 61, 2412, 1939.
69. O'Konski, C.T. and Haltner, A.J., A study of electric polarization in polyelectrolyte solutions by means of electric birefringence, *J. Am. Chem. Soc.*, 79, 5634–5649, 1957.
70. Wippler, C., Diffusion de la Lumiere par des Solutions Macromoleculaire. II. Etude Experimentale de L'Effet d'un Champ Electrique sur les Particules Rigides, *J. Chim. Phys.*, 53, 328, 1956.
71. Stoylov, S.P., Scheludko, A., and Chernev, R., Experimental investigation on the variation in the intensity of the scattered light by colloid solutions subjected to an electric field, *God. SU HF*, 58, 113–128, 1963/64.
72. Jennings, B.R. and Jerrard, H.G., Light scattering study of tobacco mosaic virus solutions when subjected to an electric field, *J. Chem. Phys.*, 44, 1291–1296, 1966.
73. Stoylov, S.P. and Petkanchin, I., Electric polarizability of the tobacco mosaic virus at the isoelectric point, *God. Sof. Uiv.*, 61, 227–231, 1966/67.
74. Stoylov, S.P. et al., Electric birefringence polytetrafluorethylene particles in agarose gels, *Biophys. Chem.*, 58, 157–164, 1996.
75. Buleva, M., Ph.D. thesis, 1981.
76. Morgan, H. and Green, N., Dielectrophoretic manipulation of rod-shaped viral particles, *J. Electrostatics*, 42, 279–293, 1997.
77. S.P. Stoylov and Petkanchin, I.B., Surface charge on the solid/water interface—heterogeneity and dynamics: an electro-optic approach, *Curr. Top. Colloid Interface Sci.*, 4, 1–13, 2001.
78. E.J. Calabrese and Baldwin, L.A., Toxicology rethinks its central belief, *Nature*, 241, 691–692, 2003.
79. A little poison can be good for you, *Fortune*, June 9, 2003, pp. 24–25.
80. N. Boyce, Is there a tonic in the toxin? *U.S. News and World Report*, October 18, 2004, pp. 74–75.
81. Buleva, M. and Stoylov, S.P., Investigation of the deformation of liquid surface in nonhomogeneous electric field, *Izv. Khim.*, 12, 4066, 1977.
82. Radeva, Ts. and Stoylov, S.P., Unusual dispersion of electro-optically determined electric polarizability of bentonite dispersed particles in the range 10 kHz–1 MHz, *Colloid Polymer Sci.*, 263, 301–505, 1985.

2 Thin Double Layer Theory of the Wide-Frequency Range Dispersion of the Polarizability of Nonconducting Spheroidal Particles

V.N. Shilov, Yu. B. Borkovskaja, and S.N. Budankova

CONTENTS

2.1 INTRODUCTION

The dispersed particle in the electrolyte solution acquires an induced dipole moment, $\vec{d}$, under the action of an external electric field, $\vec{E}$. The tensor of polarizability, $\hat{\gamma}$, relating the values of $\vec{d}$ and $\vec{E}$ ($\vec{d} = 4\pi\varepsilon_m\hat{\gamma}\,\vec{E}$, where ε_m is the dielectric permittivity of the solution), is an important parameter, determining the behavior of suspensions in an alternating electric field, in particular, the orientation of suspension particles.

The tensor of polarizability for a particle with axial symmetry has two independent components, $\gamma_\parallel$ and $\gamma_\perp$, corresponding to the longitudinal and transverse orientations of the revolution axis, respectively, with respect to the direction of the external electric field. The orientational energy, U, of such particle is determined by the following expression [1]:

$$U = \frac{4\pi\varepsilon_m}{2}\mathrm{Re}(\gamma_\parallel - \gamma_\perp)E^2\cos^2\theta \qquad (2.1)$$

where θ is the angle between the axis of revolution and the direction of the external electric field. According to the Boltzmann distribution, the angular dependence of U results in the appearance of the preferential orientation of polarized particles along the direction of the electric field.

The orientation of polarized particles manifests itself in the change of optical characteristics of the suspension under the action of the electric field, i.e., in electro-optical effects. Methods based on the measurement of these effects are widely used to study particles and macromolecules in suspensions [1–4]. It is important that the electro-optical effects are relatively easy to measure in a wide range of frequencies of the orienting electric field from

10^2 to 10^7 Hz [5]. Therefore, the investigation of the frequency dependence of electro-optic effects makes it possible to obtain information about the various mechanisms contributing to the polarization of the particles and having relaxation times within the range from 10^{-3} to 10^{-8} s. An important requirement to the theory of electro-optic phenomena, following from the Equation 2.1, is the necessity to consider the anisotropy of particle's properties, responsible for the anisotropy of its polarizability. Such a requirement makes it impossible to use the simplest and the most convenient model of spherical particles with isotropic volume and surface characteristics. Unfortunately, the abandonment of this simple model invalidates the direct employment of the method of separation of variables to the solution of the boundary-value problem of the surface contribution to particle polarizability in an alternating applied field, even for the simplest model of anisotropy as a spheroidal shape of the particle. Some remedies for such a difficulty will be considered below.

For the particle in electrolyte solutions, there are two characteristic mechanisms of electric polarization having relaxation times in the above-mentioned frequency range, which are both sensitive to the parameters of particle's electric double layer (EDL). The first mechanism is the low-frequency dispersion of EDL polarization [6], caused by the influence of the field-induced electrolyte concentration variations on the electric currents in the EDL. This mechanism, which may be characterized as the concentration polarization of the double layer, or volume diffusion mechanism of the double layer polarization is responsible for the lowest frequency range dispersion of the colloidal particle polarization in the electrolyte solution, so-called α-dispersion. The second mechanism [7] is the well-known Maxwell–Wagner dispersion of polarization, caused by the formation of the field-induced free (ionic) charge distributions near the surfaces that separate the phases with different local conductivities and dielectric constants.

Concentration fields responsible for the volume diffusion mechanism of double layer polarization are formed in the course of the electrolyte diffusion in the solution surrounding the particle, at distances closer to the characteristic linear size of the particle, l. Accordingly, the duration of such a process is determined by the time, τ_α, required for the electrolyte to diffuse through the distance corresponding to the characteristic linear size of a particle:

$$\tau_\alpha \sim \frac{2D}{l^2} \tag{2.2}$$

where D is the diffusivity of ions.

The Maxwell–Wagner dispersion of polarization may be manifested in its uncombined state within the frequency range given by the next inequality:

$$1/\tau_\alpha \ll \omega \ll 1/\tau_\varepsilon$$

when neighbor relaxation processes, namely, low-frequency dispersion of diffusion-controlled EDL polarization (characteristic time $1/\tau_\alpha$) and high-frequency dispersion of dielectric polarization of constituent phases (characteristic time $1/\tau_\varepsilon$) are not manifested. Here, ω is the cyclic frequency.

The field-induced ionic charge, responsible for Maxwell–Wagner mechanism of the interphase polarization, is formed in the diffused part of EDL near the particle's surface, when the ratio of the dielectric constants of particles to media, $\varepsilon_p/\varepsilon_m$, differs from the ratio of their conductivities, K_p/K_m and hence the continuity of the electric current, crossing the particle's surface, occurs to be incompatible with the continuity of the normal components of the electrostatic induction. The relaxation time, τ_{MW}, of Maxwell–Wagner polarization is the time necessary for the formation of field-induced ionic charge and may be represented approximately as the duration of the screening process in the electrolyte solution:

$$\tau_{MW} \sim \frac{K_m}{\varepsilon_m} \sim \frac{1}{\kappa^2 D} \tag{2.3}$$

where κ is the reciprocal of Debye screening length.

From the comparison of the above estimations of τ_α and τ_{MW}, it is evident that these characteristic times differ essentially and hence, low-frequency double layer polarization and Maxwell–Wagner polarization may be observed separately under the condition of a thin double layer:

$$\kappa l \gg 1 \tag{2.4}$$

Below, we will consider the Debye-type frequency dependence for the complex components, $\gamma_\parallel^*$ and $\gamma_\perp^*$, of polarizability of the spheroidal particle for both Maxwell–Wagner dispersion and low-frequency, α-dispersion, neglecting in that way, some deviations from Debye-type relaxation, characteristic of volume diffusion mechanism of double layer polarization [8]:

$$\gamma_{\parallel,\perp}^* = \gamma_{\parallel,\perp}^{(\infty)} + \frac{\gamma_{\parallel,\perp}^{(i)} - \gamma_{\parallel,\perp}^{(\infty)}}{1 + j\omega\tau_{MW_{\parallel,\perp}}} + \frac{\gamma_{\parallel,\perp}^{(0)} - \gamma_{\parallel,\perp}^{(i)}}{1 + j\omega\tau_{\alpha_{\parallel,\perp}}} \tag{2.5}$$

Here, $\gamma_{\parallel,\perp}^{(0)}$ is the low-frequency limit of α-dispersion, $\gamma_{\parallel,\perp}^{(i)}$ is simultaneously the high-frequency limit of α-dispersion and low-frequency limit of Maxwell–Wagner dispersion, and $\gamma_{\parallel,\perp}^{(\infty)}$ is the high-frequency limit of Maxwell–Wagner dispersion.

2.2 THEORY

As for the theory of both the mechanisms of the dispersion of a particle's polarization, it is well defined (see, Refs. [9,10]) for a homogeneously charged spherical particle. The analytical theory of the volume diffusion mechanism of the double layer polarization was developed first for a nonconducting sphere [11] with a thin EDL and was used as a basis for the theory of low-frequency dielectric dispersion of the suspensions in the electrolyte solutions [12]. The numerical theory of action of both the mechanisms—volume-diffusion controlled dispersion of EDL polarization and Maxwell–Wagner dispersion was developed by DeLacey and White [13] on the basis of the standard electrokinetic model without any restrictions on the magnitudes of the double layer thickness and ς-potential.

The theory of Maxwell–Wagner polarization was first developed for the suspension of spherical particles, characterized by the dielectric constants and the volume conductivities of the contacting phases without any specific surface parameters of the phase boundary. Maxwell–Wagner theory was generalized for the case of ellipsoidal particles by Fricke [14] and Sillars [15]. In accordance with this theory, both, $\gamma_\parallel$ and $\gamma_\perp$, components of the polarizability of spheroid, one for aligned orientation, when the revolution axis of the spheroid is parallel to the applied electric field (lower index "$\parallel$") and another for transverse orientation, when the revolution axis of the spheroid is perpendicular to the applied electric field (lower index "$\perp$"), are:

$$\gamma_{MW_{\parallel,\perp}}^* = \gamma_{\parallel,\perp}^{(\infty)} + \frac{\gamma_{\parallel,\perp}^{(i)} - \gamma_{\parallel,\perp}^{(\infty)}}{1 + j\omega\tau_{MW_{\parallel,\perp}}} \tag{2.6}$$

where

$$\gamma_{\parallel,\perp}^{(\infty)} = \frac{a^2 b}{3} \frac{\varepsilon_p - \varepsilon_m}{\varepsilon_m + (\varepsilon_p - \varepsilon_m)A_{\parallel,\perp}} \tag{2.7}$$

$$\gamma_{\parallel,\perp}^{(i)} = \frac{a^2 b}{3} \frac{K_p - K_m}{K_m + (K_p - K_m)A_{\parallel,\perp}} \tag{2.8}$$

and

$$\tau_{MW\parallel,\perp} = \frac{\varepsilon_m + (\varepsilon_P - \varepsilon_m)A_{\parallel,\perp}}{K_m + (K_P - K_m)A_{\parallel,\perp}} \tag{2.9}$$

where a is half the distance between poles and b is half the diameter of the equator.

The values, ε_p, ε_m and K_p, K_m represent dielectric constants and conductivities of particle (lower index "p") and dispersion media (lower index "m") respectively;

$A_{\parallel}$ and $A_{\perp} = \frac{1}{2}(1 - A_{\parallel})$ are the known (e.g., according to Ref. [16]) factors of depolarization of spheroid.

The expressions for $A_{\parallel}$ in the case of prolate particles (left side of the lines) and oblate particles (right side of the lines) are:

$$A_{\parallel} = \frac{ab^2}{h^3}\left[\operatorname{arctanh}\frac{h}{a} - \frac{h}{a}\right], \quad A_{\parallel} = \frac{ab^2}{h^3}\left[\frac{h}{a} - \arctan\frac{h}{a}\right] \tag{2.10}$$

$$h = \sqrt{a^2 - b^2} \quad h = \sqrt{b^2 - a^2} \tag{2.11}$$

For the solid particles in a water electrolyte solution, it is the typical situation when both the volume conductivity and the dielectric constant of the particle occur to be of negligibly small values when compared to the corresponding parameters of the water dispersion medium. For such a situation, as it is evident from Equation 2.7 and Equation 2.8 that the low- and high-frequency limits ($\gamma_{\parallel,\perp}^{(i)}$ and $\gamma_{\parallel,\perp}^{(\infty)}$) are very close values and hence, Maxwell–Wagner dispersion of polarizability is only feebly marked affinity. Nevertheless, the distinct frequency dependence of the Maxwell–Wagner polarization of nonconducting, nonpolar particle, immersed in the water electrolyte solution may arise as a manifestation of the surface conductivity conditioned by the ion transfer in the electric double layer.

The influence of the surface conductivity, K^σ, on the polarizability was first considered by O'Konski [17]. For spherical particles, the role of the surface conductivity was exactly reduced to the replacement of the particle's volume conductivity, K_p, in Equation 2.8 and Equation 2.9 by an effective value, K_p^{eff}

$$K_p^{eff} = K_p + K_p^s = K_p + \frac{2K^\sigma}{a} \tag{2.12}$$

where the factor $K_p^s = 2K^\sigma/a$ represents the surface contribution to the effective conductivity of the particle.

The surface contribution to the effective conductivity of the spheroidal particle is anisotropic and the effective volume conductivity of the spheroid, $K_{p\parallel,\perp}$, will be different, which depends on the particle's orientation:

$$K_p \to K_{p\parallel,\perp} = K_p + f_{\parallel,\perp}K^\sigma \tag{2.13}$$

where coefficients, $f_\parallel$ and $f_\perp$, have dimensions of reciprocal length and correspond to the orientation of particle's axes of revolution, parallel or perpendicular to the applied field, respectively. O'Konski suggested the procedure to assess the effective conductivity for the ellipsoidal particle. This procedure consists of the calculation of the total surface current, intersecting the contour of the particle's equatorial section, which is perpendicular to the direction of the applied electric field.

O'Konski's procedure provides a good approximation of the surface contribution to the effective conductivity in the most important limiting cases of the spheroid's shape, namely: For almost spherical particle:

$$f_{\parallel,\perp} \simeq 2/b \tag{2.14}$$

For strongly prolate (needle-shaped) spheroid:

$$f_\parallel \simeq 2/b \tag{2.15}$$

$$f_\perp \simeq 4/\pi b \tag{2.16}$$

And for a thin oblate (disk-like) spheroid with transversely oriented axes of revolution:

$$f_\perp \simeq 4/\pi a \tag{2.17}$$

The bounds of the applicability of these O'Konski's approximations of surface contribution to the polarizability of the spheroid were not yet made.

An independent approach to this problem was presented in Refs. [18–20] on the basis of Fourier method of separation of variables in the solution of the Laplace equation for the field-induced potential distribution. The main difficulty of the problem consists of the fact that the contribution of surface conductivity to the condition of the current balance at the surface cannot be taken into account, with only one spherical harmonic in spheroidal coordinates, but, contrary to the case of zero surface conductivity, requires an infinite Fourier series. An approximate procedure suggested by the authors of Refs. [18–20] makes it possible to obtain analytically the farthest reaching term (FRT) of the series expansion in spherical harmonics. This single term, which determines the dipolar coefficient of the particle, constitutes a very good approximation of the exact solution, as has been shown numerically. This approximation, which will be named below as FRT approach, was first applied by Eremova and Shilov [18,19] to the calculation of the polarizability of the spheroidal particle with a thin double layer for the cases of low-frequency limits of volume-diffusion controlled dispersion and low-frequency limits of Maxwell–Wagner dispersion. O'Brien and Rowlands [21] extend the applicability of the FRT approach to the analysis of the frequency dependence of the polarizability of spheroids for the case of Maxwell–Wagner dispersion. Later [22], it was shown that the employment of the FRT approach may be reduced to the derivation of analytical expressions for the surface contribution to the effective conductivity, K_{pef}. Applying the FRT procedure, fully similar to that used in Ref. [21], it is easy to show that the analytical expressions for K_{eff} from Ref. [22] determine not only for low-frequency limits of the components of polarizability, expressed by the Equation 2.8, but also for the relaxation time of Maxwell–Wagner dispersion, given by Equation 2.9. In this way, the FRT approach of O'Konski's factors, $f_{\parallel,\perp}$, for arbitrary shape and orientation of spheroid particle was obtained as:

For prolate particles:

$$f_\parallel = \frac{3a}{2bh}\left[\frac{a^2 - 2b^2}{h^2}\arctan\frac{h}{b} + \frac{b}{h}\right] \tag{2.18}$$

$$f_\perp = \frac{3a}{2bh}\left[\frac{a^2}{2h^2}\operatorname{arc\,cot}\frac{b}{h} + \frac{b(a^2 - 2b^2)}{2a^2h}\right] \tag{2.19}$$

For oblate particles:

$$f_\| = \frac{3a}{2bh}\left[\frac{2b^2 - a^2}{h^2}\ln\frac{h+b}{a} - \frac{b}{h}\right] \tag{2.20}$$

$$f_\perp = \frac{3h}{2ba}\left[\frac{b(2b^2 - a^2)}{2a^2h} - \frac{a^4}{2h^4}\ln\frac{b+h}{a}\right] \tag{2.21}$$

Contrary to the frequency range of Maxwell–Wagner dispersion, the surface current at low-frequency range, $\omega \le 1/\tau_\alpha$, is forced not only by the gradients of the electric potential, but also by the gradients of the electrolyte concentration (see, for example, Ref. [9]) that are formed in the liquid surrounding the particle, due to the difference in the transport numbers of the counterions between the double layer and the volume electrolyte solution. Slow process of the formation of field-induced electrolyte concentration variations gives rise to the low-frequency, volume-diffusion controlled α-dispersion of the polarizability of charged particle in the electrolyte solutions [9].

The distribution of this electrolyte concentration variation is governed by the diffusion equation. But the variables in the diffusion equation (contrary to the Laplace equation governing the potential distribution in Maxwell–Wagner frequency range) are not separable in the spheroidal coordinates. The last circumstance determines an essential obstacle for the theory of the frequency dependence of the polarizability of the spheroidal particle, in addition to the above discussed problems of the consideration of the surface conductivity contribution to the current balance near the particle's surface. However, at low-frequency limits of the α-dispersion, when $\omega \ll \omega_\alpha$, the diffusion equation is reduced to the Laplace equation. So, at such low frequencies, both distributions of the field-induced electrolyte concentration variation, δC and electric potential, $\delta\varphi$, satisfy Laplace equation and it is convenient (see Ref. [23]) to express them both through the distributions of the electrochemical potentials of cations, $\delta\mu^+$ and anions, $\delta\overline{\mu}$:

$$\frac{\delta C}{C_0} = \frac{1}{2kT}(\delta\mu^+ + \delta\mu^-), \quad \delta\varphi = \frac{1}{2e}(\delta\mu^+ - \delta\mu^-) \tag{2.22}$$

where C_0 is the equilibrium electrolyte concentration outside the double layer, k is the Boltzmann constant, T is the absolute temperature, and e is the proton's charge.

The potentials, $\delta\mu^+$ and $\delta\mu^-$, are independent of one another (as a weak influence of the factors, responsible for the interdiffusion is neglected) and satisfy the Laplace equation. Therefore, the electrochemical potentials of cations and anions in the binary electrolyte solution can be determined by solving the two independent problems, one for each ion type. Everyone of these problems are fully similar to the well-known problems of the distribution of the electric potential that forms around a conducting particle in a conductive medium under the action of an applied electric field with the only peculiarity—each of them are characterized by an appropriate particle-to-medium conductivity ratio. Similarity leads to the expressions for the polarizabilities, $\gamma_{\|,\perp}^{+(0)}$ and $\gamma_{\|,\perp}^{-(0)}$, corresponding to the field-induced distributions of the electrochemical potentials of cations, μ^+ and anions, μ^-, at low-frequency limits and having the same form as Equation 2.8:

$$\gamma_{\|,\perp}^{\pm(0)} = \frac{ab^2}{3}\frac{K_{p\|,\perp}^{\pm} - K_m^{\pm}}{K_m^{\pm} + (K_{p\|,\perp}^{\pm} - K_m^{\pm})A_{\|,\perp}} \tag{2.23}$$

In these expressions, $K_m^{\pm} = t_m^{\pm} K_m$, $K_{p\parallel,\perp}^{\pm} = t_m^{\pm} K_p + f_{\parallel,\perp} K^{\sigma\pm}$, and $K^{\sigma\pm} = t^{\sigma\pm} K^{\sigma}$ are the contributions of ions of one sign, $K_m^{\pm}$ to the conductivity of the volume solution, K_m ($t_m^{\pm}$ represents the ion transport numbers in solution); $K_{p\parallel,\perp}^{\pm}$ to the effective conductivity of the particle, $K_{p\parallel,\perp}$; $t_p^{\pm} K_p$ to the isotropic conductivity of particle's volume, K_p ($t_m^{\pm}$ is the ion transport numbers inside the particle); $K^{\sigma\pm}$ to the surface conductivity, K^{σ} ($t^{\sigma\pm}$ is the ion transport numbers in the electric double layer).

The superposition of $\gamma_{\parallel,\perp}^{+(0)}$ and $\gamma_{\parallel,\perp}^{-(0)}$ leads to the next expression for low-frequency limits of α-dispersion of electric polarizability:

$$\gamma_{\parallel,\perp}^{\alpha(0)} = \frac{ab^2}{3} \left(\frac{K_{p\parallel,\perp}^{+} - K_m^{+}}{K_m^{+} + (K_{p\parallel,\perp}^{+} - K_m^{+})A_{\parallel,\perp}} + \frac{K_{p\parallel,\perp}^{-} - K_m^{-}}{K_m^{-} + (K_{p\parallel,\perp}^{-} - K_m^{-})A_{\parallel,\perp}} \right) \tag{2.24}$$

In the following, we restricted our consideration to the typical system of nonconducting particles, $K_p = 0$, with the surface conductivity resulting mainly from the contribution of coions. Supposing, for definiteness, that the particles are charged negatively and hence that, $t^{\sigma+} \gg t^{\sigma-}$, we may have to accept for such a system:

$$K_{p\parallel,\perp}^{+} = f_{\parallel,\perp} K^{\sigma}, \quad K_{p\parallel,\perp}^{\sigma-} = 0 \tag{2.25}$$

For the case of symmetrical electrolyte solution, where the diffusion coefficients of the two types of ions are the same, the contributions of the ions of both signs to the volume conductivity are the same too, $K_m^{+} = K_m^{-} = \frac{1}{2}K_m$, Equation 2.23 for $\gamma_{\parallel}^{+(0)}$, $\gamma_{\parallel,\perp}^{+(0)}$ and $\gamma_{\parallel,\perp}^{-(0)}$, and Equation 2.24 for low-frequency limits of the α-dispersion of the particle's polarizability may be rewritten as:

$$\gamma_{\parallel,\perp}^{+(0)} = \frac{ab^2}{3} \frac{4Du_{\parallel,\perp} - 1}{1 + (4Du_{\parallel,\perp} - 1)A_{\parallel,\perp}}, \quad \gamma_{\parallel,\perp}^{-(0)} = -\frac{ab^2}{3} \frac{1}{1 - A_{\parallel,\perp}} \tag{2.26}$$

$$\gamma_{\parallel,\perp}^{(0)} = \gamma_{\parallel,\perp}^{+(0)} - \gamma_{\parallel,\perp}^{-(0)} = \frac{ab^2}{3} \frac{2(1 - 2A_{\parallel,\perp})Du_{\parallel,\perp} - (1 - A_{\parallel,\perp})}{3(1 - A_{\parallel,\perp})[1 - A_{\parallel,\perp}(1 - 4Du_{\parallel,\perp})]} \tag{2.27}$$

Here

$$Du_{\parallel,\perp} = f_{\parallel,\perp} K^{\sigma}/2K_m \tag{2.28}$$

represents the generalization of the Dukhin number for the case of the spheroidal particle. Dukhin number was accepted [24] as the criterion for the role of the double-layer polarization in the electrokinetic phenomena and its well-known expression for the spherical particle is $Du_{\parallel} = Du_{\perp} = K^{\sigma}/K_m a$. As it is follows from Equation 2.8 and Equation 2.9, the value of the generalized Dukhin number, together with the depolarization factors, $A_{\parallel}$ and $A_{\perp} = 1 - A_{\parallel}$, determines also the high-frequency limits of α-dispersion (or, what is the same—the low-frequency limits of the Maxwell–Wagner dispersion) of the particle's polarizability and the relaxation time of the Maxwell–Wagner dispersion:

$$\gamma_{\parallel,\perp}^{(i)} = \frac{a^2 b}{3} \frac{2Du_{\parallel,\perp} - 1}{1 + (2Du_{\parallel,\perp} - 1)A_{\parallel,\perp}} \tag{2.29}$$

and

$$\tau_{\parallel,\perp}^{MW} = \frac{\varepsilon_m}{K_m} \frac{1 + (\varepsilon_p/\varepsilon_m - 1)A_{\parallel,\perp}}{1 + (2Du_{\parallel,\perp} - 1)A_{\parallel,\perp}} \tag{2.30}$$

The problem concerning diffusion-controlled characteristic time of α-dispersion of the polarizability of the spheroidal particle is strongly difficult, because of the impossibility of the separation of variables for the diffusion equation in the spheroidal coordinates. To overcome this difficulty, we use the method, elaborated in Ref. [25] and consisting of the correlation between the two approaches, to the calculations of low-frequency limits of α-dispersion of the dielectric permittivity of the suspension.

The first approach is based on the calculation of the macroscopic field in diluted suspension as a superposition of a long-range (dipole) asymptotic of the fields, created by polarized particles. The application of this approach leads to the well-known expression for the particles' contribution, $\delta\varepsilon^*$, to the complex dielectric permittivity in the case of suspension of monodisperse, like-oriented, and equiform spheroids having a form:

$$\delta\varepsilon_{\parallel,\perp}^* = 4\pi n \varepsilon_m^* \gamma_{\parallel,\perp}^*$$

where n is the number of particles per unit suspension's volume,

$$\varepsilon_m^* = \varepsilon_m - j\frac{K_m}{\omega} \tag{2.31}$$

where ε_m^* is the complex dielectric permittivity of the dispersion medium.
In the low-frequency range, characterized by the inequality

$$\omega \ll \frac{K_m}{\varepsilon_m}$$

the absolute value of the second term in the right-hand side of Equation 2.28 exceeds many times the first term and due to this circumstance, the real part of the complex dielectric permittivity, $\delta\varepsilon_{\parallel,\perp}^*$, including the low-frequency limits of the α-dispersion may be approximately represented as

$$\delta\varepsilon_{\parallel,\perp}^*|_{\omega \ll K_m/\varepsilon_m} = 4\pi n \varepsilon_m \left[\frac{K_m}{\varepsilon_m \omega} \operatorname{Im}\gamma_{\parallel,\perp}^* \right]$$

Substituting the Debye-type frequency dependence of polarizability, $\gamma^*_{\parallel,\perp}$, given by Equation 2.5 to the above Equation 2.30, and considering the estimations of τ_α and τ_{MW}, given by the Equation 2.2 and Equation 2.3, under the condition (Equation 2.4), leads to the next estimation for low-frequency limits:

$$\delta\varepsilon_{\parallel,\perp}^{(0)} \equiv \lim_{\omega\tau_l \to 0} \delta\varepsilon_{\parallel,\perp}^* = 4\pi n K_m \tau_{\alpha\parallel,\perp}(\gamma_{\parallel,\perp}^{(i)} - \gamma_{\parallel,\perp}^{(0)}) \tag{2.32}$$

The second approach to the searching of $\delta\varepsilon_{\parallel,\perp}^{(0)}$, first used for the case of Maxwell–Wagner dispersion in Ref. [26] and generalized for α-dispersion in Ref. [27], is based on the well-known expression for the field-induced variations of the specific free energy, $\delta W_s^{(0)} \|\perp$, stored by the polarized suspension through its macroscopic dielectric constant, $\delta\varepsilon^{(0)}$:

$$W_{s\|,\perp}^{(0)} = \frac{1}{2}\varepsilon_{\|,\perp}^{(0)}E^2$$

This expression immediately leads to the following relation between the dielectric permeability of the dilute suspension and the contribution, $\delta W^{(0)}$, of the single dispersed particle to the stored free energy:

$$\delta W_{\|,\perp}^{(0)} = \frac{1}{2n}\delta\varepsilon_{\|,\perp}^{(0)}E^2 \tag{2.33}$$

This approach makes it possible to deduce the low-frequency limits of the dielectric dispersion by solving purely the static problem, without any information about low-frequency asymptotic of the imaginary part of the polarizability. The application of this approach, together with the above-mentioned FRT method, allowed us (see Ref. [21]) to derive the analytical expression for low-frequency limits of α-dispersion of the dielectric permeability of suspension.

The energy change, $\delta W^{(0)} \|, \perp$, includes terms related to the electrostatic energy, the energy associated with the double layer, and the energy outside the double layer. This last term, which constitutes the major contribution in the frequency range of α-dispersion of the particle's polarization, corresponds to the free energy change associated with the field-induced concentration change, $\delta C(\vec{r})$, in the electroneutral electrolyte solution around the particle:

$$\delta W_{\|,\perp}^{(0)} = nkTC_0 \int_{\text{particle surface}}^{\infty} \left(\frac{\delta C_{\|,\perp}(\vec{r})}{C_0}\right)^2 dV \tag{2.34}$$

In this expression, C_0 is the equilibrium ($E = 0$) electrolyte concentration outside the double layer.

The concentration change, $\delta C(\vec{r})$, may be expressed through the field-induced changes of the two electrochemical potentials, $\delta\mu^+(\vec{r})$ and $\delta\mu^-(\vec{r})$, considering Equation 2.22. The boundary-value problems of searching the space distributions of the electrochemical potentials under the condition of low-frequency limits of α-dispersion are similar to the pure electrostatic problem (see Ref. [17]). The approximate, analytical solutions of these problems for spheroids of arbitrary shape and orientation with homogeneous surface conductivity were obtained in Ref. [17] by means of FRT approach. The higher degree of accuracy of the analytical results was confirmed in Refs. [18,19] by their comparison with the numerical calculations. The space distributions of the field-induced variations of the electrochemical potentials of the ions, $\delta\mu^\pm(\vec{r})$, and the electrolyte concentration around the spheroidal particle, together with the corresponding analytical expressions for the integral in the right-hand side of Equation 2.34 are represented in Ref. [25]. In particular:

$$\int_{\text{particle surface}}^{\infty} \left(\frac{\delta C(\vec{r})}{C_0}\right)^2 dV = \frac{\varepsilon_m\kappa^2 E^2}{8kT}(\gamma^{(0)+} - \gamma^{(0)-})^2 I \tag{2.35}$$

To apply this equation for the oriented spheroidal particle, one should complete the symbols, $\delta C(\vec{r})$, $\gamma^\pm$, and I, with lower indexes, "$\|$" or "$\perp$", to denote their relations to parallel or perpendicular orientation of the revolution axes to the applied electric field, respectively. The expressions for $\gamma_{\|,\perp}^{(0)\pm}$ are given by Equation 2.26 and the values of the depolarization factors, $A_{\|,\perp}$ and Dukhin number, $Du_{\|,\perp}$ are specified by Equation 2.10, Equation 2.11 and Equation 2.28 respectively.

The factors, $I_{\|,\perp}$, for prolate spheroid have a form:

$$I_{\|} = \frac{3\pi}{5h^6} \left[\begin{array}{l} -a^3b^2 \ln^2 \frac{a+h}{a-h} \\ +2hb^2(a^2+b^2) \end{array} \quad \ln\frac{a+h}{a-h} + 4ah^2 \quad (a^2-2b^2) \right] \tag{2.36}$$

$$I_{\perp} = \frac{3\pi}{20h^6} \left[\begin{array}{l} -ab^4 \ln^2 \frac{a+h}{a-h} \\ +4h(a^4+b^4) \end{array} \quad \ln\frac{a+h}{a-h} + 4ah^2 \quad (3a^2-2b^2) \right] \tag{2.37}$$

and for an oblate spheroid,

$$I_{\|} = \frac{12\pi}{5h^6} \left[\begin{array}{l} -a^3b^2 \mathrm{arccot}^2 \frac{a}{h} \\ +hb^2(a^2+b^2) \end{array} \quad \mathrm{arccot}\frac{a}{h} - ah^2 \quad (h^2+b^2) \right] \tag{2.38}$$

$$I_{\perp} = \frac{3\pi}{5h^6} \left[\begin{array}{l} -ab^4 \mathrm{arccos}^2 \frac{a}{b} \\ +2h(a^4+h^4) \end{array} \quad \mathrm{arccos}\frac{a}{h} - ah^2 \quad (3a^2-2b^2) \right] \tag{2.39}$$

The comparison of Equation 2.32 with the Equation 2.33 and Equation 2.35 and considering Equation 2.36 through Equation 2.39 for the factors, $I_{\|,\perp}$, leads to the next expressions for characteristic time of low-frequency volume-diffusion controlled dispersion (α-dispersion) of polarizability of the oriented spheroidal particle:

$$\tau_{\alpha\|,\perp} = \frac{\chi^2 \left(\gamma_{\|,\perp}^{(0)+} - \gamma_{\|,\perp}^{(0)} \right)^2 I_{\|,\perp}}{16\pi \left(\gamma_{\|,\perp}^{(i)} - \gamma_{\|,\perp}^{(0)} \right)} \frac{\varepsilon_{\mathrm{m}}}{K_{\mathrm{m}}} \tag{2.40}$$

2.3   RESULTS AND DISCUSSION

Equation 2.40 completes the set of the parameters of Equation 2.5 governing the frequency dependence of the polarizability of nonconducting charged spheroidal particle in the electrolyte solution within the wide-frequency range covered in both the volume-diffusion controlled α-dispersion and the Maxwell–Wagner dispersion. The components, $\gamma_{\|}^{(0)}$ and $\gamma_{\perp}^{(0)}$, of the polarizability tensor at the low-frequency limits of α-dispersion are given by Equation 2.27; the values $\gamma_{\|}^{(i)}$ and $\gamma_{\perp}^{(i)}$, that represent simultaneously the high-frequency limits of α-dispersion and low-frequency limits of the Maxwell–Wagner dispersion are given by Equation 2.29; the high-frequency limits of the Maxwell–Wagner dispersion of the components of polarizability tensor, $\gamma_{\|}^{(\infty)}$ and $\gamma_{\perp}^{(\infty)}$ are given by Equation 2.7. The characteristic times for low-frequency, α-dispersion, $\tau_{\alpha_{\|}}$ and $\tau_{\alpha_{\perp}}$, and for Maxwell–Wagner dispersion, $\tau_{\mathrm{MW}_{\|}}$ and $\tau_{\mathrm{MW}_{\perp}}$, are represented by Equation 2.40 and Equation 2.30 respectively.

The influence of a thin double layer polarization on the above-mentioned parameters in accordance with Equation 2.28 is determined by the generalized Dukhin numbers: $Du_{\|}$ for the longitudinal and $Du_{\perp}$ for the transverse orientation of the revolution axis of the particle with respect to the applied electric field. The dependence of the Dukhin numbers on the values and orientation of the particle's axes is given by the parameters, $f_{\|}$ and $f_{\perp}$, having the dimensions of reciprocal length. These parameters, first determined by O'Konski [17] for the limiting cases of slightly and strongly prolate and oblate spheroids (see Equation 2.14 through Equation 2.17) are represented by the dotted lines at Figure 2.1a and Figure 2.1b, respectively, for longitudinal and for transverse orientations of the revolution axes (in length, $2a$).

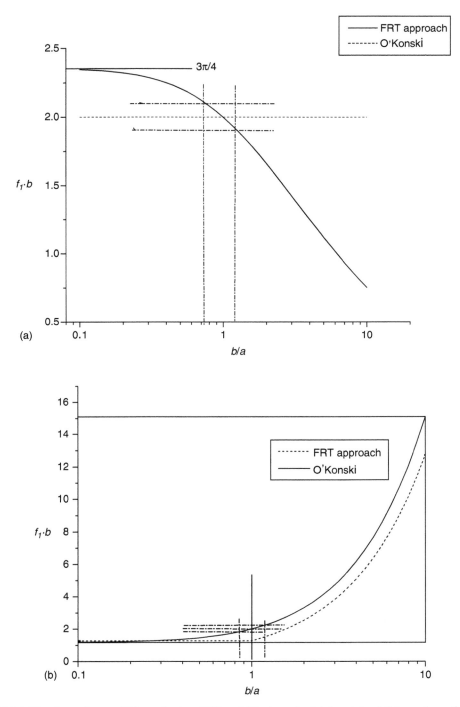

FIGURE 2.1 The dependence of dimensionless O'Konski factors, $f_{\parallel} \cdot b$ (a) and $f_{\perp} \cdot b$ (b), on the ratio b/a of the radius of the equatorial section b to the length of the semiaxes of symmetry of the spheroid a.

The solid lines on these figures represent the results of the calculations of $f_\parallel$ and $f_\perp$ by means of the FRT approach for the general case of the axes ratio in accordance with Equation 2.18 through Equation 2.21.

As follows from the data for a needle-shaped spheroid, represented by the left-hand sides of Figure 2.1a and Figure 2.1b, the results of O'Konski and FRT approaches lead to close results, especially for the case of transverse orientation, when $b/a < 0.44$, and the results practically coincide if $b/a < 0.45$. For the longitudinal orientation of a very prolate spheroid, the asymptotic of $f_\parallel$ in the frame of FRT and O'Konski approaches are, respectively, $3\pi/4b$ and $1/2b$, whereas for finite ratio, $b/a < 1$, the relative difference between the two approaches is less than 10%. Essentially, worse agreement between O'Konski and FRT approaches in the calculation of the factors, $f_\parallel$ and $f_\perp$, takes place for the disk-shaped spheroid. The factor, $f_\parallel$, calculated for the longitudinal orientation of the oblate spheroid in the framework of the FRT approach is several times lesser when compared to $1/2b$, given by O'Konski's theory. It corresponds to the note by O'Konski [17] about the inadequacy of his approach to the conversion of the surface conductivity to the effective volume conductivity for a thin oblate spheroid. A better, but not a perfect, agreement between the two approaches takes place for the case of the transverse orientation of the oblate spheroid: the relative difference between the two approaches is less than 20% only for $b/a > 5$.

For the limiting case of slightly prolate or oblate spheroid, O'Konski [17] suggests the approximation given by Equation 2.14, which is the exact value for spherical particles. It provides a good approximation (with the error less than 10%, as decorated by the dashed straight lines in Figure 2.1) within the following range of the axes ratio:
For $f_\parallel$,

$$0.75 < b/a < 1.2 \tag{2.41}$$

And for $f_\perp$,

$$0.8 < b/a < 1.1 \tag{2.42}$$

Some examples of the application of both the O'Konski and FRT approaches to the calculation of the parameters of wide-range dispersion of the anisotropy of polarizability are given in the Figure 2.2 through Figure 2.4. Figure 2.2 and Figure 2.3 represent the dependencies of the characteristic times for volume-diffusion controlled, α-dispersion, $\tau_{\alpha_\parallel}$, $\tau_{\alpha_\perp}$, and for Maxwell–Wagner dispersion, $\tau_{MW_\parallel}$, $\tau_{MW_\perp}$, on the axes ratio of the spheroidal particle of constant volume. As one can see from Figure 2.2 and Figure 2.3, the characteristic time of the polarizability of the spheroid, oriented by its longer axis along the field ($\tau_\parallel$ for oblate and $\tau_\perp$ for prolate particle) exceeds the characteristic time for the case of transverse orientation of the longer axes ($\tau_\perp$ for oblate and $\tau_\parallel$ for prolate particle) for both the dispersion ranges:

$$\tau_\parallel \geq \tau_\perp \ \text{ for } a \geq b \quad \tau_\parallel \leq \tau_\perp \ \text{ for } a \leq b \tag{2.43}$$

The dependencies of the low-frequency limits for the components of the polarizability on the particle's axes ratio are shown in the Figure 2.4 and Figure 2.5, respectively, for both the α-dispersion and Maxwell–Wagner–O'Konski dispersion.

Similar to the situation with the characteristic time, the low-frequency limits of polarizability for the spheroid oriented by its longer axes along the field exceed the low-frequency limits for the cases of transverse orientation of the longer axes:

$$\gamma_\parallel^{(0,i)} \geq \gamma_\perp^{(0,i)} \ \text{ for } a \geq b \quad \gamma_\parallel^{(0,i)} \leq \gamma_\perp^{(0,i)} \ \text{ for } a \leq b \tag{2.44}$$

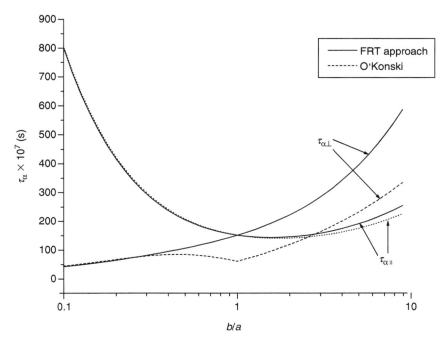

FIGURE 2.2 Relaxation time of α-dispersion of polarizability for different orientations of spheroidal particle in water KCl solution of concentration $C_m = 1.5 \times 10^{-4} N$ as a function of the axes ratio. The solution conductivity, calculated from the given value of C_m, is $K_m = 2.2 \times 10^{-4}$ 1/ohm·m. Surface conductivity $K^\sigma = 10^{-9}$ 1/ohm is caused by the counterions only. Across the curves, the particle's volume is kept constant and equal to the volume of the sphere of radius, $a_{sph} = 250$ nm.

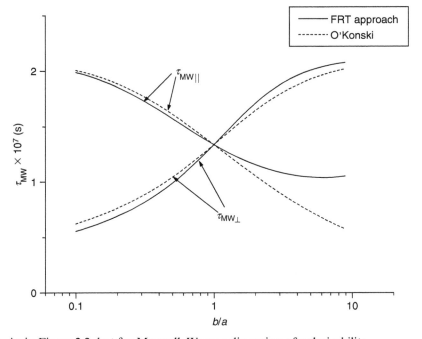

FIGURE 2.3 As in Figure 2.2, but for Maxwell–Wagner dispersion of polarizability.

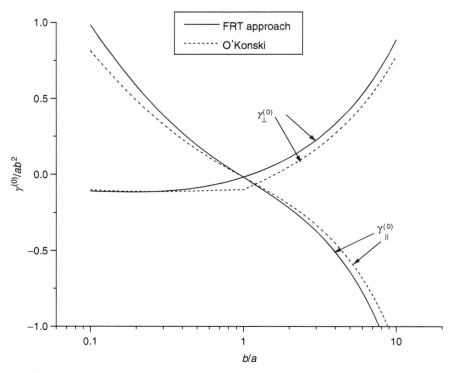

FIGURE 2.4 Dimensionless low-frequency limits of polarizability, α-dispersion of polarizability for different orientations of spheroidal particle as a function of axes ratio for the same parameters as those used in Figure 2.2.

It takes place due to both the larger value of the factor, $f_{\parallel,\perp}$, which determines (see Equation 2.13) the equivalent conductivity of the particle and the smaller value of the depolarization factor, $f_{\parallel,\perp}$, for the case of polarization along the longer axes. Another important feature of the behavior of the limiting values of polarizability, visible from the comparison of Figure 2.3 and Figure 2.4, is the somewhat larger value of the low-frequency limits of the Maxwell–Wagner dispersion, $(\gamma_{\parallel}^{(i)}, \gamma_{\perp}^{(i)})$, when compared to the low-frequency limits of $(\gamma_{\parallel}^{(0)}, \gamma_{\perp}^{(0)})$.

$$\gamma_{\parallel,\perp}^{(i)} \geq \gamma_{\parallel,\perp}^{(0)} \tag{2.45}$$

It is caused by the suppressing influence of the diffusion of the ions, responsible for the α-dispersion, on the surface current of counterions, which is the only factor leading to the increasing of low-frequency induced dipole moment above its minimal level, characteristic for the system "dielectric in conductor."

Such peculiarities of the form-dependence of the characteristic times and limiting values of polarizability are manifested in the frequency dependence of their components, $\gamma_{\parallel}(\omega)$, $\gamma_{\perp}(\omega)$, and their differences, i.e., the anisotropy of the polarizability within the wide-frequency range covering both the α- and the Maxwell–Wagner dispersions, represented in Figure 2.6 and Figure 2.7. As it is evident from these figures, in most part of the frequency ranges, we have the exceeding values of the polarizability along a longer axis, when compared to that of the#146shorter one. It takes place due to the corresponding relationship between the limiting values of the polarizabilities, reflected by the inequalities (Equation 2.44). However, within the range of the Maxwell–Wagner dispersion, a zone exists, close to the high-frequency limits of

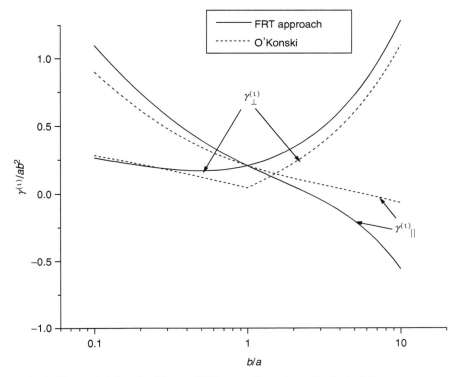

FIGURE 2.5 As in Figure 2.4, but for Maxwell–Wagner dispersion of polarizability.

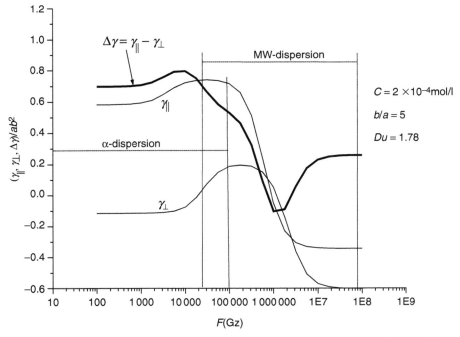

FIGURE 2.6 Wide-frequency range dispersion of polarizability for different orientations and the polarizability anisotropy with respect to the longer axes of prolate particle, $\Delta\gamma = \text{Re}\,(\gamma_\| - \gamma_\perp)$, for the same parameters as those used in Figure 2.2.

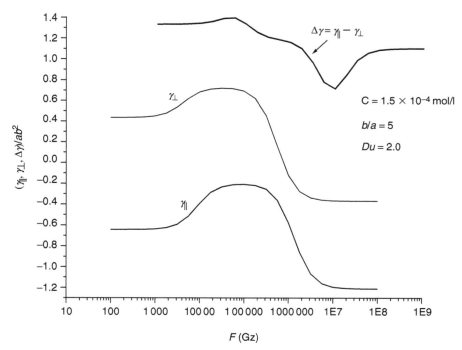

FIGURE 2.7 As in Figure 2.6, but for oblate particles; the polarizability anisotropy with respect to the longer axes of oblate particle is defined as $\Delta\gamma = \text{Re}\,(\gamma_\perp - \gamma_\parallel)$.

the dispersion of polarizability along the longer axes, where the latter value becomes smaller when compared to the polarizability along the shorter axes. The occurence of this zone is connected with the relationship between the relaxation times, expressed by the inequalities, Equation 2.43 and illustrated in Figure 2.3 and Figure 2.4. As an example, let us consider the frequency dependencies of the components of the polarizability of prolate spheroid, represented in Figure 2.6. Within the frequency zone, $1/2\pi\tau_{\text{MW}_\parallel} > F > 1/2\pi\tau_{\text{MW}_\perp}$, where the value of $\gamma_\parallel$ falls deep enough already, whereas $\gamma_\perp$ is only about the beginning of its falling, than the inversion of the sign of the difference between $\gamma_\parallel$ and $\gamma_\perp$ may occur. The existence of such an inversion within the Maxwell–Wagner dispersion gives rise to a small zone of a negative sign of polarizability anisotropy, determined as the difference between the polarizability between the long and the short axes. Another manifestation of the relationship between the relaxation times corresponding to the longitudinal and transverse orientations takes place within the range of α-dispersion. At the initial zone of the α-dispersion, characterized by the relationship, $1/2\,\pi\tau_{\alpha_\parallel} F > 1/2\pi\tau_{\alpha_\perp}$, the polarizability along the longer axes ($\gamma_\parallel$ for prolate particle) increases relatively faster, while $\gamma_\perp$ increases still slower. As a result, the polarizability anisotropy, equal to $\gamma_\parallel - \gamma_\perp$ for prolate particles, increases in this zone. In the neighbor zone of some larger frequency, where $1/2\,\pi\tau_{\alpha_\parallel} > F \simeq 1/2\,\pi\tau_{\alpha_\perp}$, the increase of $\gamma_\parallel$ falls, whereas $\gamma_\perp$ increases faster and as a result, $\gamma_\parallel - \gamma_\perp$ for prolate particles at the high-frequency part of the α-dispersion decreases. So, due to the action of the above-considered physical mechanisms, the anisotropy of polarizability within the zone of α-dispersion passes a maximum, and its Maxwell–Wagner dispersion is characterized by a deep minimum, both shown in Figure 2.6 and Figure 2.7, correspondingly for oblate and prolate spheroids.

The altitude of the maximum and the depth of the minimum, both depend on the surface conductivity (on Dukhin's number). But the dependence of the altitude maximum is stronger than the depth of the minimum. It is connected with a fact that the influence of the concentration polarization on induced dipole moment and hence the altitude of α-dispersion, is a second-order effect with respect to the surface conductivity, whereas the direct influence of the surface conductivity on the induced dipole moment in the absence of the concentration polarization responsible for the Maxwell–Wagner polarization is a first-order effect.

The influence of the surface conductivity on the polarizability and its anisotropy depends on the conductivity and hence on the concentration of the bulk solution. The frequency dependencies of the anisotropy of polarizability for set values of the concentration of KCl water solution and the same value of surface conductivity are represented in Figure 2.8 for oblate particles and in Figure 2.9 for prolate particles. As one can see from these figures, the pronounced manifestations of the above-discussed features of the α-dispersion of the polarizability anisotropy take place at lower electrolyte concentrations, $C < 10^{-4}\,N$, corresponding to the large magnitude of the Dukhin number corresponding to the polarization of the spheroid along its longer axes, $Du \geq 2/3$. Concerning the characteristic features of Maxwell–Wagner dispersion, as observed from Figure 2.6 and Figure 2.7, they are manifested pronouncedly at not so small electrolyte concentrations of about $C < 3 \times 10^{-5}\,N$ that provide not so large Dukhin numbers, $Du \geq 0.7/1$. It should be noted that similar results for the Maxwell–Wagner dispersion of the anisotropy of polarizability were obtained in Ref. [5] also by means of the FRT approach.

In conclusion, our theoretical considerations described the frequency dependence of the polarizability of the charged spheroidal particles for the wide-frequency range covering the

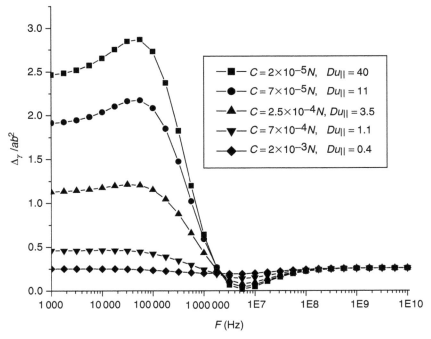

FIGURE 2.8 Wide-frequency range dispersion of the polarizability anisotropy $\Delta\gamma = \gamma_\parallel - \gamma_\perp$ of prolate particle ($a = 250\,\text{nm}$, $b = 50\,\text{nm}$) with respect to its longer axes for different concentrations of KCl displayed at the figure, together with the values of the Dukhin number. Rest of the parameters as in Figure 2.2.

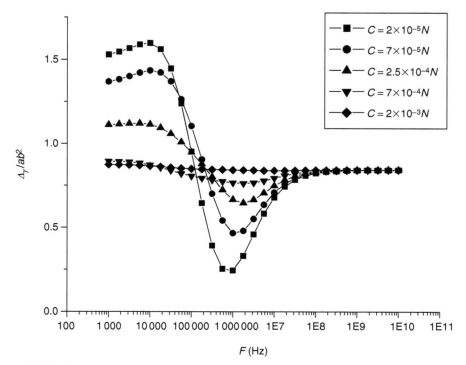

FIGURE 2.9 Wide-frequency range dispersion of the polarizability anisotropy, $\Delta\gamma = \gamma_\perp - \gamma_\parallel$, of oblate particle ($a = 50\,\text{nm}$, $b = 250\,\text{nm}$) with respect to its longer axes. Rest of the parameters as in Figure 2.8.

action of both the mechanisms that include the polarization of the EDL, namely, its concentration polarization (α-dispersion) and the storage of the ionic charge due to the electric surface current (Maxwell–Wagner–O'Konski dispersion). The theory is based on the approach used in Refs. [18–22] and allowed to convert the surface conductivity of the double layer into the effective bulk conductivity of the spheroid with the arbitrary axes ratio (FRT approach). The results obtained by the FRT approach were compared with the results of such kind of the conversion suggested first by C.T. O'Konski in Ref. [17] for the limiting cases of the axis ratio. The fields of the application of O'Konski's approach were assessed on the basis of this comparison.

REFERENCES

1. Stoylov, S.P., Shilov, V.N., Dukhin, S.S., and Petkanchin, I.B., *Elektro-optika Kolloidov (Electro-Optics of Colloids)*, Dukhin, S.S., Ed., Naukova Dumka, Kiev, 1977.
2. Stoylov, S.P., *Colloid Electro-Optics*, Academic Press, London, 1991.
3. Trusov, A. and Vojtylov, V., *Electro-Optics and Conductometry of Polydisperse Systems*, CRC Press, Boca Raton, FL, 1993.
4. Abstracts of Papers, *Symposium on Colloid and Molecular Electro-Optica*, Varna, 1991.
5. Bellini, T.T., Mantegazza, F., Degiorgio, V., Avallone, V., and Saville, D.A., *Phys. Rev. Lett.*, 82, 5160, 1999.
6. Dukhin, S.S. and Shilov, V.N., *Dielectric Phenomena and the Double Layer in Disperse Systems and Polyelectrolytes*, Wiley, Jerusalem, 1974.
7. Maxwell, J.C., *Electricity and Magnetism*, vol. 1, Dover, New York, 1954.

8. Wagner, K.W., *Arch. Electrotech.*, 2, 371, 1914.
9. Dukhin, S.S. and Shilov, V.N., Thin double layer theory of the wide-frequency range dielectric dispersion of suspensions of nonconducting spherical particles including surface conductivity, in *Interfacial Electrokinetics and Electrophoresis*, vol. 106, Delgado, A.V., Ed., Surfactant Science Series, Marcel Dekker, New York, 2002, p. 155.
10. Shilov, V.N., Delgado, A.V., Gonzalez-Caballero, F., and Grosse, C., *Colloids Surf. A*, 192, 253, 2001.
11. Shilov, V.N. and Dukhin, S.S., *Kolloidn. Zh.*, 32, 117, 1970.
12. Shilov, V.N. and Dukhin, S.S., *Kolloidn. Zh.*, 32, 293, 1970.
13. DeLacey, E.H.B. and White, L.R., *J. Chem. Soc., Faraday Trans.*, 277, 2007, 1981.
14. Fricke, H., *Phys. Rev.*, 26, 682, 1925.
15. Sillars, R.W.J., *J. Inst. Elec. Engrs.*, 80, 378, 1937.
16. Stretton, J.A., *Electromagnetic Theory*, McGraw-Hill, New York, 1941.
17. O'Konski, C.T., *J. Phys. Chem.*, 64, 605, 1960.
18. Eremova, Ju. Ja. and Shilov, V.N., *Kolloidn. Zh.*, 37, 635, 1975.
19. Eremova, Ju. Ja. and Shilov, V.N., *Kolloidn. Zh.*, 37, 1090, 1975.
20. Dukhin, S.S. and Shilov, V.N., *Adv. Colloid Interface Sci.*, 13, 153, 1980.
21. O'Brien, R.W., Rowlands, W.N., *J. Colloid Interface Sci.*, 159, 471, 1993.
22. Shilov, V.N. and Eremova, Ju. Ja., *Kolloidn. Zh.*, 57, 255, 1995.
23. Shilov, V.N., Zharkikh, and Bokovskaja, Yu.B., *Kolloidn. Zh.*, 47, 757, 1985.
24. Lyklema, J., Electrokinetics and related phenomena, in *Fundamentals of Interface and Colloid Science (FICS). Solid_/Liquid Interfaces*, vol. II, Academic Press, New York, 1995, Chap. 4.
25. Grosse, C., Pedrosa, S., and Shilov, V.N., *J. Colloid Interface Sci.*, 220, 31, 1999.
26. Grosse, C., *Ferroelectrics*, 86, 171, 1988.
27. Grosse, C. and Shilov, V.N., *J. Colloid Interface Sci.*, 193, 178, 1997.

3 Quantitative Molecular Electro-Optics: Macromolecular Structures and Their Dynamics in Solution

Dietmar Porschke and Jan M. Antosiewicz

CONTENTS

3.1   INTRODUCTION

In recent years the number of macromolecular structures, which have been determined up to high atomic resolution, has increased dramatically. This development is mainly due to technical progress and to increased numbers of applications of x-ray crystallography and of nuclear magnetic resonance (NMR) measurements. Compared to the remarkable achievements obtained by these techniques it seems that other techniques are not able to compete and moreover are hardly able to contribute. However, there are many problems, where other techniques are still useful, at least as a complement to x-ray crystallography and NMR.

Most of the available information on macromolecular structures has been provided by crystal structure analysis and this approach clearly implies some limitations. In the crystalline state the molecules are packed close to each other and, thus, interact with each other, which may lead to changes in the structure. A more important limitation is the fact that the molecules in the crystal are usually frozen into one of several possible structures. Obviously the biological function of macromolecules requires that these molecules are flexible. One of the challenges remaining for structural biology is the characterization of the dynamics of biological macromolecules during biological function.

Usually it is argued that the information on the dynamics is available from NMR measurements. Although this is correct in principle, there are clear practical limitations. First of all, extraction of information on the dynamics is not trivial in the case of large molecules. Furthermore there is an information gap between the proton exchange time range of a few milliseconds and the fluctuation time range of a few nanoseconds. In cases like DNA double helices, the NMR spectra provide information on the local structure relatively easily, but it is much more difficult to get sufficient information on the global structure. Obviously other techniques should be used to fill these gaps wherever possible. In our present review we discuss the potential of a special approach: electro-optical investigations. This potential has been increased considerably owing to recent developments both in experimental procedures and in the interpretation of experimental data.

The principle of electro-optical measurements is quite simple: electric field pulses are applied to molecules in solution and any anisotropy in the distribution induced under the electric field is analyzed by measurements of the dichroism or of some other optical parameter. The main advantages of the technique are its simplicity and high sensitivity. It has been demonstrated, for example, that the intercalation of single aromatic molecules into DNA double helices with 95 base pairs can be detected by electro-optical measurements [1,2]. Furthermore experimental data can be obtained with low amounts of material. Finally, experimental data can be obtained over a wide time range from nanoseconds to seconds and, thus, these data include information on the dynamics in the same time range.

Electro-optical investigations provide information on macromolecules and their complexes related to three different domains of physics:

1. *Hydrodynamics*: the time constants obtained from electro-optical transients are determined by coefficients of rotational diffusion and are very sensitive to global structures as well as their dynamics.
2. *Electrostatics*: the electrical parameters reflect the distribution of charges.
3. *Optics*: the dichroism (or other optical parameters) indicates the orientation of the residues contributing to the absorbance.

For macromolecules with simple structures the data can be interpreted directly without problems. In the case of macromolecules with a complex structure, however, the quantitative interpretation of the experimental data requires computational efforts based on appropriate theoretical relations.

Information on the dynamics of macromolecules may be obtained in many cases directly from electro-optical transients. For example, flexibility may be indicated by reduced time constants of rotational diffusion or by the appearance of separate relaxation processes related to the motion of parts of the molecule with respect to each other. In addition, electric field pulses may also induce conformation changes of macromolecules [3]. These field-induced changes of the structure are of particular interest with respect to bioelectricity.

A very useful compendium on the general theory and experimental procedures has been published by Fredericq and Houssier [4], describing the state of the art in 1973 (see also Ref. [5]). In the present review we describe some advances in experimental procedures. The major part of this review is devoted to the quantitative interpretation of the experimental data. The basis of an automatic procedure for the calculation of electro-optical data is discussed and some selected applications are presented. Another major part should be devoted to the analysis of various electro-optical phenomena by Brownian dynamics simulations. Because of space limitations we have discussed this method and its applications only very briefly.

3.2 ADVANCES IN EXPERIMENTAL PROCEDURES

3.2.1 DICHROISM OR BIREFRINGENCE?

As already mentioned, the principle of electro-optical measurements is very simple: electric field pulses are applied to solutions of macromolecules and the field-induced orientation of these molecules into the direction of the field vector is recorded by measurements of the optical anisotropy. The electric field strength applied to the solutions has been as high [6–9] as 80 kV/cm, but in some cases a few V/cm may already be sufficient [10] to induce easily detectable optical anisotropy effects. The technique for generation of the field pulses is partly determined by the voltage range used for the measurements and partly by the rise and decay time constants of the field pulse required for the resolution of rotational time constants of the systems under investigation.

The field-induced orientation may be recorded by different types of optical measurements, which are closely related to each other: linear dichroism and birefringence are more popular than light scattering and fluorescence. In the present review the main emphasis is on the linear dichroism, because this type of measurement provides two major advantages:

1. The existence of field-induced reactions, e.g., conformation changes or changes of ligand binding, can be checked and analyzed much more easily than by the other available procedures.
2. The electric dichroism can be calculated quantitatively and relatively easily from molecular structures (see Section 3.4).
3. The solvent in general does not contribute to the measured dichroism signals, whereas birefringence transients measured in aqueous solutions usually include a contribution from water.

The closely related technique for detecting molecular orientation by birefringence measurements has some special advantages, which result from application of lasers as light sources:

1. High light intensities in general increase the sensitivity.
2. The narrow light beam may be utilized to construct cells with long light paths and narrow electrode distances.

These advantages in cell construction lead to an extension of the experimental potential, e.g., to lower concentrations. However, the increase of sensitivity enabled by such measurements of birefringence may have the drawback that contributions of the solvent to the signal cannot be neglected anymore and must be separated out.

3.2.2   HIGH FIELD STRENGTHS AND HIGH TIME RESOLUTION

Most laboratories use commercial pulse generators for their electro-optical measurements. These instruments are constructed for generation of rectangular pulses, but in spite of the relatively large capacitors used in these instruments, there is a clear decay of the field strength in the case of long pulses. The Cober Model 606 high-power pulse generator, for example, delivers from 0 to 2500 V output at up to 12.5 A with pulse widths from 50 ns to ~1 ms; the rise and decay times are in the range of about 30 ns. These instruments may be used with transformers for an increase of the output voltage, but the amplification leads to an increase of the rise and decay times. The output voltage may also be increased by connection of two generators in series. More important is the possibility to connect two instruments for generation of pulses with reversal of the field vector, because these reverse pulses are very useful for the characterization of permanent or induced dipole moments.

A different technique, based on the discharge of cables, has been developed originally for the analysis of field-induced chemical relaxation effects [11], but the discharge of cables proves to be very useful also for the measurements of field-induced physical relaxation effects, including molecular electro-optical investigations. As described in textbooks of physics, the discharge of a cable using a load with a resistance, which is exactly matched to the cable impedance, provides pulses of ideal rectangular shape. The pulse length increases with the length of the cable, corresponding to the time required for an electric signal running back and forth along the cable. The cable technique has been used for the construction of two limit types of pulse generators, which are used for applications to solutions of low or high conductivity, corresponding to loads of high or low resistances, respectively.

In the case of solutions of low conductivity, electric field pulses may be maintained for relatively long time periods. In this limit case, the voltage applied to the sample during discharge of a cable decays with an overall shape of an exponential, but the exponential consists of small steps with a constant voltage during each step and the time of each step corresponds to the time required for an electrical signal running back and forth along the cable. Because it is useful for most investigations to apply pulses of absolutely constant voltage for relatively long time periods, cables of up to 1000-m length have been used [12]. In this case each step takes a time of 10 μs and the voltage applied to a solution during this step remains absolutely constant, irrespective of the cell conductance. The voltages processed by the cable technique are up to 100 kV. Initiation and termination of pulses requires special switching devices, because standard electric switches cannot be used in the high-kV range. Spark gaps with mobile electrodes proved to be very useful for this purpose: high voltages up to 100 kV and currents in the range of 1000 A may be switched within a few nanoseconds by reduction of the electrode distance, using a simple pneumatic construction for moving one of the electrodes toward the other one.

3.2.3 Measurements at Physiological Salt Concentrations

In the case of solutions with high salt concentrations, the extent of Joule heating resulting from application of high electric field pulses is not negligible anymore and, thus the length of the pulses has to be limited. Under these conditions it is relatively difficult to maintain constant electric field strengths even for short pulse times, if the pulses are generated by discharge of standard capacitors. This technical problem may be solved by the cable discharge technique, provided that the resistance of the sample is matched to the cable impedance. Matching of the sample resistance is possible in principle, but not convenient in practice. Because the resistance of samples with physiological salt concentrations in standard measuring cells is higher than the impedance of standard high-voltage cables, the matching problem may be solved by using a compensation cell parallel to the measuring cell, and the resistance of the compensation cell is adjusted such that the resistance of both cells in parallel connection matches the cable impedance. The first instrument using this principle has been constructed by Hoffman [13] for temperature-jump measurements. Some difficulties associated with this technique have been removed by a reconstruction of the optical detection system and of the compensation cell [14,15]. Because the electro-optical signals obtained at high salt concentrations are usually relatively small, the instrument has also been constructed for automatic sampling of large numbers of transients under the control of a PC (Figure 3.1).

3.2.4 Electric Field Pulses of Arbitrary Forms

An interesting new possibility for the generation of electric field pulses with an arbitrary shape is available owing to the development of "arbitrary waveform generators." These generators are widely used for many different industrial applications and, thus, instruments have been developed that can be easily programmed to generate pulses of any form. However, the pulse amplitudes of these generators are usually limited to about 30 V and, thus, the pulses have to be amplified for most electro-optical applications. The combination of an arbitrary waveform generator with an amplifier has been very useful for electro-optical studies of membrane proteins [10,16,17].

3.2.5 Optical Detection

The quality of the results obtained by electro-optical investigations is determined by the signal-to-noise ratio. The signal-to-noise ratio is critical for the assignment of the exponential

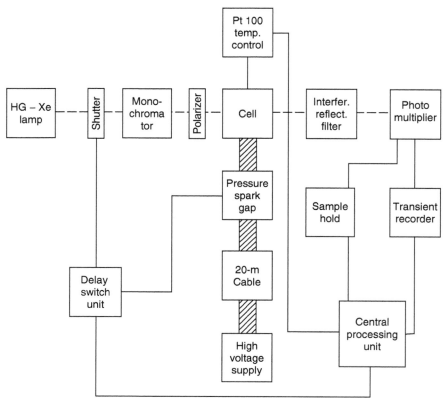

FIGURE 3.1 Block diagram of an instrument for measurements of the electric dichroism at physiological salt concentrations. The electric field pulse is generated by the discharge of a 20-m cable; the "load," corresponding to the cell together with a compensation cell (not shown), is matched to the impedance of the cable. Automatic collection of data involves the following sequence of steps: (i) when the temperature of the sample is within the specified interval, the shutter is opened for short exposure of the sample to the light beam; (ii) the sample hold units stores the total light intensity; (iii) the delay switch triggers the spark gap for initiation of the field pulse; (iv) data are collected by the transient recorder, triggered by the field pulse; (v) transfer of data to the PC and averaging. The cycle is repeated after a specified time interval. (From Porschke, D. and Obst, A., *Rev. Sci. Instrum.*, 62, 818, 1991.)

processes, which are characteristic for the given system. If the characteristic exponentials cannot be resolved from the experimental decay curves, a correct interpretation of the data may be impossible (see Section 3.5.2). A comparison of the published data clearly shows that there are large differences in the signal-to-noise ratio; in some cases these differences were the source of contradictory conclusions derived from the electro-optical measurements on the same type of system.

The signal-to-noise ratio is mainly determined by the quality of the optical detection system. Because the shot noise decreases with the square root of the light intensity, high light intensities are of advantage. Photoreactions can be avoided by using fast automatic shutters for minimal exposure of samples to the light beam. Xenon and mercury arc lamps with powers up to 600 W together with high-intensity grating monochromators and Glan crystal polarizers proved to be useful for dichroism measurements.

An important factor for measurements of unperturbed signals is the construction of the measuring cell with optical windows, which should be absolutely strain-free. Any strain in the windows may lead to artifacts in measured signals, in particular during measurements at the "magic angle" performed as a control for the existence of field-induced reactions (for a recent detailed discussion of magic angle measurements, see Ref. [18]).

The signal-to-noise ratio of experimental data is also strongly affected by the bandwidth of the detector. Usually the response time constant of the detector is adjusted to a value that is smaller by a factor of, e.g., ten than the smallest relaxation time constant of the system under investigation. This is not a necessary condition anymore, provided that deconvolution techniques are used. Deconvolution has been developed for measurements in the very short time range of, e.g., a few nanoseconds, but can be very useful also for measurements in the microseconds or milliseconds time range. If deconvolution is used, the response time of the detector may be close to the observed relaxation times, resulting in a clear increase of the signal-to-noise ratio.

3.2.6 DATA PROCESSING AND DECONVOLUTION

The potential of electro-optical measurements has been extended considerably owing to the development of fast analog–digital converters and of efficient computers. The time range accessible to investigations has been extended to a few nanoseconds. The accuracy of the measurements can be increased by accumulation of many transients. Parallel storage of the driving pulses and of the resulting optical transients at high time and amplitude resolution opened the possibility of reliable application of deconvolution procedures.

Transients obtained by electro-optical measurements should represent the response of the solution under investigation, but are often convoluted with the response of the detection system and with the "forcing function" corresponding to the electric field pulse. The ideal limit case given by an infinitely fast detector response and a rectangular forcing function is more an exception than the rule. Deconvolution has to be practiced with caution, because various factors may have to be separated out. The procedure is relatively simple if the optical signal is convoluted with the detector response exclusively. Provided that the detector is used in its linear range, we get a linear convolution product. If the detector response can be represented by a single exponential, a simple analytical procedure can be used for deconvolution [19]. When the response function of the detector is more complex, numerical procedures can be used [20]. As a test we have recorded transients for a given system with a wide range of different detector time constants. The evaluation with our deconvolution routines based on the individual detector response curves demonstrated that deconvolution is reliable. Reference transients characterizing the detector response can be easily obtained, for example by measurements of the birefringence of urea solutions.

Deconvolution of signals affected by a nonideal forcing function requires more attention, because the response function of the system to the electric field must be included. If the effect induced by the field pulse is known to increase with the square of the field strength, this can be easily implemented in the deconvolution procedure. The analysis of complex convolution products is possible, provided that the factors are sufficiently well known. For example, we have used a procedure for evaluation of a system response given as convolution product with the square of the forcing function (i.e., the electric field) and the resulting response, which is convoluted again with the detector function [21]. When the dependence of the system to this forcing function is not well characterized, deconvolution of signals folded with a forcing function is not practicable anymore.

3.2.7 STOPPED FLOW ELECTRO-OPTICS

A unique advantage of the electro-optical technique is the fact that it provides information on structures by a simple procedure within a very short time. This advantage may be exploited for the analysis of structure changes during macromolecular reactions. A major part of these reactions is relatively fast and, thus, it is not simple to get direct information on changes of structure. The most popular approach for the analysis of fast reactions, in particular in biosciences, is the stopped flow technique. For analysis of structure changes at a high time resolution, it should be useful to combine the stopped flow technique with an electro-optical procedure. The main difficulty is construction of a stopped flow cell with the possibility to apply electric field pulses. Attempts in this direction, e.g., for stopped-flow-T-jump experiments, were not really successful because of limitations in the time resolution and difficulties in getting cells without leak. Both problems were solved by novel cell construction with reduced dimensions [2]. Reduction to small dimensions of the flow system is essential for a high time resolution.

The combined stopped-flow-electric field-jump (Figure 3.2) provides the possibility to mix solutions and study reactions with a dead time of 0.2 ms. Electric field pulses can be applied at any time after mixing and resulting dichroism signals can be recorded. The time resolution of detection can be adjusted within the usual range, i.e., up to ~10 ns. The intercalation reaction of ethidium into the DNA double helix has been used as a test case. The increase in the helix length resulting from intercalation has been resolved into different relaxation processes with

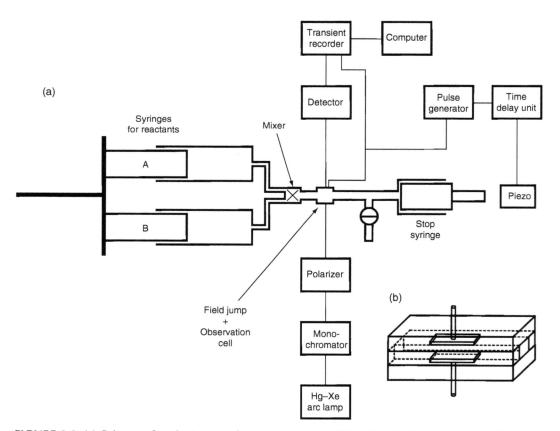

FIGURE 3.2 (a) Scheme of an instrument for measurements of the electric dichroism in combination with stopped flow mixing. (b) Observation chamber with Pt electrodes. The flow through the mixer and the observation chamber is terminated when the piston of the stop syringe hits the piezo. The piezo signal initiates the pulse generator after a prefixed delay time. The optical components serve to measure the electric dichroism as usual. (From Porschke, D., *Biophys. J.*, 75, 528, 1998. With permission of the Biophysical Society.)

time constants in the range from ~1 to ~100 ms. Obviously the new technique is very useful to distinguish relaxation processes reflecting mainly external binding without length increase from processes associated with intercalation, resulting in a length increase. The increase of length during a given relaxation process can be quantified. External and internal bindings may also be distinguished by the sign of the dichroism.

3.2.8 FLASH PHOTOLYSIS AND ELECTRO-OPTICS

The potential of electro-optical procedures to get information on macromolecular structures within a short time interval can also be utilized for the analysis of photoreactions. The photoreaction is induced by an appropriate light pulse and the response of the system is analyzed by recording electro-optical signals at given times after photo-initiation. An instrument for this type of analysis has been constructed (Figure 3.3) and used for an analysis of the purple membrane photocycle [22]. The analysis of dichroism decay curves obtained during the photocycle clearly revealed bending of the purple membrane during the M-state. For accumulation of enough M-state, natural purple membranes were analyzed at slightly increased pH values (pH 7.8 to 9.3), where the lifetime of the M-state is increased. A more convenient system, which has been studied at neutral pH, is the mutant D96N having a prolonged lifetime of the M-state.

The electro-optical results on membrane bending in the M-state are consistent with the independent conclusion obtained by electron crystallography [23] on light-induced opening of the protein channel on the cytoplasmic side of the membrane. Opening of the protein on one side leads to a wedge-like deformation, which must be amplified to a high degree of curvature

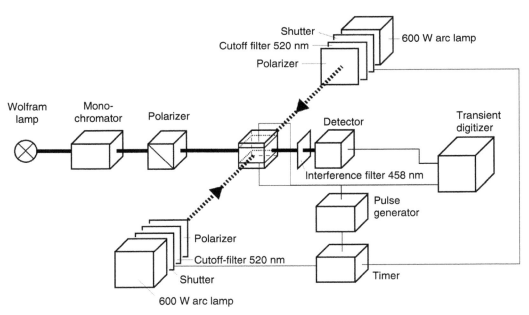

FIGURE 3.3 Scheme of an instrument for measurements of the electric dichroism combined with flash photolysis. The solution is exposed to a flash of light directed simultaneously from the front- and the back-side. After a given time interval the sample is also exposed to an electric field pulse and the electric dichroism transient is recorded. Interference of the flash light with the light used for measurements is avoided by a combination of cutoff and interference filters. (Reprinted from Porschke, D., *J. Mol. Biol.*, 331, 667, 2003. With permission from Elsevier.)

in the membrane patches. The x-ray analysis did not show this transition. Apparently "the main structural transition is inhibited or muted in the three-dimensional crystal" [23,24]. Thus, electro-optical studies are particularly useful for the characterization of structure changes *in solution*—without restrictions due to crystal packing.

3.3   BASIC THEORY

3.3.1   ELECTRIC DICHROISM

The most useful form for presentation of stationary dichroism data is the reduced electric dichroism

$$\xi(t) = \frac{\Delta A_\parallel(t) - \Delta A_\perp(t)}{\bar{A}} \tag{3.1}$$

where $\Delta A_\parallel(t)$ and $\Delta A_\perp(t)$ are the field-induced absorbance changes measured with light polarized parallel and perpendicular relative to the field vector, respectively, as a function of time t; $\bar{A}$ is the isotropic absorbance measured in the absence of an external electric field. If the absorbance changes are only due to orientation effects and do not include contributions from reactions, e.g., conformation changes, the following relation holds:

$$\bar{A} = \frac{A_\parallel + 2A_\perp}{3} \tag{3.2}$$

where $A_\parallel$ and $A_\perp$ are the absorbances of light polarized parallel and perpendicular to the field vector, respectively. In the presence of an uniform external electric field, the light absorption by the solution is characterized by cylindrical symmetry for rotations around the direction given by the field vector. Under these conditions the absorbance A^f_ϕ in the presence of an external electric field at any angle ϕ with respect to the field vector is given by

$$A^f_\phi = A^f_\parallel \cos^2\phi + A^f_\perp \sin^2\phi \tag{3.3}$$

From Equation 3.2 and Equation 3.3 we get for the absorbance change ΔA_ϕ at the angle ϕ

$$\Delta A_\phi = A^f_\phi - \bar{A} = \frac{A_\parallel - \bar{A}}{2}(3\cos^2\phi - 1) = \frac{\Delta A_\parallel}{2}(3\cos^2\phi - 1) \tag{3.4}$$

Equation 3.4 implies

$$\Delta A_\parallel = -2 \times \Delta A_\perp \tag{3.5a}$$

and

$$\Delta A_{54.8°} = 0 \tag{3.5b}$$

As mentioned above, Equation 3.2 through Equation 3.5 are valid, provided that there are no absorbance changes due to field-induced reactions. Thus, Equation 3.5a and b are used to check experimental data for the existence of such reactions. Field-induced absorbance changes observed at the "magic angle" orientation of polarized light corresponding to $\phi = 54.8°$ reflect field-induced reactions rather than orientation effects. Because the changes of light transmission due to orientation effects are often much larger than those due to chemical effects, these measurements require a very careful adjustment of the optical detection system [18,25].

3.3.2 ORIENTATION FUNCTIONS

The degree of molecular orientation and, thus, the magnitude of the electric dichroism ξ increases with the electric field strength E. Complete orientation in the direction of the electric field may be expected only in the limit of infinitely high E. The dependence of ξ on E is determined by the type and the magnitude of the electric dipole moment. The degree of orientation at a given field strength increases with the magnitude of the dipole moment. At low E values the electric dichroism ξ increases with E^2, corresponding to the Kerr law. The complete dependence of ξ on the field strength E is described by the "orientation function" Φ according to

$$\xi = \Phi \cdot \xi_\infty \tag{3.6}$$

where ξ_∞ is the limit dichroism observed at infinitely high electric field strengths. In the case of induced dipoles the orientation function [26] is given by

$$\Phi = \left\{ \frac{3}{4} \left[\left(e^\gamma / \sqrt{\gamma} \int_0^{\sqrt{\gamma}} e^{x^2} dx \right) - 1/\gamma \right] - \frac{1}{2} \right\} \tag{3.7}$$

where $\gamma = (\alpha E^2)/(2kT)$, α is the anisotrophy of the polarizability, k is the Boltzmann constant, and T is the absolute temperature.

In the case of permanent dipoles the orientation function [26] is given by

$$\Phi = 1 - \frac{3(\coth\beta - 1/\beta)}{\beta} \tag{3.8}$$

where $\beta = \mu_p \cdot E/kT$ and μ_p is the permanent dipole moment.

The orientation functions may be used to determine the limit value of the electric dichroism, corresponding to complete molecular orientation, by least squares fitting of dichroism values determined at different field strengths.

The limit value of the electric dichroism provides direct information about the orientation of the chromophores with respect to the dipole vector according to the following equation:

$$\xi_\infty = \frac{3}{2}(3\cos^2(\varphi) - 1) \tag{3.9}$$

where φ is the angle of the transition dipole moment of the chromophore relative to the dipole vector. When the transition dipole moment of the chromophore is oriented parallel to the dipole vector, corresponding to $\varphi = 0$, the limit value of the dichroism is $+3$. In the other limit case with the transition dipole moment in perpendicular direction to the long axis ($\varphi = 90$), the limit value of the dichroism is -1.5. Thus, the limit value of the electric dichroism can be used to calculate the angle φ of the optical transition dipole with respect to the direction of the electric dipole. An example for a successful application of this procedure is double helical DNA: the first evidence for the tilt of the base pairs in B-DNA double helices, i.e., the deviation of the base pair orientation from $90°$ relative to the helix axis, came from measurements of electric dichroism [27].

3.3.3 RELAXATION PROCESSES

Electro-optical experiments provide two different types of relaxation processes:

1. Application of field pulses induces relaxation of the molecules into the direction of the field vector; this process is driven by the torque resulting from some electric anisotropy of the molecules;
2. After pulse termination the molecules return to their random distribution.

The second relaxation process is determined by diffusion, whereas the first process is also determined by the torque. In the case of the first relaxation process, the observed time constants may be evaluated relatively easily only in the limit cases of either very low or very high electric field strengths. At intermediate electric field strengths the coupling of electric torque and diffusion cannot be described by simple analytical expressions and, thus, evaluation of data obtained in this range require extensive numerical calculations [28,29].

Usually the analysis of electro-optical transients is mainly restricted to the decay process, which is described by a sum of exponentials

$$\xi(t) = \sum_i \xi_i \cdot e^{-t/\tau_i} \tag{3.10}$$

where ξ_i are the amplitudes and τ_i the relaxation time constants. As discussed in Section 3.4.5, the theory predicts that the electro-optical decay curves for rigid molecules may contain up to five different relaxation processes. For molecules with structures of high symmetry, like spheres or cylinders, the dichroism decay is characterized by a single relaxation time constant

$$\tau = \frac{1}{6D} \tag{3.11}$$

where D is the coefficient for rotational diffusion. In the case of spherical molecules this diffusion coefficient is given by

$$D = \frac{kT}{8\pi\eta\sigma^3} \tag{3.12}$$

where η is the solvent viscosity and σ is the radius of the sphere. An exact analytical solution for cylinders is not available, but there are very useful approximate expressions [30,31]. Using these relations, electro-optical decay time constants may be used to derive information on molecular dimensions quite easily. More efforts are required for the quantitative analysis of molecules of arbitrary shape. Procedures for the analysis of electro-optical data, which can be applied in general without restrictions, are described in the following sections.

3.4 CALCULATION OF OPTICAL, ELECTRICAL, AND HYDRODYNAMIC PARAMETERS FROM MODEL STRUCTURES

3.4.1 MODEL STRUCTURES

Many crystal structures are deposited at the Brookhaven Protein Data Bank (PDB) [32] and, thus are easily available. These structures may be used as a basis for the generation of variant structures by molecular modeling. Tools for molecular modeling have been developed in the form of different software packages. Examples are: QUANTA and INSIGHT distributed by Accelrys/MSI; Sybyl distributed by Tripos Associates, Inc. A program for construction of different structures formed by nucleic acids, NAMOT2, has been developed by Carter and Tung [33] at the Los Alamos National Laboratory and is available free of charge. All these programs can be used to construct any desired molecular structure, either starting from an existing structure or *de novo*. For example, atoms, bonds, residues, or fragments may be added to, deleted from, and rotated within a given structure. The modeled structure may be further refined on the basis of energy calculations, for example by molecular dynamic programs.

Obviously the structures generated by these procedures should be tested by appropriate experiments. One of the experimental approaches for this purpose is the measurement of electro-optical data. Quantitative comparison of the experimental data requires simulation of these data from model structures. The procedures available for this simulation are discussed in the following sections.

3.4.2 Optics

3.4.2.1 Molecular Extinction Coefficient Tensors and Calculation of the Dichroism

It is remarkable that in the early days of quantum mechanics, it was not generally accepted that absorption of light by chromophores is anisotropic. W. Kuhn reported [34] that Max Born, one of the founders of quantum mechanics, believed in isotropic absorption of light by chromophores. In order to get experimental evidence against this view, W. Kuhn performed a new type of experiments and measured for the first time the electric dichroism of some simple chromophores. Now there is no doubt anymore that absorption of light is anisotropic. Electromagnetic radiation may induce the transition of molecules to excited states; this transition is determined by the transition dipole moment of the molecules, which describes the plane of polarization of the electromagnetic radiation required for the interaction with the molecule. The probability of light absorption in the absorption band of any molecule is maximal, when this light is polarized parallel to the transition dipole moment; the probability is zero for perpendicular polarization. In general, the absorption of light is proportional to the square of the scalar product between the electric field vector of the light and the transition dipole vector. The absorption of light by a chromophore is described quantitatively by an extinction tensor

$$\hat{\varepsilon} = \begin{pmatrix} 0 & 0 & 0 \\ 0 & 0 & 0 \\ 0 & 0 & 3\varepsilon \end{pmatrix} \qquad (3.13)$$

where ε is the measured extinction coefficient of the molecule at a given wavelength at random orientation; the factor 3 in the relation between ε_{zz} and ε describes the difference between the standard absorption at random orientation and the maximal absorption at preferential orientation of polarized light with respect to the transition dipole moment. According to convention, the zero values in all tensor positions except the zz-position means that the transition dipole is in the direction of the z-coordinate of the molecular coordinate system. The information on this direction may be derived by experimental studies on this chromophore oriented in films or crystals. It is also possible to determine the direction of transition dipole moments by quantum mechanical calculations.

Usually macromolecules contain more than one chromophore and, thus, their absorption of light is determined by the appropriate sum of all individual components. It is known that the absorbance or refractive contribution of a given chromophore does not only depend on its own transition dipole moment, but also on the distribution of transition dipole moments in its environment, i.e., on the local field [35,36]. However, in the case of the linear dichroism this type of coupling may be neglected in a first approximation.

Summation of the contributions from all different chromophores requires that first all components of the local extinction tensors of the individual chromophores have to be transformed into a common coordinate system for the whole macromolecule, which will be called the particle coordinate system (PCS). The mathematical formulation of this operation is given by

$$\hat{\varepsilon} = \sum_i \hat{R}_i \, \hat{\varepsilon}_i \, \hat{R}_i^\dagger \tag{3.14}$$

where $\hat{R}_i$ is the rotation matrix from the local coordinate system of the ith chromophore to the PCS and $\hat{R}_i^\dagger$ is the transposed form of $\hat{R}_i$. After this operation all the components of the individual tensors may be added directly to get the extinction tensor of the whole polymer. According to this procedure the extinction coefficient tensors of macromolecules may be calculated, if the atomic coordinates are known, for example from a crystal structure or from a structure obtained by molecular modeling. If several structures coexist and the relative probability of their occurrence is known, it is possible to calculate an appropriate average.

The extinction tensor $\hat{\varepsilon}$ contains the complete information on the magnitude and the directional dependence of the light absorbed and, thus can be used to calculate the electric dichroism, provided that the direction of the molecular electric dipole moment is known. For a description of the electric dichroism we define a laboratory coordinate system (LCS) the incident light is directed along its x'-axis and a uniform electric field is directed along its z'-axis. Under these conditions, $A_\parallel$ and $A_\perp$ according to Equation 3.2 are given by

$$A_\parallel = cl\langle \varepsilon_{z'z'} \rangle \quad \text{and} \quad A_\perp = cl\left\langle \frac{\varepsilon_{x'x'} + \varepsilon_{y'y'}}{2} \right\rangle$$

and the reduced dichroism is given by

$$\xi = \frac{3\langle \varepsilon_{z'z'} \rangle - \text{Tr}\,\hat{\varepsilon}}{2\bar{\varepsilon}} \tag{3.15}$$

where $\text{Tr}\,\hat{\varepsilon}$ is the sum of the diagonal elements of the tensor, which is invariant to rotations; $\bar{\varepsilon}$ is the average extinction coefficient; c is the molar concentration of particles, and l is the optical path length: the symbol $\langle \, \rangle$ denotes average values over all particles present in the solution.

In the limit of infinitely high electric field strength, the dipole moment vector of the particle is parallel to the direction of the external electric field. For this case the reduced limiting dichroism is calculated from Equation 3.15 with

$$\varepsilon_{z'z',\infty} = \mathbf{e}^\dagger \hat{\varepsilon} \, \mathbf{e} \tag{3.16}$$

where $e = \vec{m}/m$ is a unit vector in the direction of the dipole moment of the particle in the PCS and the superscript $\dagger$ indicates that the vector is written as a row.

In summary, the calculation of extinction tensors and of the reduced dichroism is straightforward from the theoretical point of view. The main difficulty for practical applications is raising of the appropriate parameters required for the calculation. Another problem may be interactions between the chromophores of the polymer constituents. Some of these problems are discussed in the following sections.

3.4.2.2 Parameters of Nucleic Acid Components

The UV absorbance of DNA double helices is determined by its four constituent bases: adenine, guanine, thymine, and cytosine. The spectral parameters of these bases have been described and discussed by several authors [36–38]. An approximate and very convenient representation of the directions of the transition dipole moments has been given by Norden [39] and Rizzo and Schellman [36] using the positions of ring atoms. This type of representation has also been used by Wada [40] for the transition dipole moments of the base pairs in DNA double helices. A summary of spectral data for the standard nucleic acid bases is compiled in Table 3.1.

The interactions between nucleic acid bases in the double helix as well as in other secondary or tertiary structures strongly affect the extinction coefficients. The stacking interactions lead to a clear reduction of the absorbance in the main UV band around 260 nm, which is known as hypochromism. Obviously it is quite difficult to determine the individual spectral parameters of the bases in helical structures containing different base residues. In a first approximation we may neglect differences in the individual contributions to the hypochromic effect. Furthermore, we assume that the directions of the transition dipole moments remain unchanged. Under these conditions the reduced dichroism may be calculated directly using the extinction coefficients of the monomer bases without specification of the hypochromic effect, because the reduced dichroism is a relative value.

As a test of this procedure we have calculated the reduced dichroism of two double helical B-DNA fragments from their x-ray structure, assuming orientation in the direction of the helix axis. For the case of the DNA dodecamer structure published by Yoon et al. [41], available under the label 1DN9 in the Brookhaven PDB, the calculated limiting reduced dichroism is -1.10; for another dodecamer structure published by Drew et al. [42], available under the label 1BNA in the PDB, the calculated limiting reduced dichroism is -1.20. These values are in the range of limiting reduced dichroisms derived from the experimental data for long DNA double helices [6]. If the bases would be stacked in exact perpendicular direction to the helix axis, the limiting reduced dichroism should be -1.5; the difference is mainly due to the "propeller twist" of the DNA base pairs.

The spectral parameters of double helical DNA are relatively simple, because DNA rarely contains any modified base residues. A much more complex composition is found in the case of tRNAs. The spectral properties of some of the modified nucleotides are closely related to

TABLE 3.1
Atoms Approximating Orientation of Transition Moments and Corresponding Extinction Coefficients in the Direction of the Transition Moment for Four Basic Nucleosides, at 248, 260, and 280 nm

$\lambda \rightarrow$	248 nm		260 nm		280 nm	
Base	Direction	$\varepsilon \times 10^3$	Direction	$\varepsilon \times 10^3$	Direction	$\varepsilon \times 10^3$
A	C4 → C5	10.8	C4 → C5	15.0	C4 → C5	2.2
G	C8 → C2	13.2	C8 → C2	12.1	C2 → C6	8.2
C	C4 → N1	6.8	C4 → N1	7.6	C4 → N1	6.8
U	N1 → C5	6.8	N1 → C5	10.0	N1 → C5	3.2

those of the standard nucleotides, because methylation of the sugar, for example, hardly affects the absorbance of the base. However, methylation at some position of the base itself may have a clearly detectable influence on the spectral parameters. For most of the modified bases, the spectral parameters have not been characterized in sufficient detail. In the absence of detailed information, a preliminary assignment of the spectral parameters based on the chemical similarity appears to be justified. According to this approximation we have used the following equivalences (the code used in the PDB is given in brackets): 1-methyladenosine (1MA) ≡ adenosine (A); 5-methylcytidin (5MC) and 2′-O-methylcytidine (OMC) ≡ cytidine (C); 1-methylguanosine (1MG), N(2)-methylguanosine (2MG), N(2)-dimethylguanosine (M2G), 7-methyguanosine (7MG), 2′-O-methylguanosine (OMG), wybutosine (YG), and inosine (I) ≡ guanosine (G); thymidine (T), dihydrouridine (H2U), ribosylthymidine (5MU), and pseudouridine (PSU) ≡ uridine (U).

When these parameters are used for the case of tRNA^{Phe} from yeast [43], the resulting molar extinction coefficient is 8.54×10^5 M^{-1} cm^{-1}. A comparison with the experimental value, 6.25×10^5 M^{-1} cm^{-1}, demonstrates an overall hypochromic reduction of the absorbance by 26.8%. This reduction is smaller than that found for DNA double helices, where the hypochromic reduction is in the range of about 40%. The difference is mainly due to the existence of single-stranded regions in the tRNA, where the bases are also stacked, but less extensively than in double helical structures. The general basis of the hypochromic effect has been described, whereas its details, e.g., the dependence on the nucleotide sequence and on the secondary–tertiary structure, have not been characterized at a sufficient accuracy. When these details are neglected in a first approximation, the calculated reduced limiting electric dichroism of tRNA^{Phe} (yeast) is ~−0.39, which is in a very satisfactory agreement with the experimental values found in the range around −0.4. The procedure used for the calculation of the electric dipole moment of tRNA^{Phe}, which is required for the calculation of the limiting electric dichroism, is discussed in Section 3.4.3.

3.4.2.3 Parameters of Protein Components

Measurements of the linear dichroism of proteins are usually restricted to the spectral range around 280 nm, where the absorbance is determined by the aromatic residues of tryptophan and tyrosine. Thus, the present discussion is restricted to the spectral parameters of these amino acids, which have been described by Yamamoto and Tanaka [44], Fasman [45], Maki et al. [46], and Philips and Levy [47]. A summary of the parameters is compiled in Table 3.2.

3.4.2.4 A Special Geometry: The Case of Curved DNA

In the cases discussed above, the orientation of each chromophore and its contribution to the extinction tensor has to be considered separately. When a given type of chromophore is

TABLE 3.2
Atoms Approximating Orientation of Transition Moments and Corresponding Extinction Coefficients in the Direction of the Transition Moment for Tryptophan and Tyrosine at 280 nm

Amino Acid	Direction	$\varepsilon \times 10^3$
Tryptophan	NE1 → CE3	4.8
Tryptophan	CG → CZ2	0.8
Tyrosine	CD1 → CD2	1.2

arranged in a repetitive ordered structure, it is possible to integrate the contributions of many chromophores and to derive analytical expressions. As an example we present the extinction tensor of curved DNA [48]. The double helix is bent at a constant curvature into a plane defined by the x- and the z-axes of the coordinate system. The bent double helix forms a circular arc, which is located symmetrically with respect to the x-axis. In this case the extinction tensor is given by

$$
\hat{\varepsilon} =
\begin{pmatrix}
\dfrac{L\varepsilon_\perp}{\rho}\left(\dfrac{\rho}{2}+\dfrac{\sin\rho}{2}\right)+\dfrac{L\varepsilon_\parallel}{\rho}\left(\dfrac{\rho}{2}-\dfrac{\sin\rho}{2}\right) & 0 & 0 \\
0 & L\varepsilon_\perp & 0 \\
0 & 0 & \dfrac{L\varepsilon_\perp}{\rho}\left(\dfrac{\rho}{2}-\dfrac{\sin\rho}{2}\right)+\dfrac{L\varepsilon_\parallel}{\rho}\left(\dfrac{\rho}{2}+\dfrac{\sin\rho}{2}\right)
\end{pmatrix}
$$

$$(3.17)$$

where ρ is the bending angle (corresponding to the arc angle); $\varepsilon_\parallel$ and $\varepsilon_\perp$ are the extinction coefficients per unit length of straight DNA for light polarized parallel and perpendicular to the helix axis, respectively; L is the length of the DNA.

3.4.2.5 A Special Chromophore: The Porphyrine Rings of Hemoglobin

Hemoglobin proteins contain porphyrine rings as prosthetic groups, which absorb light in the spectral range around 400 nm. The maximal absorbance is at 404.7 nm for oxyhemoglobins and at 435.8 nm for deoxyhemoglobins. The absorbance band is called the "Soret" or "B" band and is due to ($\pi^* \leftarrow \pi$) transitions polarized in the plane of the porphyrine rings. Thus, calculation of the extinction tensor requires determination of the planes of the 4 porphyrine chromophores contained in hemoglobin proteins. This determination was based on the coordinates obtained from the PDB [32]. For each porphyrine group its orientation was defined by calculation of the plane with the lowest sum of the squared deviations of its atoms [49]. In a local coordinate system of the porphyrine ring this plane corresponds to that formed by the x- and y-axes; the extinction tensor given in this coordinate system is diagonal with the components $\varepsilon'_{xx} = \varepsilon'_{yy} = 1$ and $\varepsilon'_{zz} = 0$; the absolute magnitude of ε'_{xx} and ε'_{yy} is not required for our calculation of the reduced dichroism (see Equation 3.1). The extinction tensor of the whole protein is the tensorial sum of all the local contributions and the final result is given in the coordinate system of the PDB crystal structure. The results, compiled for some hemoglobin species in Table 3.3, demonstrate that there are rather large differences in the orientation of the porphyrine rings with respect to the coordinate system of the crystal structure.

3.4.3 ELECTROSTATICS

3.4.3.1 Basic Definitions and Problems

As mentioned above in Section 3.3.2, there are two different types of standard dipole moments:

TABLE 3.3
Hemoglobins' Soret Absorption Ellipsoids. $\bar{\varepsilon}$ Is Mean Extinction Coefficient

Crystal Structure Data	$\varepsilon_{xx}/\bar{\varepsilon}$	$\varepsilon_{yy}/\bar{\varepsilon}$	$\varepsilon_{zz}/\bar{\varepsilon}$
2HHB	0.58	0.95	1.47
1HHO	0.63	0.88	1.49
1HBS	0.07	1.44	1.49
2DHB	0.56	0.97	1.47

1. Induced dipole moments usually reflect the polarization of electron clouds by external electric fields and, thus, are found for all types of matter. The magnitude of induced dipoles is generally assumed to be a linear function of the field strength $\vec{E}$

$$\vec{\mu}_i = \hat{\alpha}\vec{E} \tag{3.18}$$

where $\hat{\alpha}$ is the polarizability tensor.

2. Permanent dipoles are observed for molecules with a permanent anisotropy of their charge distribution. The magnitude of permanent dipoles is defined by

$$\vec{\mu} = \sum_j q_j \vec{r}_j \tag{3.19}$$

where q_j are the charges fixed in the molecular structure and $\vec{r}_j$ are the position vectors of the j-th charge in a molecular coordinate system.

The standard definition of dipole moments requires that the total charge of the polymer is zero. In this case the origin of the PCS may be selected without restrictions, because the dipole moment of particles with zero net charge is invariant to the choice of the origin of the PCS.

The application of these concepts and definitions to biopolymers raises difficulties for various reasons. Most biopolymers have a very complex structure with a large number of charged residues. The state of these residues is dependent on binding of the ligands. Because most of these ligands are charged and ligand binding is in a dynamic equilibrium, the state of each residue and its charge is subject to fluctuations. Furthermore, charged residues affect the state of their neighbors by electrostatic interactions, which are known to be far-reaching. Another problem results from the fact that binding of charged ligands is not always localized to a given site, but may be delocalized in the form of an ion atmosphere. Ion atmospheres with highly mobile ions can be displaced by electric fields of low field strengths already, resulting in extremely high polarizabilities. Compared to this particularly high sensitivity, the other parts and interactions of biopolymers are much less sensitive toward external electric fields, but in general any interactions in biopolymers with some electrostatic component may be affected by the application of external electric fields. In summary, it seems to be almost impossible to establish appropriate models for quantitative descriptions. However, a reasonable model has been developed at least for some cases.

3.4.3.2 Permanent Dipole Moments of Molecules with a Net Charge

If the standard definition would be applied without extension, dipole moments could hardly be used for biopolymers, because the numbers of positive and negative charges of most biopolymers are not equal under most solvent conditions. Although "isoelectric points" characterized by a zero net charge are known, e.g., for most proteins, these points are restricted to special pH values, which are dependent on the solvent conditions. Thus, the definition of dipole moments has to be extended to particles having a net charge. In this case the magnitude of dipole moments calculated according to Equation 3.19 is very much dependent on the origin of the coordinate system. Obviously, it is important to select the appropriate coordinate system leading to a dipole moment, which is compatible with the dipole moments obtained from the analysis of experimental data. Dipole moments are used to calculate the torque experienced under the influence of external electric fields and the interaction energy of the dipole with an external electric field for the evaluation of the orientation function (see Section 3.3.2). To get these values compatible with the experimental data, the center of the coordinate system should correspond to the center of diffusion [8,50,51]. In a coordinate system with its origin at this center, the coupling tensor within the general tensor for translational and rotational diffusion is symmetric. Moreover, the coupling tensor is zero for molecules with a symmetric shape. In this case a torque applied in a coordinate system with an origin corresponding to this center generates rotation without translation. As discussed in Section 3.5.3, special effects may result for molecules with a nonsymmetric shape. A procedure for the determination of the center of diffusion and a more detailed discussion of its physical significance is presented in Section 3.4.4.

3.4.3.3 Permanent Dipole Moments of Proteins

The problems associated with the calculation of dipole moments for biopolymers have been discussed primarily for the case of proteins. Some of these problems remained open. However, it can be demonstrated that for many proteins the dominant part of the dipole moment can be calculated by a rather simple procedure, because their dipole moments are mainly determined by residues with full charges. Most of these full charges are located on the side chains of the amino acids lysine, arginine, glutamic acid, aspartic acid, and histidine. These residues are subject to protonation–deprotonation equilibria and thus are dependent on the proton concentration given by the pH value of the solution.

A comparison with other ions interacting with biopolymers shows that protons may be considered as a special type of counterion. Under standard physiological conditions, i.e., at pH 7, their concentration is very low. Nevertheless, their effect on both structure and function is essential. The standard mode of proton interaction with proteins is "site binding" to proton acceptors, usually at side chains of amino acids. The type of "ion atmosphere" binding, encountered for the interaction of many other ions with macromolecules, is without practical relevance for protons.

3.4.3.3.1 Basic Model

The first level of approximation for a description of proton binding to the binding sites on proteins is based on the intrinsic pK values of the isolated residues. According to the law of mass action, dissociation of a proton H^+ from a site AH

$$AH \rightleftharpoons A^- + H^+ \tag{3.20}$$

is described by an equilibrium constant K_a

$$K_a = 10^{-pK_a} = \frac{[A^-][H^+]}{[AH]} \tag{3.21}$$

The equilibrium constants for protonation of all amino acid residues are well known and, thus, the state of protonation of proteins can be calculated. The equilibrium charge q_i of a residue with a pK_a at a pH value is given by

$$q_i = e(\kappa + \frac{1}{1 + 10^{(pH - pK)}}) \tag{3.22}$$

where κ is 0 for basic sites like lysine and -1 for acidic sites like glutamic acid. These charges together with their coordinates can be used via Equation 3.19 to calculate the simplest approximation to protein dipole moments. This procedure does not include any interactions between the charged residues and also does not include dipole moments resulting from α-helices. Thus, it is necessary to analyze the contributions resulting from these additional effects.

3.4.3.3.2 Advanced Model

When a titratable residue is a part of a protein, its pK_a can be significantly shifted from the value characteristic for an isolated amino acid in an aqueous environment and the basic model described above might not be valid. There are two sources for such shift. Charged residues interact with each other by Coulomb interactions, which are known to be far reaching. Thus, the protonation state of each residue is dependent on the state of its neighbors. The second possible source is a changed dielectric environment of the group in a protein: the high dielectric aqueous environment of the titratable group in isolated amino acid is, at least partially, substituted by a low dielectric environment. Moreover this low dielectric environment contains fixed atomic partial charges of other residues and interactions with these charges also play some role. Both factors can shift the group pK_a value even by a few pH units. Spectacular examples are provided by a large up shift of pK_a of Asp-26 in thioredoxin From 4.0 to 7.5 [52–54] and similarly large down shift of pK_a of the active site cysteine in protein tyrosine phosphatase from 8.3 to 4.67 [55]. Such shifts can significantly influence the dipole moment of proteins at given solvent conditions, in comparison to predictions of the basic model.

The prediction of the pK_a value of a given residue in a protein requires computation of the standard free Gibbs energy change, ΔG°, accompanying the process of proton dissociation, as

$$K_a = exp\left(\frac{-\Delta G^\circ}{RT}\right) \tag{3.23}$$

which is not an easy task. Moreover, already relatively small proteins might have large numbers of residues, which are subject to protonation, which makes the total number of possible protonation states (2^M, where M is the number of titratable residues) very large. The first difficulty is overcome by referring to a thermodynamic cycle [56] connecting proton dissociation in the protein environment to the same process for the titratable group in a reference compound environment (Figure 3.4), for which the pK_a value is assumed to be known.

As it is clear from the presented thermodynamic cycle the pK_a in the protein environment and the pK_a in the model compound environment are related by

$$pK_a^{(p)} = pK_a^{(w)} + \frac{\Delta G_{w\to p}^{A^-} - \Delta G_{w\to p}^{AH}}{2.303RT}$$

which can be rearranged to a computationally easier task

$$pK_a^{(p)} = pK_a^{(w)} + \frac{\Delta G_{protein}^{AH\to A^{--}} - \Delta G_{water}^{AH\to A^{--}}}{2.303RT} \tag{3.24}$$

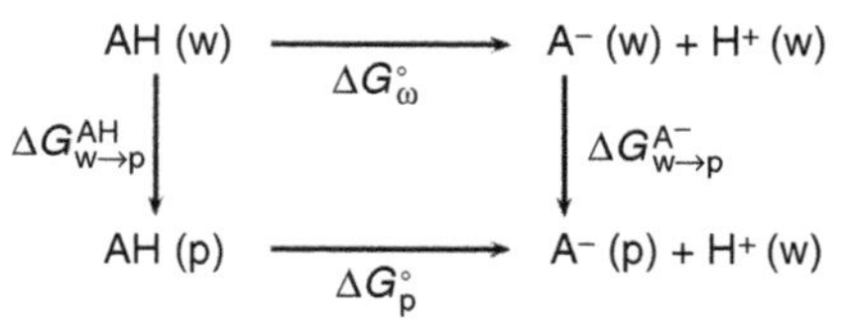

FIGURE 3.4 Thermodynamic cycle describing protonation equilibria of a titratable group in a protein and a model reference compound environment, respectively.

where $\Delta G_{\text{protein}}^{\text{AH} \to \text{A}^-}$ and $\Delta G_{\text{water}}^{\text{AH} \to \text{A}^-}$ mean the free energy change for substituting AH by A^- in protein and aqueous environment, respectively, which is relatively a small change. Equation 3.24 cannot be used directly to compute pK_a value for a given group in a protein because of the complicated protonation equilibria of all the remaining titratable groups, which renders the terms $\Delta G_{\text{protein}}^{\text{AH} \to \text{A}^-}$ and $\Delta G_{\text{water}}^{\text{AH} \to \text{A}^-}$ impossible for computation. But these terms can be computed for each particular protonation state of all remaining titratable residues of the protein. This allows one to assign each protonation state of the protein a definite energy level and compute average protonation states at given solvent conditions, and hence all pK_a values, using Boltzmann distribution law, which for one titratable group has the form

$$\frac{[\text{AH}]}{[\text{A}^-]} = \exp\left(\frac{-\Delta G}{RT}\right)$$

where ΔG is the free energy of the protonated state of a titratable group relative to the deprotonated state and dissociated proton in the solvent, which is by definition assumed to determine zero of the energy scale. Together with Equation 3.21 it gives

$$\Delta G = 2.303 RT(\text{pH} - pK_a)$$

For a macromolecule with many titratable groups it is convenient to change the reference point and instead of totally deprotonated protein take as the zero energy level the state with all titratable residues electrostatically neutral. With such a choice we can assign each protonation state of the protein the following free energy level

$$\Delta G(x_1, ..., x_M) = 2.303 RT \sum_{j=1}^{M} \gamma_j x_j (\text{pH} - pK_j^{(p)}) \tag{3.25}$$

where $(x_1, \ldots, x_M)$ is the ionization vector of M titratable sites of the protein with $x_i = 1$ when the site i is charged and $x_i = 0$ when it is neutral, and γ_i is -1 for an acidic site and $+1$ for a basic site, and subscript "a" in the pK was omitted to simplify the notation.

Because, as mentioned above, the $pK_j^{(p)}$ cannot be computed directly, the following steps are made in order to obtain computable expression for the free energy level characterizing given protonation pattern of the protein. For noninteracting sites the free energy of a protein can be expressed as a simple sum of terms analogous to that shown in Equation 3.25 for all the sites.

$$\Delta G = 2.303 RT \sum_j \gamma_j x_j (\text{pH} - pK_{\text{intrinsic}, j}) \tag{3.26}$$

where $pK_{\text{intrinsic}, j}$ is pK_a of jth site in the protein when all other sites are neutral [57].

$$pK_{\text{intrinsic}, j} = pK_j^{(w)} - \gamma_j \Delta\Delta G_{jj} / 2.303 RT$$

where $\Delta\Delta G_{jj}$ is the free energy difference for ionization of titratable site in protein environment with all other residues neutral, and ionization of the site in model compound environment

$$\Delta\Delta G_{jj} = \Delta G_{\text{protein}, j} - \Delta G_{\text{model}, j}$$

and

$$\Delta G_{\text{protein}, j} = G_{\text{protein}, j}^{\text{ionized}} - G_{\text{protein}, j}^{\text{neutral}}$$

where $G_{\text{protein}, j}^{\text{ionized}}$ is the free energy of the protein with jth site ionized and all other titratable groups neutral, and $G_{\text{protein}, j}^{\text{neutral}}$, is the free energy with all titratable groups, including the jth group, neutral. Analogous equation holds for model compound environment.

When interactions between titratable sites are taken into account, Equation 3.26 should be supplemented by an appropriate sum of interaction terms, for each pair of titratable residues when both are simultaneously ionized, hence

$$\Delta G(x_1, ..., x_M) = 2.303RT \sum_{j=1}^{M} x_j \gamma_j (\text{pH} - \text{p}K_j^{(\text{w})}) +$$
$$\sum_{j=1}^{M} x_j \Delta\Delta G_{jj} + \sum_{j=1}^{M-1} \sum_{k=j+1}^{M} x_j x_k \Delta\Delta G_{jk} \tag{3.27}$$

where

$$\Delta\Delta G_{jk} = \Delta G_{\text{protein}, j,k} - (\Delta G_{\text{protein}, j} + \Delta G_{\text{protein}, k})$$

In this equation, the meaning of $\Delta G_{\text{protein}, j,k}$ is similar to that of $\Delta G_{\text{protein}, j}$ but refers to ionization of two titratable residues, j and k, in otherwise neutral protein. Computation of $\Delta\Delta G_{jj}$ and $\Delta\Delta G_{jk}$ terms represents a separate problem. One possibility, used by the present authors [49,58,59] and by other groups, for example see references [60–65], is to model a protein as a low dielectric continuum with the shape determined by the solvent-accessible surface [66], with fixed charges at locations of the atomic nuclei, surrounded by a high dielectric continuum with the distribution of counterions determined by their energies according to Boltzmann's law. In such a case all these terms can be approximated by appropriate differences between the electrostatic energies of the corresponding charge distributions using a finite difference solution to the Poisson–Boltzmann equation [67,68]. Because of the usually huge number of possible protonation states of proteins it is not possible to compute energy levels of all the states. Instead, Monte Carlo [69] type methods may be used to generate a representative set of low-energy protonation states [8,70,71] or other approximate methods [72,73], which allow one to compute average protonation states of the residues, and hence their average charges, which can be used in permanent dipole moment computation.

Within the electrostatic approximation referenced above, each of the terms, $G_{\text{protein}, j, k}^{\text{ionized}}$, $G_{\text{protein}, j}^{\text{ionized}}$, $G_{\text{protein}, j}^{\text{neutral}}$, $G_{\text{model}, j}^{\text{ionized}}$, and $G_{\text{model}, j}^{\text{neutral}}$, is computed as

$$W = \frac{1}{2} \sum_{i=1}^{N} q_i \phi_i \tag{3.28}$$

where N is the total number of atomic partial charges corresponding to a particular situation, ϕ_i is the electrostatic potential at the location of the charge q_i and is considered to be the sum of the Coulombic potentials due to all other charges except the charge on i, plus the total reaction field at the location of charge i, resulting from existence of the dielectric boundary between the molecule and the solvent. The potential ϕ_i is a result of solving appropriate Poisson–Boltzmann equation. The electrostatic calculations are done using the UHBD program [74,75], and CHARMM22 parameter set [76] to describe atomic partial charges and

TABLE 3.4
Dipole Moments Calculated by the Basic and the Advanced Procedures

Protein	PDB	μ_7^b	ξ_7^b	pH^i	μ_i^b	ξ_i^b	μ_i^a	ξ_i^a
Lysozyme triclinic	2LZT	169	−0.530	11.1	184	−0.027	115	+0.134
Lysozyme tetragonal	1LYZ	97	−0.372	11.05	141	+0.206	85	−0.082
Subtilisin	2SBT	301	+0.251	6.6	313	+0.190	226	+0.118
Phopholipase A_2	1BP2	113	+0.786	8.2	115	+0.494	146	+0.344
Carboxypeptidase A_α	5CPA	521	−0.146	8.5	540	−0.177	554	−0.153
Concanavalin A	3CNA	237	−0.293	6.45	254	−0.435	250	−0.179
Ribonuclease A	3RN3	433	+0.618	9.65	433	+0.292	355	+0.364
Alcohol dehydrog. A	8ADH	776	−0.412	9.8	641	+0.162	554	−0.077
Myoglobin Oxy	1MBO	235	−0.498	7.5	256	−0.560	153	−0.615
Myoglobin Deoxy	5MBN	242	−0.559	7.35	256	−0.619	179	−0.690
α-chymotrypsin	4CHA	584	−0.184	9.1	555	−0.227	506	+0.214
Chymotrypsinogen A	2CGA	542	−0.083	9.75	441	−0.139	400	+0.006

Definition of symbols: $\mu \equiv$ dipole moment [D], $\xi \equiv$ limit dichroism; the subscripts 7 and i indicate values at pH 7 and the isoelectric pH^i, respectively; the superscripts b and a indicate values calculated by the basic and the advanced model, respectively; the values for the advanced model are from Ref. [71].
PDB, Protein Data Bank.

radii of amino acids in their neutral states. Ionization is represented as the addition of a ± 1 proton charge to a single atom in each group: the C atom of the main chain C-terminus; the N atom of the main chain N-terminus; CG of Asp; CD of Glu; CZ of Arg; NZ of Lys; ND1 or NE2 of His; OH of Tyr; and SG of Cys. The following reference pK_as were used: C-terminus 3.8; N-terminus 7.5; Asp 4.0; Glu 4.4; Arg 12.0; Lys 10.4; His 6.3; Tyr 9.6; Cys 8.3 (see Nozaki and Tanford [77]; Stryer [78]), and heme propionic acid 4.0 (Matthew et al. [79]). The solvent dielectric constant, ionic strength, and temperature are chosen to fit the conditions of the experimental studies. The protein and model compounds dielectric constants are set to 20.

3.4.3.3.3 Comparison of Results Obtained by the Basic and the Advanced Model
Because the known shifts of pK values for individual groups within proteins are included in the advanced model and because the advanced model is much more sophisticated, a rather large difference of the dipole moments obtained by the basic and the advanced model may be expected. However, a comparison of values computed for different proteins indicates that the difference remains relatively small (see Table 3.4). Apparently there is a compensation of pK shifts. The results in Table 3.4 demonstrate that the values obtained by the basic model, which is used in the automatic algorithm described below, are at least a reasonable approximation in most cases.

3.4.4 HYDRODYNAMICS

3.4.4.1 Macromolecule Motions in Viscous Solvents

Each macromolecule in a solution is continuously exposed to collisions and interactions with solvent molecules. In very diluted solutions, collisions between macromolecules are rare and can be neglected. Collisions and interactions with solvent molecules are a source of friction forces restricting macromolecular movement in solution on the one hand, and also generate a driving force for the random walk of the macromolecule in positional and orientational space, on the other hand. Quantitatively, frictional forces are characterized by resistance coefficients, and Brownian motion is characterized by diffusion coefficients. Both types of coefficients are related through Einstein–Stokes relations [80,81]. In what follows it is assumed that particles translate and rotate in a viscous fluid at sufficiently small Reynolds numbers, so their movements are quasisteady and the forces and torques exerted by the fluid on them are linear functions of their translational and angular velocities.

This section starts with simple scalar equations valid for molecules with spherical symmetry; subsequently a case of arbitrary shape is considered and finally practical applications for bead models of composed macromolecules are presented.

3.4.4.2 Spherical Molecules

When a spherical particle translates with the velocity $\vec{u}$ in a viscous unbounded fluid at rest, the force exerted on the molecule by the solvent is determined by Stokes formula

$$\vec{F} = -6\pi\eta\sigma\vec{u} \equiv -\zeta_t\vec{u} \tag{3.29}$$

where η is the viscosity of the solvent, σ is the radius of the sphere, and ζ_t is the translational resistance coefficient. When the particle rotates with angular velocity $\vec{\omega}$, the torque of the resistance force is

$$\vec{T} = -8\pi\eta\sigma^3\vec{\omega} \equiv -\zeta_r\vec{\omega} \tag{3.30}$$

and ζ_r is the rotational resistance coefficient. The resistance coefficients are related to the corresponding translation and rotation diffusion coefficients by the Stokes–Einstein equations

$$D_t = kT/\zeta_t \text{ and } D_r = kT/\zeta_r \tag{3.31}$$

where k is Boltzmann's constant and T is absolute temperature.

3.4.4.3 Molecules with Arbitrary Shape

When a particle of arbitrary shape moves in a viscous solvent with a linear velocity $\vec{u}_O$ and an angular velocity $\vec{\omega}$ of a given point O of this particle, then the resistance force and resistance torque experienced by the particle are not necessarily in the same directions as the velocity vectors. Moreover, translational and rotational movements are usually not independent [82]. Therefore tensors are required for connecting velocities and forces. The force $\vec{F}$ and torque $\vec{T}_O$ experienced by the particle are determined by the equation

$$\begin{pmatrix} \vec{F} \\ \vec{T}_O \end{pmatrix} = -\begin{pmatrix} \hat{R}_t & \hat{R}_{c,O}^\dagger \\ \hat{R}_{c,O} & \hat{R}_{r,O} \end{pmatrix} \cdot \begin{pmatrix} \vec{u}_O \\ \vec{\omega} \end{pmatrix} \tag{3.32}$$

where $\hat{R}$ is the translation, rotation, or coupling resistance tensor with the subscripts t, r, or c, respectively, † means transposed tensor, and subscript 'O' is used to identify quantities with a dependence of their numerical values on the location of the point O in the PCS. Tensors $\hat{R}_t$ and $\hat{R}_{r,O}$ are always symmetric, whereas tensor $\hat{R}_{c,O}$ is symmetric only, when the point O is located in the center of resistance (CR) [83,84]. These tensors describe the intrinsic geometrical properties of the body as a function of its size and shape; the solvent viscosity is included as a multiplicative factor.

It might be useful to see Equation 3.32 written for the case of a spherical particle with the PCS located at its center

$$
\begin{pmatrix} F_x \\ F_y \\ F_z \\ T_x \\ T_y \\ T_z \end{pmatrix} = - \begin{pmatrix} 6\pi\eta\sigma & 0 & 0 & 0 & 0 & 0 \\ 0 & 6\pi\eta\sigma & 0 & 0 & 0 & 0 \\ 0 & 0 & 6\pi\eta\sigma & 0 & 0 & 0 \\ 0 & 0 & 0 & 8\pi\eta\sigma^3 & 0 & 0 \\ 0 & 0 & 0 & 0 & 8\pi\eta\sigma^3 & 0 \\ 0 & 0 & 0 & 0 & 0 & 8\pi\eta\sigma^3 \end{pmatrix} \cdot \begin{pmatrix} u_x \\ u_y \\ u_z \\ \omega_x \\ \omega_y \\ \omega_z \end{pmatrix} \tag{3.32a}
$$

This equation corresponds to both Equation 3.29 and Equation 3.30. Similar to the case of spherical particles, the resistance tensors for particles of arbitrary shape can be converted to the corresponding diffusion tensors by the generalized Einstein relation [80,81])

$$
\begin{pmatrix} \hat{D}_{t,O} & \hat{D}_{c,O}^\dagger \\ \hat{D}_{c,O} & \hat{D}_r \end{pmatrix} = kT \begin{pmatrix} \hat{R}_t & \hat{R}_{c,O}^\dagger \\ \hat{R}_{c,O} & \hat{R}_{r,O} \end{pmatrix}^{-1} \tag{3.33}
$$

where $\hat{D}_{t,O}$ is translational diffusion tensor, $\hat{D}_r$ is rotational diffusion tensor, and $\hat{D}_{c,O}$ is translational–rotational coupling diffusion tensor. Again a dependence on the location of the PCS is shown by the subscript O. Translational and rotational tensors are always symmetric and the coupling tensor is symmetric only, when the coordinate system is located in the center of diffusion (CD) [85]. The CR and the CD for unsymmetric molecules are usually located at different points.

3.4.4.4 Hydrodynamic Interactions

When two or more spherical particles move in a viscous solvent, the resistance force experienced by each particle cannot be described just by Equation 3.29. The hydrodynamic interactions between these particles are described by a modified equation, e.g., for a sphere i with radius σ_i

$$
\vec{F}_i = -\zeta_i(\vec{u}_i - \vec{v}_i) \tag{3.34}
$$

where $\zeta_i = 6\pi\eta\sigma_i$, $\vec{u}_i$ is the velocity of ith sphere, and $\vec{v}_i$ is the velocity that the solvent would have at the position of the center of ith particle, if that particle were absent. This velocity is not just zero, even if the solvent is at rest at infinity, because the movement of the remaining spherical particles creates a velocity field in the solvent. This phenomenon is known as hydrodynamic interaction [86–88]. Because $\vec{F}_i$ is the frictional force exerted on the ith bead by the solvent, then $-\vec{F}_i$ is the force that this bead exerts on the medium. This force is assumed to be localized at the position $\vec{r}_i$ of the center of the bead. When a point force is exerted on a viscous incompressible fluid in steady motion at low Reynolds number, it produces an extra velocity field $\vec{v}(\vec{r})$ everywhere in the medium. This extra velocity field is a linear function of the force but not necessarily in the direction of the force and can be quantitatively characterized by an equation [87]:

$$
\vec{v}(\vec{r}) = \hat{T}(\vec{r} - \vec{r}_i) \cdot (-\vec{F}_i) \tag{3.35}
$$

with the hydrodynamic interaction tensor $\hat{T}(\vec{r})$ determined by equation [86]:

$$\hat{T}(\vec{r}) = \frac{1}{8\pi\eta r}(\hat{I} + \frac{\vec{r}\vec{r}}{r^2}) \tag{3.36}$$

where $\hat{I}$ is the unit tensor. Thus the velocity $\vec{v}_i$ can be expressed as

$$\vec{v}_i = -\sum_{j\neq i} \hat{T}_{ij} \cdot \vec{F}_j \tag{3.37}$$

where $\hat{T}_{ij} \equiv \hat{T}(\vec{r}_i - \vec{r}_j)$; $\vec{r}_i$ and $\vec{r}_j$ are position vectors for the spheres i and j. The Oseen tensor was subsequently modified by Rotne and Prager [88] and Yamakawa [89] to take into account the finite size of beads. They presented the hydrodynamic interaction tensor between the ith and jth spherical elements in the following forms:

$$\hat{T}_{ij} = \frac{2}{8\pi\eta r_{ij}}\left[\hat{I} + \frac{\vec{r}_{ij}\vec{r}_{ij}}{r_{ij}^2} + \frac{2\sigma^2}{r_{ij}^2}\left(\frac{1}{3}\hat{I} - \frac{\vec{r}_{ij}\vec{r}_{ij}}{r_{ij}^2}\right)\right] \tag{3.38}$$

when $r_{ij} > 2\sigma$, and

$$\hat{T}_{ij} = \frac{1}{6\pi\eta\sigma}\left[\left(1 - \frac{9}{32}\frac{r_{ij}}{\sigma}\right)\hat{I} + \frac{3}{32}\frac{\vec{r}_{ij}\vec{r}_{ij}}{\sigma r_{ij}^2}\right] \tag{3.39}$$

when $r_{ij} \leq 2\sigma$. Garcia de la Torre and Bloomfield [90,91] modified the Oseen tensor to take into account differences in the size of spherical subunits, and obtained the expression

$$\hat{T}_{ij} = \frac{1}{8\pi\eta r_{ij}}\left[\hat{I} + \frac{\vec{r}_{ij}\vec{r}_{ij}}{r_{ij}^2} + \frac{\sigma_i^2 + \sigma_j^2}{r_{ij}^2}\left(\frac{1}{3}\hat{I} - \frac{\vec{r}_{ij}\vec{r}_{ij}}{r_{ij}^2}\right)\right] \tag{3.40}$$

which is valid for non-overlapping beads.

3.4.4.5 Equations for Bead Models

One possibility of calculating resistance tensors for a given particle is to model the particle by spherical elements (beads) and to treat each bead as source of friction. The appropriate sum provides the resistance properties of the whole particle.

From Equation 3.34 and Equation 3.37 one gets the following system of equations for forces experienced by beads

$$\sum_{j=1}^{N} \hat{Q}_{ij}\vec{F}_j = \zeta_i\vec{u}_i \tag{3.41}$$

with

$$\hat{Q}_{ij} = \delta_{ij}\hat{I} + (1 - \delta_{ij})\zeta_i\hat{T}_{ij}$$

Assuming that all $\vec{u}_i$ are equal to (1,0,0) and solving Equation 3.41 we may find the first column of translational resistance tensor. We simply use the following relation:

$$\vec{F} = \sum_i \vec{F}_i = \hat{R}_t \vec{u}_O \tag{3.42}$$

The velocity $\vec{u}_O$ in this equation is the same as the velocity $\vec{u}_i$ assigned to all beads. Similarly, choosing unit velocities of the all beads in y and z directions, we determine the two remaining columns of the translational resistance tensor. Using calculated forces experienced by beads during translational motion with unit velocity in x, y, or z direction we calculate total torques experienced by the macromolecule and corresponding columns of translation–rotation coupling resistance tensor. For this we use the following relations:

$$\vec{T}_O := \sum_i \vec{r}_i \times \vec{F}_i = -\hat{R}_{c,O} \vec{u}_O \tag{3.43}$$

where $\vec{r}_i$ is the position vector of the ith bead with respect to O. Subsequently, we assume that the whole macromolecule rotates with unit angular velocity around x, y, or z axis and calculate the translational velocity of each bead by means of the equation

$$\vec{u}_i = \vec{\omega} \times \vec{r}_i \tag{3.44}$$

Using a corresponding procedure we get for subsequent choices of direction of three subsequent columns of transposed translation–rotation coupling resistance tensor, by means of equation

$$\vec{F} = \sum_i \vec{F}_i = -\hat{R}_{c,O}^{\dagger} \vec{\omega} \tag{3.45}$$

Finally we calculate the resulting torque on the macromolecule during rotational motions in the three directions with unit angular velocities and get the columns of the rotational resistance tensor by means of equation

$$\vec{T}_O = \sum_i \vec{r}_i \times \vec{F}_i = -\hat{R}_{r,O} \vec{\omega} \tag{3.46}$$

Exact derivations are discussed by Garcia de la Torre and Bloomfield [90,91]. The linear $3N \times 3N$ Equation 3.41 (where N is the number of spherical elements) may be solved for relatively small N by direct inversion of matrix $\hat{Q}$, but for large numbers N numerical methods are required [90–92]. The above methods were checked by construction of bead models for objects with known analytical or almost exact results and comparison of the calculated resistance coefficients [31,92,93].

The method for calculation of hydrodynamic parameters from a bead model for a macromolecule presented above leads to erroneous results in some special situations [94]. The rotational friction coefficients are obtained by calculating the translational velocity for each bead i from the rotational velocity ω of the molecule, according to

$$\vec{u}_i = \vec{\omega} \times \vec{r}_i,$$

which gives the corresponding resistance forces $\vec{F}_i$ from solving Equation 3.41, and the corresponding torque $\vec{T}_i$ experienced by the ith bead according to

$$\vec{T}_i = \vec{r}_i \times \vec{F}_i$$

Thus, a single bead located at the origin of the coordinate system is associated with $T = 0$, whereas the correct result is $T = 8\pi\eta\sigma^3\omega$. Garcia de la Torre and Rodes [94] proposed to correct the principal rotational resistance coefficients calculated according to the above-described procedures by adding to each of them the term $6\eta V$, where V is the volume of the hydrated macromolecule. This procedure leads to some overestimation of rotational resistance coefficients. If the bead model is exactly adapted to a known translational resistance coefficient, then the rotational resistance coefficient calculated for the same model is too low, when no correction is applied and too large, when corrected according to Garcia de la Torre and Rodes. A different correction term was proposed [95], which is progressively smaller, when a given object is modeled in a more detailed way by an increasing number of beads with decreasing radius. This correction is based on the "solvent-accessible volume" and is calculated in close analogy to the solvent-accessible surface of macromolecules, as described by Lee and Richards [96]. This method of correcting rotational resistance coefficients was checked by calculations for spheres and cylinders built from spherical elements. The agreement of the simulated results with the expected values for both translational and rotational coefficients is very good (for more details, see Ref. [95]).

3.4.4.6 Centers of Resistance and Diffusion

This section is ended with recalling convenient equations for transformation of resistance and diffusion tensors given in a starting coordinate system centered at an arbitrary point O to tensors in a coordinate system with its origin at the CR or CD, respectively.

When the resistance tensor is known at a certain point O, then the coordinates of the CR in this system are given by [81]

$$x_{CR} = \frac{\hat{R}_{c,32} - \hat{R}_{c,23}}{\hat{R}_{t,22} + \hat{R}_{t,33}}$$

$$y_{CR} = \frac{\hat{R}_{c,13} - \hat{R}_{c,31}}{\hat{R}_{t,11} + \hat{R}_{t,33}} \quad (3.47)$$

$$z_{CR} = \frac{\hat{R}_{c,21} - \hat{R}_{c,12}}{\hat{R}_{t,11} + \hat{R}_{t,22}}$$

Similarly, if diffusion tensors are determined in a PCS centered at O then the coordinates of the CD in this system are given by [85]

$$x_{CD} = \frac{\hat{D}_{c,23} - \hat{D}_{c,32}}{\hat{D}_{r,22} + \hat{D}_{r,33}}$$

$$y_{CD} = \frac{\hat{D}_{c,31} - \hat{D}_{c,13}}{\hat{D}_{r,11} + \hat{D}_{r,33}} \quad (3.48)$$

$$z_{CD} = \frac{\hat{D}_{c,12} - \hat{D}_{c,21}}{\hat{D}_{r,11} + \hat{D}_{r,22}}$$

3.4.5 SIMULATION OF ELECTRO-OPTICAL TRANSIENTS

In this section a procedure for the simulation of electro-optical transients for particles with arbitrary shapes is discussed. This procedure is based on the results of Wegener et al. [97]. It is limited to small electric fields and can be used for the interpretation of decay time constants.

The theory predicts for a given particle up to five individual decay time constants and their amplitudes. The procedure is very fast, because the transients are calculated by analytical expressions [97].

The PCS (see Section 3.4.2.1) is centered at the CD [85] of the particle and its axes are directed along the principal axes of rotational diffusion tensor. Thus, in this coordinate system the rotational diffusion tensor is diagonal. The permanent dipole moment, the polarizability tensor, and the extinction tensor of the particle are assumed to be known in this coordinate system. Each macromolecule is considered to be independent of the remaining macromolecules in the cell.

If $\rho(\Omega, t)$ denotes the orientational distribution at time t (which is independent of spatial position of particles because of assumed independence) with the normalization condition over all $8\pi^2$ orientations

$$\int \rho(\Omega, t)d\Omega = 1$$

and an electric field $\vec{E}$ applied to the electro-optical cell has the direction of the z'-axis of the LCS cf.section 3.4.2.1, then the reduced dichroism of the solution may be written as

$$\xi(t) = \frac{1}{\bar{\varepsilon}} \int [\varepsilon_{z'z'}(\Omega) - \varepsilon_{y'y'}(\Omega)] \rho(\Omega, t)d\Omega$$

The last equation is valid, when the z'-axis of LCS shows the direction of the applied electric field and the x'-axis the direction of the incident polarized light. However, since ρ is cylindrically symmetric about the laboratory z'-axis, $\varepsilon_{y'y'}(\Omega)$ in the last equation can be replaced by $[\varepsilon_{x'x'}(\Omega) + \varepsilon_{y'y'}(\Omega)]/2$ and the time dependence of the reduced dichroism may be presented as

$$\xi(t) = \frac{1}{2\bar{\varepsilon}} \int [3\varepsilon_{z'z'}(\Omega) - \text{Tr}\,\hat{\varepsilon}] \rho(\Omega, t)d\Omega \qquad (3.49)$$

where $\text{Tr}\,\hat{\varepsilon}$ denotes the sum of diagonal elements of the tensor and is independent of the particle orientation. Therefore the transient electro-optical signal of the investigated solution can be determined provided that the $\varepsilon_{z'z'}$ component of particle's extinction tensor in the LCS is known as a function of Ω and the orientation function $\rho(\Omega, t)$ is known as a function of time.

Wegener et al. showed that the electro-optical rise and decay signals can be reduced to sums of products of optical and orientational factors, which are functions of diffusion coefficients, and electrical and optical parameters [97,98]. For the decay curves, Wegener et al. have shown [97], that the reduced dichroism shown by Equation 3.49 can be presented as a sum of up to five different exponential terms. Their method was based on the formulation of a rotational diffusion equation for $\rho(\Omega, t)$ and solving it for given initial conditions. The existence of up to five exponential terms can be rationalized by the fact that symmetrical, second-rank tensors given in spherical coordinates have up to five independent components, which are linear combinations of the Cartesian components. Although symmetrical second-rank tensor possess up to six distinct elements, the trace of the tensor is invariant under rotation; thus there is a maximum of five independent spherical components [97,99].

For convenience, the final result of Wegener et al. for the force-free decay of electric dichroism is written as

$$\xi(t) = \frac{E^2}{15} \sum_{i=1}^{N_{\text{case}}} A_i \exp(-t/\tau_i) \qquad (3.50)$$

where N_{case} is the number of different exponential terms, A_i are the relaxation amplitudes, which are functions of components of the rotational diffusion, electric polarization and absorption coefficients tensors, and permanent dipole moment vector, and $1/\tau_i$ are the corresponding eigen values of the rotational diffusion equation, which are functions of the components of the principal values of the rotational diffusion tensor. Inverse of these eigen values determine the relaxation times for the decay curves. When all principal values of the rotational diffusion tensor are different, there are five different eigen values $1/\tau_i$ and Equation 3.50 gives five different exponential terms. When two or all three principal values of the rotational diffusion tensor are equal, the number of different eigen values $1/\tau_i$ is reduced and consequently the number of the different exponential terms in Equation 3.50 is reduced. Exact expressions for these effective amplitudes and eigen values, for different cases regarding

TABLE 3.5

Relaxation Amplitudes, A_i, and Eigen Values, $1/\tau_i$, in the Time-Dependent Electric Dichroism Decay Expression, $\xi(t) = (E^2/15) \sum_{i=1}^{N_{case}} A_i exp(-t/\tau_i)$, Valid in the Limit of Low Values of the External Electric Field E, According to Wegener et al.

i	A_i	$1/\tau_i$
	Case: all D_i are different; $N_{case} = 5$	
1	$(3/4)(c/M)^2(\varepsilon_{11} - \varepsilon_{22})(\chi_{11} - \chi_{22})+$ $(1/4)(b/M)^2(3\chi_{33} - Tr\chi)(3\varepsilon_{33} - Tr\varepsilon) - (\sqrt{3}/4)(bc/M^2)$ $[(\chi_{11} - \chi_{22})(3\varepsilon_{33} - Tr\varepsilon)+$ $(\varepsilon_{11} - \varepsilon_{22})(3\chi_{33} - Tr\chi)]$	$6D - 2\Delta$
2	$3 \in_{13} \chi_{13}$	$3(D + D_2)$
3	$3 \in_{23} \chi_{23}$	$3(D + D_1)$
4	$3\varepsilon_{12}\chi_{12}$	$3(D + D_3)$
5	$(3/4)(b/M)^2(\in_{11} - \in_{22})(\chi_{11} - \chi_{22})+$ $(1/4)(c/M)^2(3\chi_{33} - Tr\chi)(3 \in_{33} - Tr \in)+$ $(\sqrt{3}/4)(bc/M^2)[(\chi_{11} - \chi_{22})(3 \in_{33} - Tr \in)+$ $(\in_{11} - \in_{22})(3\chi_{33} - Tr\chi)]$	$6D + 2\Delta$
	Case: $D_1 = D_2 < D_3$; $N_{case} = 3$	
1	$(1/4)(3 \in_{33} - Tr \in)(3\chi_{33} - Tr\chi)$	$6D_1$
2	$3(\in_{13} \chi_{13}+ \in_{23} \chi_{23})$	$5D_1 + D_3$
3	$3 \in_{12} \chi_{12} + (3/4)(\in_{11} - \in_{22})(\chi_{11} - \chi_{22})$	$2D_1 + 4D_3$
	Case: $D_1 = D_2 = D_3$; $N_{case} = 3$	
1	$3 \in_{12} \chi_{12} + (3/4)(\in_{11} - \in_{22})(\chi_{11} - \chi_{22})$	$2D_1 + 4D_3$
2	$3(\in_{13} \chi_{13}+ \in_{23} \chi_{23})$	$5D_1 + D_3$
3	$(1/4)(3 \in_{33} - Tr \in)(3\chi_{33} - Tr\chi)$	$6D_1$
	Case: $D_1 = D_2 = D_3$; $N_{case} = 1$	
1	$(1/2)[3Tr(\in \cdot \chi) - Tr \in Tr\chi]$	$6D_1$

$D = (1/3)(D_1 + D_2 + D_3)$; $\Delta = (D_1^2 + D_2^2 + D_3^2 - D_1D_2 - D_1D_3 - D_2D_3)^{1/2}$
$\chi_{ij} = \beta^2 d_i d_j + \beta\alpha_{ij}$; $Tr\,\Xi = \Xi_{11} + \Xi_{22} + \Xi_{33}$, for $\Xi = \chi, \varepsilon$ or $\varepsilon \cdot \chi$
$b = 3(D_3 - D) + 2\Delta$; $c = \sqrt{3}(D_1 - D_2)$; $M = 2(\Delta b)^{1/2}$

Source: From Wegener, W.A., Dowben, R.M., and Koester, V.J., *J. Chem. Phys.*, 70, 622, 1979.

relations between principal values of the rotational diffusion tensor, are given in the publication of Wegener et al. [97]. Table 3.5 presents effective amplitudes and the corresponding eigen values for different possible relations between the principal values of the rotational diffusion tensors. The expressions for relaxation amplitudes shown in Table 3.5 indicate that there are some additional circumstances when the number of exponential terms in Equation 3.50 is reduced. This occurs when one axis or all three principal axes of the rotational diffusion tensor coincide with one or all three, respectively, of the principal axes of the absorption and alignment tensors. In such cases some of the off-diagonal components of these tensors, ε_{ij} or χ_{ij} with $i \neq j$, are zero and the relaxation amplitudes containing these terms might turn to zero as well.

The data presented in Table 3.5 can be used to obtain transient electric dichroism signals as functions of time. By calculation of the reduced dichroism at appropriate time intervals one gets basically the same kind of information as obtained in an electric field jump spectrometer. Such sets of data can be evaluated exactly as decay curves registered during experiments.

Wegener [98] also presented analogous equations for electro-optical rise curves. The theory for the rise curves predicts up to eight individual rise time constants and their amplitudes. In the same publication Wegener presented modified equations for the decay curves. Both amplitudes, those for the rise curves and the new amplitudes for the decay curves, include a contribution due to the hydrodynamic coupling between the translational and the rotational motions of particles. This contribution was not included in the 1979 formulation for the decay amplitudes. A very large effect of hydrodynamic coupling has been found recently [100] for the case of curved DNA fragments (see Section 3.4.2.4).

3.4.6 Algorithm for Automatic Simulation of Dichroism Data

In the preceding sections we have described the basis for the calculation of electro-optical data from macromolecular structures. For standard applications it is useful to have a simple procedure for the prediction of experimental parameters from macromolecular structures available at public data banks like the Protein Data Bank (PDB) [32]. We have developed an algorithm for automatic calculation of dichroism parameters from structures provided at the PDB [32]. Because the basis for these calculations has been described above, we simply add here some details on the calculation pathway. The first step of these calculations is extraction of all data, which are important for the evaluation of the hydrodynamic, optical, and electrical parameters.

3.4.6.1 Automatic Construction of Bead Models from Atomic Structures

Macromolecular structures are determined by x-ray analysis or NMR measurements routinely to the resolution of individual atoms. For the analysis of these structures by hydrodynamic procedures atomic resolution is not required. Usually it is sufficient to represent subunits like amino acids or nucleotides by spherical elements, the so-called beads. We practice a very simple automatic procedure for the generation of bead models from the atomic coordinates of subunits. For each subunit we calculate its center of mass, assuming the same mass for each atom. The coordinates of this center are simply generated as the average of all x-, y-, and z-coordinates in the dataset for a given subunit. The radius of a bead representing a given subunit is determined as the distance between its center and its atom with the largest distance from the center. This algorithm results in slightly different radii for different subunits. Because simulations on models with beads of a single radius are favorable, we use the average of the bead radii calculated for the subunits of a given macromolecule. The calculated bead radius may be increased to account for hydration, e.g., by the thickness of a water layer.

The selection of a proper correction for hydration may be difficult. As a demonstration that this correction need not be arbitrary, we describe here a simple example of a "calibration" procedure for the case of nucleic acids: We start from the x-ray structure for a B-DNA dodecamer and construct from this a DNA fragment with 43 base pairs. The details of our construction are described in Ref. [101]. A fragment with 43 base pairs is selected, because the dichroism decay time constant has been measured for such a fragment. After conversion into a bead model its dichroism decay time constant is calculated. By variation of the bead radius we get an approximate fit to the experimental decay time constants and an "optimal" bead radius of 6.45 Å. This radius is 1 Å larger than the value calculated by the automatic algorithm described above and, thus, provides an approximate value for a hydration correction.

The capacity of present-day computers is sufficiently large to allow for simulations of bead models with almost any number of beads. However, calculations with very high bead numbers do not really make sense. Thus, in the case of very large macromolecules we prefer to combine two or more subunits into beads of increased size, using an algorithm corresponding to that described above for single subunits.

3.4.6.2 Optical and Electrical Parameters

The optical parameters of rigid macromolecules are described by the summation of the contributions from individual chromophores in the form of the extinction tensor as discussed in Section 3.4.2. For the calculation of permanent dipole moments we use the positions of charged residues according to the basic model described in Section 3.4.3. When other contributions are expected, the advanced model may be used and the resulting dipole vector may be introduced at an intermediate step of the calculation pathway (see below). A major problem remains a reliable prediction of polarizabilities. Because a general model for predictions is not available yet, the only practical procedure is estimation on the basis of experimental data.

3.4.6.3 Sequence of Steps in the Automatic Algorithm

The automatic set of programs starts with reading the data specifying the structure of a given macromolecule from a file in the format of the PDB. When the structure includes more than a single polymer chain, the user may select a single chain or any combination of given chains. Based on the input data the program decides, whether these data are for a protein or a nucleic acid component. The decision is used for the allocation of charges. The next step is the representation of the global structure by a bead model as described above. The average bead radius evaluated by the algorithm is reported and the user may change the value to account for a hydration layer. Then the user specifies the pH value, which is required for the calculation of the net charge and of the dipole moment by the basic model. At this state the dipole moment is given with respect to the center of mass. The extinction tensor is calculated at 248, 260, and 280 nm for nucleic acids, whereas the tensor is evaluated at the single wavelength of 280 nm for proteins. A subprogram prepares the coordinates of the bead model for generation of a graphics output. The next subprogram calculates the "solvent-accessible volume" for the correction of the bead model simulations. Then a $3N \times 3N$ equation system, where N is number of beads, is solved to get the tensors for translational and rotational coupling resistances. The upper limit for the number of beads is 400 in our present implementation. The following subprogram evaluates the principal coordinate system with its origin at the CD and its axes directed along the principal axes of the rotational diffusion tensor. These data and the alignment tensor are used for calculation of individual exponential components of the dichroism decay curve according to Wegener et al. [97]. The dipole moment is given with respect to the CD and the limit value of the dichroism is evaluated at a wavelength selected by the user. Finally, the individual exponential components are combined into a dichroism decay curve, i.e., the dichroism is given as a function of time in a form that can be evaluated by routine programs for the analysis of experimental data.

3.5 THEORY AND PRACTICE

3.5.1 COMBINED ANALYSIS OF CRYSTAL AND SOLUTION STRUCTURES

Quantitative procedures for the electro-optical analysis of biological macromolecules have been practiced in this laboratory over a period of ~20 years. Among the various applications we have selected three different examples.

3.5.1.1 A Globular Protein: α-Chymotrypsin

One of the problems persisting for many years was the nature of protein dipole moments. Large dipole moments were indicated for different proteins, for example by dielectric spectroscopy [102]. However, the nature of these dipole moments remained unclear, because various interpretations of the experimental results were possible. The problem may be illustrated by the example of double helical DNA. In this case the very large dipole moments revealed by the experimental data were attributed to a polarization of its ion atmosphere. This raises the question, how to exclude such an effect in the case of proteins? Moreover, other interpretations, for example fluctuating dipole moments first proposed by Kirkwood [103], could not be excluded.

This general problem was analyzed by a combined experimental and theoretical approach. Permanent dipole moments can be distinguished from induced dipole moments by measurements of rise curves and of the response to inverted field pulses. For standard proteins the analysis requires both high-field pulses for sufficiently large signals and a high-time resolution. The comparison of rise curves for DNA restriction fragments [104] with those for a protein-like α-chymotrypsin [8] showed a clear difference. In the latter case the data demonstrated the existence of a permanent dipole moment, using the criterion established by Benoit [105], whereas in the former the rise curves showed a convolution product of a process reflecting ion binding and the process of rotational diffusion. These experiments clearly demonstrated the dominance of the permanent dipole moment in the case of α-chymotrypsin. Another experimental demonstration for the permanent dipole in α-chymotrypsin came from measurements of dichroism amplitudes as a function of the field strength [8,9]. These data were analyzed by various orientation functions, and shown to be consistent with the orientation function for permanent dipoles over the full range of field strengths. In the case of DNA fragments the stationary dichroism shows a more complex dependence on the field strength [6]. A decision on the nature of dipoles requires data over a broad range of electric field strengths. For this type of analysis the cable discharge technique with its broad voltage range proved to be very useful. The quantitative analysis of the orientation function for α-chymotrypsin provided the magnitude of the permanent dipole moment: the experimental value of 480 D (pH 8.3, after correction for the internal directing field) is remarkably high.

A final proof for the existence of large dipole moments in proteins came from the analysis of crystal structures. Because the x-ray structure of α-chymotrypsin had been determined, the experimental data could be compared with theoretical values based on the crystal structure. The basic procedure (see Section 3.4.3.3) using simple equilibrium degrees of protonation provided a value of 560 D, in satisfactory agreement with the experimental result. More extensive calculations [8,71] demonstrated that other contributions to the dipole moment of α-chymotrypsin remain relatively small, but shift the calculated dipole moment (480 D) to a perfect agreement with the experimental result [8].

TABLE 3.6
Electric Dichroism of α-Chymotrypsin

Parameter	Units	Experimental in Solution	Calculated from X-Ray Structure
Dipole moment	D	480[a]	480[b]
Limit dichroism		+0.13	+0.129
Dichroism decay time, 2°C	ns	31	29.4

[a]Corrected for the internal directing field.
[b]Advanced model.

Sources: From Antosiewicz, J. and Porschke, D., The nature of protein dipole-moments—experimental and calculated permanent dipole of α-chymotrypsin, *Biochemistry*, 28, 10072, 1989; Antosiewicz, J., Computation of the dipole-moments of proteins, *Biophys. J.*, 69, 1344, 1995.

The case of α-chymotrypsin was investigated as a model (see Table 3.6) because of its simple and relatively stable structure. For a demonstration of the reliability of the electro-optical approach, we add another example: the electric dichroism of the lac-repressor in solution was studied before any crystal structure was available. An analysis of both the rise curves and the orientation function showed the existence of a permanent dipole moment [7]. The magnitude of 1200 D evaluated from the dependence of the stationary dichroism on the electric field strength seemed to be unreasonable. Furthermore, the electro-optical results were against general expectation, because lac-repressor is a tetramer. The molecular structure was not known yet, but a tetrahedral symmetry seemed to be very likely, which should be without a permanent anisotropy of its charge distribution. A few years later the crystal structure of lac-repressor showed a clear asymmetric arrangement of the subunits [106,107], in perfect agreement with the results of the electro-optical investigation.

3.5.1.2 A Folded Nucleic Acid: tRNA[Phe] (Yeast)

Electro-optical data for structures like DNA double helices can be analyzed relatively easily, because simple basic procedures can be used for the calculation of both dichroism amplitudes and time constants. The analysis of electro-optical data for structures like tRNAs is much more difficult because of the relatively complex folding of the nucleotide chain. The problem can be solved by the simulation procedures described above. Based on the known crystal structures available for many different tRNAs, the hydrodynamic parameters can be calculated by bead models constructed, for example, by substitution of each nucleotide by a bead (Figure 3.5). The optical parameters are defined by the absorbance tensor, which is calculated from the orientation of the bases in the crystal. Finally, a permanent dipole moment is predicted from the nonsymmetric arrangement of the phosphate residues. All these parameters have been calculated [101] from the crystal structure [43] of tRNA[Phe] (yeast) using the automatic algorithm and proved to be in satisfactory agreement with experimental data (see Table 3.7).

One of the problems in the simulation of electro-optical data for molecules like tRNAs is the absence of detailed knowledge on the binding of metal ions *in solution*. For example, Mg^{2+}-ions may be bound at various sites and the degree of binding can hardly be established. In principle, the dependence of electro-optical parameters may be used as a source of information on ion binding.

The simulation implied that the structure is rigid, which is justified for its core but not for the 3 terminal nucleotides CCA, the so-called CCA-end. The deviation between the experimental and the calculated dichroism decay time constant is probably mainly due to a flexibility of the CAA-end.

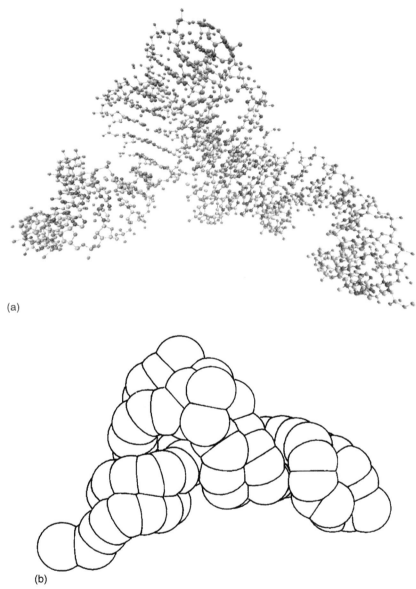

(a)

(b)

FIGURE 3.5 (a) Crystal structure of tRNA^{Phe} (yeast) and (b) bead model constructed by the automatic algorithm (Section 3.4.6). (From Ladner, J.E. et al., Atomic coordinates for yeast phenylalanine transfer-RNA, *Nucleic Acids Res.*, 2, 1629, 1975.)

3.5.1.3 A Protein–Nucleic Acid Complex: Cyclic-AMP-Receptor + Operator DNA

Protein–DNA complexes have a central function in the readout of the genetic information. This readout can be activated by protein-induced bending of specific DNA segments. Thus, the degree of DNA bending in DNA–protein complexes is of general interest. Electro-optical procedures can be very useful for the determination of bending angles *in solution*. A well-known example of DNA bending by a protein is the case of the cyclic-AMP-receptor. The crystal structure of this receptor bound to an operator DNA fragment [108] has been

TABLE 3.7
Electric Dichroism of tRNA^Phe (Yeast)

Parameter	Units	Experimental in Solution	Calculated from X-Ray Structure
Dipole moment	D	250	200[c]
Limit dichroism		−0.49[a]	−0.41[c]
Dichroism decay time	ns	41.6[b]	50.8[c]

[a] 1 mM NaCl, 1 mM MgCl$_2$, 1 mM cacodylate pH 7; 2°C.
[b] 10 mM NaCl, 1 mM MgCl$_2$, 1 mM cacodylate pH 7; 2°C.
[c] With 3 Mg^{2+}-ions in the structure; for details see Ref. [101].
Source: From Porschke, D. and Antosiewicz, J.M., *Biophys. J.*, 58, 403, 1990.

determined up to high resolution (Figure 3.6). We have analyzed the same complex in solution by measurements of the electric dichroism [109] and got the results compiled in Table 3.8. The protein–DNA complex is characterized by a positive dichroism and a permanent electric dipole moment.

The electro-optical parameters of the DNA fragment measured in the absence of the protein are completely different: its limit dichroism is negative (−1.0); the dipole of the free DNA is of the standard-induced type and does not show indications for saturation of its polarizability up to field strengths of 70 kV/cm. Thus, the type of molecular orientation is reverted completely upon protein binding. The molecular basis for this change is clearly illustrated by the structure shown in Figure 3.6. In the complex the phosphate charges are located on one side in a nonsymmetric arrangement, leading to a permanent dipole in a direction causing the positive dichroism. These qualitative expectations are verified quantitatively by the results of the simulation procedure. The almost perfect agreement of experimental and simulated parameters (Table 3.8) demonstrates that the structures in the crystal and in solution are equivalent.

An advantage of the electro-optical approach is the fact that the procedure is relatively simple and can be easily extended to the analysis of, e.g., DNA fragments of different chain lengths and sequences. Moreover the solvent conditions can be varied relatively easily. Finally, the results are not influenced by crystal-packing effects.

3.5.2 Apparent Acceleration and Retardation Effects

Dichroism decay times are considered to be a reliable source of information on the size and shape of macromolecular objects. This is certainly correct, but there are special cases, where experimental dichroism decay curves are misleading, when the individual exponential components are not resolved correctly. Correct resolution of the individual components is not always as simple as expected. The problem can be easily demonstrated for simulated dichroism transients, where difficulties resulting from a limited signal to noise ratio are not expected. An example has been reported for the case of hemoglobin mutants [49]. Although the global structure of these mutants is virtually identical and can hardly be distinguished with respect to size and shape, the dichroism decay curves are quite different (Figure 3.7).

The reason for this unexpected effect is clearly revealed by an analysis of the details. As predicted by the theory, five individual exponentials contribute to the dichroism decay. The time constants of these individual components are very similar for the different hemoglobin mutants, as expected from their very similar global structure. However, the amplitudes of these components, which are affected by the dipole vector and the extinction tensor, are very

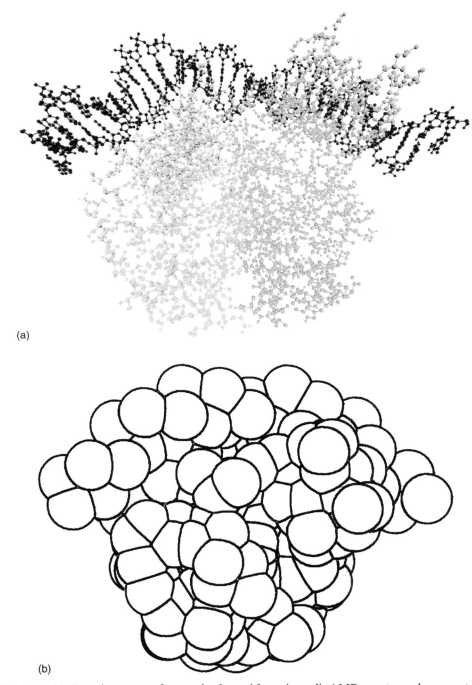

(a)

(b)

FIGURE 3.6 (a) Crystal structure of a complex formed from the cyclic AMP receptor and a promoter DNA fragment with 30 base pairs [108]; (b) bead model constructed by the automatic algorithm (Section 3.4.6).

different. The mutations lead to changes of the magnitude and the direction of the dipole. Because the overall electric dipole moment of hemoglobin is relatively small, a simple mutation may induce a relatively large change. The resulting changes of the amplitudes

TABLE 3.8
Electric Dichroism of Catabolite Gene Activator + Promoter DNA

Parameter	Units	Experimental in Solution	Calculated from X-Ray Structure
Dipole moment	D	900	1000
Limit dichroism		+0.3	+0.357
Dichroism decay time, 20°C	ns	62	64

Source: From MeyerAlmes, F.J. and Porschke, D., *J. Mol. Biol.*, 269, 842, 1997.

lead to dichroism decay curves of very different shape. This would not be a problem, if the exponential components could be separated with a sufficient reliability. It is well known that separation of exponential functions is a serious problem, when the time constants are similar. Tests on the data simulated for the hemoglobin mutants demonstrate that separation of the individual exponentials is not trivial at all—even for data resulting from computer simulations. Thus, the problem is much more serious in the case of experimental data with a limited signal-to-noise ratio.

Hemoglobin is not a singular exception. Similar effects were observed for curved DNA fragments.

The theoretical basis for a quantitative description of electro-optical transients has been presented in the literature quite some time ago. Although the factors contributing to electro-optical transients are well known, the impact of amplitudes on transients is usually neglected. The example given above is a warning that the various factors contributing to electro-optical signals should be combined for a reliable interpretation of experimental data. An isolated analysis of hydrodynamic parameters may result in serious misinterpretations.

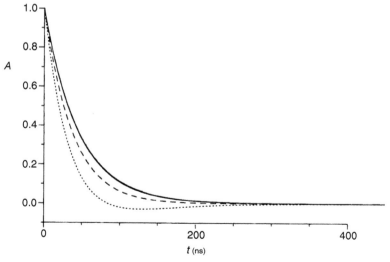

FIGURE 3.7 Dichroism decay curves calculated according to the crystal structures of hemoglobin variants: human S deoxy (1HBS) and human deoxy (2HHB) [—indistinguishable on the given scale]; human oxy (1HHO) [– – –]; horse deoxy (2DHB) [. . .]. (Reproduced from Antosiewicz, J.M. and Porschke, D., *Biophys. J.*, 68, 655, 1995. With permission of the Biophysical Society.)

3.5.3 Coupling of Translational and Rotational Motions

As mentioned in Section 3.4.4.3 already, a special electro-optical effect must be expected, when the shape of particles is not symmetric. In this case the coupling tensor within the general tensor for translational and rotational diffusion has nonzero elements, even when the center of the coordinate system is at the CD. As a result we get "hydrodynamic coupling" between translational and rotational motions of the particle.

The equations for the calculation of electro-optical transients in the presence of hydrodynamic coupling were presented by Wegener [98] in the limit of low electric field strengths. The calculation of transients at higher electric field strengths requires numerical simulations. For both types of calculations, complete information on the tensors of translational and rotational diffusions as well on the coupling tensor is required. These parameters can be calculated by bead model simulations. Furthermore, the optical parameters in the form of the extinction tensor must be available. Finally, the tensor of the electrical polarizability and the dipole vector are required. In view of the fact that some efforts are due to get reasonable values for all these parameters, it is hardly surprising that the effect of hydrodynamic coupling did not receive as much attention as necessary.

Hydrodynamic coupling effects for curved DNA have been discussed by Bertolotto et al. [110]. In a recent investigation [100] we applied detailed quantitative procedures for the calculation of hydrodynamic coupling and analyzed the effect of this coupling on the electric dichroism of smoothly bent DNA fragments. As shown by the bead model in Figure 3.8, smoothly bent DNA still has some elements of symmetry, but lost part of its symmetry with

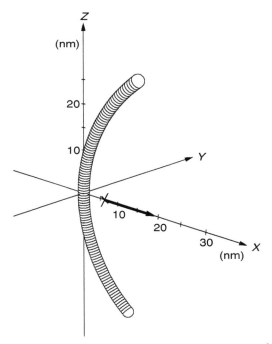

FIGURE 3.8 Model of a smoothly bent 179 bp DNA with 101 beads of 12.5 Å diameter. Bending angle 116°. The position of the center of diffusion is indicated by the x-sign. The direction of the permanent dipole is indicated by the arrow. (Reprinted with permission from Porschke, D. and Antosiewicz, J.M., *J. Phys. Chem. B*, 109, 1034, 2005; copyright 2005 American Chemical Society.)

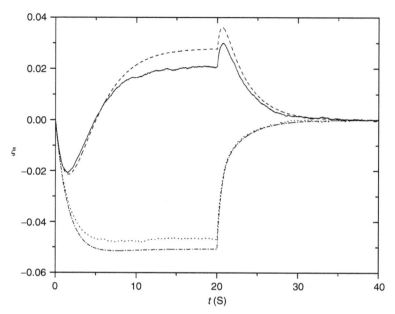

FIGURE 3.9 Reduced electric dichroism ξ as a function of time t calculated for the smoothly bent 179 bp DNA (see Figure 3.8). The electric field pulse was applied at time zero and terminated after 20 μs. The transients calculated by Brownian dynamics were calculated with and without hydrodynamic coupling (— and ..., respectively; average of 1.1×10^{10} transients). The transients were also calculated by the equations of Wegener for the limit of low electric field strengths with and without hydrodynamic coupling (- - - and ---, respectively). (Reprinted with permission from Porschke, D. and Antosiewicz, J.M., *J. Phys. Chem. B*, 109, 1034, 2005; copyright 2005 American Chemical Society.)

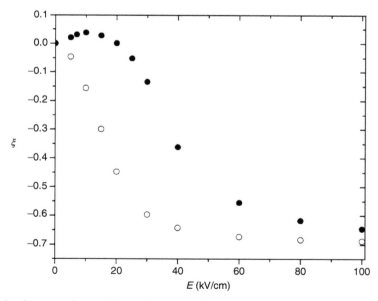

FIGURE 3.10 Stationary values of the reduced electric dichroism ξ as a function of the electric field strength E for the smoothly bent 179 bp DNA in the presence (●) and the absence (○) of hydrodynamic coupling. The parameters of the smoothly bent rod are as described in Figure 3.8. (Reprinted with permission from Porschke, D. and Antosiewicz, J.M., *J. Phys. Chem. B*, 109, 1034, 2005; copyright 2005 American Chemical Society.)

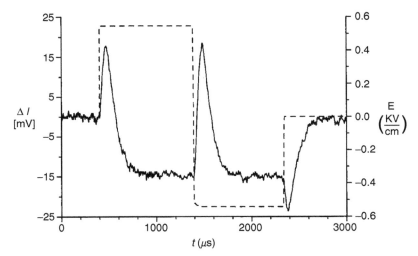

FIGURE 3.11 Intensity change of light polarized parallel to the electric field vector and transmitted through a solution of a 587 bp DNA fragment as a function of time during exposure to an electric field pulse E (dashed line); 2°C, 1 mM NaCl, 1 mM cacodylate pH 7, 0.2 mM EDTA, DNA helix concentration 0.121 μM. (Reprinted from Porschke, D., *Biophys. Chem.*, 49, 127, 1994. With permission from Elsevier.)

respect to straight DNA. The loss of symmetry leads to huge effects of hydrodynamic coupling on the electric dichroism. As shown in Figure 3.9, the transients for a bent DNA with 179 base pairs in the absence and the presence of hydrodynamic coupling are very different. A large difference is also found in the stationary values of the dichroism calculated for different electric field strengths (Figure 3.10). Obviously the case of smoothly bent DNA is special because of its dimensions, its unusual asymmetry, and its high charge density, leading to a high electrophoretic mobility in relation to its large dimensions. A much smaller influence of hydrodynamic coupling was found for tRNA^{Phe} [28], a nonsymmetric molecule with a more compact structure.

A positive stationary dichroism has been found for DNA fragments in an intermediate range of chain lengths [111] with 399, 587, and 859 base pairs (see Figure 3.11). This effect was explained by permanent dipole moments resulting from nonsymmetric charge distributions in bent conformations. The model calculation on curved DNA fragments not only supports this interpretation, but also demonstrates that a major part of the effect is due to coupling of translational and rotational motion.

Unfortunately the amount of information on the effects of hydrodynamic coupling on electro-optical effects is still very limited. Nevertheless, we may conclude that hydrodynamic coupling has huge effects in the case of molecules with a distinct nonsymmetric shape together with a high charge density. The effect of hydrodynamic coupling is clearly much smaller for molecules with a relatively compact structure, like tRNAs. More simulations are required for a sufficient knowledge of these special effects.

3.5.4 POLARIZATION DYNAMICS: EXPERIMENTS AND SIMULATIONS

Among the open problems in electro-optics is a reasonable description of the polarization of ion atmospheres around polyelectrolytes. Because of the complexity of polyelectrolytes, it has been very difficult to describe polarization quantitatively. Thus, an analysis of ion polarization by Brownian dynamics simulations should be useful. Because of space limitations we do not describe our simulation approach in detail, but simply compare some key results with experimental data obtained on the polarization dynamics.

We have simulated the response of the ion atmosphere around a fixed polyelectrolyte to electric field pulses [112]. The polyelectrolyte with a charge spacing corresponding to that of DNA and a length corresponding to 43 base pairs was fixed in the center of a box and was surrounded by mobile counterions. When the electric field is applied, the usual Brownian motion of the ions is combined with electrophoretic motion—under the restrictions imposed by electrostatic interactions. Upon these conditions the ion distribution, which is symmetric *on average* in the absence of an external electric field, loses its symmetry and a dipole is generated. One of the major advantages of the Brownian dynamics approach is the fact that simulations provide detailed information on the time response. An example for the simulated rise of a dipole induced by an electric field is shown in Figure 3.12. The dipole arrives at a stationary value after a rise period, which can be described by an exponential with a time constant of 18.6 ns.

As briefly discussed in Section 3.5.1.1, the dynamics of ion polarization around DNA has been characterized by measurements of the electric dichroism for DNA restriction fragments [104] with a high time resolution. These measurements demonstrated that the dichroism rise cannot be described by a simple exponential but is consistent with a convolution product of a process reflecting an ion dissociation reaction with a second process reflecting rotation of the

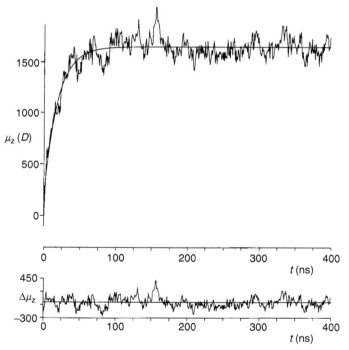

FIGURE 3.12 Rise curve of the dipole moment simulated for a linear polymer with 80 charges at a distance of 1.7 Å between these charges with an equivalent number of counterions in a cube of $40^3\,\mathrm{nm}^3$. Electric field strength $20\,\mathrm{kV/cm}$, average of 512 trajectories; the line without noise represents a least squares fit with $\tau = 18.6\,\mathrm{ns}$ and $\mu z = 1678$ D. (Reprinted with permission from Grycuk, T., Antosiewicz, J.M., and Porschke, D., *J. Phys. Chem.*, 98, 10881, 1994; copyright 1994 American Chemical Society.)

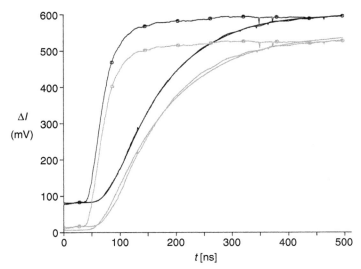

FIGURE 3.13 Dichroism rise curve for a 76 base pair fragment induced by a field pulse of 61.9 kV/cm in 0.5 mM NaCl, 0.5 sodium cacodylate, pH 7, 0.1 mM EDTA (average obtained from 10 shots). The line marked by circles is a birefringence curve of water (induced by a corresponding field pulse and recorded with the same adjustment of the detector and transient recorder) and serves as a reference for deconvolution. The data are given in black together with a fit with two exponentials ($\tau_p = 14$ ns reflecting polarization and $\tau_r = 85.4$ ns reflecting rotation) and for comparison in grey together with a fit by a single exponential (in the single-τ-fit the sum of residuals is larger by a factor of 20). (Reprinted from Porschke, D., *Biophys. Chem.*, 22, 237, 1985. With permission from Elsevier.)

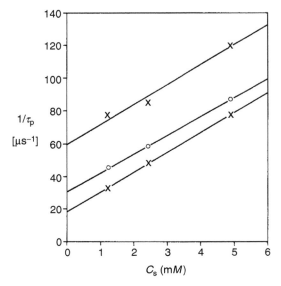

FIGURE 3.14 Dependence of the reciprocal polarization time constant $1/\tau_p$ upon the Na^+-concentration for a helix with 76 base pairs at 3 different electric field strengths: 24.6 kV/cm ($\times$); 54.1 kV/cm ($\circ$); 68.8 kV/cm ($+$). (Reprinted from Porschke, D., *Biophys. Chem.*, 22, 237, 1985. With permission from Elsevier.)

double helix (Figure 3.13). The experimental time constants for the first process at different monovalent ion concentrations were found in the range between 10 and 30 ns and revealed a concentration dependence [104], indicating an ion dissociation reaction (Figure 3.14).

The Brownian dynamics simulation of the ion polarization process is consistent with the experimental data. First of all, the time constants for the dipole rise are very similar. Moreover, the change of the ion distribution around the polymer observed in the simulations [112] indicates field-induced dissociation of ions from the polymer. The processes induced by electric fields around polyelectrolytes are very complex, but a simple form of description by an asymmetric dissociation of ions from the polymer is justified by the available data.

This very short comparison is presented to indicate the direction of further developments: an exact comparison of the experimental data obtained for well-defined systems with the results of computer simulations should be very useful to analyze the nature of complex phenomena.

3.5.5 CONFORMATIONAL DYNAMICS: DNA DOUBLE HELICES

Electro-optical methods can be used to analyze the dynamics of macromolecules by different procedures. The transients of the electric dichroism or birefringence may be used directly to get information on the internal dynamics of the system under investigation. This approach has been used, for example, to measure the persistence length of DNA [19,113] and the time constants of DNA bending [19,114].

Another very useful approach to the investigation of the dynamics of various processes, including the dynamics of biopolymers, is available by application of the magic angle technique. Electric field pulses serve to induce a shift of the equilibrium state, which is analyzed by measurements of absorbance changes under magic angle conditions (see Section 3.3.1). A recent example is the characterization of the B–A transition of DNA double helices [115,116]. It is well known that the B–A transition is induced by a reduction of the water activity, which is generated in aqueous solution by addition of ethanol. The B–A transition is observed in a narrow range of the ethanol content and is reflected by changes in the UV absorbance. Upon application of electric field pulses to DNA solutions of various ethanol content, a special relaxation process is observed under magic angle conditions within the narrow range of ethanol contents, where the B–A transition is observed by various other methods including CD spectroscopy. A detailed analysis of the transients observed for different DNA samples of various chain lengths demonstrated that electric field pulses induce a reaction in the direction from the A- to the B-form by a dipolar stretching mechanism. A spectrum of time constants was observed in the range from 2 to 100 μs. A comparison with model calculations according to a linear Ising model shows that the cooperativity of the B–A transition is relatively low. Because of the boundary conditions for induction of the B–A transition in solution, the electric field pulse method is the only currently known method for the analysis of its dynamics. The results illustrate the timescale for the internal dynamics of DNA double helices.

3.6 SUMMARY

Advances in the analysis of macromolecular structures by electro-optical procedures are described. First we discuss progress in the experimental techniques like extension to physiological salt concentrations and combinations of electro-optical techniques with stopped-flow and flash photolysis. The main part is devoted to a discussion of the theoretical background for a quantitative analysis of electro-optical data. The concept of dipole moments for polymer structures with net charges and the calculation of these dipole moments are discussed. We describe an automatic algorithm for the calculation of electro-optical parameters from macro-molecular structures. The algorithm is used to calculate hydrodynamic, electrical, and optical parameters from molecular structures deposited in the format of the PDB. In the last step of the algorithm, dichroism decay curves are calculated for the limit case of low electric field strengths. We compare theory and practice for a protein, a folded nucleic acid and a protein–nucleic acid complex. Finally, we also give examples for the analysis of more complex phenomena like hydrodynamic coupling and ion polarization by Brownian dynamics.

REFERENCES

1. Porschke, D., Geisler, N., and Hillen, W., Structure of the complex between lac repressor head-piece and operator DNA from measurements of the orientation relaxation and the electric dichroism, *Nucleic Acids Res.*, 10, 3791, 1982.
2. Porschke, D., Time-resolved analysis of macromolecular structures during reactions by stopped-flow electrooptics, *Biophys. J.*, 75, 528, 1998.
3. Porschke, D., Effects of electric fields on biopolymers, *Annu. Rev. Phys. Chem.*, 36, 159, 1985.
4. Fredericq, E. and Houssier, C., *Electric Dichroism and Electric Birefringence*, Monographs on physical biochemistry, Clarendon Press, Oxford, 1973.
5. Stoylov, S.P., *Colloid Electro-Optics: Theory, Techniques, Applications*, Academic Press, London, 1991.
6. Diekmann, S. et al., Electric properties and structure of DNA restriction fragments from measurements of the electric dichroism, *Biophys. Chem.*, 15, 157, 1982.
7. Porschke, D., Electric, optical and hydrodynamic parameters of lac repressor from measurements of the electric dichroism—high permanent dipole-moment associated with the protein, *Biophys. Chem.*, 28, 137, 1987.
8. Antosiewicz, J.M. and Porschke, D., The nature of protein dipole-moments—experimental and calculated permanent dipole of α-chymotrypsin, *Biochemistry*, 28, 10072, 1989.
9. Schonknecht, T. and Porschke, D., Electrooptical analysis of α-chymotrypsin at physiological salt concentration, *Biophys. Chem.*, 58, 21, 1996.
10. Porschke, D., Electrostatics and electrodynamics of bacteriorhodopsin, *Biophys. J.*, 71, 3381, 1996.
11. Eigen, M. and DeMaeyer, L., Relaxation methods, in *Investigation of Rates and Mechanisms of Reactions*, S. Friess, E. Lewis, and A.Weissberger, Eds., Vol. VIII, Part II of *Technique of Organic Chemistry*, Interscience, New York, 1963, pp. 895–1054.
12. Grünhagen, H.H., *Entwicklung einer E-Feldsprung-Apparatur mit optischer Detektion und ihre Anwendung auf die Assoziation amphiphiler Elektrolyte*, Ph.D. thesis, Braunschweig, 1974.
13. Hoffman, G.W., Nanosecond temperature-jump apparatus, *Rev. Sci. Instrum.*, 42, 1643, 1971.
14. Porschke, D., Cable temperature jump apparatus with improved sensitivity and time resolution, *Rev. Sci. Instrum.*, 47, 1363, 1976.
15. Porschke, D. and Obst, A., An electric-field jump apparatus with ns time resolution for electro-optical measurements at physiological salt concentrations, *Rev. Sci. Instrum.*, 62, 818, 1991.
16. Porschke, D. and Grell, E., Electric parameters of Na+/K+-Atpase by measurements of the fluorescence-detected electric dichroism, *Biochim. Biophys. Acta Bioen.*, 1231, 181, 1995.
17. Wang, G.Y. and Porschke, D., Dipole reversal in bacteriorhodopsin and separation of dipole components, *J. Phys. Chem. B*, 107, 4632, 2003.
18. Porschke, D., Analysis of chemical and physical relaxation processes of polyelectrolytes by electric field pulse methods: A comparison of critical comments with facts, *Ber. Bunsengesell. Phys. Chem.*, 100, 715, 1996.
19. Diekmann, S. et al., Orientation relaxation of DNA restriction fragments and the internal mobility of the double helix, *Biophys. Chem.*, 15, 263, 1982.
20. Porschke, D. and Jung, M., The conformation of single stranded oligonucleotides and of oligo-nucleotide-oligopeptide complexes from their rotation relaxation in the nanosecond time range, *J. Biomol. Struct. Dyn.*, 2, 1173, 1985.
21. Porschke, D. and Nolte, G., Deconvolution of electrooptical data in the frequency-domain—relaxation processes of DNA from rigid rods to coiled spheres, *Macromolecules*, 27, 590, 1994.

22. Porschke, D., Strong bending of purple membranes in the M-state, *J. Mol. Biol.*, 331, 667, 2003.
23. Subramaniam, S. and Henderson, R., Molecular mechanism of vectorial proton translocation by bacteriorhodopsin, *Nature*, 406, 653, 2000.
24. Kühlbrandt, W., Bacteriorhodopsin—the movie, *Nature*, 406, 569, 2000.
25. Porschke, D., Molecular electro-optics, in *Protein-Ligand Interactions: Hydrodynamics and Calorimetry*, S.E. Harding and B.Z. Chowdhry, Eds., Oxford University Press, Oxford, 2001, pp. 197–221.
26. Okonski, C.T., Yoshioka, K., and Orttung, W.H., Electric properties of macromolecules. IV. Determination of electric and optical parameters from saturation of electric birefringence in solutions, *J. Phys. Chem.*, 63, 1558, 1959.
27. Hogan, M., Dattagupta, N., and Crothers, D.M., Transient electric dichroism of rod-like DNA-molecules, *Proc. Natl. Acad. Sci. USA*, 75, 195, 1978.
28. Antosiewicz, J.M., Grycuk, T., and Porschke, D., Brownian dynamics simulation of electrooptical transients for solutions of rigid macromolecules, *J. Chem. Phys.*, 95, 1354, 1991.
29. Antosiewicz, J.M. and Porschke, D., Brownian dynamics simulation of electrooptical transients for complex macrodipoles, *J. Phys. Chem.*, 97, 2767, 1993.
30. Broersma, S., Rotational diffusion constant of a cylindrical particle, *J. Chem. Phys.*, 32, 1626, 1960.
31. Tirado, M.M., Martinez, C.L., and Garcia de la Torre, J.G., Comparison of theories for the translational and rotational diffusion-coefficients of rod-like macromolecules—application to short DNA fragments, *J. Chem. Phys.*, 81, 2047, 1984.
32. Berman, H.M. et al., The Protein Data Bank, *Nucleic Acids Res.*, 28, 235, 2000.
33. Carter, E.S. and Tung, C.S., NAMOT2—A redesigned nucleic acid modeling tool: construction of non-canonical DNA structures, *Comp. Appl. Biosci.*, 12, 25, 1996.
34. Kuhn, W., Dührkop, H., and Martin, H., Anisotropie der Lichtabsorption gelster Molekle im elektrischen Feld, *Z. Phys. Chem.*, B45, 121, 1939.
35. Wilson, R.W. and Schellman, J.A., Dichroic tensor of flexible helices, *Biopolymers*, 16, 2143, 1977.
36. Rizzo, V. and Schellman, J.A., Matrix-method calculation of linear and circular-dichroism spectra of nucleic-acids and polynucleotides, *Biopolymers*, 23, 435, 1984.
37. Voet, D. et al., Absorption spectra of nucleotides, polynucleotides, and nucleic acids in the far ultraviolet, *Biopolymers*, 1, 193, 1963.
38. Cech, C.L. and Tinoco, I., Circular-dichroism calculations for double-stranded polynucleotides of repeating sequence, *Biopolymers*, 16, 43, 1977.
39. Norden, B., Applications of linear dichroism spectroscopy, *App. Spectrosc. Rev.*, 14, 157, 1978.
40. Wada, A., Dichroic spectra of biopolymers oriented by flow, *Appl. Spectrosc. Rev.*, 6, 1, 1972.
41. Yoon, C. et al., Structure of an alternating-B DNA helix and its relationship to A-tract DNA, *Proc. Natl. Acad. Sci. USA*, 85, 6332, 1988.
42. Drew, H.R. et al., Structure of a B-DNA dodecamer—conformation and dynamics, *Proc. Natl. Acad. Sci. USA*, 78, 2179, 1981.
43. Ladner, J.E. et al., Atomic coordinates for yeast phenylalanine transfer-RNA, *Nucleic Acids Res.*, 2, 1629, 1975.
44. Yamamoto, Y. and Tanaka, J., Polarized absorption-spectra of crystals of indole and its related compounds, *Bull. Chem. Soc. Jpn.*, 45, 1362, 1972.
45. Fasman, G.D., *CRC Handbook of Biochemistry and Molecular Biology*, CRC Press, Cleveland, OH, 1976.
46. Maki, I., Kitaura, K., and Nishimoto, K., Effect of sigma-2 interaction on electronic-structure of molecules in excited-states. 2. Oscillator strength, *Bull. Chem. Soc. Jpn.*, 51, 401, 1978.
47. Philips, L.A. and Levy, D.H., The rotationally resolved electronic-spectrum of indole in the gas-phase, *J. Chem. Phys.*, 85, 1327, 1986.
48. Antosiewicz, J.M. and Porschke, D., Turn of promotor DNA by camp receptor protein characterized by bead model simulation of rotational diffusion, *J. Biomol. Struct. Dyn.*, 5, 819, 1988.
49. Antosiewicz, J.M. and Porschke, D., Electrostatics of hemoglobins from measurements of the electric dichroism and computer-simulations, *Biophys. J.*, 68, 655, 1995.
50. Mysels, K.J., Electric polarizability of polar ions, *J. Chem. Phys.*, 21, 201, 1953.

51. Antosiewicz, J.M. and Porschke, D., An unusual electrooptical effect observed for DNA fragments and its apparent relation to a permanent electric moment associated with bent DNA, *Biophys. Chem.*, 33, 19, 1989.
52. Jeng, M.F. and Dyson, H.J., Direct measurement of the aspartic acid 26 pK_a for reduced Escherichia coli thioredoxin by C-13 NMR, *Biochemistry*, 35, 1, 1996.
53. Langsetmo, K., Fuchs, J.A., and Woodward, C., The conserved, buried aspartic-acid in oxidized Escherichia-coli thioredoxin has a pK_a of 7.5—Its titration produces a related shift in global stability, *Biochemistry*, 30, 7603, 1991.
54. Wilson, N.A. et al., Aspartic-acid 26 in reduced Escherichia-coli thioredoxin has a pK_a greater-than-9, *Biochemistry*, 34, 8931, 1995.
55. Zhang, Z.Y. and Dixon, J.E., Active-site labeling of the Yersinia protein-tyrosine-phosphatase—the determination of the pK_a of the active-site cysteine and the function of the conserved histidine-402, *Biochemistry*, 32, 9340, 1993.
56. Warshel, A., Calculations of enzymatic-reactions—calculations of pK_a, proton-transfer reactions, and general acid catalysis reactions in enzymes, *Biochemistry*, 20, 3167, 1981.
57. Tanford, C. and Kirkwood, J.G., Theory of protein titration curves. I. General equations for impenetrable spheres, *J. Am. Chem. Soc.*, 79, 5333, 1957.
58. Antosiewicz, J.M., McCammon, J.A., and Gilson, M.K., Prediction of pH-dependent properties of proteins, *J. Mol. Biol.*, 238, 415, 1994.
59. Antosiewicz, J.M. et al., Computing ionization states of proteins with a detailed charge model, *J. Comput. Chem.*, 17, 1633, 1996.
60. Bashford, D. and Karplus, M., pK_a's of ionizable groups in proteins—atomic detail from a continuum electrostatic model, *Biochemistry*, 29, 10219, 1990.
61. Bashford, D. and Gerwert, K., Electrostatic calculations of the pK_a values of ionizable groups in bacteriorhodopsin, *J. Mol. Biol.*, 224, 473, 1992.
62. Yang, A.S. et al., On the calculation of pK_a's in proteins, *Proteins: Struct. Funct. Genet.*, 15, 252, 1993.
63. Demchuk, E. and Wade, R.C., Improving the continuum dielectric approach to calculating pK_a's of ionizable groups in proteins, *J. Phys. Chem.*, 100, 17373, 1996.
64. Alexov, E.G. and Gunner, M.R., Incorporating protein conformational flexibility into the calculation of pH-dependent protein properties, *Biophys. J.*, 72, 2075, 1997.
65. Rabenstein, B., Ullmann, G.M., and Knapp, E.W., Calculation of protonation patterns in proteins with structural relaxation and molecular ensembles—application to the photosynthetic reaction center, *Eur. Biophys. J. Biophys. Lett.*, 27, 626, 1998.
66. Richards, F.M., Areas, volumes, packing, and protein-structure, *Annu. Rev. Biophys. Bioeng.*, 6, 151, 1977.
67. Warwicker, J. and Watson, H.C., Calculation of the electric-potential in the active-site cleft due to alpha-helix dipoles, *J. Mol. Biol.*, 157, 671, 1982.
68. Klapper, I. et al., Focusing of electric fields in the active site of Cu–Zn superoxide dismutase: Effects of ionic strength and amino-acid modification, *Proteins: Struct. Funct. Genet.*, 1, 47, 1986.
69. Metropolis, N. et al., Equation of state calculations by fast computing machines, *J. Chem. Phys.*, 21, 1087, 1953.
70. Beroza, P. et al., Protonation of interacting residues in a protein by a Monte-Carlo Method—application to lysozyme and the photosynthetic reaction center of Rhodobacter-sphaeroides, *Proc. Natl. Acad. Sci. USA*, 88, 5804, 1991.
71. Antosiewicz, J.M. Computation of the dipole-moments of proteins, *Biophys. J.*, 69, 1344, 1995.
72. Bashford, D. and Karplus, M., Multiple-site titration curves of proteins—an analysis of exact and approximate methods for their calculation, *J. Phys. Chem.*, 95, 9556, 1991.
73. Gilson, M.K., Multiple-site titration and molecular modeling—2 rapid methods for computing energies and forces for ionizable groups in proteins, *Proteins: Struct. Funct. Genet.*, 15, 266, 1993.
74. Davis, M.E. et al., Electrostatics and diffusion of molecules in solution—simulations with the University of Houston Brownian Dynamics program, *Comput. Phys. Commun.*, 62, 187, 1991.
75. Madura, J.D. et al., Electrostatics and diffusion of molecules in solution—simulations with the University of Houston Brownian Dynamics program, *Comput. Phys. Commun.*, 91, 57, 1995.

76. MacKerell, A.D. et al., All-atom empirical potential for molecular modeling and dynamics studies of proteins, *J. Phys. Chem. B*, 102, 3586, 1998.

77. Nozaki, Y. and Tanford, C., Examination of titration behaviour, in *Enzyme Structure*, C. Hirs, Ed., volume XI of *Methods in Enzymology*, Academic Press, New York, 1967, pp. 715–734.

78. Stryer, L., *Biochemistry*, 4th ed., Freeman, New York, 1995.

79. Matthew, J.B., Hanania, G.I.H., and Gurd, F.R.N., Electrostatic effects in hemoglobin—hydrogen-ion equilibria in human deoxyhemoglobin and oxyhemoglobin-A, *Biochemistry*, 18, 1919, 1979.

80. Einstein, A., Über die von der molekularkinetischen Theorie der Wrme geforderte Bewegung von in ruhenden Flssigkeiten suspendierten Teilchen, *Ann. Physik*, 17, 549, 1905.

81. Brenner, H., Coupling between translational and rotational Brownian motions of rigid particles of arbitrary shape. I. Helicoidally isotropic particles, *J. Coll. Sci.*, 20, 104, 1965.

82. Happel, J. and Brenner, H., *Low Reynolds number hydrodynamics. With special applications to particulate media*, Mechanics of fluids and transport processes, 2., rev. ed edition, Noordhoff International Publishing, Leyden, 1973.

83. Brenner, H., The Stokes resistance of an arbitrary particle. II. An extension, *Chem. Eng. Sci.*, 19, 599, 1964.

84. Brenner, H., The Stokes resistance of an arbitrary particle. III. Shear fields, *Chem. Eng. Sci.*, 19, 631, 1964.

85. Harvey, S.C. and Garcia de la Torre, J.G., Coordinate systems for modeling the hydrodynamic resistance and diffusion-coefficients of irregularly shaped rigid macromolecules, *Macromolecules*, 13, 960, 1980.

86. Oseen, C.W., *Neuere Methoden und Ergebnisse in der Hydrodynamik*, Mathematik und ihre Anwendungen, Akad. Verl. Ges., Leipzig, 1927.

87. Zwanzig, R., Langevin theory of polymer dynamics in dilute solution, in *Stochastic Processes in Chemical Physics*, K. Shuler, Ed., Interscience, New York, 1969, pp. 325–331.

88. Rotne, J. and Prager, S., Variational treatment of hydrodynamic interaction in polymers, *J. Chem. Phys.*, 50, 4831, 1969.

89. Yamakawa, H., Transport properties of polymer chains in dilute solution—hydrodynamic inter-action, *J. Chem. Phys.*, 53, 436, 1970.

90. Garcia de la Torre, J.G. and Bloomfield, V.A., Hydrodynamic properties of macromolecular complexes. I. Translation, *Biopolymers*, 16, 1747, 1977.

91. Garcia de la Torre, J.G. and Bloomfield, V.A., Hydrodynamics of macromolecular complexes. II. Rotation, *Biopolymers*, 16, 1765, 1977.

92. Garcia de la Torre, J.G. and Bloomfield, V.A., Hydrodynamic properties of complex, rigid, biological macromolecules—theory and applications, *Q. Rev. Biophys.*, 14, 81, 1981.

93. Swanson, E., Teller, D.C., and Haen, C.D., Low Reynolds-number translational friction of ellipsoids, cylinders, dumbbells, and hollow spherical caps—numerical testing of validity of modi-fied Oseen tensor in computing friction of objects modeled as beads on a shell, *J. Chem. Phys.*, 68, 5097, 1978.

94. Garcia de la Torre, J.G. and Rodes, V., Effects from bead size and hydrodynamic interactions on the translational and rotational coefficients of macromolecular bead models, *J. Chem. Phys.*, 79, 2454, 1983.

95. Antosiewicz, J.M. and Porschke, D., Volume correction for bead model simulations of rotational friction coefficients of macromolecules, *J. Phys. Chem.*, 93, 5301, 1989.

96. Lee, B. and Richards, F.M., Interpretation of protein structures—estimation of static accessibility, *J. Mol. Biol.*, 55, 379, 1971.

97. Wegener, W.A., Dowben, R.M., and Koester, V.J., Time-dependent birefringence, linear dichroism, and optical-rotation resulting from rigid-body rotational diffusion, *J. Chem. Phys.*, 70, 622, 1979.

98. Wegener, W.A., Sinusoidal electric birefringence of dilute rigid-body suspensions at low field strengths, *J. Chem. Phys.*, 84, 6005, 1986.

99. Zacharias, M. and Hagerman, P.J., Influence of static and dynamic bends on the birefringence decay profile of RNA helices: Brownian dynamics simulations, *Biophys. J.*, 73, 318, 1997.

100. Porschke, D. and Antosiewicz, J.M., Strong effect of hydrodynamic coupling on the electric dichroism of bent rods, *J. Phys. Chem. B*, 109, 1034, 2005.

101. Porschke, D. and Antosiewicz, J.M., Permanent dipole-moment of transfer-RNAs and variation of their structure in solution, *Biophys. J.*, 58, 403, 1990.
102. Oncley, J., The electric moments and the relaxation times of proteins as measured from their influence upon the dielectric constants of solutions, in *Proteins, Amino Acids and Peptides*, E. Cohn and J. Edsall, Eds., Reinhold Publication Corporation, New York, 1943.
103. Kirkwood, J.G. and Shumaker, J.B., The influence of dipole moment fluctuations on the dielectric increment of proteins in solution, *Proc. Natl. Acad. Sci. USA*, 38, 855, 1952.
104. Porschke, D., The mechanism of ion polarization along DNA double helices, *Biophys. Chem.*, 22, 237, 1985.
105. Benoit, H., Contribution a l'étude de l'effet Kerr présenté par les solutions diluées de macromolécules rigides, *Ann. de Phys.*, 12e Serie, 6, 561, 1951.
106. Friedman, A.M., Fischmann, T.O., and Steitz, T.A., Crystal-structure of lac repressor core tetramer and its implications for DNA looping, *Science*, 268, 1721, 1995.
107. Lewis, M. et al., Crystal structure of the lactose operon repressor and its complexes with DNA and inducer, *Science*, 271, 1247, 1996.
108. Schultz, S.C., Shields, G.C., and Steitz, T.A., Crystal-structure of a cap-DNA complex—the DNA is bent by 90-degrees, *Science*, 253, 1001, 1991.
109. MeyerAlmes, F.J. and Porschke, D., The cyclic AMP receptor promoter DNA complex: A comparison of crystal and solution structure by quantitative molecular electrooptics, *J. Mol. Biol.*, 269, 842, 1997.
110. Bertolotto, J., Roston, G., and Ascheri, M., Electro-optical properties of DNA, *Prog. Colloid Polym. Sci.*, 128, 25, 2004.
111. Porschke, D., DNA double helices with positive electric dichroism and permanent dipole-moments—nonsymmetrical charge-distributions and frozen configurations, *Biophys. Chem.*, 49, 127, 1994.
112. Grycuk, T., Antosiewicz, J.M., and Porschke, D., Brownian dynamics of the polarization of rodlike polyelectrolytes, *J. Phys. Chem.*, 98, 10881, 1994.
113. Porschke, D., Persistence length and bending dynamics of DNA from electrooptical measurements at high salt concentrations, *Biophys. Chem.*, 40, 169, 1991.
114. Porschke, D., Electric dichroism and bending amplitudes of DNA fragments according to a simple orientation function for weakly bent rods, *Biopolymers*, 28, 1383, 1989.
115. Jose, D. and Porschke, D., Dynamics of the B-A transition of DNA double helices, *Nucleic Acids Res.*, 32, 2251, 2004.
116. Jose, D. and Porschke, D., The dynamics of the B-A transition of natural DNA double helices, *J. Am. Chem. Soc.* 127, 16120, 2005.

4 Computational Methods for Dynamic Electro-Optic Properties of Macromolecules and Nanoparticles in Solution

J. García de la Torre, F.G. Díaz Baños, and H.E. Pérez Sánchez

CONTENTS

4.1 INTRODUCTION

4.1.1 GENERALITIES

Electro-optic techniques are valuable sources of structural information about macromolecules and nanoparticles in two aspects. One of them is related to the optical or spectroscopic properties resulting from the application of the electric field. Furthermore, the response of the system is not instantaneous when the field is applied, changed, or removed, and therefore the transient, time-dependent values of those properties provide also information on dynamic aspects, which can be, in turn, related to structure.

The system considered in this chapter is a dilute solution or suspension of macromolecules or colloidal microparticles in a simple solvent. The main and always the present feature of transient electro-optics is the reorientational motions of the whole macromolecule—if it is rigid—or its component parts—if it is flexible—following the changes in the field. We shall assume that the properties that determine the molecule–field interactions are fixed, or respond instantaneously to the field. Therefore in our description the dynamics will be governed by the rotational hydrodynamics or diffusivity of a rigid solute, or the internal dynamics and energetics in the case of a flexible one. These two types of solute particles will be treated in the two sections of this chapter that follow Section 4.1, devoted to general and common aspects.

The theory of steady-state and transient electro-optic properties of macromolecular or colloidal systems was developed during the second half of the 20th century and it is well described in standard manuscripts [1–3]. The complexity of the problem limited the reach of the purely theoretical advances, posing important difficulties in the treatment, for instance, of the electro-optics in fields of moderate or greater strength, and also in the hydrodynamics of rigid particles of irregular shapes or flexible particles. Thus, the recent advances have been mainly in computational and model simulation methodologies that may permit the study of cases of arbitrary complexity. These procedures will be the major topics treated in this chapter.

4.1.2 FUNDAMENTALS

Electro-optics consider the optical (or spectroscopic) properties induced in a material by the application of an electric field. A commonly observed property is the birefringence, Δn, defined as the difference between the refractive index of the material in the direction of the applied field and that in a perpendicular direction. The material system is, in our case, a dilute solution, and Δn designates the contribution of the solute, obtained by discounting the contribution of the solvent (if appreciable) from that of the solution. Birefringence, and in general the electro-optic properties, are caused by the orientational effect of the field on the polar or polarizable (and, in the case of birefringence, optically anisotropic) solute molecules, whose orientation, which is uniformly random in the absence of the fields, becomes still random, although not uniform but biased by the external agent. Let α denote the optical polarizability of a molecule, with a given orientation, referred to a laboratory-fixed system of coordinates such that the electric field is applied along the vertical axis, z. Following the classical monograph of Frederick and Houssier [1], we write the birefringence as

$$\Delta n = K_n \nu \langle \Delta \alpha \rangle \tag{4.1}$$

where K_n is a constant related to the field-off refractive index (which will not be required for the topics in this chapter), ν is the number concentration of solute particles, $\Delta \alpha = \alpha_{zz} - \alpha_{xx}$ is the difference between the diagonal components of α in the laboratory system axes z (along the field) and x (perpendicular to the field), and $< \ldots >$ denotes an average over the many molecules, with different orientations, that composes the sample.

It may be convenient to refer the optical polarizability tensor of the molecule (either rigid, or in an instantaneous conformation of a flexible one) to some molecule-fixed system of reference; in such a case, we will use in general the notation α''. For the particular case of a rigid molecule, the theory (*vide infra*) requires the components of this tensor in the system of eigen-axes (principal axes) of the rotational diffusion tensor; for this particular case we shall employ the notation α'. For other electro-optic vectorial and tensorial quantities, the same convention (unprimed, primed, and double-primed) notation will be employed.

It should be noted that practically all the same description of electric birefringence is applicable to electric dichroism, just by replacing the optical polarizability tensor by the absorbance tensor.

The transformation from particle-fixed (primed or double-primed) vector coordinates or matrix components to laboratory-fixed ones is made through a transformation matrix, $\mathbf{M}''$, such that, for any vector μ (e.g., the dipole moment): $\mu'' = \mathbf{M}''^{\mathrm{T}} \cdot \mu$ or $\mu = \mathbf{M}'' \cdot \mu''$ (where $\mathbf{M}''^{\mathrm{T}}$ stands for the matrix transpose of $\mathbf{M}''$), and for any tensor matrix (e.g., the optical polarizability), $\alpha'' = \mathbf{M}''^{\mathrm{T}} \cdot \mu \cdot \mathbf{M}$ or $\alpha = \mathbf{M}'' \cdot \alpha'' \cdot \mathbf{M}''^{\mathrm{T}}$. This way, the birefringence can be written as

$$\Delta n = K_n \nu \left\langle (\mathbf{M}'' \cdot \alpha'' \cdot \mathbf{M}''^{\mathrm{T}})_{zz} - (\mathbf{M}'' \cdot \alpha'' \cdot \mathbf{M}''^{\mathrm{T}})_{xx} \right\rangle \qquad (4.2)$$

where α'' will have fixed components in the particle-fixed axis, and the orientational effects is represented by the matrices $\mathbf{M}''$ for each particle. Similar equations hold for the single-primed quantities. Equation 4.2 can be applied either in steady-state conditions, or in any instant, at time t, during a transient observation; in the latter case, the averaging would be over the orientations and conformations of all the molecules at that instant, and the birefringence, $\Delta n(t)$, will be a function of t.

The birefringence Δn, as written in Equation 4.1 and Equation 4.2 applies strictly to a rigid particle. A flexible particle will be considered as a set of connected rigid segments (bonds, links, segments, subunits, ...). In terms of the principle of additivity of polarizabilities, for a flexible entity these equations will be modified just by inclusion of a sum over all segments in the molecule.

4.1.3 ELECTRO-OPTICS TRANSIENTS

We shall now consider the kinetics, or time evolution of some general electro-optic property, say $p(t)$, for instance the electric birefringence, $\Delta n(t)$.

The dynamics of macromolecules and colloids in dilute solutions is determined, among other factors by a set of characteristic (reorientational or relaxation) times. As described below, the electro-optic transients of a rigid particle depend on a set of five times [4,5], while long-chain macromolecules have a full discrete spectrum of relaxation times [6,7]. The same holds for semiflexible macromolecules with segmental flexibility [8]. This set of times characteristic of the particle will be indexed as $\tau_1 \geq \tau_2 \geq \tau_3 \geq \ldots$, with τ_1 as the longest one. When an analytical result for $p(t)$ can be found, it will be a more or less complex expression involving the τ_i values. A simple and experimentally common case is the decay of birefringence or dichroism when the applied field is suddenly removed. The time function can be still complicated, but if the applied field was steady and weak enough, then the theory [6] predicts that the decay is a sum of exponentials

$$p(t) = A_1 e^{-t/\tau_1} + A_2 e^{-t/\tau_2} + \cdots \qquad (4.3)$$

where if $p(\infty) = 0$, and the A_i's are the corresponding amplitudes, with $\Sigma A_i = 1$ if $p(0) = 1$.

It is important to remark that in some particular cases the amplitudes associated to certain times may be zero or very small. The decay observed in experiments (or in computer simulations) will be still multiexponential, but we should introduce a different notation, writing

$$p(t) = A_a e^{-t/\tau_a} + A_b e^{-t/\tau_b} + \cdots \qquad (4.4)$$

where $\tau_a > \tau_b \ldots$, etc., are the observed relaxation times, and A_a, A_b, etc., are the amplitudes. It could happen, for instance, that $A_1 \cong 0$. Then $\tau_a = \tau_2$, the longest observed relaxation time would be the second (not the first) characteristic time.

The characteristic times τ_1, $\tau_2, \ldots$ are properties that depend on the size, shape, and eventually flexibility of the particle but not on the electro-optical properties. Indeed this set will be common to the various electro-optic properties and other techniques (fluorescence anisotropy decay, NMR relaxation, etc.). The dependence of the τ's on the particle's structure is usually stronger than that of other hydrodynamic or solution properties (like the sedimentation or diffusion coefficients) and therefore provide sensitive procedures for structural determination.

4.2 TRANSIENT ELECTRO-OPTICS OF A RIGID PARTICLE

4.2.1 ELECTRO-OPTICAL PROPERTIES

As mentioned above, Equation 4.2 applies directly for a rigid particle. The electro-optic vectors and tensors have fixed components in the particle-fixed systems of reference: dipole moment, μ'', optical polarizability, α'', electrical polarizability ε'', etc. For a given molecule, with a given orientation, matrix $\mathbf{M}''$ will depend on the set of angles, Ω, that gives the orientation of the molecule's axes with respect to the laboratory axes.

The average required in Equation 4.2 will be of the type

$$< \ldots >= \int f(\Omega; t) \, (\cdots) \mathrm{d}\Omega \tag{4.5}$$

where $f(\Omega; t)$ is the time-dependent distribution of orientations. The essential problem for the evaluation of electro-optic properties is the determination of this distribution, followed by the integration needed in Equation 4.5 (note that in a transient, the distribution function and, therefore, the average, are time dependent). The distribution of orientations is determined by the energy interaction of the polar or polarizable molecule and, furthermore depends on time according to the stochastic differential equation that describes rotational diffusion, and also depends on the time-dependence of the field itself. This is a problem of extraordinary difficulty that only affords simple solutions in some very simple cases.

4.2.2 HYDRODYNAMICS OF RIGID PARTICLES

The rotational diffusivity of a rigid particle in viscous solvent is governed by a rotational diffusion tensor, $\boldsymbol{D}_r$, which is a 3×3 symmetric matrix that can be diagonalized to get its three eigenvalues D_1, D_2, and D_3, and the corresponding vectors, v_1, v_2, and v_3. This tensor determines the characteristic times that enter in the time dependence of the various properties. For transient electric birefringence or dichroism, and some other properties, like fluorescence anisotropy, viscoelasticity, or NMR relaxation, there is a set of five relaxation times [4]. The dynamics of the particle after the application or removal of an electric field may be too complex to be formulated analytically. There is, however, an analytical result for the birefringence decay after steady-state application of a field of low intensity (in the so-called Kerr region): according to Wegener and coworkers, it is a sum of up to five exponentials [5]

$$\Delta n(t) = K_n \nu (E^2/15) \sum_{k=1}^{5} Q_k \exp(-t/\tau_k) \tag{4.6}$$

and the normalized birefringence is

$$\Delta n(t)/\Delta n(0) = \sum_{k=1}^{5} A_k \exp(-t/\tau_k) \tag{4.7}$$

with $A_k = Q_k/(\Sigma_{k=1,\cdots,5} Q_k)$ and with $\Sigma_{k=1,\cdots,5} A_k = 1$. For rigid particles of high symmetry the number of distinct times may be smaller, just because some of the times may be duplicated. For a particle of arbitrary shape the five τ's are listed in Table 4.1. If the particle has an axial symmetry, then two of the eigenvalues, denoted $D_\perp$, with eigenvectors perpendicular to the symmetry axis, and the distinct one, $D_\parallel$, have its eigenvector along the symmetry axis. Then, there are only three distinct relaxation times, listed in Table 4.2.

While the time constants, i.e., the τ's, like other hydrodynamic properties depend just on the size and shape of the rigid particle (and, trivially on temperature and solvent viscosity), the amplitudes depend also on the electro-optic properties. Following Wegener et al. [5], the strength of the molecule–field interaction is measured by the alignment tensor, χ, which combines the effects of a permanent dipole, μ, and an induced dipole due to the electrical polarizability tensor, ε, defined as

$$\chi = \beta^2 \mu\mu + \beta\varepsilon \tag{4.8}$$

where $\beta = 1/k_B T$ is the reciprocal of Boltzmann factor. Expressions for the amplitudes, both in the general case and in the case of axially symmetric molecules, are given, respectively in Table 4.1 and Table 4.2, in terms of the components of the alignment tensor χ' and the optical polarizability tensor, the alignment tensor α' referred to a system of axes coinciding with the eigenvectors of $\mathbf{D}_{\mathrm{rr}}$. The transformation from the components of these tensors in the particle's system of reference to the rotational diffusion matrix is $\alpha' = \mathbf{M_D}''\alpha''\mathbf{M_D}''^{\mathrm{T}}$ where $\mathbf{M_D}''$ is a matrix containing the eigenvalues of $\mathbf{D}_{\mathrm{rr}}$ in its rows.

TABLE 4.1
Fundamental Relaxation Times and Amplitudes of the Low-Field Birefringence Decay, for Rigid Particles

	Index, k	$1/\tau_k$	Q_k
0	1	$6D-2\Delta$	(a)
1	2	$3(D+D_1)$	$3\,\alpha_{23}\,\chi_{23}$
-1	3	$3(D+D_2)$	$3\,\alpha_{13}\,\chi_{13}$
-2	4	$3(D+D_3)$	$3\,\alpha_{12}\,\chi_{12}$
2	5	$6D+2\Delta$	(b)

(a) $Q_1 = (3/4)(c/M)^2(\alpha'_{11} - \alpha'_{22})(\chi'_{11} - \chi'_{22}) - (\sqrt{3}/4)(bc/M^2)[(\alpha'_{11} - \alpha'_{22})(3\chi'_{33} - Tr\chi) + (\chi'_{11} - \chi'_{22})(3\alpha'_{33}$
$\quad - Tr\alpha)] + (1/4)(b/M)^2(3\chi'_{33} - Tr\chi)(3\alpha'_{33} - Tr\alpha)$

(b) $Q_5 = (3/4)(b/M)^2(\alpha'_{11} - \alpha'_{22})(\chi'_{11} - \chi'_{22}) + (\sqrt{3}/4)(bc/M^2)[(\alpha'_{11} - \alpha'_{22})(3\chi'_{33} - Tr\chi) +$
$\quad (\chi'_{11} - \chi'_{22})(3\alpha'_{33} - Tr\alpha)] + (1/4)(c/M)^2(3\chi'_{33} - Tr\chi)(3\alpha'_{33} - Tr\alpha)$

$D = (D_1 + D_2 + D_3)/3$

$\Delta = [D_1^2 + D_2^2 + D_3^2 - D_1D_2 - D_1D_3 - D_2D_3]^{1/2}$

$b = 3(D_3 - D) + 2\Delta; c = \sqrt{3}(D_1 - D_2); M = 2(\Delta b)^{1/2}; b^2 + c^2 = M^2$

Sources: From Favro, L.D., Theory of the rotational Brownian motion of a free rigid body, *Phys. Rev.*, 119, 53, 1960; Wegener, W.A., Dowben, R.M., and Koester, V.J., Time dependent birefringence, linear dichroism and optical rotation resulting from rigid-body rotational diffusion, *J. Chem. Phys.*, 70, 622, 1979.

TABLE 4.2
Fundamental Relaxation Times and Amplitudes of the Low-Field Birefringence Decay,
for Axially Symmetric Particles

$1/\tau$	Q	$D_\perp < D_\parallel$		$D_\parallel > D_\parallel$	
		$k = 1,\ldots 5$	a,b,c	$k = 1,\ldots 5$	a,b,c
$6D_\perp$	$(1/4)(3\chi'_{33} - \mathrm{Tr}\chi)(3\alpha'_{33} - \mathrm{Tr}\alpha)$	1	a	4,5	c
$5D_\perp + D$	$3(\alpha'_{13}\chi'_{13} + \alpha'_{23}\chi'_{23})$	2,3	b	2,3	b
$2D_\perp + D$	$3\alpha'_{12}\chi'_{12} + (3/4)(\alpha'_{11} - \alpha'_{22})(\chi'_{11} - \chi'_{22})$	4,5	c	1	a

Source: From Wegener, W.A., Dowben, R.M., and Koester, V.J., *J. Chem. Phys.*, 70, 622, 1979.

Although the steady-state values are not covered here in detail, we mention that the birefringence finally reached after the application of a very weak field, which would be the starting value ($t = 0$) for the decay, is given by

$$\Delta n(0) = K_n(E^2/30)[3\mathrm{Tr}(\alpha\chi) - \mathrm{Tr}(\alpha)\mathrm{Tr}(\chi)] \tag{4.9}$$

In the above equation, $\mathrm{Tr}(\alpha)$ stands for the trace of the matrix α. A simplified—and often employed in the literature—situation is when not only the D_{rr} tensor, but also the α' and χ' tensors are axially symmetric, and the dipole moment, if it exists, is aligned with the symmetry axis. Then, two of the components in Table 4.2 vanish and the birefringence decay is monoexponential:

$$\Delta n(t)/\Delta n(0) = \exp(-6D_\perp t) \tag{4.10}$$

In contrast, amplitudes depend on both hydrodynamics and electro-optics, and may give rise to interesting effects. For example, in some special cases, the "decay" is not a monotonically decreasing function. In other special instances the birefringence may present a sign reversal. We recall that in this chapter we concentrate on the dynamic aspects; therefore we proceed to describe how the five relaxation times and the rotational diffusion coefficient can be evaluated.

4.2.3 BEAD MODEL CALCULATIONS

The calculation of hydrodynamic properties of rigid particles can be carried out using methods of fluid mechanics combined with a generalized form on Einstein's theory of Brownian motion. The most successful and popular implementation is the bead modeling methodology, based on an idea first proposed by Kirkwood [9], generalized by Bloomfield et al. [10], and extended and computationally implemented by García de la Torre and coworkers [11–14]. In the bead modeling procedure, the shape of a particle is arbitrarily represented by an array of spherical elements ("beads"), which may be of equal or unequal size. In bead modeling in strict sense, the cluster of beads must have a size and shape as close as possible to those of the particle that is modeled, as represented in Figure 4.1a. For some purposes a more appropriate version is the bead–shell model, in which it is just the surface of the particle (where friction with solvent actually takes place) by a shell of small, identical minibeads of radius σ, as illustrated in Figure 4.1b. The calculations are repeated for a series of decreasing σ_1 and extrapolated to the shell model limit of $\sigma \to 0$.

The theory and computational procedures required to calculate the properties have been described in previous reviews [13,14], so that here we will just comment on different aspects of

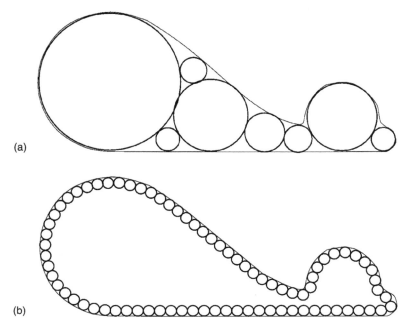

(a)

(b)

FIGURE 4.1 Example of a bead model in strict sense (a) and a bead–shell model (b).

the application of this methodology. The various types of modeling and numerical procedures have been implemented in simple, user-friendly, public domain programs, which constitute the HYDRO software suite.

HYDRO itself is the first program in the collection. It is intended for plain-bead models, constructed by the user. A main data file hydro.dat provides elementary properties of the solute and the solvent, and for each structure to be computed, a separate structural file containing the user-supplied coordinates and radii. We shall employ as a typical but simple example in which the bead model represents a structure (Figure 4.2a) with two globular domains, which are approximately spherical with radii σ_1 and σ_2, and a rodlike linker, with radius σ, which is bent at its middle, with an angle α between the two arms. The main input file for this example is in Figure 4.3. This file gives the name of the structural file, which essentially contains a list of coordinates and radii of the beads, constructed by the user (Figure 4.2a). An excerpt of the output file is in Figure 4.4. For the present purpose, the relevant quantities are the five τ's and the 6×6 generalized diffusion tensor, D, of which D_r is a 3×3 box.

With the recent growth of detailed, atomic-level structural information of biomolecules, a version of HYDRO, which would free the user of the construction of the hydrodynamic models, and *ad hoc* version of HYDRO was pertinent. Our program HYDROPRO [15] performs such a calculation using a structural file, just the PDB-formatted file containing the Cartesian coordinates. HYDROPRO avoids some problems that affect the rotational properties constructing (in an internal manner that is transparent for the user) a shell model. Figure 4.5a shows the primary hydrodynamic bead model and the shell displays a bead–shell model of lysozyme.

The HYDRO suite includes other programs with similar purpose but differing in how the structure is specified. Thus, HYDROMIC [16] does the same tasks as HYDROPRO [15], but

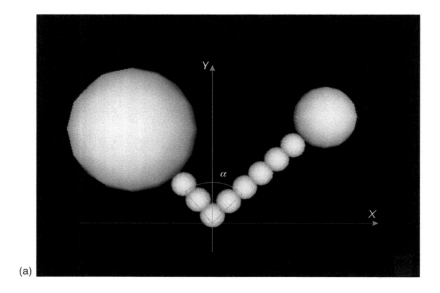

(a)

```
1.E-08   !Unit of  length  (1 Angs)
10            !Number of beads
     0.00    0.00  0. 10.0
  -14.14   14.14  0. 10.0
  -28.28   28.28  0. 10.0
  -77.78   77.78  0. 60.0
   14.14   14.14  0. 10.0
   28.28   28.28  0. 10.0
   42.42   42.42  0. 10.0
   56.56   56.56  0. 10.0
   70.71   70.71  0. 10.0
   98.99   98.99  0. 30.0
```

(b)

FIGURE 4.2 (a) Bead model, as described in the text. (b) Structural file, mystructure.txt to be used in the HYDRO calculation, corresponding to $\sigma_1 = 6.0$ nm, $\sigma_1 = 3.0$ nm, and $\gamma = 90°$.

in this program the source of structure is cryoelectron microscopy. HYDROSUB [17] works with models built from ellipsoidal and cylindrical parts, and it is suitable for multisubunit structures like antibodies, myosin, etc.

4.2.4 TIME COURSE OF BIREFRINGENCE

The quantities calculated by any of these programs are the basic hydrodynamic coefficients, the radius of gyration, and the molecular covolume. These calculations can be complemented with those of another ancillary program, SOLPRO [18,19], which computes more complex solution properties. One of them is the birefringence decay, as predicted from Equation 4.7. The HYDRO programs provide the five τ's in a file that is the input for SOLPRO, and the user has to supply the electro-optical vectors or tensors.

```
My_structure          Title
my_structure          filename
my_structure.dat      Structural file
293.                  Temperature, Kelvin
0.010                 Solvent viscosity
830000.               Molecular weight
0.760                 Solute specific volume
1.0                   Solution density
0,                    scattering: no
0,                    distances: no
0,                    covolume: no
1                     diffusion tensors: yes
*                     End of file
```

FIGURE 4.3 hydro.dat, main input data file for HYDRO.

Thus, for the bead model that we are using as an example (Figure 4.1), we present in Figure 4.6 the results for the time course of electric birefringence. We have assumed for this example that the right arm has twice the anisotropy of polarizability of the left one, and twice its dipole moment. With these assumptions, one can estimate adequate values for the optical anisotropy and alignment tensors, α and χ. Thus, with the output from HYDRO and this additional information, SOLPRO computes the time-dependency of birefringence. Figure 4.6 shows three cases: one with purely induced moment and two other cases with pure permanent moments, head-to-tail (A) and head-to-head (B). It is very interesting that we cannot always say "birefringence decay," because in some instances the birefringence may show a rise when the field is switched off. This situation had been predicted, for rigidly bent particles, by Mellado et al. [20]. At long times, in all the cases we observe a decay in which the time constant is the longest relaxation time, τ_1.

Up to this point we have considered the most common and simple case of electro-optic transient: the decay when a previously applied weak field is removed. Other potentially informative situations, like the field-on transients, or those with fields of arbitrary strength are usually avoided in the experimental work, perhaps because of the difficulty of interpreting the outcome from the experiments, due to the lack or complexity of the theoretical results.

We have devised a computer program, BROWNRIG [21], that simulates the Brownian dynamics of an arbitrary rigid body, whose hydrodynamic properties have been previously obtained from any of the HYDRO programs. The dynamics can include the interaction of an external electric field with charges located at certain points in the particle, which makes it possible to simulate the effect of electric field on particles with permanent dipole moment. Of course, electrophoretic migration can also be simulated. Actually, the modular construction of the program allows the consideration of arbitrary field–particle interactions, programmed in user-supplied subroutines that specify the interaction forces and torques. Thus, the case of particles with an induced dipole can also be treated.

BROWNRIG simulates the Brownian trajectories for each of a sufficiently large number of molecules that comprise the system sample. At regular intervals of time, the position and orientation of each molecule is recorded. From the instantaneous orientation of a molecule, the transformation matrix that appears in Equation 4.2, and determines the contribution of each molecule to the instantaneous birefringence or similar property is calculated. The observable, time-dependent property is obtained for each time, as the sample average, as formulated in Equation 4.11.

```
                        SUMMARY OF DATA AND RESULTS

                        This file:   my_structure.res
                             Case:   My_structure
                  Structural file:   my_structure.dat

                      Temperature:   293.0 K
                 Solvent viscosity:  0.01000 poise
                 Molecular weight:   8.300E+05 Da
        Specific volume of solute:  0.760 cm³/g
                 Solution density:   1.000 g/cm³
 Translational diffusion coefficient: 2.777E-07 cm²/s
      Stokes (translational) radius:  7.727E-07 cm
                Radius of gyration:   7.256E-07 cm
                           Volume:   1.051E-18 cm³
   Rotational diffusion coefficient: 2.765E+05 s-1
               Relaxation time (1):  9.134E-07 s
               Relaxation time (2):  7.315E-07 s
               Relaxation time (3):  7.211E-07 s
               Relaxation time (4):  4.499E-07 s
               Relaxation time (5):  4.498E-07 s
  Harm. mean relax.(correlation) time: 6.028E-07 s
                Intrinsic viscosity:  4.257E+00 cm³/g
          Sedimentation coefficient:  2.271E+01 svedberg

      Generalized (6x6) diffusion matrix:    (Dtt  Dtr)
                                             (Drt  Drr)


  3.012E-07  2.428E-09  0.000E+00    0.000E+00  0.000E+00  6.261E-03
  2.428E-09  2.709E-07  0.000E+00    0.000E+00  0.000E+00  2.354E-03
  0.000E+00  0.000E+00  2.610E-07    6.261E-03  2.354E-03  0.000E+00


  0.000E+00  0.000E+00  6.261E-03    4.643E+05  8.038E+03  0.000E+00
  0.000E+00  0.000E+00  2.354E-03    8.038E+03  1.860E+05  0.000E+00
  6.261E-03  2.354E-03  0.000E+00    0.000E+00  0.000E+00  1.792E+05
```

FIGURE 4.4 myresults.res, main output file from HYDRO (some selected lines only).

Next we provide an example of the use of BROWNRIG. We consider a bent rod, which has 11 beads having a diameter of 20 Å, forming two arms connected by a joint, with the arms making an angle of 120° (Figure 4.7a). The molecule has a permanent dipole moment, which is represented by a charge of $+2e$ at the joint, and charges at the two ends of $-1e$ each, which results in a zero net charge. After the trivial HYDRO calculation for this model, the diffusion tensor is given, along with the charges, field strength, etc., as input to BROWNRIG. In the simulations, the electric field may vary with time. In the present computer experiment, there is no field during the first part of the simulation, then an electric field of 5.5×10^7 N/C is applied, and when properties reach a steady state, the field is switched off.

For a particle composed of two identical, axially symmetric subunits, it can be demonstrated that the birefringence is determined solely by the degree of orientation of the subunits axes with respect to the laboratory axis along which the field is applied. If θ_1 and θ_2 are the

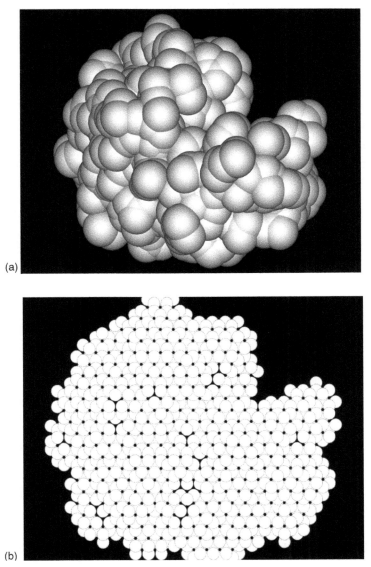

FIGURE 4.5 (a) Primary bead model of lysozyme, with "atomic" beads of radius $a = 0.31$ nm. (b) Shell model corresponding to this primary model, made of minibeads of radius $\sigma = 0.08$ nm.

instantaneous angles giving that orientation, it can be demonstrated that the time-dependent birefringence is given by

$$\frac{\Delta n(t)}{\Delta n_{\text{sat}}} = \Delta n'(t) = \frac{1}{2}(\langle P_2(\cos\theta_1)\rangle + \langle P_2(\cos\theta_2)\rangle) \tag{4.11}$$

where Δn_{sat} would be the birefringence reached in an infinitely strong field and P_2 is the second Legendre polynomial. Brownian trajectories are simulated for 1000 molecules. At any given time, $P_2(\cos\theta_1)$ and $P_2(\cos\theta_2)$ are calculated for each molecule, and from the whole

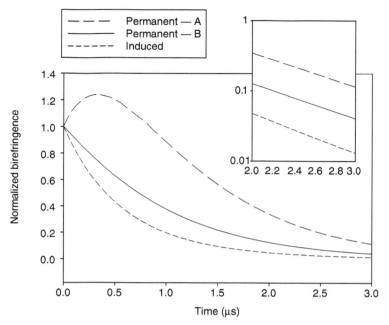

FIGURE 4.6 Normalized birefringence calculated by SOLPRO for the model in Figure 4.1. Insert shows Semilog plot at long times.

sample average we extract the value of the birefringence. The time evolution of the birefringence is displayed in Figure 4.7b. When the electric field is applied for a sufficiently long time, the birefringence reaches a plateau value, $\Delta n' = -0.10$, which agrees well with the prediction from theoretical expressions for the steady-state electric birefringence of bent rods [20]. Note that in this example the birefringence is negative because the preferential orientation of the end-to-end vector is perpendicular to the field. When the field is switched off, the birefringence returns to zero. Of course, the kinetics of the decay and rise could be analyzed in detail from the simulated trajectory; as this can be done for fields of arbitrary strength (and even for time-dependent fields), BROWNRIG enables us to predict electro-optic transients practically in any condition.

4.3 FLEXIBLE MACROMOLECULES

4.3.1 FLEXIBLE MACROMOLECULES UNDER THE INFLUENCE OF AN ELECTRIC FIELD

Rigid particles respond to the application or removal of an electric field by changing their orientation in a rotational (Brownian) motion. However, we cannot say that flexible particles rotate. Even in the absence of a field, the Brownian motion of a flexible particle consists of an overall translation (displacement of its center of mass) and a change of its instantaneous shape, as illustrated in Figure 4.8. As the flexible particle may be eventually composed of linked rigid entities (subunits, segments, bonds, etc.), it is possible to consider the rotational or reorientational motion of each of them, while the particle, as a whole executes an internal motion (but not a rigid rotation).

The application of a field induces a preferential orientation of the segments and produces changes or effects such as birefringence. As indicated previously, we concentrate in this chapter on the kinetics of those changes, which will be related to the internal motion of the flexible particle. The collective reorientation of the individual segments can also be regarded

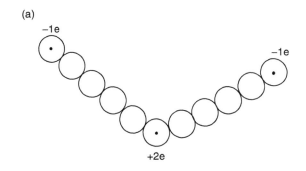

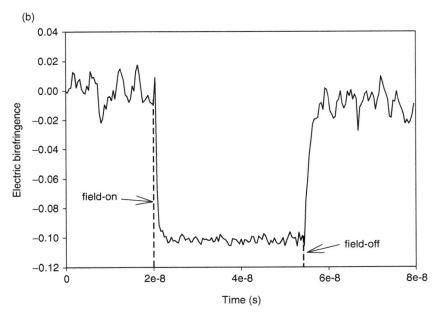

FIGURE 4.7 (a) Model for a bent-rod molecule, showing the location of the point charges. (b) Transient electric birefringence simulated for this bent rod.

as a deformation of the particle: the internal coordinates are changed and the action of the field may eventually produce a change in the mean dimensions (end-to-end distance or radius of gyration).

4.3.2 DEFORMATION AND SCATTERING

When flexible macromolecules are under the influence of an electric field, in addition to undergoing a change in the orientation, they suffer a certain deformation that can be measured and related to its structure, flexibility, and other internal characteristics.

One of the manifestations of deformation in macromolecules is the alteration of its conformational statistics and hence of its overall dimensions. Scattering of light or other electromagnetic radiation is capable of monitoring such alterations, because changes in scattering intensity reflect changes in the distribution of conformations. Indeed, the technique of electric-field light scattering has been experimentally investigated [2], although its use is not widespread. Our group has developed [22–24] a theoretical formalism for describing

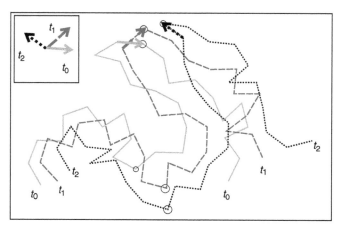

FIGURE 4.8 Schematic view of three successive snapshots of a flexible macromolecular chain. Note the reorientation of the rigid segments that compose the chain.

deformation and scattering intensities, and for the application to semiflexible macromolecules. We also use the relationship between the overall particle deformation measured by scattering and the orientation of segments measured by electric birefringence.

The average size of a macromolecule, as determined by the radius of gyration, is often obtained from the low-angle scattering of light or another type of electromagnetic radiation. When the macromolecule is deformed in an electric field, it may be possible to obtain some components of the gyration tensor from scattering with various geometries. Instrumental and theoretical aspects of the technique of electric field light scattering have been described in the literature [2]. The dependence of scattered intensity on the scattering direction is represented by the scattering form factor, $P(\mathbf{q})$, $q = (4\pi/\lambda)-\sin(\theta_s/2)$ as the modulus of the scattering vector $\mathbf{q}$. In this equation, λ is the radiation wavelength and θ_s is the angle subtended by the scattering direction and the prolongation of the incident beam. For low-angle scattering, we can write Equation 4.12, where $\mathbf{G}$ stands for the gyration tensor, which contains information about size and shape of a macromolecule

$$P(\mathbf{q}) = 1 - \mathbf{q}^{\mathbf{T}} <\mathbf{G}> \mathbf{q} \tag{4.12}$$

A particular situation of interest is that corresponding to an experimental setup in which scattering is observed in the (x,y) plane of the laboratory-fixed system of coordinates, with x as the direction of the incident beam, and the external agent acting along the z direction. In this case, we can assume that this agent produces a distribution of mass with cylindrical symmetry around the main direction. Then

$$P(\mathbf{q}) \approx 1 - \mathbf{q}^2 <\mathbf{G}>_{xx} \tag{4.13}$$

Equation 4.13 has been obtained for steady-state light scattering. If we perform experiments following the relaxation of the molecule, it is straightforward to obtain the change of the scattering form factor as a function of time

$$\frac{P(\mathbf{q},\infty) - P(\mathbf{q},t)}{1 - P(\mathbf{q},\infty)} \approx \frac{<\mathbf{G}(t)>_{xx} - <\mathbf{G}(\infty)>_{xx}}{<\mathbf{G}(\infty)>_{xx}} = \delta(t)^2_{xx} \tag{4.14}$$

This experimentally feasible combination yields the change in the component of $<\boldsymbol{G}>$, which is independent of the scattering angle. According to the previous equations, once the field is removed, the deformation produced in a TEB experiment relaxes following a certain time course. Experiments of dynamic electric field light scattering might be able to measure this evolution.

The change in the radius of gyration may be expressed in terms of a deformation ratio as the change in $<s^2>$ (where s stands for the radius of gyration) relative to its unperturbed value. This definition can be extended to all the diagonal components of the gyration tensor. Carrasco et al. [24] showed that, for a multisubunit structure, the gyration tensor and, as a consequence, $\delta_{\alpha\alpha}^2$ and δ^2 can be expressed in terms of two types of contribution, one based on the distances of the subunits to the center of mass and another on the gyration tensor of the subunits. In previous studies [22,25], we developed these expressions to obtain the deformation of a molecule in a steady-state birefringence experiment under the influence of the field and it is straightforward to extend these results to a transient experiment. In this case, the gyration tensor and, as a consequence, the deformation are functions of time, and the unperturbed value of the radius of gyration is obtained at the limit, $t = \infty$. Hence, the deformation ratio, δ^2, can be written as

$$\delta(t)_{xx}^2 = \frac{\delta(t)_{xx}'^2 < s(\infty)'^2 > + \Delta n(t)(\boldsymbol{G}^\perp - \boldsymbol{G}^\parallel)}{< s(\infty)'^2 > + (2\boldsymbol{G}^\perp - \boldsymbol{G}^\parallel)} \tag{4.15}$$

In Equation 4.15 the primed values are those obtained for a model in which the subunits were replaced by point-like elements with masses m_i positioned at the center of the subunit, while $\boldsymbol{G}^\perp$ and $\boldsymbol{G}^\parallel$ are the perpendicular and parallel components of the gyration tensor of each subunit.

4.3.3 FULLY FLEXIBLE CHAINS

Most synthetic polymers, as well as some biological macromolecules like long-chain polysaccharides, denatured proteins, and high-molecular-weight DNA behave as fully flexible random coils. The dynamics of such flexible chains can be described by the Rouse–Zimm [6,26] model. A chain with N links has a set or spectrum of many $(N–1\cong N)$ relaxation times, τ_k. Actually, there are two of such sets, $\{\tau_k\}$ and $\{\tau'_k\}$, depending on the property that is monitored, with $\tau_k = (1/2)\tau'_k$. The $\{\tau'_k\}$ set determines, for instance, dielectric relaxation, whereas transient electro-optics, and particularly birefringence, along with viscoelastic properties, depend on the $\{\tau_k\}$. Rouse [26] obtained simple theoretical formulas for τ'_k, which are not valid for the interpretation of results because this theory ignored the essential effect of hydrodynamic interaction between chain segments. Zimm [6] provided a simple numerical way to include such effects in the so-called preaveraged approximation, providing results that may be sufficient for data analysis. More accurate evaluation of hydrodynamic interaction effects, in a rigorous manner, requires computer simulations [27].

The interpretation of electro-optical transients of fully flexible macromolecules is difficult due to the complex interplay between conformational variability, mechanisms of molecule–field interaction, and internal dynamics. However, the information extracted from experimental data may be quite relevant in terms of macromolecular structure and function. An excellent example is the study of Rau and Bloomfield [28], using transient electric birefringence, of electro-optical properties, polyelectrolyte effects, and chain dynamics of long-chain DNA from T7 phage.

4.3.4 Semiflexible Macromolecular Chains

Different models have been developed to describe the flexibility of these macromolecules and to study their electro-optics (and other properties). Among them, two extreme and relatively simple models have become very popular: segmental and wormlike flexibility.

Segmentally flexible macromolecules are modeled by few rigid subunits or domains, joined by more or less flexible hinges or joints. A typical case is that of broken-rod macromolecules with two rodlike arms, a model that has been used to study the myosin rod [29,30] or some especially prepared synthetic polypeptides. Whole myosin [29] and immunoglobulins [31] are more complex examples, with more subunits and joints. Studies of the electro-optic properties of this type of molecule have been recently published by this group for steady-state [24] and transient [32,33] properties.

In wormlike macromolecules flexibility is not localized at a single or even at several joints, but is distributed along the macromolecular chain. Macromolecules with a helical structure are usually considered to present this kind of flexibility and the most paradigmatic example is probably DNA [34].

Of course most real semiflexible molecules present a mixture of both types of flexibility, although one may be more relevant in defining its characteristics. RNA [35] and DNA [36,37] are examples of this situation.

4.3.4.1 Bead and Connector Models

Bead and connector models have appeared as a versatile and plausible way to represent semiflexible macromolecular chains. First, we are going to present a general description of the methodology for constructing models of semiflexible macromolecules based on a discrete number of subunits to explain later how they can be used in the simulation of a birefringence experiment.

For a given model, we can define two contributions to the potential energy of the molecule, V_{int}, associated with the deformation of the macromolecule, i.e., to its departure from the most stable configuration. These models are chains made up of $N+1$ elements (see Figure 4.9). Their connectors define N axially symmetric subunits and we assume that unitary vectors $\mathbf{u}_i$ ($i = 1$ to N) are aligned with the symmetry axis of each subunit. The instantaneous conformation of the particle is determined by a set of $N-1$ angles, α_j, formed between two consecutive vectors. If $\alpha_{0,j}$ is the equilibrium value of these angles, the internal potential energy required for bending or deformation is given by

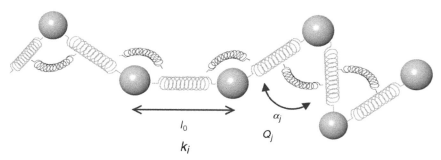

FIGURE 4.9 Generalized chain with the parameters that characterize the stretching and bending potentials.

$$\frac{V_{\text{int}}}{k_B T} = \sum_{j=1}^{N-1} Q_j (\alpha_j - \alpha_{0,j})^2 + \sum_{j=1}^{N} K_j (b_j - b_{0,j})^2 \tag{4.16}$$

In this equation $k_B T$ is Boltzmann's factor and Q_j is the flexibility parameter, with $Q_j = 0$ for the completely flexible case and $Q_j \to \infty$ for the completely rigid one.

According to the values given to parameters Q_j and the number of subunits, different types of flexibilities can be modeled. One extreme case is that whereby all Q_j's have the same value and N is sufficiently high to provide a wormlike model (with limits $Q_j = \infty$ for a rigid chain and $Q_j = 0$ for a freely jointed chain) [38]. Another extreme case is defined when all the Q's are infinity except one that has a finite (or zero) value. In this case, we have a model with two rigid arms and if the two arms are linear, we have broken-rod chain. Of course, the simplest broken-rod chain can also be modeled with two subunits and one single Q [24,30].

Although in this work the equilibrium conformation is always a straight molecule (which is defined with all $\alpha_{0,j} = 0$), cases with one or several $\alpha_{0,j} \neq 0$ can be treated in the same way. Nonlinear molecules can also be modeled using this procedure.

4.3.4.2 Longest Relaxation Time

For wormlike chains (and weakly bending rods), Hearst [39] proposed equations to obtain the longest relaxation time. Later, Hagerman and Zimm [38] assumed that the longest relaxation time could be adequately predicted as the rigid-body-treatment average of the longest of the five relaxation times of wormlike molecules. The same idea is behind the results proposed by García Molina et al. [40] for wormlike chains and randomly broken chains. In the rigid-body treatment, properties are calculated as an average of the values obtained from a set of conformations, which are treated as rigid. The appropriate set of conformations is obtained through a Monte Carlo simulation. The hydrodynamic properties of rigid macromolecules can be calculated using well-developed theoretical and computational treatments, using bead models (see above).

For segmentally flexible macromolecular formalisms to describe their diffusivity have been developed (see Ref. [41] and bibliography within), although the degree of flexibility of the joint did not enter into the treatment. Also for these molecules, Roitman and Zimm [8] were able to develop a quasianalytical description of the dynamics of the simplest broken-rod chain model, the semiflexible trumbbell, for an induced dipole-orienting mechanisms for the simplified case in which hydrodynamic interaction is neglected and the orientation is produced by very weak fields (Kerr region). Iniesta et al. [30] obtained different properties (including τ_1) using the broken-rod chain model for a semiflexible once-broken rod chain. In a different approach, Vacano and Hagerman [42] proposed the tau-ratio method, which has been applied, with certain interesting modifications, to different cases of RNA and DNA. The tau-ratio approach is based on the comparison of the longest relaxation times between one linear macromolecule and another with nonhelical elements. This ratio between relaxation times is related with the bond angle through a Monte Carlo simulation. The main result obtained is the angle that presents the structure with segmental flexibility. An improved version of this procedure, the "phased tau-ratio" has also been proposed [43].

4.3.4.3 Computer Simulation of Birefringence Experiments

Once the model that describes the internal dynamics and the interaction with the external electric field has been defined, the next step is to computationally simulate the behavior of the model in a steady state or transient electric birefringence experiment. Figure 4.10 illustrates this process.

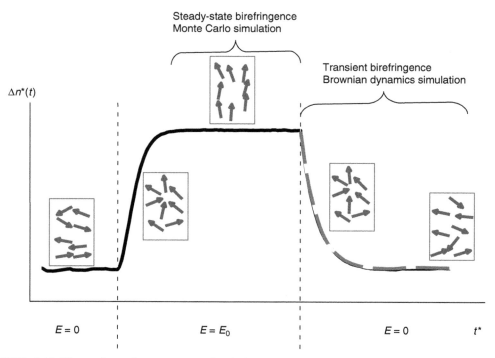

FIGURE 4.10 Illustration of a computer simulation process of a steady-state or transient electric birefringence experiment.

4.3.4.3.1   Steady-State Birefringence: Monte Carlo Simulation

In order to study the conformation of the macromolecule in the presence of an electric field, and, more specifically, to evaluate the averages needed for the steady-state molecular shape and birefringence, we employ the same Monte Carlo procedure used by Iniesta and García de la Torre [44], where total potential energy of the molecule is given by the addition of V_{int} (Equation 4.16) and V_{elect} (see Equation 4.17 below). V is a function of the set of polar angles of the arms in the laboratory system (θ_i and ϕ_i), which specify the orientations of the subunit vectors.

4.3.4.3.2   Transient Electric Birefringence: Brownian Dynamics Simulation

In recent years, it has become common to simulate the Brownian dynamics of macromolecules in solution and their orientation when an electric field is applied, or the related relaxation when this electric field is removed. Two different approaches have been used. Allison and Nambi [45] studied the electric dichroism and electric birefringence of DNA analyzing the simulated Brownian trajectories (excluding any electric field) using suitable correlation functions. One of the objectives of these authors was to approach the complex nature of polyion alignment in electric fields by comparing two orienting mechanisms, the induced dipole and the saturated induced dipole. Two correlation functions were proposed (one for each mechanism) and a low-field transient electric birefringence approximation was assumed. This method has been applied in several studies. For example, Zacharias and Hagerman [46] developed an interesting study on the influence of the static and dynamic bends in the transient electric birefringence of RNA. On the other hand, our group [24,47] studies the changes produced in the birefringence through a direct analysis of the simulated trajectories, including the presence of an electric field. In the two previous studies [32,33], we investigated the transient electric birefringence of segmentally flexible macromolecules in electric fields of arbitrary strength. In those papers, we studied the decay of the electric birefringence from the steady-state value to zero, when an orienting electric field applied on a segmentally flexible macromolecule is switched off.

4.3.4.3.3 Brownian Dynamics Simulation

The Brownian dynamics simulation of semiflexible macromolecules has been used previously by our group [48–52]. We use a simulation procedure based on Ermak and McCammon's algorithm [53], with a modification proposed by Iniesta and García de la Torre [54]. Each step is taken twice, in a predictor–corrector manner, and the position of the beads after the time step, are obtained from the previous ones. When the hydrodynamic interaction between beads is included, we use the Rotne–Prager–Yamakawa modification of the Oseen tensor [55], which corrects for the nonpoint-like nature of the frictional elements and correctly describes the possibility of overlapping (for equal-sized beads).

4.3.4.3.4 Brownian Dynamics with Electric Field

The interaction between the molecule and the field is due to permanent or induced dipoles. The contribution to the potential energy due to this interaction, V_{elect}, can be expressed as a sum of individual terms, $V_{elect,i}$, corresponding to the various subunits. These terms will depend on the permanent dipoles, μ_i, if they exist, with a possible contribution from the induced dipoles determined by the electrical polarizabilities, ε_i. Their joint effect can be expressed in terms of the so-called alignment tensor described above (Equation 4.8). This vector will be additive if expressed in a common system of reference (the lab axes), so that for the full particle, $\chi = \Sigma\chi_i$. The interaction energy with an electric field, E, is given by $V_{elect} = \chi \cdot E$. When the subunits have revolution symmetry, we assume that μ lies along the symmetry axis and that ε has parallel and perpendicular components, $\varepsilon_\parallel$ and $\varepsilon_\perp$. In this case the interaction energy is reduced to a simple form, which contains the modules of the subunit dipoles, μ_i, and polarizability differences, $\varepsilon_{i\parallel}-\varepsilon_{i\perp}$. This interaction energy depends on the angle, θ, subtended by the symmetry axis and the direction z of the electric field

$$\frac{V_{elect}}{k_B T} = -\sum_{i=1}^{N} (a_i\cos\theta_i + b_i\cos^2\theta_i) \tag{4.17}$$

In this equation $\cos\theta_i = (E \cdot u_i)/E$ and θ_i is the angle between the electric field and the symmetry axis. The parameters that describe the intensity of the molecule–field interaction are

$$a_i = \frac{\mu_i E}{k_B T} \tag{4.18}$$

$$b_i = \frac{(\varepsilon_{i\parallel} - \varepsilon_{i\perp})E^2}{2k_B T} \tag{4.19}$$

When the permanent dipoles are nonzero, there are two possibilities for describing the disposition of the dipoles: head-to-head and head-to-tail. The parameter a_i is positive if μ_i is in the same direction as u_i, and negative if it points in the opposite direction. The most relevant for broken-rod and wormlike chain models is the head-to-tail. The electric parameters are $b_i = 0$ for a purely permanent dipole moment and $a_i = 0$ for a purely induced moment. As a consequence, the field intensity, or more precisely, the strength of the molecule–field interaction is governed by the values of the a's and b's, where a is proportional to E and b proportional to E^2.

It is usual to be interested exclusively in the birefringence decay. Therefore, the field-on, birefringence-rise first part of the simulation, which has to be long to make sure that the steady state is reached, is somehow useless for our purposes. In order to save the computing time, we start the simulation with a sample of molecules generated with the Monte Carlo procedure in the presence of field. We add a period of time for the system to stabilize using the Brownian dynamics procedure. At this point, the field is removed. For the birefringence decay, we use the Brownian dynamics algorithm and individual trajectories are simulated for a large number of molecules. At any given time, t, $P_2(\cos(\theta_i))$ is evaluated for each arm ($i = 1, 2$), where $P_2(\cos\theta_i) = (3\cos^2\theta_i - 1)/2$ is the Legendre polynomial of degree 2.

For each molecule, birefringence is calculated as following:

$$\Delta n(t) = \frac{1}{\sum\limits_{i=1}^{N-1} b_{i(i+1)}} \sum_{i=1}^{N-1} b_{i(i+1)} P_2(\cos\theta_i) \tag{4.20}$$

Equation 4.20 is a weighted sum in which b_i is the instantaneous connector (spring) length. For practical purposes, in our models we can approach every pre-P_2 term equal to $1/N$. The sample averages, and, later, the final values are obtained from Equation 4.20. The duration of the decay is sufficiently long for the final birefringence to be zero within statistical error. In this way, when the field is ∞, a stationary value of saturation of birefringence is reached as a consequence $\Delta n(t) = 1$. For a more detailed description of the Brownian dynamics with electric field method, including Monte Carlo simulation, for example see Ref. [32].

4.3.4.3.5 Brownian Dynamics without Electric Field
Allison and Nambi [45] proposed two correlation functions that characterize the rotational dynamics through the corresponding analysis of the Brownian dynamics trajectories simulated in the absence of any electric field. The basis of this procedure is as follows.

We define a unitary vector **A** fixed to a coordinate system that moves with the particle. If θ_A is the angle formed between two orientations of this vector separated by a time t, then we can define $\cos\theta_A(t) = \mathbf{A}(t) \cdot \mathbf{A}(0)$. The correlation functions of interest are defined as $F(0,t) = f[P_i(\cos\theta_A)]$, where P_i is the Legendre polynomial of degree i; more specifically $P_1[\cos\theta_A(t)] = \cos\theta_A(t)$ and P_2 is as defined above.

The functions proposed by Allison and Nambi [45] are

$$H_S(lt', mt') = \frac{1}{(N-1)^2} \sum_{ik} P_2[\cos(\mathbf{u}_i(mt') \cdot \mathbf{u}_k(lt'))] \tag{4.21}$$

$$H_i(lt', mt') = \frac{1}{(N-1)^3} \sum_{ik} P_1[\cos(\mathbf{u}_i(mt') \cdot \mathbf{u}_j(mt'))] P_2[\cos(\mathbf{u}_j(mt') \cdot \mathbf{u}_k(lt'))] \tag{4.22}$$

In Equation 4.21 and Equation 4.22, $\mathbf{u}_i(mt')$ is the unit vector along the ith bond at time mt'. According to the authors, the first corresponds to an induced dipole and the second to a saturated induced dipole.

4.3.4.4 Illustrative Examples

When using models, two different approaches can be applied: the chain is described by a simple model, which reproduces the main overall characteristics of a molecule, or we use a more realistic description to gain more details about the dynamics of the molecule.

4.3.4.4.1 Simple Models

They have been used for the interpretation of experiments. Their utility is based in the idea that some electro-optical properties are related with globular properties. They can also be used for generic studies with no application to any concrete molecule.

When using simple models it is convenient to make results independent of the number of subunits and type of flexibility. The field strength must be expressed as $a^* = aN$ and $b^* = bN$, independent of the number of subunits. Regarding flexibility, the ratio between the mean square radius of gyration of a flexible model and the value for a rigid straight one can be used. When using linear-chain models of several subunits, flexibility is discretely localized in the hinges of the chain and can be defined by Q_j (see Equation 4.16). In Figure 4.11, we present plots to illustrate the correspondence between three different ways of characterizing flexibility: P/L, Q, and $<s^2>_0/<s^2>_{0,str}$.

For properties that depend only on the overall dimensions, we find the representation proposed in Figure 4.11 to be very useful. For example, it allows us to choose the most suitable model to represent DNA, because the expected values of P/L for DNA molecules of different lengths can be readily obtained. As an illustration, Figure 4.11 has been provided with an axis with the approximate number of base pairs (bp) corresponding to some values of P/L (we have assumed that $P = 50\,nm$ and that each additional base pair increases L by $0.34\,nm$). Figure 4.11 includes results for two linear-chain models of a different number of subunits ($N = 10$ and $N = 2$). We have found that $N = 10$ is a suitable number of subunits to represent wormlike flexibility within a wide range of P/L, and for example, DNA with approximately 1500 base pairs (nearly $10^6\,Da$) can be represented as a wormlike chain with $N = 10$ segments and $Q_j = 0.5$.

Using simple models, we can also consider segmentally flexible macromolecules composed of two quasirigid subunits ("arms") joined by a semiflexible swivel and interacting with an electric field. Sometimes it is desirable to study two different types of flexibility. In fact,

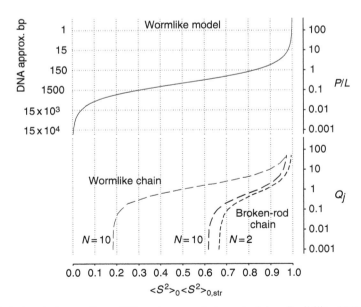

FIGURE 4.11 Relation between three different ways of characterizing flexibility (P/L, Q, and $<s^2>_0/<s^2>_{0,str}$) for a wormlike model, wormlike chain, and broken-rod chain. The approximate correspondence between the number of base pairs of DNA with $P = 50\,nm$ and P/L is included.

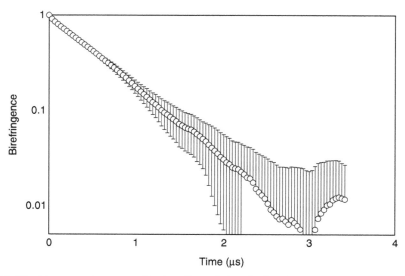

FIGURE 4.12 Birefringence decay profile (semilog plot) calculated from correlation functions for a DNA fragment of 103 base pairs.

although structural differences between wormlike and segmentally flexible molecules must be substantial, experimental results could be interpreted using different models with plausible results. Good illustrative examples are RNA [32] and DNA treated as a broken rod [33] and modeled it with three and ten beads [34].

4.3.4.4.2 More Realistic Models: DNA as an Example

Semiflexible macromolecular chains, such as nucleic acids, can be modeled with more realistic bead models. To obtain the number of beads necessary to model this molecule, we remind that DNA is regarded as a wormlike cylinder with length $L = 0.34N_{bp}$, in nm, and a hydrodynamic diameter of 2.0 nm. The diameter of the model, l, is chosen so that the chain of beads has the same hydrodynamic volume as the cylinder. As a consequence, $l = 2.45$ nm. Accordingly, the number of beads for a given contour length is $N = 0.139N_{bp}$.

In the Brownian dynamics simulation beads are joined to their neighbors by stiff springs of equilibrium length l_0 and a Hookean spring potential H. The potential energy, for an instantaneous spring length b is $V = (1/2)H(b-l_0)^2$, with a Hookean constant $H = 0.673$ erg/cm^2, whose precise value is irrelevant if it is sufficiently large to make the fluctuation around l small. The bending flexibility of DNA measured in the wormlike chain model by the persistence length P is introduced in the chain of beads by assigning to all the nonterminal beads a potential $V_{bend} = (1/2)Q(\alpha-\alpha_0)^2$, with $\alpha_0 = 0$ (usually) and where $Q = P/l_0$. If we assume P to be approximately 50 nm, $Q = 8.24\times10^{-13}$ erg/rad^2.

As an illustration of the performance and versatility of this procedure Figure 4.12 shows the birefringence decay profile calculated from correlation functions for DNA of 103 base pairs.

ACKNOWLEDGMENT

This work was supported by a grant from Ministerio de Ciencia y Tecnología (BQU2003–04517). H.E.P.S. was the recipient of a predoctoral FPU fellowship from Ministerio de Educación.

REFERENCES

1. Frederick, E., and Houssier, C., *Electric Dichroism and Electric Birefringence*, Clarendon Press, Oxford, 1973.
2. Krause, S., *Molecular Electro-Optic Properties of Macromolecules and Colloids in Solution*, Plenum Press, New York, 1981.
3. Riande, E., and Saiz, E., *Dipole Moments and Birefringence of Polymers*, Prentice Hall, New Jersey, 1992.
4. Favro, L.D., Theory of the rotational Brownian motion of a free rigid body, *Phys. Rev.*, 119, 53, 1960.
5. Wegener, W.A., Dowben, R.M., and Koester, V.J., Time dependent birefringence, linear dichroism and optical rotation resulting from rigid-body rotational diffusion, *J. Chem. Phys.*, 70, 622, 1979.
6. Zimm, B.H., Dynamics of polymer molecules in dilute solution: viscoelasticity, flow birefringence and dielectric loss, *J. Chem. Phys.*, 24, 269, 1956.
7. Doi, M., and Edwards, S.F., *The Theory of Polymer Dynamics*, Oxford University Press, Oxford, 1986, chap. 4.
8. Roitman, D.B., and Zimm, B.H., An elastic hinge model for the dynamics of stiff chains. II. Transient electro-optical properties, *J. Chem. Phys.*, 81, 6348, 1984.
9. Kirkwood, J.G., The general theory of irreversible processes in solutions of macromolecules, *J. Polym. Sci.*, 12, 1, 1954.
10. Bloomfield, V.A., Dalton, W.O., and Van Holde, K.E., Frictional coefficients of multisubunit structures. II. Application to proteins and viruses, *Biopolymers*, 5, 149, 1967.
11. García de la Torre, J., and Bloomfield, V.A., Hydrodynamic properties of macromolecular complexes. I. Translation, *Biopolymers*, 16, 1747, 1977.
12. García de la Torre, J., and Bloomfield, V.A., Hydrodynamic properties of macromolecular complexes. II. Rotation, *Biopolymers*, 16, 1765, 1977.
13. García de la Torre, J., and Bloomfield, V.A., Hydrodynamic properties of complex, rigid, biological macromolecules. Theory and applications, *Q. Rev. Biophys.*, 14, 81, 1981.
14. Carrasco, B., and García de la Torre, J., Hydrodynamic properties of rigid particles. Comparison of different modelling and computational procedures, *Biophys. J.*, 76, 3044, 1999.
15. García de la Torre, J., Huertas, M.L., and Carrasco, B., Calculation of hydrodynamic properties of globular proteins from their atomic-level structures, *Biophys. J.*, 78, 719, 2000.
16. García de la Torre, J., et al., HYDROMIC: prediction of hydrodynamic properties of rigid macromolecular structures obtained from electron microscopy, *Eur. Biophys. J.*, 30, 457, 2001.
17. García de la Torre, J., and Carrasco, B., Hydrodynamic properties of rigid macromolecules composed of ellipsoidal and cylindrical subunits, *Biopolymers*, 63, 163, 2002.
18. García de la Torre, J., Carrasco, B., and Harding, S.E., SOLPRO: theory and computer program for the prediction of SOLution PROperties of rigid macromolecules and bioparticles, *Eur. Biophys. J.*, 25, 361, 1997.
19. García de la Torre, J., Harding, S.E., and Carrasco, B., Calculation of NMR relaxation, covolume and scattering-related properties of bead models using the SOLPRO computer program, *Eur. Biophys. J.*, 28, 119, 1999.
20. Mellado, P., and García de la Torre, J., Steady state and transient electric birefringence of solutions of bent-rod macromolecules, *Biopolymers*, 21, 1857, 1982.
21. Fernandes, M.X., and García de la Torre, J., Brownian dynamics simulation of rigid particles of arbitrary shape in external fields, *Biophys. J.*, 83, 3039, 2002.
22. Pérez Sánchez, H.E., García de la Torre, J., and Díaz Baños, F.G., Birefringence, deformation and scattering of wormlike macromolecules under an external agent. Steady-state properties in an electric field, *J. Phys. Chem.*, 107, 13192, 2003.
23. Pérez Sánchez, H.E., García de la Torre, J., and Díaz Baños, F.G., Transient electric birefringence of wormlike macromolecules in electric fields of arbitrary strength: a computer simulation study, *J. Chem. Phys.*, 122, 124902, 2005.
24. Carrasco, B. et al., Birefringence, deformation, and scattering of segmentally flexible macromolecules under an external agent. Steady-state properties in an electric field, *J. Phys. Chem.*, 103, 7822, 1999.

25. Carrasco, B. et al., Birefringence, deformation, and scattering of segmentally flexible macromolecules under an external agent. Steady-state properties in an electric field, *J. Phys. Chem.*, 103, 7822, 1999.

26. Rouse, P.E., A theory of the linear viscoelastic properties of dilute solutions of coiling polymers, *J. Chem. Phys.*, 21, 1272, 1953.

27. Navarro, S., López Martínez, M.C., and García de la Torre, J., Relaxation times in transient electric birefringence and electric field light scattering of flexible polymer chains, *J. Chem. Phys.*, 103, 7631, 1995.

28. Rau, D.C., and Bloomfield, V.A., Transient electric birefringence of T7 viral DNA, *Biopolymers*, 18, 2783, 1979.

29. Harvey, S.C., and Cheung, H., *Cell and Muscle Motility*, Vol. 2, Plenum Press, New York, 1981, p. 279.

30. Iniesta, A., Díaz, F.G., and García de la Torre, J., Transport properties of rigid bent-rod macromolecules and semiflexible broken rods in the rigid body approximation, *Biophys. J.*, 54, 269, 1988.

31. Burton, D.F., *Molecular Genetics of Immunoglobulin*, Elsevier, Amsterdam, 1987, p. 1.

32. Díaz, F.G. et al., Transient electric birefringence of segmentally flexible macromolecules in electric fields of arbitrary strength, *J. Phys. Chem.*, 104, 12339, 2000.

33. Pérez Sánchez, H.E., García de la Torre, J., and Díaz Baños, F.G., Influence of field strength and flexibility in the transient electric birefringence of segmentally flexible macromolecules, *J. Phys. Chem.*, 106, 6754, 2002.

34. Bloomfield, V.A., Crothers, D.M., and Tinoco, I., *Nucleic Acids: Structures, Properties and Functions*, University Science Books, Sausalito, 2000.

35. Hagerman, P.J., Flexibility of RNA, *Annu. Rev. Biophys. Biomol. Struct.*, 26, 139, 1997.

36. Bertolotto, J.A. et al., Dependence of DNA steady-state electric birefringence on field strength, *Colloids Surf., A.*, 203, 167, 2002.

37. Lewis, R.J. et al., Brownian dynamics simulations of a three-subunit and ten-subunit wormlike chain: comparison with trumbbell theory and with experimental results from DNA, *J. Chem. Phys.*, 89, 2490, 1988.

38. Hagerman, P., and Zimm, B.H., Monte Carlo approach to the analysis of the rotational diffusion of wormlike chains, *Biopolymers*, 20, 1481, 1981.

39. Hearst, J.E., Rotatory diffusion constants of stiff-chain macromolecules, *J. Chem. Phys.*, 38, 1062, 1963.

40. García Molina, J.J., López Martínez, M.C., and García de la Torre, J., Computer simulation of hydrodynamic properties of semiflexible macromolecules. Randomly broken chains, wormlike chains, and analysis of properties of DNA, *Biopolymers*, 29, 883, 1990.

41. Mellado, P. et al., Diffusion coefficients of segmentally flexible macromolecules with two subunits. Study of broken rods, *Biopolymers*, 27, 1771, 1988.

42. Vacano, E., and Hagerman, P.J., Analysis of birefringence decay profiles for nucleic acid helices possessing bends: The tau-ratio approach, *Biophys. J.*, 73, 306, 1997.

43. Friedrich, M.W., Vacano, E., and Hagerman, P., Global flexibility of tertiary structure in RNA: yeast tRNA(Phe) as a model system, *Biochemistry*, 95, 3572, 1998.

44. Iniesta, A., and García de la Torre, J., Electric birefringence of segmentally flexible macromolecules with 2 subunits at arbitrary field strengths, *J. Chem. Phys.*, 90, 5190, 1989.

45. Allison, S.A., and Nambi, P., Electric dichroism and electric birefringence decay of short DNA restriction fragments. A. Brownian dynamics simulation, *Macromolecules*, 25, 759, 1992.

46. Zacharias, M., and Hagerman, P.J., Influence of static and dynamic bends on the birefringence decay profile of RNA helices: Brownian dynamics simulations, *Biophys. J.*, 73, 318, 1997.

47. Carrasco, B. et al., Transient orientation and electrooptical properties of axially symmetric macromolecules in an electric field of arbitrary strength, *J. Chem. Phys.*, 100, 9900, 1996.

48. Díaz, F.G., Freire, J.J., and García de la Torre, J., Viscoelastic properties of simple, flexible and semirigid models from Brownian dynamics simulation, *Macromolecules*, 23, 3144, 1990.

49. Díaz, F.G., Iniesta, A., and García de la Torre, J., Hydrodynamic interaction study of flexibility in immunoglobulin IgG1 using Brownian dynamics simulation of a simple model, *Biopolymers*, 29, 547, 1990.

50. Iniesta, A., López Martínez, M.C., and García de la Torre, J., Rotational Brownian dynamics of semiflexible broken rods, *J. Fluorescence*, 1, 129, 1991.
51. Díaz, F.G., López Cascales, J.J., and García de la Torre, J., Bead-model calculation of scattering diagrams. Brownian dynamics simulation study of flexibility in immunoglobulin IgG1, *J. Biochem. Biophys. Methods*, 26, 261, 1993.
52. Díaz, F.G., and García de la Torre, J., Viscoelastic properties of semiflexible macromolecular models, *Macromolecules*, 27, 5371, 1994.
53. Ermak, D.L., and McCammon, J.A., Brownian dynamics with hydrodynamic interactions, 69, 1352, 1978.
54. Iniesta, A., and García de la Torre, J., A second-order algorithm for the simulation of the Brownian dynamics of macromolecular models, *J. Chem. Phys.*, 92, 2015, 1990.
55. Yamakawa, H., Transport properties of polymer chains in dilute solutions. Hydrodynamic interaction, *J. Chem. Phys.*, 53, 436, 1970.

5 Simulation of Electric Polarizability of Polyelectrolytes

Kazuo Kikuchi and Hitoshi Washizu

CONTENTS

5.1 INTRODUCTION

According to the fluctuation–dissipation theorem [1], the electric polarizability of polyelectrolytes is related to the fluctuations of the dipole moment, which is spontaneously generated in the ion atmosphere around the polyions in the absence of an applied electric field [2–6]. Recently, we had started to determine anisotropy of the electric polarizability of model DNA fragments in aqueous solution by computer simulation [7–15] and had reproduced characteristics of the electric properties of polyelectrolytes in both salt-free [14,15] and salt solutions [12]. The Monte Carlo (MC) method was applied to calculate small ion distributions, electric potentials, and fluctuations of ion atmosphere polarization.

A variety of theoretical models have been proposed to describe the origin of the induced dipole moment of polyelectrolytes [2–6,16–41]. The MC Brownian dynamics method [42–46] allows us to calculate the various fields involved: concentrations of small ions; the electrical potential; and the solvent velocity as a function of space and time, which are coupled with each other through essentially nonlinear equations.

The system has prolate ellipsoidal symmetry with the foci located at both ends of the DNA polyion cylinder [47]. In salt-free solutions, at every simulation step we sort the counterions numerically in increasing order of the sum of their distances from both ends of the polyion. Two kinds of counterions are distinguished from their spatial distributions, and identified in the framework of the counterion condensation theory [3,48], which was extended to the oligomers [49,50], as condensed counterions and diffuse ion atmospheres. We define a partial polarizability tensor by calculating fluctuations of the contribution to the dipole moment from the first n counterions in the sorting list. Its introduction facilitates understanding the origin of the polarizability in close relation to the solution structure. The contribution from the condensed ions to the radial component of the polarizability tensor is very small as has often been postulated in various theories. However, contribution from a diffuse ion atmosphere is very large and cannot be neglected in the calculation of the anisotropy. In salt-free solutions the anisotropy of the electric polarizability $\Delta\alpha$ of model DNA fragments increases on dilution of the polymer concentration and is proportional to the second or higher power of the DNA molecular weight.

In salt solutions, taking into consideration the contribution of coions to the electroneutrality condition, we obtain a list of counterions at every simulation step, constituting the net charge that compensates the polyion charge. $\Delta\alpha$ decreases on addition of salt, reaching a steady value comparable to the experimental value.

5.2 MODEL

5.2.1 SALT-FREE SOLUTION

A 64/128/256 base-pair fragment of the double-stranded DNA is modeled as an impenetrable cylinder of radius 0.85 nm with 128/256/512 negative charges spaced at 0.17-nm intervals along its axis. The hydrated univalent counterions are modeled as hard spheres of radius 0.15 nm so that the radial distance of the closest approach of an ion to the axis of the DNA is 1.0 nm. The cylinder is extended 0.17 nm beyond the terminal charges at both ends. The solvent is treated as a dielectric continuum with the relative permittivity of pure water at 298 K (78.3) and no salt is added. The MC cell is a sphere of varying radius in which the DNA fragment is placed along the z-axis with its center common to that of the MC cell, or the origin of the coordinate system. The DNA concentration c_p is calculated assuming that 128/256/512 nucleotide residues occupy an effective volume equivalent to the size of the simulation sphere, whose volume is varied by changing its radius.

5.2.2 SALT SOLUTION

A 64 base-pair fragment of double-stranded DNA is modeled as an impenetrable cylinder of radius 1.04 nm. The 128 negative phosphate charges are placed along two right-handed helices embedded in the surface of the cylinder. For canonical B-DNA, positions of phosphate charges on one strand are expressed in cylindrical polar coordinates as [51]

$$r_i = 0.89$$

$$z_i = 0.208 + 0.34i$$

$$\theta_i = 95.2 + 36i, \; i = 0, \ldots, 31 \tag{5.1}$$

where radial and axial distances from and along the helix axis r_i and z_i, respectively, are given in nanometers and the angular coordinate θ_i in degrees. To generate coordinates of the phosphate charges on the other strand, z_i and θ_i ($i = 1, 2, \ldots, 32$) are negated. Equation 5.1 shows that the distance between two neighboring phosphate charges on each helix is 0.34 nm along the axis of the cylinder and that they are situated at a depth of $1.04 - 0.89 = 0.15$ nm from the surface of the DNA cylinder. The DNA cylinder is extended 0.15 nm beyond the terminal phosphate charges at both ends so that its length is

TABLE 5.1
Salt Concentrations

Salt Concentration,[a] c_s (mM)	Number of Counterions in MC Cell	Number of Coions in MC Cell	Debye Length,[b] κ^{-1} (nm)
0.0	128	0	—
0.60	256	128	12.41
1.2	384	256	8.77
2.4	640	512	6.21
4.8	1152	1024	4.38

[a]Calculated by dividing numbers of coions in the MC cell by its volume and used as approximate bulk salt concentrations.
[b]Calculated from approximate bulk salt concentrations c_s.

$L = 21.84$ nm. The hydrated univalent counter- and coions are modeled as hard spheres of radius 0.15 nm. Hence all the distances of the closest approach of small ions to the DNA phosphate charges and to each other are set to 0.30 nm. The MC cell is basically a sphere of a fixed radius of 42.86 nm from which the polymer concentration c_p is calculated to be 0.60 mM nucleotide residues. Salt concentrations c_s are calculated by dividing numbers of coions in the simulation cell by its volume and used as approximate bulk salt concentrations. They are tabulated in Table 5.1 along with the Debye lengths κ^{-1} calculated therefrom. Occasionally c_p dependence is studied by changing both the radius of the MC cell and the number of coions added.

5.3   SIMULATION

Conventional Metropolis MC procedure is used to generate a canonical ensemble. The energy of configurations is calculated as a sum of interactions of each small mobile ion in the MC cell with all the other small ions and DNA charges in the cell. A value of 0.2 nm for Ω^{-1} is used as the maximum ion displacement allowed for an MC move along each coordinate direction. The new configuration generated by a random movement of a single ion is accepted or rejected according to the probability min[1, $\exp(-\Delta U/k_B T)$], where ΔU is the change in configurational energy that would result from the move, k_B is the Boltzmann constant, and T is the absolute temperature. When a mobile ion escapes from the cell during a move in the Metropolis sampling process, another ion of the same kind is put in the symmetrical position about the center of the MC sphere. Although the Metropolis time scale Δt is fixed by the relation

$$6D\Delta t\Omega^2 = 1 \qquad (5.2)$$

where D is the diffusion coefficient for the small ion species in water at infinite dilution [42,43,45], we need not relate the Metropolis and physical time scales as far as we calculate only time averages.

The system has prolate ellipsoidal symmetry with the foci located at both ends of the DNA polyion cylinder. In salt-free solutions, at every simulation step we numerically sort counterions in increasing order of the sum of their distances from both ends of the polyion and calculate the contribution to the dipole moment from the first n counterions in the sorting list $(\mu_x(n), \mu_y(n), \mu_z(n))$ as follows:

$$
\begin{aligned}
\mu_x(n) &= \sum_{i=1}^{n} ex_i \\
\mu_y(n) &= \sum_{i=1}^{n} ey_i \\
\mu_z(n) &= \sum_{i=1}^{n} ez_i
\end{aligned}
\tag{5.3}
$$

where e is the protonic charge and (x_i, y_i, z_i) are the coordinates of the i-th counterion. We then define the partial polarizability tensor due to these n counterions with its principal components $\alpha_{xx}(n)$, $\alpha_{yy}(n)$, $\alpha_{zz}(n)$ given by

$$
\begin{aligned}
\alpha_{xx}(n) &= \left(\langle \mu_x^2(n) \rangle - \langle \mu_x(n) \rangle^2 \right)/k_B T \\
\alpha_{yy}(n) &= \left(\langle \mu_y^2(n) \rangle - \langle \mu_y(n) \rangle^2 \right)/k_B T \\
\alpha_{zz}(n) &= \left(\langle \mu_z^2(n) \rangle - \langle \mu_z(n) \rangle^2 \right)/k_B T
\end{aligned}
\tag{5.4}
$$

where $\langle\ \rangle$ denotes time or canonical ensemble average. The transverse and longitudinal partial polarizabilities, $\alpha_T(n)$ and $\alpha_L(n)$, are defined as

$$
\begin{aligned}
\alpha_T(n) &= (\alpha_{xx}(n) + \alpha_{yy}(n))/2 \\
\alpha_L(n) &= \alpha_{zz}(n)
\end{aligned}
\tag{5.5}
$$

and the partial anisotropy of the polarizability, $\Delta\alpha(n)$, as

$$
\Delta\alpha(n) = \alpha_L(n) - \alpha_T(n)
\tag{5.6}
$$

where n runs from 1 to 128/256/512 and the anisotropy of the electric polarizability, $\Delta\alpha$, to be determined is expressed as

$$
\Delta\alpha = \alpha_L(128/256/512) - \alpha_T(128/256/512).
\tag{5.7}
$$

In salt solutions, the contribution of coions to the electroneutrality condition must be taken into account. We consider that polyion phosphate charge is compensated by the same amount of net charge. At every simulation step we obtain a list of 128 counterions, which constitute the net charge as follows. First we sort all the small ions in increasing order of the sum of their distances from both ends of the polyion (more precisely, from the projections of the terminal charges at both ends of the polyion onto the DNA cylinder axis). Then if we find a coion in the list, we search the nearest consecutive counterion in the list and delete both entries. We continue this process from the beginning of the list until the first 128 consecutive entries are all occupied by counterions.

At each simulation step we calculate the contribution to the dipole moment from the first n counterions in the list constituting the net charge $(\mu_x(n), \mu_y(n), \mu_z(n))$ by Equation 5.3, define a partial polarizability tensor due to the net charge by Equation 5.4, longitudinal and transverse components of the partial polarizability $\alpha_L(n)$ and $\alpha_T(n)$ by Equation 5.5, and partial anisotropy of polarizability $\Delta\alpha(n)$ by Equation 5.6. The anisotropy of the electric polarizability $\Delta\alpha$ is determined as $\Delta\alpha = \Delta\alpha(128)$.

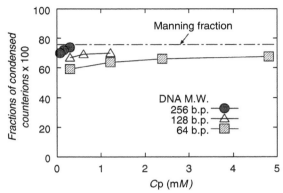

FIGURE 5.1 Numerically determined fractions of condensed counterions as functions of polymer concentration c_p for three DNA oligomers. (Modified from Washizu, H. and Kikuchi, K., *J. Phys. Chem.*, 106, 11335, 2002. Copyright 2002, American Chemical Society. With permission.)

5.4 RESULTS

5.4.1 SALT-FREE SOLUTION

The results for salt-free solutions have been described elsewhere [14,15]. We have succeeded in reproducing the characteristic features of the electric properties of polyelectrolytes. For example, the anisotropy of the electric polarizability, $\Delta\alpha$, of DNA in salt-free solutions increases on dilution of the polymer concentration, increasing more sharply at higher molecular weights, and is proportional to the second or higher power of the molecular weight consistent with the experimental value.

We have described ionic structures around charged oligomers in terms of counterion condensation theory. The figures showing distances of counterions from polyions or sum of their distances from both ends of the polyions, sorted in increasing order and averaged over a number of uncorrelated configurations, show inflection points that divide counterions into two groups. One of them is located in the vicinity of the polyion so closely to be called condensed ions and the other, a diffuse ion atmosphere. They show quite different polarization behaviors. Manning's counterion condensation theory [48] predicts for infinitely long polyelectrolyte chains that 76% of the DNA phosphate charge is neutralized by condensed ions. Figure 5.1 shows that the numerically determined fraction of condensed counterions approaches the Manning fraction as the molecular weight of the polyelectrolyte increases.

5.4.2 SALT SOLUTION

Figure 5.2a and Figure 5.2b shows counterion and coion distributions, respectively, at a salt concentration $c_s = 4.8\,\text{m}M$, which is the largest salt concentration studied. Figure 5.2a shows positions of a total of 11,520 counterions and Figure 5.2b, 10,240 coions, respectively, projected onto a z–r plane where r is the radial coordinate measured from the axis of the DNA cylinder. They are collected from ten uncorrelated configurations generated during the simulation. Besides effects of linear r dependence of the volume element of the cylindrical coordinate system, accumulation of counterions and exclusion of coions in the vicinity of the polyion can be noticed.

Counterion concentration c^+ and coion concentration c^- distributions are plotted more quantitatively in logarithmic scale (Figure 5.3). The figure is obtained by averaging over 3.49×10^7 configurations. On the surface of the polyion the very high counterion

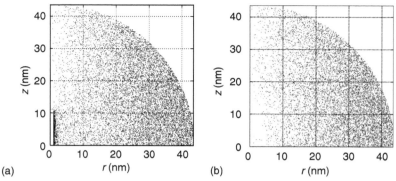

(a)   (b)

FIGURE 5.2 Projection of the (a) counterion and (b) coion distribution onto a $z - r$ plane where r is the radial coordinate measured from the axis of the DNA cylinder at a polymer concentration $c_p = 0.60$ mM nucleotide residues and a salt concentration $c_s = 4.8$ mM.

concentration (~2 M) decreases rapidly and the very low coion concentration (~10^{-4}–10^{-6} M) increases rapidly both in radial and longitudinal directions reaching their bulk values.

Ramanathan and Woodbury [50] extended the condensation theory to finite-length polymers. They showed that if the length of the oligomer L is comparable to κ^{-1} or larger, as in the present case, the fractional extent of condensation on the central portion of the polyion is the same as for infinitely long polymers. On the other hand, end effects on condensation were analyzed by Odijk [52]. He showed that the nonuniformity of the number of condensed counterions is limited to within κ^{-1} from the ends of an oligomer. Figure 5.4 shows counterion concentration c^+ profiles near the surface of a 64 base-pair DNA at a salt concentration $c_s = 4.8$ mM as the function of the radial distance r measured from the axis of the polyion at various z coordinates. It is seen that end effects are very small at this highest salt concentration studied, where L is much larger than κ^{-1}.

Manning [48] in his condensation theory made an additional assumption that the electrostatic potential ψ satisfies the condition $|e\psi/k_B T| \leq 1$ within diffuse ion atmospheres and treated them by the Debye–Hückel approximation. Figure 5.5 plots reduced electrostatic potentials at the position of the nth counterion constituting the net charge arranged in increasing order of the distance from the polyion. A horizontal line segment is drawn to indicate positions where $|e\psi/k_B T| = 1$. It is seen that the curves cross the line segment at

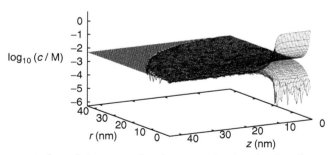

FIGURE 5.3 3D representation of the counterion (upper surface) and coion (lower surface) concentrations around a 64-base-pair DNA fragment at a polymer concentration $c_p = 0.60$ mM nucleotide residues and a salt concentration $c_s = 4.8$ mM.

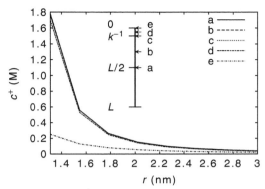

FIGURE 5.4 Counterion concentration c^+ profiles near the surface of a 64 base-pair DNA as the function of the radial distance r measured from the axis of the polyion at five representative z coordinates. The origin of the coordinate system is at one end of the DNA cylinder. Polymer concentration is $c_p = 0.60$ mM nucleotide residues, salt concentration $c_s = 4.8$ mM, and Debye length $\kappa^{-1} = 4.38$ nm. Point a: $z = L/2$, c: $z = \kappa^{-1}$, e: $z = 0$. Points b and d are the midpoints of points a and c, and c and e, respectively.

positions near the Manning fraction (corresponding to 98 counterions for a 64 base-pair DNA in water at 298 K), closer as the salt concentration increases.

According to Ramanathan and Woodbury [50], our net charge is, however, an overlap of a layer of condensed counterions and a Poisson–Boltzmann ion atmosphere, which is evaluated as if $\xi = 1$, where ξ is the linear charge density parameter of charged chains. Due to the composite nature of the ion atmosphere in salt solution, i.e., overall distribution is produced by an overlap of individually continuous components, it is not as easy to disclose the presence of a condensed counterion layer from simulation as in the salt-free solution, where nonoverlapping discontinuous distributions corresponding to the two components are glued together at the point of discontinuity [53]. Partial polarizabilities α_L (n) and α_T (n) when plotted against n will show similar behavior because a condensed layer and a Poisson–Boltzmann or a diffuse ion atmosphere must exhibit the same electric properties in the region where they are

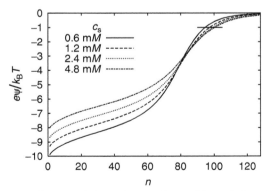

FIGURE 5.5 Reduced electrostatic potentials $e\psi/k_BT$ at the position of the nth counterion constituting the net charge arranged in increasing order of the distance from the polyion. A line segment is drawn to indicate positions where $|e\psi/k_BT| = 1$.

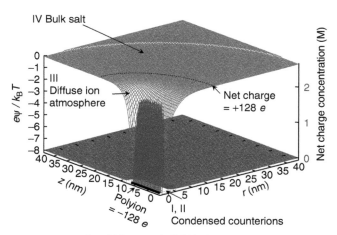

FIGURE 5.6 Ionic structure around a 64-base-pair DNA fragment, $c_p = 0.60\,mM$ nucleotide residues, $c_s = 2.4\,mM$. Reduced electrostatic potential $e\psi/k_B T$ (light) and net charge distribution (dark) are plotted in a 3D view. Truncated boundary between the ion atmosphere and the bulk salt solution is indicated by a dashed line. The solution consists of four regions, I–IV.

overlapped. We will, however, use terminology of the two-phase model [3,48] to make description simple by neglecting the minor component in each phase.

Then the ionic structure around a 64-base-pair DNA fragment is summarized in Figure 5.6 where the reduced electrostatic potential $e\psi/k_B T$ (light) and the net charge distribution (dark) are plotted in a 3D view. Numerical determination of the net charge gives us a boundary between the ion atmosphere and the bulk salt solution, whereas analytical theories do not yield a clear boundary between them. Without truncation or discretization they merge continuously at infinity.

In Figure 5.6 we visualize the solution structure as consisting of four regions I–IV. Regions I and II are composed of condensed counterions (for their distinction, see below), region III, a diffuse ion atmosphere, and region IV, the bulk salt solution. It is the net charge distribution over regions I–III that is responsible for the electric polarizability.

5.4.3 ELECTRIC POLARIZABILITY

Figure 5.7 plots partial anisotropy of polarizability $\Delta\alpha\,(n)$ determined at various salt concentrations c_s as functions of the number of contributing counterions or net charge n. We can observe three regions showing different dependence upon n: region I ($1 \leq n \leq 50$), region II ($50 \leq n \leq 70 - 100$), and region III ($70 - 100 \leq n \leq 128$). In region I, $\Delta\alpha\,(n)$ curves rise from 0 until the DNA phosphate charge is neutralized by about 50 counterions. In region II, they reach plateau values, and in region III, they grow rapidly again. It is seen that these regions correspond to those shown in Figure 5.6. As the curve for salt-free solution ($c_s = 0.0\,mM$) shows a greater curvature between regions II and III, it is possible to find a point disclosing the presence of a condensed counterion layer [15]. It is found that the difference in the phosphate charge arrangement of DNA models, linear or double-helical array of charges along the cylinder, has little effect on the calculated polarizability compared to our previous results [15].

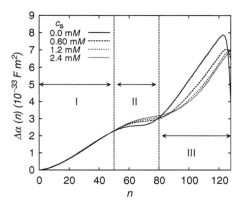

FIGURE 5.7 Partial anisotropy of polarizability $\Delta\alpha(n)$ for a 64-base-pair DNA as functions of number of contributing counterions n (net charge) at a polymer concentration $c_p = 0.60$ mM nucleotide residues and various salt concentrations c_s. (Reprinted from Washizu, H. and Kikuchi, K., *Chem. Phys. Lett.*, 320, 279, 2000. Copyright 2000, with permission from Elsevier.)

When n belongs to regions I and II, $\alpha_T(n)$ remains very small because ions are condensed in a layer over the DNA surface. Then $\Delta\alpha(n)$ is approximately equal to $\alpha_L(n)$. It increases with n in region I as build-up of the z component of the partial dipole moment $\mu_z(n)$ is regarded as a random walk starting from the origin of the coordinates. When n increases over 50, fluctuations of $\mu_z(n)$ no longer increase with n as long as n remains in region II because they distribute uniformly along the DNA. When n exits from region II, the diffuse ion atmosphere begins to contribute to the dipole moment with its spatial distribution reflected in the rapid increase of both $\alpha_L(n)$ and $\alpha_T(n)$. The $\Delta\alpha(n)$ in region III is given as a difference of two large quantities. However, the longitudinal component exhibits still larger dependence on n than the radial, resulting in a sharp increase of the anisotropy. This is because the former originates from displacements of counterions along electrostatic equipotential surfaces and the latter perpendicular.

The $\Delta\alpha(n)$ curves show no salt dependence in region I consistent with the fact that coions are strongly repelled and rarely approach the immediate vicinity of the polyion even at the highest salt concentration studied. On the other hand in region II, the presence of coions affects the curves in such a way that $\Delta\alpha(n)$ increases with the addition of the salt. Manning [35] has already explained this effect in his theory of polarizability as reflecting the decreased resistance to distortion of the condensation layer. In region III, large contribution from diffuse ion atmospheres as well as reversal of the salt concentration dependence of the $\Delta\alpha(n)$ curves is observed. Sharp drops of the curves seen in the outer part of this region should be regarded as due to the boundary conditions of our simulation cell. Linear portions of the curves are extrapolated to obtain values of $\Delta\alpha(128)$ or $\Delta\alpha$.

Figure 5.8 shows longitudinal and transverse partial polarizability contributions from all the small ions, counterions, and coions, within distance s from the polyion, $\alpha_L(s)$ and $\alpha_T(s)$, plotted in a wider region, including region IV. Partial polarizabilities as functions of s are defined in a similar manner as $\alpha_L(n)$ and $\alpha_T(n)$. A vertical broken line is drawn to indicate the boundary between regions III and IV. The bell-shaped curve is the probability distribution function of finding a 128th counterion constituting the net charge at distance s. It is seen that in region IV, polarizability components α_L and α_T increase but at the same rate, i.e., electric properties are isotropic in region IV resulting in a constant anisotropy $\Delta\alpha$.

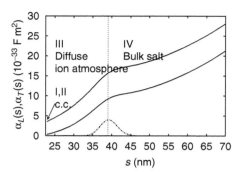

FIGURE 5.8 Longitudinal and transverse partial polarizability contributions from all the small ions, counterions, and coions, within distance s from the polyion $\alpha_L(s)$ and $\alpha_T(s)$ for a 64 base-pair DNA at a polymer concentration $c_p = 0.60 \, \text{m}M$ nucleotide residues and a salt concentration $c_s = 4.8 \, \text{m}M$. The upper curve is $\alpha_L(s)$ and the lower one $\alpha_T(s)$. A vertical broken line is drawn to indicate the boundary between region III and IV. The bell-shaped curve is the probability distribution function of finding a 128th counterion constituting the net charge at distance s.

Figure 5.9 shows the dependence of the anisotropy of the electric polarizability $\Delta\alpha$ on the polymer concentration (c_p) at a salt concentration $c_s = 4.8 \, \text{m}M$. $\Delta\alpha$ remains constant as expected from the solution structure, which preserves ionic distributions in regions I–III that are unchanged.

Figure 5.10 shows the dependence of the anisotropy of the electric polarizability $\Delta\alpha$ of a 64 base-pair DNA on the salt concentration c_s at a polymer concentration $c_p = 0.60 \, \text{m}M$ nucleotide residues. $\Delta\alpha$ decreases monotonously with the addition of salt reaching a steady value of, if the result of Figure 5.9 is combined, $\Delta\alpha = 6.5 \times 10^{-33} \, \text{F m}^2$. It is comparable to the experimental $\Delta\alpha$ value of $2.0 \times 10^{-33} \, \text{F m}^2$ determined for a 64-base-pair DNA fragment at a concentration of $20 \, \mu M$ nucleotide residues in a buffer system of $1 \, \text{m}M$ sodium cacodylate, $1 \, \text{m}M$ NaCl, and $0.2 \, \text{m}M$ EDTA at pH 7.1 by Diekmann et al. [54] despite some difference in the ionic environment.

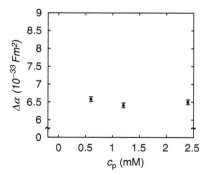

FIGURE 5.9 Effect of polymer concentration c_p on the anisotropy of electric polarizability $\Delta\alpha$ for a 64-base-pair DNA at a salt concentration $c_s = 4.8 \, \text{m}M$.

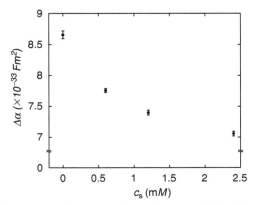

FIGURE 5.10 Effect of salt concentration c_s on the anisotropy of the electric polarizability $\Delta\alpha$ for a 64-base-pair DNA at a polymer concentration $c_p = 0.60\,\mathrm{m}M$ nucleotide residues. (Reprinted from Washizu, H. and Kikuchi, K., *Chem. Phys. Lett.*, 320, 280, 2000. Copyright 2000, with permission from Elsevier.)

5.5 CONCLUSIONS

We have succeeded in reproducing the electrical properties of polyelectrolytes in both the salt-free and the salt solutions by computer simulation. The ionic structure around charged oligomers is interpreted in the framework of the counterion condensation theory. Condensed counterions and diffuse ion atmospheres are distinguished not only by their spatial distributions but also by their fluctuation or polarization behaviors. The contribution from the condensed ions to the radial components of the electric polarizability tensors is very small as has often been postulated in various theories, whereas contribution from the diffuse ion clouds is very large and cannot be neglected in the calculation of the anisotropy. We have also shown that contribution from the condensed layer to the longitudinal component of the polarizability depends weakly on the ionic strength and in the opposite direction to that found experimentally or to the total polarizability just as the theory predicted [35].

The anisotropy $\Delta\alpha$ in salt-free solutions increases on dilution of the polymer concentration, increasing more sharply at higher molecular weights, and is proportional to the second or higher power of the molecular weight, which is consistent with the experimental value.

$\Delta\alpha$ decreases with the addition of salt, reaching a steady value comparable to the experimental value. Stronger salt concentration dependence is, however, observed in the experiment, i.e., the induced dipole moment of DNA is inversely proportional to the square root of the ionic strength of the solution tending to zero at infinite salt concentration [32,33,55,56], whereas our simulation yields linear dependence on the inverse of the square root of the ionic strength giving a finite intercept at infinite salt concentration. The discrepancy will be largely ascribed to the reduced internal fields in conducting media.

We have not taken into consideration electrophoretic motion of polyions [36,40]. Apart from these corrections, we have clarified the origin of the static electric polarizability of polyelectrolytes in the aqueous solution.

REFERENCES

1. Kubo, R., Statistical–mechanical theory of irreversible processes. I. General theory and simple applications to magnetic and conduction problems, *J. Phys. Soc. Jpn.*, 12, 570, 1957.
2. Oosawa, F., Counterion fluctuation and dielectric dispersion in linear polyelectrolytes, *Biopolymers*, 9, 677, 1970.

3. Oosawa, F., *Polyelectrolytes*, Marcel Dekker, New York, 1971.
4. Minakata, A., Imai, N., and Oosawa, F., Dielectric properties of polyelectrolytes. II. A theory of dielectric increment due to ion fluctuation by a matrix method, *Biopolymers*, 11, 347, 1972.
5. Warashina, A. and Minakata, A., Dielectric properties of polyelectrolytes. IV. Calculation of dielectric dispersion by a stochastic model, *J. Chem. Phys.*, 58, 4743, 1973.
6. Minakata, A., Dielectric dispersion of polyelectrolytes due to ion fluctuation, *Ann. NY Acad. Sci.*, 303, 107, 1977.
7. Washizu, H. and Kikuchi, K., Anisotropy of electrical polarizability of a DNA fragment, *Chem. Lett.*, 651, 1997.
8. Washizu, H. and Kikuchi, K., Electrical polarizability anisotropy of a DNA fragment, *Rep. Prog. Polym. Phys. Jpn.*, 40, 597, 1997.
9. Washizu, H. and Kikuchi, K., Concentration dependence of the anisotropy of the electrical polarizability of a model DNA fragment in salt-free aqueous solution studied by Monte Carlo simulations, *Colloids Surf. A: Physicochem. Eng. Aspects*, 148, 107, 1999.
10. Kikuchi, K. and Washizu, H., Electrical polarizability of model DNA fragments studied by Monte Carlo simulations, in *Proceedings of Yamada Conference L: Polyelectrolytes*, Noda, I. and Kokufuta, E., Eds., Yamada Science Foundation, Osaka, 1999, p. 80.
11. Washizu, H. and Kikuchi, K., Anisotropy of the electrical polarizability of a model DNA fragment in aqueous salt solution, *Rep. Prog. Polym. Phys. Jpn.*, 42, 367, 1999.
12. Washizu, H. and Kikuchi, K., Anisotropy of the electrical polarizability of a model DNA fragment in aqueous salt solution, *Chem. Phys. Lett.*, 320, 277, 2000.
13. Washizu, H. and Kikuchi, K., Anisotropy of the electrical polarizability of model DNA fragments in aqueous salt-free and salt solutions, *Rep. Prog. Polym. Phys. Jpn.*, 43, 615, 2000.
14. Kikuchi, K., Electrical polarizability of polyelectrolytes by Metropolis Monte Carlo simulation, in *Physical Chemistry of Polyelectrolytes*, Radeva, Ts., Ed., Marcel Dekker, New York, 2001, p. 223.
15. Washizu, H. and Kikuchi, K., Electrical polarizability of polyelectrolytes in salt-free aqueous solution, *J. Phys. Chem. B*, 106, 11329, 2002.
16. Schwarz, G., A theory of the low-frequency dielectric dispersion of colloidal particles in electrolyte solution, *J. Phys. Chem.*, 66, 2636, 1962.
17. Pollak, M., On the dielectric dispersion of polyelectrolytes, with application to DNA, *J. Chem. Phys.*, 43, 908, 1965.
18. Mandel, M., The electric polarization of rod-like, charged macromolecules, *Mol. Phys.*, 4, 489, 1961.
19. van der Touw, F. and Mandel, M., Dielectric increment and dielectric dispersion of solutions containing simple charged linear macromolecules I. Theory, *Biophys. Chem.*, 2, 218, 1974.
20. Mandel, M. and van der Touw, F., Dielectric properties of polyelectrolytes in solution, in *Polyelectrolytes*, volume 1 of *Charged and Reactive Polymers*, Sélégny, E., Ed., D. Reidel, Dordrecht, Holland, 1974, p. 285.
21. van Dijk, W., van der Touw, F., and Mandel, M., Influence of counterion exchange on the induced dipole moment and its relaxation for a rodlike polyion, *Macromolecules*, 14, 792, 1981.
22. Mandel, M. and Odijk, T., Dielectric properties of polyelectrolyte solutions, *Annu. Rev. Phys. Chem.*, 35, 75, 1984.
23. McTague, J.P. and Gibbs, J.H., Electric polarization of solutions of rodlike polyelectrolytes, *J. Chem. Phys.*, 44, 4295, 1966.
24. Schurr, J.M., Dielectric dispersion of linear polyelectrolytes, *Biopolymers*, 10, 1371, 1971.
25. Meyer, P.I. and Vaughan, W.E., Dielectric behavior of linear polyelectrolytes, *Biophys. Chem.*, 12, 329, 1980.
26. Meyer, P.I., Wesenberg, G.E., and Vaughan, W.E., Dielectric behavior of polyelectrolytes. II. The cylinder, *Biophys. Chem.*, 13, 265, 1981.
27. Wesenberg, G.E. and Vaughan, W.E., Dielectric behavior of polyelectrolytes III. The role of counterion interactions, *Biophys. Chem.*, 18, 381, 1983.
28. Altig, J.A., Wesenberg, G.E., and Vaughan, W.E., Dielectric behavior of polyelectrolytes IV. Electric polarizability of rigid biopolymers in electric fields, *Biophys. Chem.*, 24, 221, 1986.
29. Wesenberg, G.E. and Vaughan, W.E., Theory of the transient electric birefringence of rod-like polyions: Coupling of rotational and counterion dynamics, *J. Chem. Phys.*, 87, 4240, 1987.

30. Rau, D.C. and Charney, E., Polarization of the ion atmosphere of a charged cylinder, *Biophys. Chem.*, 14, 1, 1981.
31. Rau, D.C. and Charney, E., High-field saturation properties of the ion atmosphere polarization surrounding a rigid, immobile rod, *Macromolecules*, 16, 1653, 1983.
32. Hogan, M., Dattagupta, N., and Crothers, D.M., Transient electric dichroism of rod-like DNA molecules, *Proc. Natl. Acad. Sci. USA*, 75, 195, 1978.
33. Hornick, C. and Weill, G., Electrooptical study of the electric polarizability of rodlike fragments of DNA, *Biopolymers*, 10, 2345, 1971.
34. Weill, G. and Hornick, C., Electric polarisability of rigid polyelectrolytes, in *Polyelectrolytes*, volume 1 of *Charged and Reactive Polymers*, Sélégny E., Ed., D. Reidel, Dordrecht, Holland, 1974, p. 277.
35. Manning, G.S., Limiting laws and counterion condensation in polyelectrolyte solutions. V. Further development of the chemical model, *Biophys. Chem.*, 9, 65, 1978.
36. Manning, G.S., Linear analysis of the polarization of macroions, *J. Chem. Phys.*, 90, 5704, 1989.
37. Manning, G.S., A condensed counterion theory for polarization of polyelectrolyte solutions in high fields, *J. Chem. Phys.*, 99, 477, 1993.
38. Fixman, M., Charged macromolecules in external fields. I. The sphere, *J. Chem. Phys.*, 72, 5177, 1980.
39. Fixman, M., Charged macromolecules in external fields. 2. Preliminary remarks on the cylinder, *Macromolecules*, 13, 711, 1980.
40. Fixman, M. and Jagannathan, S., Electrical and convective polarization of the cylindrical macroions, *J. Chem. Phys.*, 75, 4048, 1981.
41. Mohanty, U. and Zhao, Y., Polarization of counterions in polyelectrolytes, *Biopolymers*, 38, 377, 1995.
42. Kikuchi, K. et al., Metropolis Monte Carlo method as a numerical technique to solve the Fokker–Planck equation, *Chem. Phys. Lett.*, 185, 335, 1991.
43. Kikuchi, K. et al., Monte Carlo method for Brownian dynamics simulation, in *Colloid and Molecular Electro-optics* 1991, Jennings, B.R. and Stoylov, S.P., Eds., IOP Publishing, Bristol, 1992, p. 7.
44. Kikuchi, K. et al., Metropolis Monte Carlo method for Brownian dynamics simulation generalized to include hydrodynamic interactions, *Chem. Phys. Lett.*, 196, 57, 1992.
45. Yoshida, M. and Kikuchi, K., Metropolis Monte Carlo Brownian dynamics simulation of the ion atmosphere polarization around a rodlike polyion, *J. Phys. Chem.*, 98, 10303, 1994.
46. Yoshida, M. and Kikuchi, K., A simple algorithm for rotational Brownian dynamics by the Metropolis Monte Carlo method, *Rep. Prog. Polym. Phys. Jpn.*, 39, 115, 1996.
47. Kellogg, O.D., *Foundations of Potential Theory*, Dover, New York, 1954, chap. 3.
48. Manning, G.S., Limiting laws and counterion condensation in polyelectrolyte solutions I. Colligative properties, *J. Chem. Phys.*, 51, 924, 1969.
49. Manning, G.S., Counterion condensation on ionic oligomers, *Physica A*, 247, 196, 1997.
50. Ramanathan, G.V. and Woodbury, C.P. Jr., Statistical mechanics of electrolytes and polyelectrolytes. II. Counterion condensation on a line charge, *J. Chem. Phys.*, 77, 4133, 1982.
51. Arnott, S. and Hukins, D.W.L., Optimized parameters for a-DNA and b-DNA, *Biochem. Biophys. Res. Commun.*, 47, 1504, 1972.
52. Odijk, T., Impact of nonuniform counterion condensation on the growth of linear charged micelles, *Physica A*, 176, 201, 1991.
53. Manning, G.S., Counterion condensation theory constructed from different models, *Physica A*, 231, 236, 1996.
54. Diekmann, S. et al., Electric properties and structure of DNA restriction fragments from measurements of the electric dichroism, *Biophys. Chem.*, 15, 157, 1982.
55. Yamaoka, K., Matsuda, K., and Takarada, K., Dependence of electric polarizability of rodlike DNA in aqueous solutions on ionic strength as studied by electric dichroism and birefringence. A comparison between experimental and theoretical results, *Bull. Chem. Soc. Jpn.*, 56, 927, 1983.
56. Rau, D.C. and Charney, E., Electric dichroism of DNA Influence of the ionic environment on the electric polarizability, *Biophys. Chem.*, 17, 35, 1983.

6 Electrokinetics of Concentrated Colloidal Dispersions

A.V. Delgado, F. Carrique, M.L. Jiménez, S. Ahualli, and F.J. Arroyo

CONTENTS

6.1   INTRODUCTION

Many colloidal suspensions of interest in industrial processes are concentrated, i.e., contain large amounts of particles. For instance, ceramic slurries usually contain ~50% of particles by volume; pharmaceutical drug suspensions are typically ~30% in particle volume fraction, and so on. Hence, it is of utmost importance to adequately evaluate the properties of such systems, including their electrokinetic behavior. In spite of this, it was not until the 1970s that Levine, Neale, and Epstein [1–3] made the first rigorous approaches to solve the electrokinetic equations to obtain the electrophoretic mobility, the electroosmotic flow or the sedimentation potential of a concentrated suspension of solid particles.

The problem is far more complex than in the case of dilute suspensions, as one must find a way to account for the effect of particle–particle interactions on the liquid and particle motions. Prior to Levine and Neale's works, several authors have solved the problem of the flow of liquid around a swarm of uncharged particles using the so-called cell models, a simple but precise enough way to take into account the interactions [4,5]. Both the free-surface model elaborated by Happel [4] and the zero vorticity model proposed by Kuwabara [5] demonstrated their validity against experimental data of liquid flow through the plugs of particles.

The basic feature of a cell model is that the problem can be reduced to that of a single particle (spherical, with radius a) immersed in a concentric shell of an electrolyte solution of external radius b, such that the particle to cell volume ratio is equal to the particle volume fraction, ϕ, throughout the whole suspension:

$$\left(\frac{a}{b}\right)^3 = \phi \tag{6.1}$$

The presence of neighbor particles is taken into account by a proper choice of boundary conditions for the velocity of the liquid, the electrical potential, and the chemical potential of ions on the outer surface of the cell, $r = b$ (Figure 6.1). In this respect, it must be mentioned that Happel's and Kuwabara's models differ in their choice of boundary conditions for the liquid velocity $\mathbf{v}$ at $r = b$. According to Happel,

$$v_r(r = a) = v_\theta(r = b) = 0 \tag{6.2}$$

$$v_r(r = b) = -v_e \cos\theta \tag{6.3}$$

$$\tau_{r\theta}(r = b) = 0 \tag{6.4}$$

where $-v_e$ is the liquid velocity (with respect to the particle) at a large distance from the particle. In an electrophoresis experiment, v_e would just be the electrophoretic velocity. Equation 6.4 indicates that the tangential shear stress on the surface of cell is zero. Kuwabara instead used the condition of zero azimuthal component of the flow vorticity on the $r = b$ sphere:

$$(\text{curl } \mathbf{v})_\phi\big|_{r=b} = 0 \tag{6.5}$$

which in fact is equivalent to assuming that curl $\mathbf{v} = 0$, as the radial and tangential components of curl $\mathbf{v}$ are automatically zero. In these expressions, r, θ, ϕ, are respectively the radial, tangential, and azimuthal spherical coordinates of a reference system centered in the particle.

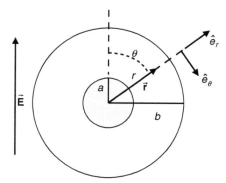

FIGURE 6.1 Schematic representation of the basic element of a cell model: the colloidal particle (radius a) is surrounded by a concentric shell of solution (thickness $b-a$). The radial ($\hat{e}_r$) and tangential ($\hat{e}_\theta$) unit vectors are also identified.

As Levine and Neale have already pointed out and as has been demonstrated by other authors [6] the results of both choices for hydrodynamic boundary conditions agree closely with each other, and can be used indistinctly for all practical purposes.

As mentioned earlier, Levine and Neale first generalized these calculations to account for the effects of the existence of a surface charge density on the particles and the subsequent electrical double layer around them. In Refs. [1,2] the problem of electrophoresis (and its reciprocal phenomenon, electroosmosis) was first solved for particles with low zeta potentials and arbitrary double layer thickness on the basis of Kuwabara cell model. Kozak and Davis [7,8] extended the model of Levine and Neale and derived a mobility expression valid for arbitrary zeta potentials with nonoverlapping double layers. More recently, Ohshima [9] presented a general mobility formula for identical spherical particles in concentrated suspensions that fulfilled the limiting cases of Levine and Neale, and Kozak and Davis. Ohshima also developed a general expression for the electrical conductivity of a concentrated suspension valid for low zeta potentials and nonoverlapping double layers [10]. Shortly after, Lee et al. [11] analyzed the electrophoresis of concentrated suspensions for arbitrary surface potentials taking into account double layer polarization and overlapping of adjacent cells, again using the Kuwabara cell model. Their study, although more general than the above-mentioned ones, is also based on the Levine–Neale electrical boundary conditions. Hsu et al. [12] have recently studied the electrophoresis of concentrated suspensions of spherical particles (Kuwabara's cell model and Levine–Neale's conditions), extending previous approaches with the inclusion of a general condition on the particle surface permitting the exchange of ions between the surface and the surrounding medium. The effects of overlapping of nearby double layers and appreciable double layer polarization are included in their model. Lin et al. [13] extended the study of Lee et al. to the case where the strength of the applied electric field is arbitrary.

Most of the relevant studies on electrokinetic phenomena in concentrated suspensions, electrophoresis, and electrical conductivity in particular, are based on Levine–Neale's boundary conditions. In a recent series of papers [14–18], a new set of boundary conditions have been checked according to the Shilov–Zharkikh cell model [19] following the suggestion of Dukhin et al. [20], who described some inconsistencies in the Levine–Neale model.

Very recently, a generalization of the theories of the electrophoretic mobility and electrical conductivity in concentrated suspensions, valid for arbitrary zeta potential, particle volume fraction, surface conductance (through a dynamic Stein layer (DSL) model), and double layer thickness (double-layer overlap is allowed), has been developed by the authors [16]. Because several contributions (surface charge, ionic strength, hydrodynamic and electrostatic interactions between particles, ...) superimpose in a complex way to yield the overall response, quantitative knowledge of the importance of overlapping effects is of great importance.

This chapter has been organized as follows: first a brief account of the basic equations of the problem is given and then their linearization and simplification are given by considering that the particles are spherical and with identical radius. Subsequently, the possibilities for specifying electrostatic and ionic boundary conditions are discussed. In Section 6.5, we will compare the results obtained with the different choices, after evaluating the dc conductivity of the suspensions. This chapter also shows how other quantities of interest, including permittivity, electrophoretic mobility, or dynamic mobility, can be calculated. Where possible, a comparison with other approaches (in particular, to the calculation of dynamic mobility) will also be discussed, as well as a view of some existing experimental data.

6.2 BASIC EQUATIONS

The full solution of the problem requires one to know, at every point $\mathbf{r}$ of the system, relevant quantities such as the electric potential $\Psi(\mathbf{r})$, the number density of each type of ions $n_i(\mathbf{r})$, the drift velocity $v_i(\mathbf{r})$ of each ionic species $(i = 1, \ldots, N)$, the fluid velocity $\mathbf{v}(\mathbf{r})$, and the pressure $p(\mathbf{r})$. The fundamental equations connecting them are [21–23]

$$\nabla^2 \Psi(\mathbf{r}) = -\frac{\rho_{el}(\mathbf{r})}{\varepsilon_m} \tag{6.6}$$

$$\rho_{el}(\mathbf{r}) = \sum_{i=1}^{N} z_i \, e \, n_i(\mathbf{r}) \tag{6.7}$$

$$\eta_m \nabla^2 \mathbf{v}(\mathbf{r}) - \nabla p(\mathbf{r}) - \rho_{el} \nabla \Psi(\mathbf{r}) = 0 \tag{6.8}$$

$$\nabla \cdot \mathbf{v}(\mathbf{r}) = 0 \tag{6.9}$$

$$\mathbf{v_i} = \mathbf{v} - \frac{D_i}{k_B T} \nabla \mu_i (i = 1, \ldots, N) \tag{6.10}$$

$$\mu_i(\mathbf{r}) = \mu_i^\infty + z_i \, e \, \Psi(\mathbf{r}) + k_B T \ln n_i(\mathbf{r})(i = 1, \ldots, N) \tag{6.11}$$

$$\nabla \cdot [n_i(\mathbf{r}) \mathbf{v}_i(\mathbf{r})] = 0 \; (i = 1, \ldots, N), \tag{6.12}$$

where e is the elementary electric charge, k_B Boltzmann's constant, T the absolute temperature, and μ_i^∞ the standard chemical potential of the ith ionic species. Equation 6.6 is Poisson's equation, where ε_m is the permittivity of the medium, and $\rho_{el}(\mathbf{r})$ the electric charge density given by Equation 6.7. Equation 6.8 and Equation 6.9 are the Navier–Stokes equations appropriate to a steady flow of incompressible fluid with viscosity η_m at low Reynolds number in the presence of an electrical body force. Equation 6.10 means that the ionic flow is the result of hydrodynamic drag, and the electrostatic and thermodynamic forces acting on each of the N ionic species, with diffusion coefficients D_i. Equation 6.12 is the continuity equation expressing the conservation of the number of each ionic species in the system.

6.3 BOUNDARY CONDITIONS

As mentioned earlier, a crucial aspect of the problem is the proper choice of boundary conditions, both in the inner $(r = |\mathbf{r}| = a)$ and outer $(r = b)$ cell surfaces. The former is effectively the solid and liquid separation, so-called slip plane: both the ions and the liquid located beneath that surface are considered immobile, strongly attached to the particle.

If $\Psi^0(r)$ is the equilibrium (i.e., in the absence of any applied field) potential distribution, its value on $r = a$ is the zeta potential, ζ:

$$\Psi^0(r = a) = \zeta \tag{6.13}$$

and the discontinuity of the normal electric displacement yields:

$$\varepsilon_m \frac{d\Psi^0}{dr}\bigg|_{r=a} = -\sigma \tag{6.14}$$

where σ is the surface charge density of the particles. The condition that the cell is electroneutral leads finally to

$$\frac{d\Psi^0}{dr}\bigg|_{r=b} = 0 \tag{6.15}$$

The key problem now is the specification of the electrostatic boundary conditions in the presence of the field. First, it is convenient to write the nonequilibrium quantities in terms of their equilibrium values plus a field-dependent perturbation:

$$n_i(\mathbf{r}) = n_i^0(r) + \delta n_i(\mathbf{r}) \ (i = 1, \ldots, N) \tag{6.16}$$

$$\mu_i(\mathbf{r}) = \mu_i^0(r) + \delta\mu_i(\mathbf{r}) \ (i = 1, \ldots, N) \tag{6.17}$$

$$\Psi(\mathbf{r}) = \Psi^0(r) + \delta\Psi(\mathbf{r}) \tag{6.18}$$

In the linear theory of electrokinetic phenomena it is proposed that the applied field strength is low enough for the perturbations to be linear in the field, and, furthermore, quadratic and higher terms will be excluded from the calculations.

On the basis of Ohshima's formula [9,10], spherical symmetry considerations permit us to introduce the radial functions $h(r)$, $\phi_i(r)$, and $\Gamma(r)$ as follows:

$$\mathbf{v}(\mathbf{r}) = \left\{ -\frac{2}{r}h(r)E^* \cos\theta, \ \frac{1}{r}\frac{d}{dr}[r\,h(r)]E^* \sin\theta, 0 \right\} \tag{6.19}$$

$$\delta\mu_i(\mathbf{r}) = -z_i e\phi_i(r)E^* \cos\theta \ (i = 1, \ldots, N) \tag{6.20}$$

$$\delta\Psi(\mathbf{r}) = -\Gamma(r)E^* \cos\theta \tag{6.21}$$

where E^* is a field yet to be defined in terms of the macroscopic electric field.

The boundary conditions described so far yield

$$h(a) = \frac{dh}{dr}\bigg|_{r=a} = 0 \tag{6.22}$$

$$L[h(r)]_{r=b} = 0 \tag{6.23}$$

$$h(b) = \frac{u_e b}{2} \tag{6.24}$$

$$\frac{d\phi_i}{dr}\bigg|_{r=a} = 0 \ (i = 1, \ldots, N) \tag{6.25}$$

where u_e is the electrophoretic mobility and L is a differential operator:

$$L \equiv \frac{d^2}{dr^2} + \frac{2}{r}\frac{d}{dr} - \frac{2}{r^2} \tag{6.26}$$

From Equation 6.6 through Equation 6.12 and Equation 6.19 through Equation 6.21, it can be found that the functions $h(r)$, $\phi_i(r)$, and $\Gamma(r)$ verify the following set of ordinary differential equations:

$$L\{L[h(r)]\} = -\frac{e}{\eta_m r}\frac{dy}{dr}\sum_{i=1}^{N} n_i^\infty z_i^2 \exp(-z_i y)\phi_i(r) \tag{6.27}$$

with

$$y = \frac{e\Psi^0(r)}{k_B T} \tag{6.28}$$

where n_i^∞ is the bulk number concentration of the ith ionic species.

$$L\phi_i(r) = \frac{dy}{dr}\left[z_i\frac{d\phi_i}{dr} - \frac{2D_i}{k_BTe}\frac{h(r)}{r}\right] \quad (i = 1,\cdots,N) \tag{6.29}$$

$$L\Gamma(r) = \frac{1}{\varepsilon_m k_B T}\sum_{i=1}^{N} z_i^2 e^2 n_i^0(r)[\Gamma(r) - \phi_i(r)] \tag{6.30}$$

Let us now consider the boundary conditions associated with $\delta\Psi$ and Γ that will also give us clues on how to define the field $\mathbf{E}^*$ in Equation 6.19 through Equation 6.21. Specifying the values of $\delta\Psi$ on the cell boundary $r = b$ is essential, and as Dukhin et al. [20] pointed out, such specification displays the connection between the macroscopic, or experimentally measured field, and the local electrical properties.

One possible choice of boundary condition is that of Levine–Neale (LN) (a Neumann-type condition):

$$\text{LN:}\quad (\nabla\delta\Psi)\cdot\hat{\mathbf{r}}|_{r=b} = -\mathbf{E}^*\cdot\hat{\mathbf{r}} = -E^*\cos\theta \tag{6.31}$$

or

$$\text{LN:}\quad \frac{d\Gamma(r)}{dr}\bigg|_{r=b} = 1 \tag{6.32}$$

Shilov et al. [19] proposed a Dirichlet-type condition (SZ hereafter):

$$\text{SZ:}\quad \delta\Psi(\mathbf{r})|_{r=b} = -\mathbf{E}^*\cdot\mathbf{r}|_{r=b} = -E^*b\cos\theta \tag{6.33}$$

and in terms of Γ,

$$\text{SZ:}\quad \Gamma(b) = b \tag{6.34}$$

Some condition must also be imposed on the ionic perturbations $\delta n_i(\mathbf{r})$ at $r = b$. Following Lee et al. [11,17], the number density of each ionic species must be equal to the corresponding equilibrium ionic density. Therefore

$$\begin{aligned}\delta_{\mu i}(r)|_{r=b} &= f_{ie}\delta\Psi(r)|_{r=b}\\ \Phi_i(b) &= \Gamma(b) \quad (i = 1,\cdots,N)\end{aligned} \tag{6.35}$$

Ding and Keh [6] suggested that an alternative condition on ionic perturbations can be

$$\begin{aligned}\frac{\partial\delta n_i(\mathbf{r})}{\partial r}\bigg|_{r=b} &= 0\\ \frac{d\phi_i}{dr}\bigg|_{r=b} &= \frac{d\Gamma}{dr}\bigg|_{r=b}\\ (i &= 1,...,N)\end{aligned} \tag{6.36}$$

In addition to the condition imposed, we must impose the constraint that in the steady state the net force acting on the cell must vanish. This condition can be expressed as [9]

$$\eta_m\frac{d}{dr}[rLh(r)]\bigg|_{r=b} = \rho_{el}^0(b)\Gamma(b), \tag{6.37}$$

where $\rho_{el}^0 (b)$ is the equilibrium charge density at $r = b$. Note that neglecting double layer overlap is equivalent to setting $\rho_{el}^0 (b) = 0$ in Equation 6.37.

6.4 LEVINE AND NEALE'S SOLUTION

The solution to Navier–Stokes equations (Equation 6.8 and Equation 6.9) adopted by Levine and Neale in his classical paper [1] is the one found by Henry [24], and hence it is strictly valid for low zeta potential:

$$p = A_0 + \int_b^r \frac{d\Psi^\circ}{dr} dr + \left[A_1 r + \frac{B_1}{r^2} - \varepsilon_m \chi \left(3 \frac{d\Psi^0}{dr} - 2\xi \right) \right] \cos \theta \tag{6.38}$$

$$v_r = \left[\frac{A_1 r^2}{10\eta_m} + \frac{B_1}{\eta_m r} + A_2 + \frac{B_2}{r^3} + \frac{2\varepsilon_m \chi}{3\eta_m} \left(\int_a^r \xi dr - \frac{1}{r^3} \int_a^r r^3 \xi dr \right) \right] \cos \theta \tag{6.39}$$

$$v_\theta = \left[-\frac{A_1 r^2}{5\eta_m} - \frac{B_1}{2\eta_m r} - A_2 + \frac{B_2}{2r^3} - \frac{2\varepsilon_m \chi}{3\eta_m} \left(\int_a^r \xi dr - \frac{1}{2r^3} \int_a^r r^3 \xi dr \right) \right] \sin \theta \tag{6.40}$$

where A_0, A_1, A_2, B_1, B_2 are constants to be determined from the fluid velocity boundary conditions, and

$$\chi = \frac{E^*}{1 - \phi} \tag{6.41}$$

$$\xi = \frac{d\Psi^0}{dr} + \frac{a^3 r}{2} \int_b^r \frac{1}{r^4} \nabla^2 \Psi^0 dr \tag{6.42}$$

In order to obtain the electrophoretic velocity, v_e, of the particles, the fact that the electrophoretic velocity will be parallel to the field if the particles are uniformly charged is made use of, and hence it is not necessary to use vector quantities. For similar symmetry considerations, both the electrical, $\mathbf{F}_{el}$, and hydrodynamic, $\mathbf{F}_h$ forces will be parallel to E^*. Their components (in the direction of the field) are given by

$$\mathbf{F}_h = 2\pi a^2 \int_0^\pi (\tau_{rr} \cos \theta + \tau_{r\theta} \sin \theta)_{r=a} \sin \theta \, d\theta \tag{6.43}$$

where the rr and $r\theta$ components of the hydrodynamic stress tensor are, respectively,

$$\tau_{rr} = -p + 2\eta_m \frac{\partial v_r}{\partial r} \tag{6.44}$$

$$\tau_{r\theta} = -\eta_m \left[r \frac{\partial}{\partial r} \left(\frac{v_\theta}{r} \right) + \frac{1}{r} \frac{\partial v_r}{\partial \theta} \right] \tag{6.45}$$

Concerning the electrical force, its value is

$$F_{el} = -\varepsilon_m E^* a^2 \left(\frac{d\Psi^0}{dr} \right)_{r=a} \tag{6.46}$$

And using the expressions of v_r and v_θ (Equation 6.39 and Equation 6.40), the total force on the particle will be

$$F_{\text{tot}} = F_{\text{h}} + F_{\text{el}} = -4\pi B_1 = -6\pi\eta_{\text{m}} v_{\text{e}} a\Omega + \varepsilon_{\text{m}} E^* a\Lambda \tag{6.47}$$

where, for the Kuwabara model,

$$\Omega = \frac{5}{5 - 9\phi^{1/3} + 5\phi - \phi^2}$$

$$\Lambda = \frac{5\phi \int_a^b \frac{r^3}{a^3}\xi dr - 5\int_a^b \xi dr}{(1-\phi)(5 - 9\phi^{1/3} + 5\phi - \phi^2)} \tag{6.48}$$

Taking into consideration that in steady motion F_{tot} must be zero, we reach the following result:

$$v_{\text{e}} = \frac{2}{3}\frac{\varepsilon_{\text{m}}}{\eta_{\text{m}}}\frac{\Lambda}{\Omega}E^* \tag{6.49}$$

The evaluation of Λ requires, as shown in Equation 6.42 and Equation 6.48, the knowledge of $\Psi^0(r)$ in the shell region. This in turn needs the integration of Poisson's equation, and this cannot be analytically achieved for spherical particles, even in equilibrium conditions, unless the potential is low everywhere in that region, that is, unless the zeta potential is low. In such a case, Equation 6.6, Equation 6.7 lead to

$$\frac{1}{r^2}\frac{d}{dr}\left(r^2\frac{d\Psi^0}{dr}\right) = \kappa^2\Psi^0 \tag{6.50}$$

where κ is the Debye–Hückel parameter (reciprocal double layer thickness):

$$\kappa = \left(\frac{\sum_{i=1}^N n_i^\infty e^2 z_i^2}{\varepsilon_{\text{m}} k_{\text{B}} T}\right)^{1/2} \tag{6.51}$$

The solution of Equation 6.50 subject to the conditions of Equation 6.13 and Equation 6.15 is

$$\Psi^0(r) = \frac{\zeta a}{r}\frac{\sinh(\kappa b - \kappa r) - \kappa b\cosh(\kappa b - \kappa r)}{\sinh(\kappa b - \kappa a) - \kappa b\cosh(\kappa b - \kappa a)} \tag{6.52}$$

and from Equation 6.49, the electrophoretic mobility, u_{e}, can be written, in the Henry approximation:

$$u_{\text{e}} = \frac{\varepsilon_{\text{m}}}{\eta_{\text{m}}}\zeta f(\kappa a, \phi) = (u_{\text{e}})_{\text{Sm}}\, f(\kappa a, \phi) \tag{6.53}$$

where $(u_{\text{e}})_{\text{Sm}}$ is the Smoluchowski expression for u_{e}, and $f(\kappa a, \phi)$ can be calculated numerically, although approximate expressions can be found in Refs. [1,9]. Figure 6.2 shows the results. Note that for large κa, $f(\kappa a,\phi)$ tends to 1 for all values of ϕ, and one finds that Smoluchowski's equation is valid irrespective of the concentration, provided the double layer

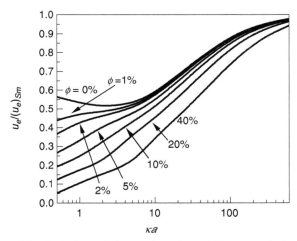

FIGURE 6.2 The cell-model calculation of electrophoretic mobility, u_e, relative to the Smoluchowski value $\varepsilon_m \zeta / \eta_m$, as a function of κa, for different volume fractions of solids, ϕ.

is sufficiently thin compared to a. If the volume fraction tends to zero, one recovers Henry's result for the mobility, as $f(\kappa a, 0)$ reads

$$f(\kappa a, 0) = \frac{2}{3}\left[1 + \frac{\kappa^2 a^2}{16} - \frac{5\kappa^3 a^3}{48} - \frac{\kappa^4 a^4}{96} + \frac{\kappa^5 a^5}{96} + \left(\frac{\kappa^4 a^4}{8} - \frac{\kappa^6 a^6}{96}\right) \exp(\kappa a) \int_{\kappa a}^{\infty} \frac{\exp(-t)}{t}\,dt\right] \tag{6.54}$$

A good approximation to this was found by Ohshima [25,26]:

$$f(\kappa a, 0) = \frac{2}{3}\left[1 + \frac{1}{2\left(1 + \dfrac{2.5}{\kappa a\{1 + 2\exp(-\kappa a)\}}\right)^3}\right] \tag{6.55}$$

6.5 ROLE OF THE CHOICE OF BOUNDARY CONDITIONS

In the above-described Levine and Neale's calculations, the boundary conditions on the external surface of the cell ($r = b$) include Kuwabara's conditions on the velocity of the fluid, as well as Equation 6.31 and Equation 6.32 for the electric potential perturbation. Also, since they follow Henry's description, the potential is just a superposition of the equilibrium distribution and a field-induced term, the electrical double layer retains its equilibrium geometry, so that Equation 6.36 is implicitly used.

We now consider how the solution of the problem is affected by the choice of boundary conditions. To do this, we will find the dc conductivity of the suspension using all the choices at hand (Dirichlet- and Neumann-type conditions for either $\delta \Psi$ or δn_i) and compare the results.

The electrical conductivity K of the suspension is defined in terms of the volume averages of the current density, $\langle \mathbf{j} \rangle$ and the electric field, $\langle \mathbf{E} \rangle$:

$$\langle \mathbf{j} \rangle = K \langle \mathbf{E} \rangle$$

$$\langle \mathbf{o} \rangle = \frac{1}{V} \int_V \mathbf{o}\, dV \tag{6.56}$$

where the integration is extended to the total volume of the cell. The average field, $\langle \mathbf{E} \rangle$, is given by

$$\langle \mathbf{E} \rangle = -\frac{1}{V} \int_V \nabla \Psi(\mathbf{r}) \mathrm{d}V = -\frac{1}{V} \int_V \nabla \delta \Psi(\mathbf{r}) \mathrm{d}V \tag{6.57}$$

where the fact that the volume average of the gradient of the equilibrium electric potential is zero has been made use of [27].

Using Equation 6.10 and Equation 6.17

$$\mathbf{j}(\mathbf{r}) = \rho_{\mathrm{el}}^0(\mathbf{r}) \mathbf{v}(\mathbf{r}) - \sum_{i=1}^{N} \frac{z_i e n_i^0(r)}{\lambda_i} \nabla \delta \mu_i(\mathbf{r}) \tag{6.58}$$

where $\lambda_i = k_{\mathrm{B}} T/D_i$ is the ionic drag coefficient. Noting that $\nabla \cdot (\mathbf{rj}) = \mathbf{j}$, and using the divergence theorem,

$$\langle \mathbf{j} \rangle = -\frac{1}{V} \sum_{t=1}^{N} z_i e \int_s \left\{ r \left(-\mathbf{n}_i^0 v + \mathbf{n}_i^0 \frac{\nabla \delta \mu_i}{\lambda_i} \right) \cdot \hat{\mathbf{r}} \right\} \mathrm{d}S \tag{6.59}$$

where S is the outer spherical surface of the cell.

Using Equation 6.20, and calling

$$C_i = -\frac{b^2}{3} \left(r \frac{\mathrm{d}\phi_i}{\mathrm{d}r} - \phi_i \right)_{r=b} \tag{6.60}$$

one obtains

$$\langle \mathbf{j} \rangle = -\frac{1}{V} \sum_{i=1}^{N} \frac{z_i^2 e^2 n_i^0(b)}{\lambda_i} \left(\frac{3C_i}{b^2} - \phi_i(b) \right) \frac{4}{3} \pi b^2 \mathbf{E}^* + \frac{1}{V} \sum_{i=1}^{N} z_i e n_i^0(b) \int (\mathbf{rv}) \cdot \hat{\mathbf{r}} \, \mathrm{d}S \tag{6.61}$$

Now, taking into account Equation 6.19 and Equation 6.24

$$\int (\mathbf{rv}) \cdot \hat{\mathbf{r}} \, \mathrm{d}S = -\frac{4}{3} \pi b^3 u_e \mathbf{E}^* \tag{6.62}$$

Also, Equation 6.21 leads to [28]

$$\langle \mathbf{E} \rangle = \frac{\Gamma(b)}{b} \mathbf{E}^* \tag{6.63}$$

and, finally, substituting Equation 6.62 and Equation 6.63 in Equation 6.61 and using Equation 6.56:

$$\frac{K}{K_{\mathrm{m}}} = \frac{\displaystyle\sum_{i=1}^{N} \left\{ \frac{z_i^2 e^2 n_i^\infty}{\lambda_i} \left[\frac{\phi_i(b)}{\Gamma(b)} - \frac{3\phi}{a^3} \left(\frac{b}{\Gamma(b)} \right) C_i \right] - 2 \left(\frac{h(b)}{\Gamma(b)} \right) (z_i e n_i^\infty) \right\} \exp\left(-\frac{z_i e \Psi^0(b)}{k_{\mathrm{B}} T} \right)}{\displaystyle\sum_{i=1}^{N} \frac{z_i^2 e^2 n_i^\infty}{\lambda_i}} \tag{6.64}$$

where $n_i^0(b)$ has been expressed in terms of the potential at b using the Boltzmann factor:

$$n_i^0(b) = n_i^\infty \exp\left(-\frac{z_i e \Psi^0(b)}{k_B T}\right) \tag{6.65}$$

and the following expression has been used to evaluate the conductivity of the medium, K_m:

$$K_m = \sum_{i=1}^{N} \frac{z_i^2 e^2 n_i^\infty}{\lambda_i} \tag{6.66}$$

If, in addition, we use the conditions in Equation 6.35 for the ionic perturbations,

$$\frac{K}{K_m} = \frac{\sum_{i=1}^{N}\left\{\frac{z_i^2 e^2 n_i^\infty}{\lambda_i}\left[1 - \frac{3\phi}{a^3}\left(\frac{b}{\Gamma(b)}\right)C_i\right] - 2\left(\frac{h(b)}{\Gamma(b)}\right)(z_i e n_i^\infty)\right\}\exp\left(-\frac{z_i e \Psi^0(b)}{k_B T}\right)}{\sum_{i=1}^{N}\frac{z_i^2 e^2 n_i^\infty}{\lambda_i}} \tag{6.67}$$

whereas if the conditions chosen for δn_i ($r = b$) are those given by Equation 6.36, the suitable equation for conductivity is Equation 6.64 with the additional constraints that $d\phi_i/dr|_{r=b} = d\Gamma/dr|_{r=b}$, $(i = 1, \ldots, N)$.

Note that if the perturbations in Equation 6.19 through Equation 6.21 were expressed in terms of $\langle E \rangle$ (that is, if $E^* = \langle E \rangle$), then Equation 6.63 would yield $\Gamma(b) = b$, which is precisely (Equation 6.34), corresponding to the SZ boundary conditions. Thus, we reach the important conclusion that if Shilov–Zharkikh boundary condition (Equation 6.34) is used, then the field E^* is necessarily equal to the macroscopic (average) one $\langle E \rangle$, and $\Gamma(b) = b$.

If, on the other hand, the Levine–Neale boundary condition (Equation 6.31, Equation 6.32) is chosen, the field E^* (that Ohshima calls applied external field, see Ref. [28]) is related to the average one as $E^* = \langle E \rangle b/\Gamma(b)$, and, in addition, $d\Gamma/dr|_{r=b} = 1$. Mixing these conditions leads to unphysical results, whereas, when properly used, Levine–Neale and Shilov–Zharkikh conditions should lead to identical evaluations of K.

6.5.1 Choice of Electrostatic Boundary Conditions

A numerical technique based on the method used by DeLacey and White [23] for solving the problem of ac conductivity was used to numerically evaluate the conductivity of the suspensions. This method has turned out to be very efficient in the integration of the electrokinetic equations in the static case [15,16]. We will first consider how the choice of the cell boundary condition for $\delta \Psi(r)$ influences the calculation of K.

Figure 6.3 shows that LN and SZ boundary conditions (with $\delta n_i(b) = 0$) lead to identical results whatever the value of zeta potential ζ (Figure 6.3a), κa (Figure 6.3b), and volume fraction ϕ (Figure 6.3c). Note that for the volume fraction involved, increasing the zeta potential ζ (Figure 6.3a) can provoke a very important raise in K, a clear consequence of the large surface conductivity contribution of such highly charged double layers.

As to the effect of κa, the results in Figure 6.3b indicate that increasing this quantity provokes a fall in $(K - K_m)/K_m$ because the influence of the medium conductivity progressively hides that of the surface conductivity. In fact, it must be recalled here that the net effect of the double layer conductivity is controlled by the Dukhin number Du, relating the surface conductivity K^σ to that of the medium, and to the particle radius:

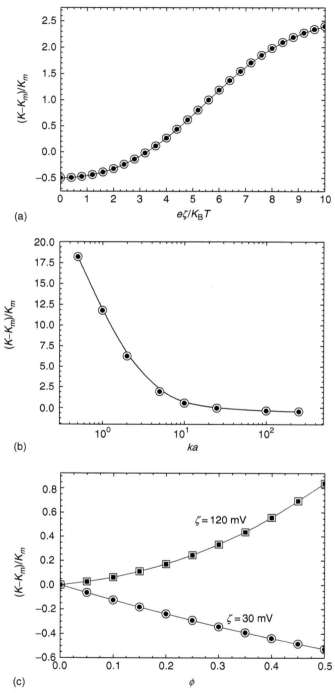

(a)

(b)

(c)

FIGURE 6.3 Ratio of the conductivity increment of the suspension $(K - K_m)$ to the conductivity of the solution (K_m) as a function of: (a) dimensionless zeta potential for $\kappa a = 10$ and $\phi = 0.4$; (b) κa for $\zeta = 120\,\text{mV}$ and $\phi = 0.4$; (c) volume fraction ϕ for $\zeta = 30\,\text{mV}$, $120\,\text{mV}$, and $\kappa a = 10$. The dispersion medium is KCl, and the particle radius is $a = 100\,\text{nm}$. Solid symbols: Levine–Neale boundary conditions. Open symbols: Shilov–Zharkikh. Ionic perturbation condition: $\delta n_i(b) = 0$.

$$Du = \frac{K^\sigma}{K_m a} \qquad (6.68)$$

Note that Du (and, as a consequence, the contribution of the particles to the conductivity K) is more important the larger ζ, the lower K_m, and the smaller the size.

Finally, Figure 6.3c is also self-explaining: increasing the volume fraction of particles with low ζ-potential leads to a reduction of the conductivity of the suspension as compared to that of the solution, whereas performing the same operation with higher ζ yields a higher conductivity.

6.5.2 BOUNDARY CONDITION FOR IONIC CONCENTRATIONS

As mentioned, we have in principle two options in establishing a boundary condition for the ionic perturbations, as specified by Equation 6.35 ($\delta n_i(b) = 0$) and Equation 6.36 ($\partial \delta n_i / \partial r|_{r=b} = 0$). Both are compatible with (and independent of) the previously discussed conditions for $\delta \Psi$.

Figure 6.4 shows the influence of the conditions chosen for δn_i on our estimations of $(K-K_m)/K_m$ as a function of the dimensionless zeta potential for different κa values and a fixed volume fraction, $\phi = 0.4$. Three features are immediately clear: (a) using $\partial \delta n_i / \partial r|_{r=b} = 0$ leads to lower conductivity increments; (b) the differences between the results obtained with $\partial \delta n_i / \partial r|_{r=b} = 0$ and with $\delta n_i(b) = 0$ increase with zeta potential; (c) the relative differences between the two approaches change with κa. In order to stress the last point, in Figure 6.5 we have plotted K/K_m as a function of κa at constant zeta potential and volume fraction.

The above features (a) and (b) can be explained by considering that forcing $\delta n_i = 0$ at $r = b$, instead of $\partial \delta n_i / \partial r|_{r=b} = 0$, opens up the possibility of the existence of diffusion fluxes of ions (mainly counterions, much more abundant at high $|\zeta|$ at the cell boundary. The increased flow of counterions will yield a higher surface conductivity and hence a larger K of the suspension, as compared to the case of no concentration gradients. That is the reason for the condition $\delta n_i(b) = 0$ leading to larger conductivities. Clearly, as the zeta potential increases, the effects associated to double layer polarization will be more intense [29,30], and the differences between the cases $\delta n_i(b) = 0$ and $\partial \delta n_i / \partial r|_{r=b} = 0$ are consistently more significant. At low $|\zeta|$, the counterion excess concentration is equally low and so will be the eventual diffusion fluxes.

The effect of κa (Figure 6.4a through figure 6.4c and Figure 6.5) appears to confirm our arguments: the two ways of calculation of K differ mainly in the medium κa regime, where double-layer polarization effects are known to be most significant for electrokinetics [31]. At low κa, the double layer is so extended that the effect of the external field on it is very small (the volume charge density is very low throughout the ionic atmosphere). On the contrary, at high κa the Dukhin number is so small that again double layer polarization effects are unimportant. In the region in between these two extremes, the effects of changes in the fluxes of counterions in the double layer will be significant, and the arguments above explained, concerning the differences between $\delta n_i(b) = 0$ and $\partial \delta n_i / \partial r = 0|_{r=b}$, apply effectively. Note (Figure 6.5) that at very high κa the conductivity ratio tends to its Maxwell value, then being only controlled by volume fraction and not by particle charge.

6.6 ELECTROPHORETIC MOBILITY

The evaluation of the conductivity of the suspensions requires in some cases (see Equation 6.64 and Equation 6.67) the previous knowledge of the electrophoretic mobility of the particles, u_e. This quantity is, according to Equation 6.24, equal to $2h(b)/b$, so that having

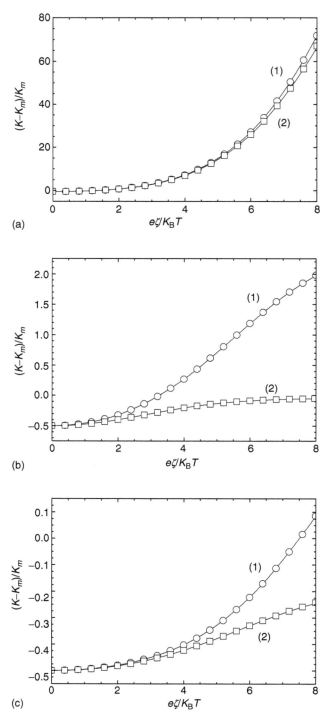

FIGURE 6.4 Relative conductivity increment, $(K-K_m)/K_m$, plotted as a function of the dimensionless zeta potential for suspensions of spheres 100 nm in radius with volume fraction $\phi = 0.4$, and two choices for ionic perturbations in the cell boundary (1: $\delta n_i(b) = 0$; 2: $\partial \delta n_i/\partial r|_{r=b} = 0$). κa values: 1 (a), 10 (b), 100 (c).

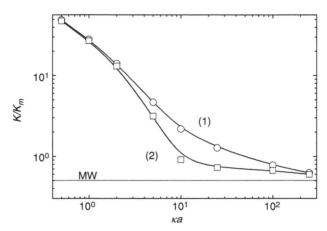

FIGURE 6.5 Relative conductivity K/K_m of suspensions of spherical particles, with volume fraction $\phi = 0.4$, dimensionless zeta potential $= 6$, as a function of κa for the choices $\delta n_i(b) = 0$ (1) and $\partial \delta n_i / \partial r|_{r=b} = 0$ (2). MW indicates the Maxwell–Wagner limit.

solved the velocity function $h(r)$ at $r = b$ it is necessary to obtain u_e. Figure 6.6 shows some representative results calculated with LN and SZ boundary conditions for $\delta \Psi$, and with $\delta n_i(r = b) = 0$ and $\partial \delta n_i / \partial r = 0$ for the ionic concentration perturbations. It is important to point out that, in this case, not only different choices for δn_i, but also different electrostatic boundary conditions for $\delta \Psi$ largely affect the estimation of u_e. Whereas the effect of δn_i boundary conditions might be expected, the fact that the choice of $\delta \Psi(b)$ influences the mobility and not the conductivity is, at a first glance, striking.

In previous papers from different authors [15,32], it was shown that calculations based on LN and SZ boundary conditions can be brought to coincidence at low $|\zeta|$ values, if the former values are multiplied by the Maxwell factor for uncharged spheres $(1-\phi)/(1+\phi/2)$. This is

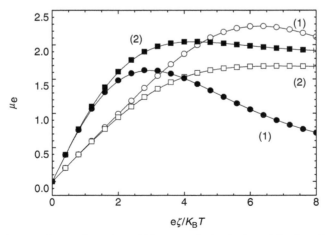

FIGURE 6.6 Dimensionless electrophoretic mobility, μ_e, of spherical particles ($a = 100$ nm) as a function of the dimensionless zeta potential, in KCl solution with $\kappa a = 10$ and for constant $\phi = 0.4$. Curves labeled (1) and (2) correspond, respectively, to setting $\delta n_i(b) = 0$ and $\partial \delta n_i / \partial r|_{r=b} = 0$ for the ionic perturbations. LN (solid symbols) and SZ (open symbols) stand for Levine–Neale and Shilov–Zharkikh boundary conditions for $\delta \Psi$. *Note:* $\mu_e = (3\eta_m e / 2\varepsilon_m k_B T) u_e$.

done in Figure 6.7 for the same set of data as in Figure 6.6: note that, indeed, both types of calculation tend to coincide in that region of zeta potentials. When $|\zeta|$ is above ~50 mV at room temperature the divergence between them is noticeable, and the Maxwell factor is clearly insufficient.

Our previous analysis of the conductivity can help us in elucidating the source of such differences: in the case of K, both approaches use the same field, $\langle \mathbf{E} \rangle$, to define the conductivity in terms of $\langle \mathbf{j} \rangle$. But when it is the electrophoretic mobility that is calculated, for the velocity of electrophoresis according to the SZ model,

$$\mathbf{v}_e = u_e^{SZ} \langle \mathbf{E} \rangle \tag{6.69}$$

whereas, according to LN approach,

$$\mathbf{v}_e = u_e^{LN} \mathbf{E}^* \tag{6.70}$$

and both the fields are related (Equation 6.63):

$$\mathbf{E}^* = \frac{b}{\Gamma(b)} \langle \mathbf{E} \rangle \tag{6.71}$$

so that u_e^{SZ} should be compared with $(u_e^{LN} b / \Gamma(b))$. However, it is advisable to define any kinetic coefficient in terms of model-independent fields; in the case of electrophoresis, this can be done by referring to the macroscopic, average electric field, $<\mathbf{E}>$, so that Equation 6.69 would be the required model-independent relationship. A detailed discussion on this subject is given by Dukhin et al. [20]. Figure 6.8 demonstrates that in this case, a perfect agreement between SZ and LN evaluations of the mobility can again be found: one can use either of them to calculate u_e, but, since they have different definitions of the field, a connection factor is needed to bring them to coincidence.

The difference between the choices $\delta n_i(b) = 0$ and $\partial \delta n_i / \partial r |_{r=b} = 0$, that was evident in conductivity calculations, still persists: the electrophoretic mobility is consistently larger in the former case, the more so the larger $|\zeta|$. Looking at the problem from the reference frame

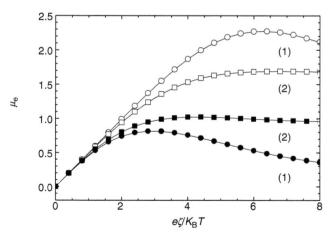

FIGURE 6.7 Same as Figure 6.6, except that LN data have been multiplied by the Maxwell factor $(1-\phi)/(1+\phi/2)$.

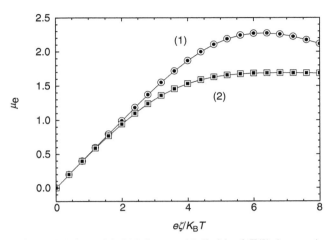

FIGURE 6.8 Same as Figure 6.6, but with LN data multiplied by $b/\Gamma(b)$ (see text).

fixed in the particle, this means that the liquid moves faster past the steady particle. Again, a qualitative explanation can be given to this fact, based on the plausible existence of nonzero diffusion fluxes of counterions if the condition $\delta n_i(b) = 0$ is chosen. These fluxes will drag liquid with them, increasing its velocity with respect to the particle. Turning to the laboratory frame, the electrophoretic velocity will be larger, as observed.

Before proceeding it may be interesting to compare our evaluations of u_e (Figure 6.8) with Levine and Neale's calculation. This can be done using the data in Figure 6.9: as observed, Levine–Neale's calculations are only valid up to $\zeta \sim 50\,\text{mV}$ at room temperature. In such conditions, all numerical models essentially agree and the original LN evaluations coincide with them. Let us mention that the Levine–Neale model has implicit the condition $\partial \delta n_i/\partial r \mid_{r=b} = 0$ for the ionic perturbation, but, as mentioned, this is not very significant for the range of zeta potential in which it is valid.

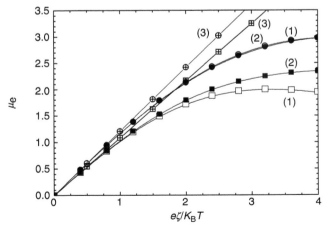

FIGURE 6.9 Dimensionless electrophoretic mobility plotted as a function of reduced zeta potential, for two volume fractions ($\phi = 10\%$ (circles) and 30% (squares)) using three methods of evaluation: (1): numerical, and $\delta n_i(b) = 0$; (2): numerical and $\partial \delta n_i/\partial r \mid_{r=b} = 0$; (3) Levine and Neale equation (Equation 6.49). In all cases $a = 100\,\text{nm}$ and KCl solution with $\kappa a = 10$.

6.7   COMPARISON WITH EXPERIMENTAL DATA

We have shown that two choices of electrostatic boundary conditions lead to essentially identical evaluations of the conductivity of the suspension, as long as the perturbing electric field is defined accordingly. However, no similar conclusion could be reached regarding ionic perturbations. One may wonder if one of the choices, namely, $\delta n_i(b) = 0$ or $\partial \delta n_i / \partial r \mid_{r=b} = 0$, can be demonstrated to be correct.

A possible way to do that is to use experimental data for comparison. The results obtained by Midmore et al. [33] on latex particles appear particularly useful. Figure 6.10 allows the comparison with our predictions: Figure 6.10a (10^{-2} M KCl) shows the experimental value obtained for the relative conductivity increment, $(K - K_m)/K_m$ as a horizontal line in a plot of this quantity versus reduced zeta potential. Two full theoretical curves obtained using the two types of ionic boundary conditions have been superimposed. Note how in the two cases one can find reasonable, similar values of ζ (~ -110 mV) that could explain the experimental findings. If KCl concentration is reduced to 5 mM, the curve corresponding to $\partial \delta n_i / \partial r$ $\mid_{r=b} = 0$ (labeled "2") requires a ζ as high in absolute value as -170 mV to explain the experimental result. On the other hand, in Figure 6.10c it is clear that it is impossible to find a ζ value, leading to the conductivity measured if Neumann conditions are used to specify the perturbations δn_i at $r = b$. This suggests that the Dirichlet specification of $\delta n_i(b)$ is the correct choice; clearly, this conclusion needs to be based on a wider range of experimental data. The next section will generalize the model to include ac fields, and calculate the complex conductivity, K^*, and permittivity, ε^*, of the suspensions.

6.8   CASE OF AC FIELDS: DIELECTRIC RESPONSE

The dielectric spectroscopy of colloidal dispersions is a powerful and promising tool for the precise characterization of the electrical properties of interfaces, providing excellent resolution in the evaluation of such properties. Several reasons can be cited for such capabilities. First, measurements are performed in a wide range of frequencies of the externally applied ac electric field, so that, contrary to the single data point obtained in dc methods (mainly electrophoresis), one can rather easily obtain hundreds of data points covering several frequency decades. Second, both the frequency dependence of the dielectric constant and the amplitude of the relaxations eventually found are extremely sensitive to even minute changes in such features as particle's surface charge, particle size and shape, colloidal stability of the system or ionic composition of the dispersion medium [34,35].

In spite of these fundamental advantages over standard dc electrokinetic methods, some aspects may be mentioned that somewhat have hindered a wider use of dielectric spectroscopy (sometimes called low-frequency dielectric dispersion, or LFDD, because it is at the low field frequencies that the most interesting phenomena are observed in suspensions). One problem is the difficulty involved in its experimental determination: in principle, measurements are easy to perform, as all that is needed is to measure the impedance of a conductivity cell immersed in the suspension and connected to an impedance meter. However, the frequency interval where dielectric relaxation takes place (typically a few kHz) can easily include the region where electrode polarization interferes with the results. Nevertheless, methods have been proposed in the literature to minimize such unwanted effects to a large extent [36–42]. Another problem can still be mentioned: the comparison between theory and experiments is often somewhat disappointing, as the former usually underestimates the latter [43–45], although consideration of a finite mobility of ions in the dense part of the double layer (so-called dynamic Stern layer, or DSL models) appears to be a promising improvement in electrokinetic theory [46–49].

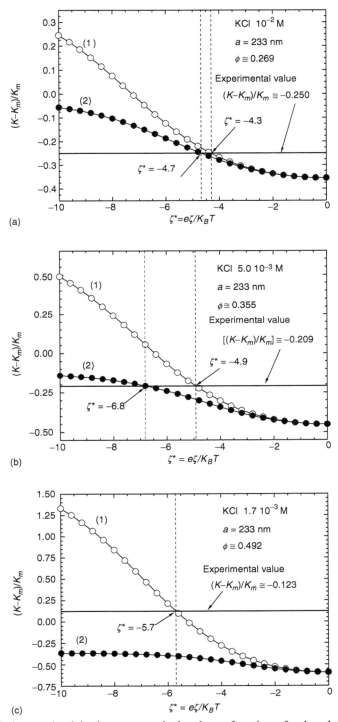

FIGURE 6.10 Relative conductivity increment calculated as a function of reduced zeta potential using the two choices of boundary conditions for ionic concentrations: (1): $\delta n_i(b) = 0$; (2): $\partial \delta n_i / \partial r|_{r=b} = 0$. Horizontal lines are experimental values of the conductivity increment found by Midmore et al. [33] for suspensions of polystyrene latex particles with the characteristics displayed on each plot.

Notwithstanding these difficulties, impedance measurements in fact inform about a collective response of the system, and so they would be an excellent tool for the analysis of the electrokinetics of concentrated suspensions if a proper theoretical treatment is available to interpret experimental data [30,50,51].

This section intends to describe the main results of the calculation of the dielectric constant of concentrated colloidal suspensions of spheres using the above-described cell model. Furthermore, taking into account that overlapping of electric double layers of neighboring particles is more likely than one could expect, our treatment also considers the possibility of double-layer overlap. The validity of the resulting calculations will finally be checked against some (rather scarce) experimental data.

It will be assumed that a harmonic field of frequency ω is acting on the suspension. Because of the finite time required for the different mechanisms responsible for the current density $\mathbf{j}(\mathbf{r})$ at each point, the relationship between the average values of $\mathbf{j}$ and $\mathbf{E}$ will involve a complex, frequency-dependent conductivity $K^*(\omega)$:

$$\langle \mathbf{j} \rangle = K^* \langle \mathbf{E} \rangle \tag{6.72}$$

Recall that the knowledge of K^* leads immediately to that of the complex permittivity, ε^*:

$$K^*(\omega) = K^*(\omega = 0) - i\omega\varepsilon^* \tag{6.73}$$

where i is the imaginary unit, and it is customary to distinguish between the real and the imaginary components of ε^*:

$$\varepsilon^*(\omega) = \varepsilon'(\omega) + i\varepsilon''(\omega) \tag{6.74}$$

For practical reasons, the relative permittivity is more frequently used,

$$\varepsilon_r^*(\omega) = \frac{\varepsilon(\omega)}{\varepsilon_0} = \varepsilon_r'(\omega) + i\varepsilon_r''(\omega) \tag{6.75}$$

where ε_0 is the vacuum permittivity. Our aim is to calculate the contribution of the particles and their double layers to the permittivity, by introducing the so-called dielectric increment, $\Delta\varepsilon^*(\omega)$ or its corresponding relative value:

$$\Delta\varepsilon_r^*(\omega) = \frac{\Delta\varepsilon^*(\omega)}{\varepsilon_0} = \frac{\varepsilon^*(\omega) - \varepsilon_m}{\varepsilon_0} = \Delta\varepsilon_r'(\omega) + i\Delta\varepsilon_r''(\omega) \tag{6.76}$$

Note that $\Delta\varepsilon_r'$ and $\Delta\varepsilon_r''$ will be linearly dependent on ϕ for dilute suspensions, whereas this will not be true for concentrated suspensions.

The equations at the beginning of the chapter that are already described in the case of constant fields must be modified to account for the time variation of the field. We simply write down the required changes: Equation 6.8 (Navier–Stokes equation) must be substituted by

$$\eta_m\nabla^2\mathbf{v}(\mathbf{r},t) - \nabla p(\mathbf{r},t) - \rho_{el}\nabla\Psi(\mathbf{r},t) = \rho_m\frac{\partial}{\partial t}\left[\mathbf{v}(\mathbf{r},t) + \mathbf{v}_e^*\exp(-i\omega t)\right] \tag{6.77}$$

where ρ_m is the liquid density, and the liquid velocity with respect to the laboratory frame of reference is written as the sum of its velocity with respect to the particle and the velocity of the particle with respect to the laboratory, $\mathbf{v}_e^*$. This complex vector is often called dynamic electrophoretic velocity of the particle.

Ionic conservation equation (Equation 6.12) changes into

$$\nabla \cdot [n_i(\mathbf{r}, t)\mathbf{v_i}(\mathbf{r}, t)] = -\frac{\partial}{\partial t}[n_i(\mathbf{r}, t)](i = 1, \ldots, N) \qquad (6.78)$$

and, in addition, all field-induced perturbations must have the same time dependence as the field itself:

$$X(\mathbf{r}, t) = X^0(\mathbf{r}, t) + \delta X(\mathbf{r})\exp(-i\omega t) \qquad (6.79)$$

where X stands here for Ψ, μ_i, n_i, or $\mathbf{v}$.

6.8.1 Overall Behavior of the Dielectric Constant of Concentrated Suspensions

Figure 6.11 gives us an idea of the main trends found in the real (ε_r') and imaginary (ε_r'') components of the dielectric constant as a function of the frequency $f(= \omega/2\pi)$ of the applied field, for different volume fractions ranging from $\phi \to 0$ (the dilute case) to $\phi = 0.4$. These calculations correspond to spheres with a radius $a = 100$ nm, zeta potential $\zeta = 100$ mV in a KCl solution with $\kappa a = 10$.

In order to interpret these data, recall that the dielectric constant of the particles ($\varepsilon_{rp} = 2$) is much lower than that of the medium ($\varepsilon_{rm} = 78.5$ at 25°C): increasing the volume fraction of particles means substituting high dielectric constant material by a material with low dielectric constant. This decreasing effect can be compensated for by the polarization of the diffuse layer, which can give rise to very high values of the dielectric constant at low frequencies. This is in fact observed in Figure 6.11 for ϕ between 0 and 0.10; for higher particle concentrations, the polarization of the double layer does not suffice to compensate for the low dielectric constant of the particles, and the low-frequency dielectric constant, $\varepsilon_r'(0)$, decreases with ϕ and is below that of the medium for $\phi > \sim 0.4$.

This plot warns us about the large errors that can take place when a theory designed for dilute systems is used with concentrated suspensions [23]. This is better appreciated in Figure 6.12a: the low-frequency dielectric constant increases with volume fraction up to $\phi \sim 0.1$, and raising ϕ above this value brings about a reduction in $\varepsilon_r'(0)$. In Figure 6.12a, the dashed line corresponds to the linear dependence $\varepsilon_r'(0) = \varepsilon_{rm} + \phi\delta\varepsilon_r'(0)$ (with $\delta\varepsilon_r'(0)$ independent of ϕ, as it was calculated for the dilute case): note that already for $\phi = 0.01$, the difference between the linear and the true volume fraction dependencies amounts to about 10%.

Finally, the relaxation (so-called α-relaxation [35]) of the double layer polarization is observed in all cases, but, as Figure 6.11a (and, more clearly, Figure 6.12b) shows, the critical frequency, f_{rel} (corresponding to the maximum in $\varepsilon_r''(f)$), is not the same for all particle concentrations: in fact, it increases very significantly with ϕ.

It is appropriate here to discuss the reasons for the two main features of Figure 6.11 and Figure 6.12: the decrease in $\varepsilon_r'(0)$ with respect to the linear behavior and the increase in f_{rel} with volume fraction. Recall that the physical origin of the high values of the dielectric constant that can be observed at low frequencies [29,34,35,52,53] is the existence of a gradient of neutral electrolyte concentration around the particle when an electric field is applied (concentration polarization). The buildup of such neutral electrolyte clouds is a slow process, as it takes a comparatively long time to transfer ions from one side of the particle to the other. This slow process, because of its phase lag with respect to the external field, causes the dipole moment of the particle to acquire an out-of-phase (imaginary) component. This in turn gives rise to an increased displacement current. Such current is sensed macroscopically as high

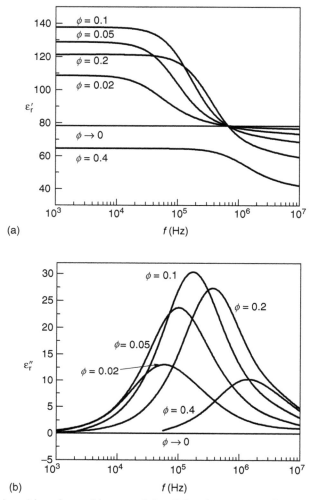

FIGURE 6.11 Real (a) and imaginary (b) parts of the dielectric constant of suspensions of spheres as a function of frequency (f) for the volume fractions indicated. Particle radius: $a = 100\,\text{nm}$; zeta potential: $\zeta = 100\,\text{mV}$; KCl solution with $\kappa a = 10$.

values of ε_r': the larger the electrolyte accumulation on both sides of the particle, the gradients, and hence the low-frequency dielectric increment, are larger. The higher the frequency of the field, the more hindered the slow diffusion processes will be, and the dielectric constant should decrease. Hence the existence of a relaxation effect in $\varepsilon_r''(f)$. Coming back to the results in Figure 6.11 and Figure 6.12, the initial increase in $\varepsilon_r'(0)$ with increase in volume fraction must be due to the accumulated effects of the polarization of all the double layers compensating for the negative effect of the low dielectric constant of the particles. As the volume fraction is increased, the presence of neighbor particles must hinder the formation of a neutral electrolyte concentration gradient: the following image can help in understanding this mechanism. Suppose that three particles are aligned with the field: if they are negatively charged, and the field is applied from left to right, then the electrolyte will be in excess on the right pole of each particle, and depleted on the left. The presence of a particle on the right of

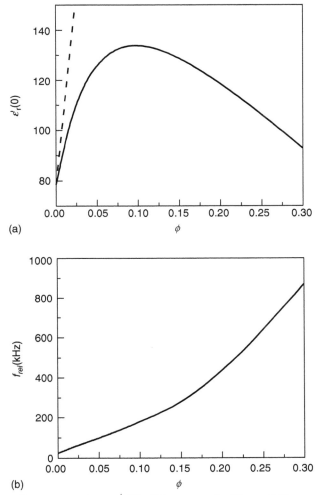

FIGURE 6.12 (a) Low-frequency values $\varepsilon_r^{'}(0)$ of the curves in Figure 6.11a as a function of volume fraction. The dashed line has been calculated assuming a linear dependence of the dielectric constant on the volume fraction (see text). (b) Characteristic or α-relaxation frequency f_{rel} of the suspensions in Figure 6.11b.

the central one will reduce the excess concentration near it (it will be partially compensated by the decrease of concentration existing on the left pole of the neighbor particle). The same applies to the left part of the central particle. Summing up, we can say that the gradient is lower in the higher concentration than in the dilute case, hence the decrease in $\varepsilon_r^{'}$ is observed at high volume fractions (Figure 6.11a and Figure 6.12a).

Concerning the increase in relaxation frequency, this can be explained on a similar basis: the diffusion length is reduced by the presence of the neighbor particles (their concentration polarizations interfere as described above), hence the time needed for the current to go from one side of the particle to the opposite one is lower, and higher frequencies are needed to freeze the mechanism and reduce $\varepsilon_r^{'}$.

In the following paragraphs we will analyze how all these physical situations are affected by the variables of interest: zeta potential, particle radius, and electrolyte concentration.

6.8.2 Zeta Potential and Dielectric Dispersion

The discussion here is limited to the effect of ζ on the main quantities $\varepsilon_r'(0)$ and f_{rel}. Figure 6.13 summarizes the results: like in dilute systems, increasing ζ yields higher dielectric constants, as the concentration polarization is also increased.

A new result that is not found in dilute systems is the increase of f_{rel} with ζ for fixed volume fraction. This effect is truly significant, as one can observe in Figure 6.13b. In order to explain this finding, note that it also occurs (although to a very low extent) at low volume fractions, in the dilute range. It is feasible that increasing ζ will bring about not only an increased electrolyte concentration on the right hand side of the particle (and its corresponding decrease on the left), but a closer approach of that ion cloud to the particle. Thus, since the effective distance that ions must travel back and forth during a field oscillation decreases with ζ, larger frequencies will be needed to fully hinder their diffusion. At high volume fractions, this effect is magnified, as the decreased diffusion length is controlled by both ζ and the geometrical limitations associated to the closer particle-to-particle vicinity.

6.8.3 Effect of κa on the Dielectric Constant

As mentioned earlier, the product κa (i.e., the ratio between the particle size and its double layer thickness) strongly affects the dielectric dispersion of dilute suspensions. Hence,

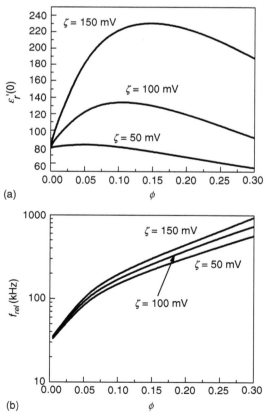

(a)

(b)

FIGURE 6.13 (a) Low-frequency dielectric constant of suspensions of spheres 100 nm in radius as a function of volume fraction, for different zeta potentials. (b) Same as (a), but for the relaxation frequency. The dispersion medium is a KCl solution, such that $\kappa a = 10$.

calculations were also performed in which ζ was maintained constant and κa varied through changes in either ionic strength or particle size. Thus, Figure 6.14a and b shows, respectively, the effect of increasing ionic strength (increasing κ or reducing the double layer thickness) on the low-frequency dielectric constant and relaxation frequency. In these plots, both the particle radius ($a = 100\,\text{nm}$) and the zeta potential ($\zeta = 100\,\text{mV}$) are maintained constant. First of all, note that for any concentration of particles the dielectric constant increases with κa. This is also well known in the case of dilute suspensions, and the same arguments apply here: increasing the ionic strength implies increasing the amount of counterions in the double layer. This will bring about a larger concentration polarization and hence a higher dielectric constant.

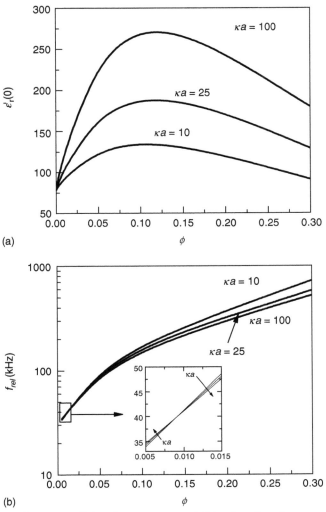

FIGURE 6.14 (a) Low-frequency dielectric constant, and (b) relaxation frequency, of suspensions of spherical particles ($a = 100\,\text{nm}$) as a function of volume fraction for different κa values, obtained by varying the concentration of KCl. The zeta potential is constant and equals $100\,\text{mV}$. Inset: detail of the low volume fraction behavior.

The effect of κa on the relaxation frequency is small for $\phi < {\sim}20\%$ if a is kept constant. Roughly, at low-volume fractions [23]

$$f_{\text{rel}} \propto \frac{D}{(a + \kappa^{-1})^2} \qquad (6.80)$$

where D, for the case of solution containing two types of ions, is

$$D = \frac{D_1 D_2 (z_1 + z_2)}{(D_1 + D_2)} \qquad (6.81)$$

(for the case of KCl, $D \approx D_1 \approx D_2$). Since for all cases shown in Figure 6.14b, $\kappa^{-1} \leq a/10$, the term $(a + \kappa^{-1})$ in Equation 6.80 is almost independent of κ. Thus the relaxation frequency changes by just a small amount when κa is increased: the direction of such a change must correspond, according to Equation 6.80, to larger f_{rel} for lower κ^{-1} (higher κa). This is in fact found true for very dilute suspensions (see inset in Figure 6.14b). However, at high volume fractions, the reverse behavior is observed: f_{rel} increases when κ^{-1} increases. This can be qualitatively explained on the basis of the obstacle that neighbor particles present to the building of the electrolyte clouds as mentioned above. At lower κa, the double layer will be thicker and counterions will be on the average closer to the surrounding particles. We can expect that the reduction in diffusion length produced by increasing ϕ will be more significant the thicker the double layer. That is, the relaxation frequency will be more increased by adding particles the larger their double layer thickness, i. e., the smaller κa, as observed in Figure 6.14b.

It may be interesting to consider suspensions at very low κa ($= 3.3$) from a different point of view: Figure 6.15 shows a Cole–Cole plot, in which the imaginary part of the dielectric increment is plotted as a function of its real part. As observed, at low to moderate volume fractions ($\phi < {\sim}15\%$), the curves display shoulders indicating the existence of two separate relaxations: the one at lower frequencies (right part of the curves) is associated with double layer polarization; at higher frequencies the Maxwell–Wagner relaxation can be observed. When the volume fractions are increased above the value previously cited the two relaxations are not distinguished, as the α-relaxation is higher with the larger ϕ whereas the Maxwell–Wagner frequency decreases with ϕ as follows [53]:

$$\omega_{\text{MW}} = \frac{K_{\text{m}}}{\varepsilon_{\text{m}}} \frac{2(1 - \phi)Du + 2 + \phi}{2 + \phi} \qquad (6.82)$$

It is a well-known fact that K^{σ} increases with ζ and is independent of volume fraction, so Equation 6.82 predicts a reduction in ω_{MW} when the volume fraction increases. We can conclude that, as

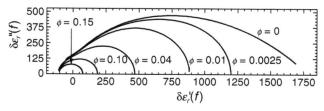

FIGURE 6.15 Cole–Cole plots of the dielectric increment (per volume fraction) of suspensions of spheres 100 nm in radius, as a function of the volume fraction. In all cases $\kappa a = 3.3$, $\zeta = 100\,\text{mV}$.

observed in Figure 6.15, the frequencies of the α- and Maxwell–Wagner relaxations will get closer the higher the volume fraction. If the latter is large enough, it will be impossible to sense them as two separate frequencies. Note also that the Cole–Cole plot acquires a semicircular shape, which is an indication of an almost pure Debye relaxation process.

6.8.4 COMPARISON WITH OTHER APPROACHES AND EXPERIMENTAL DATA

To our knowledge, no rigorous treatment has been published, other than the cell-based treatment that we have just described, concerning the permittivity of concentrated colloidal dispersions. Nevertheless, an approximate model was described in Ref. [51] based on evaluations of the characteristic diffusion length, L_D, of ions in the formation of the polarized double layer. L_D is related to the volume fraction by the following expression:

$$L_D = a\left(1 + \frac{1}{(\phi^{-1/3} - 1)^2}\right)^{-1/2} \tag{6.83}$$

There it was also shown that the frequency of the α-relaxation is related to the frequency of the dilute case, f_d, as follows:

$$f_{rel} = f_d\left(1 + \frac{1}{(\phi^{-1/3} - 1)^2}\right) \tag{6.84}$$

Similarly, the low-frequency dielectric increment, $\Delta\varepsilon_r'(0)$ depends on ϕ and on the increment for dilute suspensions $\Delta\varepsilon_{rd}'(0)$, as indicated by Equation 6.85:

$$\Delta\varepsilon_r'(0) = \Delta\varepsilon_{rd}'(0)\left(1 + \frac{1}{(\phi^{-1/3} - 1)^2}\right)^{-3/2} \tag{6.85}$$

In order to check the validity of the approximate Equation 6.84, Equation 6.85, we have plotted the ratios f_{rel}/f_d and $\delta\varepsilon_r'(0)/\delta\varepsilon_{rd}'(0)$ as a function of volume fraction in Figure 6.16 using both the approximate and full numerical evaluations of those quantities. Note that the approximate model is quite good in predicting the qualitative behaviors of both $\delta\varepsilon_r'(0)$ and f_{rel}, but the quantitative agreement is very reasonable in the calculation of the low-frequency dielectric increment, $\delta\varepsilon_r'(0)$, but worsens in the case of f_{rel}, thus limiting its applicability to moderate volume fractions, $\phi < {\sim}10\%$.

In addition, we have the possibility to compare our calculations with experimental data. Figure 6.17 and Figure 6.18 [30] include data on the determinations of permittivity of ethylcellulose latex spheres (Aquacoat latex, FMC Corp, USA) for values of ϕ between 1 and 15%. The particles had a volume-average radius $a = 93 \pm 10$ nm and the suspensions were prepared in 1 mM KCl. In each of the figures, part a refers to experimental data, and part b concerns our cell-model calculations, based on a zeta potential of -112.3 mV obtained from classical electrophoresis measurements performed on dilute suspensions. As observed, the model is capable of attaining a good qualitative explanation of the experimental data, although, as already found in dilute suspensions, calculations appear to underestimate $\varepsilon_r'(0)$ and overestimate f_{rel}. It is likely that such quantitative differences can be corrected by using a model including the possibility of ionic transport in the inner part of the double layer (a DSL model) that was found to achieve good accuracy in dilute suspensions [46,48]. This work is currently under preparation.

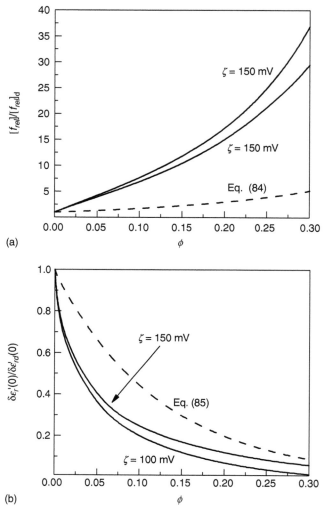

(a)

(b)

FIGURE 6.16 Relaxation frequency (a) and low-frequency dielectric increment (b) related to their corresponding values for the dilute case plotted as a function of the volume fraction ϕ of the suspensions. Full lines: numerical evaluation; Dashed lines: approximate Equation 6.84 and Equation 6.85.

6.9 DYNAMIC MOBILITY

Although electroacoustic phenomena were first studied by Debye in the 1930s [54,55], it has received renewed interest only during the last 20 to 30 years, probably as a consequence of the commercial availability of devices that, based on such phenomena, have been dedicated to the evaluation of the electrokinetic properties of concentrated suspensions.

Basically there are two such techniques. The first consists in the generation of a pressure wave when an alternating electric field is applied to the suspension. It is called electrokinetic sonic amplitude (ESA) effect. The second one is reciprocal of the ESA phenomenon: a pressure wave of a suitable frequency produces an ac field in the suspension. It is called colloid vibration potential (CVP) or colloid vibration current (CVI).

Although there were a number of previous works, particularly those by Enderby and Booth [56,57], it is widely admitted that O'Brien [58,59] set out the foundations of a theory of

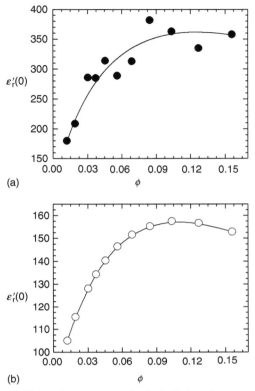

FIGURE 6.17 Low-frequency dielectric constant of ethylcellulose latex suspensions as a function of volume fraction. •: experimental points. (the full line is a guide to the eye); ○: theoretical predictions according to a cell model. (Taken from Carrique, F. et al., *J. Chem. Phys.*, 118, 1945, 2003. Copyright 2003, American Institute of Physics.)

electroacoustic phenomena. According to his derivation, the dynamic electrophoretic mobility, u_e^*, plays a fundamental role in electroacoustic phenomena, as both ESA and CVI can be related to it. This is a consequence of the linear theory of nonequilibrium thermodynamics that establishes a reciprocal relationship between ESA and CVI. If the suspension is subjected to an electric field $\mathbf{E}$ and a pressure gradient ∇p simultaneously, then the velocity of the particles, $\mathbf{v}_p$, and the average current density through the suspension, $<\mathbf{j}>$, will depend linearly on both $\mathbf{E}$ and ∇p:

$$\mathbf{v}_p = \alpha \nabla p + u_e^* \mathbf{E}$$
$$\langle \mathbf{j} \rangle = \beta \nabla p + K^* \mathbf{E}$$

(6.86)

where α and β represent the transport properties of the suspension, and K^* is its complex electrical conductivity. From Onsager's reciprocity relationship, β and u_e^* will be proportional, so that the dynamic mobility is essential in the value of the CVI. In an ESA experiment, the amplitude of the pressure wave generated by application of an ac field of frequency ω is proportional to both u_e^* and the so-called density contrast, $\Delta\rho = \rho_p - \rho_m$, where ρ_p (ρ_m) is the particle (medium) density. The great advantage of electroacoustic techniques over other classical electrokinetic methods is that they can, in principle, be applied to colloidal suspensions of any particle concentration, ϕ, as they are based on the determination of a

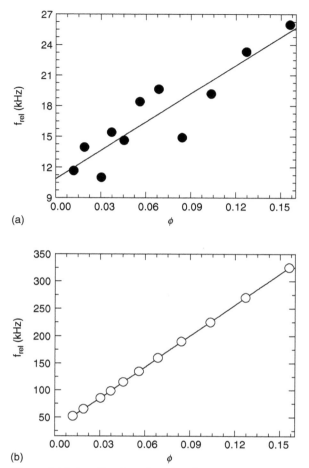

FIGURE 6.18 Same as Figure 6.17, but for the relaxation frequency. (Taken from Carrique, F. et al., *J. Chem. Phys.*, 118, 1945, 2003. Copyright 2003, American Institute of Physics.)

collective average response, and no individual particles need to be tracked. However, this implies to know the ϕ-dependence of u_e^* (or ESA, or CVI), in order to extract the zeta potential, surface conductivity, or any other parameters characterizing the electrical state of the interface [60].

6.9.1 CELL-MODEL CALCULATIONS OF DYNAMIC MOBILITY

Figure 6.19 through Figure 6.21 show, respectively, the effects of volume fraction, zeta potential and κa, on the modulus and phase of the dynamic mobility, for different frequencies of the applied field. The frequency dependence of the dynamic mobility presents a number of features that can be explained on the basis of the response of the particles and their double layers to the action of an ac field. This can be carried out by using an expression that was deduced in Ref. [29] for the case of dc fields:

$$u_e = \frac{\varepsilon_m}{\eta_m} \zeta (1 - C) \tag{6.87}$$

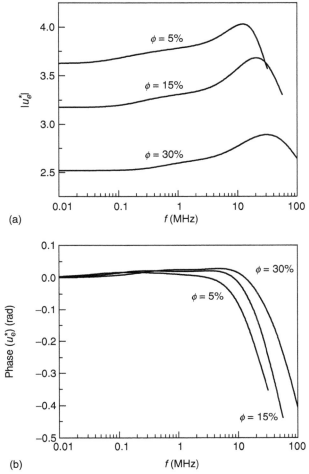

(a)

(b)

FIGURE 6.19 Modulus (a) and phase (b) of the dynamic mobility u_e^* of spherical particles 100 nm in radius in KCl solutions, plotted as a function of frequency for different volume fractions. In all cases, $\zeta = 100$ mV and $\kappa a = 18$.

where C is the induced dipole coefficient relating the dipole moment, $\mathbf{d}$, of the particle to the applied field

$$\mathbf{d} = 4\pi\varepsilon_m a^3 C\mathbf{E} \tag{6.88}$$

In the case of ac fields, the frequency dependence of the dynamic mobility u_e^* will hence be contained in the complex dipole coefficient C^*, except for an inertia term that leads to a decrease in $|u_e^*|$ when the frequency of the field is so high that the particle and liquid motions cannot follow the rapid field changes. In Figure 6.19 through Figure 6.21, it is this inertia that explains the fact that at high frequencies all double layer effects essentially disappear and all curves tend to coincide.

The frequency dependence of lower frequencies can be qualitatively explained by observing the plots of $|1-C^*|$ in Figure 6.22: note how a decrease can be expected in $|u_e^*|$ due to the α-relaxation of the double layer, when the real part of $|1-C^*|$ decreases. Below the α-relaxation frequency, the diffusion fluxes associated with the polarized double layer

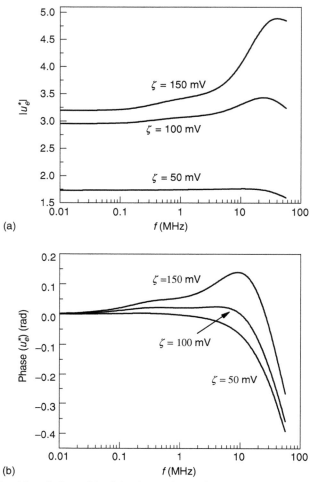

(a)

(b)

FIGURE 6.20 Modulus (a) and phase (b) of the dynamic mobility of spherical particles 100 nm in radius in KCl solutions, plotted as a function of frequency for different zeta potentials. In all cases, $\phi = 20\%$ and $\kappa a = 18$.

tend to decrease the dipole moment produced by ionic electromigration. Above that frequency, the concentration polarization of the electrical double layer cannot occur, and because those diffusion fluxes disappear, the dipole moment increases and this tends to reduce the mobility.

Concerning the maximum in the mobility around 1–10 MHz (Figure 6.19 through Figure 6.21), this must be a manifestation of the Maxwell–Wagner–O'Konski (MWO) relaxation. Recall that if the ratio between conductivity and permittivity is different for the solid particles and the medium, the application of an external field polarizes the particle by charge accumulation on the interface. For a spherical particle of radius a, the (complex) dipole moment induced by the field reads [53]:

$$\mathbf{d}^* = 4\pi\varepsilon_m \frac{\varepsilon_p^* - \varepsilon_s^*}{\varepsilon_p^* + 2\varepsilon_s^*} a^3 \mathbf{E} \qquad (6.89)$$

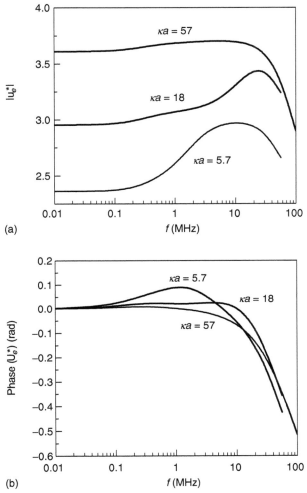

FIGURE 6.21 Modulus (a) and phase (b) of the dynamic mobility of spherical particles 100 nm in radius in KCl solutions, plotted as a function of frequency for different κa values. In all cases, $\phi = 20\%$ and $\zeta = 100\,\text{mV}$.

where

$$\varepsilon_p^* = \varepsilon_p - i\frac{K_p}{\omega} \tag{6.90}$$

$$\varepsilon_s^* = \varepsilon_m - i\frac{K}{\omega} \tag{6.91}$$

and K_p, the particle conductivity, can be ascribed to its surface conductivity, K^σ,

$$K_p = \frac{2K^\sigma}{a} \tag{6.92}$$

On substitution of Equation 6.90 and Equation 6.91 in Equation 6.89, it is easy to find that the frequency dependence of $\mathbf{d}^*$ has a characteristic or relaxation frequency (Maxwell–Wagner–O'Konski frequency, ω_{MWO}):

$$\omega_{MWO} = \frac{(1-\phi)K_p + (2+\phi)K}{(1-\phi)\varepsilon_p + (2+\phi)\varepsilon_m} \qquad (6.93)$$

For frequencies above ω_{MWO} the induced dipole moment acquires its most negative values, and a negative dipole will favor the particle's motion, thus increasing the mobility. Using Equation 6.93, it is found that, for a given ζ (i.e., for given K^σ) and ϕ, the larger the particle size the lower the ω_{MWO}, and this explains the shift in frequency of the maxima in $|u_e^*|$ when κa is reduced (Figure 6.21).

We now have the necessary tools to explain the main features of Figure 6.19 through Figure 6.21. The modulus of u_e^* in Figure 6.19a clearly shows how an increase in the volume fraction, ϕ, provokes a decrease of the mobility mainly because the liquid motion around any particle hinders the motion of its neighbors.

Another effect of the increase in ϕ is a raise in the frequency characteristic of the beginning of inertia effects: this frequency is of the order of $(\eta_m/\rho_p a^2)$ [55], so that the smaller the particle size the larger the frequency of the field that is needed to slow down the motion due to inertia. As ϕ is increased the characteristic distance changes from a to the average particle–particle distance, which clearly decreases with ϕ. Hence the increase in the inertial relaxation frequency as observed in Figure 6.19. Note that Figure 6.19b shows that the particle mobility lags behind the field (negative phase angle), and that the lag increases with frequency beyond the inertial relaxation.

Let us now consider how the zeta potential affects the mobility (Figure 6.20). The modulus of u_e^* increases with $|\zeta|$ at any frequency, as expected, but it is worthwhile to mention three features not so obvious. First is that the shoulder associated with the α-relaxation is more noticeable the higher the zeta potential, due to the growing importance of concentration polarization. The second is that the MWO increase is more abrupt if $|\zeta|$ increases, indicating that the interfacial dipole is more important at higher zeta potential because of the increased surface conductivity. Indeed, below the MWO relaxation the induced dipole strength will be larger, the more so the higher $|\zeta|$. If this dipole is hindered, as it happens above ω_{MWO}, the mobility should jump because of the $(1-C^*)$ dependence in u_e^*. The third feature is that the phase angle at moderate and high $|\zeta|$ increases with frequency (prior to the inertia characteristic frequency) and in fact can be positive for a wide f range. The explanation for this lies on the evolution of the phase of $(1-C^*)$ (Figure 6.22): as we increase f below the MWO-relaxation, $(1-C^*)$ leads the field. For higher frequencies the phase angle decreases, and, eventually, the lag associated to inertia brings about a decrease and change of sign in the phase of u_e^*.

Finally, Figure 6.21 displays the effect of the ionic strength (as represented by the product κa). Increasing κa has also significant effects on both the modulus and phase of u_e^*: the former (Figure 6.21a) raises because increasing κa at constant ζ means increased number of charged particles that will move faster for a given field. In addition, Figure 6.21b shows that the MWO raise in the phase disappears for sufficiently high κa: the effects of interfacial polarization are masked by the high values of K_m (and, correspondingly, low values of Du). Most important is the fact that the shift of ω_{MWO} towards higher frequencies as we increase the ionic strength gives raise to a hiding of the MWO raise by the inertia fall, and it is only the latter one that is observed at $\kappa a = 18$ and, mainly, $\kappa a = 57$.

6.9.2 COMPARISON WITH EXPERIMENTAL DATA

There is an increasing amount of experimental results on the electroacoustics of concentrated suspensions. Here, we will use our own results on the dynamic mobility of alumina particles

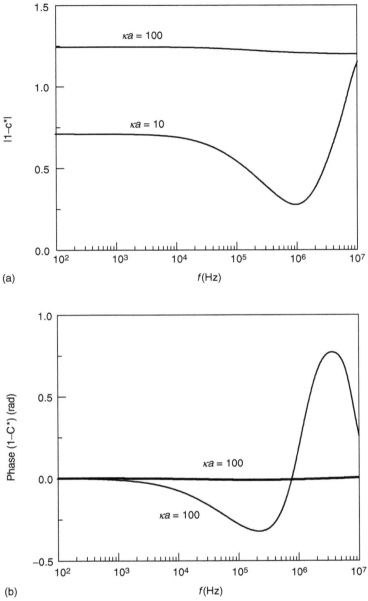

(a)

(b)

FIGURE 6.22 Modulus (a) and phase (b) of the quantity $1-C^*$ (C^*: complex dipole coefficient) appearing in the approximate expression (87), plotted as a function of frequency, for different κa values in a diluted suspension. $\zeta = 150$ mV.

deduced from ESA measurements [61]. A detailed analysis has also been given by Dukhin et al. [62]. Using the particle radius deduced from photon correlation spectroscopy (PCS) data ($a = 165$ nm), and knowing the ionic strength of the dispersion medium, we obtained κa and used the cell model to fit the frequency dependence of the mobility using ζ as the only parameter. Figure 6.23 through Figure 6.26 show the results. As observed, the cell model is capable of reproducing the experimental data using reasonable values of ζ, that, in addition,

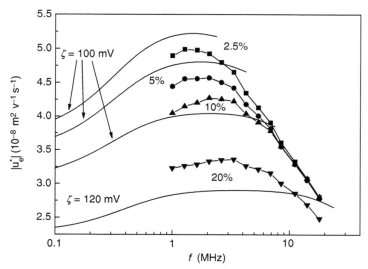

FIGURE 6.23 Modulus of the dynamic mobility of alumina particles in suspensions of different volume fractions in 5×10^{-5} mol/l KCl solutions. The points are experimental data, and the lines correspond to cell-model calculations with the zeta potentials indicated.

decrease with the concentration of the indifferent electrolyte KCl and do not change with volume fraction, as expected. Note, however, that the agreement gets worse for high volume fractions and for solutions containing the larger KCl concentrations. The differences are basically associated to the fact that experimentally the inertia relaxation starts at lower frequencies than predicted by the model. Because the frequency of hydrodynamic relaxation depends on a^{-2}, as mentioned earlier, the behavior observed may be an indication of the existence of aggregation, as the formation of aggregates would effectively correspond to an increase in a and a corresponding decrease in the inertia characteristic frequency. This

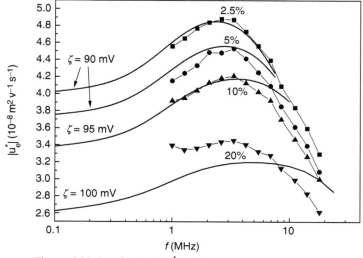

FIGURE 6.24 Same as Figure 6.23, but for 2×10^{-4} mol/l.

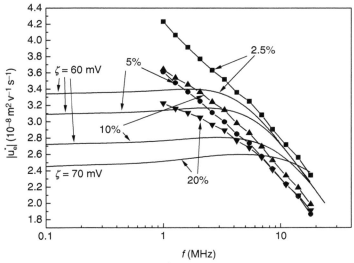

FIGURE 6.25 Same as Figure 6.23, but for 5×10^{-4} mol/l.

behavior can be demonstrated by the plots in Figure 6.27, where the cell-model calculations have been repeated after changing the size to approximately double the PCS value, i.e., using $a = 300$ nm. The calculations now come much closer to the data, thus demonstrating that electroacoustic techniques are very sensitive to aggregation and polydispersity, an additional advantage of these methods when working with concentrated systems.

6.9.3 OTHER APPROACHES

To our knowledge, there are two main independent approaches both of which contribute to enrich our understanding of electroacoustics. O'Brien and coworkers use a "first-principles"

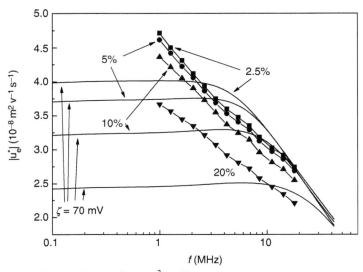

FIGURE 6.26 Same as Figure 6.23, but for 10^{-3} mol/l.

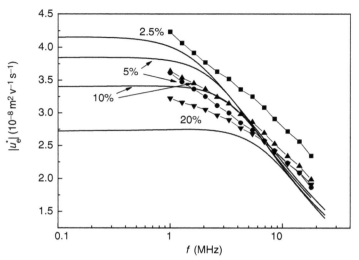

FIGURE 6.27 Like Figure 6.25, but using a particle size $a = 300\,\text{nm}$ in the cell-model calculations. (Taken from Delgado, A.V. et al., *Colloids Surf., A: Physicochem. Eng. Aspects*, 267, 95, 2005. Copyright 2005, Elsevier Science.)

approach; in Ref. [63], Rider and O'Brien calculated u_e^* for a pair of particles in mutual interaction, thus extending the range of validity of their previous calculations to volume fractions ~10%. Later, O'Brien et al. [64] gave a more complete account of particle–particle interactions assuming that only nearest-neighbor interactions take place; their formula gave excellent results for near-neutrally buoyant particles ($\Delta\rho \sim 0$). More recently, O'Brien et al. [65] found a formula for u_e^* that they claim to be valid for high volume fractions and arbitrary $\Delta\rho$. The expression is [65,66]:

$$u_e^* = \frac{\varepsilon_m \zeta}{\eta_m(2+\phi)} \frac{(2-2\phi) - 3\phi(F-1) - (2\lambda^2/(3+3\lambda+\lambda^2))}{1 + (\Delta\rho/\rho_m)\{\phi F + (2\lambda^2/3(3+3\lambda+\lambda^2))\}} \tag{6.94}$$

where F and λ are defined as

$$F = \frac{2}{3}\left[\frac{1}{2} + (4\lambda^2 I + \{1+2\lambda\}\exp(-2\lambda))J^2\right]$$

$$J = \frac{\exp(\lambda)}{1 + \lambda + (\lambda^2/3)}$$

$$I = \int_1^\infty [g(r)-1]r\exp(-2\lambda r)\,dr \tag{6.95}$$

$$\lambda = (1+i)\sqrt{\frac{\omega a^2}{2\nu}}$$

In these expressions, ν is the kinematic viscosity of the medium. The pair distribution function, $g(r)$, is calculated using the Percus–Yevick formula [67]. According to O'Brien et al., Equation 6.94 is valid if the following conditions are met: (a) the double layer is thin compared to the particle size, i.e., $\kappa a \gg 1$; (b) the zeta potential is low ($<50\,\text{mV}$, say), since no double layer polarization effects are considered. If, as usual, the experiments are conducted at high frequencies (above ~ 1 MHz), this is not a very important limitation, as double layer polarization does not have time to develop at such high frequencies; (c) the density contrast $\Delta\rho$ and the volume fraction of solid particles are arbitrary, within reasonable limits.

We will simply compare the results in the form $\mathrm{mod}(u_e^*)$, $\mathrm{phase}(u_e^*)$ data versus frequency for some situations in which both the analytical and the cell models would presumably be valid; thus we will fix $\zeta = 40\,\mathrm{mV}$ and $\kappa a = 20, 40$. Two volume fractions will be tested, one very low ($\phi = 2\%$) and the other rather high (30%).

Before proceeding with the comparison, it is necessary to take into account two corrections, one for each model, that must be applied. In O'Brien's theory, it is necessary to consider [66] that the proximity of neighbor particles alters the tangential electric field in the vicinity of each of them. The neighbors change the field because their dielectric constant, ε_p, and conductivity (represented by their surface conductivity, K^σ, in turn zeta potential- and κ-dependent) differ from those of the medium. Briefly, the formula for $|u_e^*|$ in Equation 6.94 must be multiplied by a factor

$$\frac{1}{1 + \left[\dfrac{2K^\sigma}{K^*a} + \dfrac{i\omega\varepsilon_p}{K^*}\right]\dfrac{1-\phi}{2+\phi}} \qquad (6.96)$$

Note that the surface conductance can be calculated from zeta potential using Bikerman's equation [68]. The full formula was successfully used in Ref. [65] to obtain the zeta potential and particle size distribution in silica and alumina suspensions of different volume fractions.

A subtle argument leads to the correction that must be applied to the cell model: according to Dukhin et al. [32], the dynamic mobility deduced from any cell model $(u_e^*)_c$, does not correspond to the experimentally measured one. Following O'Brien et al. [65], the mobility must be calculated subject to the condition that the macroscopic momentum per unit volume is zero. By applying that condition to the numerical calculation, it can be shown that

$$u_e^* = (u_e^*)_c \frac{1}{1 + \frac{\Delta\rho}{\rho_m}\phi} \qquad (6.97)$$

Figure 6.28 shows the results of the comparison. A good agreement is reached between the two ways of calculation, which sheds confidence on using any of them, with the only limitation that the analytical formula works best if κa is high enough (>10, say), as it is elaborated assuming locally flat double layers.

The second approach to the electroacoustics of suspensions was originally elaborated by Dukhin et al. [32,62], and was based on the application of the so-called coupled-phase model [69] to the CVI or CVP phenomena. These authors used the Shilov–Zharkihk electrostatic boundary conditions. As shown in Ref. [62] the low-frequency value of CVI can be easily compared using O'Brien's and Dukhin et al.'s approaches. The results are

$$\begin{aligned}
\text{Dukhin: } \mathrm{CVI}(\omega \to 0) &= \frac{\varepsilon_m}{\eta_m}\zeta\,|\nabla p\,|\left\{\phi\frac{(1-\phi)}{(1+0.5\phi)}\frac{(\rho_p - \rho_s)}{\rho_s}\right\} \\
\text{O'Brien: } \mathrm{CVI}(\omega \to 0) &= \frac{\varepsilon_m}{\eta_m}\zeta\,|\nabla p\,|\left\{\phi\frac{(1-\phi)}{(1+0.5\phi)}\frac{(\rho_p - \rho_m)}{\rho_m}\right\}
\end{aligned} \qquad (6.98)$$

where $\rho_s\,(= \phi\rho_p + (1-\phi)\rho_m)$ is the density of the suspension. This means that the two models would essentially differ in the density contrast (particle–liquid or particle–suspension) used.

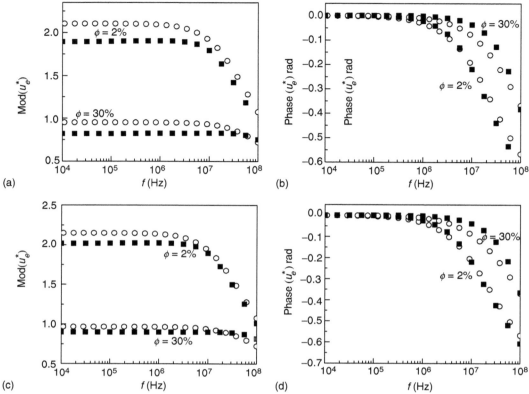

FIGURE 6.28 Modulus ($|u_e^*|$) and phase angle (θ) of the dynamic mobilities of suspensions of the volume fractions indicated, for $\zeta = 40\,\text{mV}$, and $\kappa a = 20$ (a, b) and 40 (c, d), plotted as a function of frequency. In all cases, the particle radius is 100 nm and KCl is used as electrolyte in the aqueous medium. Solid squares: cell model calculations corrected by Equation 6.97; open circles: O'Brien et al.'s Equation 6.94 and Equation 6.95 corrected by the factor Equation 6.96.

Although at moderate ϕ values such a choice is unimportant, this is not true even at moderate volume fractions: for $\rho_p = 2\rho_m$, the terms in brackets in Equation 6.98 differ by more than 10% above $\phi \approx 5\%$.

6.10 CONCLUSION

In this chapter we have tried to give an overview of the electrokinetic phenomena in concentrated systems, a field of research in colloid science that is gaining increasing attention because of the many technological applications of such suspensions. On the basis of the authors' experience, we have mainly focused on the use of cell models, which have been tested against experimental data in many different cases, and are conceptually simple. Nevertheless, we have also mentioned other approaches and discussed their applicability. Most important, it must be stressed that there is still room for new contributions, as the problem is far from fully solved, considering the variety of electrokinetic phenomena that can be used for the characterization of concentrated colloidal dispersions.

ACKNOWLEDGMENTS

Financial support from MEC, Spain, and FEDER Funds, EU (Project No. MAT2004-00866) is gratefully acknowledged.

REFERENCES

1 Levine, S., and Neale, G.H., The prediction of electrokinetic phenomena within multiparticle systems. I. Electrophoresis and electroosmosis, *J. Colloid Interface Sci.*, 47, 520, 1974.

2 Levine, S., and Neale, G.H., Electrophoretic mobility of multiparticle systems, *J. Colloid Interface Sci.*, 49, 330, 1974.

3 Levine, S., Neale, G.H., and Epstein, N., The prediction of electrokinetic phenomena within multiparticle systems. II. Sedimentation potential, *J. Colloid Interface Sci.*, 57, 424, 1976.

4 Happel, J., and Brenner, H., Viscous flow in multiparticle systems. Motion of spheres and a fluid in a cylindrical tube, *AICHE J.*, 3, 506, 1957.

5 Kuwabara, S., The forces experienced by randomly distributed parallel circular cylinders or spheres in a viscous flow at small Reynolds numbers, *J. Phys. Soc. Jpn.*, 14, 527, 1959.

6 Ding, J.M., and Keh, H.J., The electrophoretic mobility and electrical conductivity of a concentrated suspension of colloidal spheres with arbitrary double-layer thickness, *J. Colloid Interface Sci.*, 236, 180, 2001.

7 Kozak, M.W., and Davis, E.J., Electrokinetics of concentrated suspensions and porous-media. 1. Thin electrical double-layers, *J. Colloid Interface Sci.*, 127, 497, 1989.

8 Kozak, M.W., and Davis, E.J., Electrokinetics of concentrated suspensions and porous-media. 2. Moderately thick electrical double-layers, *J. Colloid Interface Sci.*, 129, 166, 1989.

9 Ohshima, H., Electrophoretic mobility of spherical colloidal particles in concentrated suspensions, *J. Colloid Interface Sci.*, 188, 481, 1997.

10 Ohshima, H., Electrical conductivity of a concentrated suspension of spherical colloidal particles, *J. Colloid Interface Sci.*, 212, 443, 1999.

11 Lee, E., Chu, J.W., and Hsu, J.P., Electrophoretic mobility of a concentrated suspension of spherical particles, *J. Colloid Interface Sci.*, 209, 240, 1999.

12 Hsu, J.P., Lee, E., and Yen, F.Y., Electrophoresis of concentrated spherical particles with a charge-regulated surface, *J. Chem. Phys.*, 112, 6404, 2000.

13 Lin, W.H., Lee, E., and Hsu, J.P., Electrophoresis of a concentrated spherical dispersion at arbitrary electrical potentials, *J. Colloid Interface Sci.*, 248, 398, 2002.

14 Carrique, F., Arroyo, F.J., and Delgado, A.V., Sedimentation velocity and potential in a concentrated colloidal suspension. Effect of a dynamic Stern layer, *Colloids Surf., A: Physicochem. Eng. Aspects*, 195, 157, 2001.

15 Carrique, F., Arroyo, F.J., and Delgado, A.V., Electrokinetics of concentrated suspensions of spherical colloidal particles: effect of a dynamic Stern layer on electrophoresis and DC conductivity, *J. Colloid Interface Sci.*, 243, 351, 2001.

16 Carrique, F., Arroyo, F.J., and Delgado, A.V., Electrokinetics of concentrated suspensions of spherical colloidal particles with surface conductance, arbitrary zeta potential, and double-layer thickness in static electric fields, *J. Colloid Interface Sci.*, 252, 126, 2002.

17 Lee, E., Yen, F.Y., and Hsu, J.P., Dynamic electrophoretic mobility of concentrated spherical dispersions, *J. Phys. Chem. B*, 105, 7239, 2001.

18 Hsu, J.P., Lee, E., and Yen, F.Y., Dynamic electrophoretic mobility in electroacoustic phenomenon: concentrated dispersions at arbitrary potentials, *J. Phys. Chem. B*, 106, 4789, 2002.

19 Shilov, V.N., Zharkikh, N.I., and Borkovskaya, Y.B., Theory of non-equilibrium electrosurface phenomena in concentrated disperse systems. 1. Application of non-equilibrium thermodynamics to cell model of concentrated dispersions, *Colloid J.*, 43, 434, 1981.

20 Dukhin, A.S., Shilov, V.N., and Borkovskaya, Y.B., Dynamic electrophoretic mobility in concentrated dispersed systems. Cell model, *Langmuir*, 15, 3452, 1999.

21 O'Brien, R.W., and White, L.R., Electrophoretic mobility of a spherical colloidal particle, *J. Chem. Soc., Faraday Trans. 2*, 74, 1607, 1978.

22 Ohshima, H. et al., Sedimentation velocity and potential in a dilute suspension of charged spherical colloidal particles, *J. Chem. Soc., Faraday Trans. 2*, 80, 1299, 1984.

23 DeLacey, E.H.B., and White, L.R., Dielectric response and conductivity of dilute suspension of colloidal particles, *J. Chem. Soc., Faraday Trans. 2*, 77, 2007, 1981.

24 Henry, D.C., The cataphoresis of suspended particles. I. The equation of cataphoresis, *Proc. R. Soc. London*, A, 183, 106, 1931.

25 Ohshima, H., Interfacial electrokinetic phenomena, in *Electrical Phenomena at Interfaces: Fundamentals, Measurements and Applications*, 2nd ed., Ohshima, H. and Furusawa, K., Eds., Marcel Dekker, New York, 1998, chap. 2.

26 Ohshima, H., Electrophoresis of charged particles and drops, in *Interfacial Electrokinetics and Electrophoresis*, Delgado, A.V., Ed., Marcel Dekker, New York, 2002, chap. 5.

27 O'Brien, R.W., The electrical conductivity of a dilute suspension of charged-particles, *J. Colloid Interface Sci.*, 81, 234, 1981.

28 Ohshima, H., Cell model calculation for electrokinetic phenomena in concentrated suspensions: an Onsager relation between sedimentation potential and electrophoretic mobility, *Adv. Colloid Interface Sci.*, 88, 1, 2000.

29 Shilov, V.N. et al., Polarization of the electrical double layer. Time evolution after application of an electric field, *J. Colloid Interface Sci.*, 232, 141, 2000.

30 Carrique, F. et al., Dielectric response of concentrated colloidal suspensions, *J. Chem. Phys.*, 118, 1945, 2003.

31 Derjaguin, B.V., and Dukhin, S.S., Nonequilibrium double layer and electrokinetic phenomena, in *Surface and Colloid Science*, Vol. 7, *Electrokinetic Phenomena*, Matijeviç, E., Ed., Wiley, New York, 1974, chap. 3.

32 Dukhin, A.S. et al., Electroacoustics for concentrated dispersions, *Langmuir*, 15, 3445, 1999.

33 Midmore, B.R., Hunter, R.J., and O'Brien, R.W., The dielectric response of concentrated lattices, *J. Colloid Interface Sci.*, 120, 210, 1987.

34 Grosse, C., Relaxation mechanisms of homogeneous particles and cells suspended in aqueous electrolyte solutions, in *Interfacial Electrokinetics and Electrophoresis*, Delgado, A.V., Ed., Marcel Dekker, New York, 2002, chap. 11.

35 Shilov, V.N. et al., Suspensions in an alternating external electric field: dielectric and electrorotation spectroscopies, in *Interfacial Electrokinetics and Electrophoresis*, Delgado, A.V., Ed., Marcel Dekker, New York, 2002, chap. 12.

36 Springer, M.M., Dielectric relaxation of dilute polystyrene lattices, Ph.D. thesis, University of Wageningen, The Netherlands, 1979.

37 Kijlstra, J., van Leeuwen, H.P., and Lyklema, J., Low-frequency dielectric-relaxation of hematite and silica sols, *Langmuir*, 9, 1625, 1993.

38 Myers, D.F., and Saville, D.A., Dielectric-spectroscopy of colloidal suspensions. 1. The dielectric spectrometer, *J. Colloid Interface Sci.*, 131, 448, 1989.

39 Rosen, L.A., and Saville, D.A., Dielectric-spectroscopy of colloidal dispersions. Comparisons between experiment and theory, *Langmuir*, 7, 36, 1991.

40 Grosse, C., and Tirado, M., Measurement of the dielectric properties of polystyrene particles in electrolyte solution, *Mater. Res. Soc. Symp. Proc.*, 430, 287, 1996.

41 Tirado, M. et al., Measurement of the low-frequency dielectric properties of colloidal suspensions: comparison between different methods, *J. Colloid Interface Sci.*, 227, 141, 2000.

42 Jiménez, M.L. et al., Analysis of the dielectric permittivity of suspensions by means of the logarithmic derivative of its real part, *J. Colloid Interface Sci.*, 249, 327, 2002.

43 Delgado, A.V. et al., Low frequency dielectric dispersion in ethylcellulose latex. Effect of pH and ionic strength, *Colloids Surf., A: Physicochem. Eng. Aspects*, 131, 95, 1998.

44 Arroyo, F.J. et al., Effect of ionic mobility on the enhanced dielectric and electro-optic susceptibility of suspensions: theory and experiments, *J. Chem. Phys.*, 116, 10973, 2002.

45 Dunstan, D.E. and White, L.R., The dielectric response of dilute polystyrene latex dispersions, *J. Colloid Interface Sci.*, 152, 308, 1992.

46 Arroyo, F.J. et al., Dielectric dispersion of colloidal suspensions in the presence of Stern layer conductance: particle size effects, *J. Colloid Interface Sci.*, 210, 194, 1999.

47 Arroyo, F.J., Carrique, F., and Delgado, A.V., Effects of temperature and polydispersity on the dielectric relaxation of dilute ethylcellulose suspensions, *J. Colloid Interface Sci.*, 217, 411, 1999.

48 Rosen, L.A., Baygents, J.C., and Saville, D.A., The interpretation of dielectric response measurements on colloidal dispersions using the dynamic Stern layer model, *J. Chem. Phys.*, 98, 4183, 1993.

49 Kijlstra, J., van Leeuwen, H.P., and Lyklema, J., Effects of surface conduction on the electrokinetic properties of colloids, *J. Chem. Soc., Faraday Trans.*, 88, 3441, 1992.

50 Shilov, V.N., and Borkovskaya, Y.B., Low-frequency dielectric-dispersion of concentrated suspensions of particles with a thin electrical double-layer, *Colloid J.*, 56, 647, 1994.

51 Delgado, A.V. et al., The effect of the concentration of dispersed particles on the mechanism of low-frequency dielectric dispersion (LFDD) in colloidal suspensions, *Colloids Surf., A: Physicochem. Eng. Aspects*, 140, 139, 1998.

52 Delgado, A.V. and Arroyo, F.J., Electrokinetic phenomena and their experimental determination: an overview, in *Interfacial Electrokinetics and Electrophoresis*, Delgado, A.V., Ed., Marcel Dekker, New York, 2002, chap. 1.

53 Dukhin, S.S., and Shilov, V.N., *Dielectric Phenomena and the Double Layer in Disperse Systems and Polyelectrolytes*, Wiley, New York, 1974.

54 Hunter, R.J., Recent developments in the electroacoustic characterisation of colloidal suspensions and emulsions, *Colloids Surf., A: Physicochem. Eng. Aspects*, 141, 37, 1998.

55 Hunter, R.J. and O'Brien, R.W., Electroacoustics, in *Encyclopedia of Surface and Colloid Science*, Hubbard, A.T., Ed., Marcel Dekker, New York, 2002, p. 1722.

56 Enderby, J.A., On electrical effects due to sound waves in colloidal suspensions, *Proc. R. Soc. London, A*, 207, 329, 1951.

57 Booth, F., and Enderby, J.A., On electrical effects due to sound waves in colloidal suspensions, *Proc. Phys. Soc.*, 65, 321, 1952.

58 O'Brien, R.W., Electro-acoustic effects in a dilute suspension of spherical-particles, *J. Fluid Mech.*, 190, 71, 1988.

59 O'Brien, R.W., The electroacoustic equations for a colloidal suspension, *J. Fluid Mech.*, 212, 81, 1990.

60 Arroyo, F.J. et al., Dynamic mobility of concentrated suspensions. Comparison between different calculations, *Phys. Chem. Chem. Phys.*, 6, 1446, 2004.

61 Delgado, A.V. et al., Dynamic electrophoretic mobility of concentrated suspensions. Comparison between experimental data and theoretical predictions, *Colloids Surf., A: Physicochem. Eng. Aspects*, 267, 95, 2005.

62 Dukhin, A.S. et al., Electroacoustic phenomena in concentrated dispersions: theory, experiment, applications, in *Interfacial Electrokinetics and Electrophoresis*, Delgado, A.V., Ed., Marcel Dekker, New York, 2002, chap. 17.

63 Rider, P.F., and O'Brien, R.W., The dynamic mobility of particles in a non-dilute suspension, *J. Fluid Mech.*, 257, 607, 1993.

64 O'Brien, R.W. et al., Electroacoustic determination of droplet size and zeta potential in concentrated emulsions, *Am. Chem. Soc. Symp. Ser.*, 693, 311, 1998.

65 O'Brien, R.W., Jones, A., and Rowlands, W.N., A new formula for the dynamic mobility in a concentrated colloid, *Colloids Surf., A: Physicochem. Eng. Aspects*, 218, 89, 2003.

66 Hunter, R.J., Measuring zeta potential in concentrated industrial slurries, *Colloids Surf., A: Physicochem. Eng. Aspects*, 195, 205, 2001.

67 Percus, J.K., and Yevick, G.J., Analysis of classical statistical mechanics by means of collective coordinates, *Phys. Rev.*, 110, 1, 1958.

68 Lyklema, J., *Fundamentals of Interface and Colloid Science*, vol. II, Solid–Liquid Interfaces, Academic Press, London, 1995, p. 3.208.

69 Dukhin, A.S., Solid–Liquid and Goetz, P.J., Acoustic spectroscopy for concentrated polydisperse colloids with high density contrast, *Langmuir*, 12, 4987, 1996.

*A.A. Spartakov, A.A. Trusov, A.V. Voitylov,
and V.V. Vojtylov*

CONTENTS

7.1 INTRODUCTION

The object of electro-optics of colloids is the study of the characteristics of colloid particles, the properties of their surfaces, and those of the adjoint double electric layer. Most issues of stability, formation, and evolution of colloid systems can be resolved by electro-optical research. The main problem of electro-optics of colloids is the establishment of the relation of geometrical, electrical, and optical properties of colloid particles and the properties of optical anisotropy of a colloid system. For heterogeneous liquid systems—lyophobic colloids, which are the subject of this chapter—some theories have been proposed that bind the properties of induced anisotropy of a colloid as a whole with the physical and chemical properties of individual colloid particles and with the macroscopic properties (viscosity, dielectric constants, etc.) of the dispersive medium. The "colloid" theories, based essentially on the same ideas as the theories of induced anisotropy in molecular liquids, have certain features that make them in some respects simpler and in others more complex than the "molecular" theories. We shall indicate a set of circumstances on which these features depend.

7.1.1 ROLE OF BROWNIAN MOTION

The influence of external electric fields in molecular liquids on orientation and motion of a molecule is considerably less than that of Brownian motion. For example, for nitrobenzene the limiting degree of orientation that can be reached in an electric field at the limit of breakdown of liquid amounts to about 1%. At the same time the influence of electric field on particle orientation and motion in a colloid can be less, equal, or bigger than that of Brownian motion. For example, the motion of rigid particles with sizes varying from tenths to tens of microns in an electric field with strength of some hundreds of V/cm can be considered as mechanical. Most colloids with such particle sizes almost do not change in such fields. On the contrary, the motion of molecules in electric fields can be described only statistically.

7.1.2 ROLE OF INTERACTION

Considering a colloid particle as small, but macroscopic, homogenous body, we can treat colloid particles in a field as independent bodies. The effect of the medium is taken into consideration through macroscopic parameters (viscosity, dielectric strength, index of refraction, etc.). Thus, from this point of view "colloid" theories have a "quasigaseous" nature.

7.1.3 QUANTITATIVE FACTORS

An important feature of the phenomena of induced anisotropy in colloid systems in comparison with molecular systems is the high magnitude of the anisotropic effect and also the relaxation time. The fact that the magnitude of anisotropy in a colloid solution can, for a given field strength, be many orders of magnitude greater than the anisotropy of a molecular liquid, explains the smaller disorienting role of Brownian motion in systems with large particles. From a practical point of view this proves the possibility of electro-optical phenomena study in colloids in comparatively weak fields (10^1–10^3 V/cm) using relatively long time periods (1–10^{-5} s).

We have indicated several factors known to be favorable to the "colloid theories" in comparison with the "molecular" ones. Now we will consider some complicating factors.

7.1.4 SECONDARY ELECTRICAL PHENOMENA

Molecular liquid anisotropy is caused directly by molecular orientation in the field (deformation of molecules can almost always be completely neglected); moreover, the electric field does not in principle cause any physical or chemical changes in the liquid. In the case of a colloid solution various complications can arise. For example, electrophoretic and electroosmotic motions can be additional factors in orientation that can be either in cooperation or in competition with the direct orientating action of the field. The polarization of electrodes that create the field in the solution can introduce constant or temporary (for alternating fields) alterations of the field. Finally, the coagulating action of the electric current passing through a colloid solution can cause irreversible changes in the system. Thus alternating electric fields are preferable when studying electro-optical properties of colloids.

7.1.5 POLYDISPERSITY AND POLYMORPHISM

In the overwhelming majority of cases, colloid particles of a given solution have various sizes (polydispersity). For nonspherical particles, apart from varying size, there is a variation in form (polymorphism). It is possible in principle to obtain monodisperse spherical particles by means of repeated fractional centrifugation, but the separation of monomorphic particles is an unsolved problem yet. Polydispersity and polymorphism complicate the quantitative comparison of theory with the experiment, making necessary an elaboration of how this comparison should be carried out.

The study of electro-optical anisotropy in colloids is associated with a set of peculiarities that favor and complicate the investigation at the same time. It is only the experiment that is able to show which of the above-mentioned circumstances (and under what conditions) can be neglected in practice and which must be taken into account in the first place. Keeping in mind the above-mentioned circumstances, the electro-optical methods were developed to study electrical, optical, and geometrical properties of colloid particles. Particle orientation in electric field is induced by their anisotropy of electric polarizability or by their permanent dipole moment; therefore it is important to elaborate the methods that would allow their determination from electro-optical research data.

7.2 THEORY

7.2.1 ELECTRO-OPTICAL EFFECT SPECIFICATION IN POLYDISPERSE AND POLYMORPHIC COLLOIDS

One of the fundamental properties of colloids is the distribution of particle parameters. Colloid particles of different sizes and shapes may also differ in some other parameters. The particles in liquid biological disperse systems include densely packed macromolecular structures. Even the widespread colloids of mineral particles have a substantial spread of geometrical, electrical, and other characteristics. The distribution of particles on their parameters always has to be considered when characterizing the particles of a colloid under study with a set of parameters $\{\xi_i\}$, $1 \leq i \leq n$. It was pointed out that the electro-optical study of colloid particle polarizability was contradictory [1] until the distribution function on particle size was introduced. Thus we do not face the problem of the determination of particle parameter set, but the one of the distribution function on these parameters instead.

The determination of the distribution of particles based on their geometrical and electrical parameters is associated with the study of colloids in external force fields. One usually observes the dependence of induced anisotropy on time, amplitude, and other field characteristics that compose the set of external parameters $\{\psi_j\}$, $1 \leq j \leq m$. The development of electro-optical methods, based on the study of optical property anisotropy induced by external electric fields allowed the determination of the distribution functions of colloid particles on sizes, shape parameters, and electrical polarizability anisotropy. Most researchers study colloids with low particle concentration; the methods and results of such studies will be presented in this paper.

In disperse systems with low particle concentration, when the interparticle interaction and second-order light scattering is negligible, the experimentally observed effect A caused by external action is usually defined as the sum of individual particles' contribution to the effect. This fact can be expressed by the equation

$$A(\vec{\psi}) = \int_{\Xi} K(\vec{\psi}, \vec{\xi}) \, \varphi(\vec{\xi}) \, d\vec{\xi} \tag{7.1}$$

where $\vec{\xi}$ is the particle parameter vector that represents the set $\{\xi_i\}$, Ξ is the integration domain that corresponds to the range of parameter values for the system under study, $\varphi(\vec{\xi})$ is the particle distribution function on their parameters, $\vec{\psi}$ stands for the external parameter set $\{\psi_j\}$, $A(\vec{\psi})$ is the generalized dependence of induced anisotropy on time or electric field parameters, $K(\vec{\psi}, \vec{\xi})$ describes $A(\vec{\psi})$ in the case of a colloid system that consists of identical particles characterized by $\vec{\xi}$.

The dependencies $A(\vec{\psi})$ can be measured experimentally for colloids under study; then relation of Equation 7.1 can be considered as a Fredholm first-kind integral equation with respect to the unknown distribution function $\varphi(\vec{\xi})$, where the kernel $K(\vec{\psi}, \vec{\xi})$ should be derived theoretically. The difficulty and the general possibility of the solution of Equation 7.1 considerably depend on the shape of $K(\vec{\psi}, \vec{\xi})$, number of particle n, and external m parameters. Usually the problem can be solved if $n \leq m$. The particle parameter set $\{\xi_i\}$, on which the observed electro-optical effect depends on, is too big, and this makes the derivation of κ a difficult problem if no additional simplifying assumptions are made and if we do not constrain to approximate dependencies in the case of simple-shaped electric fields. To reduce n one can choose appropriate electric field type so that some particle parameters have no influence on particle orientation in the field; conduct the experiment in a such a manner that some particle parameters have no influence on the observed electro-optical effect; derive a particle model that establishes a relation of particle parameters (that reduces the total number of particle parameters in the distribution function $\varphi(\vec{\xi})$).

The existing electro-optical methods allow the measurement of electro-optical effect as a dependence on one parameter ψ_j from the set $\{\psi_j\}$, $1 \leq j \leq m$, whereas other parameters are assumed to be constant. The variable parameter ψ_j usually is time, electric field strength, or frequency. Then the left part of Equation 7.1 $A(\psi_j)$ is a function of one argument and the solution is possible if and only if $n = m = 1$; the right part of 7.1 turns out to be a one-argument integral. Since $K(\vec{\psi}, \vec{\xi})$ is continuous with respect to all arguments and $\varphi(\vec{\xi}) \geq 0$ for all $\vec{\xi} \in \Xi$, then applying the mean-value theorem $n - 1$ times for all ξ_k, $k \neq i$, and considering $a \leq \xi_i \leq b$ we obtain

$$\int_{\Xi} K(\vec{\psi}, \vec{\xi}) \, \varphi(\vec{\xi}) \, d\vec{\xi} = \int_{a_i}^{b_i} K(\vec{\psi}, \vec{\xi}^{(i)}) \, \varphi(\xi_i) \, d\xi_i, \tag{7.2}$$

where

$$\varphi(\xi_i) = \int_{a_1}^{b_1} \cdots \int_{a_n}^{b_n} \varphi(\vec{\xi}) \, d\xi_1 \ldots d\xi_{i-1} d\xi_{i+1} \ldots d\xi_n. \tag{7.3}$$

The components of $\vec{\xi}^{(i)} \in \Xi$ (except ξ_i) are the set of mean values $\{\xi_k^0\}$, $k \neq i$ that in turn depends on components of $\vec{\psi}$.

The distribution function $\varphi(\xi_i)$ can be expressed by a relation

$$\varphi(\xi_i) = \frac{1}{N} \frac{dN}{d\xi_i} , \tag{7.4}$$

where dN is the amount of particles that correspond to the given interval $[\xi_i, \xi_i + d\xi_i]$ whereas other parameters are arbitrary, N is the total amount of particles in a colloid system. In the case of the distribution on one parameter Equation 7.1 reads

$$A(\psi_j) = \int_{a_i}^{b_i} \alpha(\psi_j, \xi_i)\, \varphi(\xi_i)\, d\xi_i. \tag{7.5}$$

The comparative analysis of kernels of Equation 7.1 and Equation 7.5 lets us conclude:

1. The set of particle parameters $\{\xi_k^0\}, k \neq j$ that allows, when averaging, to lower the number of arguments of the distribution function depends on $K(\vec{\psi}, \vec{\xi})$ and $\varphi(\vec{\xi})$. This set may not correspond to any real particles in the colloid system under study. Thus the kernel of Equation 7.5, $\alpha(\psi_j, \xi_i)$, may not correspond to any strict theory of electro-optical effect for identical particles, but depend on the distribution of particles in the colloid system on their parameters instead. There are two exceptions: the first is the case of infinitely narrow intervals $[a_k, b_k], k \neq i$, the second is when $K(\vec{\psi}, \vec{\xi})$ does not depend on ξ_k for $\xi_k \in [a_k, b_k]$.
2. The dependance $\alpha(\psi_j, \vec{\xi}^{(i)})$ is not the result of averaging of $K(\vec{\psi}, \vec{\xi})$ on ξ_k for $k \neq i$, but corresponds to $K(\vec{\psi}, \vec{\xi}^{(i)})$, where $\vec{\xi}^{(i)}$ varies in a narrower domain than Ξ. If light diffraction on particles considerably influences the observed electro-optical effect and $K(\vec{\psi}, \vec{\xi})$ oscillates or changes rapidly with $\vec{\xi}$, then $\alpha(\psi_j, \xi_i)$ may also oscillate or change rapidly. These diffractional oscillations can be diminished by widening and making continuous the spectrum of incident light.
3. The kernel of Equation 7.5, $\alpha(\psi_j, \xi_i)$, cannot be strictly derived from electro-optical theory, since it depends on the unknown distribution function $\varphi(\vec{\xi})$. In order to minimize the difference between $\alpha(\psi_j, \xi_i)$ and the results of a theory built for identical particles one has to choose such effect, field type, and light spectrum that $K(\vec{\psi}, \vec{\xi})$ would be smooth and monotonous as a function of particle parameters.
4. The increase in the number of particle parameters on which the unknown distribution function depends lowers the number of undefined particle parameters making the theoretical description of the kernel more precise.

Even in the case of the determination of the distribution function on one particle parameter (for example, size or rotary diffusion constant), the experimentally obtained dependence of electro-optical effect on one external parameter is not always sufficient. When considering such equations, we will furtheron for the sake of simplicity drop out subindexes of external and particle parameters and rewrite Equation 7.5 as

$$A(\psi) = \int_{a}^{b} \alpha(\psi, \xi)\, \varphi(\xi)\, d\xi. \tag{7.6}$$

This is a Fredholm first-kind integral equation. Its solution is usually unstable, since mathematically the problem is ill-conditioned and additional information about the solution $\varphi(\xi)$ is required. Even in the case of continuous smooth kernels $\alpha(\psi, \xi)$ (defined by physical theory or experiment), one has to develop numerical methods for the determination of $\varphi(\xi)$; these methods will be considered furtheron.

7.2.2 TYPES OF OPTICAL ANISOTROPY INDUCED BY ELECTRIC FIELD

Electro-optical effects in colloids were discovered by Kerr over a century ago [2], but the corresponding techniques were mainly developed by electro-optical scientific schools in the United States, Russia, Bulgaria, and Japan in the second half of the 20th century. The authors of this publication do not aspire to describe the results of electro-optical research in full entirety and will restrict to the description of the methods they took part in the development of. In 1949, Tolstoy and Feofilov, the founders of our scientific group, studied birefringence in pulse fields using the traditional scheme of Kerr cell filled with a colloid and placed between two crossed polarizers. The studied signal amplitude was the modulation of transmitted light intensity induced by the application of electric field. It was observed that in the case of two crossed polarizers, the signal amplitude was much less than in the case of one polarizer and also depended on light polarization direction [3]. The modulations of transmitted light intensity differed for light polarized along and across the electric field; the transparency of the system could increase or decrease. The signal depended on incident light polarization, light spectrum, particle type, and refractive index of dispersive medium. A colloid placed between the polarizer and the analyzer changes light polarization and partially depolarizes it. As the result some light passes through the analyzer even if the colloid in the cell does not induce birefringence. Tolstoy concluded that the optical effect is mostly caused not by birefringence, but by dichroism instead (birefringence is just a side effect). This conclusion defined the set of electro-optical effects we used furtheron.

When studying colloids with particles of different sizes and shapes, it is important that the research can be conducted using light sources with wide continuous spectrum; this is important for the solution of inverse problems of electro-optics, as the light spectrum widening diminishes the differences between strict theoretical dependence $K(\vec{\psi}, \xi)$ in Equation 7.1 and $\alpha(\psi, \xi)$ in Equation 7.6.

The size of the particles in a colloid that provide easily observed effects is comparable to the wavelength of light; this distorts the incident light wave front much more than the small optical inhomogeneities, such as fluctuations of liquid density, macromolecules, etc. Most optical effects in colloids are explained by the peculiarities of light scattering on particles.

For instance, the fact that linearly polarized light turns to elliptically polarized when transmitting through a colloid with field-oriented particles is not connected with the difference in light propagation speed along and across the applied electric field, but with the ellipticity of light waves scattered by particles. The observed electric dichroism in colloids is not connected with light absorption by a colloid, but with the difference in extinction cross section for light polarized along and across the electric field. This difference is mostly defined by the dependence of energy dissipated by colloid particles on particle orientation. Such dichroism is called "conservative."

Systematic studies of scattered light anisotropy in external electric fields started in the 1960s in scientific schools organized by Sheludko and Stoylov in Bulgaria, Jennings and Gerrard in the United Kingdom. The dependence of angular light energy dissipation on external parameters was experimentally studied, paying special attention to the study of induced dipole moment, polarizability anisotropy, and size of particles of biological as well as mineral nature.

7.2.3 LIGHT SCATTERING AND ITS ANISOTROPY INDUCED BY ELECTRIC FIELD

The study of light beam modulations as it transmits through a colloid oriented by electric field is associated with the solution of the problem of light scattering by individual particles of various orientations. Light absorption by colloid particles is impossible without partial scattering. For most colloids and suspensions, extinction light energy is mostly dissipated.

Thus, the most bright and easily observed phenomena are those connected with the peculiarities of light scattering on the particles. Moreover, some phenomena may be masked by the peculiarities of the experimentally observed light scattering. If particles are small, the solution of the light scattering problem is not so complicated, whereas in the case when particle size r is comparable to the incident light wave λ, the problem becomes especially hard to solve, if no simplifying assumptions are made [4].

However, it is the sum energy, including scattered and absorbed, that is important for most optical and electro-optical problems. In this case one deals with extinction cross section $C = C_s + C_a$. Scattering cross section C_s and absorption cross section C_a of a particle are defined as fractions of scattered and absorbed energy related to incident light energy, correspondingly. Both cross sections have the dimension of an area and are proportional to area S of the particle projection on the plane orthogonal to the direction of the propagation of incident light; that lets us write [4]

$$C_{\text{ext}} = S(k_{\text{s}} + k_{\text{a}}). \tag{7.7}$$

In this equation, k_{s} and k_{a} are the effectivity factors of scattering and absorption, respectively. One of the ways to determine the parameters of scattered light is based on the theoretical description of light scattering on an individual particle. It is connected with the solution of Maxwell equations with boundary conditions that reflect the properties of an individual particle. For monochromatic light, this solution can be represented as a superposition of partial waves with electric and magnetic vectors defined by Bessel and Hankel functions of ρ and ρ' and the adjoint Legendre polynomials as functions of scattering angle ϑ. Diffractional parameters $\rho = kr$ and $\rho' = |m|\rho$ are defined by wave vector module k, particle size r, and its complex refraction index m. Factors k_{s} and k_{a}, defined by incident and scattered light rays, depend on the angle ϑ. The problem is completely solved only for a sphere [5]; in this case size r equals its radius; for particles of other shapes, parameters r, ρ, and ρ' depend on shape; this indefiniteness leads to uncertainty in r. For nonspherical particles, partial waves and the associated C_s and C_a depend on particle orientation with respect to the directions of propagation and polarization of the incident light. Only for small particles, when $\rho \ll 1$, shape and orientation have little influence on k_{s} and k_{a}, whereas their dependence on ρ defines the dependence of scattering and absorbed light energy cross sections on particle volume v [6,7]

$$C_{\text{s}} \sim k^4 v^2, \quad C_{\text{a}} \sim k v. \tag{7.8}$$

However, even small particles of anisotropic material oriented by electric field produce observable effects. The dependence of their magnitude a on angle θ between particle orientation axis and field is defined by a well-known relation [8]

$$a = a_0 \frac{3\cos^2\theta - 1}{2}. \tag{7.9}$$

The magnitude of a_0 depends on particle optical properties. If particle orientation along the field is described by axially symmetric distribution function $w(\theta)$, one should use the orientation function

$$\Phi = \int_0^\pi \frac{3\cos^2\theta - 1}{2} w(\theta) \sin\theta \, d\theta. \tag{7.10}$$

Relation 7.9 then reads

$$a = a_0 \, \Phi. \tag{7.11}$$

When studying light scattering anisotropy in an electric field, one usually measures the magnitude

$$a = \frac{J_\mathrm{E} - J}{J}, \tag{7.12}$$

where J_E and J are light intensities scattered in a given angle ϑ for a system oriented by field E and a system with chaotic particle orientation, respectively. Then

$$a_0 \sim q^2, \quad q = \rho \sin(\vartheta/2). \tag{7.13}$$

Relations 7.11 and 7.12 are usually used when studying particle polarizability and rotary diffusion coefficient of monodisperse colloids in a sine-shaped field. The magnitude a of a stationary effect in a rapidly oscillating field of effective strength E is obtained by substituting the orientation function of the following type in 7.11 [8]

$$\Phi(E) = \frac{3}{2} \int\limits_0^1 x^2 e^{\chi x^2} \mathrm{d}x \int\limits_0^1 e^{\chi x^2} \mathrm{d}x - \frac{1}{2}, \tag{7.14}$$

where $\chi = \gamma \, E^2/k_\mathrm{B} \, T$ is the particle polarizability anisotropy. To describe the relaxation effect after terminating the electric field pulse, one should substitute

$$\Phi(t) = \Phi(E) \, \mathrm{e}^{-6Dt} \tag{7.15}$$

in 7.11. Relations 7.9, 7.10, and 7.12 hold true for $q \ll 1$, which is a less strict condition than $\rho \ll 1$. The increase in particle size will lead to an increase in C_s that will be bigger that that for C_a, and also increase ρ and ρ'.

The derivation of a strict theory for stretched or flattened particles with varying $\rho \approx 1$ is almost impossible. The use of light with wide spectrum or the study of colloid systems with a wide range of particle shapes makes statistically averaged parameters of light scattered by particles change weakly and kernels $\alpha(\psi,\xi)$ depend monotonously on volume, rotary and translational diffusion coefficients, and other particle characteristics. In these cases $\alpha(\psi,\xi)$ can be derived with the precision sufficient for the solution of inverse problems in electro-optics.

For big particles with $\rho \gg 1$ the effectivity factors k_s and k_a asymptotically tend to their limits: $k_s \to 1$, $k_a \to 1$, for particles that absorb light and $k_s \to 2$ for particles that scatter it.

7.2.4   ELECTRO-OPTICAL EFFECTS IN KERR CELLS

The intensity of light transmitted through the cell decreases with the increase in energy absorbed and scattered by particles. Transmitted light intensity includes that of forward scattering. The light scattered by a particle is coherent, but has a phase delay with respect to the incident light wave. This phase shift and particle cross section C_{ext} depend on its orientation with respect to the directions of light propagation and incident light polarization.

The natural extinction index is a magnitude convenient for electro-optical research:

$$\mathcal{K} = n \, C_{ext} L, \tag{7.16}$$

where n is the number of particles in volume unit and L is the Kerr cell length. According to the Lambert–Berr law, this index relates to the incident light intensity J_0 and transmitted though a colloid light intensity J as $\mathcal{K} = \ln(J_0/J)$ if the dispersive medium is transparent and if molecular light scattering is not taken into account. When considering a system of monodisperse colloids with particle size proportional to incident monochromatic light wavelength such that ρ and ρ' remain constant, then $\mathcal{K}$ is proportional to the cross section of the particle S. This lets us assume $\mathcal{K}(r) = \mathcal{K}_0 r^2$ for particle fraction of size r for polydisperse colloids, studied in incident light with wide spectrum. One can assume $\mathcal{K}_0$ to be constant for particle fractions with size r varying in a narrow range so that ρ does not change significantly. This assumption is justified if the deviation of other particle parameters that influence scattering considerably from their mean is big enough for the sum of effectivity parameters $k_s + k_a$ to change insignificantly with the variation of ρ.

Conservative dichroism is observed experimentally for colloids with particle size $0.1\lambda < r < 5\lambda$. The admissible range of sizes r and diffraction parameters ρ is much wider when conducting research in white light than in monochromatic light. The conservative dichroism magnitude $\mathcal{N}$ can be defined as

$$\mathcal{N} = (\mathcal{K}_{\parallel} - \mathcal{K}_{\perp})/(\mathcal{K}_{\parallel}^{sat} - \mathcal{K}_{\perp}^{sat}), \tag{7.17}$$

where $\mathcal{K}_{\parallel}$ and $\mathcal{K}_{\perp}$ are the natural indexes of extinction for light polarized along and across the electric field, $\mathcal{K}_{\parallel}^{sat}$ and $\mathcal{K}_{\perp}^{sat}$ are their magnitudes for a system totally oriented along the electric field particles. They are usually easily obtained experimentally, whereas their theoretical derivation is possible only in certain cases.

The peculiarities of electro-optical effects in molecular mediums only partially coincide with those observed in colloids. In the case of electric birefringence in liquids and gases, the refraction indexes n_P, n_S of ordinary, extraordinary rays in anisotropic medium, and n_0 in isotropic medium are related as [9]

$$\frac{n_P - n_0}{n_S - n_0} = -2. \tag{7.18}$$

In the case of electric dichroism in liquids there is a relation analogous to 7.18, which reads [10]

$$\frac{A_{\parallel} - A_0}{A_{\perp} - A_0} = -2,$$

where $A_{\parallel}$ and $A_{\perp}$ are the liquid absorption coefficients for light polarized along and across the field and A_0 is the liquid absorption coefficient in isotropic medium. This relation can be expressed in terms of transmitted light intensities by

$$\frac{\ln(J/J_{\parallel})}{\ln(J/J_{\perp})} = -2, \tag{7.19}$$

where $J_{\parallel}$ and $J_{\perp}$ are the transmitted light intensities for light polarization along and across the electric field, and J is the transmitted light intensity if no field was applied. The left part of

7.18 defines the relation of the change in refraction index for light polarized along the electric field to that polarized across for absorbing mediums with negligible scattering. For scattering systems $\ln(J/J_\parallel) = \mathcal{K}_\parallel - \mathcal{K}$ and $\ln(J/J_\perp) = \mathcal{K}_\perp - \mathcal{K}$.

Relation 7.18 in dispersive colloids cannot be experimentally justified since the notions of ordinary and extraordinary rays cannot be defined in such systems, whereas relation 7.19 does not hold true for most colloids when studying conservative dichroism.

When quantitatively describing the optical anisotropy and its relaxation in molecular mediums induced by electric filed E one can use the orientation functions 7.14 and 7.15. Notice that 7.14 does not depend on permanent dipole moment of molecules. That is true if molecules are not polar, or if the frequency of the orienting field is high enough for the dipole moment to have no influence on particle orientation. The following relations hold true for electric birefringence magnitude Δn [9]:

$$\Delta n\,(E)/\Delta n^{\text{sat}} = \Phi(E), \quad \Delta n(t)/\Delta n^{\text{sat}} = \Phi(t), \tag{7.20}$$

whereas for the dichroism magnitude [10] $\Delta A = (A_\parallel - A_\perp)/(A_\parallel + A_\perp)$

$$\Delta A\,(E)/\Delta A^{\text{sat}} = \Phi(E), \quad \Delta A(t)/\Delta A^{\text{sat}} = \Phi(t). \tag{7.21}$$

The magnitudes of Δn^{sat} and ΔA^{sat} correspond to total particle orientation in the cell. They can be calculated using the theories of optical polarizability and molecular spectrums.

The same orientation functions 7.14 and 7.15 can be used for colloid systems close to monodisperse to describe conservative dichroism [11]:

$$\mathcal{N}(E) = \Phi(E)\,, \ \mathcal{N}(t) = \Phi(t) \tag{7.22}$$

if the field frequency is high enough for the dipole moments of particles to have no influence on their orientation. Formula 7.22 is approximate, and its justification is based on assumptions concerning light diffraction by colloid particles of irregular shape and is more of experimental nature than theoretical. If the left equality in 7.22 holds true, then the right is also true, since they are both valid if 7.9 is true.

The change in light extinction by a particle as it orients in electric field can always be represented with two components. The first changes proportionally to $(3\cos^2\theta - 1)/2$, whereas the second is caused by light wave diffraction by particle not proportional to that factor. If particles were in homogenous electric field of light wave and no diffractional contribution to the natural extinction index was taken into consideration, then the influence of particle orientation on $\mathcal{K}_\parallel - \mathcal{K}$ and $\mathcal{K}_\perp - \mathcal{K}$ could be described by 7.14 and relation 7.19 would hold true. Although the peculiarities of light diffraction on colloid particles considerably influence the natural extinction indexes, one can assume that the contributions of diffraction to $\mathcal{K}_\parallel$ and $\mathcal{K}_\perp$ (measured for white light) are almost the same. If this is true, one can use 7.22 since then $\mathcal{K}_\parallel - \mathcal{K}_\perp$ does not contain the diffractional part and changes proportionally to $\Phi(E)$. Khlebtsov theoretically considered electric dichroism and birefringence in colloids [12] and obtained the dependencies of these effects on field parameters and time that are in no contradiction with this assumption. Relation 7.22 does not strictly correspond to any light scattering theory by identical particles since they describe electro-optical dependencies averaged over all particle parameters, but for γ and D. The applicability of 7.22 to the solution of inverse electro-optical problems was justified experimentally for "almost" monodisperse colloids with different optical properties.

The results of justification of 7.19 and 7.22 for aqueous colloids of graphite and diamond are presented on Figure 7.1 and Figure 7.2. Monodisperse colloids were obtained from

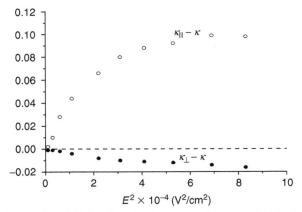

FIGURE 7.1 Results of Equation 7.19 justification for monodisperse colloid of graphite.

polydisperse colloids by centrifugation. The particles of each of the centrifugated colloids had the same sedimentation constant, whereas their sizes, diffusion constants, and polarizabilities varied in narrow ranges. Graphite particles mostly absorb light and almost do not scatter. According to electron microscopic data, they have a shape of thin plates of approximately equal thickness. Diamond particles, unlike graphite, almost do not absorb light, as they are strong scatterers. Diamond aqueous colloid has a slightly stretched and a slightly flattened particles with a variety of shapes. Experimental results, presented in Figure 7.1 and Figure 7.2, are obtained in a sine-shaped electric field with a frequency of 5 kHz. As is seen from Figure 7.1 the magnitudes of $\ln(J/J_{\parallel}) = \mathcal{K}_{\parallel} - \mathcal{K}$ and $\ln(J/J_{\perp}) = \mathcal{K}_{\perp} - \mathcal{K}$ do not agree with 7.19.

This testifies the considerable influence of light diffraction even of heavily absorbing particles on light transmittance. The results of function orientation Φ computation as well as the measured values of $\mathcal{N}$ for monodisperse and polydisperse colloids of diamond and graphite are presented in Figure 7.2. To compare these magnitudes, we plotted them as a function of dimensionless argument $\chi = \gamma E^2/(2k_B T)$ that defines particle orientation. Dependence $\Phi(\chi)$ is computed using Equation 7.14. The value of γ that provides the transition from E to χ and provides best agreement of $\Phi(\chi)$ with experimental dependencies $\mathcal{N}(E)$ was selected using a least-squares fit.

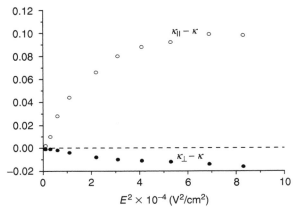

FIGURE 7.2 Results of Equation 7.22 justification. Solid curve — $\Phi(\chi)$, °— the values of $\mathcal{N}$ for monodisperse colloids, •— the values of $\mathcal{N}$ for polydisperse colloids.

In the case of monodisperse graphite, $\gamma = 8.1 \times 10^{-12}$ cm^3 and in the case of polydisperse graphite, $\gamma = 1.3 \times 10^{-12}$ cm^3. In the case of monodisperse diamond $\gamma = 4.6 \times 10^{-13}$ cm^3 and in the case of polydisperse diamond, $\gamma = 1.9 \times 10^{-14}$ cm^3. It is seen from Figure 7.2 that the dependence $\mathcal{N}(E) = \Phi\,(E)$ is justified for monodisperse colloids of diamond and graphite, whereas in the case of polydisperse systems it is impossible to find single γ values that would bring $\Phi(E)$ and $\mathcal{N}(E)$ close together.

Another electro-orientational effect, the anisotropy of electric conductivity induced by external electric fields, can be applied to the study of colloid systems. The theory of this effect is strict and also uses Equation 7.14 and Equation 7.15 to describe the orientational function. The results of the study of γ and D obtained from this technique and the electro-optical research agree [13]. This once again proves the applicability of 7.22 in the electro-optical research of polydisperse systems.

To describe the dependence $\mathcal{N}(E)$ in a polydisperse colloid, one has to use the distribution function $\varphi(\gamma)$; the ability of particles with given γ to influence the conservative dichroism magnitude can be characterized by its weight $\Delta\mathcal{K}_\gamma$. The contribution of completely oriented particles with polarizability anisotropy γ to $\mathcal{N}^{\mathrm{sat}}$ equals $f(\gamma)\mathrm{d}\gamma$ where $f(\gamma)$ is the weighted distribution function

$$f(\gamma) = \Delta\mathcal{K}_\gamma \cdot \varphi(\gamma)\,, \tag{7.23}$$

and $\mathrm{d}\gamma$ is the chosen interval of γ. Considering 7.23 as the definition of $f(\gamma)$, it can be expressed in terms of electro-optical effect $\mathcal{N}^{\mathrm{sat}}$ that corresponds to the saturated particle orientation in polydisperse colloids

$$f(\gamma) = \frac{\mathrm{d}\mathcal{N}^{\mathrm{sat}}}{\mathrm{d}\gamma}\,. \tag{7.24}$$

Since $\mathcal{N}^{\mathrm{sat}} = 1$, then, according to 7.17

$$\int_{\gamma b}^{\gamma a} f(\gamma)\,\mathrm{d}\gamma = 1. \tag{7.25}$$

In the general case, when particles are characterized by parameter ξ

$$f(\xi) = \frac{\mathrm{d}\mathcal{N}^{\mathrm{sat}}}{\mathrm{d}\xi}, \quad \Delta\mathcal{K}_\xi = \frac{\mathcal{K}_\parallel^{\mathrm{sat}}(\xi) - \mathcal{K}_\perp^{\mathrm{sat}}(\xi)}{\mathcal{K}_\parallel^{\mathrm{sat}} - \mathcal{K}_\perp^{\mathrm{sat}}}.$$

The magnitudes of $\mathcal{K}_\parallel^{\mathrm{sat}}(\xi)$ and $\mathcal{K}_\perp^{\mathrm{sat}}(\xi)$ correspond to a colloid with particles with the same parameter ξ, but the distribution of particles on other parameters and their particle number in volume unit are the same as in polydisperse colloid, for which the difference $\mathcal{K}_\parallel^{\mathrm{sat}}(\xi) - \mathcal{K}_\perp^{\mathrm{sat}}(\xi)$ was defined. The weight $\Delta\mathcal{K}_\xi$ quantitatively characterizes the contribution of particles with given ξ to the effect compared to that of other particles in a colloid.

For a diluted colloid with no completely oriented particles $\mathcal{N}(E)$ reads

$$\mathcal{N}(E) = \int_{\gamma b}^{\gamma a} \Phi\,(E, \gamma) f(\gamma)\,\mathrm{d}\gamma. \tag{7.26}$$

In this equation, the orientation function 7.14 is a function of two arguments, external field strength E and particle parameter γ. Hereafter, lower and upper integration bounds will be denoted with subindexes a and b, correspondingly.

When describing the relaxation dependence $\mathcal{N}(t)$ one can use the particle distribution function on rotary diffusion coefficients D with weight $\Delta\mathcal{K}_D$, that has the same meaning as the weight function $\Delta\mathcal{K}_\gamma$:

$$f(D) = \Delta\mathcal{K}_D \cdot \varphi(D). \tag{7.27}$$

If the orienting field is strong enough for all particles in the polydisperse system to be completely oriented, we can assume $\Phi(E) = 1$ and $\Phi(t) = e^{-6Dt}$ for all the particles. Then the relaxation dependence $\mathcal{N}(t)$ reads

$$\mathcal{N}(t) = \int_{D_a}^{D_b} e^{-6Dt} f(D)\, dD. \tag{7.28}$$

In the cases when the electric field does not provide a complete particle orientation, the shape of the relaxation curves in polydisperse colloids depends on field strength and relaxation is described by a more general relation

$$\mathcal{N}(E, t) = \int_{\gamma_a}^{\gamma_b} \int_{D_a}^{D_b} \Phi(E, \gamma)\, e^{-6Dt} f(\gamma, D)\, d\gamma\, d\, D, \tag{7.29}$$

with

$$f(\gamma, D) = \Delta\mathcal{K}_{\gamma D} \cdot \varphi(\gamma, D). \tag{7.30}$$

7.2.5 PARTICLE ORIENTATION IN SINE-SHAPED FIELDS

Electro-optical effects induced by sine-shaped electric fields are usually used to study particle polarization and to determine their shapes and sizes. The simultaneous study of permanent and induced dipole moments in polydisperse colloids is a challenging problem that does not have a uniform solution. For this reason, high-frequency fields are often used to determine γ, since the particle orientation caused by permanent dipole moments is negligible. The minimum frequency that provides orientation with negligible dipole moment contribution is determined experimentally by the absence of effect modulation in a sine-shaped field of constant amplitude. For particles with size greater than $0.1\,\mu m$ in mediums with viscosity close to that of water this minimal frequency varies from 1 to 1.5 kHz. When describing the distribution of particles on angle of orientation θ the potential energy of particles

$$u = \frac{\gamma E^2 \cos^2\theta}{2},$$

defines the angular distribution functions

$$w(\theta) = \exp(-u/k_B T) / \int_0^\pi \exp(-u/k_B T)\, \sin\theta\, d\theta$$

for dielectric equilibrium systems. In electroconductive colloids, when the period of the applied electric field is greater than the relaxation time of the double electric layer, the polarization of this double electric layer has a noticeable contribution to particle orientation. For nonequilibrium electroconductive colloid systems, potential energy u in $w(\theta)$ should be replaced by work W of the moments of all the particle orienting forces. In the frequency range from 1.5 kHz to 0.5 MHz, there is no need to consider concentrational particle polarization connected with the difference in co- and counterion transfer indexes. There is also no influence of electro-osmosis along the particle surface on their orientation. According to the theory of particle orientation in electric field [14]

$$W = \frac{\gamma E^2 \cos^2 \theta}{2},$$

where particle polarizability anisotropy is defined by its volume, specific surface conductivity κ^σ and specific volume electroconductivity of the dispersive medium $\mathcal{K}$. Unlike light scattering characteristics particle parameters γ and D for particles of specified volume are mostly defined by how flat or stretched they are, and vary insignificantly with particle shape. Thus, when describing dependencies $w(\theta)$, $\Phi(E)$, and $\Phi(t)$, particles can be represented with spheroids with semiaxes a and b, where a is the symmetry axis. The shape parameter for such particles is the relation of semiaxes $p = a/b$. Denote $\gamma_\parallel$ as the particle polarizability along a, $\gamma_\perp$ as that along b. If a and b are considerably bigger than the width of the double layer, then in the above-mentioned range of frequencies the particle polarizabilities $\gamma_\parallel$ and $\gamma_\perp$ are defined by [15,16]

$$\gamma_\parallel = \frac{\varepsilon\, ab^2}{3} \left(\frac{3I_\parallel Z}{2b} - 1 \right) \Big/ \left(1 + A_\parallel (\frac{3I_\parallel Z}{2b} - 1) \right), \tag{7.31}$$

$$\gamma_\perp = \frac{\varepsilon\, ab^2}{3} \left(\frac{3I_\perp Z}{2b} - 1 \right) \Big/ \left(1 + \frac{1 - A_\parallel}{2} (\frac{3I_\perp Z}{2b} - 1) \right), \tag{7.32}$$

where $I_\parallel$, $I_\perp$, and $A_\parallel$ depend on particle shape only. For $p > 1$

$$I_\parallel = c(2 - c^2)\, \arcsin\frac{1}{c} + \frac{c^2}{p}, \qquad I_\perp = \frac{p^2 + 1}{p} + \frac{c^2}{2p} - pc^2 + \frac{c^3}{2}\, \arcsin\frac{1}{c},$$

$$A_\parallel = \frac{c^3}{2p}\, \ln\frac{c + 1}{c - 1} - \frac{c^2}{p}, \qquad c = \frac{p}{\sqrt{p^2 - 1}},$$

whereas for $p < 1$

$$I_\parallel = c^3 \left(\frac{1}{p} - \frac{p}{2} \right) - \frac{c^2 p^4}{2}\, \ln\left(\frac{1}{c} + \frac{1}{p} \right), \qquad I_\perp = c^3 \left(\frac{2}{p} - p \right)\, \ln\left(\frac{1}{c} + \frac{1}{p} \right) - c^2,$$

$$A_\parallel = (c^2 + 1) \left(1 - c\, \arctan\frac{1}{c} \right), \qquad c = \frac{p}{\sqrt{1 - p^2}}.$$

The magnitudes of $\gamma = \gamma_\parallel - \gamma_\perp$ as well as the orientation functions $\Phi(E)$ considerably and nonmonotonously depend on the relation of specific surface and volume electroconductivity $Z = \kappa^\sigma / \mathcal{K}$. In colloids with low electroconductivity $\mathcal{K}$, it is possible to use strong electric fields that change particle orientation from weak to saturated and $\mathcal{N}$ from 0 to 1. In colloids with high ion concentration K is high, thus they can only be studied in weak fields, as the application of strong fields heats the system and changes its properties. Let us note that the electro-optical research methods in strong and weak electric fields are still under development.

7.2.5.1 Strong Field

In many colloid systems, particle orientation can vary from weak when the observed effect is proportional to E^2 (the Kerr law), to saturated, when the further increase of field strength does not change the observed effect. For such systems one can study the nonlinear electro-optical effects; the electric fields that are used in such studies are called strong fields.

To study particle polarizability and the influence of double electric layer characteristics one can use relations 7.26 and 7.28. To simplify the task it is expedient to consider one function $f(r)$ instead of two functions $f(\gamma)$ and $f(D)$. Particle parameter r equals the radius of a sphere with rotary diffusion constant D. The dependence $D(r)$ is defined by relation $D(r) = k_{\mathrm{B}}T/(8\pi\eta r^3)$ in dispersive mediums with viscosity η. The dependence of r on spheroid semiaxes a and b can be obtained by expressing D in terms of a and b [17].

For $p > 1$ and $c = \mathrm{p} / \sqrt{p^2 - 1}$

$$D = \frac{3 k_{\mathrm{B}} T}{16\pi\eta ab^2} \left(-\frac{c^2}{p^2 + 1} + \frac{c(c^2 + 1)}{2(p^2 + 1)} \cdot \ln\frac{c + 1}{c - 1} \right),$$

whereas for $p < 1$ and $c = p/\sqrt{1 - p^2}$

$$D = \frac{3 k_{\mathrm{B}} T}{16\pi\eta ab^2} \left(\frac{c^2}{p^2 + 1} + \frac{c^3(1 - p^2)}{p^2(p^2 + 1)} \arctan\frac{1}{c} \right).$$

The dependence of polarizability anisotropy $\gamma(r)$ can be computed for spheroids using 7.31, 7.32, and the interrelation of a, b, and r. Calculations showed that if $p > 1.5$ (prolate ellipsoids) or $p < 0.8$ (oblate ellipsoids) then

1. Product $\gamma \cdot D$ for given Z depends weakly on a and b. For most real systems $\gamma \cdot D$ can be considered constant.
2. The magnitude of γ remains almost constant for given r and Z [11,18].

Such a choice of r allows us not to consider the distribution of particles on shape when studying their polarizability. Substituting $\gamma = \gamma(r)$ in 7.14, we can derive the orientation function as a function of external field strength E, particle size r, and electric surface parameter Z. Let us define this orientation function as $\Phi(E, \gamma(r))$. Equation 7.26 and Equation 7.28 now read

$$\mathcal{N}(E) = \int_{r_a}^{r_b} \Phi(E, \gamma(r)) f(r) \, \mathrm{d}r \tag{7.33}$$

$$\mathcal{N}(t) = \int_{r_a}^{r_b} \mathrm{e}^{-6D(r)t} f(r) \mathrm{d}r. \tag{7.34}$$

These integral dependencies can be used when studying colloid particle properties, if supplemented with experimental dependencies $N(t)$ and $N(E)$ for the colloid under study. If $\mathcal{N}(t)$ is known from the experiment, then 7.34 is a Fredholm first kind integral equation with respect to the unknown function $f(r)$. Using the dependence of $f(r)$ and formulas 7.31, 7.32, the right part of 7.33 allows one to calculate $\mathcal{N}(E)$ for arbitrary Z. Varying Z, one can derive the dependence $\mathcal{N}(E)$ and the corresponding value Z_0 that provide the best agreement with the experimental dependence $\mathcal{N}(E)$ and correspond to the particles in a colloid. If K is measured, then the specific surface electroconductivity of particles is defined by a simple relation $\kappa^\sigma = Z_0 \cdot K$.

The function $f(r)$ allows the determination of $\varphi(r)$ that characterizes the particle distribution on sizes; if the particles are not too stretched, r is close to their real size. Otherwise additional information on particle shape that allows to compute real, geometrical sizes, is required. The determination of $\varphi(r)$ is based on the analysis of weight function $\Delta \mathcal{K}_r$ in

$$f(r) = \Delta \mathcal{K}_r \cdot \varphi(r), \tag{7.35}$$

which, according to the definition of weight functions, can be represented as

$$\Delta \mathcal{K}_r = \frac{\mathcal{K}_\parallel^{sat}(r) - \mathcal{K}_\perp^{sat}(r)}{\mathcal{K}_\parallel^{sat} - \mathcal{K}_\perp^{sat}}.$$

When conducting experiments in white light we can consider the numerator to change proportionally to r^2 (see 7.7), whereas the denominator does not depend on r. As a result

$$\varphi(r) = c \cdot r^{-2} f(r), \quad c \cdot \int_{r_a}^{r_b} r^{-2} f(r) \, dr = 1, \tag{7.36}$$

where c is defined from the normalization of $\varphi(r)$.

7.2.5.2 Weak Field

Weak fields are the electric fields that induce the observed electro-optical effect proportional to E^2.

The research in weak fields is expedient if the colloid system is electroconductive and the application of strong electric fields is impossible. Experimental research of polarizability in polydisperse colloids in weak fields is possible only if the difference $\mathcal{K}_\parallel^{sat} - \mathcal{K}_\perp^{sat}$, required for the derivation of $\mathcal{N}$, is known *a priori*.

When particle orientation is weak, $\gamma E^2 \ll 2k_B T$ and relation 7.14 reads [19]

$$\Phi(E) = \frac{\gamma E^2}{15 \, k_B T} .$$

Substituting this function to 7.26 we derive

$$\beta = \frac{1}{15 \, k_B T} \int_{\gamma_a}^{\gamma_b} \gamma f(\gamma) \, d\gamma \tag{7.37}$$

where $\beta = d\mathcal{N}/dE^2 |_{E=0}$. If the dependence of γ on D or r is known (formulas 7.31 and 7.32 can be used to derive it) then, instead of 7.37, β reads

$$\beta = \frac{1}{15 k_B T} \int\limits_{D_a}^{D_b} \gamma(D) f(D) \, dD, \quad \beta = \frac{1}{15 k_B T} \int\limits_{r_a}^{r_b} \gamma(r) f(r) \, dr. \tag{7.38}$$

Remind that for a sine-shaped field $E = E_0 \sin(\Omega t)$ with frequency Ω, $E = E_0 / \sqrt{2}$, and formula 7.37 is valid if $\gamma E^2 \ll (2 k_B T)$ for all particles in the polydisperse system under study.

Equation 7.37 and Equation 7.38 cannot be used to determine the distribution functions $f(r)$, $f(\gamma)$, and $f(D)$ from the weak-field experimental data, but one can determine $f(D)$ and $f(r)$ if conducting the experiment in weak fields of a special kind described below.

To solve this problem we suggest to use weak fields of the following type:

$$E = E_0 \sin(\Omega t) \sin(\omega t) \tag{7.39}$$

and study the dependence of electro-optical effect N on ω at fixed Ω. Particle polarization, induced by such field, has a spectrum with frequencies $\Omega + \omega$ and $\Omega - \omega$. If $\Omega \gg \omega$, the dispersion of polarization can be neglected and particle orientation is caused by polarizability anisotropy γ on frequency Ω. The analysis of the solution of diffusion equation in field 7.39 shows [20] that the influence of permanent dipole moment μ on their orientation is negligible if

$$\Omega \gg \frac{\mu E_0^2 D}{4\sqrt{2} k_B T} \left(1 + \left(\frac{3D}{\omega} \right)^2 \right)^{-1/4},$$

then the dependence of the orientation functions on frequency ω reads

$$\Phi(\omega, D) = \frac{\gamma E_0^2}{60 k_B T} \left(\left(1 + \left(\frac{\omega}{3D} \right)^2 \right)^{-1/2} \sin\left(2\omega t + \arctan\left(\frac{3D}{\omega} \right) + 1 \right). \tag{7.40}$$

The dispersion relation $\mathcal{N}$ in a polydisperse colloid reads

$$\mathcal{N}(\omega) = \int\limits_{D_a}^{D_b} \Phi(\omega, D) f(D) \, dD. \tag{7.41}$$

After substituting 7.40 in 7.41 $\mathcal{N}(\omega)$ reads

$$\mathcal{N}(\omega) = \mathcal{N}_0 + \mathcal{N}_s(\omega) \sin 2\omega t - \mathcal{N}_c(\omega) \cos 2\omega t, \tag{7.42}$$

where

$$\mathcal{N}_0 = \frac{E_0^2}{60 k_B T} \int\limits_{D_a}^{D_b} \gamma f(D) \, dD, \tag{7.43}$$

$$\mathcal{N}_s(\omega) = \frac{E_0^2}{60 k_B T} \int\limits_{D_a}^{D_b} \gamma D \cdot \frac{\omega}{3D} \left(1 + \frac{\omega^2}{9D^2}\right)^{-1} \frac{f(D)}{D}\, dD, \tag{7.44}$$

$$\mathcal{N}_c(\omega) = \frac{E_0^2}{60 k_B T} \int\limits_{D_a}^{D_b} \gamma D \left(1 + \frac{\omega^2}{9D^2}\right)^{-1} \frac{f(D)}{D}\, dD. \tag{7.45}$$

The dependence 7.43 is analogous to 7.33, but 7.43 depends on the distribution functions $f(D)$ and field amplitude E_0, whereas 7.33 depends on $f(D)$ and effective field strength E. The dependencies 7.44 and 7.45 can be reduced to integral equations with respect to the unknown $f(D)$, if $\gamma \cdot D$ is computed. The results of such computations show that $\gamma \times D$ changes insignificantly when varying particle semiaxes if $a/b > 1.5$ or $a/b < 0.8$ and all particles in a colloid have equal surface conductivity. This lets us assume that γD is constant for most real colloids (this fact should be accounted for when applying 7.44 and 7.45 to describe real colloids).

When determining the degree of polydispersity in systems with high electroconductivity there is no need to determine $K_\parallel^{\text{sat}} - K_\perp^{\text{sat}}$ as we did with γD. Taking into consideration the dependence 7.17, 7.42 reads

$$ln(J_\perp/J_\parallel) = (\mathcal{K}_\parallel^{\text{sat}} - \mathcal{K}_\perp^{\text{sat}}) \cdot (\mathcal{N}_0 + \mathcal{N}_s(\omega) \sin 2\omega t - \mathcal{N}_c(\omega) \cos 2\omega t).$$

In weak fields of type 7.39, it is sufficient to experimentally study the dependence of $ln(J_\perp/J_\parallel)$ on field frequency ω and time. This allows to calculate the dependencies $\alpha_s(\omega) = C_1(\mathcal{K}_\parallel^{\text{sat}} - \mathcal{K}_\perp^{\text{sat}}) \cdot \mathcal{N}_s(\omega)$ and $\alpha_c(\omega) = C_1(\mathcal{K}_\parallel^{\text{sat}} - \mathcal{K}_\perp^{\text{sat}}) \cdot \mathcal{N}_c(\omega)$ for the colloid under study and determine $f_{-1}(D)$, where

$$f_{-1}(D) = \frac{f(D)}{D}, \qquad C_1 = \frac{E_0^2(\mathcal{K}_\parallel^{\text{sat}} - \mathcal{K}_\perp^{\text{sat}})}{60 k_B T} \gamma D. \tag{7.46}$$

Taking into consideration 7.44 and 7.45, $f_{-1}(D)$ can be defined by solving any of the following integral equations

$$\alpha_s(\omega) = \int\limits_{D_a}^{D_b} \frac{\omega}{3D} \left(1 + \frac{\omega^2}{9D^2}\right)^{-1} f_{-1}(D)\, dD, \tag{7.47}$$

$$\alpha_c(\omega) = \int\limits_{D_a}^{D_b} \left(1 + \frac{\omega^2}{9D^2}\right)^{-1} f_{-1}(D)\, dD. \tag{7.48}$$

It is not hard to compute the distribution functions $f(D)$ and $f(r)$ with the unknown multiplier C_1. The normalization of $f(D)$ and $f(r)$ can be used to determine C_1:

$$\int\limits_{D_a}^{D_b} f(D)\, dD = 1, \qquad \int\limits_{r_a}^{r_b} f(r)\, dr = 1. \tag{7.49}$$

7.2.6 EFFECTS IN ROTATING FIELDS

Rotating electric fields were at first used to study the permanent dipole moment μ of colloid particles, but then were applied to the study of polarizability anisotropy γ. The rotating field technique is applicable to the solution of inverse problems in polydisperse systems; the idea to use rotating fields belongs to Tolstoy. When the role of permanent dipole moment in particle orientation is noticeable, low-frequency electric fields can be used to conduct electro-optical research of μ. Special attention should be paid to the condition that electrophoretic particle orientation and electrode polarization should not distort particle orientation. Rotating fields do provide such conditions; there are three techniques of such studies available [21]. The first, R_1-field technique, employs electric fields with strength E, rotating with angular velocity ω. The torque moments of forces acting on a particle in that field are the same as in the case of constant field with the same effective magnitude. The difference is that in constant field the torque moment changes with time, whereas when using a rotating field it does not. The second, R_2-field technique, employs the fields that provide particle rotation due to γ only. This can be achieved making the field vector E, slowly rotating with angular velocity ω, oscillate fast, alternating its direction. The third, R_3-field technique, employs such fields that particles rotate only due to their permanent dipole moments μ. This can be achieved by making the field vector E rotate slowly with angular velocity ω, oscillate fast, and simultaneously change orientation in the rotation plane on $\pi/2$. Then the influence of particle polarization on their orientation disappears (when particles rotate in field plane, the fast field orientation change on $\pi/2$ would change the sign of the torque momentum caused by γ, but will not change its absolute magnitude. Thus the particle will not "be able to follow" the sign change of the torque momentum, and will rotate because of μ only). The description of particle orientation in rotating fields of arbitrary strength takes into consideration the electric moment that rotates the particle, viscous force momentum, and the influence of Brownian motion. This description is a complicated problem that has approximate solutions; the R_3-field technique is in principle inapplicable in the case of weak fields. If the influence of electric field on particle movement is greater than that of Brownian motion, the particle movement becomes mechanical and follows the vector of electric field rotation.

Here we will consider the case when the orientation of permanent particle dipole moment and its higher polarizability axis coincide. This fact is easily checked by comparing $\mathcal{K}_\parallel - \mathcal{K}_\perp$ for the colloid under study, measured in low-frequency rotating fields of various types. According to R_1-field technique, the torque caused by electric field E is caused by two components of the general equation that characterizes stationary particle rotation

$$\frac{1}{2}\gamma E^2 \sin 2\alpha + \mu E \sin \alpha = V[p]\eta\omega. \tag{7.50}$$

In this equation, ω is the frequency of field rotation, α is the lag angle (the angle between μ and E if $\gamma \parallel \mu$), V is the particle volume, $[p]$ is the shape factor [17], and η is the viscosity of the dispersive medium. The viscous friction coefficient $V[p]\eta$ equals k_BT/D.

If a particle orients due to induced dipole moment only, $\gamma E \gg \mu$ (this is often the case of particle orientation in nonpolar mediums), then the angle α depends on ω/E^2:

$$\sin 2\alpha = \frac{2k_BT}{D\gamma} \cdot \frac{\omega}{E^2}. \tag{7.51}$$

If it is only the permanent dipole moment that causes particle orientation, $\gamma E \ll \mu$ (this is sometimes the case of particle orientation in polar mediums), then the angle α depends on ω/E:

$$\sin \alpha = \frac{k_B T}{D\mu} \cdot \frac{\omega}{E}. \tag{7.52}$$

The inverse is also true. If α and ω/E remain constant when E and ω change, then particle orientation is caused by their permanent dipole moment only, and if α does not change when ω/E^2 remains constant, then particle orientation is caused by their induced dipole moment only. When describing R_2-field technique and R_3-field technique, the particle motion is defined by 7.50 with averaged force momentum $<M>$ created by oscillating fields. In the case of R_2-field technique

$$\langle M \rangle = \frac{1}{2} \gamma E^2 \sin 2\alpha,$$

where E is the effective field strength. Particle motion is described by Equation 7.52; in strong fields α does not change if ω/E^2 remains constant. In the case of R_3-field technique

$$\langle M \rangle = \frac{1}{\sqrt{2}} \mu E \sin \alpha,$$

where α is the angle between μ and the bisectrix of the field oscillations angle. Particle motion is described by Equation 7.52; in strong fields α does not change if ω/E remains constant.

Field-oriented particles form an imperfect polaroid that rotates with frequency ω, placed on the optical path of the incident light. As a result transmitted light intensity J_ω is modulated with frequency 2ω and modulation depth ΔJ

$$J(\omega) = J + \Delta J \sin 2\omega t.$$

The angles $\tilde{\alpha}$ between field and the axis of polaroid when transmitted light intensity is the smallest can be measured experimentally. Turning the polarization plane of incident light, one can always find such an angle that $\tilde{\alpha} = \alpha$ for monodisperse colloids. In polydisperse colloids angles α differ for different particles. Assuming they all depend on particle parameter D angle $\tilde{\alpha}$ is defined by

$$\tan 2\tilde{\alpha} = \int_{D_b}^{D_\omega} \sin 2\alpha(D) f(D) \, dD \bigg/ \int_{D_b}^{D_\omega} \cos 2\alpha(D) f(D) \, dD \tag{7.53}$$

where D_ω is the smallest D value among all particles in real polydisperse system that still participate in regular rotation, following the electric field vector E at frequency ω. In R_1-field technique for $\gamma E \ll \mu$ and in R_3-field technique Equation 7.52 is used. For $D = D_\omega \sin \alpha$, then

$$D_\omega = \frac{k_B T}{\mu} \frac{\omega}{E},$$

and the relation 7.53 reads

$$\tan 2\tilde{\alpha} = 2k_{\mathrm{B}}T \cdot \frac{\int\limits_{D_b}^{D_\omega} \sqrt{E^2/\omega^2 - k_{\mathrm{B}}^2 T^2/(\mu D)^2}/.(\mu D) \cdot f(D)\, \mathrm{d}D}{\int\limits_{D_b}^{D_\omega} (E^2/\omega^2 - 2k_{\mathrm{B}}^2 T^2/(\mu D)^2) \cdot f(D)\, \mathrm{d}D}. \tag{7.54}$$

In R_1-field technique for $\gamma E \gg \mu$ and in R_2-field technique, Equation 7.51 is used. For $D = D_\omega \times \sin 2\alpha = 1$, then

$$D_\omega = \frac{2k_{\mathrm{B}}T}{\gamma} \frac{\omega}{E^2},$$

and relation 7.53 reads

$$\tan 2\tilde{\alpha} = 2k_{\mathrm{B}}T \frac{\int\limits_{D_b}^{D_\omega} 1/(\gamma D) \times f(D)\, \mathrm{d}D}{\int\limits_{D_b}^{D_\omega} \sqrt{E^4/\omega^2 - 4k_{\mathrm{B}}^2 T^2/(\gamma D)^2}f(D)\, \mathrm{d}D}. \tag{7.55}$$

If particle orientation is caused by permanent dipole moment only, then according to 7.54 $\tilde{\alpha}$ does not change if E/ω remains constant. But, in order to state that in polydisperse colloids particle orientation is defined by the integration of μ and E only, particle orientation should not change for given E/ω, that is $\tilde{\alpha}$ and light modulation amplitude ΔJ should remain constant. Analogously, if particle orientation is only caused by induced dipole moment, then for given E^2/ω, $\tilde{\alpha}$ and ΔJ remain constant. Using R_2-field technique, when concentrational polarization can be neglected and formulas 7.31 and 7.32 are applicable to describe γ, the product γD remains almost constant for particles of different size and shape in polydisperse colloid. Its magnitude depends on $Z = k^\sigma / \mathcal{K}$ considerably. Assuming $\gamma \cdot D$ constant 7.55 can be simplified as

$$\tan 2\tilde{\alpha} = \frac{2k_{\mathrm{B}}T/(\gamma D)}{\sqrt{E^4/\omega^2 - 4k_{\mathrm{B}}^2 T^2/(\gamma D)^2}} \tag{7.56}$$

and made more suitable for the study of particle surface conductivity in polydisperse colloids, because in this case there is no need to determine particle distribution functions on their parameters.

The relations 7.54, 7.55, and 7.56 can be used to study permanent and induced dipole moments as well as particle surface conductivity, using dependencies of $\tilde{\alpha}$ on E and ω in rotating fields.

7.3 MATHEMATICAL TECHNIQUES FOR THE DETERMINATION OF DISTRIBUTION FUNCTIONS AND NUMERICAL MODELING

7.3.1 PROBLEM SPECIFICATION

Equation 7.26, Equation 7.28, Equation 7.34, Equation 7.44, Equation 7.45, Equation 7.47 and Equation 7.48 contain the weighted distribution functions $f(\xi) = \Delta\mathcal{K}_\xi \cdot \varphi(\xi)$; their left parts can be determined experimentally. As already noticed, these relations can be considered as integral equations of type 7.6 with the difference that $\varphi(\xi)$ is replaced by $f(\xi)$:

$$A(\psi) = \int_a^b \alpha(\psi,\xi) f(\xi)\, d\xi, \quad \psi \in [c,d]. \tag{7.57}$$

Notice that 7.57 can describe the observed effect A only in the case if weight functions ΔK_ξ do not change their sign for all the parameters that influence the observed effect of all particles in a colloid. Otherwise the transition from 7.1 to 7.57 would be impossible. For this condition to hold true one should conduct experiment in white light, or the wavelength of the incident light has to be such that particle diffraction parameter ρ would differ from 1 considerably and the particle cross section C_{ext} would not oscillate as ξ varies.

Expression 7.29 is more complex than 7.57:

$$A(\psi_1, \psi_2) = \int_{a_1}^{b_1} \int_{a_2}^{b_2} \alpha(\psi_1, \psi_2, \xi_1, \xi_2) f(\xi_1, \xi_2)\, d\xi_1\, d\xi_2. \tag{7.58}$$

There exists yet another problem with the formulation of the mathematical solution of 7.1, 7.6, 7.57, and 7.58. It is associated with the definition of integration bounds for these equations. The solution of these equations becomes more precise with the increase in the precision of integration bounds determination (the bounds where the unknown distribution functions are not zero): Ξ in the case of 7.1 and $[a,b]$ in the cases of 7.6 and 7.57. So, the integration bounds determination is in part the mathematical problem statement. They can be determined from some additional information about the solution or the analysis of left parts of equations. For example, for Equation 7.26, Equation 7.28, Equation 7.34, Equation 7.44, and Equation 7.45, where γ, D, and r stand for ξ and E, t and ω stand for ψ, the integration bounds can be approximately determined analyzing experimental dependencies $N(E)$, $N(t)$, $\alpha_s(\omega)$ and $\alpha_c(\omega)$, by assuming that for big and small ψ values the change in these dependencies corresponds to ξ close to a and b. The precise values for a and b can be determined with iterative method, that is, having found $f(\xi)$ for some a and b, compute $A(\psi)$, and using (7.57) to compare it with the obtained experimental dependence $A(\psi)$, refine a and b.

Equation 7.6 and Equation 7.57 are Fredholm first kind integral equations; their solution belongs to a class of ill-conditioned problems [22]. Even small perturbations in the experimentally obtained dependence $A(\psi)$ may cause considerable errors in the recovered distribution function. The choice of solution method depends on kernel-type $\alpha(\psi,\xi)$ and supplementary additional information on function $\varphi(\xi)$ or $f(\xi)$. There are several ways to use additional information that makes the specification of problems 7.6 and 7.57 mathematically correct and functions $\varphi(\xi)$ and $f(\xi)$ weakly dependent on the experimental error in $A(\psi)$ and small inexactness in $\alpha(\psi,\xi)$.

The easiest and most disseminated way is the representation of the unknown distribution function with an analytic dependence with two to three unknown parameters. Usually this is how the distribution function on size $\varphi(r)$ is determined; it is usually sought as a log-normal distribution function [23]

$$\Phi(r) = e^{-(\ln r - \ln r_0)2/\sigma^2},$$

if the particles were obtained by fine crushing of a substance. If particles are aggregated, consisting of smaller particles, then the normal distribution function for $\varphi(r)$ would work better:

$$\varphi(r) = e^{-(r-r_0)^2/\sigma^2}.$$

Parameters r_0 and σ are determined when combining the experimental dependencies $A(\psi)$ with the analytical $\varphi(r)$. Such a method is also quite applicable for other analytical functions $\varphi(r)$, though *a priori* it is hard to predict the shape of the sought distribution function and find some suitable analytical representation for $\varphi(r)$.

There is another additional information about the sought distribution function: that is the continuosity of the solution with its derivatives on $[a,b]$. Perturbations in $A(\psi)$ should also be taken into consideration. In this case the regularization method [24] can be used to solve 7.6 and 7.57. Since these equations and the additional conditions are similar, let us restrict to considering Equation 7.57. According to the theory of regularization method the solution $f(\xi)$ of Equation 7.57 depends on undefined parameter, which is called the regularization parameter. It is determined using methods of conjugate gradients, various residues, and other techniques that employ some peculiarities of the solution. There are numerical algorithms based on these regularization techniques by Tikhonov [25] developed to solve Equation 7.26, Equation 7.28, Equation 7.34, Equation 7.44 and Equation 7.45 quite reliably [26]. These variants of this method are universal, but have some shortages, for example the choice of regularization parameter is badly substantiated and not so effective. Some inexactness in the choice of interval $[a,b]$ considerably reduces the precision of the solution; even small systematic errors in $A(\psi)$ have great influence on the solution.

Equation 7.6 and Equation 7.57 can be modified by changing the kernels and the left parts. Various techniques, such as scaling [22], differentiation, integration with respect to ψ [27] can be used. These procedures can smooth or mark out the peculiarities in the left part of the original equation and its solution. Kernels of 7.26, 7.28, and 7.34 depend on the product of their arguments $\alpha(\psi, \xi) = \alpha(\psi \cdot \xi)$, and those of 7.44 and 7.45 on the ratio of their arguments $\alpha(\psi, \xi) = \alpha(\psi / \xi)$. This peculiarity allows the derivation of integral equations with respect to moments of derivatives of the distribution function $\xi f(\xi)'$ [27]. Some modified equations can be solved more precisely with the regularization method than the original Equation 7.57.

A more general method than the above-mentioned regularization technique has been elaborated: the method of penalty functions. The main concepts of this method belong to Babadzanjanz [28]. According to this method the solution is sought in the class of polynomial functions that satisfy the original equation, modified equations, and other peculiarities of the solution in the least square sense. Each equation or other additional condition is assigned to some penalty function, which is positive valued and decreases as the condition is better fulfilled. The solution of the problem corresponds to the minimum of the objective functional that consists of these penalty functions. It was justified that in the inverse electro-optical problems the penalty functions method is preferable: the precision is higher than that of the regularization method, there is no undefined parameter, the initial determination of interval $[a,b]$ may be imprecise; this method is also applicable in the case of two-dimensional integral equation 7.58.

7.3.2 NUMERICAL ANALYSIS

The described numerical techniques of the solution of Fredholm first kind integral equations were implemented in Fortran 90. Computer modeling of experimental data can provide good estimation of applicability and reliability of a method, since it is hard to estimate the shape of the unknown distribution function based on experimental data; computer modeling also lets us estimate the possibility to resolve several peaks in the unknown distribution function. It is also important to estimate the influence of natural distortions in A on the recovered distribution function f.

The effectiveness of application of regularization technique to the solution of 7.26, 7.28, 7.34, 7.44, and 7.45 was considered earlier [20,26,30]. Here we only provide the comparison of two implementations of the regularization technique, a standard program [29] PTIPR with the choice of regularization parameter by the method of conjugate gradients and a ZERT program by Zernova T.Yu. that was developed for the processing of electro-optical data, and a penalty function method-based program ICESolver.

The comparison was carried out by comparing the solutions of Equation 7.28 and Equation 7.46 obtained by different programs. To compare the multiple peaks resolution we picked one- and two-peaked functions as model. To compare the sensitivity of programs on the precision of the determination of integration bounds $[a,b]$ they were chosen to be at least 30% off the real function variability domain. The comparison of techniques of $f(\xi)$ recovery was carried out by the following scheme:

1. We used the model functions $f(\xi) = \sum_{k=1}^{l} e^{-(a_k\xi + b_k)^2}$, $l = 1,2$ to compute $A(\psi)$.

2. For various values of ψ, the "experimental data" obtained according to Equation 7.28 and Equation 7.47 was corrupted with 5% multiplicative white noise ε in some numerical experiments. We considered kernels $K_{rel} = e^{-6\psi y}$ and $K_\omega = (\psi/3\xi)/(1 + (\psi/3\xi)^2)$.

3. The three codes mentioned above were used to solve Fredholm equation of the first kind to recover $f(\xi)$ from $A(\psi)$. The obtained $f(\xi)$ was compared to the model function.

The results of $f(\xi)$ recovery are presented in Figure 7.3 and Figure 7.4.

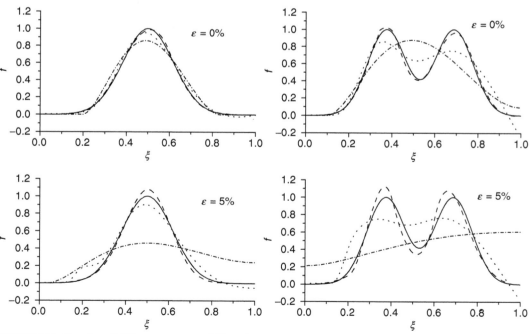

FIGURE 7.3 Results of $f(\xi)$ recovery modelling for Equation 7.28. Solid curve is the model function, dotted curve is the solution obtained by ZERT, dash-dotted curve is obtained by PTIPR, dashed curve is obtained by ICESolver.

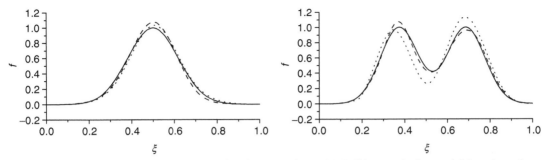

FIGURE 7.4 Results of $f(\xi)$ recovery modeling for Equation 7.47. Solid curve is the model function, the solutions obtained by `ICESolver` with $\varepsilon = 0\%$ (dotted curve), $\varepsilon = 5\%$ (dashed curve).

It is seen that if integration bounds are not determined precisely the regularization techniques solve these equations with low precision even in the case of one-peaked $f(\xi)$. In the case of two peaks the regularization techniques are unsuitable, whereas the penalty functions method recovers $f(\xi)$ quite reliably. It is worth noticing that the precision of the regularization techniques improves greatly if the integration bounds $[a,b]$ are defined precisely and the regularization parameter is tweaked accordingly.

The modeling of two-dimensional equation solution was carried out in the same way for Equation 7.58 with the kernel $\alpha\,(\psi_1\,\psi_2,\,\xi_1,\,\xi_2) = \exp\,(-6\psi_1\,\xi_1)\,\Phi(\psi_2\,\xi_2)$ that corresponds to Equation 7.29. The model distribution function

$$f(\xi_1,\xi_2) = \sum_i a_i \exp\left(-\frac{b_i(\xi_1 - c_i)^2}{c_i^2 - (\xi_1 - c_i)^2} - \frac{d_i(\xi_2 - f_i)^2}{f_i^2 - (\xi_2 - f_i)^2}\right).$$

was recovered from the modeled experimental data alternated with 1% white noise error. The result is presented in Figure 7.5.

The comparison of these dependencies proves the applicability of penalty functions technique to the determination of two-dimensional distribution functions $f(\xi_1, \xi_2)$. This gives us the hope that the further development of electro-optics of polydisperse colloids and the methods of solution of Fredholm first kind integral equations will allow a precise correlational analysis of particle parameters.

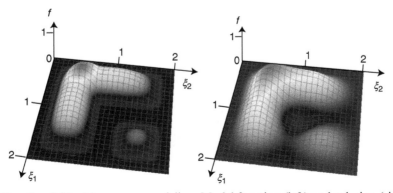

FIGURE 7.5 Results of $f(\xi_1, \xi_2)$ recovery modeling. Model function (left) and solution (right).

7.4 RESULTS

Now we will consider the results of research of geometrical and electrical characteristics of several water polydisperse colloids that can demonstrate the peculiarities and possibilities of electro-optics of polydisperse colloids. Here we will consider conservative dichroism as the electro-optical effect $\mathcal{N}$; the research was conducted in white light.

7.4.1 PARTICLE SIZE AND DISTRIBUTION OF PARTICLES

When studying geometrical characteristics of real particles in colloid systems with electro-optical methods one usually uses transition processes induced by electric fields. Transition time depends on particle rotary diffusion constants D. Equation 7.28 and Equation 7.34, considered earlier, contain the dependence $\mathcal{N}(t)$ that describes electro-optical effect relaxation from the state of saturated orientation of all particles in a colloid. Such relaxational dependence can be obtained experimentally when studying colloids with low electric conductivity that do not change their properties in strong electric fields. The distribution function $f_{-1}(D)$ can be determined experimentally using weak electric fields by studying electro-optical effect dispersion in sine-shaped fields of type 7.39. It is worth noticing that such measurements can be made for colloids with low and high electric conductivities and also colloids that do not change in weak electric fields only. The function $f_{-1}(D)$ is the solution of Equation 7.47 and Equation 7.48 that contains experimental data that describe electro-optical effect dispersion $\mathcal{N}$. The distribution functions $f(D)$, $f_{-1}(D)$, and $f(r)$ are related and we will use only the function $f(r)$ to compare the results of particle size study obtained by different methods. (Note that transition from D to r can be accomplished using formula $D = kT/8\pi\eta r^3$.)

Let us compare the distribution functions $f(r)$ of aqueous colloid of palygorskite obtained by studying effect $\mathcal{N}$ in weak and strong electric fields. Palygorskite particles are needle-shaped particles of clay with length 20–25 times bigger than width. The interparticle distance in the prepared colloid was five times bigger than their length; electroconductivity K of the colloid equaled 1.9×10^{-4} Ohm^{-1} cm^{-1}. The research showed that sine-shaped field pulses with frequency equal 5 kHz, when $E = 500$ V/cm, ranging from tenth of a second to seconds applied to the colloid of palygorskite bring the system to the state of saturated particle orientation whereas the colloid itself does not change in such an electric field. That field was used to determine the relaxation curve $\mathcal{N}(t)$. The study in weak field of type 7.39 was conducted with $\Omega = 5$ kHz (ω varied in the experiment) and $E_0 = 75$ V/cm. Functions $f(r)$ were determined by solving Equation 7.34 and Equation 7.47. They are presented in Figure 7.6 [20].

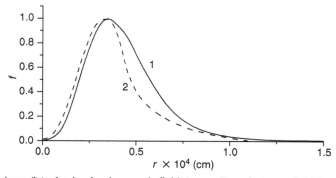

FIGURE 7.6 Functions $f(r)$ obtained using weak-field (curve 1) and strong-field (curve 2) techniques.

As is seen from this figure the obtained dependencies $f(r)$ from these equations for palygorskite particles are close; this is also the case for a whole set of other water colloids that we studied (diamond, graphite, anisaldazine, etc.). The distribution functions, $f(\xi)$, recovered from electro-optical data, contain weight $\Delta\mathcal{K}_\xi$; to compare these results with the results obtained by other techniques one has to determine $\varphi(\xi)$. Relation 7.36 allows this transition from $f(r)$ to $\varphi(r)$. To demonstrate the possibilities of electro-optical methods in the determination of these distribution functions, we picked a suspension of bacteria cells *Pseudomonas fluorescens*. First, the cell size was defined using a biological microscope. The cell length varied from 1.5 to 5.0 μm, the width from 0.5 to 1.5 μm. Values of r were computed for all the cells and the corresponding histogram was built. Before conducting the electro-optical research, the suspension was diluted with distilled water. In the field of type 7.39 with $\Omega = 5\,\text{kHz}$ and $E_0 = 45\,\text{V/cm}$ the following dependence was obtained [20]

$$\mathcal{N}(\omega) = \mathcal{N}_0\sin(2\omega t + \Delta). \tag{7.59}$$

The experimentally obtained dependencies $\mathcal{N}_0(\omega)$, $\Delta(\omega)$, and the computed $\mathcal{N}_s(\omega)$ that occur in Equation 7.44 are presented in Figure 7.7 (remember that the left parts of Equation 7.44 and Equation 7.47 are similar). After the solution of 7.47 functions $f_{-1}(D)$, $f(r)$, and $\varphi(r)$ were determined. Dependencies $f(r)$, $\varphi(r)$, and the histogram are presented in Figure 7.7.

As is seen from the figure the electro-optical technique reflects the most probable cell sizes correctly, but the histogram is narrower than the distribution function $\varphi(r)$. The difference between the histogram and $\varphi(r)$ can be explained by big error in the experimental data (~ 10%) for this suspension, since bacteria are too big for electro-optical research: the effect of N is small, and its dispersion can be observed only in small-frequency range ($0.01 < \omega/2\pi < 1.0\,\text{Hz}$). If the research of such systems like *Ps. fluorescens* was conducted using statistical data accumulation (unlike direct measurement of $\mathcal{N}$ and Δ, which was the case for the experimental data presented in Figure 7.7) in weaker fields, then the electro-optical technique would give more precise results.

7.4.2　ANISOTROPY OF PARTICLE ELECTRICAL POLARIZABILITY

In this section, we will restrict to the study of polarizability anisotropy γ of colloid particles at frequencies greater than the upper limit of their polarizability α-dispersion. At these

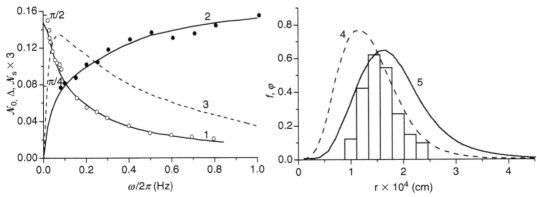

FIGURE 7.7 The results research of size distribution for bacteria *Ps.fluorescens*. The dependencies $\mathcal{N}_0(\omega)$–1, $\Delta(\omega)$–2, $\mathcal{N}_s(\omega)$–3, $j\phi(r)$–4, $f(r)$–5.

frequencies, as shown experimentally, a permanent dipole moment influencing their orientation can be neglected, making the study of γ less complicated than in the low-frequency range. Since particle polarizability and its anisotropy γ considerably depend on particle size, polydisperse colloids with a size distribution also have a distribution of γ that has to be considered. A direct technique of the determination of this distribution is the study of particle orientation in sine-shaped electric fields of radiotechnical frequency and varying amplitude. The study of the electro-optical effect dependence on applied electric field amplitude allows the determination of this distribution without considering any model for particle polarizability. Measuring the dependence $\mathcal{N}(E)$ and using Equation 7.26 one can determine the distribution function $f(\gamma)$. This dependence can be determined without considering any model for particle polarizability. The transition from $f(\gamma)$ to $\varphi(\gamma)$ can be accomplished if weight function $\Delta\mathcal{K}_\gamma$ for the particle polarizability anisotropy in the colloid under study is known.

The magnitude of $\mathcal{K}$ and its alterations do not depend on the wavelength of light λ [11]. This lets us conclude that $\Delta\mathcal{K}_\gamma$ is proportional to particle cross section S when the particles are oriented along the electric field.

Let us consider the transition from $f(\gamma)$ to $\varphi(\gamma)$ for particles of an aqueous colloid of graphite. Graphite particles are flat, thin, and all have almost equal width; thus their cross section S and $\Delta\mathcal{K}_\gamma$ are proportional to their volume V. The electrical structure of these particles is complex: they are electroconductive along their flat surface and do not conduct in perpendicular direction. Formula 7.33 is not suitable for the description of such particle polarizability. The components of particle polarizability tensor are proportional to particle volume V; that means $\gamma \sim V$. This allowed us to assume that for particles with properties like those just described $\Delta\mathcal{K}_\gamma \sim \gamma$. Thus we do not face any difficulties with the transition from $f(\gamma)$ to $\varphi(r)$ ($f(\gamma) \sim \gamma\phi(\gamma)$). Dependencies $\mathcal{N}(E^2)$, $f(\gamma)$, and $\varphi(r)$ for an aqueous polydisperse colloid of graphite [30] are presented in Figure 7.8.

For the same colloid we studied the relaxation of effect of $\mathcal{N}$ induced by fields of various amplitudes. This allowed us to use Equation 7.29 to determine the distribution function $f(\gamma,D)$ for aqueous colloid of graphite. The experimental dependence $\mathcal{N}(E,t)$ and the recovered distribution function $f(\gamma,D)$ are presented in Figure 7.9.

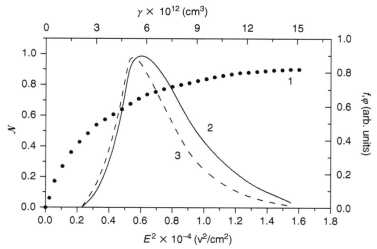

FIGURE 7.8 Graphite particle distribution functions on polarizability anisotropy in water. Curve 1: experimentally obtained dependency $\mathcal{N}(E^2)$; curve 2: calculated dependencies from this experimental data $f(\gamma)$; curve 3: $j(\gamma)$.

Electro-optical research of the distribution of colloid particles on two parameters was conducted for the first time. The presented function $f(\gamma,D)$ agrees with that determined for this colloid distribution functions of one argument $f(\gamma)$ and $f(D)$. As is seen from the plot the relation of γ and D for most particles is close to dependencies $\gamma \sim 1/D$, but there are some deviations. The range of polarizability anisotropy is considerably lower for big rotary diffusion constants D than for small constants. It is possible that on an equal footing with big particles, there are aggregates of small particles with D close to that of big particles. Since aggregates do not have any particular structure, they may have a considerably wider range of γ than particles of the same size.

The comparison of experimental data with theory that describe real nonspherical particles polarization is of great interest since it will help us determine the bounds of applicability of this theory and let us better determine the possibilities of electro-optical method when studying electrical characteristics of particles.

The polarization theory of spheroids can be used when describing the dependence of electro-optical effect magnitude on the amplitude of field applied to polydisperse colloid, if the distribution function $f(r)$ is known. A given r corresponds to only one D, but can correspond to a number of volume and shape combinations, if we do not restrict to the geometry of a spheroid or define p. Using electronic microscope one can determine the bounds of p for the colloid particles under study and the value of p that best fits the theory with the experimental data. Magneto optical method can also be used to determine the distribution of particles on p[31]; this distribution can be used in electro-optical effect calculations. As we noticed, for particles with given r, γ depends weakly on p if its value is greater than 1.5 for prolate or less than 0.8 for oblate particles.

Relation 7.33 lets us calculate $\mathcal{N}(E^2)$ if $f(r)$ is given and use formulas 7.31 and 7.32 to describe particle polarizability. To determine $f(r)$ for real colloids, one can use the experimentally obtained dependence $\mathcal{N}(t)$ and Equation 7.34. The computed and experimentally obtained dependencies $\mathcal{N}(E^2)$ can be compared to check the applicability of the model for particle polarization for real particles. Dependence $\mathcal{N}(E^2)$ computed using 7.34 depends on functional dependence $\gamma = \gamma(r)$ that can be obtained using formulas 7.31 and 7.32. This dependence will be a function of $Z = \kappa^\sigma / \mathcal{K}$; this parameter usually is the same for all particles in the colloid under study. Magnitudes Z and the shape of the function $f(r)$ considerably influence the theoretical dependence $\mathcal{N}(E^2)$, whereas p considerably influences it only when p is close to 1.

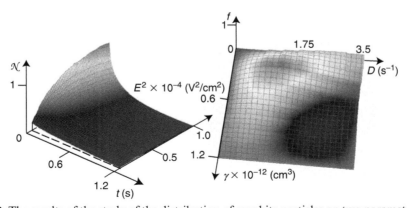

FIGURE 7.9 The results of the study of the distribution of graphite particles on two parameters γ and D.

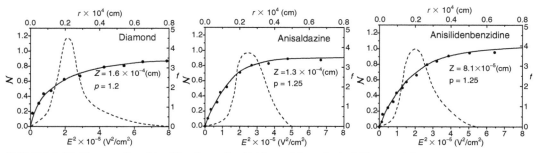

FIGURE 7.10 The results of justification of colloid particle polarizability model.

The comparison of theoretical and experimental dependencies $\mathcal{N}(E^2)$ was conducted for aqueous colloids of diamond, anisaldazine, anisilidenbenzidine, benzopurpurine, palygorskite, and some others. Figure 7.10 shows the experimentally obtained and theoretically computed dependencies $\mathcal{N}(E^2)$ for aqueous colloids of diamond, anisaldazine, and anisilidenbenzidine [11].

Here the experimentally obtained functions $f(r)$ that are required for computations are also presented. The particles of these colloid systems are slightly prolate, thus p influences the shape of the computed curves $\mathcal{N}(E^2)$. Values of Z and p that provide the best correspondence of theoretical and experimental dependencies $\mathcal{N}(E^2)$ were determined using least squares method. Z equals 1.6×10^{-4}, 1.3×10^{-4}, and 8.2×10^{-5} cm for colloid particles of diamond, anisaldazine, and anisilidenbenzidine in water, respectively. Parameter p influences $\mathcal{N}(E^2)$ considerably weaker than Z. The best-fit estimates for p are 1.2 for diamond particles and 1.25 for anisaldazine and anisilidenbenzidine. The obtained values correspond to the electromicroscopic data. For these Z and p the maximum deviation of experimental dependence $\mathcal{N}(E^2)$ from the corresponding theoretical value equaled 2, 5, and 6% for colloids of diamond, anisaldazine, and anisilidenbenzidine, respectively. This deviation is within the experimental error bounds. Such an agreement of theoretical and experimental dependencies testifies the applicability of the polarization model of spheroid with surface conductivity for qualitative and quantitative descriptions of polarization of real particles with irregular shape and to the applicability of formulas 7.31 and 7.32 in electro-optical research of colloids. The electroconductivity magnitudes of dispersive mediums for these colloids equaled several MOhm^{-1} cm^{-1}, magnitudes of κ^σ, computed from the determined Z and K equaled several tens of GOhm^{-1}. The obtained values of Z and p let us compute γ using formulas 7.31 and 7.32 for spheroids of given size r. Table 7.1 presents the corresponding results.

TABLE 7.1
The Values of $\gamma \times 10^{15}$cm^3 for Particles of Various Sizes

$r \times 10^5$ cm of Particles	Particles in Water		
	Diamond	Anisaldazine	Anisilidenbenzidine
0.5	2.7	2.6	2.1
2.5	260	245	165
5.0	1600	1430	895

Instead of analyzing dependence $\mathcal{N}(E^2)$ to study particle polarizability one can analyze dependence $\tilde{\alpha}(\omega)$. Equation 7.31, Equation 7.32, Equation 7.55, and functions $f(r)$ or $f(D)$ let us compute $\alpha(\omega)$ for aqueous colloids of diamond, anisaldazine, and anisilidenbenzidine (relation 7.56 that eliminates the need for taking into account polydispersity and polymorphism of colloids is inapplicable to colloids of diamond, anisaldazine, and anisilidenbenzidine since the particles in these colloids are not prolate enough). Experimental dependencies $\alpha(\omega)$ for these colloids were determined using R_2-field technique. For the magnitudes of Z and p, determined from $\mathcal{N}(E^2)$, there is an agreement of experimental and theoretical dependencies $\alpha(\omega)$ for all three colloids with a precision within the experimental error bounds.

The study of aqueous colloid of benzopurpurine with prolate particles ($p \approx 2.5$) and palygorskite with needle-like particles ($p \approx 25$) showed that dependencies $\mathcal{N}(E^2)$ and $\alpha(\omega)$ change weakly when varying p in the case of benzopurpurine and almost do not change for palygorskite. The obtained magnitudes of Z from the experimental data equal 1.0×10^{-4} and 2.1×10^{-4} cm for benzopurpurine and palygorskite in water, respectively [32,33].

The next point of the theory proposed by Shilov that has to be tested experimentally is whether the minimum of the theoretical dependence $\gamma(Z)$ reveals itself in the experiment. The value of Z decreases with the increase of electrolyte concentration in a colloid. Thus the answer can be obtained when comparing the theoretical and experimentally measured values of $\beta = d\mathcal{N}/dE^2|_{E=0}$. According to 7.31 and 7.32, polarizability anisotropy γ changes considerably with Z, which changes with ion concentration in the dispersive medium. The increase in ion concentration slightly changes κ^σ and considerably increases K. As is seen from 7.37, β changes proportionally to γ of the particles. It is evident that when changing ion concentration of a colloid we will also change κ^σ and γ; this means β also reflects the changes of γ. The increase of electrolyte amount in the solution lowers the electric charge and polarizability of particle, which may induce the formation of particle aggregates and make the study of ion influence on particle polarizability anisotropy more difficult.

Relation 7.38 lets us conclude that alternations in $f(D)$ and $f(r)$ can considerably change β. Thus the comparison of β is possible for systems with different electrolyte concentrations only if their distribution functions $f(r)$ are equal. Even pair aggregates orient in weak fields and influence the effect $\mathcal{N}$ (and β) considerably stronger than individual particles. The absence of aggregates can be controlled by studying $\mathcal{N}(t)$. Pair aggregates have D six to eight times smaller than real particles. For this reason even the formation of a small amount of aggregates will change $\mathcal{N}(t)$. If electrolyte addition lowered γ (and thus increased K) in such a way that it is impossible to obtain full particle orientation because of colloid heating in electric field, then the experimentally controlled dependence $\mathcal{N}(t)$ would change even if there were no aggregates formed. In this case weak fields of type 7.39 should be used, and the effect should be analyzed using Equation 7.59. When particle polarization changes in the absence of aggregates in a colloid function, $f(D)$ does not change and as is seen from 7.40, 7.41, and 7.59 the dependencies $\mathcal{N}_0(\omega)$ will be similar for different electrolyte concentrations, whereas $\Delta(\omega)$ will remain unchanged.

We conducted the study of aqueous colloid of diamond with electrolyte additives of $Th(NO_3)_4$ and $AlBr_3$ and aqueous colloid of palygorskite with electrolyte additives of $NaCl$ [34]. The number of particles per volume unit was low (it equaled 1×10^6 cm^{-3} for diamond particles and 1×10^7 cm^{-3} for palygorskite particles), and electrolyte additives did not induce any aggregate formation. An exception was the isoelectric point, where the ultrasound-cured colloids could be studied as systems with isolated particles and no aggregates. The frequency of the electric field equaled 1.5 kHz; electroconductivity was studied at the same frequency.

The influence of electrolyte concentration c for $Th(NO_3)_4$ and $AlBr_3$ on β is presented on Figure 7.11 for an aqueous colloid of diamond. The same figure shows dependence $\beta(Z)$, that was calculated using formula 7.31, formula 7.32, and formula 7.38 (it is required for the calculation of the distribution function $f(r)$).

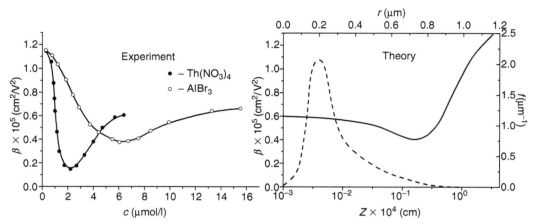

FIGURE 7.11 The dependencies of β on electrolyte concentration c for aqueous colloid of diamond in $Th(NO_3)_4$ and $AlBr_3$ electrolytes (left) and on Z (right). Distribution function $f(r)$ for aqueous colloid of diamond (right, dashed curve).

Dependencies $\beta(c)$ have a minimum, their nonmonotonous behavior can be explained in terms of the theory of polarization of a spheroid with surface conductivity. The increase in electrolyte concentration, as it was already noticed, increases K and decreases Z; moreover, Z decreases with the increase in counterion valence. For diamond colloid Z should be smaller than 1.6×10^{-4} cm^{-1} with the addition of electrolyte, while K should be greater than $1.0 \cdot \times 10^{-6}$ Ohm^{-1} cm^{-1}. Dependence $\beta(Z)$ also has a minimum that corresponds to $Z = 1.5 \times 10^{-5}$ cm. The minimums of dependencies $\beta(c)$ correspond to $c = 2.0 \times 10^{-6}$ mol/l, $K = 5.9 \times 10^{-6}$ Ohm^{-1} cm^{-1} for Th(NO$_3$)$_4$ and $c = 6.1 \times 10^{-6}$ mol/l, $K = 5.9 \times 10^{-6}$ Ohm^{-1} cm^{-1} for AlBr$_3$. Comparing $\beta(Z)$ with $\beta(c)$ in their minimum values, we can determine the corresponding values of κ^{σ}. They equal 7.5×10^{-11} Ohm^{-1} and 9.0×10^{-11} Ohm^{-1} for diamond particles with additives of aqueous electrolytes Th(NO$_3$)$_4$ and AlBr$_3$, respectively. For diamond particles in water with no additives $\kappa^{\sigma} = 1.6 \times 10^{-10}$ Ohm^{-1}. As we see, when changing Th(NO$_3$)$_4$ electrolyte concentration from 0 to 2.0×10^{-6} mol/l, the electroconductivity K increases 5 times, whereas κ^{σ} decreases 1.8 times. When changing AlBr$_3$ electrolyte concentration from 0 to 6.1×10^{-6} mol/l electroconductivity K increases 6.1 times, whereas κ^{σ} decreases 2.1 times. We conducted electrophoretic research of these colloids too. The concentrations of electrolytes Th(NO$_3$)$_4$ and AlBr$_3$ in isoelectric point equaled 1.3×10^{-6} mol/l and 4.5×10^{-6} mol/l, respectively. These values are approximately 1.5 times smaller than the concentrations that correspond to the minimum of $\beta(c)$. Such a distinct difference in concentrations is typical for isoelectric and isodipole points for aqueous colloids.

One-valent electrolytes have much less influence on Z and κ^{σ}. The values of surface conductivity of colloid particles of palygorskite for various concentrations of NaCl electrolyte studied in the electric fields of the same frequency are presented in Table 7.2

TABLE 7.2
Influence of Aqueous Electrolyte NaCl on Surface Conductivity of Palygorskite Particles

Method	Conductometrical			Electro-Optical		
$c \times 10^3$ mol/l of NaCl	0.1	0.77	3.9	0.1	0.4	0.8
$\kappa^{\sigma} \times 10^9$, Ohm^{-1}	5.7	5.7	1.6	4.0	3.8	3.6

The values of κ^{σ} presented in [35] and measured without taking polydispersity into account when studying the increment of electroconductivity (conductometrical method) of aqueous colloids of palygorskite are also provided.

The nonmonotonous character of $\gamma(c)$ can be experimentally studied using R_2–field technique. The increase and decrease of particle γ corresponds to an increase and decrease in lag angles between the electric field and the major polarizability axis; this corresponds to increase and decrease of experimentally measured angles $\tilde{\alpha}$. Formula 7.56 shows that for all particles orienting in electric field for a given relation E^2/ω the minimum of $\gamma(Z)$ (and the minimum of γD) corresponds to maximum values of $2\tilde{\alpha}$ and $\tilde{\alpha}$ (during the experiment $0 < \tilde{\alpha} < \pi / 4$). We conducted the study of dependencies $\tilde{\alpha}$ on concentration c of electrolyte $Th(NO_3)_4$ for an aqueous colloid of diamond, palygorskite, anisaldazine, and benzopurpurine when varying particle concentration c_{pt}. The R_2–field technique [36] is connected with application of strong field to the colloid system under study and thus is only applicable to colloids with low electroconductivity. Th^{4+} ions change γ considerably even at low $Th(NO_3)_4$ electrolyte concentrations, which allow us to conduct the study without taking into account the negligible (in this case) system heating caused by electric field application. Figure 7.12 shows the dependencies $\tilde{\alpha}(c)$ measured at $E = 300\,V/cm$ and $\omega = 45.8$.

As seen from the figure, for all four-particle types, equal relations c_{pt}/c corresponded to approximately equal $\tilde{\alpha}$ see Equation 7.55. This testifies to the fact that at low electrolyte concentrations of $Th(NO_3)_4$ all its counterions are absorbed by the particle surface and thus their density of surface charge is defined as c_{pt}/c.

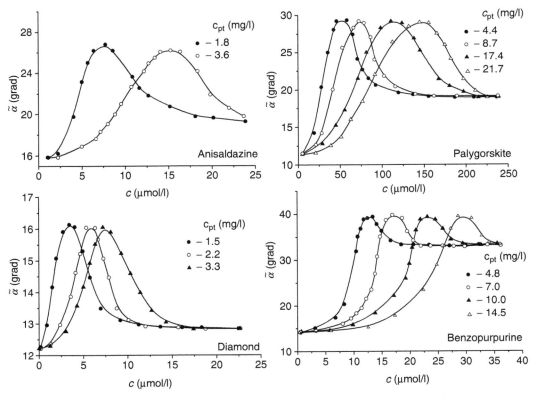

FIGURE 7.12 Results of the study of particles of aqueous colloids using R-field technique.

7.5 CONCLUSION

We can conclude that the electro-optics of polydisperse colloids is a prospective direction of the study of colloids and other nanodisperse systems. Using the techniques described above one can obtain the data inaccessible by other methods. These techniques include the study of electric conductivity and surface polarization of the particles in real systems with particles of different sizes and shapes. The analysis of the recovered particle distribution functions shows that they have some similarities and also dissimilarities. The further experimental data accumulation is essential to build a general theory that would describe the particle distribution on their parameters in heterogenous systems. The techniques of the particle distribution functions recovery require further development and modification, though, as it was shown, this complicated problem is already partially solved. The recovery of two-dimensional distribution functions allows to study particle parameter correlation in complex systems. The considered methods of permanent and induced particle polarizability study can be successfully used to study its influence on colloid stability, kinetic coagulation processes, and electric particle interaction.

The combined study of polydisperse colloid systems with different methods, such as electro-, magneto-optical, and conductivity anisotropy techniques, allows one to obtain more complete and detailed results.

ACKNOWLEDGMENTS

The authors thank the Russian Fund of Basic Research (grant 05-0332292) and the Government of Sankt-Petersburg (grant of Sankt-Petersburg in Science and Scientific Technique Work, N 37/05) for financial support.

REFERENCES

1. Sokerov, S., Petkanchin, I., and Stoylov, S., in *Surface Forces in Thin Films and Dispersions*, Deriagin, B.V., Ed., Nauka, Moskow, 96, 1972 (in Russian).
2. Kerr, J., *Phil. Mag.*, 50, 337, 1875.
3. Tolstoy, N.A., and Feofilov, P.P., *Doklady Acad. Nauk SSSR*, 66, 617, 1949.
4. Van de Hulst H.C., *Light Scattering by Small Particles*, Wiley, New York, 1952.
5. Mie, G., *Ann. Phys., Leipzig*, 25, 377, 1908.
6. Kerker, M., *The Scattering of Light and Other Electromagnetic Radiations*, Academic Press, New York, 1969.
7. Newton, P.G., *Scattering Theory of Wave and Particles*, McGraw-Hill, New York, 1966.
8. Stoylov, S.P., *Adv. Colloid. Interface Sci.*, 3, 45, 1971.
9. Fredericq, E. and Houssier, C., *Electric Dichroism and Electric Birefringence*, Clarender Press, Oxford, 1973.
10. O'Konski, C.T., in *Colloid and Molecular Electrooptics*, Jennings, B.R. and Stoylov, S.P., Eds., Institute of Physics Publishing, Bristol, 1992, p. 245.
11. Trusov, A. and Vojtylov, V.V., *Electrooptics and Conductometry of Polydisperse Systems*, CRC Press, Boca Raton, 1993.
12. Khlebtsov, N.G., Melnikov, A.G., and Bogatyrev, V.A., *J. Colloid Interface Sci.*, 146, 463, 1991.
13. Vojtylov, V.V., Kakorin, S.A., and Trusov, A.A., *Colloid J. USSR*, 48(1), 1986. Translated from *Kolloidnyi Zhurnal*, 48, 139, 1986.
14. Shilov, V.N., *Doklady Acad. Nauk SSSR*, 200, 1161, 1971.
15. Eriomova, Yu.Ya. and Shilov, V.N., *Colloid J. USSR*, 37(3), 1975. Translated from *Kolloidnyi Zhurnal*, 37(3), 535, 1975.
16. Eriomova, Yu.Ya. and Shilov, V.N., *Colloid J. USSR*, 37(6), 1975. Translated from *Kolloidnyi Zhurnal*, 37(6), 1090, 1975.
17. Perrin, F., *J. Phys. Rad.*, 5, 497, 1934.

18. Vojtylov V.V. and Trusov A.A., *Colloid J. USSR*, 47(4), 1985. Translated from *Kolloidnyi Zhurnal*, 47(4), 672, 1985.
19. O'Konski, C.T., Yoshioka, K., and Orttung, W.H., *J. Phys. Chem.*, 63, 1558, 1959.
20. Trusov, A.A., Vojtylov, V.V., and Zernova, T.Yu., *Colloids Surf. A. Physicochem. Eng. Aspects*, 201, 31, 2002.
21. Tolstoy, N.A., and Spartakov, A.A., in *Colloid and Molecular Electrooptics*, Jennings B.R. and Stoylov, S.P., Eds., Institute of Physics Publishing, Bristol, 1992, p. 37.
22. Pogorzelski, W., *Integral Equation and Their Applications*, Vol. 1, Pergamon Press, Oxford, 1966.
23. Waterman, D.R., Golley, C.R., and Jennings, B.R., in *Colloid and Molecular Electro-optics*, Jennings B.R. and Stoylov S.P., Eds., Institute of Physics Publishing, Bristol, 1992, p. 135.
24. Tikhonov, A.N. and Arsenin, V.Y., *Solution of Ill-Posed Problems.* W.H. Winston, Ed., Washington, D.C., 1977.
25. Tikhonov, A.N., et al., *Numerical Methods for the Solution of Ill-Posed Problems. Mathematics and its Applications*, Kluwer Academic Publishers, Dordrecht, 1995.
26. Vojtylov, V.V., Spartakov, A.A., and Trusov, A.A., *Opt. Spectrosc. (USSR)*, 44(3), 351, 1978. Translated from *Opt. Spektrosk.*, 44, 604, 1978.
27. Babadzanjanz, L.K. et al., *Colloids Surf. A: Physicochem. Eng. Aspects*, 148, 29, 1999.
28. Babadzanjanz L.K. and Voitylov, A.V., Recovery of distribution functions of colloid particles and macromolecules that satisfy Fredholm first kind integral equations, in *Processes of Management and Stability*, St. Petersburg State University Press, Saint Petersburg, 2001, p. 125 (in Russian).
29. Yagola, A.G. http://foroff.phys.msu.ru/illposed/eng/index.html
30. Vojtylov, V.V. et al., *Colloids Surf. A: Physicochem. Eng. Aspects*, 209, 123, 2002.
31. Spartakov, A.A. et al., *Colloids Surf. A: Physicochem. Eng. Aspects*, 209, 131, 2002.
32. Vojtylov, V.V. et al., *Colloid J. USSR*, 43(3), 1981. Translated from *Kolloidnyi Zhurnal*, 43(3), 440, 1981.
33. Vojtylov, V.V. and Trusov, A.A., *Colloid J. USSR*, 47(3), 1985. Translated from *Kolloidnyi Zhurnal*, 47(3), 455, 1985.
34. Vojtylov, V.V. et al., *Colloid J. USSR*, 52(6), 1990. Translated from *Kolloidnyi Zhurnal*, 52(6), 1057, 1990.
35. Maliarenko, V.V., Ovcharenko, D.F., and Dukhin, S.S., *Colloid J. USSR*, 36(3), 1974. Translated from *Kolloidnyi Zhurnal*, 36(3), 484, 1985.
36. Vojtylov, V.V. et al., *Colloid J. USSR*, 52(4), 1990. Translated from *Kolloidnyi Zhurnal*, 52(4), 637, 1990.

8 Application of the Scaling Method in Electro-Optics

Maria Stoimenova

CONTENTS

8.1 INTRODUCTION

Dielectric and electro-optic low-frequency analyses are the basic source of information on the electric polarization of charged colloids [1,2]. Dielectric measurements are made at weak electric fields. The advantage of electro-optic techniques is the possibility to measure the complete field intensity dependence of the effects. In electro-optic investigations, the saturation curves are used to apply fitting procedures to test theoretical constructions [3,4] or to obtain the saturation value at complete particle orientation related to particle dimensions [2]. As for the frequency analysis used for separation of the polarization mechanisms and the contributions of the induced and permanent dipoles, electro-optic procedures are restricted mainly to the low-field regime. Conventionally, this is realized at low degrees of orientation (where the method is grounded theoretically), following variations of the initial slopes of the field intensity dependences, or measuring directly the frequency dependence at constant field intensity. The reversing-pulse technique is an alternative method for separation of slow and fast processes, using fitting procedures for testing the polarization theories [5].

In a series of recent papers [6–10], we demonstrate the possibility of using the complete electro-optic saturation curves (or arbitrary part of them) to follow the relative variations of the particle electric parameters upon varying the field frequency or the ionic content of the medium. Several advantages of the proposed method are evident.

First, in many cases (small particle axial ratios, significant multiple scattering, etc.) the linear regime is not available. For particles of higher refractive indices and larger dimensions or for concentrations involving multiple scattering, the effects can display nonlinear behavior (even sign reversal) at low degrees of particle orientation [11,12]. Due to the oscillating character of the optical functions of wavelength-size particles and their complicated theory, the problem has been successfully solved for a limited number of model samples, strongly restricted in their refractive indices, dimensions, forms, etc. [2].

Second, the data obtained by the complete field intensity curves in polydisperse systems is closer to the parameters of the average-sized particles. Even a small portion of aggregates or large particles can influence the results significantly at low fields.

And third (the most important), the changes in the functional dependence of the effects could hardly be recognized in the low-field regime. The conventionally measured frequency curves depend on the choice of field intensity even at low degrees of particle orientation [8]. This explains the contradictory results for permanent dipole moments presented in literature. The conventional procedure for determination of particle electric parameters presumes that low-frequency (Hertz range) effects are independent of electric polarizability. Obtained at kilohertz frequency, the value of electric polarizability is used in the analysis of data for dc pulses to determine the permanent dipoles. Such procedure is not consistent with the clearly revealed dependence of the shape of the frequency curves on the field intensity [4,8]. The proposed approach helps to analyze the anomalous low-frequency range and to test the dependence of the slow effects on particle surface electric polarizability (relaxing in the kilohertz domain).

8.2 SAMPLES AND METHODS USED FOR ILLUSTRATION OF SCALING PROCEDURES

In the present paper we illustrate the application of the proposed method on a series of colloidal systems containing particles of different forms, dimensions, refractive indices, structures, and mechanisms of surface charging—aqueous suspensions of palygorskite [7,13–15], β-FeOOH [8,9], γ-Al$_2$O$_3$ [16,17], SiO$_2$ [12] particles. The aim is to demonstrate similar tendencies in the low-frequency electro-optic behavior of the charged particles and the applicability of the proposed procedure for analysis of the effects in this range.

The clay mineral palygorskite (attapulgate) is an aqueous silicate of a shistous-banded structure. The particles are rigid polydisperse rods of average length and diameter 500 nm and 20 nm, respectively. Particle refractive index is ca. 1.5, and particle surface charge in the aqueous suspensions is ca. $0.7\,\mu C/cm^2$.

The β-FeOOH are highly monodisperse ellipsoids of major and minor axes 285 ± 11 nm and 71 ± 4 nm, respectively (refractive index ca. 2.2). They are prepared by acid hydrolysis of $1.8 \times 10^{-2}\,M$ FeCl$_3$ solution containing $10^{-5}\,M$ HCl over a period of 3 weeks at room temperature [18]. The extraneous ions are removed by repeated centrifugation in distilled water till the supernatant conductivity reaches the value of 10^{-4} S/cm. The investigated samples are obtained by dilution of the stock suspension in bidistilled water.

The aluminum and silicium oxide particles are obtained from Degussa: The electro-optic effect is displayed by aggregates of small spherical particles (diameter ~20–40 nm) present in aqueous suspensions of the commercial products [16,12]. According to electro-optic data the aggregates are of average size about 300 nm (refractive index ca. 1.5).

Figure 8.1 shows schematically the relative variations of surface charge σ, ζ-potential, and electric polarizability γ of the β-FeOOH particles with pH of the medium [8]. Similar pH dependence of the surface characteristics is reported for the other oxides [16,19]. They differ in the position of the point of zero charge (pzc). Aqueous suspensions of β-FeOOH (pzc ≈ 6.5) are investigated in the pH interval 4–5.5; suspensions of γ-Al$_2$O$_3$ (pzc ≈ 8), in the pH interval 4–7.5, where the particles are positively charged. The pH values are obtained by addition of HCl. The silica particles (pzc ≈ 2) are negatively charged in the investigated pH interval 5.5–9 (the pH values are obtained by addition of NaOH under inert atmosphere). Ionic strength is supported by NaCl. Values of pH corresponding to the steep branch of the γ vs. pH dependence are chosen for illustration of the charge dependence of the electro-optic effects.

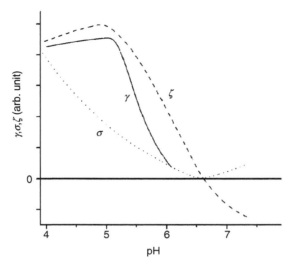

FIGURE 8.1 Relative variation of electric polarizability γ (solid line), surface charge σ (dotted line), and ζ-potential (dashed line) of β-FeOOH particles with pH of the medium at ionic strength $10^{-4}\,M$. (Adapted from Miteva, S. and Stoimenova, M., *J. Colloid Interface Sci.*, 273, 490, 2004.)

The electro-optic data presented in this chapter is measured by the electric light-scattering method. The experimental setup (described in Ref. [2]) is used in numerous studies of colloidal systems. The light beam is incident on the sample, which is placed in a measuring cell of volume 8 ml and diameter 20 mm (presented results are obtained by the use of white unpolarized light). Platinum plate electrodes of radius 5.5 mm and distanced at 2.6 mm are incorporated in the cell. The electric pulses are applied perpendicularly to the observation plane. The scattered light intensity is measured at observation angle 90° using a photomultiplier or an oscilloscope registration. The scattered intensity of the unperturbed system (I_0) is compensated and the field-induced light-scattering variations ($\Delta I = I_e - I_o$) with time are directly observed on the oscilloscope. The electro-optic effect is defined as a relative quantity $\alpha = \Delta I / I_o$. The steady component of the effect is determined taking the middle of the modulated steady state signal as I_e. The alternating component is determined by the amplitude of the steady state modulated response, defined as a relative value $(\alpha_{max} - \alpha_{min})/\alpha_{max}$.

8.3 CONVENTIONAL PROCEDURES FOR ELECTRO-OPTIC FREQUENCY ANALYSIS

The theoretical basis of the conventional electro-optic method for determination of the relative variations of electric polarizability is the generalized Kerr law, presented by the equation $d\alpha/dE^2 = \Omega\gamma$ (α is an arbitrary electro-optic effect, γ is the (anisotropy of) particle electric polarizability, E is the electric field intensity, Ω is the optical function). The equation is valid at low degrees of particle orientation. For particles of similar optical function, it gives the ratio $\gamma_1/\gamma_2 = (d\alpha/dE^2)_1/(d\alpha/dE^2)_2$. In practice, this ratio is usually obtained by taking the ratio of effects of the compared samples at constant field intensity. Such procedure is adequate for similar polarization mechanisms of the compared effects. As will be demonstrated in this paper, the low-field data is not sufficient to distinguish between different orientation mechanisms.

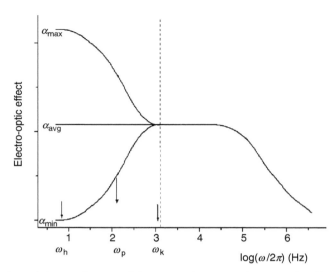

FIGURE 8.2 Variation of the maximum, minimum, and average electro-optic effects with electric field frequency. Dashed vertical line separates the range of the alternating component of the effect, the arrows indicate the characteristic frequencies of its relaxation interval.

Figure 8.2 shows schematically the frequency range where charged colloid-sized particles display the gigantic-low-frequency dispersion. For most aqueous colloids it is below 1 MHz. The dotted vertical line in the figure separates two frequency domains related to different relaxations. In the hertz domain the relaxation frequency of particle rotation is detected by the relaxation of the alternating component of the electro-optic response. The relaxation interval of the alternating component is determined by three frequencies, two of them marking the beginning (ω_h) and the end (ω_k) of the interval, and the third one is the relaxation frequency of particle rotation (ω_p). The characteristic frequencies depend on particle geometry and polydispersity. In the kilohertz domain only a steady component remains, related to particle electric polarization and (as generally accepted) the relaxation of particle surface polarizability. The boundary separates domains with and without alternating component of the responses, its frequency depending upon particle dimensions and form.

The behavior of the effects in the kilohertz domain is consistent with the Kerr law—at low fields the effects are linear with the square of electric field intensity [2]. The corresponding initial slopes are proportional to the value of electric polarizability (γ) with a coefficient depending on the particle optical properties. The determination of the absolute value of γ requires determination of the optical coefficient. As no explicit expressions for the latter are available in the majority of cases, the value of γ is determined with precision to a constant. At constant dispersity of the system the optical coefficient is constant and the relative variation of the initial slopes follows the relative variation of electric polarizability.

Figure 8.3 demonstrates the essentially different electro-optic behavior of the samples in the hertz domain. Electro-optic data is presented for the palygorskite particles, used in many studies as model electro-optic object. At low degrees of orientation the form of the frequency curves changes drastically with field intensity, precisely in the domain of relaxation of the alternating component of the responses. It can indicate the interference of a longitudinal or a transverse permanent dipole depending on the choice of field intensity. The different form of the saturation curves measured at kilohertz frequency and at dc pulses shows different functional dependence of the effect at low frequency.

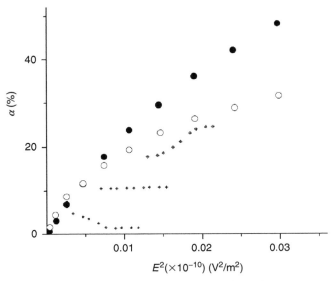

FIGURE 8.3 Field intensity dependence of the electro-optic effect for aqueous suspension of palygorskite: solid symbols, 1 kHz; open symbols, dc pulses. Insetted: the form of the frequency curves in the interval 10 Hz–10 kHz for three values of field intensity (7.7, 15, 30 kV/m). (Adapted from Stoimenova, M., Radeva, Ts., and Stoylov, S., *Colloid Polym. Sci.*, 257, 1226, 1979.)

8.4 SCALING OF ELECTRO-OPTIC SATURATION CURVES—PHYSICAL BASIS OF THE METHOD

The dependence of the electro-optic effects on the electric field intensity is determined by particle orientation mechanism and can be different in the different frequency domains. Scaling of the electro-optic saturation curves by field intensity permits to find the frequency range in which the functional dependence of the effects, hence the orientation mechanism, remains unchanged. Similar procedure can be applied to verify the changes of this dependence with variation of particle electric parameters.

As mentioned above, at low degrees of particle orientation Kerr law determines the ratio γ_1/γ_2 as the ratio of the initial slopes $(d\alpha/dE^2)_1/(d\alpha/dE^2)_2$. Due to the linearity of the effects it can be obtained by taking either the ratio of effects of the compared samples at constant field intensity or the ratio of field intensities at constant effect: $\gamma_1/\gamma_2 = (d\alpha/dE^2)_1/(d\alpha/dE^2)_2 = (\alpha_1/\alpha_2)_{E\text{const}} = (E_2^2/E_1^2)\alpha_{\text{const}}$. In the linear regime both procedures give identical results. However, the second procedure can be applied for arbitrary degrees of orientation.

The method is grounded on the following considerations: the electro-optic effect (i.e., the degree of particle orientation) is determined by particle orientation energy. The induced dipole orientation ($U = f(\gamma E^2)$) can be varied through changes in either electric polarizability or electric field intensity. For systems of equal degree of orientation $\gamma_1 E_1^2 = \gamma_2 E_2^2$, and hence $\gamma_1/\gamma_2 = E_2^2 / E_1^2$. This equality is valid for arbitrary degree of orientation and is independent of the optical properties of the system (if they remain unchanged). As optical interactions introduce oscillating optical functions and could lead to equal electro-optic effects at different degrees of orientation, a direct comparison of single effects could be erroneous. The polarizability ratio can be determined by searching for the best fit of the saturation curves, scaling the effects with field intensity. A good fit of the curves is expected for a monodisperse system if particle electric polarizability is independent of the field intensity.

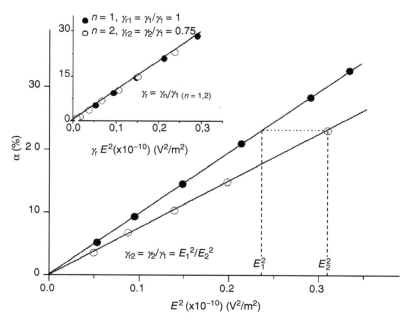

FIGURE 8.4 Field intensity dependence of the electro-optic effect (frequency 3 kHz) for FeOOH particles of concentration 5×10^{-3} g/l, washed up to different conductivity of the suspension: solid symbols, $\kappa = 8 \times 10^{-6}$ S/cm; open symbols, $\kappa = 4 \times 10^{-6}$ S/cm. Insetted: the same curves on the scale ($\gamma_r E^2$). (Adapted from Miteva, S. and Stoimenova, M., *J. Colloid Interface Sci.*, 273, 490, 2004.)

In Figure 8.4 and Figure 8.5 the method is illustrated by experimental data for suspensions of monodisperse ellipsoidal β-FeOOH particles. The effects are presented for two suspensions of different surface charge (achieved by washing of the stock suspension up to different conductivity before dilution). The ratio of the initial slopes of their field intensity curves ($\gamma_2/\gamma_1 = 0.75$) equals the reciprocal ratio of the squared field intensities at constant effect (Figure 8.4). We can introduce a scale $\alpha/(\gamma_r E^2)$, where γ_r is the ratio of electric polarizability of each sample towards a reference sample. The value of γ_r is different for each sample. Choosing for a reference sample the suspension of higher conductivity, we obtain the figure shown in the inset of Figure 8.4. On this scale the initial slopes of the two systems coincide. Figure 8.5 shows on the same scale the saturation curves of the suspensions (the original curves α/E^2 are seen in the inset of the figure). They coincide up to complete particle orientation. The essential point is that we do not need the linear part of the curves to obtain the ratio γ_2/γ_1. Arbitrary portions of the curves can be used for the purpose. The obtained value can be corrected by searching for the best fit of the curves. For the suspensions presented on Figure 8.4 and Figure 8.5 both methods (initial slopes and $\alpha/(\gamma_r E^2)$ scaling) give identical results. Any field intensity dependence of the scaling coefficient is to be related to field intensity dependence of the electric polarizability ratio (due to interparticle interactions, anisotropic saturation effects, polydispersity effects, etc.) or to side effects. The latter cases can be distinguished by appropriate additional experiments.

An evident advantage of the proposed method is that it is not restricted to the linear portion of the field intensity curves. This helps the analysis of experimental data in cases where the linear part cannot be determined experimentally or might be affected by optics and polydispersity, thus extending the comparative electro-optic studies to a wide range of real colloids. Since the optical functions do not influence the scaling coefficients, the method is

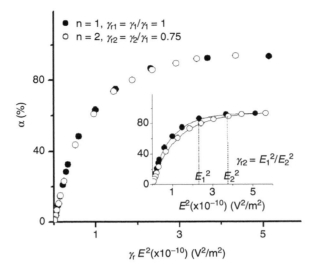

FIGURE 8.5 Complete saturation curves for the suspensions presented in Figure 8.4 on the same scales. (Modified from Miteva, S. and Stoimenova, M., *J. Colloid Interface Sci.*, 273, 490, 2004.)

convenient for the separation of optical and electrical effects in optically complicated systems. It is particularly advantageous for the study of semidilute dispersions. The significant difference in their electro-optic behavior could originate in both optical (multiple scattering) and electrical interactions. The polarization process in semidilute dispersions could display anisotropic polarizability changes. The scaling method helps to separate them from optical effects.

8.5   APPLICATION OF THE SCALING METHOD

The above described scale $\alpha/(\gamma_r E^2)$ is used for different purposes. If the functional dependence of the electro-optic effect on field intensity is known, we can use it for determination of the scaling coefficients, i.e., the relative variations of electric polarizability. This approach is applied in the kilohertz domain (see Figure 8.1). If the mechanism of particle orientation is not clarified, the proposed scale is used for the analysis of slow effects accounting for their dependence on the kilohertz-induced dipole. We apply this approach for the study of the slow effects in the hertz domain.

In some cases the presented field intensity curves are measured up to saturation value of the effects, in others they are restricted to intermediate degrees of orientation. This is due to either the restricted range of the available generators or the different purpose of the paper from which the experimental data is exported.

8.5.1   KILOHERTZ DOMAIN

The method was first applied in the kilohertz domain [6,7]. The experimentally established quadratic dependence of the effects on field intensity (at low fields) indicates induced dipole orientation. It is generally accepted that for colloid-sized particles, mainly the polarization processes in the particle surface electric layer determine their orientation in the kilohertz frequency domain. By scaling of the field intensity curves we can verify if their functional dependence is altered upon variation of particle surface electric state.

It was demonstrated on a variety of samples [6] that the field intensity curves in the kilohertz range coincide on the scale $\alpha/(\gamma_r E^2)$. Then the scaling coefficients present

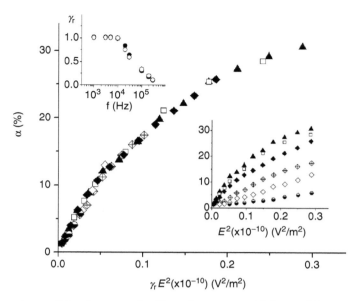

FIGURE 8.6 Dependence of the electro-optic effect of aqueous γ-Al$_2$O$_3$ particles on $\gamma_r E^2$ for different field frequencies in the kilohertz range: triangles, 10 kHz; squares, 20 kHz; solid diamonds, 30 kHz; crossed diamonds, 100 kHz; open diamonds, 200 kHz; circles, 300 kHz. Insetted: right, the same curves on the scale α/E^2; left, relative variation of electric polarizability with frequency—solid symbols, scaling coefficients; open symbols, constant field intensity (23 kV/m) values. (Adapted from Stoimenova, M. and Peikov, V., *J. Colloid Interface Sci.*, 164, 285, 1994.)

the frequency variation of electric polarizability. Figure 8.6 shows data for the aqueous aluminium oxide suspension. The inset in the figure presents the scaling coefficients, i.e., the normalized frequency dependence of electric polarizability. In parallel, a conventional frequency curve measured at constant low field is given for comparison. The scaling method and the conventional method yield similar results for the relaxation interval of γ. The fitting of the curves is better for narrow particle-size distribution. Polydispersity worsens the fitting in the low-field domain [6].

Several figures illustrate the application of the method to follow the variation of γ with particle surface electric parameters. Variations of electric polarizability with particle surface charge (varied by pH of the suspension) are demonstrated on the oxide samples. Figure 8.7 shows the results for three suspensions (of different pHs) of the aluminum oxide particles. The logarithmic scale is used to make visible the fitting of the curves at low fields. Figure 8.8 demonstrates the surface charge dependence of γ for the silica particles. Field intensity curves of the compared suspensions at frequency 100 kHz are included. It is seen that for the low-charge particles this frequency is beyond the relaxation frequency of γ, while for the higher-charge particles it is almost in the plateau region. Hence, the relaxation frequency of kilohertz polarizability displays well-expressed surface charge dependence. Similar dependence was reported for the aluminum oxide particles [16].

The relative variation of electric polarizability with the ionic strength of the medium is shown for the palygorskite particles. They are of constant surface charge and the electrolyte concentration influences mainly the thickness of the Debye atmosphere. Figure 8.9 presents data for three palygorskite suspensions of different electrolyte concentration. The inset in the figure presents the scaling coefficients, i.e., the normalized dependence of electric

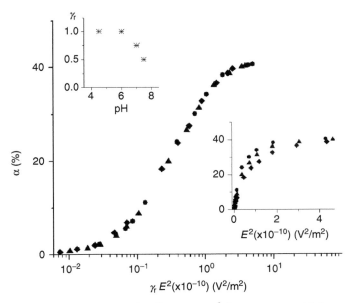

FIGURE 8.7 Dependence of the electro-optic effect on $\gamma_r E^2$ for aqueous γ-Al$_2$O$_3$ particles of different surface charges: circles, pH = 4.5–6; triangles, pH = 7; diamonds, pH = 7.5. Insetted: right, the same curves on the scale α/E^2; left, scaling coefficients (relative variation of electric polarizability with surface charge). (Modified from Buleva, M. and Stoimenova, M., *J. Colloid Interface Sci.*, 141, 426, 1991.)

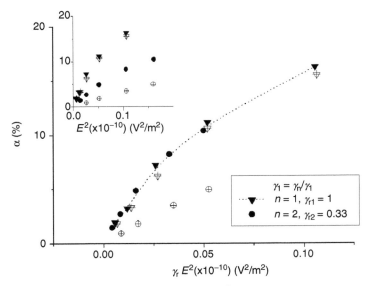

FIGURE 8.8 Dependence of the electro-optic effect on $\gamma_r E^2$ for aqueous SiO$_2$ particles (concentration 0.1 g/dm^3) of different surface charges: triangles, pH = 9; circles, pH = 5; solid symbols, 1 kHz; crossed symbols, 100 kHz. Insetted: right, scaling coefficients; left, the same curves on the scale α/E^2. (Adapted from Stoimenova, M. and Buleva, M., *J. Colloid Interface Sci.*, 152, 483, 1992.)

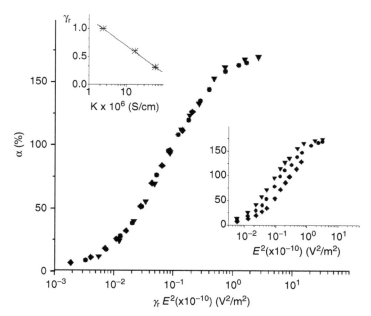

FIGURE 8.9 Dependence of the electro-optic effect (frequency 5 kHz) on $\gamma_r E^2$ for palygorskite suspensions (concentration 0.1 g/dm^3) of different NaCl concentrations: triangles, $\kappa = 2.4 \times 10^{-6}$ S/cm; circles, $\kappa = 1.7 \times 10^{-5}$ S/cm; diamonds, $\kappa = 6 \times 10^{-5}$ S/cm. Insetted: left, scaling coefficients (relative variation of electric polarizability with ionic strength of the medium); right, the same curves on the scale α/E^2. (Adapted from Peikov, V., Radeva, Ts., and Stoimenova, M., *J. Colloid Interface Sci.*, 168, 1, 1994.)

polarizability on the ionic strength of the medium. Unfortunately our experimental setup is not adapted for investigations at higher electrolyte concentrations and we have reliable data only for the presented intervals. But it is worth mentioning that the fitting gets worse with further increase of the ionic strength, particularly at higher fields.

Figure 8.10 shows results for the β-FeOOH samples. The figure includes variation of both surface charge and ionic strength.

8.5.2 Low-Frequency Domain

In the hertz domain the electro-optic effects show different field intensity dependence (see Figure 8.3).

Figure 8.11 shows field intensity curves in this domain for the samples presented in the chapter. Two types of frequency changes are seen in the figure. For some samples (e.g., SiO$_2$, Figure 8.11a) the electro-optic effect decreases systematically with the decrease of frequency, precisely in the relaxation interval of the alternating component of the effect (in cases reaching to sign reversal). For other samples (e.g., β-FeOOH, Figure 8.11b) the same tendency is seen at higher-field intensity, but it is reversed at low fields. The aqueous palygorskite suspension can be related to either groups depending on the applied scale (Figure 8.11c and Figure 8.11d), as the reversal for this sample is realized at very low fields.

Figure 8.12 through Figure 8.14 concern the first type of samples. We can consider the frequency variation of the effect resulting from the superposition of the kilohertz effect and a negative effect (reducing the kilohertz one) relaxing in the relaxation interval of the alternating component [10]. Assuming additivity of the effects we present the negative effect as: $\Delta = \alpha(\omega_k) - \alpha(\omega_h)$ (ω_h and ω_k are the frequencies limiting the relaxation

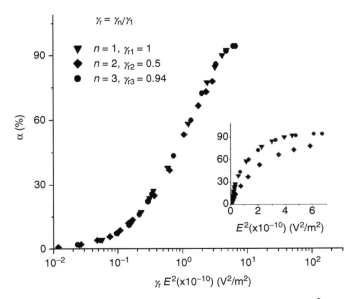

FIGURE 8.10 Dependence of the electro-optic effect (frequency 3 kHz) on $\gamma_r E^2$ for aqueous β-FeOOH suspensions (concentration 5×10^{-3} g/dm³) of different pH and NaCl concentrations: triangles, 3×10^{-5} M HCl, pH = 4.5; diamonds, 3×10^{-5} M NaCl, pH = 5.5; circles, aqueous, pH = 5.5. Insetted: left, scaling coefficients; right, the same curves on the scale α/E^2. (Adapted from Miteva, S. and Stoimenova, M., *J. Colloid Interface Sci.*, 273, 490, 2004.)

interval) (Figure 8.12). The assumption for additivity of the effects is not important for our further considerations. We use the difference Δ only to visualize the dependence of the negative effect on the field intensity and its parallel increase with the kilohertz-induced dipole. It is seen in Figure 8.12 that for the aqueous suspensions of palygorskite, γ-Al_2O_3, and SiO_2 particles the negative effect linearizes with the square of the field intensity and is comparable in value with the kilohertz effect. Further, we use the proposed scale $\alpha/(\gamma_r E^2)$ (which equalizes the kilohertz-induced dipole moments of the compared samples) to check the dependence of the negative effect on the kilohertz-induced moment through variation of particle surface electric parameters. Figure 8.13 shows the variation of the negative effect with particle surface charge for the silica particles presented in Figure 8.8. The negative effect is shown directly by the low-frequency curves $\alpha(\omega_k)$, $\alpha(\omega_p)$, and $\alpha(\omega_h)$, corresponding to the characteristic frequencies of the alternating components. No dependence of the slow effect on particle surface charge is displayed on the $(\gamma_r E^2)$ scale, which signifies that its ratio to the kilohertz-induced moment remains unchanged. Similar behavior can be seen in the data reported for the γ-Al_2O_3 particles [16]. Figure 8.14 shows on the same scale the variation of the negative effect with the ionic strength of the medium for the palygorskite suspensions presented in Figure 8.9. Again the ratio of the negative effect to the kilohertz-induced moment remains unchanged.

The aqueous β-FeOOH particles display different low-frequency behavior. On a larger scale the difference is not significant, but in the low-field regime (usually exploited for frequency analysis) it is clearly seen. In the kilohertz range the effects are linear with the square of the field intensity. Deviation from linearity is observed in the low-frequency relaxation interval, showing different tendency above and below particle relaxation frequency. Similar tendencies are observed for the palygorskite at very low degrees of particle orientation [13]. The complicated variation of the field curves in this domain can be obtained by superposition of two slow effects (linear and quadratic with field intensity), relaxing in the same frequency interval, upon the

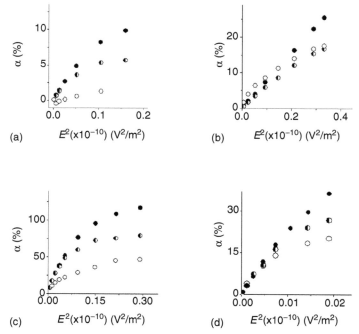

FIGURE 8.11 Field intensity dependence of the electro-optic effect at the characteristic frequencies in the low-frequency domain for (a) SiO_2: solid circles, 1 kHz; crossed circles, 100 Hz; open circles, 10 Hz. (Modified from Stoimenova, M. and Buleva, M., *J. Colloid Interface Sci.*, 152, 483, 1992.) (b) β-FeOOH: solid circles, 3 kHz; crossed circles, 200 Hz; open circles, 10 Hz. (Adapted from Miteva, S. and Stoimenova, M., *J. Colloid Interface Sci.*, 273, 490, 2004.) (c) Palygorskite: solid circles, 5 kHz; crossed circles, 100 Hz; open circles, 20 Hz. (Modified from Peikov, V., Radeva, Ts., and Stoimenova, M., *J. Colloid Interface Sci.*, 168, 1, 1994; Stoimenova, M., Radeva, Ts., and Stoylov, S., *Colloid Polym. Sci.*, 257, 1226, 1979.) (d) Low-field range of (c).

kilohertz effect. One of them is negative (reducing the kilohertz effect) and better displayed at high fields, the other is positive (increasing the effect) and better displayed at low fields. Three of the curves, $\alpha(\omega_h)$, $\alpha(\omega_p)$, and $\alpha(\omega_k)$, corresponding to the characteristic frequencies of the alternating component, are sufficient to describe the frequency behavior in this interval. In electro-optic experiments dc pulses are usually preferred to low-frequency pulses. The dc pulses of appropriate duration can replace the lowest characteristic frequency, but this is correct when square reversing pulses are used for frequency analysis [8].

Figure 8.15 illustrates the application of the characteristic field intensity curves for analysis of the slow effects. We apply the above proposed scaling to follow the dependence of the slow effects on particle induced dipole relaxing in the kilohertz domain. The dependence of the slow effects on particle surface charge and Debye layer thickness is plotted for the β-FeOOH particles on the scale (γ_r E^2) (where γ_r is the ratio of the kilohertz electric polarizabilities of the compared samples to a reference sample, see Figure 8.10). The comparison is demonstrated on the linear scale $\gamma_r^{1/2}$ E, to show the linear dependence of the positive effect at low fields (where the kilohertz and the negative effect are sufficiently small). The positive effect diminishes at higher surface charge (and particularly in the presence of ferric ions [8]), while the negative effect does not change significantly. The latter effect is similar to the effect observed by the previous samples (Figure 8.12 through Figure 8.14).

Significant changes of the negative effect are observed in the presence of specific adsorption of surfactants [20], polymers [21], and polyelectrolytes [22]. Figure 8.16 shows the relative

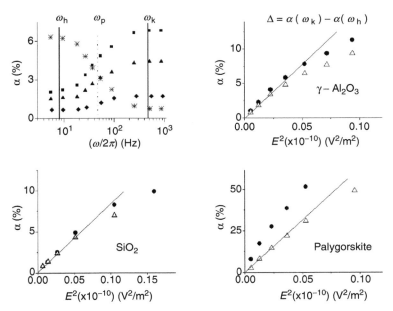

FIGURE 8.12 Frequency dependence of the electro-optic effect for aqueous γ-Al_2O_3 at different field intensities: diamonds, 15 kV/m; triangles, 25 kV/m; squares, 30 kV/m; stars, alternating component. (Modified from Stoimenova, M. and Radeva, Ts., *J. Colloid Interface Sci.*, 141, 433, 1991.) Field intensity dependence of the kilohertz effect and the negative low-frequency effect for aqueous γ-Al_2O_3, SiO_2, palygorskite.

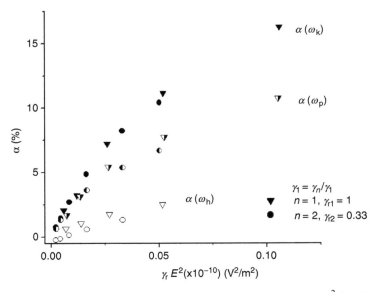

FIGURE 8.13 Dependence of the characteristic low-frequency curves on $\gamma_r E^2$ for the aqueous SiO_2 particles presented in Figure 8.8: triangles, pH = 9; circles, pH = 5; solid symbols, 1 kHz; crossed symbols, 100 Hz; open symbols, 10 Hz.

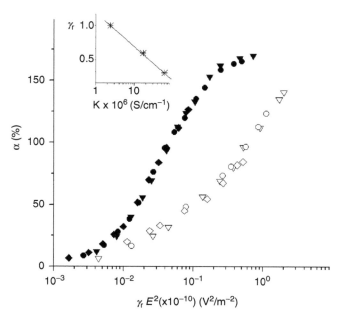

FIGURE 8.14 Dependence of the characteristic low-frequency curves $\alpha(\omega_k)$ and $\alpha(\omega_h)$ on $\gamma_r\, E^2$ for the palygorskite suspensions presented in Figure 8.9: triangles, $\kappa = 2.4 \times 10^{-6}$ S/cm; circles, $\kappa = 1.7 \times 10^{-5}$ S/cm; diamonds, $\kappa = 6 \times 10^{-5}$ S/cm; solid symbols, 1 kHz; open symbols, dc pulses. (Modified from Peikov, V., Radeva, Ts., and Stoimenova, M., *J. Colloid Interface Sci.*, 168, 1, 1994.)

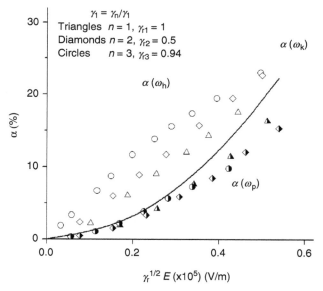

FIGURE 8.15 Dependence of the characteristic low-frequency curves on $\gamma_r^{1/2}\, E$ for the aqueous β-FeOOH suspensions presented in Figure 8.10: triangles, 3×10^{-5} *M* HCl, pH = 4.5; diamonds, 3×10^{-5} *M* NaCl, pH = 5.5; circles, aqueous, pH = 5.5; solid line, 3 kHz; crossed symbols, 200 Hz; open symbols, dc pulses. (Adapted from Miteva, S. and Stoimenova, M., *J. Colloid Interface Sci.*, 273, 490, 2004.)

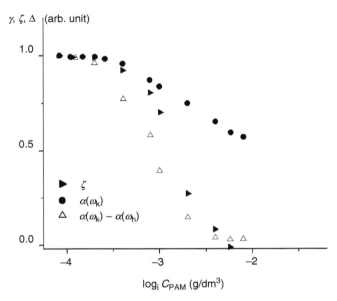

FIGURE 8.16 Relative variation of ζ-potential (solid triangles), kilohertz electric polarizability γ (circles), and negative low-frequency effect (open triangles) of γ-Al$_2$O$_3$ particles with the concentration of polyacrylamide. (Adapted from Radeva, Ts. and Stoimenova, M., *Colloids Surf.*, A, 54, 235, 1991.)

variation of the kilohertz and the negative effect with polymer adsorbtion on the surface of the γ-Al$_2$O$_3$ particles. Their ratio changes drastically and the slow effect follows the variation of electrokinetic potential [21].

The results in Figure 8.15 show that the two types of samples are not essentially different. The two figures related to palygorskite particles also demonstrate this (Figure 8.11c and Figure 8.11d). When polyelectrolyte molecules are adsorbed on the surface of the γ-Al$_2$O$_3$ particles the positive effect appears [22]. Hence, both the quadratic and the linear effect are related to particle surface electric state.

Summarizing, in the general case the superposition of two slow effects (linear and quadratic with field intensity) can explain the complicated frequency curves in the relaxation interval of particle rotation. One of the slow effects is observed for all polarizable particles. It is of negative sign and displays the features of an induced dipole effect dependent on the kilohertz-induced moment. It is displayed through electrokinetic rotation and seems to be a slow stage of the surface polarization process, related to electrokinetic charge. The linear slow effect shows permanent dipole-like behavior and is relatively independent of the kilohertz-induced moment. It depends on the ionic content of the medium. As we have no grounds to expect significant changes in the counterion mobility of the compared samples (Figure 8.15), it is probably due to asymmetric particle surface charge.

8.5.3 OPTICALLY INTERACTING PARTICLES

The electro-optic characteristics obtained by the scaling procedure are independent of the optical properties of the colloidal system (including multiple scattering) and the method can be applied to optically interacting systems. The application of the method to follow variations of electric polarizability was demonstrated in Figure 8.4 and Figure 8.5 for two β-FeOOH suspensions of different particle surface charge and equal particle concentration (0.005 g/l). Suspensions of similar surface charge and higher particle concentration (0.06 g/l) are

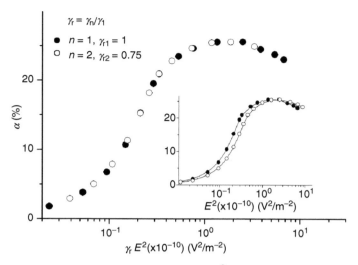

FIGURE 8.17 Dependence of the electro-optic effect on $\gamma_r E^2$ (frequency 3 kHz) for FeOOH particles of concentration 6×10^{-2} g/dm^3, washed up to different conductivities of the suspension: solid symbols, $\kappa = 8 \times 10^{-6}$ S/cm; open symbols, $\kappa = 4 \times 10^{-6}$ S/cm. Insetted: the same curves on the scale α/E^2. (Adapted from Miteva, S. and Stoimenova, M., *J. Colloid Interface Sci.*, 274, 531, 2004.)

presented in Figure 8.17. Despite the drastic changes in the saturation curves due to multiple scattering, the polarizability ratio is the same ($\gamma_2/\gamma_1 = 0.75$).

Figure 8.18 shows for several samples the variation of scattered light intensity by the unperturbed system (I_0) with particle concentration. In parallel the electro-optic effect (saturation value) is presented for the β-FeOOH particles. At very low concentrations the

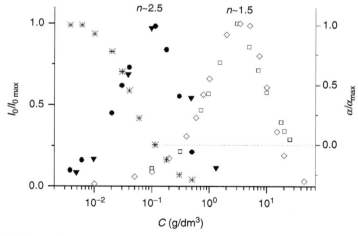

FIGURE 8.18 Variation of normalized scattered light intensity with particle concentration (left axes): Circles, β-FeOOH. (Adapted from Miteva, S. and Stoimenova, M., *J. Colloid Interface Sci.*, 274, 531, 2004.) Down triangles, TiO$_2$. (Adapted from Stoimenova, M. and Buleva, M., *J. Colloid Interface Sci.*, 152, 483, 1992.) Up triangles, palygorskite. (Adapted from Peikov, V., "Electro-optic Investigation of Semidilute Disperse Systems", Ph.D. thesis, Bulg. Acad. Sci., Sofia, 1994.) Squares, SiO$_2$. (Adapted from Stoimenova, M. and Buleva, M., *J. Colloid Interface Sci.*, 152, 483, 1992.) Stars, the normalized electro-optic effect for aqueous β-FeOOH (right axes). (Adapted from Miteva, S. and Stoimenova, M., *J. Colloid Interface Sci.*, 274, 531, 2004.)

intensity of scattered light increases linearly (the particles are scattering light independently), and the electro-optic effect is constant. At higher concentrations multiple scattering produces attenuation and the curve of I_0 displays a maximum and a consequent decrease. The precise form of the curve depends on the particle dimensions and on the parameters of the detecting setup. Since particle orientation increases light scattering, the electro-optic effect is very sensitive to optical interaction and decreases drastically reaching a sign reversal. Samples of concentrations in the vicinity of the electro-optic sign reversal display strongly deformed field intensity curves (even reversal from positive to negative effect) and peculiar form of the responses. The samples are grouped by particle refractive indices. The effects of optically dense particles (e.g., β-FeOOH) show considerable decrease at concentrations far below the critical concentration of particle electric interaction. The scaling procedure is applicable for any volume fraction but the precision is reduced if the amplitude of the response is small, as is the case near the electro-optic sign reversal.

Figure 8.19 presents the electro-optic saturation curves for four concentrations of aqueous β-FeOOH particles in the interval 0.005–0.5 g/l. Two of the suspensions are obtained by 10- and 100-fold dilution of the concentrated sample and washing by centrifugation to similar low conductivity (2×10^{-6} S/cm). The fourth suspension is obtained by the supernatant of the concentrated sample thus assuring 100-fold dilution of particles and similar ionic content of the medium. Due to the optical interactions the effect is changing drastically reaching sign reversal. Both the relaxation frequency of kilohertz electric polarizability and the

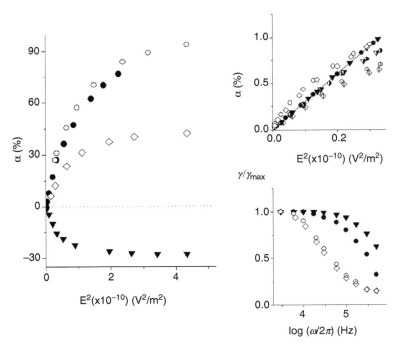

FIGURE 8.19 Electro-optic saturation curves (frequency 3 kHz) for β-FeOOH suspensions of different particle concentrations: circles, $C \approx 5 \times 10^{-3}$ g/dm^3; diamonds, $C \approx 4 \times 10^{-2}$ g/dm^3; triangles, $C \approx 5 \times 10^{-1}$ g/dm^3 and ionic content of the medium: open symbols, $\kappa = 2.5 \times 10^{-6}$ S/cm; solid symbols, $\kappa \approx 3 \times 10^{-5}$ S/cm. Right up—dependence of the electro-optic effect on $\gamma_r^{1/2} E$ at the characteristic frequencies: solid line, 3 kHz; crossed symbols, 200 Hz; open and solid symbols, dc pulses. Right down—relative variation of the kilohertz electric polarizability with field frequency. (Modified from Miteva, S. and Stoimenova, M., *J. Colloid Interface Sci.*, 274, 531, 2004.)

low-frequency characteristic curves (Figure 8.19, right) show well-expressed dependence on the ionic content of the suspensions and no dependence on particle volume fraction.

8.6 CONCLUSIONS

The scaling method facilitates the electro-optic analysis of the real colloidal systems. The investigation demonstrates that the electro-optic parameters determined for a highly dilute colloidal system in the absence of multiple scattering can be obtained for systems of higher volume fractions displaying significant optical interactions. Precision is reduced in the vicinity of the electro-optic sign reversal where the effects decrease strongly. The method is applicable at constant dispersity of the samples (controlled by the saturation value of the effect and particle relaxation time). It permits to follow precisely the relative variations of particle electric polarizability and to determine its relaxation frequency. Only the determination of the absolute value of electric polarizability requires knowledge of the optical coefficients.

We apply the proposed method to a series of colloidal systems containing particles of different geometries, structures, and mechanisms of surface charging. The aim is to demonstrate similar tendencies in the frequency variation of the electro-optic responses in the low-frequency range and the applicability of the proposed procedure for analysis of the effects in this range. As generally accepted, the kilohertz effect is determined by the polarization of the particle Debye atmosphere. Comparing the results obtained for the studied suspensions (and other literature data) in the hertz domain, we come to the conclusion that the superposition of two slow effects (linear and quadratic with field intensity) explains the complicated frequency curves in this domain. The two slow effects are related to surface processes and are displayed through electrokinetic rotation. The negative slow effect is observed for all polarizable particles (at sufficiently high fields) and displays the features of an induced dipole effect dependent on the kilohertz-induced moment. It seems to be a slow stage of the surface polarization process, related to electrokinetic charge.

The positive slow effect shows permanent dipole-like behavior. It is displayed only at certain conditions and is not directly dependent on the kilohertz-induced moment. The positive effect is probably due to surface charge nonuniformity. For disk-shaped particles the asymmetric charge is expected to be in transverse direction (on the disk surfaces) and the positive effect would change sign to negative. The superposition of two negative effects in the same domain leads to sign reversal of the responses at low field intensity and significant negative effect at high fields. This could explain the low-frequency behavior of disk-shaped particles like bentonite, purple membrane fragments, etc.

Although the obtained results are sufficiently general (at least for the low-electrolyte regime) our attempts to correlate them with existing theories of surface polarization of charged particles [23–28] meet difficulties. Relaxation frequency of the kilohertz-induced dipole shows expressed dependence on particle surface charge, as predicted by Maxwell–Wagner–O'Konski theory [23] or an extended Maxwell–Wagner theory [28]. The negative effect in the hertz domain shows some of the features of the electrokinetic model [24–27]. The discouraging point is that our data for relaxation frequencies is two orders of magnitude lower than that predicted by the theories. Only the positive slow effect finds natural explanation in the theoretical schemes describing surface charge nonuniformity [29–31].

REFERENCES

1. Lyklema, J., *Fundamentals of Interface and Colloid Science*, Vol. II, *Solid–Liquid Interfaces*, Academic Press, San Diego, CA, 1995.
2. Stoylov, S.P., *Colloid Electro-Optics*, Academic Press, London, 1991.

3. O'Konski, C.T., Yoshioka, K., and Orttung, W.H., Determination of electric and optical parameters from saturation of electric birefringence on solutions, *J. Phys. Chem.*, 63, 1558, 1959.

4. Wang, G.Y. and Porschke, D., Dipole reversal in bacteriorhodopsin and separation of dipole components, *J. Phys. Chem.*, B, 107, 4632, 2003; Porschke D., Slow modes of polarization in purple membranes, *Phys. Chem. Chem. Phys.*, 6, 165, 2004.

5. Yamaoka, K., Tanigawa, M., and Sasai, R., *J. Chem. Phys.*, 101, 1625, 1994.

6. Stoimenova, M. and Peikov, V., Electro-optic investigation of oxide suspensions. Scaling of the electro-optic effects by field intensity, *J. Colloid Interface Sci.*, 164, 285, 1994.

7. Peikov, V., Radeva, Ts., and Stoimenova, M., Electro-optics of semidilute dispersions: separation of electrical and optical effects, *J. Colloid Interface Sci.*, 168, 1, 1994.

8. Miteva, S. and Stoimenova, M., Electro-optic characteristics of aqueous β-FeOOH particles, *J. Colloid Interface Sci.*, 273, 490, 2004.

9. Miteva, S. and Stoimenova, M., Electro-optic characteristics of optically interacting particles, *J. Colloid Interface Sci.*, 274, 531, 2004.

10. Stoimenova, M. and Miteva, S., Steady electro-optic parameters of charged colloidal particles, *J. Colloid Interface Sci.*, inpress.

11. Stoimenova, M. and Stoylov, S., Orientational optic effects from above Rayleigh particles, *J. Colloid Interface Sci.*, 76, 502, 1980.

12. Stoimenova, M. and Buleva, M., Electro-optic investigation of oxide suspensions. Optic effects in semidilute dispersions, *J. Colloid Interface Sci.*, 152, 483, 1992.

13. Stoimenova, M., Radeva, Ts., and Stoylov, S., On the low frequency dispersion of the electro-optic effects, *Colloid Polym. Sci.*, 257, 1226, 1979.

14. Radeva, Ts. and Stoimenova, M., Low frequency dispersion of the electro-optic effects from monocationic forms of palygorskite, *J. Colloid Interface Sci.*, 76, 315, 1980.

15. Peikov, V., "Electro-optic Investigation of Semidilute Disperse Systems", Ph.D. thesis, Bulg. Acad. Sci., Sofia, 1994.

16. Buleva, M. and Stoimenova, M., Electro-optic investigation of oxide suspensions. Mechanism of formation of the induced dipole moment, *J. Colloid Interface Sci.*, 141, 426, 1991.

17. Stoimenova, M. and Radeva, Ts., Electro-optic investigation of oxide suspensions. On the nature of "permanent" dipole moment, *J. Colloid Interface Sci.*, 141, 433, 1991.

18. Radeva, Ts., Overcharging of ellipsoidal particles by oppositely charged polyelectrolytes, *Colloids Surf.*, A, 209, 219, 2002.

19. Buleva, M., "Electro Moments of Colloid Particles and Structure of the Electric Double Layer", Ph.D. thesis, Bulg. Acad. Sci., Sofia, 1990.

20. Stoimenova, M. and Zhivkov, A., Electro-optic investigation of oxide suspensions. Surface charge sign reversal through specific adsorption, *J. Colloid Interface Sci.*, 142, 92, 1991.

21. Radeva, Ts. and Stoimenova, M., Electro-optic investigation on polymer adsorption and its influence on the electric double layer of colloid particles, *Colloids Surf.*, A, 54, 235, 1991.

22. Radeva, Ts. and Stoimenova, M., Electro-optic study of colloid dispersions stabilized by polyelectrolyte adsorption, *J. Colloid Interface Sci.*, 160, 475, 1993.

23. O'Konski, C. and Krause, S., Theory of the Kerr constant of rigid conducting dipolar macromolecules, *J. Phys. Chem.*, 74, 3243, 1970.

24. Dukhin, S.S. and Shilov, V.N. *Dielectric Phenomena and the Double Layer in Disperse Systems and Polyelectrolytes*, Wiley Interscience, New York, 1974.

25. Fixman, M., Charged macromolecules in external fields, *J. Chem. Phys.*, 72, 5177, 1980.

26. DeLacey, E.H.B. and White, L.R., Dielectric response and conductivity of dilute suspension of colloidal particles, *J. Chem. Soc., Faraday Trans. 2*, 77, 2007, 1981.

27. Russel, W., Saville, D., and Schowalter, W., *Colloidal Dispersions*, Cambridge University Press, Cambridge, 1989.

28. Saville, D., Bellini, T., Degiorgio, V., and Mantegazza, F., *J. Chem. Phys.*, 113, 6974, 2000.

29. Anderson, J.L., Effect of nonuniform zeta potential on particle movement in electric fields, *J. Colloid Interface Sci.*, 105, 45, 1985.

30. Velegol, D., Electrophoresis of randomly charged particles, *Electrophoresis*, 23, 2023, 2002.

31. Jones, J.F. et al., Charge nonuniformity light scattering, *Colloids Surf.*, A, 267, 79, 2005.

Part II

New Applications

*A. Rigid Particles,
Polyelectrolytes, Biosystems*

9 Counterions Dynamics as Studied by Electric Light Scattering

Ivana B. Petkanchin

CONTENTS

9.1 INTRODUCTION

The electric double layer (EDL) around a colloid particle, in aqueous media, in the absence of an external field is in equilibrium. This state is characterized by a minimum of free surface energy, and according to Lyklema this layer is "relaxed" [1]. The application of an electric field will deform or polarize the EDL, giving rise to a nonequilibrium state, characterized by an induced dipole moment. The nonequilibrium state is manifested in electrophoresis and in dielectric, dielectrophoretic (dipolophoretic), and electro-optical methods of investigation. In electrophoresis, the polarization of EDL will produce an additional impeding force superimposed with the Stockes viscous force and electrophoretic retardation. This phenomenon was first introduced and theoretically described by Overbeek [2] and later considered in the theory of electrophoresis as well [3,4]. It has to be noted that the theoretical works of Dukhin [4,5] on the nonequilibrium EDL and its influence on electrophoresis were initiated on a larger scale from electric light scattering (ELS) studies on electric polarizability of various colloidal particles at changed electrolyte concentrations or by adsorption of surface-active substances [6,7]. The longstanding discussions with a member of the Bulgarian electro-optic group could be considered as stimulating the development of nonequilibrium EDL theory.

In dielectric and electro-optic methods, the appearance of a particle-induced dipole moment leads to changes in orientation of the particles, and subsequent changes in dielectric and electro-optical properties of the disperse system [8,9]. According to Lyklema, the induced dipole moment, i.e., particle electric polarizability, is the "most basic characteristic of the polarized nonequilibrium double layer" [1]. All this determines the importance of the electro-optical methods as a main source of the electric polarizability, its value, and dynamics.

Recently, a great variety of colloidal particles: oxy hydroxide, clay minerals (natural and synthetic), bio-objects (purple membranes, fd virus, chloroplasts) were studied extensively by ELS in our electro-optical group. The results collected allow to draw an important conclusion that the electric polarizabilities of most of the colloid particles have an interfacial nature [8,9].

One basic feature of EDL is its dynamics, i.e., the time for EDL polarization or the time to restore the equilibrium distribution of the charges in a polarized EDL when the external field is switched off. Unfortunately, studies on this problem are not widespread. Considering electrophoresis, which is performed mainly in dc fields, to study the time dependence of electrophoretic mobility at short times ($10^{-9}-10^{-5}$ s) after the field is switched on is a complicated task from an experimental point of view. On the contrary, electro-optic and dielectric methods offer the possibility to follow EDL dynamics, when an ac electric field is applied, by studying the frequency dependence of the electro-optical effect, or of the dielectric increment.

From the frequency dependence of the electro-optical effect, one can distinguish different polarization mechanisms to obtain values of electric polarizability (induced dipole moment) and its relaxation time (τ_γ).

Recently, the dynamics of EDL was analyzed and a network simulation method was used to solve the differential equations, describing the polarization processes in EDL [7–13]. This method is applied to obtain the frequency response of the real and imaginary part of the dielectric increment at different κa (κ^{-1}, thickness of EDL, a, particle radius) for different ζ-potentials. The results obtained coincide well with those of De Lacey and White by using the standard electrokinetic model [14]. The increase of ζ-potential at constant κa leads to an increase in the dielectric increment. The critical relaxation frequency of the latter, with κa increase, at $\zeta =$ constant, is shifted to higher frequencies.

This chapter reviews the obtained results from ELS with an emphasis on counterions dynamics. The influence of ionic conditions, of adsorption (surfactants, neutral polymers, and polyelectrolytes) on relaxation time of electric polarizability will also be considered. As ELS is basically used, a brief description of the method will be given. The process connected with the polarization of EDL in external electric fields will be described within the framework of the model developed in Ref. [10].

9.2 POLARIZATION OF COLLOID PARTICLES WITH EDL IN EXTERNAL ELECTRIC FIELDS

The application of an electric field of strength E, on anisometric particle of length l and dielectric constant ε_p, in aqueous solution of dielectric constant ε_m, gives rise to accumulation of charges at particle interface. This process is known as Maxwell–Wagner (MW) interfacial polarization mechanism [15,16] and builds the noncharged (not connected with particle's charge) component of the induced dipole moment, $\bar{p}^{(1)}$. The latter moment is directed opposite to electric field direction when $\varepsilon_p<\varepsilon_m$, and is determined by the difference in the dielectric constant and conductivity of particle and media. In Figure 9.1, this process is shown schematically for an ellipsoidal particle. The tangential component (with respect to the particle surface) of the applied electric field (E_0), E_t causes movement of charges and their accumulation at particle surface. The normal component of E_0, E_n exerts a torque leading to particle orientation [17].

The relaxation time of MW polarization process of spherical, charged particle with radius a, is given by [8]

$$\tau_{MW} = \frac{(2\varepsilon_m + \varepsilon_p)\varepsilon_0}{2K_m + \left(K_p + \frac{2K^\sigma}{a}\right)} \tag{9.1}$$

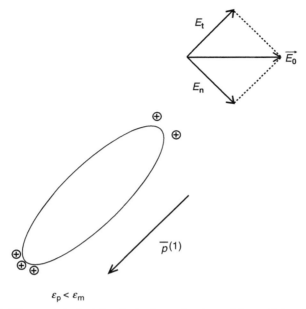

FIGURE 9.1 Maxwell–Wagner interfacial polarization of anisometric particle with long axis 1 (bound charges at solid–liquid interface). Applied electric field direction—$\bar{E}_0$, tangential component—E_t, normal one-E_n.

where K^σ is the surface conductivity, introduced first in electro-optics by O'Konski [18], K_p and K_m are the conductivity of particle and media, respectively. The independence of τ_{MW} on particle dimensions and electric charge has to be noted. If the particle is nonconducting ($K_p = 0$) and if K^σ is not very large, τ_{MW} coincides with the electrolyte relaxation time

$$\tau_{MW} \approx \frac{\epsilon_m \epsilon_0}{K_m} = \frac{\kappa^{-2}}{D^+ + D^-} \tag{9.2}$$

where κ^{-1} is the thickness of EDL and D^+, D^- are diffusion coefficients of cations and anions, respectively. The induced dipole moment $\bar{p}^{(1)}$ of a spherical particle with radius a is given by

$$\bar{p}^{(1)} = \gamma^{(1)} \bar{E}_0 = C 4\pi \epsilon_0 \epsilon_m a^3 E_0 \tag{9.3}$$

where, $\gamma^{(1)}$ is the particle interfacial electric polarizability, E_0 is the strength of the applied electric field, and C is the so-called dipole coefficient (or dipole strength). De Lacey and White [14] introduce the latter coefficient in their theoretical paper, considering the influence of EDL polarization on dielectric increment within the framework of standard electrokinetic model. In our opinion, it is more correct to use electric polarizability instead of particle dipole coefficient, having in mind the importance of the polarizability as a basic characteristic of the nonequilibrium double layer [1]. Generally speaking, the electric polarizability is proportional to the charge (σ_s), whose movement in applied electric field creates the induced dipole moment, and to charge mobility, u_i ($\gamma \sim \sigma u_i$). The surface conductivity K^σ is related to surface charge density, σ_s through the ions mobility u_i [19]:

$$K^\sigma = \sigma_s . u_i \tag{9.4}$$

As we are interested in EDL dynamics, an inspection of Equation 9.2 shows that an increase in the electrolyte concentration (compression of the EDL) leads to the decrease in relaxation time, which means an increase of the relaxation frequency. Typical relaxation times for MW interfacial polarization are in megahertz frequency range (10^{-7}–10^{-8} s) in 1–1 valent indifferent electrolyte with $10^{-3} M$ concentration.

When the time $t > \tau_{MW}$, a new induced dipole moment, $\bar{p}^{(2)}$, is created by the process of electromigration of counterions along the particle surface (the process could be called surface diffusion) by the tangential component of the applied field. This new dipole moment is determined mainly by the conductivities instead of difference in dielectric constants (MW interfacial polarization mechanism).

For nonconducting spherical particle with EDL, the surface electric polarizability $\gamma^{(2)}$ is given by [1]

$$\gamma^{(2)} = 4\pi\epsilon_0\epsilon_m a^3 \left(\frac{1 - Du}{1 + 2Du}\right) \tag{9.5}$$

where, $Du = K^\sigma / K_m.a$ is the Dukhin number, reflecting the relative contribution of surface conductivity for the polarization process. The relaxation time of this process is $\tau = (a^2)/D$, a is the particle radius, and D is counterions translational diffusion coefficient [1]. The electric polarizability $\gamma^{(2)}$ due to the electromigration predominantly of counterions along particle surface building the respective ionic fluxes, and presents the charge-dependent ionic component of the induced dipole moment, $p^{(2)}$. In Equation 9.4, σ_s is the particle surface charge. Schwarz [20] used a similar mechanism when considering the dielectric dispersion of colloid particles. But he neglected any fluxes of charges into or out of the bulk solution, assuming that the solution charge near to the particle responds to the applied electric field only by surface diffusion.

In Figure 9.2, the polarization of negatively charged anisometric particle is shown schematically, when $Du \ll 1$ (small value of K^σ) by the tangential electric field component, E_t. The

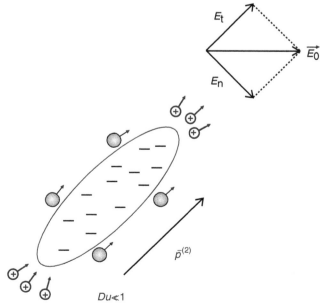

FIGURE 9.2 Polarization of negatively charged anisometric particle by electromigration (surface electric diffusion) $Du \ll 1$.

interaction of polarized charge, created by E_t, with the normal field component, E_n would lead to a particle orientation in field direction.

In summary, the electromigration of the excess ions in EDL is the reason for double layer polarization and its shift from equilibrium to a nonequilibrium state. The additional electric surface current proportional to the surface conductivity, K^σ, redistributes the electric charges in EDL, so creating an induced dipole moment. The EDL polarization will change the hydro-dynamic and electrolyte concentration fields around the particle. The tangential movement of counterions, i.e., along particle surface, increases their accumulation at particle top and depletes them at the particle bottom. Some of the ions diffuse into the bulk electrolyte and others will encounter cations coming from the bulk solution. These processes build up an excess and deficit of neutral electrolyte, at particle top and bottom, respectively, i.e., the appearance of an effect known as concentration polarization, or volume diffusion (VD) [1,5,21]. The concentration gradient gives rise to diffusion fluxes, $\bar{j}_d$ of counterions, opposite to the tangential electromigra-tion fluxes j_E, decreasing the polarization charge at both ends of an anisometric particle. This is shown in Figure 9.3 for a negatively charged particle. The electrolyte concentration increase at the top of the particle and respective decrease at the bottom means a decrease and increase of EDL thickness, respectively. The concentration polarization creates a third induced dipole moment $\bar{p}^{(3)}$, directed opposite to field direction when $Du \ll 1$.

The relaxation time, τ_{VD}, of the concentration polarization in the case of ellipsoidal particle scales as [10]

$$\tau_{VD} = \frac{l^2}{2D_{eff}} \qquad (9.6)$$

where, l is the particle long dimension and $D_{eff} = (2D^+ D^-)/(D^+ + D^-)$. On the basis of the above, briefly considered physical picture and applying network-simulation method developed in Refs.

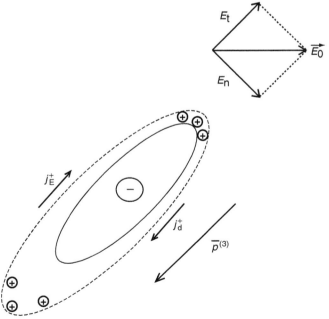

FIGURE 9.3 Polarization of a negatively charged anisometric particle by volume diffusion. j_E^+—electromigration counterions fluxes, j_d^+—diffusion counterions fluxes.

[11,12], the induced dipole coefficient C is calculated as a function of the time t for different ζ-potentials [10], showing two relaxation regions: one of MW polarization and second of VD, $\tau_{\mathrm{MW}} < \tau_{\mathrm{VD}}$. The opposite inequality is valid for critical relaxation frequency of the electro-optical effect ($\nu_{\mathrm{cr,MW}} > \nu_{\mathrm{cr,VD}}$). The dipole coefficient increases with increase in the ζ-potential.

9.3   POLARIZATION OF POLYELECTROLYTES IN EXTERNAL ELECTRIC FIELDS

Recently, many theoretical and experimental works on electro-optical phenomena in poly-electrolyte systems are published [22–26]. It is shown that electric birefringence in suspensions from nonspherical polyelectrolytes has an electrokinetic nature and the Kerr constant B gives information about the electric polarizability. The experimental studies on electric birefringence of nearly spherical and rodlike monodisperse polyelectrolyte particles have shown that B is a decreasing function of field frequency with non-Debey behavior. The function is described well with Fixman function and is similar to those of the real part of the dielectric increment ($\Delta \varepsilon'$) of spherical polyelectrolytes [27]. The relaxation time of spherically charged particle of radius R in the approximation of thin double layer ($\kappa R \gg 1$) is $\tau_{\mathrm{r}} = R^2/2D_i$, where D_i –is the diffusion coefficient of counterions.

The similarity between electric birefringence and dielectric phenomena probably is due to the fact that the time for EDL polarization is important in both studies. This time is determined by the time for which the small counterions diffuse over a distance comparable with particle dimensions. Aqueous dispersion of rodlike charged polytetrafluorethylene (PTFE) particles are studied at different ionic strengths and ζ-potentials [23,24]. The latter is independent of the ionic strength but strongly dependent on the surface density of ionizable sites [25]. The characteristic relaxation times are obtained by fitting the Fixman function of electric birefringence and dielectric increment data. It is shown that relaxation time $\tau_{\mathrm{s,B}}$ (from birefringence data) strongly depends on the ionic strength. On the contrary $\tau_{\mathrm{s,\varepsilon}}$ (from dielectric dispersion) does not depend on ionic strength and ζ-potential. The relaxation frequency of Kerr constant of PTFE particles at lowest surface charge is around 10^4–10^5 Hz, which was two orders of magnitude smaller than those for particles with higher surface charge. A second relaxation (undershoot) occurs above 10^7 Hz showing a dependence on the ionic strength [26]. The second relaxation is explained with the anisotropy of particles' electric polarizability.

In summary, the polarization of polyelectrolytes in external electric fields could be regarded with a model combining MW and electrokinetic approach. The relaxation of Kerr constant below 1 mHz is explained with electric and hydrodynamic processes in EDL. The experimental results obtained, concerning the counterions dynamics, refer to polyelectrolyte particle whose charge is regulated through surfactant adsorption. In our opinion it is better to study the model particles like oxy hydroxide because they could be obtained with high monodispersity and the nature of their EDL is well studied and understood.

9.4   FREQUENCY-RESOLVED ELECTRIC LIGHT SCATTERING

The relative change in scattered light intensity (α) of colloidal dispersion from anisometric particles in aqueous media, at low degrees of orientation ($U/kT \ll 1$, where U is the potential energy of particle in applied electric field of strength E and kT is the thermal energy), can be written as

$$\alpha = \frac{I_{\mathrm{E}} - I_0}{I_0} = \frac{\int_{\omega} \left(3\cos^2\theta - 1\right) i\mathrm{d}\omega}{2\int_{\omega} i\mathrm{d}\omega}(\mu^2 + \delta)E^2 \tag{9.7}$$

where I_E and I_0 are intensities of scattered light in the presence of electric field and without the presence of electric field respectively; θ is the angle between particle symmetry axis and the direction of the electric field; $\mu = p/kT$, where p is the permanent dipole moment along particle symmetry axis; $\delta = \Delta\gamma/kT$, where $\Delta\gamma$ is the particle electric polarizability anisotropy, $\Delta\gamma = \gamma_\parallel - \gamma_\perp$, $\gamma_\parallel, \gamma_\perp$ electric polarizability along symmetry and transverse axes of the particle, respectively; $d\omega = 2\pi \sin\theta\, d\theta$ is the space element; i is the intensity of the scattered light by a single particle with certain orientation. The relative quantity α is called electro-optical effect [8,9].

Disperse particles having a structural charge asymmetry means that they possess a permanent dipole moment (p), independent of the electric field. The electric polarizability depends on field-induced asymmetry of surface charge distribution. In other words, the applied electric field creates an interfacial induced dipole moment ($\bar{p}$) by perturbation of the symmetry of equilibrium EDL. The induced dipole moment is a linear function on E. The interaction of the particles' electric moments with the applied electric field results in a rotation torque M, expressed as $M = [E_n \times E_t]\Delta\gamma$ (see Figure 9.2). The theory of ELS refers to elastic light scattering (no energy exchange) from independent scatterers (small particles' concentration) and excludes interparticle interactions. The monodisperse particles should be in Rayleigh–Debey–Gans (RDG) approximation (relative refractive index of the particles close to 1; small difference in the phase of electromagnetic wave in the particle and out of it [8,9]. The steady-state electro-optical effect at the above condition for elongated particles at low degrees of orientation, is given as

$$\alpha = \frac{A(K_l, K_b)}{I_0(K_l, K_b)}(\mu^2 + \delta)E^2 \qquad (9.8)$$

where $A(K_l, K_b)$ and $I_0(K_l, K_b)$ are optical functions depending on particle form and dimensions; l and b are long and transverse particle axes, respectively, $K = (2\pi)/\lambda_n \sin(\theta/2)$, where λ_n is the wavelength of the incident light. From the field-strength dependence of α, (Equation 9.8), knowing particle form and dimensions, the electric moments of the particles could be calculated.

When an alternating electric field is applied to suspension from anisometric particles, the magnitude of α depends on field frequency:

$$\alpha_\omega = \alpha_{st} + \alpha_{osc} \qquad (9.9)$$

where α_ω is the electro-optical effect at field frequency ν ($\omega = 2\pi\nu$), α_{st} is the steady-state component of α, and α_{osc} is the oscillating with double frequency component of α, because maximum value of α_{osc} corresponds to maximum of E, and E diminishes in the first half-cycle and again repeats in the second half-cycle. So α_{osc} oscillates with double frequency. If the particles have only a permanent dipole moment, the particle rotation in the second half-cycle of the field is in opposite direction, which means changes of electro-optical effect sign. If the particles have only an induced dipole moment, changes in the sign of α are not observed. So the sign of the electro-optical effect at changing field frequency allows to distinguish orientation due to permanent or induced dipole moments.

Theoretical expressions for α_ω are obtained based on the solution of Brownian motion equation, giving orientational distribution function. The latter depends on ω and time t. The expressions for α_ω of particles with permanent and induced dipole moments describing the frequency dispersion of α could be found in Ref. [8] and refer to effects related only to particle's rotational diffusion. This means that field frequency is low enough to obtain an equilibrium polarization of particle EDL, i.e., polarization charges could follow the frequency changes of E. The time needed for the development of an equilibrium EDL polarization is called relaxation time of electric polarizability, τ_γ. The latter is related with translational mobility of charges, whose displacement in an external electric field creates an induced dipole moment.

The dispersion dependence of the electro-optic effect α gives the possibility to distinguish different polarization mechanisms, i.e., orientation due to permanent or "permanent-like" and induced dipole moment (slow or fast). When the inequality $D_r \ll \nu \gg 1/\tau_\gamma$ is fulfilled, a dispersion of electric polarizability could be observed (in the expression given, D_r is the rotational diffusion coefficient of the particle and ν is the field frequency). The nonrelaxed induced dipole moment of colloid particles, as studied by ELS, acts in a frequency range between several kilohertz and several tens of kilohertz [8]. This frequency range is called "plateau" of the dispersion dependence of the electro-optical effect. After these frequencies, a relaxation of induced dipole moment should be expected.

Another possibility to study polarization mechanism in the time domain offers the application of pulses of reversing polarity. Theoretically this case was considered first by Tinoco and Yamaoka [28] for electric birefringence at low degrees of orientation. Further, theoretical and experimental developments could be found in the papers of Yamaoka et al. [29,30]. Very often the observed minima or maxima of electric birefringence at field reversal are interpreted as a manifestation of a slow induced dipole moment, related to the lower translational mobility of counterions, responsible for the formation of the induced dipole. Electric dichroism investigations of purple membrane (disc-like particles) have demonstrated the existence of slow polarization modes in the range of microseconds, probably due to counterions bound to the disc and less mobile than the ions in the diffuse double layer [31].

9.5 ELECTRIC LIGHT SCATTERING STUDIES OF COUNTERIONS DYNAMICS

Some experimental results obtained with ELS method (frequency dependence of α) for different particles–clay minerals (palygorskite and kaolinite) oxyhydroxide particles (β-FeOOH, SiO$_2$-aerosil) are used to show the influence of the ionic conditions (ionic strength, type of counterions, pH), of particle surface modification through neutral polymers or polyelectrolytes adsorption on the counterions dynamics.

In Figure 9.4, the calculated electric polarizability γ (Equation 9.8) of rodlike palygorskite particles (the particles do not posses a permanent dipole moment) is presented as a function on field frequency at different KCl concentrations. The decrease of γ could be seen clearly, and a shift of the relaxation frequency of γ toward higher frequencies with increase of KCl concentration

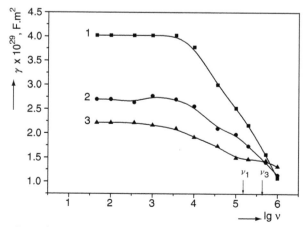

FIGURE 9.4 Frequency dependence of electric polarizability γ of palygorskite particles at different KCl concentration 9.05×10^{-5} (curve 1, ■), 2.73×10^{-4} (curve 2, ●), and 4.6×10^{-4} mol/L (curve 3, ▲).

[32]. The same trend is observed for aerosil particles (fractal aggregates) [33] and purple membrane fragments (disc-like particles) at increasing NaCl (indifferent electrolyte) concentration [34]. In Figure 9.4, the two arrows mark the frequencies where γ decreases twice from its plateau value, indicating the critical frequency of electric polarizability [32]. Because our ELS device is limited to frequencies below $500 \, \text{kHz}$, the marked values of ν_1 and ν_3 are approximate, but a trend of an increase in ν_{crit} with an increase in electrolyte concentration is seen. The approximate relaxation time of γ, in the studied case, decreases twice, and in the range $(1-0.5) \times 10^{-6} \, \text{s}$, for a fivefold increase in KCl concentration. The decrease of τ_γ means that the counterions appear to be more quicker when the EDL thickness becomes smaller, i.e., increased counterions mobility. The reason for this enchanced mobility is not clear. Some cooperative effects, arising in the compressed EDL (or increase of the charge near to particle surface), should be taken into account. Suppose, if the mobility of the counterions is not changed, the decrease of τ_γ means displacement of counterions at a smaller distance ($L = l_p + 1/\kappa$) with the electrolyte concentration increase.

In another ELS study, the influence of counterions type (F^-, Cl^-, Br^-, and I^-) at constant ionic strength ($10^{-4} \, M$) on the frequency dependence of β-FeOOH suspensions was followed [35]. The relaxation times of electric polarizability follow the order $\tau_{\gamma, I^-} > \tau_{\gamma, Cl^-, Br^-} > \tau_{\gamma, F^-}$, and approximately in the range $3 \times 10^{-6} - 4 \times 10^{-7} \, \text{s}$. The greater $\tau_{\gamma, \, I^-}$ compared with F^- ions at constant particle dimensions and constant ionic strength means a decreased I^- mobility. The latter fact could be explained with the larger I^- radius and greater adsorption ability, and some interactions with the surface groups.

The increase of fd-virus (optically "soft"—in RDG approximation, and rodlike form) particle concentration shifts the high-frequency relaxation of the electro-optical effect α toward higher frequencies [36], i.e., the relaxation time of the polarizability decreases. This effect is a manifestation of electrostatic interactions when the diffuse parts of EDL are overlapped at increased particle concentration. In the overlapping zone, the ionic concentration is locally increased, which is equivalent to ionic strength increase. In both cases (concentration of the particles or increase in ionic strength), the systems respond with a frequency shift of α in the same direction (to higher frequencies).

The study on the influence of Al_2O_3 (fractal aggregates) dimension on their electro-optical behavior has shown two high relaxation ranges of α, one relaxing in the kilohertz range and second one appearing in hundreds of kilohertz [37]. One possible explanation could be related to the aggregates structure and the existence of two counterions types, one located in "inner" pores EDL and other in "outer" EDL. The first-type counterions are with decreased mobility whereas the "outer" counterions are more free than the "inner" one.

The study of the electro-optical behavior of β-FeOOH or kaolinite particles when neutral polymer polyacrylamide (PAM) is adsorbed has revealed that the high-frequency relaxation of α is shifted toward higher frequencies with an increase in PAM concentration [38,39]. A similar effect was observed for mica particles in the presence of adsorbed PAM [40]. The shift of high-frequency relaxation i.e., the decrease of polarizability relaxation time, τ_γ is explained in Ref. [41] with conformational transition in the adsorbed polymer layer. In the region of transition the formation of loops and tails and the change of layer dielectric constant would influence τ_γ. It has to be noted that the adsorption of PAM on β-FeOOH or kaolinite particles at PAM concentration ($10^{-3} \, \text{g dm}^{-3}$) causes stabilization of the suspensions. Secondly, a similar effect, the start of an electro-optical effect relaxation at higher frequency (more longer "plateau" region) was observed with a decrease of diffuse layer thickness (case considered before).

In Figure 9.5, the dispersion dependence of α for β-FeOOH particles (curve 1) and in the presence of adsorbed sodium polystyrene sulfonate (NaPSS) (strongly charged polyelectrolyte), at a concentration of 10^{-2}g dm^{-3} for low molecular weight ($M = 7 \times 10^4 \, \text{Da}$) (curve 2) and high molecular weight ($M = 1 \times 10^6 \, \text{Da}$) (curve 3) are presented. At this polymer concentration the particles are recharged and the dispersions are stabilized [42]. As can be seen, the kilohertz

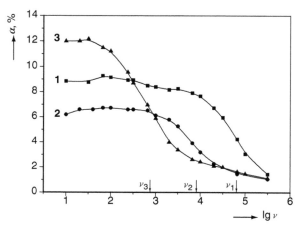

FIGURE 9.5 Frequency dependence of electro-optical effect α of β-FeOOH particles (curve 1, ■) and in the presence of NaPSS, MW $= 7 \times 10^4$ Da (curve 2, ●), and MW $= 1 \times 10^6$ Da (curve 3, ▲). Field strength 2.4×10^4 Vm^{-1}, pH $= 5.4$, suspensions conductivity 3.3×10^{-6}–6.5×10^{-6} Sm^{-1}.

electro-optical effect related with the movement of free counterions decreases. Secondly at low frequencies an additional electro-optical effect appears, whose relaxation frequency (ν_2, ν_3) is shifted toward lower frequencies compared to the relaxation frequencies of free counterions (ν_1). The relaxation frequency ν_1, ν_2, and ν_3 are determined as the frequency at which the plateau electro-optical effect decreases twice, and is used to show the tendency and not used to calculate the electric polarizability relaxation time, τ_γ. This frequency shift is an indication for decreased counterions mobility. One possible explanation could be the binding (or condensation) of counterions to the adsorbed polyions. The fact that $\nu_3 < \nu_2$ is explicable with the higher charge of NaPSS with higher molecular weight. The same trend of relaxation frequency shift is observed for recharged β-FeOOH particles in presence of adsorbed hydrolyzed polyacrylamide (HPAM 23). In Figure 9.6, the obtained results are shown for β-FeOOH without polymer (curve 1) and in the presence of 10^{-2} g dm^{-3} HPAM 23 (curve 2) [43]. The electro-optical effect at given frequency, α_ν is scaled to the electro-optical effect in kilohertz range (α_{pl}). With arrows are marked the high relaxations frequencies for both curves. Again the low-frequency additional electro-optical effect appears and the relaxation frequency of α when HPAM 23 is adsorbed is shifted in the low-frequency range. In the absence of HPAM, only high-frequency relaxation is observed (5–500 kHz) (curve 1). The relaxation of α (100 Hz–10 kHz) could be explained with a decrease of mobility of the counterions as a result of their "condensation" or "bounding" on the polyions, nevertheless, on the Manning charge parameter $\zeta < 1$ for the studied HPAM 23 [44]. Some conformational transition could lead to an increase in effective polymer charge, allowing "condensation" or "binding" of counterions.

The adsorption of low-charged polyelectrolyte, polyethyleneimine (PEI) with pH-dependent charge on α-Fe$_2$O$_3$ particles have shown again the decrease of the relaxation frequency of the electro-optical kilohertz effect [45]. The same trend is observed when polyacrylic acid is adsorbed on α-Al$_2$O$_3$ particles and the particles are recharged [46]. A very interesting dispersion dependencies were obtained, when low-charged PEI is adsorbed on aerosil particles (fractal aggregates) after the isoelectric point of the particles at 10^{-6} monomol/L PEI. [47]. In Figure 9.7, the dispersion dependencies of aerosil (SiO$_2$) suspensions at increasing PEI concentration are presented. Here we observe the opposite trend to that shown previously; the relaxation frequency of the kilohertz electro-optical effect shifts toward higher frequencies. It has to be noted that the studied suspensions are stabilized (no changes in particles relaxation

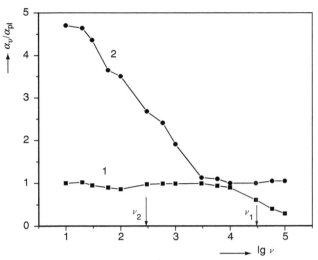

FIGURE 9.6 Frequency dependence of the scaled electro-optical effect ($\alpha_\nu \alpha_{pl}$) for β-FeOOH particles (curve 1, ■) and in presence of 10^{-2} g dm^{-3} HPAM 23 (curve 2, ●). Field strength 3.1×10^4 Vm^{-1}, pH = 5.5.

times of disorientation are observed) and secondly the frequency dependence for bare aerosil particles at pH 6 and same ionic strength is similar to that shown in Figure 9.7 (curve 1). An increase of the polarizing charge [48] responsible for the orientation of the particles when PEI is adsorbed (the kilohertz effect increases) is obvious from Figure 9.7.

The influence of adsorption of Na-humate (natural polyelectrolyte) on aerosil particles (the polymer and particle surfaces have the same sign of surface charge) is shown in Figure 9.8. The frequency dependence of (α_ν/α_{pl}) is presented. Again a decrease of the relaxation frequency is observed ($\nu_2 < \nu_1$). The increase of Na–humate concentration leads to an increase of kilohertz electro-optical effect (results not shown). This fact shows that negatively charged polyions are

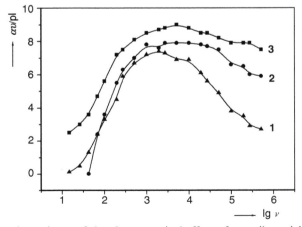

FIGURE 9.7 Frequency dependence of the electro-optical effect of aerosil particles in presence of PEI. 2×10^{-5} (curve 1, ▲), 2×10^{-4} (curve 2, ●), and 2×10^{-3} monomol/L (curve 3, ■). Field strength 3.3×10^4 Vm^{-1}, pH = 6, ionic strength 5×10^{-4}.

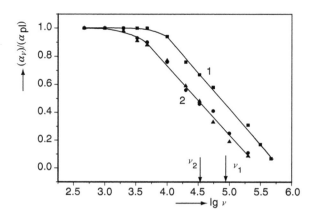

FIGURE 9.8 Frequency dependence of scaled electro-optical effect of aerosil particles (curve 1, ■) and in presence of 5×10^{-2} g dm^{-3} NaHS (▲), and 0.1 g dm^{-3} NaHS (●) (curve 2), pH $= 6$. Field strength 5.4×10^{4} Vm^{-1}.

adsorbed on the same charged surface because of strong specific interactions [49]. Regarding the frequency shift of relaxation frequency, a decreased counterions mobility, connected with their "binding" on the polyions could be supposed. We should take into account that the complicated and branched structures of NaHS (and when adsorbed) do not make possible any estimate of Manning charge parameter responsible for counterions "condensation" [50]. On the basis of the above data' it is difficult to present the reasons for a decrease in counterion mobility.

ELS studies of model anisometric particles, coated with polyelectrolyte multilayers, have shown that the polarization of "condensed" counterions is responsible for the electro-optical effect. The displacement of these counterions occurs on distances in the order of polymer contour length, not on the particle long dimension. A significant result obtained is that the counterions dynamics of the last adsorbed polyelectrolyte determines the electro-optical behavior of the whole multilayers [51].

9.6   CONCLUSIONS

The relaxation frequency of the induced dipole moment, i.e., the counterions dynamics, is shown to be dependent on EDL charge and thickness, on the conformation of the adsorbed neutral and charged polymers, on interparticle and specific interactions. These indicate the possibility for the application of electro-optical methods for studying counterions dynamics. When adsorbed charged polymers are studied, an important conclusion on "condensation" of counterions could be drawn.

The theoretical models of EDL polarization processes and their relaxation times should be compared with the results obtained from ELS studies (and also from other investigations, e.g., dielectric measurements, dipolophoresis). Such a comparison would help to better understand these processes. In this respect, the study of model colloid particles with a known nature of EDL, monodispersity, and form will be very helpful.

REFERENCES

1. Lyklema, J., *Fundamentals of Interface and Colloid Science*, Vol. II, Academic Press, London, 1995.
2. Overbeek, J.Th.G., Theorie der Electrophorese. Der Relaxations effect, *Koll. Beih*, 54, 287–364, 1943.

3. Booth, F., The cataphoresis of spherical, solid non-conducting particles in a symmetrical electrolyte, *Proc. R. Soc. London A*, 203, 514–533, 1950.

4. Dukhin, S.S., Theory of polarization of the thin double layer of colloidal particles, in *Investigation in the Field of Surface Forces*, Derjaguin, B.V., Ed., Nauka, Moskow, 1967, pp. 335–356.

5. Dukhin, S.S. and Shilov, N.V., *Dielectric Phenomena and Double Layer in Disperse Systems and Polyelectrolytes*, John Wiley and Sons, New York, 1974, p. 174.

6. Stoylov, S. and Petkanchin, I., Influence of 1–1 valent electrolyte on the electric polarizability of colloid particles in water solution, *Bulg. Acad. Nauk., Izv. Inst. Physikokhim.*, 5, 73–85, 1965.

7. Petkanchin, I., Sokerov, S., and Stoylov, S.P., Influence of surface active substances on the interfacial electric properties and stability of clay suspensions, in *Proceedings of the Sixth International Congress on Surface Activity*, Carl Hansen Verlag, Muenchen, 1973, pp. 671–677.

8. Stoylov, S., Sokerov, S., Shilov, N.V., Dukhin, S.S., and Petkanchin, I., *Electro-optics of Colloids*, Naukova Dumka, Kiev, 1977, p. 200.

9. Stoylov, S.P., *Colloids Electro-optics: Theory, Techniques, Applications*, Academic Press, London, 1991, p. 280.

10. Shilov, N.V., Delgado, A.V., González-Caballero, F., Horno, J., Lópes-Garcia, J.J., and Grosse, C., Polarization of the electrical double layer. Time evolution after application of an electric field, *J. Colloid Interface Sci.*, 232, 141–148, 2000.

11. Lópes-Garcia, J.J., Moya, A.A., Horno, J., Delgado, A.V., and González-Caballero, F., A network model of the electrical double layer around a spherical colloidal particle, *J. Colloid Interface Sci.*, 183, 124–132, 1996.

12. Lópes-Garcia, J.J., Horno, J., Delgado, A.V., and González Caballero, F., *J. Phys. Chem.*, 51, 103, 1999.

13. Lópes-Garcia, J.J., Horno, J., González-Caballero, F., Grosse, C., and Delgado, A.V., Dynamics of the electric double layer: analysis in the frequency and time domain, *J. Colloid Interface Sci.*, 228, 95–104, 2000.

14. De Lacey, E.H.B. and White, L.R., Dielectric response and conductivity of dilute suspensions of colloidal particles, *J. Chem. Soc. Faraday Trans. 2*, 77, 2007–2039, 1981.

15. Maxwell, J.C., *A Treatise on Electricity and Magnetism*, Clarendon Press, Oxford, 1892.

16. Wagner, K., Zur Theorie der unvollkommenem Dielectrica, *Ann. Phys.*, 40, 817, 1913.

17. Fixman, M., Charged macromolecules in external fields. 2. Preliminary remarks on the cylinder, *Macromolecules*, 13, 711–716, 1980.

18. O'Konski, C.T., Electric Properties of macromolecules. Theory of ionic polarization in polyelectrolytes, *J. Phys. Chem.*, 64, 605–619, 1960.

19. Ermolina, I., Morgan, H., Green, N.G., Milner, I.I., and Feldman, Yu., Dielectric spectroscopy of tobacco mosaic virus, *BBA*, 1622, 57–63, 2003.

20. Schwarz, G., A theory of frequency dielectric dispersion of colloid particles in electrolyte solution, *J. Phys. Chem.*, 66, 2636–2642, 1962.

21. Derjaguin, B.V. and Dukhin, S.S., Nonequilibrium double layer and electrokinetic phenomena, in *Surface and Colloid Science*, Matijevic, E., Ed., Vol. 7, John Wiley and Sons, New York, 1974, chap. 3.

22. Bellini, T., Degiorgio, V., and Mantegazza, F., The electric birefringence of polyelectrolytes: an electrokinetic approach, *Colloids Surf. A*, 140, 103–117, 1998.

23. Delgado, A.V. et al., Frequency dependence of the dielectric and electro-optic response in suspensions of charged rod-like colloidal particles, *Colloids Surf. A*, 140, 157–167, 1998.

24. Mantegazza, F. et al., Electrokinetic properties of colloids of variable charge. II. Electric birefringence versus dielectric properties, *J. Chem. Phys.*, 109, 6905–6910, 1998.

25. Bellini, T. et al., Electrokinetic properties of colloids of variable charge. I. Electrophoretic and electro-optic characterization, *J. Chem. Phys.*, 103, 8228–8237, 1995.

26. Bellini, T. et al., Electric polarizability of polyelectrolytes: Maxwell–Wagner and Electrokinetic relaxation, *Phys. Rev., Lett.*, 82, 5160–5163, 1999.

27. Fixman, M., Charged macromolecules in external fields. I. The sphere, *J. Chem. Phys.*, 72, 5177–5186, 1980.

28. Tinoco, I. and Yamaoka, K., The reversing pulse technique in electric birefringence, *Phys. Chem.*, 63, 423–427, 1959.

29. Yamaoka, K., Tanigawa, M., and Sasay, R., Reversing-pulse electric birefringence of disclike suspension in the low electric field region: an extension of the ion-fluctuation model, *J. Chem. Phys.*, 101, 1625–1631, 1994.
30. Sasai, R. and Yamaoka, K., Electro-optics of dispersed systems. 4 Steady-state electric birefringence of disk-shaped particles with various electric moments and orientation function: case of montmorillonite in aqueous media, *J. Phys. Chem.*, 99, 17754–17762, 1995.
31. Poerschke, D., Slow modes of polarization in purple membranes, *Phys. Chem. Chem. Phys.*, 6, 165–171, 2004.
32. Petkanchin, I., Electrical characterization of colloid particles, Ph.D. thesis, Institute of Physical Chemistry, Sofia, 1973.
33. Buleva, M. and Stoimenova, M., Electro-optic investigation of oxide suspensions. Mechanism of formation of the induced dipole moment, *J. Colloid Interface Sci.*, 141, 426–432, 1991.
34. Petkanchin, I., Taneva, S.G., and Todorov, G., Influence of surface active substances on the electric surface properties of purple membrane fragments, *Colloids Surf. A*, 52, 257–267, 1991.
35. Dimitrova, A., Peikov, V., and Petkanchin, I.B., Electric surface properties of model β-FeOOH particles in the presence of different counterions, *J. Colloid Interface Sci.*, 185, 548–550, 1997.
36. Peikov, V., Stoylov, S., and Petkanchin, I., Interactions and dimensions of fd-virus particles determined from electric light scattering, *J. Colloid Interface Sci.*, 171, 173–178, 1995.
37. Petkanchin, I. and Dimitrova, A., Electric surface properties of fractal aggregates—rotational and translational electrokinetics, *Colloids Surf. A*, 148, 35–41, 1999.
38. Radeva, Ts., Petkanchin, I., and Varoqui R., Electric surface properties in presence of neutral polymer, in *Colloid and Molecular Electro-optics*, Jennings, B.R. and Stoylov, S.P. Eds., IOP Publishing Ltd, Bristol, 1991, pp. 111–114.
39. Radeva, Ts., Petkanchin, I., Varoqui, R., and Stoylov, S.P., Electro-optical studies of Kaolinite /Polyacrylamide suspensions, *Colloids Surf. A*, 41, 353–361, 1989.
40. Radeva, Ts., Netzel, J., Petkanchin, I., and Stoylov, S.P., Electro-optic study of neutral polymer adsorption on mica, *J. Colloid Interface Sci.*, 160, 493–495, 1993.
41. Radeva, Ts. and Stoimenova, M., Electro-optic information on polymer adsorption, *Colloids Surf. A*, 54, 235–244, 1991.
42. Radeva, Ts. and Petkanchin, I., Electric properties of adsorbed polystyrenesulfonate. I. Dependence on the polyelectrolyte molecular weight, *J. Colloid Interface Sci.*, 220, 112–117, 1997.
43. Radeva, Ts. and Petkanchin, I., Electro-optic study of oxide particles in hydrolyzed polyacrylamide solutions, *J. Colloid Interface Sci.*, 182, 1–5, 1996.
44. Radeva, Ts., Widmaier, J., and Petkanchin, I., Adsorption of hydrolyzed polyacrylamides on ferric oxide particles. Counterions mobility in stabilized suspensions, *J. Colloid Interface Sci.*, 189, 23–26, 1997.
45. Radeva, Ts. and Petkanchin, I., Electric properties and conformation of polyethyleneimine at the hematite–aqueous solution interface, *J. Colloid Interface Sci.*, 196, 87–91, 1997.
46. Buleva M., Peikov, V., Pefferkorn, E., and Petkanchin, I., Adsorption of polyacrylic acid on α-Al_2O_3 colloid particles as studied by electro-optics, *Colloids Surf. A*, 186, 155–161, 2001.
47. Buleva, M. and Petkanchin, I., unpublished data, 2003.
48. Stoylov, S.P. and Petkanchin, I.B., Surface charge on the solid/water interface-heterogeneity and dynamics: an electro-optic approach, *Current Topics in Colloid and Interface Science*, 4, 1–13, 2001.
49. Buleva, M. and Petkanchin, I., Interaction of humic substances with silica and alumina colloids: adsorption and stability. Electro-optical study, *Colloids Surf. A*, 151, 225–231, 1999.
50. Manning, G., Limiting laws and counterion condensation in polyelectrolyte solutions. I. Colligative properties, *J. Chem. Phys.*, 51, 924–933, 1969.
51. Radeva, Ts., Milkova, V., and Petkanchin, I., Electro-optics of colloids coated with multilayers from strong polyelectrolytes: surface charge relaxation, *J. Colloid Interface Sci.*, 266, 141–147, 2003.

10 Optical Kerr Effect of DNA Oligomers and tRNA

Sergio Aragon, Martin Perez, Kerwin Ng, and Don Eden*

CONTENTS

10.1 INTRODUCTION

The structure of DNA is thought to play an important role in its function. In particular, it is known that some protein–DNA interactions cause large structural distortions in the bound DNA. The long-term goal of this study is to develop the optical Kerr effect experiment to detect these structural changes. Like the closely related transient electric birefringence experiment, which may also be termed as the electric Kerr effect, the birefringence of a solution is used to monitor changes in the structure of the molecules in the solution. In both the methods, an electric field is used to induce birefringence in a sample solution.

The early work in electric birefringence has been described in the classic book by Fredericq and Houssier [1], and Benoit's historical contribution at the beginning of this volume should

---

*In Memoriam.

also be consulted. The optical Kerr effect (OKE) has not been studied much for biomolecules and our exposition emphasizes this technique and compares it to the standard transient electric birefringence (TEB) on the same systems

In 1956, several years before the invention of the laser, Buckingham [2] theoretically predicted the possibility of inducing birefringence in an isotropic medium using a high intensity optical field. This effect is known as the optical Kerr effect, OKE. In 1964, Mayer and Gires [3] were the first to experimentally use the laser method to induce birefringence in a pure liquid. Paillette [4] then proceeded to study several different liquids. These early studies confirmed that the primary observed effect was due to molecular reorientation. In 1968, Kielich [5] provided an exposition of the theory of molecular interactions.

The technique was then used to study organic liquids, aqueous salt solutions, liquid crystals, and finally in 1974, Coles and Jennings observed tobacco mosaic virus (TMV), DNA, bentonite, and hyaluronic acid solutions [6]. They reported birefringence and relaxation results, but used long pulse lengths (around 200 μs) to reach an equilibrium orientation. Dobek et al. [7] studied the tRNA solutions and observed a change in birefringence, and presumably tertiary structures upon the addition of Mg^{2+} to the tRNA solutions. They used pulse lengths at half width of tens of nanoseconds. In an effort to determine what effects contribute to the OKE, they then studied the pulse lengths of the order of picoseconds [8]. The latest published works in this field did not have such convincing results. In 1987, Norden and coworkers [9] published an account of their efforts to reproduce the DNA orientation, observed by Coles and Jennings [6], and found no OKE. This work [9] predicts theoretically that the OKE of DNA would be unmeasurable. Our work shows that this is not the case.

In this chapter we demonstrate the use of the optical Kerr effect to provide quantitative structural information on very short double helical DNA fragments and a yeast tRNA.

In 1991, Eimer and Pecora [10] performed a study on fragments 8, 12, and 20 base-pairs in length; dynamic light scattering (DLS) was used to study the translational diffusion and dynamic depolarized light scattering (DPS) was used to study the rotational diffusion. These studies were performed in the high-Na buffer described below. We used identical solution conditions and samples and performed an OKE experiment for each of these fragments. Our results are compared to those published previously.

Transfer RNA provides the crucial link between the triple-base sequence codon and the insertion of a specific amino acid into a growing protein chain. This process involves a series of complex events. It is assumed that the tRNA is not a passive adapter between mRNA and the peptidyl transferase site. Many experiments have been performed on tRNA to support the fact that tRNA changes its conformation according to its environment. The experiments performed till 1977 were reviewed by Crothers and Cole [11]. Reeves et al. [12] examined $tRNA^{Ala}$ in various Mg^{2+} concentrations. They performed experiments involving melting curves, ultraviolet absorption spectra, circular dichroism spectra, hydrodynamic behavior (gel filtration and ultracentrifugation), and denaturation. Their data were interpreted to indicate that above $70 Mg^{2+}$ ions per tRNA molecule, the tRNA became much more compact. Thomas et al. [13] used fluorescence polarization anisotropy (FPA) to measure the tRNA rotational time as a function of Mg^{2+} concentration and found no change in the rotational time up to $40 Mg^{2+}$ per tRNA molecule. Electric dichroism measurements have offered evidence that supports Reeves' experiments. Porschke et al. [14] studied six different tRNA molecules, including $tRNA^{phe}$, by electric dichroism decays and by bead model simulations. In these experiments, a strong electric field is applied to the tRNA molecules to induce orientation. The decay rate of dichroism after removal of the field can be used to characterize the rotational diffusion time. Porschke et al. found a difference of nearly a factor of two in the rotational relaxation time in the ethylenediaminetetra-acetic acid (EDTA), which is slower, than in the $MgCl_2$-containing buffers. At about the same time, Patkowski et al. [15] performed a polarized and depolarized DLS study of the tRNA^{phe} in the absence of Mg^{2+}. The depolarized Rayleigh spectrum with a high-resolution confocal Fabry–Perot interferometer provided the rotational diffusion time, and polarized photon correlation spectroscopy provided the translational diffusion time. Their rotational relaxation measurements agree with the work of Thomas et al. [13]. A less direct technique to extract structural information is to utilize electro-optics to study tRNA without using dyes or fluorescence compounds. Dobek et al., using optical induced birefringence techniques [7], also observed a structural change in tRNA^{phe} induced by Mg^{2+}. The induction light was from an Yttrium Aluminum Garnet (YAG) laser, and the birefringence was observed by a weak He–Ne laser. The structural change was indicated by the change in optical Kerr constant only because they did not measure relaxation times.

TEB experiments [16] can yield the Kerr constant and rotational time of tRNA^{phe} simultaneously. However, because the applied electric field is from a constant electric current pulse, sample heating from ion conduction is a major limitation on ionic concentration. In our OKE experiments, the electric field that orients the molecule is from an infrared laser that oscillates rapidly, and the ions cannot respond to the rapid oscillation. Therefore, heating from ionic conduction is not a factor and that allows the ionic concentration to be much higher, including actual physiological conditions. In addition, no motion will result from a rapidly oscillating electric field interacting with the permanent dipole moment of the macromolecule. The mechanism that is responsible for the orientation is due to the induced electronic dipole moment [1]. As the probe source and the inducing electric field are in the optical frequency range, the observed birefringence is always positive. Thus, the data analysis is simpler without the coupling of the induced ionic dipole moment and the permanent dipole moment. OKE measurements are compared with TEB measurements of tRNA to show the differences in the two techniques.

10.2 THEORY

In both the OKE and TEB experiments, birefringence is transiently induced in an aqueous solution by an externally applied electric field. Birefringence is defined as the difference in the index of refraction along two perpendicular coordinate axes in the laboratory frame of reference. The probe beam propagation direction defines the z-axis, while the externally applied field defines the x-axis. Birefringence is the difference in index of refraction between the directions parallel (along x) and perpendicular (along y) to the applied field:

$$\Delta n = n_x - n_y = \frac{(\delta_x - \delta_y)\lambda}{2\pi\ell} \qquad (10.1)$$

where λ is the radiation wavelength, δ is the phase, and ℓ is the path length of the cell.

10.2.1 TEB Response

In the Kerr region, for an orienting field at constant intensity, at equilibrium we will reach the steady-state birefringence

$$\Delta n_{\text{ss}} = n\,K_{\text{sp}}\,C_{\text{v}}\varepsilon|E|^2 \qquad (10.2)$$

where K_{sp} is the specific Kerr constant, n is the solution index of refraction, and C_{v} is the volume fraction of the solute. In an ideal TEB experiment, the field can be turned off instantaneously so that temporal profile of its intensity is a perfect step function. The observed signal will then decay exponentially from Δn_{ss} to zero with multiple exponential components (depending on the five eigenvalues of the tensorial rotational diffusion operator). The decay transients for such a step-off are described by

$$\Delta n(t) = \Delta n_{ss} \sum_{i=1}^{5} A_i e^{-t/\tau_i}, \quad \text{where } \sum_{i=1}^{5} A_i = 1 \tag{10.3}$$

The mechanism of orientation of DNA in the TEB experiment is dominated by the ion-induced dipole, thus we expect the DNA axis to orient along the direction of the applied electric field.

10.2.2 OKE RESPONSE

We have derived the birefringence response of the noninteracting anisotropic macromolecules experiencing rotational relaxation in dilute solution [17]. The solvent is usually birefringent and adds to the total response function. The macromolecules are assumed to be dilute enough so that the bulk properties of the solvent are unaltered. Thus, we must linearly add the birefringent response of the solvent to that of the macromolecules. However, the solvent usually consists of small molecules that move very quickly, thus, the bulk response function is identical to its laser pulse function. Written out explicitly, the response function is

$$R_{\text{total}}(t) = n_{\text{bulk}} K_{\text{B}} f(t) + n_{\text{bulk}} K_{\text{sp}} C_{\text{v}} \int_{-\infty}^{t} f(t') \sum_{i=1}^{5} A_i \exp\left[-\frac{(t-t')}{\tau_i}\right] dt' \tag{10.4}$$

where the first term corresponds to the solvent, $f(t)$ is the laser pulse shape, n_{bulk} is the solution refractive index, K_{B} is the solvent Kerr constant, K_{sp} is the macromolecule Kerr constant, A_i is the relaxation amplitude, and τ_i is the relaxation time.

If $f(t)$ is a unit area Gaussian, the convolution integral may be solved analytically.

$$R_{\text{total}}(t) = n_{\text{bulk}} K_{\text{B}} f(t) + n_{\text{bulk}} K_{\text{sp}} C_{\text{v}}$$
$$\times \sum_{i=1}^{5} \frac{A_i \sigma \sqrt{\pi}}{\tau_i} \text{erf}\left[\sqrt{2}\left(\frac{t}{\sigma} - \frac{\sigma}{2\tau_i}\right)\right] \exp\left[-\left(\frac{t}{\tau_i} - \frac{\sigma^2}{4\tau_i^2}\right)\right] \tag{10.5}$$

This expression agrees with that found by Ho and Alfano [18].

In terms of the polarizability tensor components, the specific Kerr constant is given by [1]

$$K_{\text{sp}} = \frac{1}{n C_{\text{v}}} \lim_{E \to 0} \left(\frac{\Delta n}{E^2}\right) = \frac{\pi N}{15 k T n^2 C_{\text{v}}} \left[(\alpha_{11} - \alpha_{22})^2 + (\alpha_{11} - \alpha_{33})^2 + (\alpha_{22} - \alpha_{33})^2\right] \tag{10.6}$$

The components of the polarizability tensor are those in the body-centered reference frame. The two wavelengths in use in our OKE experiment, 488 and 532 nm, are close enough that setting α^i and α^o to be equal is reasonable for the DNA oligomer experiments. For DNA, the polarizability is nearly axially symmetric and is largest perpendicular to the longitudinal axis. DNA will tend to orient perpendicular to the applied electric field in this case.

10.3 MATERIALS AND METHODS

10.3.1 DNA OLIGOMERS

DNA oligomers were purchased from Operon Technologies, Alameda, CA, and CSUPERB's Microchemical Core Facility, San Diego State University, San Diego, CA. All samples were delivered, desalted, and lyophilized. The DNA from Operon was not pure and might

TABLE 10.1
The DNA Oligomer Sequences Used

Source	Sequence
SDSU-MCF	CGCGCGCG
SDSU-MCF	CGCGCGCGCGCG
SDSU-MCF	CGTACTAGTTAACTAGTACG
Operon	GCGAATTCGC
Operon	CGCGAATTCGCG

therefore be contaminated with an unknown quantity of failure sequences. These would consist of a range of molecular sizes and would thus make the measured data questionable. The DNA from San Diego is purified by polyacrylimide gel electrophoresis (PAGE), a procedure that is expected to remove most of the failure sequences. The DNA oligomer sequences used are shown in Table 10.1.

The following buffers were used:

Sodium cacodylate [7]: 0.1 M NaCl, 0.01 M Na-cacodylate, pH 6.5
High salt [10]: 50 mM phosphate, 100 mM NaCl, 2 mM EDTA, 0.1% NaN$_3$, pH $= 7$
Low salt:1 mM sodium ions, ions from Phosphate buffer made using monobasic and
  dibasic sodium phosphate salts, pH $= 7$

OKE was used to measure the rotational transport properties of B-DNA using the same sequences and buffers used by in the polarized and depolarized DLS of Eimer and Pecora [10]. Low salt buffers are required in TEB to avoid Joule heating in the sample and electrode polarization. For TEB measurements, the removal of the salt from the system was accomplished with commercially available dialysis tubes. The samples were equilibrated with the low salt buffer.

The DNA sample concentration is determined by UV and VIS spectrometry. The sample is diluted such that the absorbance may be measured in the range 0.05–1.5 molar extinction. The mass concentration is commonly determined by the absorbance at 260 nm, with 1 molar extinction indicating $C_{\text{molar extinction}} = 1 = A_{\text{molar extinction}} = 1/\epsilon\, l = 50 \mu g/mL$ [19]. The concentration of DNA is then given by

$$C = \left(50\, \frac{\mu g}{mL}\right) A_{\text{molar extinction}} \tag{10.7}$$

The theory to be applied is valid only in the dilute limit, so we check this condition by considering the value of NL^3, where N is the number density and L is the characteristic length. If NL^3 is less than 1, each of the N boxes the solvent volume is divided into is less than singly occupied and the particles are taken to be noninteracting. In a dynamic light-scattering study by Russo et. al. [20], it was shown that, for a solution of stiff-chain polymers, interparticle effects were not observed with $NL^3 < 100$. The DNA oligomer solution data are shown in Table 10.2.

The DNA arrives single-stranded in a lyophilized pellet contained in a small tube and must be resuspended and annealed. This must be done in buffer that is at least 100 μM sodium and pH near 7 [21]. The appropriate buffer was added to the tube, it was shaken and heated to 10°C below its melting point, and then was allowed to cool. All strands used are self-complementary, so it was not necessary to remove any unmatched strands. It was found that with the 12 mer and 20 mer purchased from the SDSU facility, a jellylike white solid

TABLE 10.2
DNA Oligomer Solution Data

Sequence	A260	BP	Dilution	μg/ml	μmol/ml	NL^3
GCGAATTCGC	1.207	10	100	6035	0.997	0.024
CGCGAATTCGCG	0.5779	12	100	2890	0.396	0.016
CGCGCGCG	0.9164	8	163.3	7482	1.552	0.019
CGCGCGCGCGCG	1.528	12	70	5347	0.733	0.030
CGTACTAGTTAACTAGTACG	1.095	20	140	7663	0.627	0.119

settled in the bottom of the tube. The clear solution was carefully drawn off the surface with a syringe and used for the experiments.

10.3.2 tRNA PREPARATION

The Mg^{2+} free configuration of tRNA was obtained from a suspension of the tRNA in EDTA buffer, and the Mg^{2+} dependent configuration was brought about by the addition of $MgCl_2$ buffers. H_2O base buffers were used in TEB experiments; D_2O based buffers were used in OKE experiments because they absorbed little infrared light at 1064 nm, allowing greater field strength to induce orientation in the sample. H_2O base buffers were prepared as follows. The $MgCl_2$ buffer consisted of 20 mM trishydroxymethylaminomethane (Tris), 18 mM NaCl, 0.5 mM $MgCl_2$. The EDTA buffer consisted of 20 mM Tris, 18 mM NaCl, 0.5 mM EDTA. The pH of the buffers was adjusted to 7.2. The D_2O buffer was prepared by lyophilizing the H_2O buffer, and dissolving the solids in an equivalent volume of D_2O.

5 mg of yeast tRNA^{phe} was purchased from Sigma. Half of the tRNA was transferred to EDTA H_2O buffer in a green Centricon tube, and the other half was transferred to $MgCl_2$ buffer. To perform the OKE experiments, the H_2O buffers were displaced by repeated washings with the D_2O buffers. The final volumes were about 200 μl each. After the OKE experiments, the tRNA was washed in H_2O buffers in green Centricon tubes and diluted to about 1 ml.

10.4 BIREFRINGENCE EXPERIMENTAL SETUP

10.4.1 TEB EXPERIMENTAL SETUP: RNA

The TEB experiments were described earlier in Highsmith and Eden [16] and the RNA work by Ng[22]. An Ar^+ probe laser was set at 100 mW at 514.5 nm. The incident polarization was set at 45° to the applied electric field in the sample cell. The sample cell was 2.8 cm in length and had a volume of 1 ml. Since the experiment was generally run at about 4°C, a plastic case, with holes for the laser beam, enclosed the cell, with nitrogen blowing constantly to prevent condensation of water. The electrodes were made of Platinum to avoid chemical reactions.

A schematic of the TEB experimental setup is shown in Figure 10.1. For the RNA measurements, the electric field in the cell was generated by a Spice SPI-5 high-voltage pulse generator. The applied voltage was typically at 800 V to generate an electric field of approximately 3000 V/cm across the gap of 2.7 mm for durations from 250 to 500 ns. A quarter-wave plate was used to convert the elliptically polarized argon light to linearly polarized light, and the light intensity after the analyzer changed with the birefringence of the sample. The optical signal was recorded in the same manner as in the OKE experiment. Instead of using the photodiode as a trigger for the digitizer, a trigger output from the pulse generator was used.

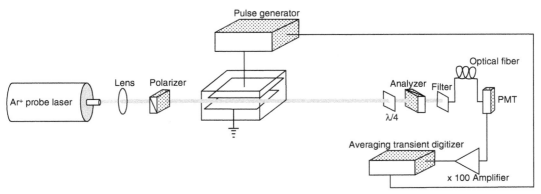

FIGURE 10.1 TEB experimental setup.

The birefringence signal was recorded by a LeCroy CAMAC Model 6880A digitizer. It had amplitude range of 500 mV, peak to peak with error less than $\pm 2\%$ and temperature coefficient of $\pm 0.1\%$ after calibration. The sampling interval was set at 0.742 ns. The step response of the overall detection circuitry was equal to or less than 1.5 ns rise time with less than 5% overshoot using a 1 ns rise time input pulse. The digitizer was controlled by a LeCroy CAMAC Model 6010 Magic Controller. The controller communicated commands and data to a personal computer via an IEEE 488 interface. A count of 1000 pulses were used for the background, and 10,000 pulses were typically used for each data set.

The data from TEB experiment differed from that of the OKE experiment in two respects. First, the inducing electric field was a square pulse, and the exact magnitude was known. Second, the birefringence arose from both the permanent and the induced dipole moments, whereas only the induced dipole moment contributed in the OKE experiments. Birefringence contributed by the permanent dipole moment was a first order process; therefore, the rotational times for the rise and decay were different. Despite the differences, the analysis procedure was similar. The acquired data were converted into birefringence units from the voltages in the digitizer, using a Fortran program written by K. Ng. Each data set was divided into sections of birefringence buildup and decay. The rise time and the decay were determined separately. The final results were average of the experimental data sets, and the 95% error limit was also determined by the Student T-test.

Figure 10.2 shows a typical birefringence response of the tRNA in a TEB experiment. The birefringence response of tRNA was significantly slower than that of the equipment and

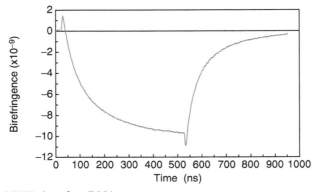

FIGURE 10.2 Typical TEB data for tRNA.

the buffer. Therefore, deconvolution of the tRNA and equipment response was not important. The signal in Figure 10.2 was broken into two parts, the buildup and the decay.

When the electric field was first applied, the birefringence buildup was positive. This is the contribution from the buffer. Experiments on the buffer yielded the system response time of 3.5 ns. The response of the buffer was expected to be significantly faster than 3.5 ns. However, this response was a convolution of the buffer signal and the response of the excitation and detection electronics. As time progressed, the total birefringence was dominated by the tRNA birefringence due to a slower rotational time. As the field strength was known, the Kerr constant was computed from the sum of the tRNA component amplitudes and the electric field squared. Pulses of 500 ns durations were used.

10.4.2 OKE Experimental Setup: RNA

The experimental setup of the OKE, shown in Figure 10.3, shared the same argon laser, photomultiplier tube (PMT), amplifier, and digitizer with the TEB experimental setup. The argon ion probe laser was typically set at 100 mW in the experiments. The Q-switched YAG laser was a General Photonic Model T002-44 with wavelength of 1064 nm with 90 MW pulses at a 2 Hz repetition frequency. The half amplitude full width was 10 ns. The polarization of the YAG laser was set at 45° to the polarization of the argon laser. The YAG laser was combined collinearly with the probe laser by dichroic mirror m_1. A small fraction of the YAG laser power was transmitted through the mirror and hit a photodiode that sent a trigger signal to the digitizer. The sample cell was made of a capillary tube with an inside diameter of 2 mm inside diameter and a length of 5 cm. Teflon heat shrink tubing held the round windows on both ends of the cell. The birefringence of the windows was negligible. The half-wave plate rotated the polarization of the YAG beam to create an opposite birefringence in the reference cell to cancel the buffer signal when required. The half-wave plate for the YAG beam was designed so that it was a full-wave plate for the argon beam. Thus, the polarization of the argon beam was unaltered by the plate. Mirror m_2 diverted the YAG beam to a beam dump. The power of the YAG pulse could be monitored at this location using a Coherent J25 pyroelectric power meter. The 514 nm quarter-wave plate was used to convert the elliptically polarized light of the probe beam into a linearly polarized light. Thus the light intensity after passing through the analyzer became a function of the sample birefringence. The optical signal was transmitted to the PMT by an optical fiber and was converted into an electrical signal, which was then amplified 100 times by two Comlinear CLC-100 video amplifiers; the −3dB bandwidth was up to 500 MHz. The second one was modified to provide ac coupling. The birefringence signal was recorded in the same manner as in the TEB experiments.

Nulling was performed before the sample birefringence recordings. To null the system, the analyzer and the quarter-wave plate were adjusted to minimize the current magnitude to the

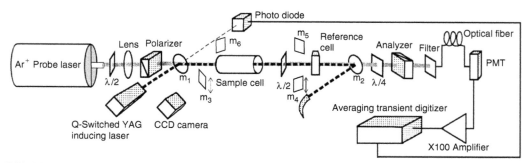

FIGURE 10.3 OKE experimental setup.

anode of PMT. Then the anode current was set to twice the minimum current by adjusting the analyzer. The position of the analyzer at these two current settings gave the extinction ratio. Then the analyzer was adjusted to $50\,\mu A$ for birefringence recording.

To determine the birefringence of the sample, the power of the YAG beam needed to be characterized. The Q-switched YAG beam had a Gaussian shape temporally, and the beam profile was called the lamp function. The parameters of the lamp function were determined by the birefringence of D_2O, because D_2O had a very fast response time, much less than 1 ns. D_2O was selected because it had a lower absorption in the infrared range than does H_2O.

After the excitation behavior was determined, the buffer signal was canceled using the reference cell so that the birefringence of the sample was more dominant. The peak buffer birefringence could be as high as 1.5×10^{-8}, while the sample signal could be as low as 0.3×10^{-8}. The amount of opposite retardation can be adjusted by rotating the half-wave plate. The half-wave plate was to be adjusted until the buffer signal peak was gone.

The background was collected with the probe beam blocked at the source, and the YAG laser was set pulsing. Immediately after the background was collected, the probe beam was unblocked to collect the data. The number of background pulses was determined by the signal quality. Two hundred pulses were generally enough for D_2O data. Several sets of data were collected for an accurate description of the YAG laser temporal profile. Typically, 200 pulses for the background and 1000 pulses for the data were collected. After each data set was collected, the system was nulled, and after the OKE experiments were done, the sample was diluted to just over 1 ml for TEB measurements.

10.4.3   OKE Experimental Setup: DNA Oligomers

The experimental setup for the DNA oligomer measurements was as shown in Figure 10.3 above, with two important differences. The Lecroy CAMAC was replaced by a Lecroy 334A digital oscilloscope (DSO), and the frequency of the beam from the General Photonics YAG laser was doubled to 532 nm to avoid infrared absorption in water. This required dichroic mirrors (m_1, m_2) for 532 and 488 nm operation. A Hamamatsu PMT, custom modified for exceptionally rapid response, was powered by a high-voltage source that is itself powered by an isolated transformer [23]. The PMT output was fed to a high-speed 10X video amplifier. The resulting current can be routed to the DSO when a temporal profile is required or to an ammeter (microampere range). The ammeter does not respond to rapid transients, making it useful for the measurement of stable baselines.

The DSO collects the temporal profile of the transmitted light during a birefringence inducing pulse. The output of the video amplifier is sampled at 2 gigasamples per second, that is, 1 sample per 0.5 ns. The length of the record, or trace, may be varied to capture extended transients. Traces are averaged to increase the signal to noise ratio. For OKE measurements, the scope trigger is provided by a photodiode that responds to green light that leaks past the 532 and 488 nm dichroic mirror. A Labview program, written by a previous group member [24], was used to automate the function of electronic shutters and complete data acquisition. The data were transferred to PC using a GPIB utility provided by the DSO manufacturer [25].

To enhance the sensitivity of the OKE detection, we use a solvent signal suppression system. Although DNA is highly anisotropic, most of the volume of the solution is made up of optically anisotropic water. The observed response will be the sum of the birefringence responses from water and DNA molecules. DNA is oriented more slowly than water, so for a short pulse (relative to the DNA relaxation times), the DNA signal will be small relative to the buffer signal. Because of the large difference in reorientation times, it is possible to selectively remove the signal component due to the buffer and leave only that due to the DNA.

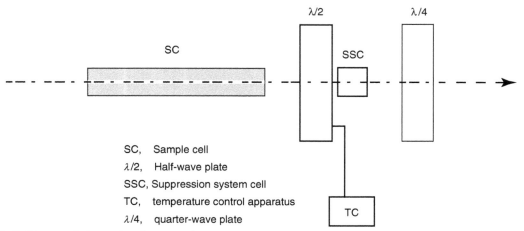

SC,   Sample cell
$\lambda/2$,   Half-wave plate
SSC, Suppression system cell
TC,   temperature control apparatus
$\lambda/4$,   quarter-wave plate

FIGURE 10.4 Solvent signal suppression system.

We place an optically anisotropic fluid (benzene) that can reorient as quickly as water in a quartz cuvette after the sample cell to create birefringence in the second cell but in the opposite sense from the sample cell. The polarization of the exciting beam can be rotated using a half-wave retardation plate for 532 nm. A retardation plate has been custom-made so that it acts as a half-wave plate at 532 nm, but a full-wave plate at 488 nm so that the probe beam is transmitted unaffected. The exciting polarization vector is turned by $\pi/2$ and then allowed to pass through the second cell, turning the second fluid into a negatively birefringent material. If the amplitude of the birefringent response were the same in both cells, and the temporal profiles were the same, then they would perfectly cancel out. There are two free parameters in the suppression system. The first is the orientation through which exciting polarization vector is turned by the half-wave plate, and the second is the phase difference introduced by the second cell's contents. The first is easily adjusted by turning a half-wave plate by θ turns and the polarization vector by 2θ. The second parameter is varied by the choice of the liquid (although varying the cell length, if this were practical, could also serve as a control). Ho and Alfano [18] have compiled a list of birefringent liquids that may be considered.

The solvent signal suppression system is utilized immediately after the sample cell and the quarter-wave plate as shown in more detail in Figure 10.4. The retardation plate has been custom manufactured to be a full-wave plate for 488 nm, but a half-wave plate for 532 nm. It is used to rotate the polarization vector of the inducing beam while allowing the probe beam to pass unaltered. Its behavior depends on the precise thickness of the plate, so great care is taken to prevent thermal expansion. The plate is held in a rotary mount whose metal body is temperature controlled using a flat adhesively attached resistive element. This mount allows the positioning of the plate precisely to $0.02°$.

The beam alignment procedures require great care and the use of a Pulnix 7M CCD camera, a BeamView Analyzer, a PCI card, and a software. The process is described in detail by Perez [17].

10.5 OKE AND TEB EXPERIMENTAL RESULTS

10.5.1 OKE OF DNA OLIGOMERS

The expressions introduced in the Section 10.2 apply to an arbitrarily shaped body. In these experiments the body has a high degree of symmetry that simplifies the expected response. We expect to see only the time constant associated with end-over-end tumbling of our rigid rod.

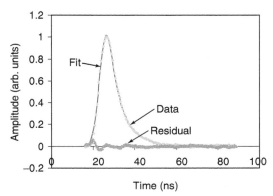

FIGURE 10.5 8 mer with buffer suppression.

There will thus be only three parameters in our fit: the fast solvent amplitude, $n_{bulk}K_B$; the DNA amplitude, $n_{bulk}K_{sp}C_v$; and the tumbling associated time constant, τ. The suppression system was used to reduce the fast solvent component. Although the fast component was significantly reduced, it was still present in the total response. The signal was indeed robust; a typical response for an average of 200 pulses is displayed below. When the response data were plotted alongside the lamp function, the Gaussian pulse that drives it, a long decay tail was clearly present. This long time decay was found to be due to a coupling of the YAG laser power supply to the photomultiplier and was eliminated by suitable electrical isolation. Sample data are shown in Figure 10.5 and Figure 10.6.

A Mathematica Package, Globals Optimization 4.0, implements the fit to the integrated Equation 10.6 of Section 10.2 [26]. The lamp function was taken to be a unit area Gaussian, and a single exponential was used to describe the decay. With these assumptions, the data obtained are shown in the Table 10.3. The percent error refers to the difference with the DLS data of Eimer and Pecora [10]. We note that the agreement with previous measurements is excellent.

10.5.2 OKE OF RNA

The rotational times and the specific Kerr constants of tRNA were derived from the deconvolution of the observed birefringence and the lamp function. The data were fitted according to the following equation for the multicomponent rise and decay of birefringence.

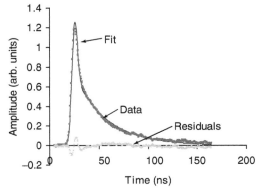

FIGURE 10.6 20 mer with buffer suppression.

TABLE 10.3
Average Relaxation Times over 4–6 Runs ($T = 12.1 \pm 0.2°C$)

mer	τ (ns)	σ	% Error
8	3.21	0.10	−0.08
12	6.44	0.12	0.81
20	16.18	0.17	−0.01

$$\Delta n_i(t_j) = K \tau T (t_{j-1}) (1 - e^{-\Delta t/\tau}) + \Delta n_{i-1} (t_{j-1}) e^{-\Delta t/\tau} \tag{10.8}$$

where Δn_i is the birefringence from the ith component of the birefringence decay, j is the jth time channel of the square pulse approximation of the data and lamp function, and $t = j\Delta t$. In the experimental analysis, the amplitude, a_i, of the component was substituted for $K\tau_i$, and $T(t_{j-1})$ was the magnitude of the buffer birefringence. By doing so, a_i became the magnitude ratio of the component to the buffer birefringence. The advantage of this method was that the loss in birefringence amplitude and excitation efficiency became less important and the shape of the lamp function is arbitrary.

Figure 10.7 displays a typical OKE response of the average of 3 data sets. Each set was an average of 1000 YAG pulses from 15.93 mg/ml tRNA in D_2O EDTA buffer at 20°C. The time constants were $\tau_1 = 9.5$ ns and $\tau_2 = 35.5$ ns, and the relative magnitudes were $a_1 = 0.36$ and $a_2 = 0.64$.

The rotational times and the birefringence magnitudes in the following tables are the average of individual experiments with the error range of 95% confidence derived from the Student T-test [27].

The rotational time constants in Table 10.4 do not show a significant change with concentration of tRNA. Within the error bars, one can see that the rotational time constant of the $MgCl_2$-containing samples was independent of tRNA concentration. The solution contained at most $4 Mg^{2+}$ ions per tRNA molecule, and the average rotational time was 27 ± 1 ns at 20°C in D_2O for this case. There is no statistically significant difference with the samples containing EDTA (at most 3 EDTA/tRNA), yielding an average rotational time of 29 ± 4 ns at 20°C in D_2O. At the highest concentration of tRNA, however, the OKE experiments showed two distinct time constants (with the slow one representing 60 to 53% of the decay). A short time constant around 7 ns was measured in addition to the overall tumbling

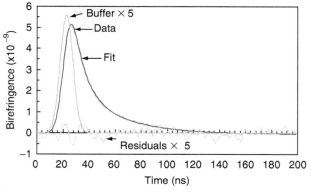

FIGURE 10.7 Typical OKE response and excitation.

TABLE 10.4
OKE tRNA^phe Rotational Time Constants and Birefringence

MgCl$_2$ Buffer					EDTA Buffer				
Conc. (mg/ml)	τ (ns)	95% (ns)	Biref. (10^{-9})	95% (10^{-9})	Conc. (mg/ml)	τ (ns)	95% (ns)	Biref. (10^{-9})	95% (10^{-9})
13.80	25.27	±1.50	0.85	±0.07	15.93[a]	7.71	±1.36	0.41	±0.06
12.23	26.45	±0.83	0.97	±0.10		32.76	±3.31	0.70	±0.07
6.60	27.58	±3.80	0.62	±0.08	15.80[a]	6.82	±0.84	0.52	±0.04
3.23	28.54	±3.22	0.67	±0.05		26.71	±3.56	0.78	±0.07
					8.30	24.71	±2.12	0.54	±0.03
					4.06	33.15	±4.20	0.34	±0.02

[a]Consists of two components.

time constant. This time constant may be interpreted as due to an internal flexing motion of the tRNA.

K_{sp} for the OKE measurements was calculated using the following equation:

$$K_{sp} = \frac{\lambda}{nC_v} aB(D_2O, \lambda)$$

where λ is the probe laser wavelength at 514 nm and $B(D_2O, \lambda)$, the Kerr constant of D$_2$O at 514 nm, is 2.649×10^{-16} m/V^2 determined in this laboratory [28]. The precision of this number, ±1.5%, is much greater than that of OKE experiments. C_v is the volume fraction (tRNA concentration divided by 1.82 g/ml, the RNA density), and a is the relative magnitude.

The data in Table 10.5 show that EDTA keeps the Kerr constant nearly unchanged, allowing tRNA to remain flexible. On the other hand, the MgCl$_2$-containing sample can be interpreted as Mg^{2+} condensation on the tRNA stiffening the structure at low tRNA concentration (at fixed Mg concentration). The stiffer structure is more anisotropic, increasing the K_{sp}.

10.5.3 TEB OF RNA

The signal strength in the OKE experiment depends exclusively on an induced moment via the molecular polarizability. Thus, relatively large concentrations of tRNA were used and only overall average processes were detected. In the TEB case, there is a significant dipole moment

TABLE 10.5
OKE Specific Kerr Constant

MgCl$_2$ Buffer			EDTA Buffer		
Conc. (mg/ml)	K_{sp} (10^{-20})	95% Conf. (10^{-20})	Conc. (mg/ml)	K_{sp} (10^{-20})	95% Conf. (10^{-20})
13.8	1.14	±0.09	15.9	1.30	±0.12
12.2	1.47	±0.15	15.8	1.53	±0.68
6.60	1.75	±0.22	8.30	1.21	±0.68
3.23	3.89	±0.28	4.06	1.56	±0.10

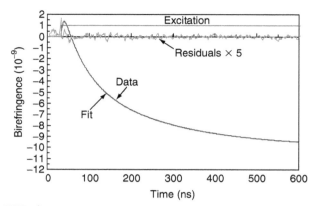

FIGURE 10.8 tRNA TEB rise.

that will contribute to the orientation mechanism, giving rise to a large signal even for small concentrations. We expect to see more detail in this case, with more challenges in interpretation. The typical TEB experiment data set shown in Figure 10.2 was separated into the rise and the decay of the birefringence signal. The analysis of the rise of the negative birefringence (composed of data from the first point rise that was above the noise to 3 points before the sudden birefringence drop due to cessation of the electric field), is shown in Figure 10.8.

The decay of birefringence after the electric field was removed is shown in Figure 10.9. The analyzed data started after contribution from the buffer became insignificant 3.7 ns after the electric field was deactivated. The data ended at the last data point collected. By not fitting the fast component from the buffer, the data fitting would have too fewer fitting parameters. Thus, the rotational time of tRNA could be obtained more accurately and certainly, because the buffer birefringence would be small after 3.7 ns. Often due to thermal shock or other phenomena, the signal did not return to zero in the recorded time interval. Therefore, a baseline was added to the fit. The baseline was entered as a component with a very long time constant of 10^{10} ns.

All TEB experiments were performed at 4°C. The pulse length is 500 ns at 800 V, and the field strength is 3000 V/cm. The following rotational times and the birefringence magnitudes are averages of individual experiments, and the error range is 95% confidence derived from the Student T-test. We begin with an analysis of the birefringence decay curve.

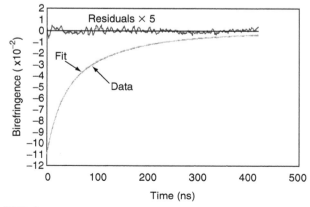

FIGURE 10.9 tRNA TEB decay.

TABLE 10.6
TEB Rotational Decay Time Constants at 4°C

MgCl$_2$ Buffer					EDTA Buffer				
Conc. (mg/ml)	τ (ns)	95% (ns)	Biref. (10^{-9})	95% (10^{-9})	Conc. (mg/ml)	τ (ns)	95% (ns)	Biref. (10^{-9})	95% (10^{-9})
1.93	31	±3	−7.0	±0.4	2.10	25	±4	−6.3	±0.2
	100	±19	−2.4	±0.6		117	±13	−5.7	±0.4
0.61	33	±2	−2.4	±0.2	0.93	29	±3	−3.1	±0.2
	114	±16	−1.0	±0.1		127	±10	−3.9	±0.3
0.27	26	±19	−1.8	±2	0.26	29	±10	−2.2	±0.9
	106	±203	−1.6	±3		110	±24	−4.5	±0.9
0.27[a]	43	±11	−3.2	±0.2	[a]0.26	78	±1	−6.2	±0.1

[a]Using a single relaxation model.

In the TEB measurements, as shown in Table 10.6, the signal to noise was good enough to detect two relaxation times. There was no statistically significant variation with concentration in either buffer. The average fast relaxation time is 30 ± 3 ns for the Mgcl$_2$ buffer, and an equivalent 27 ± 2 ns for the EDTA buffer. Evidently, the conformation in the EDTA buffer is not significantly different (at most 46 EDTA/tRNA) from that in the Mgcl$_2$ buffer (at most 45 Mg^{2+}/tRNA). The slow relaxations show similar behavior. The average slow relaxation time is 106 ± 20 ns for the Mgcl$_2$ buffer, and an equivalent 118 ± 15 ns for the EDTA buffer. The statistics of the fit show no significant difference between these values. The slow relaxations do not correspond to possible rigid body rotational diffusion eigenvalues of the very asymmetric geometry of tRNA. Accurate hydrodynamic computations referred to in the next section demonstrate that the rigid body relaxations are all clustered around 21 ns (20°C, water). Thus the slow relaxation detected in this work may arise from aggregates or other contaminants in the sample, or are artifacts of the fitting process. It is noteworthy that the fast relaxation times corrected to 20°C yield times of 16 ns or less, in disagreement with all other measurements for the rotational diffusion time. This indicates that it may have been unwarranted to fit the data to more than one exponential.

The data showing a fit to only a single relaxation give an average value over the rigid body motions of the tRNA. Converting the Mgcl$_2$ buffer time of 43 ± 11 ns at 20°C, we obtain 24 ± 6 ns, in complete agreement with the OKE data. Unlike the OKE data, on the other hand, the amount of EDTA per tRNA is much larger (46 versus 3) for the smallest concentration of tRNA in the TEB buffer. The larger relaxation time of 78 ns at 4°C, equivalent to 44 ns at 20°C may be indicative of a more extended conformation for tRNA under those conditions.

The rise of the birefringence tells a much more complex story. In this case also, the curves can be fit to two relaxation times as shown in Table 10.7. In the case of the MgCl$_2$ buffer, we see values that are fairly independent of tRNA concentration. The two exponential fit to the smallest concentration has large error and is not statistically different from those of the other two concentrations. An average weighted by the uncertainties in the quantities yields for the MgCl$_2$ buffer fast relaxation time value of 58 ± 6 ns, statistically indistinguishable from the value obtained from a single exponential fit to the lowest concentration. Similarly, a weighted average of the slow relaxation yields 182 ± 28 ns. The fits yield relative amplitudes of 40–60% for the slow relaxation, so both relaxations appear similarly weighted. Thus, it appears that the low concentration value in the Mgcl$_2$ buffer has poor signal-to-noise ratio

TABLE 10.7
TEB Birefringence Rise Time Constants at 4°C

In MgCl$_2$ Buffer			In EDTA Buffer		
Conc. (mg/ml)	τ (ns)	95% Conf. (ns)	Conc. (mg/ml)	τ (ns)	95% Conf. (ns)
1.93	58	±7	2.10	36	±3
	179	±28		155	±7
0.61	58	±5	0.93	38	±3
	195	±28		164	±7
0.27	47	±32	0.26	30	±14
	114	±175		102	±14
0.27[a]	63	±12	[a]0.26	89	±3

[a]Using a Single relaxation fit.

and cannot be fully trusted. However the same caveats with respect to fitting to two exponentials compared to one are also applicable here.

As there are no other data in the literature with this level of detail in the rise, we discuss these results here. Because there is practically a factor of three between slow and fast relaxations, and tRNA is expected to be reasonably rigid in the presence of Mg^{2+}, one is quite tempted to conclude that these values correspond to the relaxations from the orientation of the permanent dipole moment (the slow value), and that of the polarizability induced dipole moment (the fast value). The relative weights would indicate that the permanent and induced dipole both contribute significantly to the observed signal. If this was the case, how is these data consistent with the observed decay time constants? If we were to take the decay scenario at face value, we would predict that there should be four relaxation times in the rise, two for each detected orientation mechanism, and the values would be about 30, 90, 106, and 318 ns. Clearly, we cannot resolve four relaxation times. It is quite possible that the previous spectrum is simply represented by the average of the pairs, 60 and 212 ns, which is quite close to what is observed. Thus, we propose that the data are consistent but the lack of resolution makes the analysis ambiguous.

The EDTA buffer relaxations show some statistically significant differences with respect to those of the Mgcl$_2$ buffer. The weighted average of the fast relaxations is 36 ± 4 ns, while that of the slow relaxations is 157 ± 10 ns. In this case the ratio between relaxations is closer to five. Evidently, there are contributions from flexibility that further mix relaxations making this case quite difficult to disentangle.

10.6   DISCUSSION

The OKE measurements on the DNA oligomers clearly demonstrate that the equipment developed in this laboratory (by the late Don Eden) is capable of measuring, with good sensitivity and accuracy, the molecular reorientation after the perturbation from the YAG pulse. We have shown that careful instrument design, including solvent signal suppression, can surmount the difficulties previously expressed by Wirth et. al. [9]. Thus, we can apply this method with confidence to the study of a more complex system, tRNA. The TEB and OKE results reveal different information about tRNA. The experimental data for OKE and TEB of tRNA presented here corroborates measurements performed by previous authors.

Measurements of the OKE rotational correlation time at 20°C in our laboratory gave average values of 23 ± 1 ns in the presence of $MgCl_2$, and 25 ± 3 ns in the presence of EDTA (both values corrected to the viscosity of water from that of D_2O). At the lowest concentrations of tRNA used, these data correspond to at most $4 Mg^{2+}$/tRNA, and at most 3 EDTA/tRNA. Thomas et al. [13] used FPA measurements to determine the rotation times of tRNA[phe] in various Mg^{2+} concentrations at 20°C. The tRNA was in 10 mM, pH 7.0 phosphate buffer and 100 mM NaCl. Ethidium bromide was added in the quantity of 1 ethidium bromide per 20 tRNA molecules. Thomas reported two tRNA time components in experiments ranging from 0 to 40 Mg^{2+} per tRNA molecule, and Mg^{2+} concentration did not affect the decay times in the fluorescence and anisotropy experiments. The average two component fluorescence time constants are 3.6 and 24.9 ns, and the relative amplitudes are 0.055 and 0.945, respectively. The anisotropy relaxation is identified with the rotational motion of tRNA. The tumbling relaxation time increased slightly from 21.35 ns to 24.54 ns as Mg^{2+} concentration increases from 0 to 40 Mg^{2+}/tRNA. Patkowski et al. [15] used depolarized DLS to measure the rotational time of tRNA at concentrations from 2.25 to 75.0 mg/ml in Tris buffer containing 200 mM NaCl at pH 7.2. They found that the rotational times decrease only by 0.7 ns over the stated tRNA concentration range. The average value is 17.2 ns at 24.1°C or 19.7 ns when corrected to 20°C, a value consistent with the 21.4 ns Mg^{2+}-free value of Thomas. Our data are consistent with all of these measurements.

Hydrodynamic computations [29] based on our boundary element program BEST, using a uniform hydration model of 1.1A (which approximates the hydration for nucleic acids), yields the following $J = 2$ five relaxation times for the very asymmetric tRNA structure: 16.93, 17.08, 21.10, 23.89, and 25.35 ns. The average of these values, 21 ns, is in excellent agreement with the measured values quoted above (these relaxations will be almost equally weighted due to the anisotropy of the rotational diffusion coefficient). These values were computed from the crystal structure of yeast tRNA[phe] (1EVV.pdb). This demonstrates that the solution's native structure is well represented by the crystal structure in the presence of Mg^{2+}.

Furthermore, Porschke and Antosiewicz [14] used electric dichroism to determine the rotational time of various tRNA, including tRNA[phe]. They used two Na-cacodylate base buffers at pH 7 with 10 mM NaCl, with 200 μM EDTA and with 1 mM $MgCl_2$. The sample with EDTA has a rotational time of 102 ns at 2°C, or 57.4 ns at 20°C, and the sample with 1 mM $MgCl_2$ had the rotational time of 41.6 ns at 2°C, or 23.4 ns at 20°C. The $MgCl_2$-containing sample has a relaxation time quite consistent with those reported here from the OKE and consistent with measurements of other groups quoted. The EDTA value, however, is widely discrepant. This discrepancy can be understood if one takes into account the small concentration of tRNA used by Porschke (1 μM), for this yields about $50 Mg^{2+}$/tRNA and 200 EDTA/tRNA. Evidently, a large amount of EDTA is needed in order to completely sequester all divalent cations away from the tRNA and allow the molecule to assume its extended, nonnative conformation. In our OKE work, only 3 EDTA/tRNA was used and this is insufficient to remove divalent cations already complexed with the tRNA. The samples used by other workers in the absence of EDTA can also be assumed to be in the native state even when no additional Mg^{2+} was added.

The TEB data reported here for tRNA[phe] show a level of detail not present in the experiments reported by other authors, including a slow relaxation time. One could wonder if the low temperature sample was fully at equilibrium and contained some molecules in a nonnative conformation. For example, the other closest technique applied to tRNA is the work of Porschke and Antosiewicz [14] cited earlier. In their electric dichroism experiments, they applied pulses of 1.6 μs in duration, which is quite a bit longer than the 500 ns pulses used in our TEB experiments. However, the relaxation time of 41.6 ns obtained in these measurements, which is essentially the same as the single relaxation fit of 43 ns to our TEB decay data (lowest concentration) at very similar temperatures, indicates that there was no significant contribution from a slow relaxation. If a slow relaxation were actually present, then a single exponential fit would yield a weighted average relaxation time larger than 41.6 ns if the problem were one of lack of resolution. This relaxation time agrees with the measurements done at 20°C and the hydrodynamics computations. Thus, we conclude that the slow relaxation time detected in our TEB, but not in our OKE measurements, is probably contamination of some sort or an artifact of over fitting the data. This latter possibility is bolstered by the large statistical errors in the double exponential fits.

In conclusion, we have shown that robust signals are observable in the OKE from nucleic acids and that the data are essentially consistent with TEB data obtained on the same samples in our laboratory. Furthermore, there is a good agreement between our measurements and the measurements made with other techniques by other authors. Further computational work is in progress in this laboratory in which more accurate hydrodynamic and polarizability calculations are carried out using the boundary element method [30,31,32].

ACKNOWLEDGMENTS

The authors wish to thank J. Michael Schurr for helpful discussion on this work. This work was supported by the National Institutes of Health, MBRS SCORE Program — Grant No. S06 GM52588-3 to D. Eden and Grant No. S06 GM52588-10 to S. Aragon. M. Perez was supported by an NIH Rise Fellowship during his M.S. work on DNA oligomers at SFSU.

REFERENCES

1. Fredericq, E. and Houssier, C., *Electric Dichroism and Electric Birefringence*, Clarendon Press, Oxford, 1973.
2. Buckingham, A.D., Birefringence resulting from the application of an intense beam of light to an isotropic medium, *Proc.Phys. Soc.*, B69, 344, 1956.
3. Mayer, G. and Gires, F., *C.R. Acad. Sci.*, 258, 2039, 1964.
4. Paillette, M., *C.R. Acad. Sci.*, 262, 264, 1966.
5. Kielich, S., Molecular interaction in optically induced nonlinearities, *IEEE J. Quantum Electron.*, 4, 744, 1968.
6. Coles, H.J. and Jennings, B.R., Optical Kerr effect in biopolymer solutions, *Biopolymers*, 14, 2567, 1975.
7. Dobek, A., Klimecki, M., Patkowski, A., and Labuda, D., Optically induced birefringence in tRNA solutions, *Acta Phys. Pol.*, A, 62(5–6), 431, 1982.
8. Dobek, A. and Deprez, J., Optical Kerr effect induced by picosecond light pulses in tRNA solutions, *J. Colloid Interface Sci.*, 111, 75, 1986.
9. Wirth, M., Eriksson, T., and Norden, B., Critical aspects on optical-Kerr effects on macromolecules. Lack of measurable orientation of DNA, *J. Phys. Chem.*, 91, 1957, 1987.
10. Eimer, W. and Pecora, R., Rotational and translational diffusion of short rod-like molecule in solution: Oligonucleotides, *J. Chem. Phys.*, 94, 2324, 1991.
11. Crothers, D.M. and Cole, P.E., Conformational changes in tRNA, in *Transfer tRNA*, Altman, S., Ed., MIT press, Cambridge, 1978, p. 196.
12. Reeves, R.H. , Cantor, C.R., and Chambers, R.W., Effect of magnesium ions on the conformation of two highly purified yeast alanine transfer ribonucleic acids, *Biochemistry*, 9, 3993, 1970.
13. Thomas, J.C., Schurr, J.M., Reid, B.R., Ribeiro, N.S., and Hare, D.R., Effect of magnesium(2+) on the solution conformation of two different transfer ribonucleic acids, *Biochemistry*, 23, 5414, 1984.
14. Porschke, D. and Antosiewicz, J., Permanent dipole moment of tRNAs and variation of their structure in solution, *Biophys. J.*, 58, 403, 1990.
15. Patkowski, A., Eimer, W., and Dorfmüller, T., A polarized and depolarized dynamic light scattering study of the tRNA^{Phe} conformation in solution, *Biopolymers*, 30, 93, 1990.

16. Highsmith, S., and Eden, D., Transient electrical birefringence characterization of heavy meromyosin, *Biochemistry*, 24, 4917, 1985; Highsmith, S. and Eden, D., Myosin subfragment 1 has tertiary structural domains, *Biochemistry*, 25, 2237, 1986; Highsmith, S. and Eden, D., Limited trypsinolysis changes the structural dynamics of myosin subfragment-1, *Biochemistry*, 26, 2747, 1987.

17. Perez, M., Optically induced birefringence in short DNA oligomers, M.S. thesis, San Francisco State University, 2003.

18. Ho, P. and Alfano, R., Optical Kerr effect in liquids, *Phys. Rev. A: At., Mol., Opt. Phys.*, 20, 2170, 1979.

19. Maniatis, T., *Molecular Cloning: A Laboratory Manual*, Cold Spring Harbor Press, Cold Spring Harbor, 1982.

20. Russo, P.S., Karasz, F.E., and Langley, K.H., Dynamic light scattering study of semi-dilute solutions of a stiff-chain polymer, *J. Chem. Phys.*, 80, 5312, 1984.

21. Schurr, J.M., Personal communication, 2002.

22. Ng, K., Transient Electric Birefringence and Optical Kerr Effect of Yeast Phenylalanine tRNA, M.S. thesis, San Francisco State University, San Francisco, CA, 1997.

23. Eden, D., Personal communication, 2000.

24. Armata, D., LabView Program, San Francisco State University, San Francisco, CA, 1998.

25. Scope Explorer v1.1, Lecroy Instruments, 2001.

26. Loehle Enterprises , *Globals Optimization 4.0*, Naperville, IL, 2001.

27. Ingram, J.A., *Introductory Statistics*, Cummings Publishing Company, Menlo Park, CA, 1974.

28. Fuchs, M. and Ng, K., San Francisco State University, unpublished data.

29. Aragon, S.R., A precise boundary element method for macromolecular transport properties, *J. Comput. Chem.* 25, 1191, 2004.

30. Aragon, S.R. and Hahn, D.K., "The polarizability and capacitance of platonic solids and the Kerr constant of Proteins", Lecture series on computer and computational sciences vol. 4, Brill: Leiden, 25, 2005.

31. Aragon, S.R. and Hahn, D.K, "Accurate Transport Properties of Proteins from high precision boundary element calculations. I Diffusion and specific volume", *Biophysical J.* (in press, 2006).

32. Hahn, D.K. and Aragon, S,R, "The intrinsic viscosity of proteins from high precision boundary element calculations", *J. Chem. Theory Comput.* (in press, 2006).

11 The Use of Transient Electric Birefringence to Analyze Curvature in Naturally Occurring, Mixed-Sequence DNA Molecules

Nancy C. Stellwagen and Yongjun Lu

CONTENTS

11.1   INTRODUCTION

Transient electric birefringence (TEB) and the related technique, transient electric dichroism, are very useful methods for studying DNA conformation in solution [1–8], the flexibility of the helix backbone [9–14], the optical factor and electrical polarizability of DNA [11,15–18], the mechanism of DNA gel electrophoresis [19–25], and the effect of pulsed electric fields on the internal structure of the agarose gel matrix [26–29]. Because several of these topics have been reviewed in detail previously [30–34], this chapter will focus on the use of TEB to analyze the curvature of naturally occurring, random-sequence DNA molecules, with an emphasis on recent experimental results.

    Normal, random-sequence DNA molecules in free solution have a conformation that is best described as a wormlike coil, due to the rotational flexibility of the bonds in the helix backbone. The relative stiffness of a wormlike coil is described by its persistence length, the distance over which the direction of orientation of two successive chain segments is correlated [30,35]. Some DNA molecules are also anisotropically flexible, and bend preferentially in one direction upon the application of an external force [7]. Other DNA molecules contain sequences that are intrinsically bent or curved [1,2,5]. Such sequence-dependent differences in DNA conformation and flexibility are thought to play an important role in many processes in the cell, including transcription, replication, and recombination, and also may be involved in the packaging of DNA into viral capsids and higher order nucleosome structures [36–40].

Much of what is known about DNA curvature has come from polyacrylamide gel electrophoresis and ligase-catalyzed cyclization studies of oligomers containing various target sequences repeated every 10 base pairs (bp), approximately in phase with the helix repeat. The results indicate that phased A-tracts, runs of 4–6 adenine (or thymine) residues in a row, are primarily responsible for stable curvature of the DNA helix backbone [41–44]. The observed bend angle ranges from $13°$ to $22°$ per A-tract, depending on the number of adenine residues in the A-tract and the method of measurement [2,44–51]. The most probable bend angle appears to be $20° \pm 2°$ per A-tract [44].

Relatively little is known about curvature in naturally occurring, mixed-sequence DNA molecules, because phased A-tracts are relatively rare in genomic DNA. Curved restriction fragments are often identified adventitiously by their anomalously slow electrophoretic mobilities in polyacrylamide gels [52,53]. A more systematic method of identifying curvature in mixed-sequence DNA molecules is the circular permutation assay [54], which can be applied to kilobase-sized DNA molecules by electrophoresing the various permuted sequence isomers in large-pore polyacrylamide gels [55–58]. DNA molecules containing regions of stable curvature or anisotropic (hinge-like) flexibility near their centers migrate more slowly than permuted sequence isomers that have their regions of curvature or flexibility located near one end. Hence, the fastest migrating, permuted sequence isomer has a region of curvature located at or near one end of the molecule. Restriction fragments containing the putative bend centers can then be isolated and analyzed by more quantitative techniques, such as TEB.

TEB is particularly suitable for analyzing DNA curvature, because the measured rotational relaxation times (τ) are approximately proportional to the third power of macromolecular length [59]. Hence, the difference in relaxation times between normal and curved (or anisotropically flexible) DNA molecules can be accurately determined under a given set of experimental conditions [1,2,5–8]. Measuring the relaxation times as a function of ionic strength allows a distinction to be made between stable curvature and anisotropic flexibility [5,7]. If the ratio of the relaxation times (τ-ratio) of the target DNA molecule and a normal restriction fragment of the same size is independent of ionic strength, the target DNA is stably curved; if the τ-ratios vary significantly with ionic strength, the target DNA is anisotropically flexible. The τ-ratios can also be used to estimate the apparent bend angle, if the location of the site of curvature is known [60–62]. Global fitting of the τ-ratios of a series of overlapping restriction fragments allows the apparent bend angle and its sequence position to be determined separately [6–8].

This chapter describes the use of TEB to analyze the curvature of three naturally occurring, mixed-sequence DNAs: (1) an electrophoretically anomalous 153-bp restriction fragment (called 12A) taken from an intergenic region in plasmid pBR322; (2) restriction fragments taken from the origin of replication of bacteriophage M13, which has been cloned into the plasmid Litmus 28; and (3) restriction fragments taken from the VP1 gene in the simian virus 40 (SV40) minichromosome.

11.2 MATERIALS AND METHODS

11.2.1 DNA Samples

The methods used to prepare the relatively large quantities of DNA needed for TEB studies have been described previously [7,63]. To summarize briefly, fragment 12A and a normal control fragment of the same size (called 12B) were obtained by *Msp*I digestion of plasmid pBR322, isolated by polyacrylamide gel electrophoresis, eluted by the crush and soak method, and recovered by ethanol precipitation. Multimers of each fragment were then cloned into separate plasmids [63] and amplified using standard methods [64]. After exhaustive restriction enzyme digestion, the 12A and 12B fragments were isolated by agarose or polyacrylamide gel electrophoresis, eluted by the crush and soak method, purified and concentrated on small diethylaminoethylcellulose (DEAE) columns, eluted into the desired buffer, and stored at −20°C until needed. The identities of the fragments were verified by suitable restriction enzyme digestion and polyacrylamide gel electrophoresis.

Restriction fragments derived from the M13 origin and the VP1 gene in SV40 were obtained by digesting plasmid Litmus 28 (New England Biolabs) or the SV40 genome with suitable restriction enzymes. The desired restriction fragments were isolated by agarose gel electrophoresis, subcloned into the polylinker of plasmid pUC19, and amplified by standard methods [64]. After restriction enzyme digestion, the desired fragments were isolated by electrophoresis in 1% agarose gels, excised, and the agarose dissolved with a chaotropic salt (QIAquick gel extraction kit, Qiagen). The DNA fragments were then concentrated and desalted by adsorption on small DEAE columns, eluted into the desired buffer and stored at −20°C until needed. All fragments were sequenced to verify their identities.

The normal DNA fragments used to generate the standard curves of birefringence relaxation times as a function of molecular weight were isolated from plasmids pUC19, pBR322, or Litmus 28 by restriction enzyme digestion, and amplified and purified as described above. The normal fragments contained 171–543 bp, have no or only a few isolated A- or T-tracts, exhibit normal electrophoretic mobilities in polyacrylamide gels, have normal birefringence relaxation times, and can be described by the same persistence length [14].

The DNA samples were dissolved in three different buffers, called B, C, and E for brevity. All buffers contained $1\,\mathrm{m}M$ Tris-Cl, pH 8.0, plus $0.1\,\mathrm{m}M$ EDTA (buffer B), $0.01\,\mathrm{m}M$ EDTA (buffer C), or $1\,\mathrm{m}M$ NaCl plus $0.01\,\mathrm{m}M$ EDTA (buffer E).

11.2.2 TEB Measurements

The apparatus and procedures used to carry out the TEB measurements have been described in detail previously [5,14]. Briefly, the birefringence instrument includes a Spectra Physics model 117A helium neon laser, quartz Glan-Thompson polarizer and analyzer, a mica quarter-wave plate typically rotated −3° from the crossed position to enable linear detection, and a photodiode-based optical amplifier with zero supression circuit. The time constant of the detector is ∼$0.23 \pm 0.06\,\mu\mathrm{s}$. Square wave pulses ranging from $1\,\mu\mathrm{s}$ to several milliseconds in duration and 100 V to 2 kV in amplitude are generated by a Model 605P Cober high-power pulse amplifier. The Kerr cell is a shortened 1-cm path length quartz spectrophotometer cell, chosen for its negligible strain birefringence and thermostatted at 20°C, unless otherwise indicated. The electrodes are parallel platinum plates with a 1.5-mm separation mounted on a Teflon support of standard design [65]. A 400-Ω low-impedance resistor is connected in parallel with the cell to keep the current constant. The stray light constant of the optical system [5], with the DNA solution and electrodes in the cell, is typically $(0.5\text{--}1.0) \times 10^{-5}$.

Each DNA solution was pulsed 15 times in the single-shot mode, and the field-free decay of each birefringence signal was analyzed using a nonlinear least squares fitting program (CURVEFIT) adapted from the program designed by Bevington [66]. This method of analysis was adopted because small but detectable decreases in the relaxation times were noted if the decay curves were averaged before analysis. The decay curves of fragments containing ≤250 bp could usually be fitted with a single relaxation time; larger fragments required two relaxation times, as described previously [14,15]. In such cases, the slower relaxation time, typically comprising 80% or more of the decay curve, was assigned to the end-over-end rotation of the DNA molecules. The single relaxation time observed for the smaller DNA fragments and the slower relaxation time observed for fragments larger than ~250 bp are described collectively as the terminal relaxation times.

The average standard deviation of the terminal relaxation times measured for any DNA sample under any given set of buffer conditions was less than $\pm 3\%$, even when different, independently prepared DNA stock solutions were used. The terminal relaxation times measured for the various DNA samples were independent of DNA concentration from 5–25 μg/μL, independent of pulse length from 3–30 μs and independent of electric field strength from 1–9 kV/cm [14]. Therefore, a standard set of operating conditions (DNA concentration = 14 μg/μL, electric field strength = 5 kV/cm, and pulse length = 8 μs) was used for the measurements.

The terminal relaxation times of curved and normal DNA fragments of the same size were used to calculate τ-ratios, τ_R, defined as $\tau_{curved}/\tau_{normal}$, and τ-decrements, defined as $100(\tau_R - 1)$. Here, τ_{curved} is the relaxation time of a target, possibly curved fragment and τ_{normal} is the relaxation time of a normal fragment containing the same number of base pairs. If the target fragment is curved or anisotropically flexible, the τ-decrements are negative, as $\tau_{curved} < \tau_{normal}$. If $\tau_R \sim 1$, the target fragment has the conformation of normal DNA.

To distinguish between stable curvature and anisotropic (hinge-like) flexibility, the relaxation times of the target fragments are measured in a series of different buffers. If the τ-ratios and τ-decrements are essentially independent of buffer composition, the target fragment is stably curved. On the other hand, if the τ-ratios and τ-decrements depend significantly on buffer composition, the target fragment is anisotropically flexible.

As small DNA molecules in low ionic strength solutions can be approximated as rigid rods, the terminal relaxation times, τ, can be related to end-to-end length, L, by Equation 11.1

$$\tau = \frac{\pi \eta L^3}{18kT[\ln p - 0.662 + 0.92/p]} \tag{11.1}$$

where η is the viscosity of the solvent, k is Boltzmann's constant, T is the absolute temperature, and p is the axial ratio [67]. Curved and normal DNA molecules containing the same number of base pairs will have different end-to-end lengths.

More quantitative estimates of DNA curvature can be obtained by comparing the τ-ratios of curved and normal fragments of the same size to theoretical equations relating the τ-ratios to apparent bend angles and fractional distances of the bend from the end of the molecule [60–62]. Theoretical curves have been derived for rigid, once-broken rods [60,61] and wormlike chains containing a flexible hinge [62]. Both models predict similar apparent bend angles for a given τ-decrement, within $\pm 4\%$ [7]. Therefore, either theoretical curve can be used to estimate the apparent bend angles from the observed τ-ratios or τ-decrements.

If the location of the bend in a target fragment is known, the apparent bend angle can be estimated directly from the theoretical curves of the τ-ratios as a function of bend angle. If the location of the bend is not known, the apparent bend angle and its sequence position can be separately determined by the method of overlapping fragments [6–8]. Briefly, several overlapping restriction fragments are generated with sequences somewhat offset from the center of the target fragment. The terminal relaxation times of all the fragments are measured and compared with the relaxation times of normal controls of the same size. The τ-ratios and τ-decrements of the overlapping fragments are then compared with theoretical curves calculated as a function of apparent bend angle and fractional distance, S, of the bend from one end of the molecule [60–62], making an initial assumption about the magnitude or location of the bend. The calculation is repeated iteratively, with different assumptions about the magnitude or location of the bend, until the τ-ratios or τ-decrements lead to a consistent set of bend angles and sequence locations for the various overlapping fragments [7,8]. The validity of the calculation depends only on the assumption that the magnitude of the bend does not depend on its distance from the end of the fragment.

For the curved fragment 12A taken from pBR322, fragment 12B, obtained from the same restriction enzyme digest of pBR322 and containing the same number of base pairs, was used as the normal control. For curved fragments obtained from the M13 origin and the VP1 gene of SV40, the relaxation times of normal fragments containing the same number of base pairs were taken from log–log plots of the relaxation times of normal DNA restriction fragments as a function of base pair size, measured in the same buffers used to characterize the target fragments [40]. This procedure has the advantage that control fragments of exactly the same size as the target fragment(s) do not need to be created. More importantly, this procedure ensures that the relaxation times of the target fragments are compared with the relaxation times of "average" normal fragments of the same size.

11.3 RESULTS AND DISCUSSION

11.3.1 FRAGMENT 12A FROM PBR322

The relaxation times observed for the curved (12A) and normal (12B) 153-bp fragments from pBR322, measured in buffers B, C, and E, are summarized in Table 11.1. The relaxation times of fragments 12A and 12B vary somewhat in the different buffers, because of the dependence of DNA persistence length on ionic strength [13,14,68–70]. However, the τ-ratios are essentially independent of buffer composition, indicating that fragment 12A is stably curved, not aniso-tropically flexible. The results are in agreement with electron microscopy studies, which have shown that plasmid pBR322 is stably curved at or near the location of fragment 12A [71].

The differences in the relaxation times observed for fragments 12A and 12B in the various buffers correspond to about a 3% difference in end-to-end length, according to Equation 11.1. If fragment 12A is assumed to be a bent rod with arms of equal length [60], the average τ-ratio corresponds to a deflection of one arm by about 30° from the straight-rod limit. A more detailed analysis of the curvature of fragment 12A is not possible without more information about the precise location of the apparent bend center.

11.3.2 M13 ORIGIN OF REPLICATION

Five overlapping restriction fragments were isolated from the M13 origin of replication, containing 207, 219, 224, 471, and 489 bp. The relationship between the various fragments is shown schematically in Figure 11.1A. For ease of reference, A- and T-tracts containing five or more residues are highlighted by small boxes. All five restriction fragments migrated

---

TABLE 11.1
Relaxation Times (τ), τ-Ratios, and τ-Decrements Observed for Fragments 12A and 12B in Different Buffers

	Buffer B	Buffer C	Buffer E
τ (12A), μs	$2.0 \pm 0.1_5$	$2.2 \pm 0.0_5$	1.9 ± 0.1
τ (12B), μs	$2.2_5 \pm 0.1_5$	$2.5 \pm 0.0_5$	$2.1 \pm 0.1_5$
τ-ratio	0.89	0.88	0.90
τ-decrement	−11.1	−12.0	−9.5

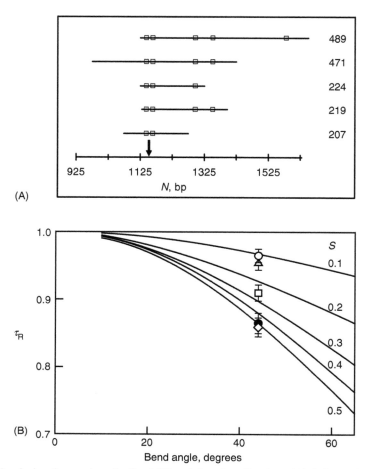

(A)

(B)

FIGURE 11.1 Analysis of curvature in the M13 origin of replication. (A) Schematic diagram of the overlapping restriction fragments used for the analysis. For ease of reference, A_n- and T_n-tracts ($n \geq 5$) are indicated by square boxes; the location of the apparent bend center is indicated by the arrow on the horizontal axis. (B) Comparison of the τ-ratios observed for the various M13 fragments in buffers B and C with theoretical curves calculated for various bend angles and fractional distances, S, of the bends from one end of the fragment [62]. From top to bottom, the symbols correspond to restriction fragments containing: (○), 219 bp; (△), 489 bp; (□), 224 bp; (•), 207 bp; and (◇), 471 bp. The error bars correspond to the uncertainties in the measured values of the τ-ratios.

anomalously slowly in polyacrylamide gels, suggesting that they are curved or anisotropically flexible. The τ-ratios and τ-decrements measured by TEB in buffers B and C are summarized in Table 11.2. As the τ-ratios and τ-decrements are reasonably constant in the two buffers, the restriction fragments are intrinsically curved, rather than anisotropically flexible.

To avoid concentrating on small buffer-to-buffer differences in conformation, the τ-ratios observed for the various M13 fragments in buffers B and C were averaged, giving the values reported in the last column of Table 11.2. The average τ-ratios are compared with theoretical curves calculated as a function of bend angle and fractional distance of the bend from one end of the molecule [62] in Figure 11.1B. Global fitting of the τ-ratios to the theoretical curves indicates that an apparent bend of $44° \pm 2°$ is located at fractional distances of 0.50, 0.10, 0.25, 0.50, and 0.15 from one end of the 207-, 219-, 224-, 471-, and 489-bp fragments, respectively.

TABLE 11.2
The τ-Ratios and τ-Decrements Observed for Restriction Fragments Derived from the M13 Origin

DNA size, bp	Buffer B		Buffer C		Average τ-Ratio
	τ-Ratio	τ-Decrement	τ-Ratio	τ-Decrement	
207	0.87	−13.0	0.86	−14.0	0.86_5
219	1.00	0	0.95	−4.8	0.97_5
224	0.94	−5.8	0.99	−1.1	0.96_5
471	0.83	−16.8	0.87	−12.8	0.85
489	0.96	−3.5	0.98	−2.1	0.97

Hence, the apparent bend center in the M13 origin is located at Litmus 28 sequence position 1153 ± 2 bp [7], close to the −35 promoter of the gene II protein mRNA [72]. Intrinsic curvature of the helix backbone at this site may facilitate wrapping the DNA around RNA polymerase [7], which is the first step in rolling circle replication [73].

The magnitude of the apparent bend angle in the M13 origin is approximately equal to that observed for two phased A-tracts. The sequence surrounding the apparent bend center contains four A- and T-tracts within ~40 bp, as shown in Scheme 11.1. Replacing a segment of normal DNA with this sequence causes the modified DNA to exhibit approximately the same τ-ratios as observed for the original 207-bp M13 fragment. Hence, the sequence shown in Scheme 11.1 is necessary and sufficient to cause stable curvature of the DNA helix backbone. The A- and T-tracts shown in bold type contribute importantly to the observed curvature; replacing any of them by site-directed mutagenesis substantially reduces the τ-ratios, and hence the apparent curvature, of the mutated fragments [7].

... **AAAA**TTCGCGTT**AAATTTTT**GTT**AAA**‖TCAGCTCA**TTTTT** ...

SCHEME 11.1 Sequence surrounding the apparent bend center (‖) in the M13 origin of replication. The A- and T-tracts in bold type contribute importantly to the observed curvature.

Because the four A- and T-tracts in the sequence shown in Scheme 11.1 are of different lengths, the relative spacing between them will depend on whether their 5′- or 3′-ends are compared. Previous studies have shown that the relative phasing of the residues at the 3′-ends of unsymmetrical A-tracts is more important than the phasing of the 5′-residues [8,74], most likely because A-tract-induced bending increases in the 5′ → 3′ direction [75]. The two T_5-tracts flanking the apparent bend center are separated by 19 bp, approximately two turns of the double helix. The terminal residue of the A_4-tract at the beginning of the sequence is approximately out-of-phase with the terminal residue of the first T_5-tract, causing the bends due to these two sequence elements to occur in the same direction. In addition, the terminal residue of the central A_3-tract is out-of-phase with the terminal residues of the two flanking T_5-tracts. Hence, all the bends caused by the A- and T-tracts surrounding the apparent bend center occur in the same direction, contributing additively to the observed curvature.

The thermal stability of the curvature in the M13 origin was investigated by measuring the τ-ratios of the 207- and 471-bp fragments in buffer C as a function of temperature. As the location of the apparent bend centers in the two fragments is known, the τ-ratios can be used to calculate the apparent bend angle as a function of temperature. The results are illustrated in Figure 11.2. The average apparent bend angle was $44.6° \pm 1.1°$ for the 207-bp fragment and $46.3° \pm 2.1°$ for the 471-bp fragment, at temperatures ranging from 4° to 53°C. Hence, the

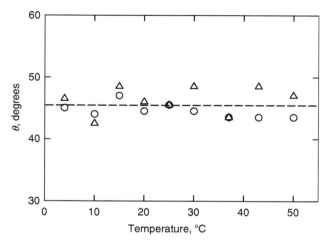

FIGURE 11.2 The temperature dependence of the apparent bend angles, θ, observed for the 207- and 471-bp fragments in buffer C. The solid line represents the average bend angle observed for both fragments at various temperatures, $45 \pm 2°$.

intrinsic curvature in the M13 origin of replication is thermally stable at the biologically relevant temperature of 37°C.

The apparent bend angle in the 471-bp fragment was also measured by atomic force microscopy (AFM) [72], to verify the accuracy of the bend angle determined by the TEB method of overlapping fragments. A normal control fragment of nearly the same size, 476 bp, was also analyzed by AFM. The bend angle in an AFM experiment is defined as the angle formed between lines tangent to the two ends of individual molecules chosen randomly for statistical analysis. As shown in Figure 11.3A, the normal 476-bp fragment exhibited a Gaussian distribution of curvature angles centered at 0°. Hence, the 476-bp fragment is not intrinsically curved, in agreement with the TEB results [14,72]. The most probable angle of curvature of the curved 471-bp fragment was 40°–50°, as shown in Figure 11.3B, in good agreement with the apparent bend angle determined for this fragment by TEB, $46° \pm 2°$.

11.3.3 VP1 GENE IN THE SV40 MINICHROMOSOME

Four overlapping restriction fragments, containing 199-, 204-, 206-, and 340- bp, were isolated from the VP1 gene in SV40. The relationship of these four fragments to each other is shown schematically in Figure 11.4A; for ease of reference, A- and T-tracts containing five or more residues are indicated by square boxes. The terminal relaxation times of the four restriction fragments were measured in buffers B and C and compared with the relaxation times of normal fragments containing the same number of base pairs, taken from standard curves measured in the same two buffers [8,14]. The τ-ratios observed for each of the target fragments in each of the buffers were then averaged and compared with theoretical curves calculated as a function of bend angle and fractional distance, S, of the bend from one end of the fragment [60,61], as shown in Figure 11.4B. Global fitting of the experimental τ-ratios to the theoretical curves indicates that an apparent bend of $46° \pm 2°$ is located at fractional distances of 0.50, 0.18, 0.33, and 0.41 from one end of the 199-, 204-, 206-, and 340-bp fragments, respectively. These fractional distances correspond to sequence position 1922 ± 2 bp, near the center of the VP1 gene. The apparent bend angle is similar in magnitude to that found in the M13 origin.

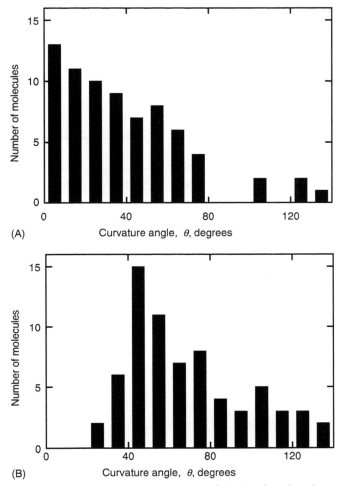

FIGURE 11.3 Histograms of the frequency of occurrence of the bend angles observed by atomic force microscopy for (A) the normal, 476-bp control fragment ($N = 73$); and (B) the curved 471-bp fragment from the M13 origin of replication ($N = 69$). (Adapted from Lu, Y.J., Weers, B.D., and Stellwagen, N.C., *Biophys. J.*, 85, 409, 2003.)

The sequence surrounding the apparent bend center in the VP1 gene includes four A_n- or T_n-tracts ($n \geq 4$) and one mixed A_3T_4 element within a span of 60 bp, as shown in Scheme 11.2. Replacing this sequence by a segment of normal DNA of the same size causes the resulting construct to exhibit normal birefringence relaxation times, τ-ratios and τ-decrements [8]. Hence, the sequence shown in Scheme 11.2 is necessary and sufficient to cause stable curvature of the DNA helix backbone.

...**AAAAAAA**CTCATG**AAAAT**GGTGCTGG**AAAA**‖
CCCATTCAAGGGTCAAATTTTCATTTTTTT...

SCHEME 11.2 Sequence surrounding the apparent bend center (‖) in the VP1 gene of SV40. The A-tracts shown in bold contribute the most importantly to the observed curvature.

To determine whether the five unphased A- and T-tracts surrounding the apparent bend center are responsible for the observed curvature, a 199-bp fragment containing the sequence

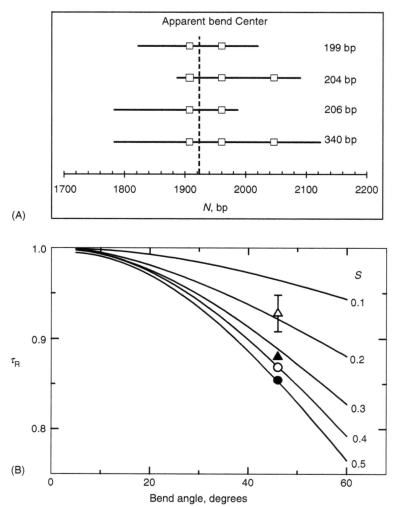

FIGURE 11.4 (A) Schematic diagram of the relationship between the four overlapping restriction fragments used to analyze the curvature of the VP1 gene in the SV40 minichromosome. The square boxes indicate the positions of A- and T-tracts containing five or more residues; the vertical dashed line indicates the location of the apparent bend center. (B) Comparison of the average τ-ratios observed for the four overlapping fragments in buffers B and C with theoretical curves calculated as a function of the apparent bend angle and relative position, S, with respect to the end of the fragment [60,61]. The error bar bracketing the symbol for the 204-bp fragment represents the variation in the τ-ratios observed in the two buffers. From top to bottom, the symbols correspond to fragments containing: ($\triangle$), 204 bp; ($\blacktriangle$), 206 bp; ($\bigcirc$), 340 bp; and ($\bullet$), 199 bp. (Adapted from Lu, Y.J., Weers, B.D., and Stellwagen, N.C., *Biophys. J.*, 88, 1191, 2005.)

shown in Scheme 11.2 was modified by site-directed mutagenesis, keeping the total number of base pairs constant. Each of the A- and T-tracts was mutated individually; a construct was also prepared in which all five A- and T-tracts were mutated simultaneously. The names of the various sequence mutants, the A- and T-tracts eliminated, and averages of the τ-ratios and τ-decrements observed in buffers B, C, and E are summarized in Table 11.3.

The absolute values of the average τ-decrements observed for mutants 199F, 199B, 199C, 198A, and 199E, each missing one of the A- or T-tracts surrounding the apparent bend center,

TABLE 11.3
Characterization of SV40 Sequence Mutants Lacking One or More A-Tracts

Fragment Name	Mutated A-Tract(s)	Average τ-Ratio[a]	Average τ-Decrement[b]
199	None	0.83	-17 ± 3
199F	A_6	0.88	-12 ± 1
199B	A_4T	0.90	-10 ± 2
199C	A_4	0.89	-11 ± 4
198A	A_3T_4	0.86	-14[c]
199E	T_7	0.86	-14 ± 2
199K	All five	1.00	-1 ± 2

[a]Average of the τ-ratios observed in buffers B, C, and E.
[b]Average of the τ-decrements observed in buffers B, C, and E and the standard deviation.
[c]Measured in buffer E only.

were lower than observed for the parent 199-bp fragment. Hence, all five A- and T-tracts contribute to the observed curvature. However, the first three A-tracts, shown in bold in Scheme 11.2, contribute the most importantly; eliminating any of them caused the average τ-decrement to decrease by 30–40%. Eliminating all five A- and T-tracts surrounding the apparent bend center (fragment 199K) caused the average τ-ratio to become equal to 1.00 and the average τ-decrement to decrease essentially to zero, indicating that fragment 199K has a normal conformation in free solution. Hence, A-tract bending is the primary cause of the curvature observed in the VP1 gene of SV40.

The biological function of curvature in the middle of the VP1 gene is not clear. The apparent bend center forms the boundary of a strong nucleosome positioning signal *in vitro* [76], suggesting that wrapping the DNA around histones might be facilitated by curvature of the helix backbone. However, nucleosomes appear to be preferentially excluded from this site *in vivo*, suggesting that this site may be involved in the formation of higher order nucleosome structure [77].

The thermal stability of the curvature in the VP1 gene was analyzed by determining the τ-decrements of the 199-bp fragment in buffer C as a function of temperature. As shown in Figure 11.5, the apparent bend angle was independent of temperature from 4° to 50°C; the

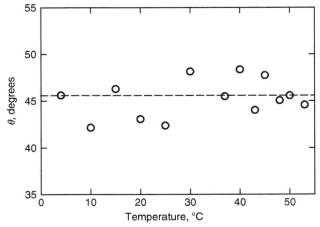

FIGURE 11.5 The temperature dependence of the apparent bend angles, θ, observed for the 199-bp SV40 fragment in buffer C. The dashed line represents the average bend angle, $46 \pm 6°$.

average bend angle observed over this temperature range was $46° \pm 6°$. Hence, as observed for the M13 origin, curvature in the SV40 VP1 gene is stable at the biologically important temperature of 37°C.

11.4 CONCLUDING REMARKS

TEB is a powerful technique for determining the intrinsic curvature of naturally occurring, random sequence DNA molecules. In this chapter, TEB relaxation times have been used to analyze the curvature of restriction fragments taken from an intergenic region in plasmid pBR322, the M13 origin of replication, and the VP1 gene in the SV40 minichromosome. In the last two cases, the apparent bend angles and sequence locations were determined independently by measuring the τ-ratios of a series of overlapping restriction fragments and comparing the observed τ-ratios with a series of theoretical curves calculated as a function of bend angle and distance of the bend from one end of the molecule. Both the M13 origin of replication and the VP1 gene in SV40 contain intrinsic bends of ~45°, which are stable over a wide range of temperatures. Hence, the curvature of these two fragments is apt to be biologically important.

The sequences surrounding the apparent bend centers in the M13 origin and the VP1 gene contain 4 or 5 A- and T-tracts within a span of 40–60 residues. The A- and T-tracts are absolutely required to observe stable curvature; their sequential removal by site-directed mutagenesis causes the modified DNA fragments to gradually adopt a normal conformation. Therefore, the apparent bends in the M13 origin and the VP1 gene in SV40 are not sharp kinks but are delocalized regions of curvature distributed over 40–60 bp.

The magnitude of the apparent bend angle observed in the M13 origin of replication was confirmed by AFM, thus validating both methods of measuring DNA curvature. In a way, the accuracy of the apparent bend angles determined by the TEB method of overlapping fragments is somewhat surprising, since the theoretical equations were derived for once-broken rods [60,61] or small loci of flexibility [62], not for distributed bends comprising up to 30% of the length of the restriction fragment. Hence, the theoretical equations appear to have a greater validity than expected from the relatively limited cases for which they were derived.

ACKNOWLEDGMENTS

We gratefully acknowledge the contribution of Brock D. Weers, who made and characterized most of the restriction fragments used in this study. Financial support by grant GM071464 from the National Institute of General Medical Sciences is also acknowledged.

REFERENCES

1. Hagerman, P.J., Evidence for the existence of stable curvature of DNA in solution, *Proc. Natl. Acad. Sci. USA*, 81, 4632, 1984.
2. Levene, S.D., Wu, H.-M., and Crothers, D.M., Bending and flexibility of kinetoplast DNA, *Biochemistry*, 25, 3988, 1986.
3. Lewis, R.J., Pecora, R., and Eden, E., Transient electric birefringence measurements of the rotational and internal bending modes in monodisperse DNA fragments, *Macromolecules*, 19, 124, 1986.
4. Diekmann, S. and Pörschke, D., Electro-optical analysis of 'curved' DNA fragments, *Biophys. Chem.*, 26, 207, 1987.

5. Stellwagen, N.C., Transient electric birefringence of two small DNA restriction fragments of the same molecular weight, *Biopolymers*, 31, 1651, 1991.
6. Nickol, J. and Rau, D.C., Zinc induces a bend within the transcription III-A-binding region of the 5 S RNA gene, *J. Mol. Biol.*, 228, 1115, 1992.
7. Lu, Y.J., Weers, B., and Stellwagen, N.C., Analysis of DNA bending by transient electric birefringence, *Biopolymers*, 70, 270, 2003.
8. Lu, Y.J., Weers, B., and Stellwagen, N.C., Intrinsic curvature in the VP1 gene of SV40: comparison of solution and gel results, *Biophys. J.*, 88, 1191, 2005.
9. Hagerman, P.J., Investigation of the flexibility of DNA using transient electric birefringence, *Biopolymers*, 20, 1503, 1981.
10. Elias, J.G. and Eden, D., Transient electric birefringence study of the persistence length and electrical polarizability of restriction fragments of DNA, *Macromolecules*, 14, 410, 1981.
11. Elias, J.G. and Eden, D., Transient electric birefringence study of the length and stiffness of short DNA restriction fragments, *Biopolymers*, 20, 2369, 1981.
12. Chen, H.H., Rau, D.C., and Charney, E., The flexibility of alternating dA-dT sequences, *J. Biomol. Struct. Dyn.*, 2, 709, 1985.
13. Charney, E., Chen, H.-H., and Rau, D.C., The flexibility of A-form DNA, *J. Biomol. Struct. Dyn.*, 9, 353, 1991.
14. Lu, Y.J., Weers, B., and Stellwagen, N.C., DNA persistence length revisited, *Biopolymers*, 61, 261, 2002.
15. Stellwagen, N.C., Electric birefringence of restriction enzyme fragments of DNA: optical factor and electric polarizability as a function of molecular weight, *Biopolymers*, 20, 399, 1981.
16. Diekmann, S. et al., Electric properties and structure of DNA restriction fragments from measurements of the electric dichroism, *Biophys. Chem.*, 15, 157, 1982.
17. Rau, D.C. and Charney, E., Electric dichroism of DNA. Influence of the ionic environment on the electric polarizability, *Biophys. Chem.*, 17, 35, 1983.
18. Pörschke, D., DNA double helices with positive electric dichroism and permanent dipole moments: non-symmetric charge distributions and 'frozen' configurations, *Biophys. Chem.*, 49, 127, 1994.
19. Stellwagen, N.C., Orientation of DNA molecules in agarose gels by pulsed electric fields, *J. Biomol. Struct. Dyn.*, 3, 299, 1985.
20. Holzwarth, G. et al., Transient orientation of linear DNA molecules during pulsed-field gel electrophoresis, *Nucleic Acids Res.*, 23, 10031, 1987.
21. Stellwagen, N.C., Effect of pulsed and reversing electric fields on the orientation of linear and supercoiled DNA molecules in agarose gels, *Biochemistry*, 17, 6417, 1988.
22. Jonsson, M., Åkerman, B., and Nordén, B., Orientation of DNA during gel electrophoresis studied with linear dichroism spectroscopy, *Biopolymers*, 27, 381, 1988.
23. Mayer, P., Sturm, J., and Weill, G., Stretching and overstretching of DNA in pulsed field electrophoresis. I. A quantitative study from the steady state birefringence decay, *Biopolymers*, 33, 1347, 1993.
24. Mayer, P., Sturm, J., and Weill, G., Stretching and overstretching of DNA in pulsed field gel electrophoresis. II. Coupling of orientation and transport in initial response to the field, *Biopolymers*, 33, 1259, 1993.
25. Åkerman, B., Tube leakage during electrophoresis retards reptating DNA in unmodified and hydroxyethylated agarose gels, *Phys. Rev. E*, 54, 6697, 1996.
26. Stellwagen, N.C. and Stellwagen, J., "Flip-flop" orientation of agarose gel fibers in pulsed alternating electric fields, *Electrophoresis*, 14, 355, 1993.
27. Stellwagen, J. and Stellwagen, N.C., Transient electric birefringence of agarose gels. I. Unidirectional electric fields, *Biopolymers*, 34, 187, 1994.
28. Stellwagen, J. and Stellwagen, N.C., Transient electric birefringence of agarose gels. II. Reversing electric fields and comparison with other polymer gels, *Biopolymers*, 34, 1259, 1994.
29. Stellwagen, J. and Stellwagen, N.C., Internal structure of the agarose gel matrix, *J. Phys. Chem.*, 99, 4247, 1995.
30. Hagerman, P.J., Flexibility of DNA, *Annu. Rev. Biophys. Biophys. Chem.*, 17, 265, 1988.
31. Charney, E., Electric linear dichroism and birefringence of biological polyelectrolytes, *Q. Rev. Biophys.*, 21, 1, 1988.

32. Nordén, B. et al., Microscopic behaviour of DNA during electrophoresis: electrophoretic orientation, *Q. Rev. Biophys.*, 24, 103, 1991.

33. Hagerman, P.J., Sometimes a great motion: the application of transient electric birefringence to the study of macromolecular structure, *Curr. Opin. Struct. Biol.*, 6, 643, 1996.

34. Stellwagen, N.C., The use of transient electric birefringence to characterize the conformation of DNA in solution, the mechanism of DNA gel electrophoresis, and the structure of agarose gels, *Colloid Surf. A*, 209, 107, 2002.

35. Bloomfield, V.A., Crothers, D.M., and Tinoco, I., Jr., *Physical Chemistry of Nucleic Acids*, Harper & Row, New York, 1974, chap. 5.

36. Perez-Martin, J. and de Lorenzo, V., Clues and consequences of DNA bending in transcription, *Annu. Rev. Microbiol.*, 51, 593, 1997.

37. Allemann, R.D. and Egli, M., DNA recognition and bending, *Chem. Biol.*, 4, 643, 1997.

38. McGill, G. and Fisher, D.E., DNA bending and the curious case of Fos/Jun, *Chem. Biol.*, 5, R29, 1998.

39. Widom, J., Role of DNA sequence in nucleosome stability and dynamics, *Q. Rev. Biophys.*, 34, 3, 2001.

40. Minsky, A., Information content and complexity in the high-order organization of DNA, *Annu. Rev. Biophys. Biomol. Struct.*, 33, 317, 2004.

41. Olson, W.K. and Zhurkin, V., Twenty years of DNA bending, in *Biological Structure and Dynamics*, Sarma, R.H. and Sarma, M.H., Eds., Adenine Press, Schenectady, 1996, p. 341.

42. Haran, T.E. et al., Dynamics of curved DNA molecules: prediction and experiment, *J. Am. Chem. Soc.*, 125, 11160, 2003.

43. Beveridge, D.L. et al., Molecular dynamics of DNA and protein-DNA complexes: progress on sequence effects, conformational stability, axis curvature, and structural bioinformatics, in *Nucleic Acids: Curvature and Deformation*, Stellwagen, N.C. and Mohanty, U., Eds., American Chemical Society, Washington, D.C., 2004, chap. 2.

44. Zhurkin, V.B. et al., Sequence-dependent variability of B-DNA, in *DNA Conformation and Transcription*, Ohyama, T., Ed., Springer, New York, ISBN, 158706281x, http://www.eurekah.com, 2005, chap. 2.

45. Koo, H.-S., Drak, J., Rice, J.A., and Crothers, D.M., Determination of the extent of DNA bending by an adenine-thymine tract, *Biochemistry*, 29, 4227, 1990.

46. Rivetti, C., Walker, C., and Bustamante, C., Polymer chain statistics and conformational analysis of DNA molecules with bends or sections of different flexibility, *J. Mol. Biol.*, 280, 41, 1998.

47. MacDonald, D. et al., Solution structure of an A-tract bend, *J. Mol. Biol.*, 306, 1081, 2001.

48. Barbic, A., Zimmer, D.P., and Crothers, D.M., Structural origins of adenine-tract bending, *Proc. Natl. Acad. Sci. USA*, 100, 2369, 2003.

49. Wu, Z. et al., Overall structure and sugar dynamics of a DNA dodecamer from homo- and heternuclear dipolar couplings and ^{31}P chemical shift anisotropy, *J. Biomol. NMR*, 26, 297, 2003.

50. Tchernaenko, V., Halvorson, H.R., and Lutter, L.C., Topological measurement of an A-tract bend angle: variation of duplex winding, *J. Mol. Biol.*, 326, 751, 2003.

51. Tchernaenko, V. et al., Topological measurement of an A-tract bend angle: comparison of the bent and straightened states, *J. Mol. Biol.*, 326, 737, 2003.

52. Marini, J.C. et al., Bent helical structure in kinetoplast DNA, *Proc. Natl. Acad. Sci. USA*, 79, 7664, 1982.

53. Stellwagen, N.C., Anomalous electrophoresis of deoxyribonucleic acid restriction fragments on polyacrylamide gels, *Biochemistry*, 22, 6186, 1983.

54. Wu, H.-M. and Crothers, D.M., The locus of sequence-directed and protein-induced DNA bending, *Nature*, 308, 509, 1984.

55. Holmes, D.L. and Stellwagen, N.C., Estimation of polyacrylamide gel pore size from Ferguson plots of linear DNA fragments. II. Comparison of gels with different crosslinker concentrations, added agarose and added linear polyacrylamide, *Electrophoresis*, 12, 612, 1991.

56. Stellwagen, N.C., Use of polyacrylamide gel electrophoresis to detect structural variations in kilobase-sized DNAs, *Electrophoresis*, 16, 691, 1995.

57. Strutz, K. and Stellwagen, N.C., Intrinsic curvature of plasmid DNAs analyzed by polyacrylamide gel electrophoresis, *Electrophoresis*, 17, 989, 1996.

58. Stellwagen, N.C., Sequence-dependent bending in plasmid pUC19, *Electrophoresis*, 24, 3467, 2003.

59. Fredericq, E. and Houssier, C., *Electric Dichroism and Electric Birefringence*, Clarendon Press, Oxford, 1973, chap. 1.
60. García Bernal, J.M. and García de la Torre, J., Transport properties and hydrodynamic centers of rigid macromolecules with arbitrary shapes, *Biopolymers*, 19, 751, 1980.
61. Mellado, P. and García de la Torre, J., Steady-state and transient electric birefringence of solutions of bent-rod molecules, *Biopolymers*, 21, 1857, 1982.
62. Vacano, E. and Hagerman, P.J., Analysis of birefringence decay profiles for nucleic acid helices possessing bends: the τ-ratio approach, *Biophys. J.*, 73, 306, 1997.
63. Stellwagen, N.C., Circular dichroism and thermal melting of two small DNA restriction fragments of the same molecular weight, *Biochemistry*, 23, 6311, 1984.
64. Sambrook, J. and Russell, D.W., *Molecular Cloning*, 3rd ed., Cold Spring Harbor Laboratory Press, Cold Spring Harbor, New York, 2001.
65. Pytkowicz, R. and O'Konski, C.T., Characterization of *Helix Pomatia* hemocyanin by transient electric birefringence, *Biochim. Biophys. Acta*, 36, 466, 1959.
66. Bevington, P.R., *Data Reduction and Error Analysis for the Physical Sciences*, McGraw-Hill, New York, 1969.
67. Tirado, M.M., López Martínez, M.C., and García de la Torre, J., Comparison of theories for the translational and rotational diffusion coefficients of rod-like macromolecules. Application to short DNA fragments, *J. Chem. Phys.*, 81, 2047, 1984.
68. Pörschke, D., Persistence length and bending dynamics of DNA from electrooptical measurements at high salt concentrations, *Biophys. Chem.*, 40, 169, 1991.
69. Baumann, C.G. et al., Ionic effects on the elasticity of single DNA molecules, *Proc. Natl. Acad. Sci. USA*, 94, 6185, 1997.
70. Wenner, J.R. et al., Salt dependence of the elasticity and overstretching transition of single DNA molecules, *Biophys. J.*, 82, 3160, 2002.
71. Muzard, G., Théveny, B., and Revét, B., Electron microscopy mapping of pBR322 DNA curvature. Comparison with theoretical models, *EMBO J.*, 9, 1289, 1990.
72. Lu, Y.J., Weers, B.D., and Stellwagen, N.C., Analysis of the intrinsic bend in the M13 origin of replication by atomic force microscopy, *Biophys. J.*, 85, 409, 2003.
73. Higashitani, A. et al., Multiple DNA conformational changes induced by an initiator protein precede the nicking reaction in a rolling circle replication origin, *J. Mol. Biol.*, 237, 388, 1994.
74. Koo, H.-S., Wu, H.-M., and Crothers, D.M., DNA bending at adenine-thymine tracts, *Nature*, 320, 501, 1986.
75. Burkoff, A.M. and Tullius, T.D., The unusual conformation adopted by the adenine tracts in kinetoplast DNA, *Cell*, 48, 935, 1987.
76. Stein, A., Unique positioning of reconstituted nucleosomes occurs in one region of simian virus 40 DNA, *J. Biol. Chem.*, 262, 3872, 1987.
77. Ambrose, R.K. et al., Location of nucleosomes in simian virus 40 chromatin, *J. Mol. Biol.*, 214, 875, 1990.

12 Electro-Optics and Electroporation of Synthetic Lipid Vesicles

Zoltan A. Schelly, Niloofar Asgharian, N. Mariano Correa, Viktor Peikov, Sixin Wu, Hongxia Zeng, and Hongguang Zhang

CONTENTS

12.1 INTRODUCTION

Electro-optical phenomena encompass changes in all possible optical properties of matter caused by the application of an electric field E, including changes in energies of quantum states. The most commonly studied effects in solutions and suspensions, however, are birefringence, light scattering, and dichroism. Our particular focus will be on the course of time-dependent changes in birefringence $\Delta n(t)$ and light scattering $\Delta S(t)$ in aqueous solution of unilamellar bilayer vesicles. Such studies can provide useful information about morphology, structure, electrical and mechanical properties, and dynamic behavior of the particles involved [1]. As vesicles are deformable by the applied field, the ensuing structural changes can be drastic, which may lead to pore formation in the bilayer (electroporation, EP) and fusion of the vesicles (electrofusion). Thus, although not directly observable through birefringence and light scattering experiments, the mechanism of electroporation will also be addressed by means of additional trans-membrane transport studies.

12.2 EXPERIMENTAL SECTION

12.2.1 INSTRUMENTATION

For transient birefringence studies of liquid-phase samples, the major components of the experimental setup are a high-voltage pulse generator and the sample cell (Kerr cell) within the standard [2] optical detection system. The schematic top view of the latter is depicted in Figure 12.1. Looking from the detector (PMT), the polarizer (P) and the slow axis of the quarter-wave plate ($\lambda/4$) are set at $\pi/4$ and $3\pi/4$ angles, respectively, relative to the direction of the applied field E, and the analyzer is adjusted to an (optimized) small angle $\pm\alpha$ relative to the $\lambda/4$ plate. Angles are measured in the trigonometric direction (anticlockwise). The light intensity $I(t)$ incident on the detector is recorded as a function of time t, from which—by assuming $I(t)$ is modulated solely by the evolving optical anisotropy—the birefringence signal $\Delta n(t)$ is computed.

However, any other concurrent process in the sample that may affect the detected light intensity $I(t)$ will invalidate the preceding assumption, i.e., the process will contribute to or reduce the actual birefringence [3,4]. The most common of such processes are the clustering of induced dipolar particles [4–6] and phase separation or percolation [3,7]. Each of these processes may cause transient turbidity $\Delta I_t(t)$, i.e., loss in transmitted light intensity due to increased light scattering integrated over the full solid angle. It is absolutely necessary to know whether transient turbidity occurs. In the event it does, the turbidity must first be extracted from the detected $I(t)$ before computing the true birefringence signal [6]. The pure turbidity component $\Delta I_t(t)$ of the composite signal $I(t)$ can be obtained independently, for instance, through particular reorientation of the optical components of the detection system, which renders it blind to sensing birefringence [3]. In contrast, if induced birefringence and light scattering (turbidity) occur simultaneously, the pure birefringence component of the signal cannot be obtained independently.

For studying induced light scattering alone, the $\lambda/4$ plate and analyzer are removed, the polarizer is set either parallel or perpendicular to the external field E, and the time-dependent change of the corresponding transmitted light intensity, $\Delta S_\parallel(t)$ or $\Delta S_\perp(t)$, is monitored [4].

Further details of the instrumentation and the use of reversing polarity pulses for perturbation have been described previously [3,4,8,9].

12.2.2 PREPARATION, CHARACTERIZATION, AND LOADING OF VESICLES

Large unilamellar bilayer vesicles of narrow size distribution were prepared through multiple extrusion (Extruder, Lipex Biomembranes) from the zwitterionic synthetic phospholipid dioleoylphospahatidylcholine (DOPC; Figure 12.2; from Avanti Polar Lipids). Their unilamellar nature and size distribution were established by ^{31}P NMR (Bruker MSL-300) and dynamic light scattering (Brookhaven Model BI-200 SM) measurements, respectively [4].

Only for (i) establishing the mechanism of electroporation through transmembrane transport studies [10,11], (ii) determining the effect of electrolytes [4] and the mechanism of global polarization of the vesicles [9], and (iii) utilizing electroporation for the creation of subnanometer size quantum dots (QDs) [12–15], were the interior compartment of the vesicles and

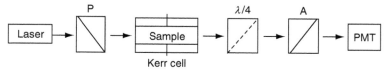

FIGURE 12.1 Schematic top view of the detection system. P, polarizer; Kerr Cell with electrodes; $\lambda/4$, quarter-wave plate; A, analyzer; PMT, photomultiplier tube detector.

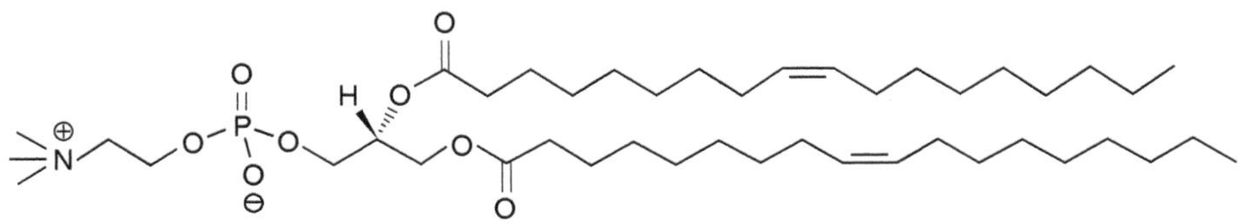

FIGURE 12.2 Structure of the zwitterionic phospholipid dioleoylphospahatidylcholine (DOPC) used for the preparation of the vesicles.

the bulk medium loaded with suitably chosen different ions during preparation of the vesicles. Otherwise, the electric pulse experiments were performed at zero ionic strength.

12.2.3 Experimental Variables

In a large number of experiments, the domain of experimental variables investigated are listed below:

Vesicle size in terms of mean hydrodynamic diameter $<D_h>$: 95, 120, and 190 nm (which correspond to mean lipid aggregation numbers N of $(N/10^5)$: 0.8, 1.3, and 3.2, respectively.

Lipid concentration C (mg/ml): 0.031, 0.063, 0.125, 0.25, 1, 2, and 3 (which result in the mean center-to-center distance d between vesicles of $(d/10^3 \, \text{nm})$: 2.34, 1.85, 1.47, 1.17, 0.74, 0.59, and 0.51, respectively.

Ionic strength: 0 to $5 \times 10^{-4} \, M$ (NaCl inside or outside the vesicles).

Electric field strength E: 0.5 to 7.2 kV/cm

Pulse length Δt: 10 μs to 10 ms

Temperature (°C): 15, 25, and 35. The liquid-to-crystalline phase transition temperature of the DOPC bilayer is -17.3°C.

12.3 RESULTS AND DISCUSSION

12.3.1 Observed Effects

As a function of the experimental variables used, three distinct effects are observed:

(a) Induced transient positive birefringence $\Delta n(t)$ *only*

Conditions: 0.8 kV/cm $\leq E \leq E^*$ $\Delta t \leq \Delta t^*$ *and* C < 0.5 mg/ml (for E^* and Δt^* see (b))

Forward relaxation: $\Delta n(t) = \Delta n_0 - \Sigma_i A_i \exp(-t/\tau_i)$ (double exponential) with

$\Delta n_0 = \Sigma_i A_i = \Sigma_{-i} A_{-i}$

Reverse relaxation: $\Delta n(t) = \Sigma_{-i} A_{-i} \exp(-t/\tau_{-i})$ (triple exponential)

(b) Concomitant induced birefringence and induced turbidity (components can be separated)

Conditions: $E \geq 2$ kV/cm *and* C ≥ 0.5 mg/mL. E, C, and Δt must exceed certain threshold values (*) that are functions of other variables:

$C^*(E, \Delta t)$, $E^*(C, \Delta t)$, and $\Delta t^*(C, E)$

Induced transient light scattering (measured as turbidity) $\Delta S_{\parallel}(t)$ and $\Delta S_{\perp}(t)$. The signals exhibit nonexponential relaxations.

(c) Electroporation (electric field-induced pore formation)

Conditions: $E > 4$ kV/cm *and* $\Delta t \geq 30$ μs

12.3.2 Discussion

The large data set obtained allows the description of the sequence and nature of physical events (Figure 12.3) that lead to the observed optical response of vesicles ensuing the perturbation by a single high-voltage rectangular electric pulse.

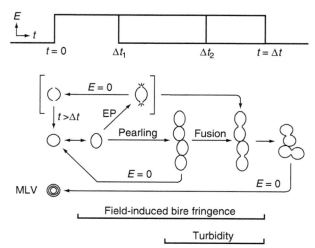

FIGURE 12.3 Schematic summary of the electric field-induced processes in DOPC vesicles in relation to the rectangular electric pulse of different possible lengths (Δt). The double arrow ($\leftrightarrow$) indicates the sole step that is reversible with respect to the presence and absence of the field E. For electroporation (EP) to occur Δt_1 must be greater than 30 μs, and the resealing of the pores takes longer than the length of the pulse, which created them ($t > \Delta t$). Notice the simultaneous occurrence of birefringence and turbidity caused by pearling and fusion. The latter requires a pulse length of Δt_2. Reequilibration takes place at zero field strength ($E = 0$) after termination of the pulse. If fusion did occur, reequilibration leads to multilamellar vesicles (MLV) instead of the original unilamellar vesicles.

Prior to application of the pulse, the equilibrium system is characterized by the incessant thermal shape fluctuation of the time-average spherical shells. Whereas in a symmetrical instantaneous deformed shape (such as a prolate or oblate ellipsoid) the vesicle possesses no dipole moment, whereas in an asymmetrical shape (such as an egg or a pear) it has a multipole moment or an equivalent *instantaneous* dipole moment. Upon turning the field E on, the system is polarized, which renders the vesicle with an induced dipole moment. Our reversing-pulse electric birefringence studies have shown that the polarization occurs by a "slow ion polarization" mechanism [9]. Due to the Maxwell stress during polarization, the DOPC vesicle is slightly deformed to a prolate ellipsoid with its major axis parallel to E. (Alternatively, in the case of anisotropic electric polarizability of the membrane, and depending on the relative magnitudes of the dielectric constants of the medium (water, ε_w) and the radial (ε_β) and tangential (ε_γ) components of the dielectric constant of the membrane, the deformation may lead to an oblate ellipsoid or the extent of deformation may be zero [16].) The resultant of the instantaneous and induced dipole moments tends to reorient parallel to E, which may occur either through a solid rotation or a peristaltic rotation [17] of the vesicle. The ensuing structural anisotropy of the solution is the origin of the observed transient birefringence. If the applied pulse is long enough, the induced dipolar prolate ellipsoids aggregate to linear chains (pearling) parallel to E, which further increases the birefringence and, simultaneously, causes turbidity. As mentioned before, the two components of the light intensity signal can be separated, or the turbidity component can be studied independently. If the field is maintained even longer, the vesicles within the chain fuse (electrofusion).

Provided the applied field strength E and pulse length Δt exceed 4 kV/cm and 30 μs [10,11], respectively, the early events take a route different from that outlined in the preceding paragraph. i.e., the elongation of the spherical membrane shell to a dipolar prolate ellipsoid with its major axis parallel to E entails: (i) an increase of pressure in the interior compartment of the vesicle and (ii) the increase of curvature of the bilayer (i.e., destabilization of membrane structure) at the polar cap regions of the vesicle which face the electrodes, where also the induced transmembrane potential $\Delta V(\theta,t)$ has its maximum—with θ being the polar coordinate of locus on the membrane with respect to E. Due to the combination of these effects, pores open at the polar cap regions of the vesicle (electroporation) through which part of the content (water, or entrapped solution) of the vesicle's compartment is ejected into the bulk [10,11]. This circumstance can be utilized for the creation of QDs as will be described below. While the field is on, the diameter of the holes grow, but they rapidly reseal after turning the field off.

Termination of the pulse initiates the reverse relaxations from the steady state reached under the field. However, due to the relative rates of the distinct steps involved and the fact that all vesicle states (except the original spherical state) of the overall forward process are existent only in an electric field, removal of the field $(E=0)$ prevents the system from back tracing the forward path. Thus, depending on the pulse length Δt applied, return to equilibrium occurs through alternate paths (Figure 12.3), which create several thermodynamic cycles.

12.3.3 Quantum Dots Via Electroporation

The interest in nanoscopic metal and semiconductor particles (QDs) arise from their unusual size-dependent optical, electronic, and magnetic properties. Their synthesis and investigation in solution represent experimental challenges because of the spontaneous, rapid self-aggregation of the constituent atoms or molecules. Customarily, to arrest particle growth at the smallest possible size, the QDs are covalently covered ("capped") with a protective layer of surfactants or polymers. By using such methods, QDs with narrow distribution of size down to 1-nm diameter have been achieved. Capping, however, has deleterious effects: it alters the surface electronic properties of the particles and prevents the evolution of optical and electronic properties to be observed as a function of increasing particle size.

The preparation of even subnanometer size, uncapped QDs has been developed in our laboratory by utilizing the electroporation of vesicles [12]. The essence of this novel method is the initial segregation of the two reactants of the synthesis reaction (one inside the vesicles and the other in the bulk). During electroporation—using the example of creating PbS QDs [14] by the reaction Pb^{2+} (inside) $+ S^{2-}$ (outside) $\rightarrow$ PbS (outside) (Figure 12.4)—a part of the Pb^{2+} ions is ejected into the bulk where the reaction occurs. Such metered admission of Pb^{2+} throughout into the S^{2-} solution prevents high local concentrations that would be unavoidable in macroscopic scale mixing. An additional crucial feature of the method is the adsorption of the nascent PbS molecules on the exterior surface of the vesicle, which drastically slows down their self-aggregation from the otherwise diffusion-controlled rate to the hour and day timescales. The slow growth of the QDs in the subnanometer (molecular) size range can be conveniently monitored through the previously unobserved alternating red and blue shifts of the UV absorption band associated with distinct sizes of the particles. Above the molecular size range, the particles exhibit a familiar monotonic red shift of the absorption band with increasing size. The alternating spectral shifts we observed are in agreement with density functional calculations of the structure and transition energy of such tiny clusters [18].

FIGURE 12.4 Diagram of the reaction used for preparation of PbS quantum dots (QDs) via electroporation (EP) of vesicles.

Uncapped QDs of molecular size of AgBr [12], CdS [13], and gold [15] have also been obtained via electroporation of vesicles. Clearly, the method opens up the possibility for discovering novel properties of particles in solution in the unexplored subnanometer size domain.

ACKNOWLEDGMENT

This work was supported in part by The Welch Foundation, the National Science Foundation, the ACS Petroleum Research Fund, and the Texas Advanced Research Program.

REFERENCES

1. Schelly, Z.A., Transient electro-optics of organized assemblies, *Colloids Surf.*, *A*, 209, 305, 2002.
2. Frederiq, E., and Houssier, C., *Electric Dichroism and Electric Birefringence*, Clarendon Press, Oxford, 1973.
3. Tekle, E., Ueda, M., and Schelly, Z.A., Dynamics of electric field induced transient phase separation in water-in-oil microemulsion, *J. Phys. Chem.*, 93, 5966, 1989.
4. Asgharian, N., and Schelly, Z.A., Electric field-induced transient birefringence and light scattering of synthetic liposomes, *Biochim. Biophys. Acta*, 1418, 295, 1999.
5. Chen, H.M., and Schelly, Z.A., Dynamics of laser-induced transient light scattering in pure liquids and reverse micelles, *Chem. Phys. Lett.*, 224, 69, 1994.
6. Chen, H.M. and Schelly, Z.A., Laser-induced transient electric birefringence and light scattering in Aerosol-OT/CCl$_4$ reverse micelles, *Langmuir*, 11, 758, 1995.
7. Tekle, E. and Schelly, Z.A., Modeling the electric birefringence relaxations of AOT/isooctane/H$_2$O water-in-oil microemulsions, *J. Phys. Chem.*, 98, 7657, 1994.
8. Feng, K.-I., and Schelly, Z.A., Electric birefringence dynamics of crystallites and reverse micelles of sodium bis(2-ethylhexyl) phosphate in benzene, *J. Phys. Chem.*, 99, 17212, 1995.
9. Peikov, V., and Schelly, Z.A., Reversing-pulse electric birefringence study of unilamellar DOPC vesicles, *J. Phys. Chem. B*, 105, 5568, 2001.
10. Correa, N.M., and Schelly, Z.A., Dynamics of electroporation of synthetic liposomes studied using a pore-mediated reaction, $Ag^+ + Br^- \rightarrow AgBr$, *J. Phys. Chem. B*, 102, 9319, 1998.
11. Correa, N.M., and Schelly, Z.A., Electroporation of unilamellar vesicles studied by using a pore-mediated electron-transfer reaction, *Langmuir*, 14, 5802, 1998.
12. Correa, N.M., Zhang, H., and Schelly, Z.A., Preparation of AgBr quantum dots via electroporation of vesicles, *J. Am. Chem. Soc.*, 122, 6432, 2000.
13. Zeng, H.X., Marynick, D.S., and Schelly, Z.A., CdS quantum dots prepared via electroporation of vesicles: experimental and computational results, *Abstract of Papers, 225th ACS National Meeting*, New Orleans, LA, March 23–27, 2003; American Chemical Society, Washington, DC, 2003; COLL-188.
14. Wu, S.X., Zeng, H.X, and Schelly, Z.A., Ultra-small, uncapped PbS quantum dots prepared via electroporation of vesicles, *Langmuir*, 21, 686, 2005.
15. Wu, S.X., Zeng, H.X., and Schelly, Z.A., Growth of uncapped, subnanometer size gold clusters prepared via electroporation of vesicles, *J. Phys. Chem. B*, 109, 18715, 2005.
16. Peikov, V., and Schelly, Z.A., Modeling of the electric field-induced birefringence of unilamellar vesicles, *J. Phys. Chem. B*, 108, 9357, 2004.
17. Schelly, Z.A., Dynamics in water-in-oil microemulsions, *Curr. Opin. Colloid Interface Sci.*, 2, 37, 1997.
18. Zeng, H.X., Schelly, Z.A., Ueno-Noto, K., and Marynick, D.S., Density functional study of the structures of lead sulfide clusters (PbS)$_n$, $n = 1$–9, *J. Phys. Chem. A*, 109, 1616, 2005.

13 Electro-Optical Analysis of Bacterial Cells

Alexander Angersbach, Victor Bunin, and Oleg Ignatov

CONTENTS

13.1 INTRODUCTION

The electrophysical methods for analysis of bioobjects, in general, and electro-optical methods, in particular, have a long history of evolution. First electro-optical experiments were conducted already in the nineteenth century. However, until present, the application of the method for bacterium analysis had a phenomenological character. In the majority of the publications on bacterium analysis, illustrations of the electro-optical method potentially substituted its wide practical application for solving specific biotechnological problems. As a result, there are still no techniques developed for the application of the electro-optical methods or tools to utilize them.

The electro-optical method helps in obtaining unique information on structural, electrical polarization of the investigation objects that is inaccessible when using any other nondestructive method for cell analysis. Electro-optical technique used in the wide frequency range of electric field gives information about dielectric parameters of an averaged cell of the assembly. These data comprise the information on the parameters and properties of the cell structures with different complex dielectrical permeability. At the same time, optical analysis provides actual study of distribution of the refractive index throughout a cell. This component is a real part of complex dielectric permeability measured asymptotically at the optical boundary of electromagnetic field frequencies [1,2].

This chapter is not a review; it is an attempt to outline the spectrum of the tasks in which electro-optical method could provide new information. The experimental data obtained were used to demonstrate the areas where electro-optics is applied and the results were compared with other similar methods.

13.2 THEORY, MEASUREMENT, AND CALCULATED PARAMETERS

The basic properties of bacterial cells as the objects of analysis are size, its comparable with wavelength of light, shape anisotropy, morphometric and electrophysics heterogeneity, and ability to rapidly change their properties depending on physicochemical parameters of the environment. The selection of optimal variant of electro-optical method for cell analysis was carried out taking into account the indicated properties of the test object. The most suitable was recognizing the method in which alternative electric field with fixed position of vector in the measurement chamber acts on the cells, and photometric registration of the changes of suspension optical density is conducted. The Maxwell–Wagner bulk polarizability is used as the main type of polarizability for experimental data analysis. This type of polarizability occurs on the boundary of two media with different dielectric parameters, $\varepsilon = \varepsilon' + \sigma/j\omega$, where the real ($\varepsilon'$) and the imaginary ($\sigma/j\omega$) parts of complex permeability of a medium-structure and σ represents the conductivity of the structure (medium) [3–5].

The advantages of the above-mentioned combined method for cell monitoring in microbiology, biotechnology, and biodetection are measurement of heterogeneous system of suspended cells, rapid analysis and simple algorithms for calculation of cell parameters on the basis of experimental data.

Between the action of electric field on the cells and the changes of the optical density, a number of processes take place. The starting process is generation of induced charges on the boundaries of the cell structures that have different dielectric properties. Tensor of cell polarizability α is a quantitative parameter that describes spatial distribution of its charges. Its components are expressed in absolute values not dependent on the strength of electric field E and is the sum of partial function $F_i(\varepsilon_i/\varepsilon_{i+1})$ for adjacent cell structures i and $i+1$ [6].

The charges interact with electrical field and cause a rotation moment. The cell distribution is changed from chaotic to oriented. For fixed position of electrical-field vector, the process of orientation takes a few to ten seconds. Cell sizes and viscosity of the medium influence the duration of the process.

The stationary distribution of cells in orientation angles is described by a Bolzman function. Argument for this exponential function is the value $M = \Delta\alpha E^2/kT$, referred to as orientation factor. The variable M includes dα, which is the difference between longitudinal and orthogonal components of the polarizability vector (below the anisotropy of polarizability); kT is the energy temperature [7].

The changes of nonspherical cell orientation are accompanied by changes of medium optical density. It is explained by modification of cell light dispersion after orientation. Linear dependence between the stationary value of optical density variation and the value of polarizability anisotropy dα is observed when orientation factor M and optical factor $Q = n * a * 2\pi/\lambda$ are not more than a few units (a is size of the cell, n is relative refractive index, λ is wave length) [8–10]. When electric-field strength is less then 50 V/cm, these conditions are effective for most of bacterial cells.

When the orientation factor $M < 2$ to 5, the shapes of relaxation curves for cell orientation and cell disorientation are same at first approximation, and it is exponential function for the same-size cells. For suspension of different-size cells, relaxation curve is the sum of exponents that have different exponential factors and different weight coefficients. The exponential factor is equal to the coefficient of cell rotary diffusion (CRD) that at first approximation is proportionate to cell volume. Weight coefficient is proportionate to partial concentration of the cell.

The data presented above are sufficient for general understanding on the specificity of the electro-optical method. More detailed information on the theory of bulk polarization and optical effects, as well as specificity of solution of direct and inverse electro-optical tasks, is presented in a number of scientific papers [11–13].

Therefore, electro-optical measurements connected with action of alternative electric field on cells allow measuring two signals at each frequency of orienting field—stationary value of optical density variation and shape of relaxation curve. First signal at $M < 2-5$ is proportional to polarizability anisotropy of an averaged cell of assembly. The second signal is connected with morphometric parameters of cells. The area under relaxation curve may be used in the majority of applications as a parameter that reflects average size of cells. Algorithm of strict reconstruction of cell size distribution according to the relaxation curve form is unstable and makes the routine analysis difficult [14,15].

During electro-optical measurements, a series of impulses, where electric field has the fixed value, and variable frequencies are generated. Registration of photometric signal allows calculating a frequency disperse of polarizability anisotropy (FDPA), however, they do not provide additional information on morphometric parameters. The number of experimental FDPA points is normally from 2 to 10–15. For the full manifestation of FDPA variations, at polarizability of Maxwell–Vagner type frequency the range from 40 kHz to 40 MHz is used. However, for the major practical tasks this range may be decreased.

Electrophysical cell model allows determining electrophysical parameters of cell structures by the form of FDPA diagram. A cell is described as ellipsoids of rotation comprising enclosed structures in the form of ellipsoids of rotation. These structures are the cytoplasm, membrane, and cell wall [6]. Qualitative model charts of FDPA depending on dielectric parameters of cell structures are shown in Figure 13.1.

These charts may be useful for choosing frequency range of electrical field and experimental data interpretation.

13.2.1 General Restriction

Linear connection between the measured value of optical density variation and FDPA is realized only at restriction for orientation of M and optical factors of cell parameters. Obliteration of this fact brings to nonlinear distortion of FDPA shape in the area of maximal values of the function. Linearity of experimental function can be proved if dependence of optical density variation on square power of electric field is presented by linear regression dependence.

Another factor that distorts results of electro-optical measurements is the influence of dielectric parameters of supporting medium on FDPA. The FDPA changes as a function of the electric conductivity of the medium; the experimental diagrams of various bacterial species, e.g., *Bacillus subtilis*, are shown in Figure 13.2. The reproducibility of experimental data in this figure is confirmed by several curves formed at constant value of electric conductivity [16].

For comparing the data from different electro-optical measurements, electric conductivity of medium must be maintained at approximately 4–6 μS/sm. In other cases, the influence of electroconductivity on the form and absolute value of FDPA should be taken into account.

The FDPA of cell is a strong physical parameter determining distribution and absolute values induced by electric-field charges. Experimental value of optical density variation must be normalized on E^2 and optical density of suspension to avoid the influence of cell concentration and electric-field power.

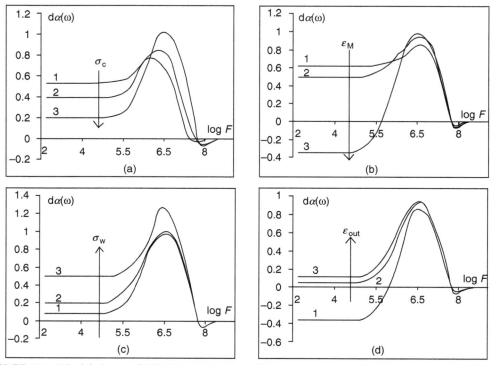

FIGURE 13.1 Model charts of DPAF calculated at different parameters of cell structures. (a) Effect of cytoplasmic electroconductivity on FDPA (1: $\sigma_c = 0.1\,\text{Sm/m}$, $\varepsilon_c = 60$; 2: $\sigma_c = 0.2\,\text{Sm/m}$, $\varepsilon_c = 60$; and 3: $\sigma_c = 0.4\,\text{Sm/m}$, $\varepsilon_c = 20$). (b) Effect of membrane dielectric permeability on FDPA (1: $\sigma_m = 0.2\,\text{mSm/m}$, $\varepsilon_m = 10$; 2: $\sigma_m = 0.4\,\text{mSm/m}$, $\varepsilon_m = 10$; and 3: $\sigma_m = 0.3\,\text{mSm/m}$, $\varepsilon_m = 5$). (c) Effect of cell wall electroconductivity on FDPA (1: $\sigma_w = 0.2\,\text{Sm/m}$, $\varepsilon_w = 60$; 2: $\sigma_w = 0.3\,\text{Sm/m}$, $\varepsilon_w = 20$; and 3: $\sigma_w = 0.6\,\text{Sm/m}$, $\varepsilon_w = 60$). (d) Effect of medium dielectric permeability on FDPA (1: $\sigma_{out} = 0.015\,\text{Sm/m}$, $\varepsilon_{out} = 81$; 2: $\sigma_{out} = 0.01\,\text{Sm/m}$, $\varepsilon_{out} = 70$; and 3: $\sigma_{out} = 0.01\,\text{Sm/m}$, $\varepsilon_{out} = 75$).

The linearity of optical response at normalization to optical density is broken at high cell concentrations. That is why performing measurements in the cell suspension with optical density over 0.4 U (optical length 10 mm) is undesirable.

On the other hand, the accuracy of FDPA measurement in transparent media is decreased at low value of optical density due to the measurement error. Compromise value of optical density of test suspension lies within 0.05–0.2 U.

At first approximation, variation of light wavelength is insignificant. Adsorption of visible light by bacteria is practically absent, and light dispersion does not have specific areas of absorption [17].

The transfer of cells to the medium with low electroconductivity causes excessive osmotic pressure in them. The pressure breaks off the basic transport processes in the cell membrane. Osmotic pressure is a positive factor at a short of time between cell transfer and start of the measurement procedure. There are no changes in polarization parameters in this case because this state of cell looks like frozen. After 10 min between the preparation of samples and electro-optical measurements, osmosis may change cell polarization parameters. For resistant osmotic pressure inside the cell, some nondissociated components may be added in the outer

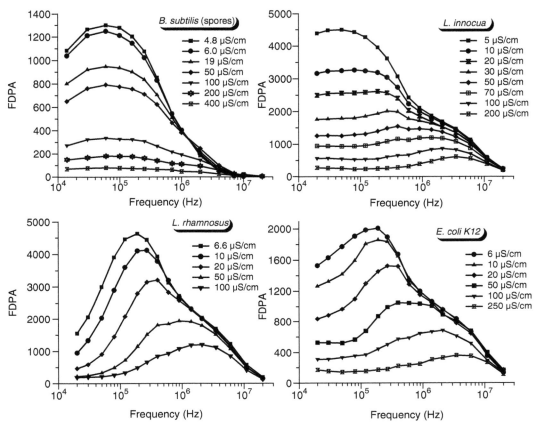

FIGURE 13.2 FDPA dependency of external conductivity by various bacterial species: spores of *Bacillus subtilis*; *Listeria innocua* NCTC11289; *Lactobacills rhamnosus* (LGG); *Escherichia coli K12*.

medium, for instance, some kinds of saccharides. Glucose or saccharose cannot be used for this purpose because most bacteria have enzyme systems for direct utilization of these substances. At the same time, application of manitol or inozitol for producing the compensating osmotic pressure causes no side effects. The majority of microorganisms has no enzymes for immediate utilization of these saccharides and is not able to synthesize these enzymes even after preliminary adaptation.

13.2.2 Area of Electro-Optical Analysis Application

The electro-optical method occupies a special place among the methods for rapid control of cell parameters. Electro-optical analysis makes possible a direct real-time measurement. The method provides monitoring of cell structural transformation during evolution processes and changes of environmental parameters. Usually, for experiments only distilled water is used. Key question in electro-optical experiments is how structural, biochemical, and physiological characteristics are reflected in the polarized parameters and how unambiguously reverse transition of experimental data interpretation can be realized.

There are several ways of using electro-optical analysis in biotechnology and microbiology, which make it advantageous compared to the other competitive techniques:

1. Electro-optical analysis can be used in industrial monitoring of cell cultivation as whole objects on the optimal growing media. The purpose of the monitoring is to produce maximal amount of viable cells at minimal expenses for their production. This purpose is of importance in production of vaccines in pharmacies, ferments, and starting cultures in the food industry. Measurements of FDPA may be used for controlling cultivation process and for forecasting the process duration.
2. The analysis can also be used in industrial monitoring of cell cultivation as whole objects on the optimal growing media for biosynthesis of genetically modified products, for instance, insulin, immunoglobulins, homopoethin, etc. The purpose of this analysis is to achieve maximal outcome. This application is of importance for optimization of production of the protein synthesized by recombinant strains of bacteria and used as a new drug substance.
3. It can be used in scientific and industrial monitoring of cell cultivation on media with toxic compounds. The purpose is to optimize the utilization of toxic compounds in processes. Cells in this case act as small plants for transforming the toxic compounds into neutral chemical compounds. This method is applied in environmental microbiology. The measurement of FDPA may be used for optimizing the experimental conditions.
4. It also finds use in the electrophysical analysis of cells under the influence of different physical and chemical factors. The purpose of the analysis is to investigate the transformation of the cell structures. FDPA and size variation are used as tools for rapid analysis of the modifications in the cell structure.
5. It can be used to express the identification of microorganisms by using selective markers. In contrast to the well-known methods, electro-optical analysis provides rapid detection on the first step of binding of selective marker with surface antigen by variation of the FDPA function.

13.2.3 CELL CULTIVATION

Different kinds of cells have different forms of FDPA functions. Most of these measurements were carried out 10–30 years ago on phenomenological stage of investigation [11,18]. The experimental data on FDPA for different kinds of cells at stationary stage of growth are shown in Figure 13.3.

Differences in FDPA for different kinds of cells are mainly caused by morphometric parameters of cells, i.e., size and axial ratio (a/b), where a and b are long and short sizes of cells, respectively, but not by genotype. Differences in the dielectric properties of similar structures of different kinds of cells are actually insignificant. Noticeable differences are observed when vegetative forms of cells, spores or cells with different cytoplasmatic inclusions, are compared.

During growing of one-type bacteria, more dramatic and significant variation of FDPA function, mean size, and axial ratio was observed. Figure 13.4 illustrates variation of FDPA function during cell growing for *Listeria innocua* and *Escherichia coli K12*. In Figure 13.5, change of FDPA function during long time as 3D picture is demonstrated.

Variations of FDPA functions include at least two processes with different timescales. Fast oscillations reflect cell division, biosynthesis, and metabolic activity. Slow component reflects changing of total bacteria activity, limitation of process by deficiency of growth-stimulating substances, and other processes. Separation and interpretation of each process is a complex problem, however, this is a direct way for optimization of biotechnology process in general.

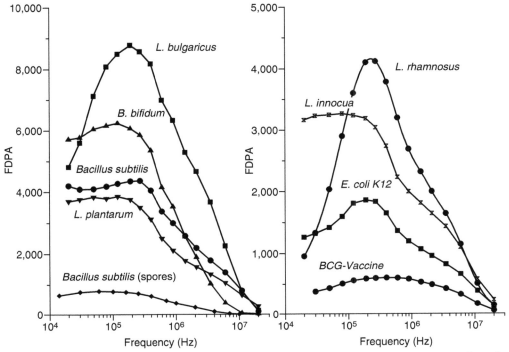

FIGURE 13.3 FDPA of various bacterial species in stationary phase. Measured by external medium conductivity 10–14 µS/cm and different growth conditions. *Lactobacillus bulgaricus, Bifidobacterium bifidum, Bacillus subtilis* (vegetative and spores), *Lactobacillus plantarum, Lactobacillus rhamnosus, Listeria innocua, Escherichia coli K12 DH 5α*, BCG-Vaccine (*Bacillus Calmette-Guerin-Vaccine*) against *Mycobacterium tuberculosis*.

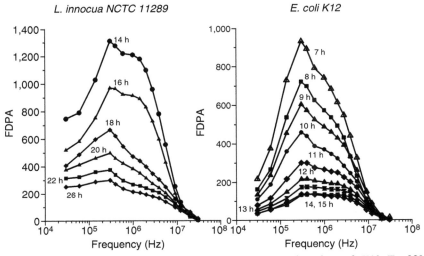

FIGURE 13.4 Series of FDPA functions for *Listeria innocua* and *Escherichia coli K12*. $T = 32°C$, aerobic growth.

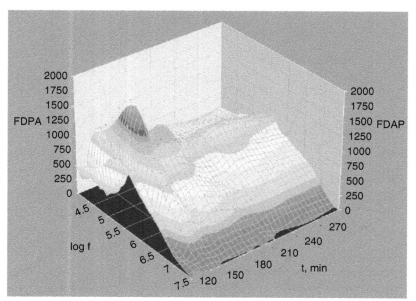

FIGURE 13.5 Time dependence of AP of *Escherichia coli K12*. Long cultivation $T = 36°C$, aerobic condition, not corrected pH, MRS culture medium.

13.2.4 RELATION BETWEEN THE RESULTS OF ELECTRO-OPTICAL MEASUREMENTS AND PHYSIOLOGICAL AND MORPHOLOGICAL CELL PARAMETERS

There are three kinds of experimental data that may be obtained through electro-optical measurements: FDPA function, relaxation curve, and changing of these parameters with time.

The execution of the experiments at standard conditions, considering the above-mentioned limitations, allows avoiding different artifacts and concentrating directly on electrophysical interpretation of the cell status.

FDPA function is the most complicated parameter to analyze because it reflects an abstract parameter, i.e., summarized distribution of inducible charges in the cell volume on the boundaries of different media—cell structures with heterogeneous dielectric parameters. The purpose of such analysis is to establish the relation between polarized parameters of each cell structures and their chemical composition and physiological function. This relation is based on the change of phase bound to free water ratio, change or redistribution of proteins in cell volume, and formation of new cell structures, e.g., cytoplasm body inclusions, spores. Conductivity variation may be connected with change of the concentrations of different movable ions during biosynthesis processes, e.g., H^+, K^+, Na^+ [19,20].

Dielectric permeability is more sensitive to phase transitions. The electric conductivity of cell structures is more sensitive to chemical and energy interaction with the environment.

The electrophysical modeling of a cell helps in calculating complex dielectric parameters of cell structures on conditions that the restrictions for their range are considered. However, no clear picture is obtained in this case. Static FDPA function cannot be measured faster than 10–15 min. This time is comparable with lifetime of one bacteria generation. Within this time, many changes take place in the cell. Every event is reflected in the cell structure, current state, and ion composition of the cell substance. Obviously, 2–4 pictures of cell population for lifetime of one generation cannot reflect all characteristics of the running processes.

Time analysis of polarized parameters carried out in 3–8 min provides principally new viewpoint to vital activity of cells. With this time measurement scale, the reasons for change of polarized parameters may be differentiated and connected with special cell structure and specific metabolic process.

Cell size and its axial ratio influence the form of FDPA. Measurement of relaxation curve and calculation of the cell size may be used for the estimation of this dependence. In some cases, analysis of mean size variation during cell growing may be used for identification of cell division and assessment of population synchronization.

13.2.5 DEVICES AND SOFTWARE

Devices for analysis of electro-optical bacterial cells should automatically execute the following operations:

1. Primary dilution of the cell suspension to achieve fixed optical density
2. Replacing cells from the initial support medium by fixed outer medium having low electroconductivity
3. Synthesis in the sample electrical field with programmable intensity, frequency, and position of electrical-field vector
4. Measuring the optical density variation and calculating the FDPA and morphometrical parameters of cells

The time required for performing each of these processes is limited. The time required for the diffusion of passive ions from the double electrical layer of the cells to the outer medium determines the duration of the medium replacing process; usually, it takes 3–5 min. Duration of electro-optical measurement s at each frequency is proportional to the cell size. Total time of measurements is proportional to the number of registered FDPA points.

Therefore, the instruments for electro-optical analysis of bacterial cells should be capable of executing complicated measurement protocol with some additional operations. Manufacture of electro-optical cell analyzer for scientific and commercial applications by Biotronix GmbH Company is the first attempt to provide the new area of bacterial cell electro-optical monitoring with first-class devices.

13.2.6 ELECTRO-OPTICAL MONITORING OF PERIODICAL CELL CULTIVATION

Industrial monitoring of the cell cultivation is based on the FDPA function analysis presented during fermentation by one or several points. In other words, this is a time cross section of experimental data demonstrated in Figure 13.4 and Figure 13.5. Time dependence of FDPA function at a definite frequency F_t is named as AP function.

Production of maximal quantity of viable cells is the main purpose of such process of cultivation. During the cultivation, cells are selected from fermenter and brought to measurement medium with low electroconductivity. The cells are used for electro-optical measurements and simultaneously are checked for growth on petri dishes. The cell growth on the petri dish is a generally accepted method of viability control. The main disadvantage of this method is long time of incubation. However, in this case this method is used only once to confirm the connection between the polarized parameters and viability of cells.

Results of FDPA measurements for *E. coli* cells and simultaneously optical density and concentration of viable cells after 24 h are presented in Figure 13.6.

The main reason for variations in cell polarization parameters in this case is the changes in the cytoplasm conductivity, especially in the area adjacent to the cell membrane. Moreover,

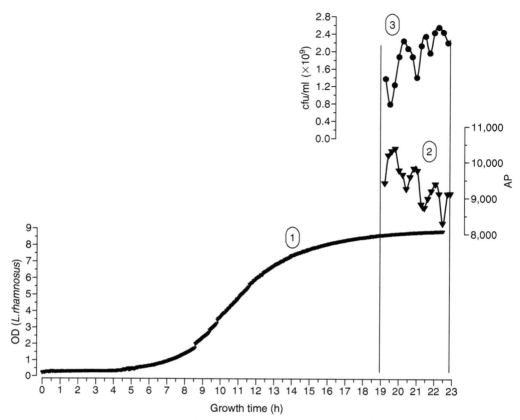

FIGURE 13.6 Growth curve (1), FDPA (2), and counters on petri dishes (3) of *Lactobacillus rhamnosus*. $T = 32°C$, anaerobic growth, MRS culture medium.

there are a considerable number of reasons for polarized parameters changes. However, more important here is that polarized parameters and cell viability have a stable correlation index of approximately equal to -1. Therefore, electro-optical monitoring of cell cultivation can forecast the time of the process completion with maximal concentration of viable cells.

For other kinds of microorganisms, correlation between polarized parameters and viability may be more complicated. It depends on different genetically dependent ways of metabolic activity used by cells in different conditions.

13.2.7 Electro-Optical Monitoring of Recombinant Strains Cultivation

Recombinant strains are widely used in biotechnology as producers of various proteins. The main procedure to provide maximum yield of final protein is the cultivation of the strain producer. In view of the lack of methods for appropriate evaluation of protein accumulation during cell growth, an efficient and precise technique to control the fermentation process is needed.

The cultivation of recombinant strain may be divided into two stages. The first stage includes cell growth and accumulation of living cells without synthesis of target protein. The second stage includes activation of biosynthesis inductor and accumulation of target protein in the soluble form or in the inclusion bodies. Different substances, variation of the medium temperature, or pH may serve as inductor [21].

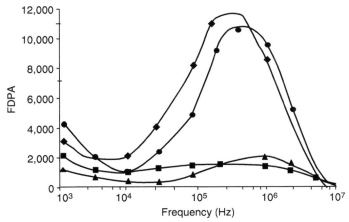

FIGURE 13.7 FDPA of recombinant strain *E. coli strain SG20050* (a) plasmid free (1 h (●), 10 h (■)); (b) pThy-315 for producer of tumornecrosisfactor + Thimosin −α (1 h (◇), 10 h (▲)). Feed-batch cultivation, $T = 37°C$, aerobic condition.

Critical point of this process is time of the cells introduction into culture medium, time of the inductor activation, and time of the process completion. All these points of the process may be optimized by the current electro-optical analysis.

Results of FDPA measurements during cultivation of *E. coli* with plasmids pThy-315 for synthesis of fusion protein tumornecrosisfactor + thimosin − α as inclusion body is shown in Figure 13.7 [22].

Variation of FDPA function reflects the changes of dielectric parameters of cell structures, synthesis of inclusion bodies, and redistribution of cell cytoplasm components.

Analysis of AP functions during cell cultivation allows determining the optimal time of the process. More detailed data on the relation between AP functions and concentration of target protein were obtained for insulin biosynthesis by another *E. coli* strain. Figure 13.8 shows one of the AP function areas illustrating the optimal process of protein biosynthesis. Additionally, Figure 13.8 includes the diagram of insulin concentration measured using high-performance liquid chromatography (HPLC) technique after 30 min.

The maximal concentration of insulin production is corresponding to local minimum of cell polarizability. From the viewpoint of cells metabolic activity, the main cycle of biosynthesis is completed and the cycle of recombinant protein utilization is initiated.

Cultivation of recombinant strain has a number of specific problems because this process includes several different metabolic pathways for specific protein synthesis. Electro-optical monitoring of cell cultivation provides identification of the milestones of this process and outlining the basic strategy of online control.

13.2.8 ELECTRO-OPTICAL MONITORING OF CELL CULTIVATION IN THE CULTURE MEDIUM WITH TOXIC COMPONENTS

Specific strains of bacteria are used for decomposition of toxic components. These microorganisms produce enzymes capable of degrading toxic components and use the resulting products as sources of carbon or other building materials.

By using electro-optical technique, the process of development of several cultures was investigated. Microorganisms *Acinetobacter calcoaceticum A-122* and *Brevibacterium* sp. *13PA* were provided by the "Biocatalysis" Science Research Institute's Laboratory

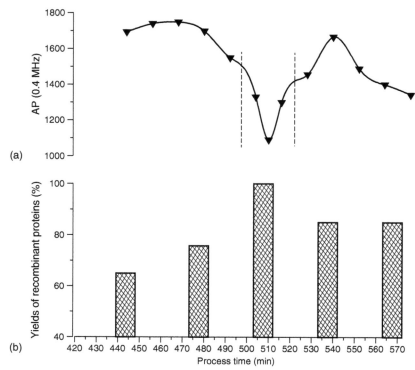

FIGURE 13.8 Cultivation of recombinant strain *E. coli* as producer of human insulin. Feed-batch fermentation, $T = 37°C$, aerobic condition, correction of pH. (a) AP function in time range of maximal production. (b) Yields of recombinant human insulin that is measured by high-performance liquid chromatography.

of Microbiological Transformation (Saratov, Russia). *Acinetobacter calcoaceticum A-122* use p-nitrophenol (PNP) as its sole carbon source. *Brevibacterium* sp. *13PA* use acrylamide or acrylic acid also as the source of carbon with the help of the inducible enzyme amidase.

Results of FDPA measurements for *Brevibacterium* sp. *13PA* after incubation with acrylamide are presented in Figure 13.9 [23].

Oxidation of acrylamide carried out by amidase is accompanied by active oxygen consumption. Parallel measurement of respiratory activity of cells has confirmed the change of cell metabolic activity also observed using electro-optical technique.

In Figure 13.10, the results of FDPA measurement for cells *A. calcoaceticum A-122* after incubation with PNP are demonstrated [24].

Dependence of FDPA on PNF concentration in the range from 0.1 to 1 mM PNF has been experimentally proved. According to the principle of catabolite repression, bacteria degraders synthesize the enzyme. Its concentration is proportional to concentration of utilized substance. Linear character of FDPA dependence as function of PNF concentration confirms both the fact of existence of such type of kinetics of the process and linear dependence between metabolic activity of cells and its FDPA function.

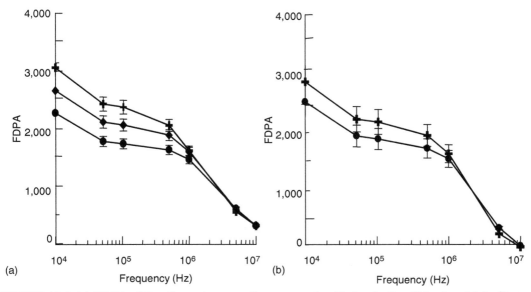

FIGURE 13.9 (a) FDPA of *Brevibacterium* sp. cells upon acrylamide incubation (0.1 g/l and 0.5 g/l): +, cells incubated in deionized water; ♦, acrylamide-incubated cells (0.1 g/l); •, acrylamide-incubated cells (0.5 g/l). (b) FDPA of *Brevibacterium* sp. cells without amidase activity upon acrylamide incubation (0.5 g/l): +, cells incubated in deionized water; •, acrylamide-incubated cells.

13.2.9 ELECTRO-OPTICAL ANALYSIS OF CELL MEMBRANE FUNCTIONS

This type of analysis provides assessment of membrane functions that are affected by some substance or physical factor. Membrane status and its mechanical stability can be evaluated by analysis of passive diffusion process. Operating of different types of pumps can be checked during active transport of ions with low molecular weight. Ionic streams are caused mainly by ions (K^+, N^+, Mg^{2+}, H^+, etc.) and their movement is natural for the cell. Diffusion and active

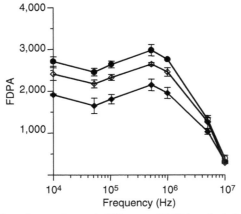

FIGURE 13.10 FDPA of *A. calcoaceticum A-122* upon PNP-incubated: •, control; ♦, 1.0; ◇, 2.5 (m*M* PNP).

transport evidently change the ion concentration from both sides of membrane. Passive diffusion of ions occurs in direction from cell to outer medium with low electroconductivity. As a result, concentration of ions is decreased in cytoplasm and accordingly FDPA function is also decreased. Active transport of ions may change its own direction and this process is typical for a live cell.

Figure 13.11 illustrates FDPA function of *E. coli K12* after action of ethanol in different concentrations.

Membrane can be damaged by ethanol and this process leads to death of bacteria sometime later. For live cell passive diffusion across the membrane is insufficient. Ethanol makes pores in the membrane and ions move from the cell to outer medium. FDPA variation is largely conditioned by changing of cytoplasm conductivity and the area of these pores may be calculated. The volume of the outer medium is tens of thousand times larger than the volume of cells. In spite of the different electroconductivity of cells and outer medium, even complete destruction of cells is unable to essentially change the parameter of outer medium. If the membrane is resistant, ethanol cannot damage it. Such membrane testing allows predicting parameters of cells during their long storage.

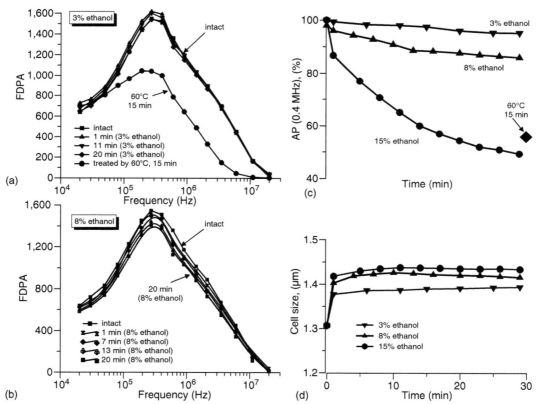

FIGURE 13.11 Kinetic of ion concentration in the cytoplasm and change in cell size in *E. coli K12* induced by membrane destruction after adding of ethanol. (a) FDPA change kinetic after adding of 3% ethanol and heat treatment on 60°C for 15 min; (b) FDPA change kinetic after adding of 8% ethanol; (c) decrease of AP on 400 kHz as a indicator of the loss ion concentration in the cytoplasm; (d) change in cell size after ethanol addition.

Active transport of glucose into the cell may be used for testing of membrane pumps. Results of FDPA analysis of this process for *E. coli Nissle 1917* and *L Plantarum* cells are shown in Figure 13.12.

Active transport includes several processes. The first is active transport of glucose by pumps or due to passive diffusion inside the cell. It is accompanied by increase of cell size and AP value as may be observed on the curve. After some time, excessive concentration of glucose inside the cell leads to the termination of its transport. The level of metabolic activity is typically reduced and FDPA decreased.

These examples are illustrations of the results of cells electro-optical analysis carried out after simple types of influences. However, they demonstrate how it is possible by means of physical or chemical factors variation to test cell membrane and its functional activity.

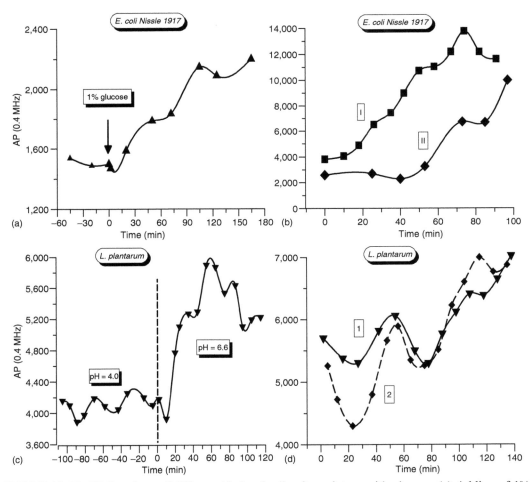

FIGURE 13.12 AP functions of different kinds of cells after mixture with glucose. (a) Adding of 1% glucose to *E. coli* strain *Nissle 1917* in stationary phase: pH = 6.0, $T = 36°C$, aerobic conditions. (b) Inoculation of *E. coli* strain *Nissle 1917* in fresh growth medium: aerobic, $T = 36°C$, pH = 6.6, experiment I, II (inoculum-overnight culture). (c) Adjustment of pH = 6.6 to a culture of *Lactobacillus plantarum* strain ATCC 8014 in stationary phase by pH = 4.0, anaerobic condition. (d) Inoculation of *Lactobacillus plantarum* strain ATCC 8014 in fresh growth medium: $T = 36°C$, anaerobic condition, pH = 6.0, experiment 1, 2 (inoculum-overnight culture).

13.2.10 Cell Type Identification by Electro-Optical Technique

Detection and identification of pathogenic bacteria have some specific features. Usually pathogenic microorganisms are present in nature simultaneously with other ballast cultures that have the same electrophysical properties, shapes, and sizes. Of practical interest is detection of small bacteria concentration. The time of analysis is limited and should not be longer than 30 minutes.

The electro-optical method of pathogenic bacteria detection and identification provides results in an automode each 8–25 min. Long time of cell identification corresponds to low concentration and accumulation of experimental data and vice versa.

Identification of bacteria in the sample by electro-optical technique can be performed only if selective tag is used. Monoclonal (mAb) or polyclonal (pAb) antibodies to cell surface epitops or selective phage to cell are two possible types of selective tags.

Binding of selective tags with cell surface leads to variation of electrophysical properties of cells. Comparison of FDPA function before and after binding gives an answer regarding the presence of target cells. mAb and pAb antibodies give similar results. However, the risk of false-positive result for pAb may arise because of nonpecific linkage with a cell surface.

Selective moderate phages also may be used for cell detection. Binding phage with surface and its penetration into cell is accompanied by FDPA variation. Usage of moderate phages as selective tags reduces the cost of analysis, but increases its duration. Application of phages allows to evaluate target cells viability since the phage is capable of changing the structures of only viable cells.

Samples for detection and identification of target bacteria usually include low concentration of such bacteria among the other types of cells. Fortunately, FDPA measurement immediately does not exclude biological particles and phytoparticles. These particles have greater sizes and low value of bulk polarizability and are selected by measurement procedure. Nevertheless, the problem of electro-optical signal reception with low noise remains. The rough evaluation of the system sensitivity gives the following result: the optical density of cell suspension optical density $= 0.1$, concentration of cells with average size $2\,\mu m$ is approximately 10^8 cells/ml. It leads to the change of intensity of the light passing through the sample by 10%. Target cell concentrations in the sample used for detection are not higher than 10^4 cell/ml. It corresponds to relative light intensity variation $10^{-4}-10^{-5}$. The change of optical density after cell orientation is not higher than 1% and relative variation of its value after binding with selective tag is no more than 30%. Signal to noise ratio >5 is used for measuring experimental signal. Consequently, Fotometric signal to noise ratio (F/N) must be no less than 2.5×10^8. This value may be increased if higher concentrations of target cells are detected. High values of F/N may be reached successfully by two ways: narrow-band filtration of experimental signal and its accumulation.

The electro-optical technique for target cells detection and identification was checked on *Listeria Monocytogene*, *Azosprila Br.*, and vaccine strain ST1 of *B. Antrax* using mAb selective tag. The following characteristics of interactions were investigated:

1. Variation of FDPA as function of mAb concentration
2. Selective properties of the tag–cell interaction in terms of FDPA variation
3. Influence of other type cells on the selective binding
4. Optimization of physical conditions for binding and reproducibility of measurement results

Experimental results for all types of microorganisms were similar. That is why, in Figure 13.13 only the data for binding mAb and *Listeria Monocytogenos* with different concentrations of mAb are shown [25].

Dependence of FDPA on mAb concentration has linear character at low concentrations and saturation at definite concentrations of mAb. Decreasing of signal for extremely high mAb concentration describes the kinetics of process at excessive mAb. Analysis of FDPA has established that the presence of nonspecific cells was not detected by specific mAb and did not influence selective binding of mAb and bacteria. Further refining of binding process has led to mixture incubation at temperature 22–24°C during 3–5 min. These improvements do not influence the difference between experimental FDPA functions before and after binding.

Interaction of cells and the phage as selective tag has been investigated for *E. coli XL-1* and moderated phage M13K07. Experimental FDPA function of this process and AP function is shown in Figure 13.14 [26].

Selective infection of cells by phage has two main stages. During the first stage the phage interacts with the cell surface and prepares it for penetration into the cell. FDPA function at this moment is slightly decreased. After 20 min, new stage begins. The phage penetrates across membrane into the cell and transforms all system of host for reproduction of phage copies. FDPA function is further decreased down to low level and describes the change of cell structure and passive diffusion of ions into outer medium. During the last stage, FDPA is increased and at this time the cell is producing new copies of the phage using all accumulated substances.

Therefore, electro-optical detection and identification of cells by means of mAb can be carried out during 10–15 min without manual operations.

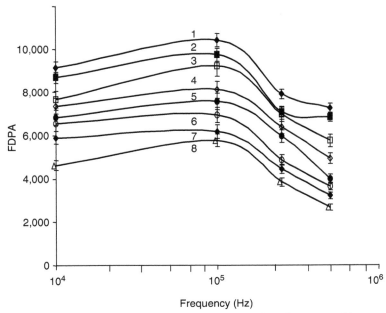

FIGURE 13.13 FDPA of *L. monocytogenes* after incubation with different mAb concentrations: (1) control without mAb; (2) 0.475 μg/ml mAb; (3) 1.425 μg/ml mAb; (4) 1.9 μg/ml mAb; (5) 2.3 μg/ml mAb; (6) 4.75 μg/ml mAb; (7) 7.1 μg/ml mAb; (8) 9.5 μg/ml mAb.

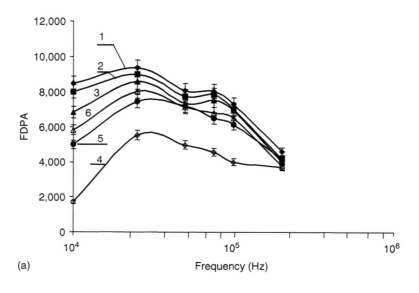

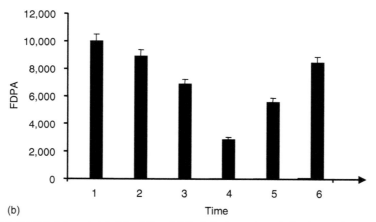

FIGURE 13.14 FDPA (a) and AP (b) of *E.coli* XL-1 during phage infection cells suspended in distilled water (conductivity, $1.8\,\mu S/m$), obtained after incubation with 20 M13K07 phages per bacterium: (1) control without phages; (2) after 1 min; (3) after 10 min; (4) after 30 min; (5) after 60 min; (6) after 90 min.

Analysis of data reproducibility has confirmed stability of the measuring procedure when one sample was used for all measurements and when a new sample was used for every new measurement.

13.3 CONCLUSION

In this paper, electro-optical analysis of bacteria is considered from two viewpoints. Physical aspect describes the advantages, deficiencies, and limitations of this technique. Applied aspect underlines specific features of bacteria during electro-optical analysis and confirmed by experimental results. The work does not apply for the profound description of theoretical bases of the method and specific (genetic, biochemical power) properties of investigated microorganisms. It solves the other problem and gives an answer to the question on what should be taken into account when conducting electro-optical analysis and where this type of the analysis is competitive.

The main advantage of electro-optical technique is direct measurement of bacteria parameters in real time and obtaining information on their structures and properties. The measuring procedures do not cause irreversible changes of their properties and do not demand additional reagents. Transfer of bacteria to other medium with low electroconductivity does not cause any changes in their viability since the results of electric-field action on the cells are commensurable with forces causing thermal movement.

The above-mentioned examples prove that the electro-optical method solves analytical problems better than the competitive methods used for this purpose. Automode of electro-optical analysis allows optimizing a number of biotechnological process that was impossible when using the known alternative methods. This allows making an optimistic conclusion on broadening the area of electro-optical method applications in the near future.

REFERENCES

1. Kerker, M., *The Scattering of Light*, Academic Press, London, 1969, p. 347.
2. Stoylov, S.P., *Colloid Electro-Optics: Theory and Applications*, Academic Press, London, 1991, p. 156.
3. Fricke, H., The Maxwell–Wagner dispersion of ellipsoids, *J. Phys. Chem.*, 57, 934–937, 1953.
4. Bottcher, G.P.F., *Theory of Electrical Polarizability*, Academic Press, New York, 1982, p. 480.
5. Landau, L.D., and Liphshitz, E.M., *Electrodynamics of Continuous Medium*, Nauka, Moscow, 1957, p. 43.
6. Styopin, A.A., Calculation of polarizability parameters of complex particles, *J. Exp. Theor. Phys.*, 8, 1230–1238, 1972.
7. Dukhin, S.S., *Electrooptic of Colloids*, Naukova Dumka, Kiev, 1973, p. 562.
8. Van de Hulst, H.C., *Light Scattering by Small Particles*, Wiley Interscience, New York, 1957, p. 321.
9. Lattimer, P., Light scattering by ellipsoids, *J. Colloid Interface Sci.*, 53, 102–109, 1975.
10. Shchyogolev, Yu. et al., Inverse Problems of spectroturbidimetry of biological disperse systems with random and ordered particle orientation, *Proc. SPIE*, 2082, 167–176, 1994.
11. Pohl, H.A., *Dielectrophoresis*, Cambridge University Press, Cambridge, 1978, p. 48.
12. Miroshnikov, A.I., and Fomchenkov, V.M., *Electrophysical Analysis and Separation of Cells*, Nauka, Moscow, 1986, p. 184.
13. Bunin, V.D., Electrooptical analysis of a suspension of cells and its structure, in *Encyclopedia of Surface and Colloid Science*, Marcel Dekker, New York, 2002, pp. 2032–2043.
14. Tikhonov, A.N., and Arsenin, V.J., *Techniques of Solving Incorrect Tasks*, Nauka, Moscow, 1979, p. 380.
15. De la Torre, G., and Blumfield, V.A., Hydrodynamic properties of complex, rigid biological macromolecules; theory and application, *Q. Rev. Biophys.*, 14, 81–139, 1981.
16. Angersbach, A., Heinz, V., Schlüter, O., Ananta, E., Knorr, D., and Bunin, V., Sicherung der Reproduzierbarkeit von Populationszuständen bei der Untersuchungen von Mikroorganismen unter Nutzung einer elektro-optischen Messmethode. GDL-Kongress "Lebensmitteltechnologie 2001" vom 8. bis 10. November 2001 in Berlin, GDL Bonn, 2001, pp. 233–238.
17. Jennings, B.R., and Morris, V.J., Light scattering by bacteria: size and electrical properties of *E. coli*, *J. Colloid Interface Sci.*, 27, 377–382, 1968.
18. Bunin, V.D., and Voloshin, A.G., Determination of cell structures, electrophysical parameter sand cell population heterogeneity, *J. Colloid Interface Sci.*, 180, 122–126, 1996.
19. Carstensen, E.L., Marquis, R.E., and Gerhardt, P., Dielectric properties of native and decoated spores of *Bacillus megaterium*, *J. Bacteriol.*, 140, 917–928, 1979.
20. Schwan, H.P., Electrical properties of cells: priciples, some recent results, and some unresolved problems, in *The Biophysical Approach to Excitable Systems*, Plenum Press, New York, 1981, pp. 3–24.

21. Glick, B.R., and Pasternak, J.J., *Molecular Biotechnology*, ASM Press, Washington, D.C., 1998, p. 589.
22. Bunin, V.D. et al., Electrophysical monitoring of cultivation process of recombinant *Esherichiae coli* Strains, *Biotechnol. Bioeng.*, 51, 720–724, 1996.
23. Ignatov, O.V., Khorkina, N.A., Singirtsev, I.N., Bunin, V.D., and Ignatov, V.V., Exploitation of the electro-optical characteristics of microbial suspensions for determining *p*-nitrophenol and microbial degradative activity, *FEMS Microbiol. Lett.*, 165, 301–304, 1998.
24. Ignatov, O.V. et al., Comparison of the electrooptical properties and specific respiratory activity of Acinetobacter calcoaceticum A-122, *FEMS Microbiol, Lett.*, 173, 453–457, 1999.
25. Bunin, V.D. et al., Studies of *Listeria monocytogenes*—antibody binding using electro-orientation, *Biosens. Bioelectron.*, 19, 1759–1761, 2004.
26. Bunin, V.D. et al., Electro-optical analysis of the *Escherichia coli*–phase interaction, *Anal. Biochem.*, 328, 181–186, 2004.

14 Geometry of Purple Membranes in Aqueous Medium

Alexandar M. Zhivkov

CONTENTS

14.1 INTRODUCTION

Purple membranes (PMs) are violet distinct areas of the cytoplasmic membrane of the extremely halophilic bacteria *Halobacterium salinarium* (*halobium*) [1]. They are isolated by bacteria lysis with an osmotic shock followed by centrifuging in density gradient [2]. In aqueous medium, PMs are not closed in vesicles [3] and as a result the membrane electric asymmetry manifests itself as a permanent dipole moment. PMs are built only by a single protein—bacteriorhodopsin (bR) [4], functioning as a photoinduced proton pump, converting the solar energy in a transmembrane proton gradient [5]. bR is a chromoproteid whose chromophore (retinal residue) is involved in a photochemical cycle [6,7]: it results in a proton transfer to the outer membrane surface.

In order to understand the mechanism of the proton pump, the structure of bR and photocycle have been the object of intensive investigations during the last 30 years. Now the three-dimensional bR structure and the groups of H^+-transfer path have been established [8–11]. This makes the mechanism of H^+-pump almost clear from the structure point of view, but bR conformational changes remain uncertain. The difficulties come from the fact that PMs are studied in vacuum, when the pump is inactive or the methods used are not sensitive enough as they register the changes in independent bR macromolecules. PM's unique structure gives the possibility to study the conformational changes summed in all excited bR by studying membrane curvature, because it depends on bR macromolecular shape. In fact, after exciting bR by a laser impulse, the light scattering of PM suspension shows changes, which can be interpreted as conformational macromolecule alterations or as PM curving, because of asymmetric change of local pH [12]. The second possibility can be tested by changing the surface charge (by the pH of the medium and ion adsorption) in dark when the photocycle is inactive.

No less important is the question of comparing bR dipole moment p_{bR} calculated on the basis of structure models with that determined electro-optically. As a rule, PMs are considered as rigid planar plates. That allows presenting their dipole moment p as a sum of the dipole moments of bR macromolecules in the membrane and asymmetrically located charged lipids. If PMs in aqueous medium are not planar, it is necessary to determine their curvature and its influence on p. Also, a question arises whether PMs behave like rigid particles during the electric impulse.

These problems require the establishment of the shape of PMs in aqueous medium and whether it changes under the influence of transmembrane and external electric fields. In order to solve these tasks the electro-optical and static light scattering methods have been used here. They offer advantages of sensitivity to particle shape over alternative methods and are capable of providing complex information on the optical, electric, and geometric properties of the particles.

In Section 14.2, a concise literature review on the problem of PM geometry is done. The facts published in literature are discussed and arguments for and against a planar shape for PMs are discussed.

In Section 14.3, the light scattering theory in Rayleigh–Debye–Gans (RDG) approximation and the electro-optical theory have been presented. The exposition is limited to the minimum necessary for the interpretation of the results given later.

In Section 14.4, the application of the RDG theory to PMs' water suspension is provided. It is shown that PMs scatter light as nonabsorbing, optically isotropic, and independent particles. New formulas have been derived (Section 14.4.2) on light scattering under small and large angles for a segment of spherical shell and cylindrically bent elliptical disk as well as their permanent dipole moment.

In Section 14.5, the geometry of undisturbed PM in aqueous medium is determined. In order to solve the problem the number of possible shapes has been restricted to two. The reasons are given for using the rotational diffusion coefficient for electro-optical determination of the area of ellipsoids with unknown axes ratio. Then PM area is determined by the relaxation time of electric light scattering, the radius of gyration—by small angle scattering and the membrane curvature—with the aid of the formulas deduced in Section 14.4.2.1. The comparison of the experimental light scattering indicatrix with those calculated by the formulas derived in Section 14.4.2.2 for large angles scattering shows that the undisturbed PMs in aqueous medium have the shape of a cylindrically bent disk. That conclusion is supported by the existence of two relaxation times in a practically monodisperse suspension.

In Section 14.6, the change in PM geometry under the effect of pH, macroions absorption, and external electric field is studied. It is shown that the change in surface charge leads to alteration in PM curvature and area. The quantitative analysis of this effect described by the formulas from Section 14.4.2.2 shows that PMs are cylindrically bent. Besides, the angle of curvature practically coincides with that obtained by small-angle scattering (Section 14.5.3.2). It has been established that the electro-optical effect at high field intensity has an orientation–deformation nature, whereas at low intensity PMs behave as rigid particles.

In Section 14.2 and Section 14.3, data from the literature are given. The results in Section 14.4 and Section 14.5 are new and so far unpublished. The greater part of the results in Section 14.6 has already been published [13–17] except for the quantitative interpretation of pH-induced changes in curvature (Section 14.6.1).

14.2 LITERATURE REVIEW

14.2.1 PURPLE MEMBRANES: STRUCTURE AND PROPERTIES

bR is a chromoproteid with a molecular mass of 26,534. Its polypeptide chain forms seven α-spiral regions, which are oriented normal to the membrane [18]. The chromophore is a retinol residue connected by protonated Schiff base with a lysine residue [4]. The retinols in light-adapted PMs have a *trans*-conformation and are oriented at $\approx 20°$ to the membrane plane [19].

bR macromolecules are grouped into three and every trimer has two circles of α-spiral segments: an inner one from 9 and an outer from 12. The trimers are closely packed and their centers form a two-dimensional hexagonal lattice with a distance of 6.2 nm between the lattice points [20]. The trimers are almost cylindrically shaped, which does not allow PMs to close in vesicles.

bR amounts to 75% of the PM mass, and the rest of the 25% consists of lipids [2]. At neutral pH about 90% of the lipids are negatively charged [21]. The lipid molecules fill in the space among the trimers and can be partly extracted without destroying the PM structure [22].

The density of the PM is $1.18 \, g/ \, cm^3$ [2]. They are stable over a wide temperature and pH ranges of the medium [4]. Structure changes occur at temperatures $>90°C$ or $pH > 11$ [23]. At $pH < 3.2$ [24] bR is transformed into a blue spectral form but without destroying bR structure.

The retinal determines the absorption in the visible region: a broad band with a maximum at 570 nm and molar absorption coefficient $63,000 \, ^{-1} \cdot cm^{-1}$ for light-adapted PM [25]. In the range 350–470 nm PM absorption does not exceed 20% from the maximum at 570 nm and over 650 nm it drops to zero [7]. The refractive index of PM determined by light scattering of water–glycerol suspension is 1.51 at 450 nm and 1.53 at 650 nm [26].

The surface charge of PMs is due chiefly to bR and lipids and partly to bound cations [27]. The isoelectric point is at pH 1.5 [28] and zeta-potential is 32 mV at neutral pH in the dark [29]. The charge is asymmetrically distributed [30,31], with the cytoplasmic surface more electronegative [32,33]. The surface potential shows titrable groups with pK 3.5 and pK 10 [9,34].

As in aqueous medium PMs are not closed in vesicles, the electrical asymmetry of the membrane is manifested as a transverse oriented permanent dipole moment [35]. PMs are oriented in an electric field as a result of the competition between permanent and induced dipole moments [36]. The permanent moment prevails at frequencies below 100 Hz [37] and intensity below 140 V/cm [38]. In this case, PMs are oriented transversely to the electric field. At kilohertz frequencies, only to third electric polarizability is manifested and PMs are oriented parallel to the field. The permanent dipole moment is around 100 D for a bR [39]. It is proportional to the square of PM diameter and has a maximum at pH 5–6 [37,40,41]. The electric polarizability is proportional to the—third or fourth power of the diameter [39], has a maximum at pH 5–6 [40], and depends linearly on Debye–Hückel parameter [42]. PMs are considered as planar and rigid when their electro-optical properties are determined [43].

14.2.2 PM GEOMETRY

In vivo PMs are cylindrically bent as the bacteria cell has the shape of a cylinder with a diameter 0.7 μm and height 4–10 μm [44]. In electron micrographs, PMs have the shape of a flat elliptical disk with eccentricity about 1.2, thickness 5 nm, and diameter 0.3–1.2 μm, with an average size of 0.55 μm [40].

It is considered that in aqueous medium PMs are planar and rigid. The assumption is based on PM's quasicrystalline structure and their electron micrographic image. But trimers' hexagonal packaging does not mean that the whole PM can be considered as a two-dimensional crystal as its diameter is two orders of magnitude greater than its thickness. The planar shape of the PMs in electron microscopic preparations are probably due to an interaction with the substrate when the adhesion forces exceed the module of elasticity at bending.

The authors who do not share the assumption that PMs are planar and rigid are a few. The first experimental evidence against the generally accepted assumption is given by quasielastic light scattering—the wave vector has an abnormal behavior that cannot be described by the formulas for a plane and rigid disks [45]. Contradictions that could not be explained by the conception of rigid particles have been obtained earlier as well—pH-induced decrease of electric polarizability γ when the surface charge grows [40], however, no suggestion of PM bending has been made in this work.

Czégé was the first to suggest bending of PM induced by the pH of the medium [12,46]. His conclusions are based on change in light scattering of PM suspensions during the photocycle at varying pH and viscosity of the medium. According to him PMs are bent with their cytoplasmic surface inwards at pH<5 or outwards at pH>5, and at pH≈5 they are planar because of surface charge symmetry. He explains the photoinduced curvature change with conformational changes in bR and the local change of pH because of proton pump. Czégé has calculated the form-factor as a function of curvature for spherically and cylindrically bent membranes with a fixed orientation [12,47]. It is to be regretted that these calculations are erroneous, as one can see when comparing his form-factor at the two extreme curvature values with the analytical formulas for plane disks and a spherical shell. Besides, Czégé's experiment is complicated to interpret and, probably, that is why his studies have not received due appreciation. The later studies of Czégé et al. [48,49] describe the kinetics of photoinduced change in light scattering but do not contribute to the solution of the problem with the curvature in dark.

A far more reliable interpretation is possible at pH change when the photocycle is not operating. Electric dichroism studies show that the pH value at which the permanent dipole moment p has a maximum is at pH 5–6 for the wild strain [37,40,41] and varies at bR mutants [41]. The authors accept that at this pH the PMs are planar and they bend when moving away from it. The drawback of this approach is that the alteration of p can be due to a change in charge asymmetry at invariable PM geometry.

In spite of the fact that all studies of p and polarizability γ show a similar course with pH, it is impossible to determine quantitatively the curvature from this dependence because the contributions of charge and curvature to p and γ must be separated. The presence of pH-maximum of γ allows an unambiguous interpretation only on a qualitative level as a change in geometry. Consequently, the investigation of PM electrical characteristics is not enough for a quantitative interpretation. Especially, pH-maximum of p and γ does not mean that PMs are planar at this particular pH but only that the curvature is minimal.

In order to avoid these difficulties, independent information on PM geometry is needed. Light scattering can provide such knowledge because of its high sensitivity to particle geometry. The dependence of light scattering in electric field on pH has been studied [13,16]. The steady-state values of the electro-optical effect at full orientation at two wavelengths, the γ and the rotational diffusion coefficient unambiguously show that at 3.6 and 5.8 PM curvature is minimal. The influence of the surface charge has been confirmed by the adsorption of polyvalent macroions that bind to PM lipids [14]. The PM behavior at field intensity $5\,\text{kV/cm}$ refutes the conception of PM as rigid particles in spite of the fact that they behave as such at low fields [15,17].

A common drawback of all published works is that they do not determine the curvature quantitatively and even do not give a reliable answer to the question whether at pH 5–6 the PMs are planar or the curvature is minimal. That problem has been solved by the results presented in paragraph in Section 14.5.3 ... as well as about pH-induced change in curvature (Section 14.6.1) and Section 14.6.1 about undisturbed PMs in aqueous medium, as well as about pH-induced changes in curvature. The problem of a quantitative estimation of curvature changes at operating photocycle remains unsolved.

14.3 THEORY

14.3.1 LIGHT SCATTERING IN RAYLEIGH–DEBYE–GANS APPROXIMATION

The theory presented below describes the single elastic scattering of linearly polarized monochromatic light by optically independent particles in the range of linear polarization of matter. This means that the scattered light has the same frequency as the incident light, the refractive index n and the absorption index κ do not depend on light intensity and particles have a random distribution in space. The surplus polarizability $\alpha = \alpha_1 - \alpha_0$ is determined from the polarizabilities of the particle α_1 and the medium α_0. Similarly the intensity of the surplus scattering is determined as: $I = I_1 - I_0$.

In the general case the particles can be absorbing and optically anisotropic. Absorbing particles have a complex refractive index $m = n - i\kappa$, where n and κ are real and positive. Besides n determines light phase velocity and κ—the distance $l = \lambda_0/2\pi\kappa$, at which the wave amplitude decreases e times [50]. The absorption index κ is defined by eq. (14.1) for a dielectric layer with thickness Z and absorbance $A = log(I_0/I)$ at wavelength in vaccum λ_0:

$$\kappa = (1/4\pi \log e)(\lambda_0/z)A. \tag{14.1}$$

14.3.1.1 Light Scattering Coefficient

If the linear dimensions of the scattering volume are small compared to the distance r to the photodetector, it can be considered as a point source with force J, radiating light with intensity $I = J/r^2$ in z direction. The scattered light flux $\Phi = \omega J$, incident on a photodetector area s normally orientated to z, is proportional to the space angle $\omega = s/r^2$, the scattering coefficient R_θ, the scattering volume Ω, and the intensity of the incident light I_0. The scattering coefficient (Rayleigh ratio) R_θ is determined as

$$R_\theta = \frac{\Phi}{\omega \, \Omega \, I_0} = \frac{r^2 I}{\Omega I_0}. \tag{14.2}$$

The space distribution of scattered light is determined by the polarization angle φ (the angle between electric vector of the incident light and the observation plane) and the scattering angle θ (the angle between the incident and the scattered beam). The observation plane where the incident and scattered beams lie is assumed to be horizontal, and the light is polarized vertically at $\varphi = \pi/2$. Light scattering theory relates R_θ at angles θ and φ with the shape, the relative size B/λ, and the relative refractive index $n = n_1/n_0$ (where n_1 and n_0 are respective refractive indices of the particles and the medium, $\lambda = \lambda_0/n_0$ is the wavelength in the medium) of particles with diameter B.

Rayleigh–Debye–Gans approximation [51] can be applied if the refractive indices of particle and medium are close:

$$|m - 1| \ll 1 \tag{14.3}$$

and the wave phase in the particle or outside it vary insignificantly:

$$2\pi(B/\lambda)|m - 1| \ll 1, \tag{14.4}$$

where: B is the maximal particle size and $m = m_1/m_0$ is the relative complex refractive index, where $m_1 = n_1 - i\kappa_1$.

These conditions limit n, as well as κ. If

$$2\pi(B/\lambda_0)\kappa \ll 1, \tag{14.5}$$

the wave amplitude does not decrease notably within the particle and the polarization does not depend on absorption—this allows to replace m with n. If condition 14.3 is met, the effective electric field in the particle is almost the same as the active field of the incident wave, so particle polarizability does not depend on its shape. The surplus polarizability of a particle with a volume V_1 is equal to

$$\alpha = \frac{3n_0^2}{4\pi} \left(\frac{n_1^2 - n_0^2}{n_1^2 + 2n_0^2} \right) V_1 = \frac{3n_0^2}{4\pi} \left(\frac{n^2 - 1}{n^2 + 2} \right) V_1. \tag{14.6}$$

Because of condition 14.4 the amplitude and the phase of the light electric field, acting on every element of particle volume, do not differ substantially from the amplitude and the phase of the field of the incident wave. If the linear dimension b of the greatest of these elements is small as compared to the wavelength in the particle $\lambda_1 = \lambda_0/n_1$:

$$2\pi(b/\lambda_1) \ll 1, \tag{14.7}$$

the particle can be considered as a system of independently radiating dipole oscillators and at that the scattered wave amplitude is a sum of the complex amplitudes of the elementary waves of the oscillators. To account for the interference of these waves the function $P(\theta)$ is introduced, which depends on particle shape, its relative size, optical anisotropy and orientation, as well as on the scattering angle and polarization of the incident and scattered light. The light intensity, scattered by one particle (when $r \gg \lambda$) is equal to

$$I_1 = k_0^4 (1/r)^2 \left(1 - \sin^2 \theta \cos^2 \varphi\right) \alpha^2 P(\theta) I_0, \tag{14.8}$$

where $k_0 = 2\pi/\lambda_0$ is the wave number.

The intensity of light, scattered by a volume Ω, containing N independently scattering particles is $I = I_1 N$. Combining Equation 14.2, Equation 14.6, and Equation 14.8 at vertical incident light polarization, the following expression is obtained for the scattering coefficient of a disperse system with concentration $N_0 = N/\Omega$:

$$R_{\theta v} = r^2 I_1 N_0 / I_0 = \frac{9\pi^2 n_0^4}{\lambda_0^4} \left(\frac{n^2 - 1}{n^2 + 2}\right)^2 V_1^2 N_0 P(\theta). \tag{14.9}$$

In order to solve the inverse scattering problem [52] it is convenient to express N_0 by the weight concentration c of the dispersed substance and particle molar mass M, as well as to express α by the increment of the refractive index dn/dc at $c \rightarrow 0$:

$$\alpha = \frac{n_s^2 - n_0^2}{4\pi N_0} \cong \frac{n_0 M}{2\pi N_A} \left(\frac{n_s - n_0}{c}\right) = \frac{n_0}{2\pi N_A} \left(\frac{dn}{dc}\right) M, \tag{14.10}$$

where $N_0 = c N_A / M$, and n_s and n_0 are the refractive indices of the disperse system and the medium at λ_0 respectively. Then

$$R_{\theta v} = \frac{4\pi^2 n_0^2}{\lambda_0^4 N_A} \left(\frac{dn}{dc}\right)^2 cMP(\theta) = cHMP(\theta), \tag{14.11}$$

where H is the optical constant of the suspension at wavelength λ_0.

The refractive index increment dn/dc is determined by the refractive indices of the medium n_0 and the particles n_1, having a specific partial volume $\bar{v}$

$$\frac{dn}{dc} = \frac{\bar{v}\left(n_0^2 + 2\right)\left(n_1^2 - n_0^2\right)}{2n_0 \left(n_1^2 + 2\right)}. \tag{14.12}$$

14.3.1.2 Internal Interference Function

The function $P(\theta)$ (form-factor) for particles of optically isotropic substance, satisfying condition 14.3 and condition 14.4, does not depend on incident light polarization. In that case $P(\theta)$ for particles with arbitrary shape and chaotic orientation can be calculated according to the Debye formula [51]: the particle is considered as built by N discrete scattering centers with distance d_{ij} between the i-th and j-th center:

$$P(\theta) = \frac{1}{N^2} \sum_{i=1}^{N} \sum_{j=1}^{N} \frac{\sin\left(2Kd_{ij}\right)}{2Kd_{ij}}, \tag{14.13}$$

where K is determined by the wave number in the particle $k = 2\pi/\lambda$ and the scattering angle θ:

$$K = k\sin(\theta/2) = 2\pi(n_0/\lambda_0)\sin(\theta/2). \tag{14.14}$$

At small KB, confining to the first two terms of the sine expansion in series at $Kd_{ij} \ll 1$, Equation 14.13 is reduced to Guinier formula [52] for the small-angle scattering of chaotic oriented optically isotropic particles:

$$P(\theta) \underset{\theta\to0}{=} 1 - (2/3)K^2R_g^2 + ... \approx \exp\left(-2K^2R_g^2/3\right), \tag{14.15}$$

where R_g is the radius of gyration, determined by particle size and shape:

$$R_g^2 = \sum_i m_i r_i^2 \Big/ \sum_i m_i, \tag{14.16}$$

where r_{ij} is the distance between the mass center and a volume element with mass m_i.

For a particle of an isotropic substance with the shape of [53]:

– concentric hollow sphere with radii $R > r$ (shell at $|r/R-1| \ll 1$):

$$R_{gs}^2 = \frac{(3/5)\left(R^5 - r^5\right)}{R^3 - r^3}, \tag{14.17}$$

– thin elliptical disc with axes $2a > 2B$:

$$R_{ge}^2 = (1/4)\left(a^2 + b^2\right). \tag{14.18}$$

In particular, R_g^2 is equal to $3R^2/5$ for a sphere and $B^2/8$ for a circular disk.

The linear dependence of $P(\theta)$ function on $\sin^2(\theta/2)$ at small scattering angles θ allows to extrapolate R_θ to $R_{0°}$, where $P(0)° \equiv 1$. The combination of Equation 14.11 and Equation 14.15 gives

$$\ln(cH/R_\theta) = \ln(1/M) + \left(16\pi^2/3\right)\left(n_0^2/\lambda_0^2\right)R_g^2\sin^2(\theta/2). \tag{14.19}$$

That equation allows determining M and R_g measuring R_θ, c, dn/dc, λ_0, and n_0. Besides preliminary data on particle shape are not necessary.

14.3.2 Electro-Optical Methods

The application of an electric field to the suspension induces a change of its optical properties because of the orientation of the particles and, possibly, because of their deformation and aggregation. The last two effects can be removed by choosing appropriate experimental conditions. Light scattering and dichroism in electric field are considered below; the methods are based on the measurement of the electro-orientational effect in a monodisperse suspension. It is assumed that there are no electric and optical interactions between the particles, and the axes of symmetry of the geometric, optical, and electrical properties of the particles coincide [54].

The magnitude of the electro-optical effect (EOE) is determined by the change in absorbance $\Delta A = A_E - A$ or scattered light intensity $\Delta I = I_E - I$, where the index E denotes the presence of an electric field. The relative EOE ($\Delta A/A$ or $\Delta I/I$) can be presented as a product of two functions—orientational and optical.

14.3.2.1 Electro-Orientational Function

The orientation function $F(p, \gamma, \omega, E, T, D, t)$ expresses the degree of statistical orientation of particle with a permanent dipole moment p, electric polarizability γ, and rotational diffusion coefficient D at temperature T in moment t after applying electric field with frequency ω and intensity E.

14.3.2.1.1 Steady-State Effects
The orientation function $F(E)$ at steady-state EOE ($Dt \gg 1$) depends only on the ratio of the orientation energy $(p + \gamma E)E$ and the chaotic movement energy kT. Here are the equations about disk-like particles with a permanent dipole moment, oriented along the normal and polarizability, parallel to the surface [55].
   At low degree of orientation ($pE \ll kT$ and/or $\gamma E^2 \ll 2kT$) $F(E)$ depends on the square of the electric field:

$$F(p, \gamma, E, T) = \frac{1}{15kT}\left(-\gamma + \frac{p^2}{kT}\right)E^2. \tag{14.20}$$

At high degrees of orientation ($pE \gg kT$ and/or $\gamma E^2 \gg 2kT$) $F(E)$ asymptotically increases with the electric field. At full orientation ($E \to \infty$), when the particle symmetry axis is directed in parallel or normally to the field, the function $F(E)$ obtains respectively the values 1 and $-1/2$.

14.3.2.1.2 Transition Processes
The orientation process with only the induced dipole moment at low degree of orientation in steady-state ($pE \ll \gamma E^2 \ll kT$) is an exponential function of the time t after field switch on and does not depend on the particle's electric properties [56]:

$$F(D, t) = F(\gamma, E, T)[1 - \exp(-6Dt)]. \tag{14.21}$$

At high degree of orientation in steady-state ($pE \ll \gamma E^2 \gg kT$), when only the induced dipole moment γE is expressed, the orientation rise is nonexponential and its increment from 0 to 63% for time τ_E depends on orientation moment $\gamma E \times E$ [57]:

$$\tau_E = 1.63(kT/\gamma E^2 D). \tag{14.22}$$

After orientation degree $F(E,0)$ is achieved, the decay of the orientation effect after field switch off is determined by the relaxation time $\tau_0 = 1/6D$ (the time taken for the effect to decrease e times its steady-state value) [56]:

$$F(D, t) = F(E, 0)\exp(-6Dt) = F(E, 0)\exp(-t/\tau_0). \tag{14.23}$$

14.3.2.1.3 Rotational Diffusion Coefficient
The rotational diffusion coefficient $D = kT/f$ of particles with an arbitrary shape at temperature T is determined by the friction coefficient f when the particle is rotated around one of its

axes. EOE from axis-symmetrical particles is determined at rotation around the axis, which is orthogonal to the axis of symmetry. For particles with wettable surface, f is determined by their size, shape, and medium viscosity η at temperature T.

The rotation diffusion coefficient of a sphere with a diameter B_s is

$$D_s = \frac{kT}{\pi \eta B_s^3}. \tag{14.24}$$

The friction coefficients of an ellipsoid with semiaxes $a>B$ and axes ratio a/b are given by Perrin's equations as a ratio f_e/f_s of the friction coefficients of an ellipsoid f_e and a sphere f_s with an equal volume [58]. For an oblate ellipsoid ($a=c>B$) with volume $V=(4\pi/3)a^2b$ at rotation around the big axis the combination with Equation 14.24 gives

$$D = \frac{kT}{4\eta V}\left(\frac{q^2}{1-q^4}\right)\left[1 + \frac{1-2q^2}{q\sqrt{1-q^2}}\,\text{arctg}\,\frac{\sqrt{1-q^2}}{q}\right], \tag{14.25}$$

where $q=b/a$ is the reciprocal axes ratio. For a strongly oblate ellipsoid ($a \gg B$) with a big axis $B_e = 2a$ Equation 14.25 is simplified:

$$D_e = \frac{3kT}{4\eta B_e^3}. \tag{14.26}$$

The inaccuracy of this formula as compared to 14.25 does not exceed $+1\%$ at $a/b \geq 12$ (at axis ratio $a/b = 5, 10, 15$ the error is respectively 5.5%, 1.4%, and 0.6%).

14.3.2.2 Orientation-Optical Function

The optical function determines the magnitude of the EOE for a certain degree of orientation. In contrast to the orientation function it is specific for every electro-optical method. The relative EOE can be represented as a product of the orientational and optical functions.

14.3.2.2.1 Electric Dichroism

In this calculation, only consumptive dichroism [59] is considered: scattering is negligibly small and the optical density is determined by the absorption of linearly polarized light. In this case the relation $\Delta A^{\parallel} = -2\Delta A^{\perp}$ is observed, where $\Delta A = A_E - A$ are EOE values for parallel $\Delta A^{\parallel}$ and orthogonal $\Delta A^{\perp}$ orientation of the light polarization plane to the electric field and $A = \log(I_0/I)$ is the absorbance.

At full orientation when particle's axis of symmetry is oriented parallel to the electric field, the optical function is [60]:

$$(\Delta A_\infty/A)^{\parallel} = -2(\Delta A_\infty/A)^{\perp} = 3\cos^2\Psi - 1, \tag{14.27}$$

where Ψ is the angle between the dipole moment of the optical transition and the axis of symmetry of the particle.

EOE is a linear function of the orientation degree $F(E)$. In particular, EOE decay is obtained by a combination of Equation 14.23 and Equation 14.27:

$$(\Delta A/A)_t^{\parallel} = (3\cos^2\Psi - 1)\cdot F(E,0)\exp(-6Dt) = (\Delta A/A)_{t=0}^{\parallel}\exp(-6Dt). \tag{14.28}$$

14.3.2.2.2 Electric Light Scattering

At light scattering in Rayleigh–Debye–Gans approximation the magnitude of the electro-optical effect $\Delta I = I_E - I$ is determined by the difference between the functions of internal interference (form-factor) at a certain degree of orientation $P(\theta,F)$ and chaotic orientation $P(\theta)$:

$$\Delta R_\theta = cHM[P(\theta, F) - P(\theta)], \tag{14.29}$$

where: c is the weight concentration of the dispersed substance, H the the optical constant of the suspension at wavelength λ_0, and M the mass of the particles.

The relative EOE does not depend on c, M, H, as well as on r, Ω, I_0 in Equation 14.2 and Equation 14.11:

$$\frac{\Delta R_{\theta,F}}{R_\theta} = \frac{\Delta I}{I} = \frac{P(\theta, F)}{P(\theta)} - 1. \tag{14.30}$$

As $P(\theta, F)$ and $P(\theta)$ depend in a complex way on shape, size, and optical anisotropy, the optical function is specific for particles with a certain geometry. In order to calculate $P(\theta, F)$ it is necessary to know the function of internal interference for a fixed orientation $P(\theta, \beta)$ and the orientational distribution of particles. Formula 14.30 is convenient for EOE calculation at full orientation $\Delta I_\infty / I$, when $P(\theta, F) = P(\theta, \beta)$. At incomplete orientation it is necessary to determine $P(\theta, F)$ dependence on the degree of orientation F. For this aim the function $A(KB)$ is introduced [61]. It depends on particle geometry and looks differently at low and high degree of orientation. Then

$$P(\theta, F) = P(\theta) + A(KB) \cdot F(p, \gamma, \omega, E, T, D, t). \tag{14.31}$$

EOE magnitude can be expressed as a product of the optical and orientational functions:

$$\frac{\Delta I}{I} = \frac{A(KB)}{P(\theta)} \cdot F(p, \gamma, \omega, E, T, D, t). \tag{14.32}$$

In the general case $A(KB)$ depends on the orientation degree F. At low degrees of orientation this dependence could be neglected and EOE is a linear function of $F(E)$. In this case EOE decay is monoexponential and is described by a combination of Equation 14.23 and Equation 14.32:

$$(\Delta I/I)_t = \frac{A(KB)}{P(\theta)} \cdot F(p, \gamma, E) \exp(-6Dt) = (\Delta I/I)_{t=0} \exp(-t/\tau_0), \tag{14.33}$$

where $(\Delta I/I)_t$ and $(\Delta I/I)_{t=0}$ are EOE values at moment t and in the moment of electric field switch off, $t = 0$.

At high degrees of orientation $A(KB)$ is a nonlinear function of $F(E)$ and the relaxation process is described by a polyexponential dependence. For cylindrical particles, EOE decay after achieved full orientation can be expanded in the following series [62]:

$$\left(\frac{\Delta I}{I}\right)_t = \frac{A_1(KB)}{P(\theta)}.F(E)\exp(-6Dt) + \frac{A_2(KB)}{P(\theta)}.F(E)\exp(-20Dt)$$

$$\tag{14.34}$$

$$+ A_3 \frac{(KB)}{P(\theta)}.F(E)\exp(-42Dt) + \dots$$

14.4 RDG LIGHT SCATTERING AND APPLICABILITY TO PURPLE MEMBRANES

14.4.1 Applicability of the Theory to Purple Membrane Suspension

The formulas given in Section 14.3.1 and derived in Section 14.4.2 describe single scattering from independent, nonabsorbing, and optically isotropic particles. As PMs are absorbing and anisotropic and their size is commensurate with the wavelength in the medium, the applicability of the RDG light scattering theory presented to PM water suspensions is needed to be study.

14.4.1.1 Influence of Purple Membrane's Optical Properties

14.4.1.1.1 *Refractive Index*

The averaged overall orientations PM refractive index n_1 has been determined from light scattering coefficient R_θ in water-glycerin medium with refractive index n_0 [26]. The extrapolation to $R_\theta = 0$ gives $n_1 = n_0$, where n_0 has been determined refractometrically. The result is $n_1 = 1.51$ at $\lambda_0 = 496$ nm and $n_1 = 1.53$ at $\lambda_0 = 647$ nm. The larger n_1 at larger λ_0 is connected with anomalous dispersion of n_1 in the main absorption band of bR with a maximum at $\lambda_0 = 570$ nm. These n_1 values determine the relative refractive index $n = n_1/n_0 \approx 1.13$–1.15 in aqueous medium.

When condition 14.3 is satisfied the surplus polarizability α is uniform in all particle's volume. Then the quantity of α and the polarization of the scattered light do no depend on the form and orientation of a particle from optically isotropic substance. The absence of measurable scattering light depolarization at vertical polarization of incident light (see the next paragraph) is an experimental evidence of the validity of condition 14.3 for a PM water suspension.

A theoretical assessment of the condition $|n-1| \ll 1$ can be done by calculating the α for varying shapes of the particle. For this purpose the main elements of the polarizability tensor [51] of a rotational ellipsoid from an optically isotropic substance is calculated. The results show that at $n = 1.14$ the α of a strongly oblate ellipsoid with a chaotic orientation differs from α of a sphere with the same volume by only 0.5%. Further, at small n the particle polarization nonuniformity can be ignored and one can assume that α of a disk and an ellipsoid are equal. This presents the grounds to consider that condition 14.3 is fulfilled for PMs with sufficient accuracy and Equation 14.6 can be used.

14.4.1.1.2 *Anisotropy of Refractive Index*

As far as condition 14.3 is satisfied for PM in aqua, the optical anisotropy δ is determined only by membranes' refractive index anisotropy. The molecular structure of PMs allows assuming that refractive index $n_1^{\parallel}$ differs from the $n_1^{\perp}$ (parallel and normal to the membrane plane respectively), which guarantees $\delta \neq 0$.

The δ can be determined with the aid of $P(\theta)$ equations for an anisotropic disk [63], using the experimentally determined depolarization coefficient $\Delta_v = I_v^H/I_v^V$ from the intensity of the horizontal I_v^H and the vertical I_v^V components of scattered light at vertical polarization of the incident light. The measurement in the range $\theta = 15$–$160°$ at $\lambda_0 = 415$ nm did not show measurable horizontal component, but it was evaluated in view of apparatus sensitivity as Δ_v does not exceed 1×10^{-3}. This value corresponds to $\delta \leq 0.06$ for a disk with a relative diameter $B/\lambda = 2.5$ at $\theta = 30°$ and $\lambda_0 = 415$ nm.

At $\delta = 0.06$ the ratio of $P_v^{V}(\theta)_\delta$ for an anisotropic disk to $P(\theta)$ for an isotropic disk with size $B/\lambda = 2.5$ is 1.01 at $\theta = 0°$. It decreases with θ between $10°$ and $30°$ and oscillates around 0.9 in the range $\theta \approx 30$–$150°$. As far as an error around 10% at large angles can be deemed acceptable, PMs can be considered as optically isotropic particles.

14.4.1.1.3 *Absorption Index*

The attenuation of light intensity upon passing through an absorbing particle influences the surplus polarizability α, as well as the internal interference function $P(\theta)$. That influence is negligible if condition 14.3 and condition 14.4 are fulfilled, which limits the value of κ according to Equation 14.5.

For calculating the absorption index κ the absorbance A and the thickness z of the absorbing layer can be expressed with the specific partial volume $\bar{v}$ of PM and the molar mass M' of membrane elements. The latter are composed of bR macromolecules and lipids in mass ratio 3:1. Combining with Equation 14.1 we get

$$\kappa = \left(\frac{\ln 10}{40\pi}\right)\left(\frac{\varepsilon\lambda_0}{\bar{v}M'}\right) \qquad (14.35)$$

where $[\varepsilon]$ is expressed in $dm^3/(mol.cm)$, $[\bar{v}]$ in cm^3/g, $[\lambda_0]$ in μm, and $[M']$ in g/mol.

As bR with mass $M_{bR} = 26,500$ and $\varepsilon = 63,000$ at 570 nm amounts to 75% of PMs with $\bar{v} = 1/1.18$, then κ, averaged over all orientations, is equal to $= 2.2 \times 10^{-2}$. For κ in membrane plane and normal to it at chromophore orientation $\chi = 70°$ we get $\kappa^{\parallel} = 7.7 \times 10^{-3}$ and $\kappa^{\perp} = 2.9 \times 10^{-2}$.

If the electromagnetic wave ($\lambda_0 = 570$ nm) propagates in continuous medium with $\kappa = 2.2 \times 10^{-2}$, its amplitude decreases e times at a distance $l = 4.1$ μm, which is 5 greater than PM diameter $B = 0.8$ μm. As particle absorbance is proportional to its volume V_1, and the disk volume with a diameter B and thickness h is $(2B/3h)$ times smaller than the volume of a sphere with the same diameter, PM with $B = 800$ nm and $h = 5$ nm absorbs energy, which is two orders less than in the case of a dense sphere with equal B and κ. The comparison of the scattering coefficients for absorbing and nonabsorbing spheres according to Mie's theory at $\kappa = 1 \times 10^{-4}$ and $B/\lambda = 2.5$ shows an insignificant difference [64]. This equation confirms the validity of condition 14.5. This gives the premise to assume that PMs scatter light as nonabsorbing particles even when λ_0 coincides with the absorption maximum at 570 nm, which is due to the low value of κV_1.

14.4.1.2 Influence of Particle Size

For PMs with $n \approx 1.14$ and size $B/\lambda \geq 1$ condition 14.4 is not fulfilled, which questions the applicability of RDG theory to PM water suspensions in the visible spectral range.

14.4.1.2.1 Spherical Particles

A quantitative evaluation was done for spherical particles with $n = 1.15$ and size $B/\lambda \approx 0.2$–2 by comparing of R_θ calculated in RDG-approximation with R_θ according to Mie's theory about unpolarized light at aperture angle $\Delta\theta = 10°$ [65]. For particles with $B \approx \lambda$ RDG deviation from the exact theory reaches 15% at small θ, but in that range the error does not depend on θ, which allows using small angle scattering for accurate R_g determination. This conclusion strongly holds at vertical polarization, as the radiation of the electric quadrupole is not manifested in the absence of a horizontal component of the incident light.

14.4.1.2.2 Disk-Like Particles

RDG theory's inaccuracy is due to the fact that it takes into account only dipole radiation and ignores multiple radiation. In the case of disk-like particles with a diameter $B/\lambda \approx 1$ and thickness $h/\lambda \ll 1$ the main contribution in R_θ comes from dipole and quadrupole radiation. Quadrupole moment is at its maximum at horizontal polarization of the incident and scattered light and horizontal disk orientation. For that reason I used vertically polarized light and kilohertz electric field, when the disk surface is predominantly vertically oriented.

At chaotic orientation the quadrupole contribution can be estimated from depolarization at $\theta = 90°$, where its radiation is at its maximum. The measurements at unpolarized and vertically polarized incident light showed that in both cases the horizontal component of scattered light is negligibly small, which showed insignificant optical anisotropy and absence of substantial quadrupole scattering. That appears as an experimental evidence of the application of RDG approximation to PM water suspensions.

14.4.1.3   Influence of Particle Concentration

RDG theory describes single scattering by optically independent particles. As suspension concentration increases, interparticle interference grows and the contribution of multiple scattering is increased. This changes the intensity and the polarization of the scattered light.

14.4.1.3.1   Independent Particles

A sufficient condition of independent scattering is the average distance between neighboring particles to be more of three times larger than their radius [51].

For calculating l we will assume that the mass center of every particle coincides with the center of a sphere with a radius $l/2$, and these spheres are close packed in a three-dimensional hexagonal structure with a coordination number of 12. As the volume of the elementary cell of the hexagonal lattice is $(3/4)l^3$, the number of particles in volume V is equal to $4V/3l^3$.

bR macromolecules in PMs form trimers, the mass center of which are situated at the nodes of a two-dimensional hexagonal lattice with a period d. As the area of the elementary cell is $(3/4)^{1/2}d^2$, a membrane with a diameter B has $(\pi/2\sqrt{3})B^2/d^2$ trimers. The volume V contains $CN_A V/3$ trimers or $(2N_A\sqrt{3}/3\pi)\,CVd^2/B^2$ membranes at bR molar volume concentration C. For the average distance between mass centers of neighboring PM (at absence of aggregates) we get:

$$l = \frac{1}{10}\left[\frac{2\pi B^2}{\sqrt{3}N_A d^2 C}\right]^{1/3} = \frac{1}{10}\left[\frac{2\pi\varepsilon b B^2}{\sqrt{3}N_A d^2 A}\right]^{1/3}, \tag{14.36}$$

where $A = \varepsilon bC$ is the absorbance at optical path b and molar absorption coefficient ε at measuring units: $[\varepsilon] = \mathrm{dm^3 \cdot mol^{-1} \cdot cm^{-1}}$; $[b] = \mathrm{cm}$; $[C] = \mathrm{mol/dm^3}$, $[l] = \mathrm{m}$. The average distance is $l = 1.7\,\mu\mathrm{m}$ at $\varepsilon = 63{,}000$, $d = 6.2\,\mathrm{nm}$, $B = 500\,\mathrm{nm}$, and $A = 0.5$.

Accounting for typical PM sizes, it can be assumed that particles scatter independently at $l \geq 1\,\mu\mathrm{m}$, which corresponds to bR concentration $c = 1\,\mathrm{mg/ml}$. In the experiments described in Section 14.5 and Section 14.6, the concentration is with an order lower, which fully satisfies the requirements of independent scattering at chaotic orientation. In particular when PM geometry is determined by small-angle scattering and pH dependence ($B = 800\,\mathrm{nm}$, $A = 0.05$) the distance between the particles is $l \approx 5\,\mu\mathrm{m}$.

14.4.1.3.2   Single Scattering

The contribution of multiple scattering, growing with suspension concentration c and the optical path, is manifested as a deviation from linearity of $R_\theta = f(c)$ dependence. To avoid multiple scattering, c was limited to $0.02\,\mathrm{mg/ml}$ bR at optical path $10\,\mathrm{mm}$. In the experiments described in Section 14.5 and section 14.6, a linear concentration dependence $R_\theta = f(c)$ was observed, which is an evidence of single scattering. However this is not a guarantee for absence of multiple scattering at positive EOE, when $R_{\theta F}$ increases up to three times at full PM orientation. EOE concentration dependence $\Delta R_{\theta F} = f(c)$ at high degrees of orientation in a sinusoidal field of 1 kHz also shows a linear behavior, which is an evidence that there is no multiple scattering in this least favorable case.

14.4.2   Spherically and Cylindrically Bent Disks

In the general case the surplus polarizability α and the function of internal interference $P(\theta)$, which determine the light scattering intensity, depend on the distribution of complex refractive index $m = n - i\kappa$ in particle volume.

For particles satisfying condition 14.3 and condition 14.4, light scattering intensity is determined by Equation 14.8, where $P(\theta)$ depends on the components of the tensor m in each particle element with volume ΔV_i. Its linear dimensions are limited by condition 14.7. Because of 14.3 (which limits n, as well as κ), m can be substituted with n. In the case of homogenous particles $\alpha = f(n_1, n_0) \cdot V_1$ is determined by volume-averaged n_1 in accordance with 14.6. That allows considering a particle as consisting of N elements with an equal volume ΔV_i. So, in the case of homogeneous, nonabsorbing, and optically isotropic particles, when $P(\theta)$ is calculated, only the disposition of the volume elements should be considered. For such particles $P(\theta)$ averaged overall orientations is determined by Debye's formula 14.13, which at small angles of scattering is degenerated to Guinier formula 14.15.

14.4.2.1 Small-Angle Scattering

If one of the particle's three dimensions is much smaller than the two others, the particle can be considered as a material surface with a constant density and surface area S, and in that case ΔV_i can be substituted by a surface element ΔS_i. Then (assuming that the distance to the element is a continuous function of the coordinates) for the mass coordinate x_c and the radius of gyration R_g (Equation 14.16) one can write:

$$x_c = \int_S x \, dS \bigg/ \int_S dS \tag{14.37}$$

$$R_g^2 = \int_S r^2 \, dS \bigg/ \int_S dS. \tag{14.38}$$

14.4.2.1.1 Segment of a Thin Spherical Shell

The size of the circular spherical segment will be determined by the spherical radius R and the central angle 2α in the spherical diametrical plane. These parameters unequivocally determine the segment geometry: area of the spherical surface $S = 2\pi R^2(1 - \cos \alpha)$, space angle $\omega = 2\pi(1 - \cos \alpha)$, base circumference diameter $d = 2R\sin\alpha$ and height $h = R(1 - \cos \alpha)$.

Fitting the axis of symmetry of the segment with axis Oz and expressing in spherical coordinates (R, θ, φ) the surface $dS = R^2 \sin\theta d\theta d\varphi$ of a surface element with coordinate $z = R\cos\theta$, for the coordinate z_c of the mass center we get

$$z_c = \frac{\int_S z \, dS}{\int_S dS} = \frac{R \int_0^\alpha \int_0^{2\pi} \sin\theta \cos\theta \, d\varphi \, d\theta}{\int_0^\alpha \int_0^{2\pi} \sin\theta \, d\varphi \, d\theta} = \frac{R(1 - \cos 2\alpha)}{4(1 - \cos \alpha)}. \tag{14.39}$$

Taking into account that the distance between the mass center and the i-th element is $r = [R^2 + z_c^2 - 2z_c R \cos \theta]^{1/2}$, for R_g we get

$$R_g^2 = \frac{\int_S r^2 \, dS}{\int_S dS} = R^2 + z_c^2 - \frac{2z_c R \int_0^\alpha \int_0^{2\pi} \sin\theta \cos\theta d\varphi d\theta}{\int_0^\alpha \int_0^{2\pi} \sin\theta d\varphi d\theta} = \left[1 - \left(\frac{1 + \cos\alpha}{2}\right)^2\right] R^2. \tag{14.40}$$

14.4.2.1.2 Cylindrically Bent Thin Elliptical Disk
Let us admit that an ellipse with semiaxes ($a>b$) is a part of the lateral surface of a circular cylinder with a radius R, the minor axis $2b$ parallel to the generating line of the cylindrical surface, and the major axis $2a$ coincides with the directing circumference with central angle $\alpha = a/R$. Putting the circumference center in the origin of cylindrical coordinates (R, φ, z) and the ellipse center on the polar axis Ox (the ellipse axis $2b$ is in the zero meridian plane), I can write the equation of a cylindrically bent ellipse as $\varphi^2 R^2/a^2 + z^2/b^2 = 1$; hence $z = R(b/a)[\alpha^2-\varphi^2]^{1/2}$. A surface element with area $dS = R\,d\varphi\,dz$ has a coordinate $R\cos\varphi$ on axis $2a$, and the ellipse surface is $S = \pi ab$.

The mass center lies on polar axis Ox and its coordinate is

$$x_c = \frac{4R^2}{S} \int\limits_0^\alpha \int\limits_0^z \cos\varphi\,dz\,d\varphi = 2\frac{R^2}{a} J_1(a/R) = 2R\frac{J_1(\alpha)}{\alpha} \qquad (14.41)$$

where: J_1 is a first-order Bessel function of the first kind.

For R_g (the distance from the mass center to the element of the surface with an applicate z is equal to $r = [R^2 + x_c^2 + z^2 - 2x_o\,R\cos\varphi]^{1/2}$) using Equation 14.38 we obtain:

$$R_g^2 = R^2 + x_c^2 + \frac{4R}{S} \int\limits_0^\alpha \int\limits_0^z z^2 dz d\varphi - \frac{8x_c R^2}{S} \int\limits_0^\alpha \int\limits_0^z \cos\varphi dz d\varphi \qquad (14.42)$$

where the third term is equal to $b^2/4$, and the last is $-2x_c^2$. Finally:

$$R_g^2 = R^2 + \frac{b^2}{4} - 4\left(\frac{R^4}{a^2}\right) J_1^2(a/R). \qquad (14.43)$$

14.4.2.2 Large-Angle Scattering

At arbitrary angles of scattering the function $P(\theta)$ can be calculated with Debye's formula 14.13 for homogenous nonabsorbing and optically isotropic particles, which are with chaotic orientation and satisfy condition 14.3 and condition 14.4. It is assumed that the particle is built of N discrete scattering centers (elementary oscillators). In this case the problem is reduced to determining the coordinates of all the centers and the distances a_{ij} between each center couple. As the double sum in Equation 14.13 forms a symmetrical matrix, the diagonal elements of which are 1 (for them $a_{ij}=0$), for the calculation of $P(\theta)$ it is enough to calculate $(N^2-N)/2$ terms of the double sum.

Particles with inhomogeneous composition, biological membranes in particular, can be considered as optically homogeneous, if all their elements (with linear dimensions limited by the condition 14.7) have the same refractive index. In this case one can assume that the density of the elementary oscillators is constant throughout the particle volume and their coordinates are a continuous function. That allows substituting the summing up over discrete centers with integration over the volume (or over the surface in cases, where the particle can be considered as a material surface).

14.4.2.2.1 Segment of a Thin Spherical Shell
A segment of a spherical surface with radius R and central angle in the diametrical plane 2α could be considered as composed of N discrete scattering centers, distributed uniformly along its surface S. Then $N = 2\pi(1-\cos\alpha)/(\Delta\theta\Delta\varphi \sin\theta)$ and Debye's formula 14.13 can be written as

$$P(\theta) = \left[\frac{\Delta\varphi\,\Delta\theta}{2\pi(1-\cos\alpha)}\right]^2 \sum_{i=1}^{N}\sum_{j=1}^{N}\frac{\sin(2Kd_{ij})}{2Kd_{ij}}\sin\theta_i\sin\theta_j. \tag{14.44}$$

At continuous distribution and at $n=$ const we get

$$P(\theta) = \int_0^\alpha \int_0^{2\pi} \int_0^\alpha \int_0^{2\pi} \frac{\sin\theta_1\sin\theta_2\sin[2Kd_{ij}(\theta_1,\theta_2,\varphi_1,\varphi_2)]}{8\pi^2(1-\cos\alpha)^2 Kd_{ij}(\theta_1,\theta_2,\varphi_1,\varphi_2)}d\varphi_2 d\theta_2 d\varphi_1 d\theta_1, \tag{14.45}$$

where

$$d_{ij} = R\sqrt{2[1-\cos\theta_1\cos\theta_2 - \sin\theta_1\sin\theta_2\cos(\varphi_1-\varphi_2)]}.$$

Because of the symmetry this expression can be simplified as

$$P(\theta) = \frac{1}{\pi(1-\cos\alpha)^2} \int_0^\alpha \int_0^\alpha \int_0^\pi \frac{\sin[2Kd_{ij}(\theta_1,\theta_2,\varphi_2)]}{2Kd_{ij}(\theta_1,\theta_2,\varphi_2)}\sin\theta_1\sin\theta_2 d\varphi_2 d\theta_2 d\theta_1, \tag{14.46}$$

where

$$d_{ij} = R\sqrt{2(1-\cos\theta_1\cos\theta_2 - \sin\theta_1\sin\theta_2\cos\varphi_2)}.$$

The numerical solution of this integral can be reduced almost twice, if I take into account that at equal integration steps (α/k) over θ_1 and θ_2, they form a square matrix θ_{ij}. In this matrix the elements symmetric to the main diagonal are equal. In such a way, for $P(\theta)$ calculation, it is enough to calculate $(k^2+k)/2$ values of the integral over $d\varphi_2$.

14.4.2.2.2 Cylindrically Bent Thin Elliptical Disk
A thin elliptical disk with semiaxes $a>b$, bent around a circular cylinder with radius R (the minor ellipse axis is parallel to the cylinder axis and the ellipse center is on the polar axis), consists of $N=\pi ab/R\Delta\varphi\Delta z$ discrete scattering centers. Using cylinder coordinates (R,φ,z) at continuous distribution of the elementary oscillators, accounting for ellipse symmetry, we get for internal interference function

$$P(\theta) = \frac{4}{(\pi ab)^2} \int_0^\alpha \int_0^z \int_{-\alpha}^\alpha \int_{-z}^z \frac{\sin[2Kd_{ij}(\varphi_1,\varphi_2,z_1,z_2)]}{2Kd_{ij}(\varphi_1,\varphi_2,z_1,z_2)}dz_2 d\varphi_2 dz_1 d\varphi_1, \tag{14.47}$$

where

$$d_{ij} = \sqrt{2R^2[1-\cos(\varphi_2-\varphi_1)] + (z_2-z_1)^2},$$

and the upper limit for integration over the applicate z is

$$z = (b/\alpha)\sqrt{\alpha^2-\varphi^2}.$$

14.4.2.3 Permanent Dipole Moment

A particle, which is a neutral, asymmetrically charged surface with area S and meets condition 14.4 has the size of the order of micrometer and that is why it behaves like a point dipole p in a macroscopic electric field. If every electrically neutral element of the surface s_i (small enough to be considered as flat), has a dipole moment p_i directed along the plane normal and when $\mu_s \equiv p_i/s_i = \text{const}$ all over the surface $S = \Sigma s_i$, the particle's dipole moment $p = \Sigma \mu_s s_i$ is directed along the symmetry axis and its module is

$$p = \mu_s \int_S \cos \beta dS, \tag{14.48}$$

where β is the angle between the symmetry axis and the normal to s_i.

14.4.2.3.1 Segment of a Thin Spherical Shell

For the permanent dipole moment of a spherical segment with a radius R and a central angle in the diametrical plane 2α, we get according to Equation 14.48 (the angle $\beta \equiv \theta$):

$$p = \mu_s R^2 \int_0^\alpha \int_0^{2\pi} \sin \theta \cos \theta d\varphi d\theta = \frac{\pi}{2} R^2 (1 - \cos 2\alpha)\mu_s, \tag{14.49}$$

or

$$p = \pi R^2 \mu_s \sin^2 \alpha = \frac{1 + \cos \alpha}{2} \mu_s S = \left(1 - \frac{S}{4\pi R^2}\right)\mu_s S.$$

14.4.2.3.2 Cylindrically Bent Thin Elliptical Disk

The dipole moment of an elliptical disk with semiaxes $a>b$ and surface $S = \pi ab$, bent around circular cylinder with radius R (the minor ellipse axis is parallel to the cylinder axis and the major axis determines central angle $\alpha = a/R$) is directed along the normal to the ellipse center. Its module according to Equation 14.48 (the angle $\beta \equiv \varphi$) is

$$p = 4R\mu_s \int_0^\alpha \int_0^z \cos \varphi \, dz \, d\varphi = 2\pi Rb J_1(a/R)\mu_s = \frac{2J_1(\alpha)}{\alpha} \mu_s S. \tag{14.50}$$

where: J_1 is a first order Bessel function of the first kind.

It is interesting to note that the dipole moment and the position of the mass center of a cylindrically bent elliptical disk do not depend on axis ratio a/b at constant a and R (in spite of the fact that the particle's mass depends on b). That can be explained with the constancy of z/b ratio at fixed angle φ.

14.5 UNDISTURBED PURPLE MEMBRANE GEOMETRY IN AQUEOUS MEDIUM

The geometry (shape and size) of the particles unequivocally determines the radius of gyration R_g and the rotation diffusion coefficient D. But in order to determine the size from the experimental values of R_g and D independent information on PM shape is necessary. The task is greatly simplified if the number of PM shapes is restricted, when their molecular structure is accounted for. This allows using a limited number of equations, relating R_g to the size at the respective particle's shape. Further in this section, the possibility of electro-optical determination of the size of a particle of unknown shape is validated. PM curvature has been determined from R_g (measured by small-angle scattering), using the equations, deduced in Section 14.4.2.1. At the end, a choice was made between two alternative PM shapes employing a large-angle light scattering and a conclusion was reached: PMs are cylindrically bent. Additional arguments about the shape of the PMs are presented and the complicative influence of suspension polydispersity and light nonmonochromaticity is discussed.

14.5.1 Possible Purple Membrane Shapes

The PMs make part of a cylindrical surface with a diameter 0.7 μm of the cytoplasmic membrane of the bacterial cell [44]. In this case they are shaped as a cylindrically bent elliptical disk. After lysis the bacterial cell, the freed PMs take a shape, which is determined at first place by the geometry of PM structure elements (trimer bR macromolecules with lipid molecules associated with them). A precondition for that is the conformational stability of bR and the dense trimer packaging, which do not allow a PM closure in vesicle. In particular the membrane will be flat when the elements take the shape of a cylinder or bent when they have the shape of a cut cone. In the last case, the shape most favorable energetically is that of a sector of a spherical shell, but for such a transition, the membrane's structure elements should have a lateral mobility. The electrical asymmetry may play a certain influence on PM curvature but that is a factor of second order, as even a large change of the surface charge density does not induce a significant alteration of PM curvature (Section 14.6.1).

As the trimers are packaged in hexagonal lattices, it can be assumed that their lateral diffusion is strongly encumbered because of overcoming a high potential barrier. In spite of this, the time elapsed from bacterial lysis to the start of the electro-optical experiment can be sufficient for the transition of the PMs from cylindrical to spherical bending—predominantly if lateral diffusion is carried out through bR molecules passing from one trimer to another. The transition from cylindrical to spherical symmetry is accompanied by a change in particles' optical and electrical properties, which are estimate here by the formulas, deduced in Section 14.3.2.

At our experimental conditions ($\theta = 90°$, $\lambda_0 = 427$ nm and PM surface area equal to that of a disk with diameter $B = 0.8$ μm), the transition from a cylindrically bent elliptic disk with $R_{cyl} = 205$ nm to a segment of a spherical shell with $R_{sph} = 318$ nm, should be accompanied by R_θ increase with 70% (Equation 14.47 and Equation 14.46). This is also accompanied by a 18% raise of p, according to Equation 14.50 at $\alpha_{cyl} = 125°$ and Equation 14.49 at $\alpha_{sph} = 83°$. However the control measurements of R_θ, γ, p, and D, carried out in 6 months' time did not show any changes in these parameters. As the time from bacterial lysis to the experiment's start is much smaller than this period, the existence of intermediate shapes can be ruled out and the number of possible PM shapes can be reduced to two: a cylindrically bent elliptical disk and a spherical shell segment.

As the structural elements have the shape of cylinder and their axis of symmetry is perpendicular to the membrane's plane, it can be concluded that the elasticity coefficient is isotropic in that plane. That is why the bending force will be proportional to the membrane's diameter in a given direction. Therefore at cylindrical bending of an elliptical disk the minimal free energy corresponds to the case of the small elliptical axis parallel to the cylinder axis. Because of PM homogeneity, the module of elasticity at transverse bending is constant along the disk surface (the compressive and tensile stresses on the two sides of the membrane being in equilibrium) and the free energy will be at its lowest when the disk is a part of a circular cylinder. The same conclusion can be reached if PM is considered as a thin, structureless plate, bent by external forces.

14.5.2 Rotational Diffusion Coefficient

As PM axis ratio $B/h \approx 100$ and their surface is hydrophilic, if they were flat, Perrin's equation 14.26 for strongly oblate ellipsoid with wettable surface is a good approximation. However if PMs are bent, applying this equation leads to a systematic error, which increases with the degree of bending. That makes the assessment of particles' shape necessary.

14.5.2.1 Influence of Particle Shape

The analysis of the rotational motion of anisodiametrical particles shows that the friction coefficient is determined by two factors: particle's surface area S and the perturbation of the liquid. The influence of the particle's shape is particularly connected with the perturbation of the disperse medium, which grows with the transition from sphere to ellipsoid. These two factors can be divided by comparing the friction coefficients of a strongly oblate ellipsoid and a sphere with equal surface areas, as in the first case the perturbation is at its highest, whereas it is totally absent in the second.

For the ratio of the friction coefficients of an oblate ellipsoid f_e and a sphere f_s with diameters B_e and B_s Equation 14.26 and Equation 14.24 give

$$f_e/f_s = (4/3\pi)(B_e/B_s)^3. \tag{14.51}$$

At equal areas ($S_e = S_s$ at $B_e/B_s = \sqrt{2}$) the ellipsoid's friction coefficient is only 1.2 times greater than that of the sphere. That means that the perturbation contribution is only 20% and the area makes the greatest contribution in the rotational friction. This result gives the basis to use electro-optical methods for determining the diameter of particles, shaped as oblate ellipsoid with an unknown axis ratio from the relaxation time $\tau_0 = 1/6D$ with the aid of Equation 14.26, which at 20°C in water medium becomes

$$B_e = [(9kT/2\eta)\tau_0]^{1/3} = 263 \cdot \sqrt[3]{\tau_0}, \tag{14.52}$$

where $[\tau_0]$ is expressed in ms and $[B]$ in nm. The numerical coefficient at 15, 25, and 30°C is, respectively, 252, 273, and 284.

For a flat thin disk with diameter B_d and surface S_d ($S_d \approx S_e$ at $B_d = B_e$ with error $\leq 3\%$ when $a/b \geq 10$) at 20°C in water, Equation 14.52 becomes

$$S_d = (\pi/2)B_d^2 \approx 0.109 \cdot \tau_0^{2/3}. \tag{14.53}$$

where $[S]$ is expressed in μm^2.

The weak dependence of D on the shape of the particles allows using τ_0 for determining PM bilateral area as a total area of a flat disk by Equation 14.53, neglecting their curvature.

14.5.2.2 Membrane Surface Area

The relaxation time is determined from electric light scattering measurements. This method allows reducing PM concentration with an order as compared to dichroism and in this way to avoid collective effects and aggregation. At that, as in the experiments described in Section 14.5 and Section 14.6.1, the average distance between particles is $l \approx 5 \ \mu m$. This concentration has been established, accounting for τ_0, I_0, and ΔI independence on PM concentration, which are criteria for absence of electrical and optical interactions. Low degrees of orientation were used, where EOE decay is monoexponential in a monodisperse suspension, in contrast with high orientation degrees. The linear dependence of steady-state EOE on E^2 was used as a criterion of low orientation degree.

Figure 14.1 shows EOE decay curve after switching off the electric field with intensity $E = 100 \ B/cm$, which corresponds to a low degree of orientation. The relaxation time τ_0 (the time taken for ΔI_s to decrease e times) measured at these conditions is 28 ms, which according to Equation 14.53 at $20°C$ corresponds to $S = 1.0 \ \mu m^2$ (area of both membrane surfaces), equal to that of a disk with diameter $B = 0.80 \ \mu m$.

14.5.2.3 Polydispersity or Asymmetric Particles

The polydispersity of the suspensions strongly complicates the interpretation of results. So a threefold differential centrifuging is applied after ultrasound disaggregation to remove the small size (supernatant at 2×10^4 rpm $\times 20$ min) and the large size (sediment at 1×10^4 rpm $\times 10$ min) fractions. The polydispersity of the suspension obtained is estimated from EOE decay, as the cubic dependence of τ_0 from B is highly sensitive to particle size distribution.

Figure 14.1 shows a semilogarithmic graph of EOE decay after low degree steady-state orientation (25% of EOE at full orientation) has been reached. The average relaxation time is $\tau_0 = 28$ ms. The nonlinear dependence of $\ln(\Delta I_t)$ on time t could be interpreted as a manifestation of polydispersity. Function expansion with peeling method [60] shows two exponents with relaxation times $\tau_1 = 75$ ms and $\tau_2 = 13$ ms (Figure 14.1, lines 3 and 4). This can be interpreted as a presence of particles from two fractions, corresponding to disks with diameters $B_1 = 1.11 \ \mu m$ and $B_2 = 0.61 \ \mu m$ respectively: the large size fraction is responsible for

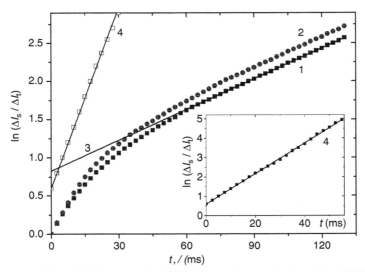

FIGURE 14.1 Decay of electric light scattering EOE at low degree of orientation (100 V/cm, 1 kHz) in semilogarithmic scale (curve 1, $\tau_0 = 28$ ms), its two components with relaxation times $\tau_1 = 75$ ms (line 3) and $\tau_2 = 13$ ms (line 4 in base figure and in the inset), and at high degree of orientation (500 V/cm, 1 KHz) (curve 2, $\tau_0 = 23$ ms).

44% of the steady-state EOE, whereas the small size contributes 55%. The availability of only two fractions is suspicious, as the precision of the method allows separating still other components.

In spite of that we will assume that after the sonication and centrifugation the dispersion phase consists of two types of particles: single membranes with a diameter $B = 0.61$ μm and aggregates with areas corresponding to a disk with $B = 1.11$ μm. As aggregate surface area is 3.2 times greater than that of a single membrane, the stability of the aggregate can be achieved if it is built of at least 10 to 15 PMs. The latter is confirmed by electron microscopic investigation of PM–gelatin gels, where the average number of membranes is 10–30 [66]. The calculation of particles' sedimentation coefficient shows that at 1×10^4 rpm $\times 10$ min (6 cm distance from the axis and 13 mm sedimentary path) such aggregates should have an almost full sedimentation and could not be responsible for 44% of the total EOE. As for the single PM with $B = 0.61$ μm, a greater part of them should not be precipitated at 2×10^4 rpm $\times 20$ min, but the optical density of the supernatant shows that almost all PMs are in the sediment. In this way the hypothesis of presence of two types of particles in the suspension is erroneous.

The alternative assumption is that the suspension is monodisperse with a relaxation time $\langle \tau_0 \rangle = 28$ ms and $\tau_1 = 75$ ms and $\tau_2 = 13$ ms correspond to two different rotational friction coefficients. As $\tau_0 = 28$ ms corresponds to a disk with a diameter $B = 0.80$ μm, the sedimentation coefficient of such membranes allows the sedimentation of the greater part of them at 2×10^4 rpm $\times 20$ min centrifuging. The latter is in accordance with the experiment. In such a way it can be inferred that the two monoexponential relaxation processes are connected with a simultaneous revolution of monodisperse particles around their two axes. This means that the particles do not have an axis of symmetry and therefore they cannot be shaped like a flat circular disk or a segment of spherical shall.

An alternative way of evaluating polydispersity is measuring the dependence $\tau_0 = f(E)$ at low degrees of orientation, as τ_0 decreases with E proportionally to the scatter of size of the particles. At field intensities ($E = 100$ B/cm and 450 B/cm at 1 kHz), which correspond to low and high degrees of orientation, the steady-state EOE $\Delta I_s/I$ are respectively 25 and 91% of $\Delta I_\infty/I$ at full orientation and the relaxation times are $\tau_0 = 28$ ms and 23 ms. These results show that the suspension investigated is practically monodisperse. The fact that τ_0 diminishes with 20% at high degree of orientation is connected with the contribution of the faster than $6Dt$ terms in Equation 14.34. That conclusion is based one side on theoretical evaluation of cylindrical particles [62] and on the other on experimental studies of betonite [67] (its disk-like particles are similar to PM).

The considerations presented up to now lead to the conclusion that the PM suspension under study is monodisperse but the particles show two coefficients of rotational diffusion. This means that the particles do not have the shape of a flat circular disk or a spherical shell segment. In the case of cylindrically bent elliptical disk, the optical effect caused by disorientation should be due to three diffusion coefficients, corresponding to rotation around the three coordinate axes. The cause for presence of only two relaxation times is that at turning round the axis parallel to the incident beam, the change of inner interference (function $P(\theta, \beta)$ at orientation angle β) is insignificant, and does not lead to measurable optical effect. At rotation around the other two axes, the $P(\theta, \beta)$ has close values, which can be seen from the approximately equal contribution of the two "fractions" that differ six times in relaxation times.

14.5.3 Membrane Size

As it was already shown in Section 14.5.2.1, the membrane surface can be determined with satisfactory precision from the rotational diffusion coefficient and the curvature—with the formulas obtained in Section 14.4.2.1. To do the latter it is necessary to determine in an independent way the radius of gyration, which is done here with small-angle light scattering.

14.5.3.1 Radius of Gyration

As it has been shown in Section 14.4.1.1, PMs scatter the light as optically isotropic non-absorbing particles, which allows the use of Equation 14.19 to determine R_g. Besides instead of determining R_θ according Equation 14.2, it is sufficient to measure I_θ at different θ.

The experimental dependence $\ln(1/I_\theta) = f[\sin^2(\theta/2)]$ is linear up to $\theta = 26°$ and its slope is equal to 33.7 (the inset in Figure 14.3). At $\lambda_0 = 400$ nm and $n_0 = 1.343$ [68], Equation 14.19 gives $R_g^2 = 6.41 \times 10^{-2}$ μm^2, i.e., $R_g = 253$ nm for PM in aqueous medium.

14.5.3.2 Angle of Curvature

The curvature of spherically or cylindrically bent membrane is determined by sphere's (circular cylinder's) radius R and the central angle 2α. As R and α determine radius of gyration R_g, the curvature can be calculated, comparing R_g of a bent and a flat thin disks with equal surface areas S.

The radius of gyration of a flat elliptical disk $R_{g, fed}$ with axis ratio $q = a/b$ and bilateral surface area $S = 2\pi ab$ is determined from Equation 14.18:

$$R_{g,fed}^2 = (1/8\pi)(q + 1/q)S \tag{14.54}$$

For the R_g ratio of a segment of spherical shell $R_{g\ sss}$ and a flat circular disk $R_{g\ fd}$ with equal areas combining with Equation 14.40 gives

$$\frac{R_{g,sss}^2}{R_{g,fd}^2} = \frac{1}{(1 - \cos\alpha)}\left[1 - \left(\frac{1 + \cos\alpha}{2}\right)^2\right]. \tag{14.55}$$

Figure 14.2 shows the dependence of $R_{g\ sss}/R_{g\ fd}$ ratio on the curvature angle α.

For R_g-ratio of a cylindrically bent $R_{g\ cbed}$ and flat $R_{g\ fed}$ elliptical disks with axis ratio $q > 0$, when the minor axis $2b$ is parallel to the cylinder axis, combining Equation 14.43 and Equation 14.54 gives

$$\frac{R_{g,cbed}^2}{R_{g,fed}^2} = \frac{1}{1 + q^2}\left[1 + \frac{4q^2}{\alpha^2}\left(1 - \frac{4}{\alpha^2}J_1^2(\alpha)\right)\right]. \tag{14.56}$$

Figure 14.2 shows the dependence of $R_{g,cbcd}^2/R_{g,fed}^2$ ratio vs. curvature angle α. At measured $\tau_0 = 28$ ms the bilateral membrane surface area according to Equation 14.53 is equal to 1.03 μm^2. For a flat elliptical disk that corresponds to: $R_{g,fed} = 286$, 289, and 297 nm for axis ratio $q = 1.0$, 1.25, and 1.5, respectively (Equation 14.54). As R_g measured in a small-angle scattering experiment is equal to 253 nm, the ratio $R_{g,cbed}/R_{g,fed}$ in the case of cylindrically bent elliptical disk ($R_{g,cbed} = R_g$) corresponds to the angle of curvature $\alpha = 137$, 125, and 129° for $q = 1.0$, 1.25, and 1.5, respectively. Besides the cylinder radius R is respectively equal to 178, 207, and 220 nm.

In the case of a segment of spherical shell ($R_{g,sss} = R_g = 253$ nm) and a flat circular disk $R_{g,fd} = 286$ nm) the ratio $R_{g,sss}/R_{g,fd}$ corresponds to $\alpha = 83°$ (Figure 14.2) and spherical radius $R = 280$ nm.

Figure 14.2 shows the dependence of dipole moment ratio of a bent disk (spherically p_{sss} or cylindrically p_{cbed}, calculated from Equation 14.49 and Equation 14.50 and a flat disk $p_{fd} = \mu_s S$

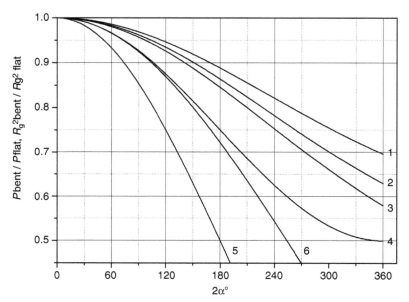

FIGURE 14.2 Ratio of squares of radii of gyration (curves 1–4) and permanent dipole moments (curves 5 and 6) of a cylindrically bent to flat elliptical disk with axis ratio $q = 1$ (curve 1), 1.25 (curve 2), 1.5 (curve 3), and of a spherical shell segment (curve 4) to those of flat circular disk vs. curvature angle α.

vs. the curvature angle α. At $\alpha = 83°$ for a spherical shell segment and $\alpha = 125°$ for cylindrically bent disk, the dipole moment p is respectively equal to 56 and 51% of that of a flat disk.

So, PM shape in water medium corresponds to a part of the surface of circular cylinder with a diameter 415 nm or a sphere with a diameter 560 nm. In both cases the bending reduces twice the permanent dipole moment of PM, which should be reckoned when calculating the dipole moment of a bP macromolecule.

14.5.4 CYLINDRICAL OR SPHERICAL BENDING

PM shape can be defined, comparing the experimental indicatrix $R_\theta = f(\theta)$ of light scattering with the theoretical dependencies $R_\theta = \text{const} \cdot P(\theta)$ at large θ, calculated for particles with fixed geometry. This task can be carried out in practice after the sizes of the two possible PM shapes have been determined.

Figure 14.3 shows $P(\theta)$ functions for chaotically oriented membranes with a surface area equal to an area of circular disk with diameter $B = 800$ nm in the cases of a spherical shell segment ($R = 280$ nm, $\alpha = 83°$) and elliptical disk with axes ratio $q = 1.25$—flat and cylindrically bent ($R = 205$ nm, $\alpha = 125°$). The points are calculated according to Equation 14.46 and Equation 14.47 in the range $K = 5$–20 μm^{-1}, where K corresponds to scattering angles $\theta = 30$–150°. In this range $P(\theta)$ functions are characterized by two maxima in the case of cylindrically bent elliptical disk and two shoulders in the case of a spherical shell segment. At $\lambda_0 = 400$ nm and $n_0 = 1.343$ the maxima at $K = 8.6 \cdot 10^{-3}$ nm^{-1} and $K = 1.7 \cdot 10^{-2}$ nm^{-1} correspond to $\theta = 48°$ and $\theta = 105°$ (Figure 14.3).

The forms of $P(\theta)$ and the corresponding indicatrix $R(\theta)$ enable to discern convincing between the three geometrical shapes. However one cannot expect such clearly expressed extrema in the experimental indicatrix, because of particles' size polydispersity (membrane area ΔS and curvature $\Delta\alpha$) and dispersion of scattering parameter $\Delta K(\Delta\lambda, \Delta\theta)$ determined by light polychromaticity $\Delta\lambda$ and aperture angle $\Delta\theta$.

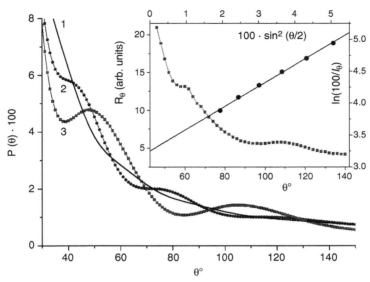

FIGURE 14.3 Dependence of form-factor $P(\theta)$ on scattering angle θ for a flat elliptical disk ($q = 1.25$, curve 1), spherical shell segment ($R = 280$ nm, $\alpha = 83°$, curve 2) and cylindrically bent elliptical disk ($R = 205$ nm, $\alpha = 125°$, $q = 1.25$, curve 3) at $\lambda_0 = 400$ nm and $n_0 = 1.343$ ($K = 5.5–20\ \mu m^{-1}$). The surface area is equal to area the of a flat circular disk with diameter $B = 800$ nm ($KB = 4.4–16$). *Inset*: Experimental dependence of scattering coefficient R_θ vs. scattering angle θ (the curve, the left ordinate and bottom abscissa) and low-angle scattering (the straight line, the top abscissa and right ordinate) for aqueous PM suspension.

The function $\langle P(\theta)\rangle_{\lambda,\theta}$, averaged on λ_0 and θ accounting for the spectral sensitivity of the apparatus $i(\lambda)$ and $\Delta\theta = \theta_2 - \theta_1$, is

$$\langle P(\theta)\rangle_{\lambda,\theta} = \int_{\theta_1}^{\theta_2}\int_0^{\infty} i(\theta)i(\lambda)P(\theta,\lambda)d\lambda d\theta \Big/ \int_{\theta_1}^{\theta_2}\int_0^{\infty} i(\theta)i(\lambda)d\lambda d\theta. \tag{14.57}$$

The solid lines in Figure 14.4 were obtained through numerical integration of Equation 14.57 at parameters for the apparatus used: $i(\lambda)$ (product of spectral intensity of lamp's radiation, transmission coefficient of light filters, and the photocathode spectral sensitivity) and $i(\theta)$ at circular aperture of the photodetector and vertical light polarization. $P(\theta, \lambda)$ was calculated from Equation 14.46 and Equation 14.47 at fixed angle θ and n_0, corresponding to λ_0. As it is clear from the comparison of Figure 14.3 and Figure 14.4, the light polychromaticity and nonzero aperture angle slightly decrease the amplitudes of the shoulders and extrema but practically do not shift them with regard to θ. So, the imperfection in the apparatus leads to a smoothing of indicatrix but its influence is insignificant.

Figure 14.3 (the inset) shows the experimental PM suspension's indicatrix at vertical polarization of the incident and scattering light. The presence of shoulders is evidence that PMs are not flat. The positions of the small shoulder at $\theta = 62°$ and the larger one at $\theta = 108°$ are close to those expected for cylindrically bent membranes with surface area $S = 1.0\ \mu m^2$, axis ratio $q = 1.25$, and curvature angle $\alpha = 125°$. Rendering an account of the inaccuracy in determining these averaged values as well as the influence of their polydispersity, the agreement between the positions of calculated and experimental maxima should be considered satisfactory. The shift of the small-angle maximum could be explained with the slightly smaller real

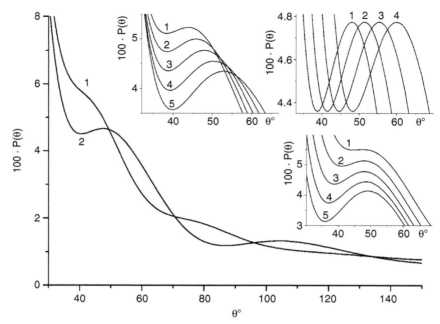

FIGURE 14.4 Form-factor $P(\theta)$ vs. scattering angle θ for a spherical shell segment (curve 1) and cylindrically bent elliptical disk (curve 2) at polychromatic light ($\lambda_{0max} = 400$ nm, semiwidth $\Delta\lambda_{1/2} = 62$ nm at $i(\lambda_{max})/2$) and circular aperture ($\Delta\theta = 7°$). The particles, parameters are the same as those in Figure 14.3. *Left inset*: Form-factor $P(\theta)$ vs. scattering angle θ for spherically bent elliptical disk ($q = 1.25$) at curvature angle $\alpha = 115°$ (1), $120°$ (2), $125°$ (3), $130°$ (4), and $135°$ (5) at $\lambda_0 = 400$ nm and $n_0 = 1.343$. The surface area is equal to area of a flat circular disk with diameter $B = 800$ nm. *Right upper inset*: Form-factor $P(\theta)$ vs. scattering angle θ for spherically bent elliptical disk (axes ratio $q = 1.25$, curvature radius $R = 205$ nm) with area equal to the area of a flat disk with diameter $B = 700$ nm (1), 750 nm (2), 800 nm (3), 850 nm (4), and 900 nm (5) at $\lambda_0 = 400$ nm and $n_0 = 1.343$. *Right lower insert*: Form-factor $P(\theta)$ vs. scattering angle θ for spherically bent elliptical disk (axes ratio $q = 1.25$, curvature angle $\alpha = 125°$) with area equal to area the of a flat disk with diameter $B = 800$ nm (1), 750 nm (2), 700 nm (3), and 650 nm (4) at $\lambda_0 = 400$ nm and $n_0 = 1.343$.

membranes' size as compared to that calculated from the mean relaxation time $\tau_0 = 28$ ms (for calculating S in Equation 14.53 the formula for a flat disk was used but PMs are not flat and the suspension has two relaxation times). The graphics of $P(\theta)$ function for different particles' area at permanent radius $R = 205$ nm (Figure 14.4, the upper right inset) or permanent angle $\alpha = 125°$ of curvature (Figure 14.4, the lower right inset) suggest this possibility.

The influence of membrane curvature on the indicatrix is presented in Figure 14.4 (left inset), where $P(\theta)$ functions are given for cylindrically bent membrane with an area equal to the area of a disk with diameter $B = 800$ nm. It can be seen that the extrema's position depends on the curvature, which leads to smoothing of the indicatrix with increasing α polydispersity even when all particles have the same area.

In the PM suspension studied the polydispersity on area ΔS is strongly restricted by differential centrifuging (Section 14.5.2.3), but it is possible to observe a considerable polydispersity on curvature $\Delta\alpha$. The centrifuging allows separating fractions, which are monodisperse in area, as the translational diffusion coefficient depends weakly on particle shape at one and the same particle area. This polydispersity of shape is very difficult to detect analyzing EOE decay, because the rotational diffusion coefficient depends weakly on particle shape (in contrast to its strong sensitivity to particle area).

The membrane curvature can be steady state (structurally determined by lipid composition and electric charge) or dynamic (curvature fluctuations). In the second case the polydispersity will manifest itself at cylindrically bent PMs (the curvature change at spherically bent membranes is strongly encumbered because structure elements in such membranes have to be displaced). The small elasticity module at cylindrical bending favors the curvature fluctuations, and hence the bending of the membrane occurs concurrently with its rotational and translational movement.

So the comparison of the experimental indicatrix with the theoretically calculated ones leads to the conclusion that PMs are cylindrically bent with mean radius of curvature $R \cong 205$ nm. This conclusion is supported by the presence of two rotation diffusion coefficients, determined in Section 14.5.2.3, which rules out the probability for PM to be shaped as a flat disk or a spherical segment.

14.6 CHANGES IN PURPLE MEMBRANE GEOMETRY

It can be suggested that PM curvature is determined by the form of structure elements and the its electrical asymmetry. In the absence of other forces the membrane will be flat at cylindrical trimers or spherical at cone-like trimers. The tangent components of electrostatic forces on the two-membrane surfaces could change the membrane curvature. However it is obvious that the charge asymmetry is not the main factor, which determines the curvature, as in the opposite case PM would have closed in a sphere (with a greater charge density on the outer surface). The influence of the electrical asymmetry could be established by pH-induced change of the surface charge, polyvalent ions adsorption, or application of a high-voltage field. Because of PM structure asymmetry, these factors lead to asymmetrical change of electrostatic interaction of both membrane surfaces and can change the curvature.

14.6.1 pH-INDUCED ALTERATIONS

When pH of the medium changes, the total membrane charge changes, as well as its asymmetry. The change of the electrostatic interactions can shift the balance of the forces of attraction and repulsion between membrane subunits, which is accompanied by a change in PM geometry. If the charge growth should lead to an increase of the membrane surface, the asymmetric change of the area of the two membrane surfaces leads to a change of PM curvature.

14.6.1.1 Membrane Curvature

For particles, commensurate with λ, R_θ is most sensitive to particles' geometry and so the method of light scattering was used. It is more favorable than dichroism and birefringence because at this method optical function does not depend (for particles light scattering as optically isotropic, see Section 14.4.1.1.2) on their inner structure. That allows studying pH-induced changes in PM curvature and size notwithstanding the possible structure changes. Besides, the higher sensitivity of light scattering allows decreasing the particles' concentration with an order as compared to dichroism and in this way avoiding aggregation and optical interactions.

The equations for $P(\theta)$ from Section 14.4.2 show that at a relative size $B/\lambda \approx 1$–3 light scattering is very sensitive to membrane shape at large θ (Figure 14.6), which is not so at small-angle light scattering (Figure 14.2). This fact is the reason to choose $\theta = 90°$ when studying the changes in PM geometry. Blue ($\lambda_{0max} = 400$ nm) and red ($\lambda_{0max} = 750$ nm) light have been used. The possible pH-induced changes in PM geometry were investigated in the pH range 3–6, where pH can be upheld without a buffer. Figure 14.5 shows the dependence of

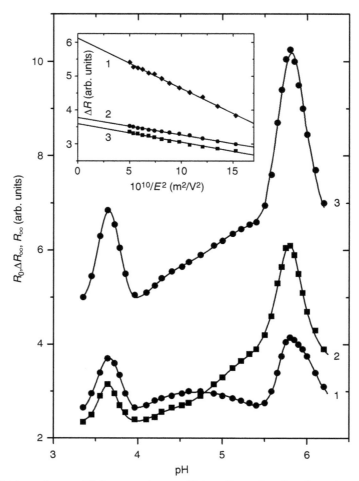

FIGURE 14.5 pH dependence of light scattering coefficient R_θ at chaotic orientation (R_0, curve 1) and full orientation (R_∞, curve 3) and EOE at full orientation (ΔR, curve 2) of PM at wave length $\lambda_{0max} = 400$ nm and scattering angle $\theta = 90°$. *Inset*: Electric light scattering steady-state EOE at high degree of orientation, $\lambda_{0max} = 400$ nm, $\theta = 90°$ at pH 5.81 (line 1), pH 5.35 (line 2), and pH 6.21 (line 3).

$R_{90°}$ vs. pH at blue light ($K = 1.4 \times 10^{-2}$ nm^{-1}) and a constant electric conductivity. The changes observed at $R_{90°}$ can be due to changes in membrane geometry (area and curvature), the refractive index or (in the case of aggregation) to changes in the particles' mass and size.

An evidence for the absence of aggregation is the fact that $R_{90°}$ and the relaxation time τ_0 do not change during the experiment in the pH range studied.

The 1.5-fold increase of $R_{90°}$ observed in the maxima at pH 3.6 and 5.8 could be due to an increase of $n1$ from 1.51 to 1.55 according to Equation 14.9 at $n_0 = 1.338$ or a raise of dn/dc increment with 22% (Equation 14.11). However direct refractometric measurements of PM suspension show that dn/dc does not change (with an accuracy $\pm 1\%$) in the range of pH 4–7.

The presence of peaks in $R_{90°}$ dependence on pH can be due to a nonmonotonic course of $P(\theta)$ at gradual changes in curvature and area. That possibility could be checked by EOE dependence on pH, as $P(\theta)$ is strongly dependent on particles' orientation. EOE field dependence at high degrees of orientation (the inset in Figure 14.5) shows that ΔR_∞ value (EOE extrapolated to $E = \infty$) outlines a peak at pH 5.8. Figure 14.5 shows the light scattering coefficient at full orientation $R_\infty = R_\theta + \Delta R_\infty$, obtained as a sum of the scattering coefficient $R_{90°}$ at chaotic orientation and EOE at full orientation. The fact that at full orientation peaks are observed at the same pH as the peaks at chaotic orientation shows that these maxima are connected with abrupt changes in PM geometry. In addition this result shows the applicability of RDG approximation, as PM orientation at 1 kHz (the membrane surface is parallel to the field and perpendicular to the scattering plane) decreases the incorrectness of the theory (Section 14.4.1.2.2).

As the dependence of τ_0 vs. pH shows (Figure 14.7), PM area does not change observably at pH 3.6 and 5.8. The only factor that could explain the 50% $R_{90°}$ increase in the region of the pH-induced peaks (Figure 14.5) is the change of PM curvature α. The $P(\theta)$ dependence vs. α (Figure 14.6) shows that at $\theta = 90°$ and $B = 800$ nm, the change in curvature has a significant influence on R_θ only at strongly bent membranes with curvature semiangle $\alpha \geq 90°$ (half a sphere or more). Therefore the $R_{90°}$ changes observed in Figure 14.5 show that PMs are not flat or weakly bent ($\alpha \leq 60°$).

As shown in Section 14.5.1, possible PM shapes can be reduced to cylindrically bent elliptical disk or a segment of spherical shell. According to Equation 14.53, PM area with $\tau_0 = 28$ ms is 1.0 μm^2 and corresponds to a circular disk with a diameter of $B = 0.8$ μm. $P(\theta)$ dependence vs. α calculated from the Equation 14.46 and Equation 14.47 for PM with this size is shown in Figure 14.6. The 50% $R_{90°}$ increase observed in the region of the pH-induced peaks at blue light ($K = 14.9$ μm^{-1}) corresponds to a decrease of PM curvature in the range $\alpha \approx 110–140°$ or to its increase at $\alpha > 140°$ in the case of a cylindrically bent disk (curve 3). For a segment of a spherical shell 50% $R_{90°}$ increase corresponds to a decrease of curvature in the range $\alpha \approx 100–145°$ or to its increase at $\alpha > 150°$ (curve 1).

As it is clearly seen from Figure 14.6, at blue light it is not possible to establish whether $R_{90°}$ maxima (Figure 14.5) corresponds to a less or larger curvature. But the choice between

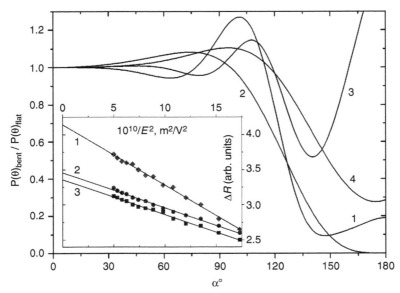

FIGURE 14.6 Relation of form-factor of bent to flat membrane at scattering angle $\theta = 90°$ and wave length $\lambda_0 = 400$ nm (curves 1 and 3) and $\lambda_0 = 750$ nm (curves 2 and 4) in the cases of: spherical shell segment to flat circular disk (curves 1 and 2) and cylindrically bent to flat elliptical disk ($q = 1.25$) (curves 3 and 4) vs. curvature semiangle α. *Inset*: Electric light scattering steady-state EOE at high degree of orientation, $\lambda_{0max} = 400$ nm, $\theta = 90°$ at pH 5.81 (line 1), pH 5.35 (line 2), and pH 6.22 (line 3).

the two possibilities can be made by comparing the theoretical and the experimental curves at another wavelength. Figure 14.6 demonstrates that if the curvature increases over $\alpha \approx 140°$, a raise of $R_{90°}$ should be observed at blue light ($\lambda_0 = 400$ nm, curves 1 and 3) and a decrease at red light ($\lambda_0 = 750$ nm, curves 2 and 4). However, field dependencies show that at red light, as well as at blue, a maximum at pH 5.8 is observed (the Inset in Figure 14.6). This allows concluding unambiguously that the 50% $R_{90°}$ increment, observed in the pH dependence, is induced by a decrease in membrane curvature starting from $\alpha = 127$–$140°$ for cylindrically bent or $\alpha = 119$–$147°$ for spherically bent PM.

It is to be regretted, that the similar dependencies of $P(\theta)$ on α for cylindrically and spherically bent membranes do not allow to choose between the two possible PM shapes only on the basis of pH dependence. But a definite conclusion can be drawn comparing the results with that from small-angle light scattering (Section 14.5.3.2). The curvature $\alpha \approx 130°$, obtained from pH dependence, practically coincides with the result of a cylindrically bent disk, described in Section 14.5.3.2 ($\alpha \approx 125°$). In the case of a segment of a spherical shell the minimal curvature, corresponding to a 50% $R_{90°}$ increase in pH dependence is $\alpha \approx 120°$, whereas the small-angle scattering gives $\alpha = 83°$. That discrepancy is an indirect evidence that PMs are cylindrically bent.

14.6.1.2 Membrane Area

The most sensitive parameter to PM area changes is the relaxation time $\tau_0 \propto B^3$ because of the strong dependence on size of the particles B. As shown in Section 14.5.2.1, $D = 1/6\tau_0$ depends weakly on particles' shape and that is why τ_0 is a measure of the total membrane area (at the absence of aggregation). In order to study pH-induced changes, τ_0 has been measured from EOE decay at light scattering. This method allows diminishing PM concentration with an order (as compared to dichroism) and gives such a reliable indicator of aggregation as R_θ.

At low concentration, monodisperse suspension was studied. The average distance between PMs is about 5 μm (Section 14.4.1.3). The absence of aggregation is evidenced by the constancy of $R_{90°}$, γ, and τ_0 with time. Besides, when pH diminishes, τ_0 also decreases (Figure 14.7, curve 1), in spite of the fact that approaching the isoelectric point (pH 1.5) is favorable for aggregation.

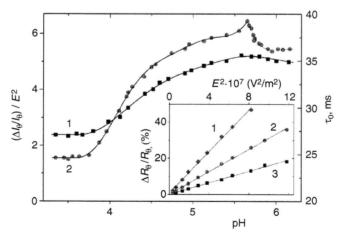

FIGURE 14.7 pH dependence of EOE in PM suspension: relaxation τ_0 (curve 1, the right ordinate) and initial slope of field dependence (curve 2, the left ordinate). *Inset*: Field dependence at low degree of orientation at pH 5.71 (line 1), pH 4.11 (line 2), and pH 3.75 (line 3).

With the pH-induced growth of the surface potential the difference between the bulk and surface pH increases. In order to reduce the difference in local pH an indifferent electrolyte was used, which could influence the electrostatic interactions on the membrane surface. However the experiment has shown that in the range from 0 to 1.6 mM NaCl at pH 4.0 $R_{90°}$ and τ_0 do not change, which means that PM geometry does not depend on ionic strength at these conditions. This allows to study pH dependence at constant electroconductivity instead of at constant ionic strength (when pH increases from 3.35 to 6.25, the ionic strength increases 3.3 times), ensuring better reproducibility.

The relaxation time (curve 1 in Figure 14.7) was measured from EOE decay after a steady-state orientation has been reached, when $\Delta I/I$ is in the linear part of the field dependence $\Delta I/I = f(E^2)$ at the corresponding pH. This means that in the pH range studied, only low degrees of orientation have been reached in spite of the fact that when pH increases, PM polarizability increases four times (curve 2 in Figure 14.7). At these conditions τ_0 does not depend on the orientation degree and it is a quantitative measure of the diffusion coefficient D.

From τ_0 dependence vs. pH at constant medium electroconductivity (Figure 14.7, curve 1) it can be concluded that in the regions of the $R_{90°}$ peaks at pH 3.6 and 5.8 (Figure 14.5) the total area of the membrane does not change. This fact can be explained as follows: when the area of one of the surfaces increases, that is compensated by area decrease of the other. The general course of pH dependence, however, shows a tendency to a rise of the total PM area when pH of the medium increases. The observed raise of τ_0 from 27 ms to 35 ms corresponds to an increase of a flat disk diameter with 9%.

As it has been shown in Section 14.6.1.1, PM curvature changes insignificantly, which allows using Equation 14.53 to estimate the change in the total membrane area. Thus the results in Figure 14.7 show that when pH increases from 3.5 to 5.5 the PM surface area increases by 19%. A similar effect has been observed on a lipid monolayer; in that monolayer, the average area of one molecule increases from 0.55 to 0.66 nm^2 [69].

14.6.1.3 Electric Polarizability

As interfacial polarizability γ is determined by the charge and the geometry of the particles, the dependence $\gamma = f(\text{pH})$ can give additional information about the curvature. When moving away from the isoelectric point the surface charge increases, which is accompanied by an increase in γ. Besides, the increase of curvature diminishes γ because the difference between longitudinal and transverse polarizability components decreases. When the influence of the curvature predominates, the opposite course of $\gamma = f(\text{pH})$ can be observed as compared to the same dependence at constant shape. So if γ decreases when moving away from the isoelectric point (pH 1.5 for PM), this fact is an unequivocal indication of a curvature rise.

At electric light scattering the situation is complicated by the fact, that when particles' geometry is changed (the change in R_θ indicates that), the optical function $A(KB)/P(\theta)$ also changes. This makes it impossible to determine quantitatively the polarizability change $\Delta\gamma$ at $R_{90°}$ peaks' region at pH 3.6 and 5.8 (Figure 14.5). The semiquantitative $\Delta\gamma$ determination is possible only in the range pH 4.0–5.5, where $R_{90°}$ alters gradually. The initial slope $(\Delta I/I)/E^2$ of the curves at 1 kHz is proportional (within a constant) to γ for particles with a permanent geometry according to Equation 14.20 and Equation 14.32. At this frequency only the induced dipole moment of PM is manifested, which is almost fully related to the interfacial polarizability.

In Figure 14.7 one can see that $(\Delta I/I)/E^2$ increases with pH about four times. $R_{90°}$ at chaotic orientation has the same value at pH 3.9 and 5.4 (Figure 14.5, curve 1) but the electric polarizability increases about three times in this range. This is mainly due to the total charge growth, whereas the contribution of the size increment is an order smaller (30% τ_0 increase in

Figure 14.7 corresponds to $\Delta B \approx 9\%$ and $\gamma \propto B^3$). The anomalous course of $(\Delta I/I)/E^2$ around pH 3.5 and 5.7 suggests that PM curvature changes. The decrease of $(\Delta I/I)/E^2$ in the pH range 5.7–6.0 is an evidence that the curvature increase has a greater impact on γ than the charge growth.

In such a way, pH dependence of $(\Delta I/I)/E^2$ shows an increase of interfacial electric polarizability because of total charge increase in pH range 3.8–5.6 and change in PM curvature in the regions of pH 3.6 and 5.7. The comparison with the typical pK values of the proteins and membrane lipids allows to conclude that at pH 5.8 imidazole groups of histidine are ionized, whereas at pH 3.6, the phosphate groups of acidic lipids [70] and carboxylic groups of the asparagine acid residues of bR [9].

The change in PM curvature in the pH range 3.6–5.7 could be explained by the different pK values of chemically identical ionizable groups, which are situated on both membrane surfaces. The differences in pK-values can be due to the local electric field of a neighboring charge, as well as to the averaged potential on each of the surfaces. Because of differences in dissociation constants of groups of the same type, pH-induced charge changes occur asymmetrically on both membrane surfaces. Since the surface charge density affects the monolayer area, the membrane itself, similarly to a bimetallic plate upon heating, will bend at pH-induced asymmetric charge change.

14.6.2 POLYVALENT MACROION ADSORPTION

PM changes, induced by the pH of the medium, could be due to changes in the charge density, as well as conformational changes in bR macromolecules, which lead to an alteration of the trimers' geometry. The last possibility could be ruled out if the electrostatic interactions are changed by adsorption of globular proteins as polyvalent ions. Very appropriate for that purpose is lectin, a positively charged glycoprotein, binding with membrane's glycolipids [71]. As lipids are situated among bR trimers, the polyions are not adsorbed on bR macromolecules and thus the possibility of conformational bR changes is avoided.

The results presented here show the adsorption of phytohemagglutinin (PhHA, a positively charged plant lectin, extracted from Phaseolus vul.). Microelectrophoretic measurements showed that the electrophoretic mobility u of PM at pH 6 decreases about 9 times (from -1.4×10^{-8} to -1.6×10^{-9} m$^2 \cdot$V$^{-1} \cdot$s^{-1}) when PhHA concentration increases up to 1 μg/ml (Figure 14.8). In the range 1–10 μg/ml, u practically does not change, which indicates that all binding sites are occupied and 1 μg PM binds 0.1 μg PhHA. In spite of the big PhHA charge, the isoelectric point is not attained, which supports literature data that PhHA binding is determined by specific interactions with the lipids.

As the sign of lectin's electric charge is opposite to that of PM, its adsorption should reduce interfacial polarizability γ. For studying the change in γ, electrical dichroism was chosen, as the modification in particles' geometry has a weaker influence on optical function as compared to light scattering. This allows a more reliable determination of γ changes from the slope of the field dependence $\Delta A/A = f(E^2)$. On the other hand, dichroism requires higher PM concentrations, which together with the lowering of the total charge at PhHA adsorption, create favorable conditions for PM aggregation. In spite of that, the absence of observable EOE changes (at 1 μg/ml PhHA) with time shows that no aggregation occurs. The reason might lie in the large size of lectin macromolecules, which makes impossible the approach of the two membranes at a distance, where van der Waals interactions exceed the electrostatic interactions. This makes this case different from the adsorption of inorganic ions on a smooth surface.

As the interfacial polarizability is approximately proportional to the charge density, the reduction of u with an order should be accompanied by an abrupt decrease of γ. The experiment, however, proves the opposite: not only that γ does not decrease, but also on

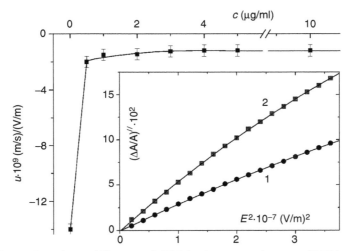

FIGURE 14.8 Electrophoretic mobility u of PM–lectin suspension vs. PhHA concentration c. *Inset*: Relative EOE of electric dichroism at parallel polarization $(\Delta A/A)^{\|}$ vs. square of electric field intensity E^2.

the contrary, it increases (the inset in Figure 14.8). This result looks paradoxical on the background of literature data on solid particles (for instance, Ref. [72]). PhHA-macroions, bound to membrane's lipids are immobile at 1 kHz and do not take part in the induced dipole moment. There is no basis to assume that the optical function $(\Delta A/A)_{\infty}$ changes because of chromophore reorientation in bR macromolecule. In this way the only reasonable explanation of EOE increase at lectin adsorption on PM remains the change of the membrane curvature.

The PM shape influences the slope of the field dependence $(\Delta A/A)^{\|} = f(E^2)$ in two ways: at first place, $\gamma = \gamma^{\|} - \gamma^{\perp}$ is determined by the difference between the longitudinal $\gamma^{\|}$ and the transverse component $\gamma^{\perp}$, and, at second place, the membrane curvature changes $(\Delta A/A)_{\infty}$ in accordance with Equation 14.27. Both effects lead to increase of $(\Delta A/A)/E^2$ only in the case when the curvature decreases, hence PMs have been initially bent and PhHA adsorption has straightened them.

The result obtained indicates that electrostatic forces play a considerable role in determining PM curvature. The probable reason for that is an asymmetric change of electrostatic interactions on both membrane surfaces. The preconditions for this are the asymmetric distribution of the lipids in both PM monolayers [73] and the specific bonding of lectins with glycolipids [71].

Taking into account that the glycolipids are located on the extracellular surface of cytoplasmic membranes [74] it can be concluded that PhHA magnifies PM electric asymmetry because of the adsorption of the positively charged lectin macromolecules on the less negatively charged membrane surface. From here it can be concluded that the free PM in water medium are bent in the same manner as in *in vivo*—the more negatively charged intracellular surface is concave. In this case, PhHA adsorption on the convex extracellular surface causes a straightening of PM because of growing of the asymmetry of tangential electrostatic interactions on the both membrane surfaces.

14.6.3 EXTERNAL HIGH-VOLTAGE ELECTRIC FIELD

The fact that electrostatic interactions in the membrane can change PM curvature allows assuming that a similar effect could be observed in an external field also. The mechanism of these changes would depend on field intensity and on the membrane module of elasticity at bending. High-voltage field can change the dissociation constants of ionizable groups, which will induce changes in PM geometry, similar to those at pH-induced change of the surface charge (Section 14.6.1). Conformational changes in bR macromolecules at very high fields are possible because of breaking of hydrogen bonds. PMs are subjected to hydrodynamic forces upon their orientation and translation in the field, which should be added to the intramembrane deformational force.

If the external electric field is applied in water suspension the possible changes in PM geometry and structure should be observed on the background of electro-orientational effect. The separation of the effects of deformation and orientation is possible at parallel application of some electro-optic methods with a different optical function. The results presented below have been obtained by electric light scattering and dichroism in PM monodisperse suspension. While dichroism is sensitive to chromophore orientation, light scattering depends only on PM geometry regardless of their internal structure (as far as PMs scatter as optically isotropic particles, see Section 14.4.1.1). That allows separating the effects of changes in PM structure and geometry. But at high degrees of orientation EOE at light scattering is not a linear function of the orientation degree F (in contrast to electric dichroism), which might lead to the wrong impression of geometry changes in the absence of those. That difficulty might be avoided by comparing the transition processes at light scattering with those at dichroism.

Figure 14.9 (curve 4) shows EOE decay at light scattering in monodisperse suspension after orientation in an electric field with intensity $E = 4.8\,\text{kV/cm}$. PM sizes correspond to a flat disk with a diameter $B = 0.5\,\mu\text{m}$ (calculated according to Equation 14.53 from $\tau_0 = 7\,\mu\text{s}$, measured by dichroism at $E = 1\,\text{kV/cm}$) [17]. The decay has two components: microsecond and millisecond; the slower one is exponential and it expresses the disorientation process. The fast component may be due to: (1) a presence of small particles; (2) changes in membranes' geometry; and (3) the nonlinearity of the optical function at high degrees of orientation. In the last case the contribution of the fast terms in Equation 14.34 should be (at the same degree of

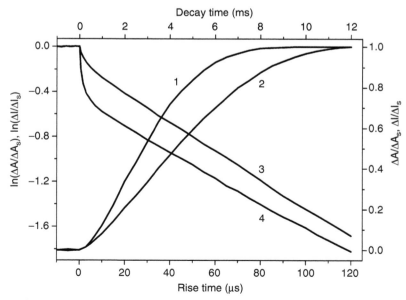

FIGURE 14.9 Rise (curves 1 and 2, right ordinate) and decay (curves 3 and 4, left ordinate) of EOE at electric dichroism (curves 1 and 3) and electric light scattering (curves 2 and 4) at $E = 5\,\text{kV/cm}$.

orientation) equal at orientation and disorientation of the particles. Consequently, both transient processes should have a similar fast component at high degree of orientation, but EOE rise does not show it (curve 2 in Figure 14.9). Besides, such a fast process of optical nature should be also observed at moderate intensity of electric field, enough to achieve a high degree of orientation. However EOE decay at light scattering upon orientation in an electric field of 1 kV/cm shows an absence of a fast component. These two facts allow concluding that the fast relaxation process (curve 4 in Figure 14.9) is not a consequence of optical function nonlinearity, but rather expresses a real process.

In case when the microsecond process is connected to the presence of small PM fragments (possibly obtained during the ultrasound disaggregation of the suspension) they should form the basic fraction with regard to particle numbers. This conclusion follows from the ratio of the fast to slow process at light scattering (curve 4), as at this method the contribution of small size particles B/λ is weak. At dichroism the relative portion of the fast process should be prevailing, as EOE magnitude does not depend on particles' size at a certain degree of orientation. The experiment however showed that the microsecond process at dichroism (curve 3 in Figure 14.9) is three times smaller than at light scattering (curve 4). Besides at moderate fields intensity (1 kV/cm) dichroism does not show a fast relaxation process [17]. These two facts rule out the presence of small PM fragments. It can be assumed that the presumed small particles are nonabsorbing (parts of the bacterial wall, metal particles of the ultrasound disintegrator tip and others). In this case EOE increase at light scattering should be much faster than that at dichroism, reflecting the contribution of the hypothetical small particles. However the experiment shows that τ_E at light scattering (curve 2 in Figure 14.9) is not smaller than τ_E at dichroism (curve 1). This rules out the presence of small particles in the suspension.

Thus the conclusion obtrudes itself that the fast relaxation process is connected with changes in PM geometry, induced by the external electric field. The difference in EOE decay in the two methods is due to the high sensitivity of light scattering to particles' geometry as opposed to the low sensitivity of dichroism. The smaller relative share of the fast process at dichroism suggests a change in curvature but not a change in chromophore orientation.

If the particles do not change their geometry during orientation, EOE rise should satisfy Equation 14.22. This suggests an independent criterion of particles' deformability—the deviation from linearity of the dependence $\tau_E = f(1/E^2)$. As particles' shape does not affect the rotational diffusion coefficient strongly (see Section 14.5.2.1), the deviation from linearity is mainly due to a change in the optical function. So light scattering was the method chosen: it is very sensitive to the geometry of particles at commensurate B and λ (Figure 14.6 and Figure 14.10).

The inset in Figure 14.10 shows the dependence of τ_E on E^{-2}. The linear dependence shows that at $E < 2.5$ kV/cm PMs behave as rigid particles. That is the electric intensity range most widely used because at these low fields, PM size and polarizability are determined [43]. At $E = 4.8$ V/cm one should observe $\tau_E = 0.45$ μs, but the experiment gives $\tau_E = 0.05$ μs. This difference amounting to an order could not be explained otherwise but with appearance of an additional process, which accelerates the EOE effect.

A possible reason for that is a field-induced increase of the polarizability, for instance because of additional ionization or counterions desorption (the normally bonded Mg^{2+} and Ca^{2+}). This would bring about an increase of the induced dipole moment and an actual acceleration of orientation but it should not affect particles' disorientation. However the experiment shows that EOE decay has a fast relaxation component, which is clearly discerned on the background of the slow disorientation (curve 4 in Figure 14.9).

The results obtained allow concluding that at high-intensity fields ($E \geq 5$ kV/cm) EOE has an orientation–deformation nature. The deformation is expressed mainly as a change of PM

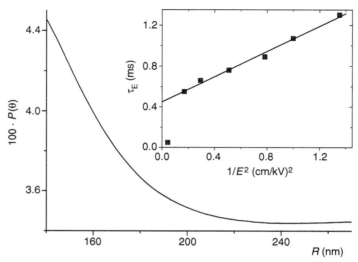

FIGURE 14.10 Form-factor $P(\theta)$ for a cylindrically bent elliptical disk at $\theta = 90°$, $\lambda_0 = 400$ nm and $n_0 = 1.343$ $(K = 14.9 \, \mu m^{-1})$ vs. cylinder radius R. The surface area is equal to area of a flat circular disk with diameter $B = 500$ nm $(KB = 7.45)$. *Inset:* The rise time τ_E of electric light scattering EOE at high degree of orientation vs. reciprocal value of square of field intensity $1/E^2$.

curvature and manifests itself on the background of the orientation effect. The PM flexion, induced by the electric field affects the friction coefficient and the light scattering intensity. The first factor leads to an actual acceleration of disorientation but its contribution is not great as D is chiefly determined by particles' area (Section 14.5.2.1). The predominant effect is due to light scattering increase in steady-state of EOE. That creates an illusion of a faster disorientation because both erection and disorientation decrease the scattering intensity. The comparison of the relaxation curves by the two methods (Figure 14.9) shows that the optical effect from curvature change at light scattering is of greatest importance. The smaller effect at dichroism is due to the change of mean chromophore orientation to axis of symmetry (average angle Ψ in Equation 14.28).

The reasons for the deformation remain hypothetical. In Refs. [75,76] curvature change in electric field is explained by an inverse flexoelectric effect connected with the asymmetry of charged lipids in the two monolayers of the membrane. Membrane curvature is changed because of an electric-field induced flip-flop of lipid molecules from one monolayer to the other. But here this mechanism looks hardly probable as PM lipids are associated with bR macromolecules, which makes them not readily mobile.

The second hypothesis is based on field induced changes of the dissociation constants of H^+ and of adsorbed counterions. That leads to an asymmetrical change of the surface charge density, resulting in changes of membrane total area and curvature similarly to the effect of pH (Section 14.6.1). As a result, the electric polarizability, friction coefficient and light scattering intensity are changed. As it was concluded, the change in curvature and the accompanying change of light scattering are of primary importance.

The third hypothesis is based on the capability of the electric field to induce reversible conformational changes in bR macromolecule [77–79], which alters the trimer' geometry and hence the PM curvature. If the changes do not lead to chromophore reorientation to local normal (direction of retinal in bR macromolecule), the change in geometry should manifest itself weakly at dichroism, but strongly at light scattering, which is in fact observed in the experiments described here.

REFERENCES

1. Stoeckenius, W., and Rowen, R., A morphological study of *Halobacterium halobium* and its lysis in media of low salt concentration, *J. Cell. Biol.*, 34, 365, 1967.
2. Oesterhelt, D., and Stoeckenius, W., Isolation of the cell membrane of *Halobacterium halobium* and its fractionation into red and purple membrane, *Methods Enzymol.*, 31, 667, 1974.
3. Blaurock, A., and Stoeckenius, W., Structure of the purple membrane *Halobacterium halobium*, *Nature New Biol.*, 233, 152, 1971.
4. Oesterhelt, D., and Stoeckenius, W., A rhodopsin-like protein from the purple membrane of *Halobacterium halobium*, *Nature New Biol.*, 233, 149, 1971.
5. Oesterhelt, D., and Stoeckenius, W., Function of a new photoreceptor membrane, *Poc. Natl. Acad. Sci. USA*, 70, 2838, 1973.
6. Lozier, R.H., and Niederberger, W., The photochemical cycle of bacteriorhodopsin, *Fed. Proc.*, 36, 1805, 1977.
7. Balashov, S.P., and Litvin, F.F., *Photochemical conversions of bacteriorhodopsin*, Moscow State University, Moscow, 1985, chap. 2.
8. Haupts, U., Tittor J., and Oesterhelt, D., Closing in on bacteriorhodopsin: progress in understanding the molecule, *Annu. Rev. Biophys. Biomol. Struct.*, 28, 367, 1999.
9. Balashov, S.P., Protonation reactions and their coupling in bacteriorhodopsin, *Biochim. Biophys. Acta*, 1460, 75, 2000.
10. Heberle, J. et al., Bacteriorhodopsin: the functional details of a molecular machine are being resolved, *Biophys. Chem.*, 85, 229, 2000.
11. Lanyi, J.K., and Luecke, H., Bacteriorhodopsin, *Curr. Opin. Struct. Biol.*, 11, 415, 2001.
12. Czégé, J., Changes in lights scattering and conformation of bacteriorhodopsin during the photocycle, *FEBS Lett.*, 242, 89, 1988.
13. Zhivkov, A.M., pH-induced changes in the geometry of the purple membranes, *Compt. Rend. Acad. Bulg. Sci.*, 49, 17, 1996.
14. Zhivkov, A.M., Changes in purple membrane curvature caused by lectin adsorption, *Compt. Rend. Acad. Bulg. Sci.*, 50, 19, 1997.
15. Zhivkov, A.M., Changes in purple membrane geometry induced by high voltage electric field, *Compt. Rend. Acad. Bulg. Sci.*, 50, 25, 1997.
16. Zhivkov, A.M., pH-dependence of electric light scattering by water suspension of purple membranes, *Colloids Surf. A*, 209, 319, 2002.
17. Zhivkov, A.M., Orientation-deformational electro-optical effect in water suspension of purple membranes, *Colloids Surf. A*, 209, 327, 2002.
18. Henderson, R., and Unwin, P.N.T., Three-dimensional model of purple membrane obtained by electron microscopy, *Nature*, 257, 28, 1975.
19. Heyn, M.P., Cherry, R.J., and Muller, U., Transient and linear dichroism studies on bacteriorhodopsin: determination of the orientation of the 568 nm all-trans retinal chromophore, *J. Mol. Biol.*, 117, 607, 1977.
20. Fisher, K.A., and Stoeckenius, W., Freeze-fractured purple membrane particles: protein content, *Science*, 197, 72, 1977.
21. Kates, M., Kushwaha, S.C., and Sprott, G.D., Lipids of purple membrane from extreme halophilic and of methanogenic bacteria, *Methods Enzymol.*, 88, 98, 1982.
22. Henderson, R., Jubb, J.S., and Rossmann M.G., A contracted form of the trigonal purple membrane of *Halobacterium halobium*, *J. Mol. Biol.*, 154, 501, 1982.
23. Shnyrov, V.L., Tarahovsky, Y.S., and Borovyagin, V.L., Study of structural changes in purple membranes of *Halobacterium halobium* at thermal and alkali denaturation, *Bioorganic Chem.* (Russia), 7, 1054, 1981.
24. Moore, T.A. et al., Studies of an acid-induced species of purple membrane from *Halobacterium halobium*, *Biochem. J.*, 171, 469, 1978.
25. Stoeckenius, W., Lozier, R.H., and Bogomolni, R.A., Bacteriorhodopsin and purple membrane *Halobacterium halobium*, *Biochem. Biophys. Acta*, 505, 215, 1979.

26. Zhivkov, A., and Pechatnicov, V.A., Photometric determination of refractive index of purple membrane, *Biophyzika* (Russia), 36, 1004, 1991.

27. Jonas, R., Koutalos, Y., and Ebrey, T.G., Purple membrane: surface charge density and the multiple effects of pH and cations, *Photochem. Photobiol.*, 52, 1163, 1990.

28. Kantcheva, M.R., Popdimitrova, N., and Stoylov, S., Electrophoretic mobility of purple membrane from *Halobacterium halobium, Stud. Biophys.*, 90, 125, 1982.

29. Popdimitrova, N. et al., Surface charge density of purple membrane fragments, *Cell Electrophoresis*, Schütt, W., and Klinkmann, H., Eds., Walter de Gruyter & Co., Berlin–New York, 1985, 167.

30. Neugebauer, D.C., Oesterhelt, D., and Zingcheeim, H.P., The two faces of the purple membrane. II. Differences in surface charge properties revealed by ferritin binding, *J. Mol. Biol.*, 125, 123, 1978.

31. Renthal, R., and Chang, C.H., Charge asymmetry of the purple membrane measured by uranyl quenching of dansyl fluorescence, *Biophys. J.*, 45, 1001, 1984.

32. Fisher, K.A., Yanagimoto, K., and Stoeckenius, W., Orientated absorption of purple membrane to cationic surfaces, *J. Cell. Biol.*, 77, 611, 1978.

33. Nachliel, E. et al., Protonation dynamics of the extracellular and cytoplasmic surface of bacteriorhodopsin in the purple membrane, *Proc. Natl. Acad. Sci. USA*, 93, 10747, 1996.

34. Carmeli, C., Quintanilha, A.T., and Packer, L., Surface charge changes in purple membrane and photoreaction cycle of bacteriorhodopsin, *Proc. Natl. Acad. Sci. USA*, 77, 4707, 1980.

35. Keszthelyi L., Orientation of membrane fragments by electric field, *Biochem. Biophys. Acta*, 598, 429, 1980.

36. Todorov, G., Sokerov, S., and Stoylov, S.P., Interfacial electric polarizability of purple membrane, *Biophys. J.*, 40, 1, 1982.

37. Kimura, Y. et al., Electric dichroism of purple membrane suspension, *Photochem. Photobiol.*, 33, 435, 1981.

38. Todorov, G., Sokerov, S., and Stoylov, S.P., Birefringence of purple membrane, *J. Coll. Interf. Sci.*, 165, 154, 1994.

39. Kimura, Y., Fujiwara, M., and Ikegami, A., Anisotropic electric properties of purple membrane and their change during the photoreaction cycle, *Biophys. J.*, 45, 615, 1984.

40. Barabas, K. et al., Electro-optical measurements on aqueous suspension of purple membrane from *Halobacterium halobium, Biophys. J.*, 43, 5, 1983.

41. Mostafa, H.I.A. et al., Electrooptical Measurements on purple membrane containing bacteriorhodopsin mutants, *Biophys. J.*, 70, 468, 1996.

42. Papp, E., Fricsovszky, G., and Meszéna, G., Electrodichroism of purple membrane, *Biophys. J.*, 49, 1089, 1986.

43. Taneva, S.G., and Petkanchin, I.B., Surface electric properties of biological systems, *Trends Photochem. Photobiol.*, 6, 113, 1999.

44. Larsen, H., Family Halobacteriaceae, in *The Procariots*, Star, M. et al., Eds., Springer Verlag, Berlin, 1981, p. 985.

45. Kubota, K. et al., Study on dynamic light scattering of the purple membrane suspension, *Biophys. Chem.*, 23, 15, 1985.

46. Czégé, J., Light scattering changes during the photocycle of bacteriorhodopsin, *Acta Biochim. Biophys. Hung.*, 22, 463, 1987.

47. Czégé, J., Bent membrane of the purple membrane. Theoretical details and further experimental data, *Acta Biochim. Biophys. Hung.*, 22, 479, 1987.

48. Czégé, J., and Reinisch, L., Cross-correlated photon scattering during photocycle in bacteriorhodopsin, *Biophys. J.*, 58, 721, 1990.

49. Czégé, J., and Reinisch, L., The pH-dependence of transient changes in the curvature of the purple membrane, *Photochem. Photobiol.*, 54, 923, 1991.

50. Shifrin, K.S., *Scattering of Light in a Turbid Medium*, Gostehizdat, Moscow, 1951, chap. 1.

51. van de Hulst, H.C., *Light Scattering by Small Particles*, John Wiley, New York, 1957, chap. 7.

52. Eskin, B.E., *Light Scattering by Polymer Solutions*, Nauka, Moscow, 1973, chaps. 1, 2.

53. Svergun, D.I., and Feigin, L.A., *X-ray and Neutron Small-Angle Scattering*, Nauka, Moscow, 1986, 62.

54. Stoylov, S.P., *Colloid Electro-optics*, Academic press, London, 1991, chaps. 1–3.

55. Shah, M., Electric birefringence of bentonite. II. An extension of saturation birefringence theory, *J. Phys. Chem.*, 67, 2215, 1963.
56. Benoit, H., Contribution à l'étude de l'effet Kerr présenté par les solutions diluées de macromolécules rigides. *Ann. Phys.*, 6, 561, 1951.
57. Schwarz G., Zur theorie der leitfahigkeitsanisotropie von polyelektrolyten in losung, *Z. Phys.*, 145, 563, 1956.
58. Perrin, F., Mouvement brownien d'un ellipsoïde. I. Dispersion diélectrique pour des molecules ellipsoïdales, *J. Phys. Radium*, 5, 497–511, 1934.
59. Volkenstein, M.V., *Molecular Optics*, Gostehizdat, Moscow, 1951, chap. 7.
60. Fredericq, E., and Houssier, C., *Electric Dichroism and Electric Birefringence*, Clarendon Press, Oxford, 1973.
61. Wippler, C., Étude théorique de la diffusion de la lumière par des sols de bâtonnets orientés, *J. Chim. Phys.*, 51, 122, 1954.
62. Stoimenova, M., Electric light scattering by cylinder symmetrical particles, *J. Colloid Interface Sci.*, 53, 42, 1975.
63. Picot, C., Weill, G., and Benoit, H., Light scattering by anisotropic disks. Application to polymer single crystals. *J. Colloid Interface Sci.*, 27, 360, 1968.
64. Shifrin, K.S., and Zelmanovich, I.L., *Tables of Light Scattering. II. Coefficients of Angular Scattering*, Hydrometeorogical Publishing House, Leningrad, 1968.
65. Sidko, F.Ya. and Lopatin, V.N., *Introduction to optics of cell suspensions*, Nauka, Novosibirsk, 1988.
66. Vsevolodov, N.N., *Biopigments—Photoregisters*, Nauka, Moscow, 1988, p. 82.
67. Suong, T.T. and Stoylov, S.P., Comparative determination of the size of disk-shape colloids particles by conventional and electric light scattering. *Izvestia Khimia*, 15, 542, 1982.
68. Ioffe, B.V., *Refractometric Methods in Chemistry*, Himia, Leningrad, 1983, p. 307.
69. Tocanne, J.F. et al., A monolayer and freeze-etching study of charged phospholipids. I. Effect of ions and pH on the ionic properties of phosphatidylglycerol, *Chem. Phys. Lipids*, 12, 201, 1974.
70. Ermakov, Yu.A., Ion equilibrium near lipid membranes: empirical analysis of the simplest model, *Colloid J.*, 62, 389, 2000.
71. Lutsik, M.D., Panasyuk, E.N., and Lutsik A.D., *Lectins*, Vissha shkola, Lvov, 1981.
72. Stoimenova, M. and Zhivkov, A., Electro-optic investigation of oxide suspensions: surface charge sign reversal through specific adsorption, *J. Colloid Interface Sci.*, 142, 92, 1991.
73. Blaurock, A.E. and King, G.I., Asymmetric structure of the purple membrane, *Science*, 196, 1101, 1977.
74. Yamakawa, T., and Nagai, Y., Glycolipids at the cell surface and their biological functions, *Trends Biochem. Sci.*, 3, 128, 1978.
75. Petrov, A.G., *The Lyotropic States of Matter: Molecular Physics and Living Matter Physics*, Gordon and Breach Publishing, New York, 1999.
76. Petrov, A.G., Flexoelectricity of model and living membranes, *Biochim. Biophys. Acta*, 1561, 1, 2001.
77. Tsuji, K., and Neumann, E., Electric-field induced pK-changes in bacteriorhodopsin, *FEBS Lett.*, 128, 265, 1981.
78. Tsuji, K., and Neumann, E., Structural changes induced by electric fields in membrane bound bacteriorhodopsin, *Biophys. Struct. Mech.*, 7, 284, 1981.
79. Tsuji, K., and Neumann, E., Conformational flexibility of membrane proteins in electric fields. I. Ultraviolet absorbance and light scattering of bacteriorhodopsin in purple membrane, *Biophys. Chem.*, 17, 153, 1983.

Part II

New Applications

B. Adsorption on Particles

15 Nanoscale Charge Nonuniformity on Colloidal Particles

Darrell Velegol

CONTENTS

15.1 INTRODUCTION TO CHARGE NONUNIFORMITY ON COLLOIDAL SYSTEMS

Two critical challenges with colloidal systems are stability and assembly. Analyses of particle stability commonly follow a diffusion-type model, which predict the time (t_p) required for two particles to approach each other and aggregate [1]. The original result due to Smoluchowski (i.e., the "rapid flocculation rate") is the diffusion-limited result, but it was Fuchs who accounted for arbitrary interaction potentials and introduced the concept of a "stability ratio" (W) into the aggregation time (t_p) for spherical colloids:

$$t_p = \frac{\pi \eta a^3}{\phi k T} W \tag{15.1}$$

where η is the fluid viscosity, a is the sphere radius, ϕ is the particle volume fraction, and kT is the thermal energy. W gives the ratio of the number of times two particles would collide in the absence of an interparticle potential energy of mean force ($\Phi = 0$), compared with the number of times the two particles would collide in the presence of the actual Φ.

A rough relationship between W and Φ is given by [1]

$$W = 0.25 \exp\left(\frac{\Phi_{max}}{kT}\right) \tag{15.2}$$

where Φ_{max} is the height of the energy barrier to flocculation. This barrier could be due to electrostatic repulsion or other forces between particles. Estimates from Equation 15.2 show that to stabilize for a year 100-nm-diameter colloids in water at $T = 300$ K, $\Phi_{\mathrm{max}}/kT = 27$. If we want to stabilize the particles for a month or 10 years, the result for Φ_{max}/kT changes little because the relationship in Equation 15.2 is exponential.

The interparticle potential of mean force (Φ) is commonly given by the DLVO model, which accounts for electrostatic forces (usually repulsive) and van der Waals forces (usually attractive). A model is developed for the potential of mean force between two flat plates, and then the Derjaguin approximation is used to evaluate the potential of mean force. Hogg et al. showed that for low potentials one obtains approximately the following for the electrostatic potential (Φ^{ES}) between identical spheres

$$\Phi^{\mathrm{ES}} = 2\pi\varepsilon a\psi_0^2 e^{-\kappa\delta} \tag{15.3}$$

where ψ_0 is the surface potential of the spheres, ε is the fluid permittivity ($80\varepsilon_0 = 7.1 \times 10^{-10}\,\mathrm{C^2/Nm^2}$ for water), κ^{-1} is the Debye length of the solution, and δ is the gap distance between the spheres at the point of closest approach. A quick evaluation using Equation 15.3 in water with the parameters $\psi_0 = -50$ mV, $a = 50$ nm, and $\kappa\delta = 1$ (a typical value for where the maximum Φ occurs, when van der Waals forces are included) gives $\Phi^{\mathrm{ES}}/kT = 50$. This value is well beyond what is needed to stabilize the particles for centuries.

And here is the heart of the problem addressed in this chapter. Even though the classical models often predict that particles should be stable to aggregation for centuries due to electrostatic repulsion under given conditions, experiments reveal that the particles might be stable for only minutes to hours under the given conditions [2,3]. What has gone wrong?

There are a number of assumptions in Equation 15.3 [4]. Forces like depletion, solvation, and steric are ignored. The particles are assumed to be geometrically smooth. The gap is assumed to be small, such that $\delta/a \ll 1$, and the charge distribution is assumed to be uniform. Our lab group has been examining this last assumption in detail for colloidal surfaces (Figure 15.1) [4–13].

This chapter has a number of purposes: (1) to review several known or hypothesized origins of charge nonuniformity on oxide and polymer particles; (2) to discuss techniques for measuring charge nonuniformity on particles; (3) to review the literature on charge nonuniformity for various particulate systems; (4) to describe some ramifications of charge

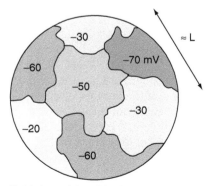

FIGURE 15.1 Schematic of a colloidal particle with charge nonuniformity (Figure originally found in Velegol et al., *J. Colloid Interface Sci.*, 230, 114, 2000.) The average surface potential on the particle is roughly -45 mV, but depending on the location on the surface, the surface potential can be from -20 to -70 mV.

nonuniformity for colloidal stability and assembly; (5) to predict the future opportunities in studying and controlling charge nonuniformity for colloidal systems. My aim is not to provide every reference on the subject of charge nonuniformity—a more comprehensive list can be obtained from the references in this chapter. Rather, my aim is to guide those readers who are new to the topic in examining the theoretical and experimental evidence in the literature for the five purposes listed above.

15.2 ORIGINS OF CHARGE NONUNIFORMITY

For micron-size particles, it is important to recognize that random Poisson charge placement of univalent charges does not account for any significant amount of charge nonuniformity [8]. An expression giving the charge nonuniformity (σ_ζ^2/N) on a particle with random charge placement is

$$\frac{\sigma_\zeta^2}{N} = \frac{Z_0^2 e^2 q}{A^2 \kappa^2 \varepsilon^2 \cosh^2\left(\frac{Ze\zeta_0}{2kT}\right)} = \frac{4Z_0 e\rho_0/(A\kappa^2\varepsilon^2)}{4 + \left(\frac{Ze\rho_0}{\kappa\varepsilon kT}\right)^2} \tag{15.4}$$

where N is the number of regions on the sphere, σ_ζ is the standard deviation of zeta potential over those N regions, Z_0 is the valence of the "point charges" on the particle surface, e is the proton charge, q is the number of charges on the surface, A is the particle surface area, κ^{-1} is the Debye length, ε is the fluid permittivity, Z is the valence of the $Z{:}Z$ symmetric ions in solution, ζ_0 is the area-averaged zeta potential on the particle surface, ρ_0 is the average surface charge density, and kT is the thermal energy. The use of σ_ζ^2/N might seem nonintuitive at first, but one can visualize that the smaller the region examined, the more the variation might be in surface charge density.

It is helpful to see an example with Equation 15.4. Say we have a 1-μm-diameter particle with negative univalent charges on the surface (i.e., $Z_0 = 1$). Each charge occupies 20 nm^2 (i.e., surface charge density is 0.8 μC/cm^2), and so there are 157,000 charges on the particle. The particle is in 1 mM KCl, so that $\kappa^{-1} = 10$ nm. The average $\zeta_0 = -80$ mV. If we mentally divide the surface into 100-nm-diameter circular regions, then $N = 400$. Equation 15.4 then predicts that if the charges are placed randomly (by a Poisson charging process), then $\sigma_\zeta = 2.3$ mV. This is a significant but small number. To emphasize, random Poisson charging with univalent charge groups will not give large charge nonuniformity on colloidal particles.

Charge nonuniformity has different origins on oxide particles, polymer particles, bacteria, and viruses. For mineral oxide surfaces, surface heterogeneity is the norm due to different surface oxygen groups on a crystal plane, different crystal planes exposed (e.g., the 001 face of gibbsite is unreactive over a particular pH range), plane imperfections, disordered lattices, selective adsorption of various ions, or surface impurities [14]. One description of surface heterogeneity is given by the MUti-SIte Complexation (MUSIC) model [15,16]. This model accounts for the presence of different types of surface groups, such as differences in surface oxygens arising from different numbers of coordinating metal ions. By examining given surface protonation reactions, the model predicts the equilibrium constants of singly, doubly, and triply metal-coordinated oxygen surface groups. The model can be applied to both random heterogeneity and patchwise heterogeneity, and this has been done for gibbsite, goethite, alumina, hematite, rutile, and other minerals [14]. Currently the models do not predict variations in surface potential or the length scale over which these variations occur, but perhaps they will have utility in doing so in the future.

Charge nonuniformity on polymer particles might arise from the emulsion polymerization process used to synthesize them. The final polymer colloid is synthesized by the coagulation of primary particles from the emulsion polymerization [17]. When the primary particles have different numbers of charge groups on their surface, the final particle will be nonuniformly charged. Returning to the use of Equation 15.4, one might imagine the surface to be dotted with "charge groups" (i.e., primary particles) of different valence, from $Z_0 = 1$ to $Z_0 = 100$. To simplify our calculation, we will assume that $Z_0 = 100$ dominates the surface charge, and so assume that the $Z_0 = 100$ dot the surface randomly, even though in fact primary particles form the entire surface. For the same charge density and ionic strength as before, we now see that Equation 15.4 predicts $\sigma_\zeta = 23$ mV, a much more significant number. Even using $Z_0 = 10$ gives $\sigma_\zeta = 7.3$ mV. I do not intend to indicate that random charging due to different primary particles is definitely the mechanism for charge nonuniformity in polymer colloids; rather, I use these arguments to make this hypothesis, which must be tested experimentally. Another potential cause of charge nonuniformity in hydrophobic polymer colloids is the "pearling effect," which causes charge groups to bundle [18].

Living systems can also have charge nonuniformity. For bacterial systems this can occur due to cell polarity [10,19]. For example, in *Escherichia coli* this bacterial polarity is caused by the localization of peptidoglycan, proteins, and phospholipids on the surface of the bacterium. Furthermore, the polarity can be dynamic within the cell.

15.3 TECHNIQUES FOR MEASURING CHARGE NONUNIFORMITY

A number of techniques exist for measuring charge nonuniformity and other heterogeneity on surfaces (Table 15.1), but few techniques exist for measuring charge nonuniformity on colloidal particles. The techniques listed in Table 15.1 have two primary challenges: either the particles must be examined under vacuum, not in their native aqueous environment, or

TABLE 15.1
Surface Characterization Techniques for Mapping Surface Composition, and Therefore Charge Nonuniformity

Technique	Capability	Best Spatial Resolution (nm)	Vacuum Required
AFM/SPM	Surface imaging, force imaging	10	No
ATR-FTIR	Chemical bonding, molecular structure	10,000	No
Auger	Surface analysis, high resolution depth profiling	100	Yes
FE-SAM	Similar to standard Auger	10	Yes
SEM/TEM	Imaging and elemental analysis	5	Yes
FE-SEM	High-resolution SEM	1.5	Yes
environmental SEM	Imaging in aqueous environments	10	No
ToF-SIMS	Surface analysis of polymer, organics	100	Yes
XPS/ESCA	Surface analysis of organic/inorganic molecules	10,000	Yes

In analyzing charge nonuniformity of aqueous colloids, the technique of rotational electrophoresis can examine with resolutions of less than 100 nm. Most of the techniques listed—atomic force microscopy (AFM), attenuated total reflectance fourier transform infrared spectroscopy (ATR-FTIR), field-emission scanning Auger microprobe (FE-SAM), scanning (or transmission) electron microscopy (SEM/TEM), time-of-flight secondary ion mass spectrometry (ToF-SIMS), x-ray photoelectron spectroscopy (XPS), or electron spectroscopy for chemical analysis (ESCA)—give specific surface chemical information, but cannot resolve less than 100 nm or work in aqueous environments.

the resolution of the technique is insufficient to examine nanoscale nonuniformity on the particle.

Atomic force microscopy (AFM) provides one possible route for measuring charge nonuniformity. Early AFM measurements suggested charge variations over the surface of purple membranes, although it could not be ascertained whether the variations were due to charge nonuniformity or random measurement error [20]. This is perhaps because AFM has several difficulties in measuring charge nonuniformity: (1) the particle is fixed, not Brownian, meaning that only a few orientations of the particle can be examined; (2) the electrostatic forces can be difficult to distinguish from other surface heterogeneities, including hydrophobic heterogeneity; (3) the effective size of the tip due to electrostatics often limits the spatial resolution of the technique. Nevertheless, AFM has been used to map polysaccharide nonuniformity on *Saccharomyces cerevisiae* yeast cells using a concanavalin A functionalized tip [21], as well as examining charge nonuniformity on very large (>500 μm) glass spheres [22]. Perhaps the most successful AFM mapping came recently when investigators used AFM to find 50–100-nm domains on micron-size particles, although the particles were dried [23]. It is likely that as skills and techniques improve, AFM will become a more viable method for measuring charge nonuniformity on colloidal particles.

Direct imaging of particles also provides a method for measuring charge nonuniformity. Direct elemental maps of individual particles [24] have used electron spectroscopy imaging (ESI) to determine the location of individual atoms in poly[styrene-*co*-(2-hydroxyethyl methacrylate)] latex. Whereas carbon and oxygen atoms were distributed fairly uniformly throughout the particle, sulfur (i.e., from sulfate or sulfonate groups) was missing from particle–particle contact areas, and potassium was mostly found in a thin outer particle layer. In a separate study, domains with either positive or negative potentials reaching a few hundred nanometers in size were found with scanning electric potential microscopy (SEPM) together with noncontact AFM images. Scanning electron microscopy (SEM) has been used to examine the adhesion of small particles to larger ellipsoids [25]. The method did not reveal any charge nonuniformity on the ellipsoids. Cryo-transmission electron microscopy has been used to examine nonuniformity in bacterial samples [26].

Perhaps the most popular method for measuring charge nonuniformity (induced or permanent dipoles) is the electro-optic method, for which the book written by Stoylov is an excellent introduction [27]. Often the particles examined are Rayleigh particles (i.e., small compared with the wavelength of light), and cylinders [28], and other shapes have been examined. Permanent dipole electro-optic effects (i.e., due to the charge nonuniformity examined in this chapter) get larger as the frequency of the applied electric field gets smaller, and so the permanent dipole is distinguishable from the induced. The electric field due to optical frequencies tends not only to be small in magnitude compared with the applied electric field in the system, but also for the optical frequencies that are usually higher than what a colloidal particle can effectively respond to in an electro-optic manner. When one knows the expression for the electro-optic effect (e.g., electric birefringence, electric dichroism, electric light scattering) and the steady-state orientation function of the particles, one can compute all electro-optic light intensities.

Electrophoretic rotation [29–31] or rotational electrophoresis is another useful method for measuring charge nonuniformity. In its original form, the method examined doublets consisting of two spheres (1 and 2) with different (but each uniform) zeta potentials. The angular velocity (Ω) for such a doublet can be predicted precisely using the electrokinetic equations as

$$\Omega = -\frac{\varepsilon(\zeta_2 - \zeta_1)}{\eta L} N E_0 \times \mathbf{e} \qquad (15.5)$$

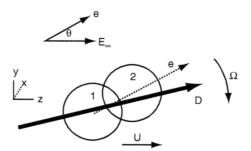

FIGURE 15.2 Schematic of doublet in an applied electric field. Note that the dipole (D) need not align with the doublet axis (e). (Figure originally found in Feick and Velegol, *Langmuir*, 18, 3454, 2002.)

where L is the center-to-center distance between the spheres, N is the electrophoretic rotation coefficient ($N = 0.64$ for rigid doublets consisting of touching, equal-sized spheres), E_0 is the applied external field (sometimes called E_∞), and **e** is the unit vector pointing from the center of sphere 1 to the center of sphere 2 see Figure 15.2.

Experimentally, one records the rotation of the doublet in an applied electric field (Figure 15.3). Standard image processing software finds the angles of the doublets as a function of time. For doublets consisting of two spheres with different (but uniform) zeta potentials, the analysis proceeds by fitting the angle versus time data to Equation 15.5. As experimental variables like the applied electric field, the solution viscosity and permittivity, and the particle sizes are known, along with $N = 0.64$ for rigid doublets, one can measure $\zeta_2 - \zeta_1$ for the doublet.

If $\zeta_2 - \zeta_1$ is fairly large (e.g., 20 mV), one could more easily take differences from translational electrophoresis. For example, if particle 1 has $\zeta = -50$ mV and particle 2 has $\zeta = -80$ mV, this difference could be taken from ZetaPals or similar measurements. However, this becomes difficult when the difference starts to approach the error in the measurement of a zeta potential, which is typically several millivolts. Then one can take small differences of large numbers. For example, if the two zeta-potentials were -50 and -51 mV, taking differences of measurements from translational electrophoresis measurements would likely give inaccurate results.

For particles with random charge nonuniformity, it is nearly impossible to get useful information from translational electrophoresis. A very useful method is to use a modification of the electrophoretic rotation technique described above. The modified technique, which solves the electrokinetic equations for a random charge distribution, is often called rotational electrophoresis. Limitations of the interpretation are that the variation of charge nonuniformity must occur over a length scale greater than the Debye length, and the electrical double layer must be thin when compared with the particle dimensions. The experiments proceed exactly as shown in Figure 15.3, and the interpretations used for random nonuniformity are the following equations instead of Equation 15.5:

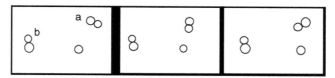

FIGURE 15.3 Schematic of doublet in an applied electric field. Note that the dipole (**D**) need not align with the doublet axis (**e**). (Figure originally found in Feick et al., *Langmuir*, 20, 3090, 2004.)

$$\langle U_i U_j \rangle = U_i^{\text{Smol}} U_j^{\text{Smol}} + \left(\frac{\varepsilon}{\eta}\right)^2 \frac{\sigma_\zeta^2}{N} E_k E_m T_{ikjm} \tag{15.6}$$

$$\langle \Omega_i \Omega_j \rangle = \left(\frac{\varepsilon}{\eta b}\right)^2 \frac{\sigma_\zeta^2}{N} E_k E_m W_{ikjm} \tag{15.7}$$

Indicial notation has been used to represent the vector translational (U) and rotational (Ω) velocities, and the fourth order tensors T and W depend on the geometry of the system and have been calculated [6]. The dimension (b) is of the semiaxis of the equivalent spheroid to a doublet.

The Smoluchowski velocity (U^{Smol}) gives the translational velocity due to the average zeta potential of the doublet. Equation 15.6 reveals why it is so difficult to observe charge nonuniformity from translational electrophoresis measurements. The first term completely dominates the second term (which is due to charge nonuniformity), and therefore the second term gets lost in the noise. But for the angular velocity the average zeta potential makes no contribution, regardless of the shape [32]. Thus, the leading order term is due to the charge nonuniformity, as given by Equation 15.7. This gives the angular velocity the sensitivity needed to measure charge nonuniformity at high resolution. It should be noted that for doublets with e perpendicular to E_0, Equation 15.7 has been simplified for the case of touching spheres of equal size:

$$\langle \Omega_x^2 \rangle = 0.884 \left(\frac{\varepsilon}{\eta a}\right)^2 \frac{\sigma_\zeta^2}{N} E_0^2 \tag{15.8}$$

where a is the sphere radius and Ω_x is the angular velocity visible in a microscope (see Figure 15.3).

One further enhancement to the rotational electrophoresis is related to the electro-optic technique. Charge nonuniformity light scattering (CNLS) examines doublet orientation with light scattering [13]. This method is currently under development, but it has simplified the rotational electrophoresis. Whereas rotational electrophoresis experiments and their interpretation can consume a day, CNLS can be done in minutes.

15.4 CHARGE NONUNIFORMITY FOR PARTICULAR COLLOIDAL SYSTEMS

Charge nonuniformity has been measured on polystyrene latex spheres, bacteria, viruses, and other particles. A summary of useful results from the electro-optics literature exists in the book by Stoylov in Table 8.4 (clay particles, typical values 10×10^{-25} C-m), 8.5 (dyes, typical values 10 to 1000×10^{-25} C-m), 9.1 (tobacco mosaic virus, typical values $< 1 \times 10^{-25}$ C-m), 9.2 (E. coli bacteria, typical values 1 to 100×10^{-25} C-m), 9.3 (purple membranes, typical values 1 to 10×10^{-25} C-m), and 9.4 (DNA, typical values $< 1 \times 10^{-25}$ C-m) [27]. Many of these measurements were taken with the electric light scattering method, although some were found with electric birefringence or electric dichroism.

Our own laboratory has used rotational electrophoresis to examine random charge nonuniformity on colloidal spheres (Figure 15.1). Of the hundreds of doublets consisting of 4.5-μm PS latex spheres that we examined, almost all rotated by electrophoresis, meaning that they were nonuniformly charged. For these particles in 1 mM KCl, the electrostatic interaction takes place primarily in a circular region about 300 nm in diameter near the point of closest approach, and there were about 1800 such regions on a particle. We estimated that the standard deviation in zeta potential among those regions was often as high as 60 mV (average for the whole particle was usually above 100 mM), but could be as high as 100 mV [7].

Obviously, when the standard deviations get so large, it is possible that our model breaks down as the charge is not likely to be truly random at that point. Nevertheless, the point is that the charge nonuniformity was significant for those particles.

We sought a method for altering the charge nonuniformity, and our first hypothesis was that surfactants or polyelectrolytes would reduce the charge nonuniformity [12]. This is because on those regions that are already highly charged, a like-signed surfactant would be unlikely to adsorb. To those regions that were not charged on the polymer, the surfactant would adsorb tail-down, leaving its charge outward and thus smoothing out the charge distribution. Figure 15.4 shows that this indeed happened. Those systems with no surfactant or polyelectrolyte remained nonuniformly charged, as did those with neutral (Triton X100) or submonolayer (1.6 nM PEI). However, all the others showed very little angular velocity, meaning that charge nonuniformity had been greatly reduced.

Two further questions were answered in subsequent research. First, could the charge nonuniformity on the polystyrene latex be reduced by raising the temperature to 130°C, well above the glass transition temperature (90°C) of PS latex? Upon conducting these experiments, we found that indeed the charge nonuniformity did decrease after 30 min of the high temperature, perhaps due to rearrangements of the charge groups at the surface [11]. Although the charge nonuniformity decreased further after 60 min of high temperature, most of the reduction occurred in the first 30 min.

Another important question is whether the doublets rotate due to random charge nonuniformity, or do the doublets simply have two uniform but different spheres together? This question is answered by examining the stationary angle to which doublets reach. If the zeta potentials on sphere 1 and sphere 2 are uniform but different, the doublet will rotate until $\mathbf{e}$ is parallel with E_0. However, if the charge is random, the dipole ($\mathbf{D}$) will not be parallel with $\mathbf{e}$, and so $\mathbf{e}$ will point to some other direction than E_0 (see Figure 15.2). Experiments revealed that many doublets did in fact rotate such that $\mathbf{e}$ was not parallel with E_0; however, those doublets that rotated very fast did have $\mathbf{e}$ parallel with E_0. Thus, the assumption of randomness is often but not always true.

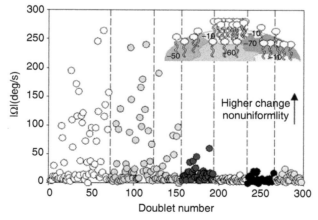

FIGURE 15.4 Charge nonuniformity with various additives in solution. (A) bare (IDC batch 1314,1), (B) with 1 mM Triton X100, (C) 0.16 nM PEI, (D) 1.6 nM PEI, (E) 10 μM NaPSS, (F) 10 mM SDBS, (G) 10 μM CTAB. Those letters for which large angular velocities were measured show significant charge nonuniformity. Notice that surfactants and polyelectrolytes often reduce the charge nonuniformity greatly. The inset shows our picture of how surfactants reduce charge nonuniformity. (Figure originally in Feick et al., *Langmuir*, 20, 3090, 2004.)

It is useful to compare the dipole moments listed in Stoylov's book [27] with the zeta potential nonuniformity measurements from our own studies. In order to do this, I will relate everything back to a common expression for zeta potential on a doublet (I could also have used a sphere [33]). Equation 15.5 gives the angular velocity of a doublet, given the two zeta-potentials of the constituent spheres. Equation 15.8 gives a similar expression for a randomly charged doublet. Equating the square root of Equation 15.8 with Equation 15.5 gives the "equivalent doublet" that would give the same angular velocity as the random case, if in fact the doublet consisted of two different (but each uniform) spheres:

$$(\zeta_2 - \zeta_1)_{\text{eff}}^{\text{rot-ep}} = 2.938 \frac{\sigma_\zeta}{\sqrt{N}} \tag{15.9}$$

As the electro-optic measurements give the dipole (p, in units C-m), we can calculate a torque by multiplying by E_0, to obtain $T = pE_0$. The resulting angular velocity can then be computed using the well-known result for torque on a rotating sphere ($T = \pi \eta a^3 \, \Omega$) to obtain

$$\Omega = \frac{pE_0}{8\pi\eta a^3} \tag{15.10}$$

Equating to Equation 15.5 gives

$$(\zeta_2 - \zeta_1)_{\text{eff}}^{\text{electro-optic}} = \frac{0.124p}{\varepsilon a^2} \tag{15.11}$$

The rotational electrophoresis experiments gave typical values of $\sigma_\zeta/\sqrt{N} = 0.3$ to $3\,\text{mV}$ for polystyrene latex spheres, giving $(\zeta_2 - \zeta_1)_{\text{eff}}^{\text{rot-ep}} = 1$ to $10\,\text{mV}$. The values listed in Ref. [27, Table 8.4] for clays are dipoles of 10^{-25} to $10^{-24}\,\text{C-m}$, for $a = 0.5\,\mu\text{m}$. The resulting $(\zeta_2 - \zeta_1)_{\text{eff}}^{\text{electro-optic}} = 0.07$ to $0.7\,\text{mV}$. Thus, the two measurements give values at least within an order of magnitude.

15.5 CONSEQUENCES OF CHARGE NONUNIFORMITY

Charge nonuniformity has a significant impact on colloidal interactions, and this has been studied mostly through theory [34–37]. These different papers had different conclusions about the importance of charge nonuniformity on interparticle forces and stability. One difficulty was that they used different parameters to quantify the charge nonuniformity.

Our lab group related interparticle forces, in a statistical manner, to charge nonuniformity. We assumed that the Derjaguin approximation held, such that the gap between the spheres was much smaller than the size of the nonuniformly charged regions on the sphere. For two interacting spheres of radius (a) and zeta potential (ζ), separated by a gap (δ), we found for $\kappa\delta > 2$ that [4]

$$\langle\Phi\rangle \approx 2\pi\varepsilon a\zeta_0^2 e^{-\kappa\delta} \tag{15.12}$$

$$\left(\frac{\sigma_\Phi}{\langle\Phi\rangle}\right)^2 = \frac{4\kappa a}{N}\left[2\left(\frac{\sigma_\zeta}{\zeta_0}\right)^2 + \left(\frac{\sigma_\zeta}{\zeta_0}\right)^4\right] \tag{15.13}$$

where ζ_0 is the area-averaged zeta potential on the particles, $<\Phi>$ is the ensemble average potential of mean force between the particles, σ_ζ is the standard deviation of zeta potentials over the N regions on each sphere, and σ_Φ is the ensemble standard deviation of potential of mean force between the particles. If $\sigma_\zeta \ll \zeta_0$, then we can ignore the second term of Equation 15.13. What remains has two interesting aspects: (1) we are left with an expression that contains σ_ζ^2/N, which is exactly the parameter that we obtain from rotational electrophoresis measurements; (2) the ratio $|\sigma_\Phi/<\Phi>|$ increases with κa. That is, for larger spheres or smaller Debye lengths, the impact of charge nonuniformity is more important for instability of the colloidal systems. Whereas some orientations of the two spheres will give larger repulsive forces, other regions will give lower repulsive forces. It is these latter, low-repulsive valleys that can allow two spheres to slip into a permanent primary energy minimum between the particles (due to van der Waals attractive forces).

15.6 OPPORTUNITIES IN CONTROLLED CHARGE NONUNIFORMITY

So far we have mostly examined random charge nonuniformity on particles. As previously shown, random charge nonuniformity rarely provides any benefit. We have developed two applications that can benefit from controlling the charge nonuniformity on particles. One was shown in Figure 15.4, where charge nonuniformity was reduced in the presence of various surfactants and polyelectrolytes. However, if we could reduce charge nonuniformity during the synthesis stage—even if the process costs a bit more—we might reduce or eliminate the need for steric or electrosteric dispersants [38] in controlling the colloidal stability after the synthesis.

A second application is the development of "colloidal molecules" [39]. Colloidal molecules are a bottom-up assembly set of particles with designed shape, size, and material properties. For example, whereas water consists of an oxygen atom and two hydrogen atoms—with a fixed bond angle, fixed bond energies, fixed molecular mass, fixed properties in general—we might imagine creating a colloidal assembly out of one silica particle with two polymer particles hanging off it in a designed orientation (Figure 15.5b).

In order to create an assembly like that shown in Figure 15.5a, one can follow the method briefly shown in Figure 15.5c, which we call "particle lithography." Positively charged particles are brought to the surface (e.g., by gravity), where it adheres to a negatively charged glass Petri dish. Now a polyelectrolyte is introduced, which covers the entire surface of the

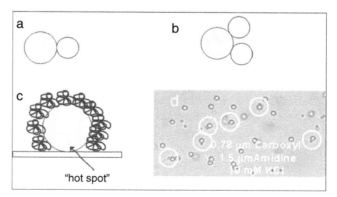

FIGURE 15.5 A schematic of "colloidal molecules." (a) A simple dimer "molecule," with an image from our lab shown in (d). (b) Schematic of a "water molecule." (c) The essential "particle lithography" step of creating these assemblies. A particle (here positive) is brought to a substrate (usually negative). A polyelectrolyte is introduced into the system to cover the entire sphere, except for the region where the substrate geometrically excludes the polyelectrolyte. That region gives a "hot spot." While the rest of the particle is now negative due to the polyelectrolyte, the hot spot remains positive.

particles, except for the region where the Petri dish excludes the polyelectrolyte from reaching. When the polyelectrolyte in solution is then washed out and the particles are sonicated off the surface, one has a particle covered by 98% negative polyelectrolyte, with 2% of the area (i.e., the "hot spot") having the original particle charge (here, positive). That precursor particle can be now readily aggregated with another negatively -charged particle, which will stick only to the hot spot. It should be noted that we almost never see large or random aggregates during these steps, due to the particle lithography method.

15.7 CONCLUSION

We have examined the measurement and consequences of charge nonuniformity in colloidal systems. Two key techniques for measuring charge nonuniformity are the electro-optic technique and the rotational electrophoresis technique. Random charge nonuniformity on particles is almost always detrimental to a process, and it has been shown that one way to reduce random charge nonuniformity is by introducing surfactants or polyelectrolytes into the sample. Indeed, it often happens that while these additives do not change the average zeta potential significantly, which is often thought to be one way that these molecular dispersants operate, the additives do greatly reduce the charge nonuniformity. A further examination of charge nonuniformity might lead not only to better synthesis methods that avoid it, but also to the creation of sophisticated, bottom-up assembled "colloidal molecules."

ACKNOWLEDGMENT

The author thanks many students who have worked with him to measure charge nonuniformity in his laboratory and to develop the techniques. Most especially he thanks Dr. Jason Feick. In addition, the author thanks the National Science Foundation for funding this work through CAREER Grant # CTS-9984443 and CRAEMS Grant CH3-0089156. Finally, the author thanks Professor John L. Anderson for many exciting discussions concerning this topic.

REFERENCES

1. Russel, W.B., Saville, D.A., and Schowalter, W.R. *Colloidal Dispersions*, Cambridge University Press, New York, 1989 (corrections 1991), chap. 8. [Much of the introduction to this chapter can be found in this now classic text, and the original references can be found there.]
2. Behrens, S.H., Borkovec, M., and Schurtenberger, P. Aggregation in charge-stabilized colloidal suspensions revisited, *Langmuir*, 14, 1951, 1998.
3. Behrens, S.H., Christl, D.I., Emmerzael, R., Schurtenberger, P., and Borkovec, M. Charging and aggregation properties of carboxyl latex particles: experiments versus DLVO theory, *Langmuir*, 16, 2566, 2000.
4. Velegol, D. and Thwar, P.K. Analytical model for the effect of surface charge nonuniformity on colloidal interactions, *Langmuir*, 17, 7687, 2001.
5. Velegol, D., Feick, J.D., and Collins, L. Electrophoresis of spherical particles with a random distribution of zeta potential or surface charge, *J. Colloid Interface Sci.*, 230, 114, 2000.
6. Feick, J.D. and Velegol, D. The electrophoresis of spheroidal particles having a random distribution of zeta potential, *Langmuir*, 16, 10315, 2000.
7. Feick, J.D. and Velegol, D. Measurements of charge nonuniformity on polystyrene latex particles, *Langmuir*, 18, 3454, 2002.
8. Velegol, D. Electrophoresis of randomly-charged particles, *Electrophoresis*, 23, 2023, 2002.
9. Velegol, D and Feick, J.D. Evaluating randomness of charge distribution on colloidal particles using stationary electrophoresis angles, *Langmuir*, 19, 4592, 2003.

10. Jones, J.F., Feick, J.D., Imoudu, D., Chukwumah, N., Vigeant, M., and Velegol, D. Oriented adhesion of *E. coli* to polystyrene particles, *Applied Environ. Microbiol.*, 69, 6515, 2003.

11. Feick, J.D. and Velegol, D. Reducing surface charge nonuniformity on individual particles through annealing, *Ind. Eng. Chem. Res.*, 43, 3478, 2004.

12. Feick, J.D., Chukwumah, N., Noel, A.E., and Velegol, D. Altering surface charge nonuniformity on individual colloidal particles, *Langmuir*, 20, 3090, 2004.

13. Jones, J.F., Holtzer, G.L., Snyder, C., Yake, A.M., and Velegol, D. Charge nonuniformity light scattering, *Colloids Surf. A*, 267, 79, 2005.

14. Koopal, L.K. Mineral hydroxides: from homogeneous to heterogeneous modeling, *Electrochima Acta*, 41, 2293 (1996).

15. Hiemstra, T., Van Riemsdijk, W.H., and Bolt, G.H. Multisite proton adsorption modeling at the solid/solution interface of (hydr)oxides: a new approach. I. Model description and evaluation of intrinsic reaction constants, *J. Colloid Interface Sci.*, 133, 91, 1989.

16. Hiemstra, T., De Wit, J.C.M., and Van Riemsdijk, W.H. Multisite proton adsorption modeling at the solid/solution interface of (hydr)oxides: a new approach. II. Application to various important (hydr)oxides, *J. Colloid Interface Sci.*, 133, 105, 1989.

17. Fitch, R.M. *Polymer Colloids*, Academic Press, New York, 1997.

18. Dobrynin, A.V., Rubenstein, M., and Obukov, S.P. Cascade of transitions of polyelectrolytes in poor solvents, *Macromolecules*, 29, 2974, 1996.

19. Shapiro, L., McAdams, H.H., and R. Losick. Generating and exploiting polarity in bacteria, *Science*, 298, 1942, 2002.

20. Butt, H-J. Measuring local surface charge densities in electrolyte solutions with a scanning force microscope, *Biophys. J.*, 63, 578, 1992.

21. Gad, M., Itoh, A., and Ikai, A. Mapping cell wall polysaccharides of living microbial cells using atomic force microscopy, *Cell Biol. Int.*, 21, 697, 1997.

22. Schellenberger, K and Logan, B.E. Effect of molecular scale roughness of glass beads on colloidal and bacterial deposition, *Environ. Sci. Technol.*, 36, 184, 2002.

23. Tan, S., Sherman, R.L. Jr., Qin, D., and Ford, W.T. Surface heterogeneity of polystyrene latex particles determined by dynamic force microscopy, *Langmuir*, 21, 43, 2005.

24. Cardoso, A.H., Leite, C.A.P., and Galembeck, F. Elemental distribution within single latex particles: determination by electron spectroscopy imaging. *Langmuir*, 14, 3187, 1998.

25. Ho, C.C. and Ottewill, R.H. Investigation of the charge distribution of ellipsoidal particles, *Colloids Surf. A*, 141, 29, 1998.

26. Matias, V.R.F., Al-Amoudi, A., Dubochet, J., and Beveridge, T.J. Cryo-transmission electron microscopy of frozen-hydrated sections of *Escherichia coli* and *Pseudomonas aeruginosa*, *J. Bacteriol.*, 185, 6112, 2003.

27. Stoylov, S.P. *Colloid Electro-Optics*, Academic Press, New York, 1991.

28. Stoimenova, M.V. Electric light scattering by cylinder-symmetrical particles, *J. Colloid Interface Sci.*, 53, 42, 1975.

29. Fair, M.C. and Anderson, J.L. Electrophoresis of dumbbell-like colloidal particles, *Int. J. Multiphase Flow*, 16, 663, 1990; corrigenda *Int. J. Multiphase Flow*, 16, 1131, 1990.

30. Fair, M.C. and Anderson, J.L. Electrophoresis of heterogeneous colloids: doublets of dissimilar particles, *Langmuir*, 8, 2850, 1992.

31. Velegol, D., Anderson, J.L., and Garoff, S. Probing the structure of colloidal doublets by electrophoretic rotation, *Langmuir*, 12, 675, 1996.

32. Morrison, F.A. Jr. Electrophoresis of a particle of arbitrary shape, *J. Colloid Interface Sci.*, 34, 210, 1970.

33. Anderson, J.L. Effect of nonuniform zeta potential on particle movement in electric fields, *J. Colloid Interface Sci.*, 105, 45, 1985.

34. Miklavic, S.J., Chan, D.Y.C., White, L.R., and Healy, T.W. Double layer forces between heterogeneous charged surfaces, *J. Phys. Chem.*, 98, 9022, 1994.

35. Grant, M.L. and Saville, D.A. Electrostatic interactions between a nonuniformly charged sphere and a charged surface, *J. Colloid Interface Sci.*, 171, 35, 1995.

36. Holt, W.J.C. and Chan, D.Y.C. Pair interactions between heterogeneous plates and spheres, *Langmuir*, 13, 1577, 1997.
37. Stankovich, J. and Carnie, S.L. *J. Colloid Interface Sci.*, 216, 329, 1999.
38. Napper, D.H. *Polymeric Stabilization of Colloidal Dispersions*, Academic Press, New York, 1984.
39. Snyder, C.E., Yake, A.M., Feick, J.D., and Velegol, D. Nanoscale functionalization and site-specific assembly of colloids by particle lithography, *Langmuir*, 21, 4813, 2005.

16 A Study of the Anomalous Kerr Effect on Dispersions of Clays in the Presence of Excess Salt and Water-Soluble Amphiphilic Additives

P. Schmiedel, S. Holzheu, and H. Hoffmann

CONTENTS

Transient electric birefringence measurements were performed with the normal- and reversing-pulse method. The measurements were carried out on dispersions of saponite, a synthetic clay mineral, in order to obtain new information about the anomalous Kerr effect. Both the dilute and the semidilute concentration ranges were investigated without and with added salt and with water-soluble amphiphilic additives. In the dilute dispersion the anomaly is not observable neither in the buildup of the birefringence nor in the decay with the single-pulse method. Signals with the reversing-pulse method, however, reveal that even in this concentration regime a second process with opposite sign exists besides the main process.

Two processes with opposite signs are found in the semidilute regime without added salt. The positive one, which is attributed to the alignment of single particles with an induced dipole moment, prevails at higher field strengths. It remains unchanged upon reversing the field. At lower field strength the negative process predominates. It is attributed to the alignment of the clay particles perpendicular to the electric field. This negative process decays and rebuilds, however, provoking a dip or a hump in the birefringence signal upon reversing the field. This process can be explained according to a theory of Yamaoka [14] with the presence of an ion-fluctuation dipole moment. On addition of excess salt, the anomaly first increases, and then decreases and disappears at higher ionic strengths. Here a third process with the sign of the first one appears in the buildup and the decay of birefringence. This process is also very sensitive to the reversing of the field. It decays rapidly after changing polarity, and rebuilds.

The anomalous signal in the semidilute concentration region was also studied as a function of amphiphilic polymers that adsorb on the clay particles. For transient measurements with the single-pulse method the anomaly disappeared and, on saturation of the clays with adsorbed polymers, a single exponential signal is observed that is due to the alignment of the particles parallel to the electric field. Measurements with the double-pulse method showed, however, that the total amplitude of the signal consists of two parts, one coming from an induced dipole and one from a seemingly permanent dipole. On field reversal, the birefringence signal therefore dips down and then recovers to the same value that it had reached during the first pulse. On turning off the field, the total signal relaxes with a single time constant back to zero. It is concluded that the rotation of the apparently permanent dipole of the platelets from perpendicular to parallel with increasing polymer adsorption is due to the immobilization of the counterions on the clay surface by the adsorption of the polymers.

In summary it can be concluded that the total signal that is observed in single-pulse measurements has two contributions. One that is due to an induced dipole moment, and one that is due to an apparent dipole moment. The apparent dipole in the salt-free solution is perpendicular to the clay particles, although with the adsorbed polymers or surfactants, the dipole is parallel to the clay particle surface. It is likely that the different signals come from the particles in the sample, which monitor different microenvironments.

16.1 INTRODUCTION

For several years the anomalous Kerr effect is a matter of controversial discussions. The effect is observed in the colloidal dispersions at the overlap concentration of the particles. In a transient electric birefringence experiment, the simultaneous existence of at least two processes with opposite signs is observed. In the case of the saponite-system dealt with in this study, the first process has a positive sign and the second process, a negative sign, which remains constant for a longer time. The second effect is competing with the first one. Under certain conditions, a third and even a fourth process with the sign of the first one can be found [1,2]. The superposition of the birefringence of all processes can lead to very complicated signals.

The anomalous Kerr effect was observed a long time ago on bentonite dispersions [3]. Other investigations exist on solutions of polymers of biological origin [4], on surfactant systems [1,2,5], on polyelectrolyte solutions [6], on polytetrafluoroethylene (PTFE) fibers [7], and also on saponite dispersions [8] as in the present study, but these investigations were based on the single-pulse method only.

Different theoretical approaches for the molecular interpretations of the effect have been made [9–11]. Thurston, for instance, assumed the presence of a permanent dipole perpendicular to the induced one. With this hypothesis, it is possible to explain the experimental data of the buildup of birefringence in a consistent way. Nevertheless this theory cannot explain the fact that the two processes with different signs also appear in the field-off decay of the birefringence. In this case one would have to expect a simple monoexponential decay because both processes are brought about by the same type of particle (one-particle process). This theory cannot explain changes in the permanent dipole due to the addition of salt or adsorbing amphiphiles. Moreover, the molecular reason for a permanent dipole of a highly symmetric aggregate, e.g., rodlike micelles, is not clear. For this reason, Cates proposed in a recent publication a theoretical model that gets along without the requirement of a permanent dipole [12]. Cates postulates the presence of clusters of particles with a local disklike order. These clusters are aligned transverse to the electric field because their polarizability perpendicular to the particles' axis is larger than the one in the direction of the axis.

The reversing-pulse electric birefringence (RPEB) technique seems to be a suitable experimental technique to clarify this controversy. In this method a rectangular positive electric field pulse is applied to the sample, and, without any delay a negative one is applied. The birefringence signal is detected during this time. The first pulse aligns the particles regardless of the type of dipole moment and builds up the birefringence. The birefringence that is due to the particles with an induced dipole should not be affected by the change of polarity because the particles do not have to turn around with the field. Only their induced dipole moment is reversed and this can occur by a small movement of the counterions and the resulting displacement of the charge centers. A permanent dipole, however, has to turn around with the field and has to rotate 180°. Therefore, one would expect more or less a decay and a rebuild of the birefringence after the reversing of polarity of the field. If the two types of dipole moments appear simultaneously—as it probably happens in the case of the anomalous Kerr effect—the two processes can be separated by suitable fit functions and the contributions of the permanent and induced dipoles can be determined.

The theory of the reversing-pulse method is well developed. On the basis of the theory developed by Szabo et al. [13], Yamaoka et al. [14] succeeded in deriving analytical expressions for the signals of an RPEB experiment. These authors postulate the presence of two electric dipole moments of the particle. One, the ion-fluctuation dipole, is due to ion-atmosphere distribution along the longitudinal axis of the particle caused by diffusion motion of the ions in the ion-atmosphere and consequently the displacement of the charge centers of the particles with the charge centers of the counterions. The other is due to instantaneously field-induced polarizability anisotropy.

Resolving the rotational diffusion equation, they calculate the signals that depend strongly on a parameter q. This parameter denotes the ratio of the ion-atmosphere polarizability, which is responsible for the ion-fluctuation dipole moment, to the polarizability anisotropy ($q = 0$: pure induced dipole; $q = \infty$: pure ion-fluctuation dipole). At different values of q one gets either a hump or a dip at the instant of reversing polarity. Recently, this theory was further extended to the particles with permanent dipole moments in the direction of the symmetric axis in addition to the previously considered dipole moments [15].

Unfortunately these theories failed to explain both the two processes in the decay of the birefringence and the third process that occurs under certain conditions. Also the application of these theories is strongly limited to the Kerr law region at very dilute systems.

Experimental investigations with the reversing-pulse method exist on polystyrene latex dispersions [16].

16.2 EXPERIMENTAL

For reversing-pulse measurements, we used an electric birefringence apparatus that has already been described in previous publications [17]. He–Ne laser was used as a light source. The light beam—the polarization of which is improved by a Glan–Thompson prism—passes the measuring cell between two steel electrodes. Another Glan–Thompson prism serves as an analyzer. The light passing through the analyzer is detected by a photomultiplier. Most of the measurements were performed with the so-called square detection. To determine the sign of the birefringence a quarter wave-device is used. This component is also necessary for the de Senarmont optic, which is more sensitive and allows the detection of the sign. Some signals have been monitored using this geometry.

For the generation of the electric field pulses we use two Cober high-voltage pulse generators. The outputs of these devices are not grounded. Therefore it is possible to connect the positive pole of one Cober with the negative pole of the other. One of these connections is grounded and attached to one electrode of the cell; the other connection is attached to the other electrode. If the falling flank of the trigger output of the first (positive) Cober is used to trigger the second (negative) Cober we get our reversing pulse. The change of polarity requires less than 1 μs. In order to check the voltage, the symmetry and the correct form of the pulses can be observed on an oscilloscope equipped with 1:1000 probes.

Our saponite samples are a gift from the Hoechst AG. Saponite is a synthetic layered silicate with the formula $[Mg_3(Si_{3.7},Al_{0.3})O_{10}(OH)_2]Na_{0.3}$. In solution it disperses into disklike particles with a diameter of ca. 500 Å and a thickness of 10 Å. Due to the dissociation of the Na+ ions the particles are negatively charged with a surface charge density of $e/150\,Å^2$. We investigated solutions with two different concentrations: a dilute one with 0.01 wt% and a semidilute one with 0.7 wt%. At these concentrations the rotational volumes of the particles begin to overlap. To study the dependence of the ionic strength on the signals different amounts of NaBr were added.

16.3 RESULTS

16.3.1 DILUTE REGIME

16.3.1.1 Samples without Addition of Salt

Generally the anomalous Kerr effect is found only in a very small concentration range around the overlap concentration where the ion-atmospheres of the particles begin to interact. Our results demonstrate, however, that even in the dilute regime two processes with different signs can be detected with the reversed-pulse technique because of its high sensitivity to the values of the electric parameters of the particle. Looking only at the first pulse and the decay of birefringence, we cannot identify these two processes.

In Figure 16.1, a sequence of birefringence measurements on the dilute solution of 0.01 wt% without salt at different field strengths is represented. The thin solid lines connect the measuring points whereas the bold lines are the fit functions. The dashed vertical lines mark the reversing of polarity.

The measurements show a buildup of the birefringence during the positive pulse. After the reversing of polarity the birefringence does not decay as one would expect, but it rises, passes over a maximum, and levels out to the same value as for the first pulse. This is an interesting result. Looking only at the signal of the first pulse one cannot see that a process with different sign occurs. The reversing, however, clearly indicates the presence of the negative process. After the change of polarity, the negative process decays whereas the positive one remains almost constant. Then the negative process rebuilds, competing with the positive one it leads

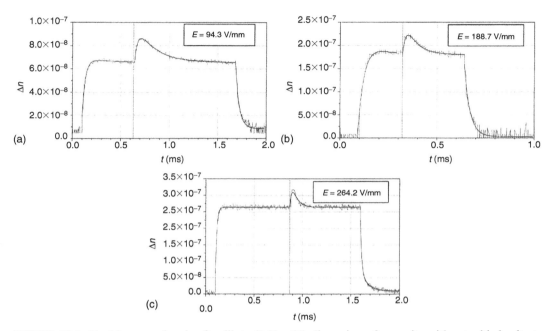

FIGURE 16.1 Birefringence signals of a dilute (0.01 wt%) dispersion of saponite without added salt at different field strengths. The dashed lines in the graphs (a)–(c) mark the time of reversing of polarity. The signals are monitored with square detection. In the buildup and in the decay only a positive process can be seen, but the reversing reveals the presence of a negative process.

to the same value of birefringence as before. This qualitative description is supported by the fit functions. In principle, it is possible to fit the buildup of the birefringence of the first pulse with a single relaxation time, but the fit with two times τ_1 and τ_2 and two amplitudes A_1 and A_2

$$\Delta n = A_1 \exp(t/\tau_1) - A_2 \exp(t/\tau_2) \tag{16.1}$$

with opposite sign is considerably better for the 94.3 V/mm signal. At higher field strengths the A_2-process fades and the 264.2-V/mm signal can be fitted monoexponentially. (In the following we call the fast positive process related with the amplitude A_1 as the A_1-process and the slower negative process, the A_2-process.)

The part of the signal on reversing the field is fitted with the following function:

$$\Delta n = A_1 + A_2 \exp(t/\tau_{2d}) - A_2 \exp(t/\tau_{2r}) \tag{16.2}$$

This means, as described above, that the A_1-process remains constant; the A_2-process, however, decays with the time constant τ_{2d} and rises again with τ_{2r}, causing a maximum of the birefringence.

The time constants τ_1 and τ_2, represented in Figure 16.1b, decrease with the increasing field strength. This is not surprising because the torque, and therefore the aligning effect, enlarges with increasing field strength, whereby the decay time τ_{2d} is less affected than the rise time τ_{2r}.

In the case of the lower field strengths the decay of birefringence can be fitted with a monoexponential function with time constants of 0.035 ± 0.005 ms. For the highest field strength only a biexponential function with $\tau_1 = 0.013 \pm 0.002$ ms and $\tau_2 = 0.064 \pm 0.006$ ms function yields a satisfactory fit. With the decay times of the signals, which are independent of field strength if not chosen too high, the length of the particles can be estimated according to the equation

$$l \approx \frac{18 k_B T \cdot \tau_d^3}{\pi \eta} \tag{16.3}$$

where τ_d denotes the decay time and η denotes the solvent's viscosity. One obtains $l = 940 \pm 50\,\text{Å}$.

Obviously, both the positive and the negative processes decay with the same time constant. Therefore they cannot be distinguished any more. This result suggests that the same particles are responsible for both the positive and the negative processes (one-particle process). We will discuss this aspect later (see section 16.4).

16.3.1.2 Samples with Excess Salt

On adding salt in the dilute solutions the anomaly can even be found in the first pulse. In Figure 16.2 the birefringence signal of a 0.01% solution with 10 mM NaBr is represented. The signal is monitored with the de Senarmont geometry, which has the advantage that the sign of the birefringence can directly be seen. Owing to its higher sensitivity the noise also is amplified.

In the first part of the signal the positive process and subsequently the negative process can be seen. In this case the amplitude of the negative process is notably larger than the amplitude of the positive one. Consequently the birefringence gets negative. On reversing the polarity the negative process decays causing a positive peak and thereafter rebuilds to the same value as in the first pulse. Analogous to the solutions without salt, the two processes cannot be distinguished in the decay of the birefringence. In comparison to the samples without salt, the total amplitude of the birefringence is remarkably smaller.

16.3.2 SEMIDILUTE REGIME

In the following section we present measurements on solutions at the overlap concentration. Overlap concentration can be defined as the concentration at which the smallest spheres,

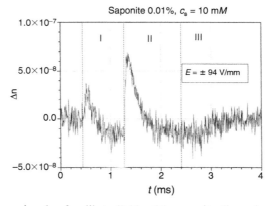

FIGURE 16.2 Birefringence signals of a dilute (0.01 wt%) saponite dispersion with 10 mM added salt. The signal is monitored with the de Senarmont optic. Here even in the buildup of the birefringence a negative process can be detected. The roman numbers in the plot mean: I, positive pulse; II, negative pulse; III, field-off decay.

which can be drawn around a clay particle to contain the particle, begin to overlap. The particles can still more or less freely rotate at this concentration.

Generally in this range the anomalous Kerr effect is most evident. This leads to the complicated looking signals, under certain conditions, that we would like to elucidate; a third process appears in addition to the two effects that are always present. This complicates the evaluation of the signals because at least six parameters are necessary to fit the results. The obtained fits under these conditions are not as accurate as in the dilute regime.

16.3.2.1 Samples without Added Salt

A sequence of birefringence signals of a 0.7% saponite solution at different field strengths (18.9, 56.6, 105.7, 179.2 V/mm) is represented in Figure 16.3. The signals a–c in the figure are monitored with the square detection, which does not allow the discrimination of the sign of the birefringence. Hence the signs are written into the plots. The roman numbers in the figure have the following meaning: I, positive pulse; II, negative pulse; III, decay of birefringence.

In graph (a), one can see only the negative A_2-process. The A_1-process is still too weak to be detected and a monoexponential fit is possible. In this context it is noteworthy that the buildup of the birefringence does not start immediately when the positive pulse is applied but with a certain delay in a kind of sigmoidal curve. This fact suggests that permanent dipoles are responsible for the A_2-process. Even on reversing the polarity, the A_1-process is hardly visible. The birefringence decays to zero before the negative process rebuilds even though the best fit is obtained, if a small amplitude A_1 is already taken into account. The decay of birefringence in graph (a) can be fitted monoexponentially. In graph (b), the fast

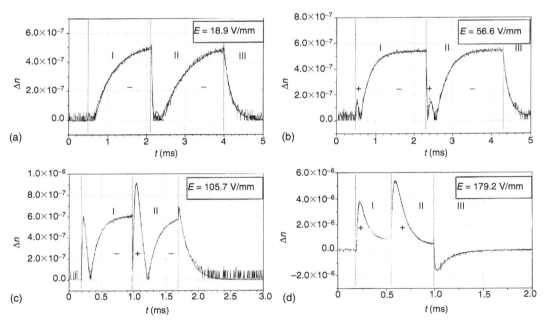

FIGURE 16.3 A series of birefringence signals of a semidiluted (0.7 wt%) dispersion of saponite without added salt at different field strengths. The signals (a)–(c) are monitored with the square detection; signal (d) is registered with the de Senarmont geometry. The signs of the signals are written in the graphs. At low field strengths the negative process dominates, on increasing the field strength, however, the positive process overtakes the negative one.

A_1-process is clearly visible even in the first pulse. Upon reversing polarity the positive peak emerges. This means that the A_1-process grows relative to the A_2-process. The decay can still be fitted monoexponentially. Consequently both processes still decay with the same time constant and are probably brought about by the same type of particles as in the case of the dilute solutions. This changes with higher field strength. In graph (c), a negative peak can be seen at the beginning of the decay interval. The positive A_1-process decays faster than the negative A_2-process. The magnitude of the signal therefore increases for a moment before the A_2-process vanishes too and the birefringence fades away. This fact can be considered as a proof that two different coexisting species are responsible for the processes with different signs (two-particle process). Without this peak in the decay, it could be assumed that during the buildup of birefringence, in the beginning the particles are aligned in field direction, producing a positive birefringence and then collectively turn around perpendicular to the field direction, causing the negative birefringence. In this case, however, the decay would have to be monoexponential because only one orientation of particles would exist in stationary conditions at the end of the pulse. In our case, however, at the beginning of the decay, two more or less independent species with different orientations exist and decay with different time constants causing the peak in the beginning of the decay interval.

The buildup of the birefringence in graph (c) is qualitatively equal to that in graph (b). The difference is, however, that the A_1-process is further increased. This tendency is continued in graph d where the A_1-process has grown so much that the birefringence does not become negative anymore. Also the decay evidences the two processes, which show quite different time constants.

16.3.2.2   Samples with Excess Salt

In the following section, we will study the influence of excess salt on the RPEB signals. Salt causes a reduction of the Debye length and therefore a weakening of electrostatic repulsion forces. This has, as demonstrated below, a significant influence on the electro-optical behavior.

At low salt concentrations up to $7\,mM$ one does not find any important change in the signals. Only the values of amplitudes and time constants slightly differ. In Figure 16.4, a series of birefringence signals of the 0.7% solution with $7\,mM$ of added NaBr is represented. Qualitatively the signals have the same form as those without salt. Both the amplitude of the positive process A_1 and the negative process A_2 are bigger than in the case without salt. As mentioned above, this is surprising, bearing in mind that in micellar systems the A_2-process decreases on the addition of salt. A further notable difference can be found in the decay of birefringence. Whereas at low field strengths the decay is monoexponential (graphs a and b) and at intermediate field strengths the peak at the beginning of the decay can be found (graphs c–f), contrary to the solutions without salt this peak vanishes again at high field strengths (graphs g and h). Here the decay proceeds with a very short time constant.

At a salt concentration above $11.5\,mM$ the signals begin to differ significantly from those without salt. In Figure 16.5, a series of signals on a 0.7% saponite solution with $11.5\,mM$ NaBr is represented. At low voltages (graphs a and b), the negative process still prevails, but contrary to the solutions with less salt the positive one can also be found and the two processes decay with different time constants provoking the decay peak. The amplitude A_2 has become much smaller than in Figure 16.3. In graph (c), an interesting development takes place. Contrary to the solutions without salt the birefringence is positive during the whole time and in the moment of reversing polarity the birefringence does not rise and pass over a maximum, but it decreases and passes through a minimum. In the case of the solutions with less salt, we observed a hump always at this position (see Figure 16.4g with $7\,mM$ salt or Figure 16.3d without salt) because we assumed that the negative A_2-process decays whereas the positive A_1-process remains constant. In the current case, obviously one part of the

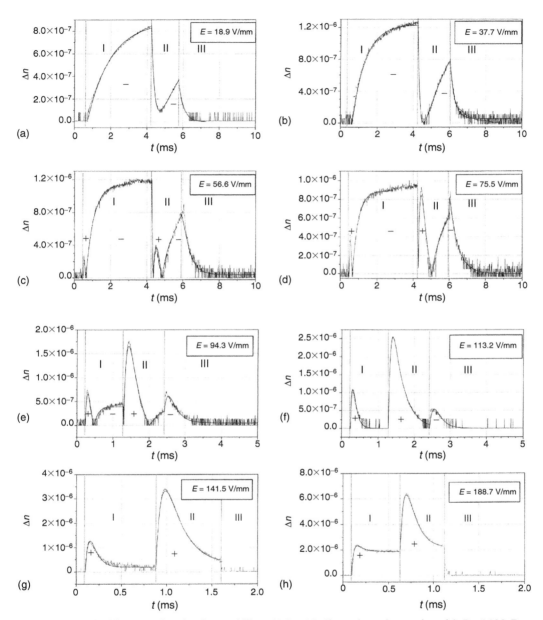

FIGURE 16.4 Birefringence signals of a semidilute (0.7 wt%) dispersion of saponite with 7 mM NaBr at various field strengths.

positive birefringence also decays. It decays even faster than the A_2-process. This means that now another positive process appears, even though it is not yet visible in the first part of the signal that can still be fitted with two amplitudes and two time constants. The dip after the reversing of polarity, however, proves its presence. We will call this third process the A_3-process. It also appears on surfactant systems on the addition of salt [3]. In graph (d), the A_3-process is clearly visible, both before and after the reversing. Obviously the A_3-process is sensitive to the reversing of polarity and decays before rebuilding, whereas the A_1-process is still not affected.

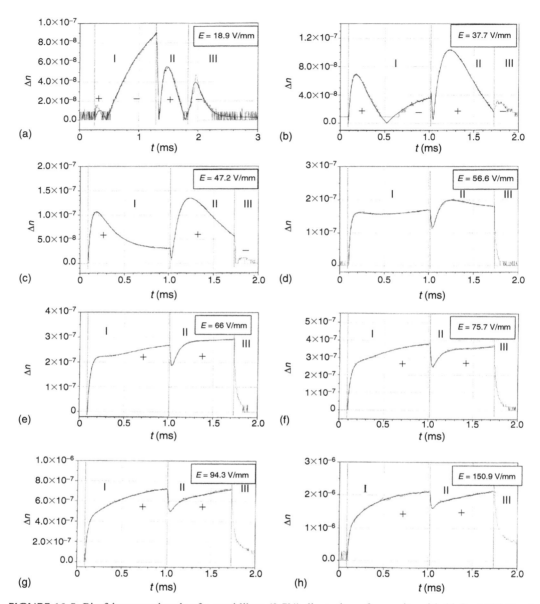

FIGURE 16.5 Birefringence signals of a semidilute (0.7%) dispersion of saponite with 11.5 mM NaBr at various field strengths.

On reversing the polarity one finds at intermediate and stronger fields always the dip in the birefringence (Figure 16.5, graphs c–h) because here the A_2-process, which causes the hump in the birefringence of solutions with little salt, loses importance whereas the A_3-process grows at fields above 40 V/mm. At fields stronger than 80 V/mm A_2 process does no longer appear in the build-up of the birefringence during the first pulse.

For further illustration of the salt dependence of the birefringence in Figure 16.6, a series of measurements is represented at a constant field strength of 56.5 V/mm and with various salt concentrations. The field strength has been chosen in a way that all processes can be found. Below

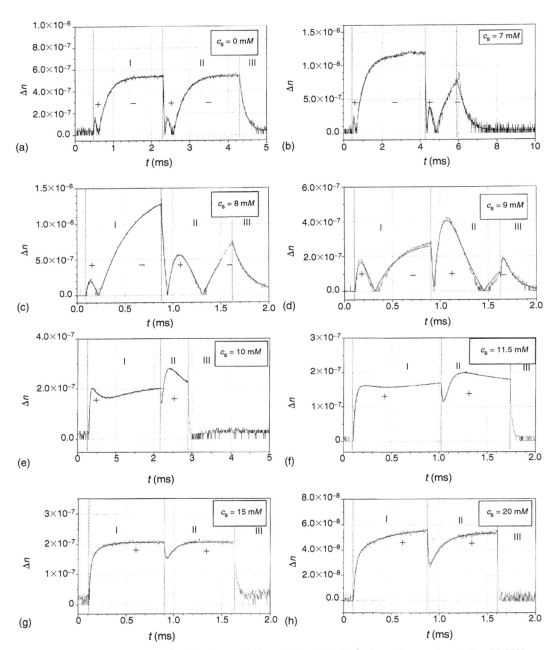

FIGURE 16.6 Birefringence signals of a semidilute (0.7 wt%) dispersion of saponite at $E = 56.5$ V/mm with various amounts of added NaBr.

9 mM A_2 is always bigger than A_1, causing a negative plateau value of birefringence. At a salt concentration of 10 mM the A_3-process comes into play whereas the A_2-process fades away. This provokes a positive plateau value of birefringence. The A_3-process always begins close to the concentration of the sol–gel transition. Addition of salt moves this transition to lower concentrations. The amplitudes of the processes become smaller at this transition. This is an indication that some of the clay particles have already formed gel clusters.

16.4 DISCUSSION

In the last section, we have seen the appearances of the birefringence signals as functions of the variables concentration, electric field strength, and ionic strength. Now we can analyze and interpret these signals with regard to the different mechanisms that we have already partially presented in the introduction.

Most easy to understand in the semidilute solutions is the A_1-process. This process is always present and it is certainly attributed to the free rotation of single particles in the electric field due to their induced dipole moments. It is observable both in the buildup of birefringence and in the decay. Upon reversing the polarity of the external field, the particles do not have to turn around; only the induced dipole moments do so. Thus, the process is not sensitive to the reversing. This agrees with the theory of Yamaoka et al. [14]. For a pure induced dipole moment the parameter q in this theory is equal to zero and in this case the calculated signal remains constant upon reversing.

The slower A_2-process with the inverse sign gives more reasons for a controversial discussion. It is generally accepted that in this process the particles align with their longitudinal axis perpendicular to the electric field. This has also been proven experimentally by electric light scattering and neutron scattering in electric fields on the above mentioned PTFE fibers. It remains to be clarified whether all particles are involved collectively in the A_2-process, i.e., all particles follow the A_1-process and afterwards the A_2-process, or if one part of the particles participates in the A_1-process aligning in field direction and the other part follows the A_2-process orienting transverse to the field direction. This question has been answered partially in the previous section. There are two observations that favor the second mechanism the first one is the peak at the beginning of the field-off decay. If the orientation of all particles would be homogeneous, one would expect a single monoexponential decay as in the dilute solution. The peak demonstrates that the degree of order in the solution increases once again before the final decay occurs. This could not be understood if the two different orientations would not have coexisted previously.

The second argument in favor of coexisting domains of particles is that Kerr's law holds for both single processes under certain conditions but not for the plateau value of birefringence, i.e., the sum of the A_1- and the A_2-process. See, for instance, the series of signals in Figure 16.3 ($c_s = 0$). The plateau value of the birefringence of the signals in graphs (a)–(c) is always around 6×10^{-7} and almost independent of the field strength. The single amplitudes, however, grow notably. This is only understandable when the two orientations coexist.

Another problem regarding the A_2-process is its negative sign, which means that the species are aligned perpendicular to the field. Are single particles responsible for the A_2-process or probably clusters of particles, as proposed by Cates? Unfortunately our measurements provide arguments for both models. The existence of clusters, i.e., a kind of domains with a liquid crystalline order in the solution, has already been tried to check in a previous investigation [18]. It is known that the A_2-process cannot be found at very high field strengths. Therefore, it could be imagined that the assumed clusters are destroyed by high fields. The idea in this study was to destroy existing clusters with a high-field pulse and to test whether the A_2-process reappears in a subsequent weaker-field pulse of the same polarity. Illner found that the anomaly reappears immediately and concluded that there are no clusters. This conclusion may, however, not be correct and the existence of aggregates cannot be completely ruled out on the basis of these results. Rather, it might be that the clusters form again very quickly in the electric field. Furthermore it is not sure that the A_2-process does not appear in a strong field. Rather, it is conceivable that it cannot be found anymore by a fitting procedure because its saturation value is very much lower than that of the A_1-process. Consequently the A_2-process is hidden by the much bigger A_1-process.

The third (A_3) process appears chiefly at high ionic strengths and at high fields. In solutions of entangled wormlike micelles, this process is commonly attributed to the alignment of chain segments of the persistence length in their tubes or the orientation and relaxation of liquid crystalline domains [1,21]. In solutions of short rodlike micelles this effect is explained by a cooperative motion of correlated micelles [1].

The characteristic features of the A_3-process are its long time constants and its sensitivity to reversing. Upon reversing the field the A_3-process decays even faster than the A_2-process. Therefore, we get the dips in the signals. Furthermore it appears also in the decay. (Unfortunately the decays in Figure 16.5 are not registered long enough but the tendency is observable: after the quick decay of the A_1-process the birefringence relaxes with a long time constant.) Considering this long time constants and the fact that the A_3-process appears at high ionic strength and at high fields, it is likely that the process has to do with bigger aggregates. Under these conditions the electrostatic repulsion forces between the particles are small and the attraction by local dipole forces is strong. Probably a kind of particle chaining starts similarly as in an electrorheological fluid. This would require both a rotational and a translational diffusion process to explain the long time constants.

16.5 INFLUENCE OF ADSORPTION ON BIREFRINGENCE ANOMALY OF CLAY DISPERSIONS

16.5.1 ADSORPTION OF POLYMERS

Clays can adsorb a large amount of surfactants or amphiphilic polymers and block copolymers. In the dispersed form they actually can bind the multifold weight of additives of their own weight [19]. The adsorption of surface-active additives can most easily be studied with surface tension measurements. The clays and the additive covered clays are hydrophilic and not surface-active and only the free additive concentrations determine the surface tension of the systems. The adsorption of amphiphilic substances has a dramatic influence on the electric birefringence signal on clay dispersions [20].

Typical signals which were obtained with the single-pulse method when increasing amounts of additives were added to the clays are shown in Figure 16.7, when the applied field and the other conditions were kept constant. A clay concentration was used that was high enough to show the large birefringence anomaly.

As the signals indicate, one observed that with increasing amount of polymers the anomaly disappears completely and a normal signal appears with the same sign as the signal in the dilute concentration range. The buildup and the decay of the signal are controlled by

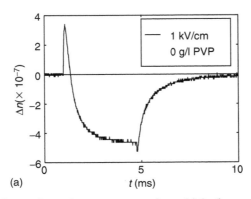

(a)

FIGURE 16.7 Signals with increasing polymer concentrations: (a) 0 g/l;

(Continued)

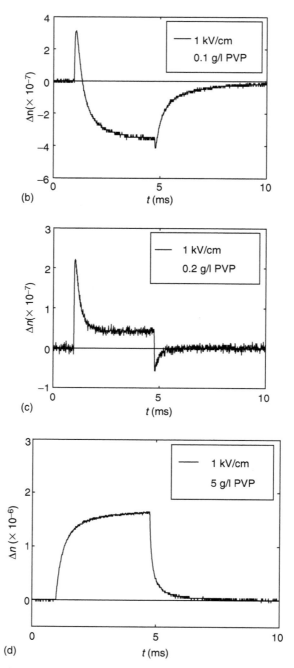

(b)

(c)

(d)

FIGURE 16.7 (**Continued**) (b) 0.1 g/l. (c) 0.2 g/l; and (d) 5 g/l.

about the same time constant. This time constant is approximately the same as the time constant that is observed in the dilute concentration range without adsorption. The signal is therefore due to the alignment of the single particles.

However, the situation is more complicated as it looks. The simple signal is deceiving and is actually hiding a process that is not shown in the simple transient signal but shows up with the pulse reversal method as shown in Figure 16.8. Now the signal shows a dip on reversal of the electric field and then recovers to the same height as during the first pulse. This is a clear indication that the total signal has two contributions, one that comes from the alignment of a permanent dipole and the other that comes from an induced dipole. However, in contrast to the situation in the dilute solution without additives where the signal had a positive hump, it has a negative dip. This indicates that the contribution from the dipole comes from a dipole that is parallel to the surface of the clay particle. We therefore should explain the 90° rotation of the apparently permanent dipole.

It is likely that the largest fraction of the counterions is located in the Stern layer of the clay particles. There is a constant exchange of these condensed counterions with the free counterions in the diffuse double layer. The condensed counterions can also migrate easily in the applied electric field, probably with a similar velocity as in the bulk phase. This means that the dipole moment parallel to the clay surface can actually change its sign very quickly, at least in a time much shorter than the rotation time of the platelet. A permanent dipole in the wrong direction is therefore quickly replaced by an induced dipole in field direction.

This situation changes dramatically with the adsorption of additives. Now the condensed counterions are trapped by the adsorbed polymers and can no longer diffuse or migrate freely. Now the permanent dipole can no longer be transformed into an induced dipole with opposite sign. The consequence of this slow down of mobility of the counterions is reflected in the response of the platelets in the electric field.

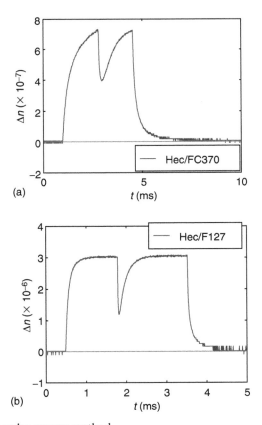

(a)

(b)

FIGURE 16.8 Signal with pulse reverse method.

With the adsorption of the polymer on the clay surface the mobility of counterions on the clay surface is reduced. The bound polymers are obstacles for the free-surface diffusion of the counterions. The permanent dipole along the clay surface can therefore no longer quickly disappear when the electric field is turned on. This dipole moment seems to be larger than the one in the perpendicular direction. With increasing adsorption of the polymer on the clay, the permanent dipole of the clay therefore aligns the particles from perpendicular to parallel. The signal that is monitored by the pulse reversal method changes from a positive to a negative hump.

Several models exist that describe the origin of the A_2-process. Owing to the decay and the rebuild of the A_2-process after reversing polarity, the existence of a permanent dipole moment or an ion-fluctuation dipole moment has to be assumed. The latter one results from a displacement of the charge centers of a particle and its ion-atmosphere. This would be a one-particle process because the same species of particles provokes the A_1- and the A_2-process. On further increasing both the salinity and the field strengths, a third process with the sign of the first one finally comes into view. This A_3-process seems to quench the A_2-process that disappears under these conditions. The A_3-process might supposedly be a chaining of particles oriented in field direction.

Finally, it should be mentioned that there is a new and completely different approach for the explanation of the electric birefringence anomaly by Netz [21].

16.6 CONCLUSIONS

In this study we investigated dilute and semidilute dispersions of saponite with the RPEB technique in order to get new information about the anomalous Kerr effect. It is shown that even in dilute systems two processes with opposite signs occur that cannot be detected with a single-pulse but only with the reversed-pulse method.

In semidilute solutions, we find the positive A_1-process and the negative A_2-process. As it is already known from older investigations the A_2-process prevails at lower fields but saturates at intermediate field strengths, whereas the A_1-process grows with increasing field and predominates at high fields. The A_1-process is caused by the alignment of single particles due to their induced dipole moment and it is not notably affected by the reversing of the external field.

The anomalous transient electric birefringence signals from dilute and semidilute clay dispersions are the result of a superposition of two separate signals, one that is caused from an induced electric dipole and the other that is caused from an apparently permanent dipole. In samples that contain no excess salt the two signals have opposite sign. In dilute solutions the positive signal is larger than the negative one, while in semidilute solutions the negative signal can be larger than the positive one. The positive signal is due to the alignment of the clay platelets parallel to the electric field. The buildup (τ_1) is controlled by the rotational diffusion coefficient of the particles. This process occurs faster than the buildup of the second process (τ_2) that is due to the alignment of the permanent dipole in the electric field. As a consequence of the superposition of the two signals one observes a complicated signal with a positive maximum, a crossover from positive to negative and a stationary negative signal with the single-pulse method.

In dilute solutions, the negative signal is smaller than the positive one and is therefore not visible in the total signal if the transient method is used. With the pulse reversal method, the negative signal that is hidden in the total signal reveals itself as a positive hump on reversal of the field because the permanent dipole from which the signal comes has to completely turn around. It is assumed that the permanent dipole is a result of the counterion fluctuation around the ionically charged clay particle. At any moment the center of the counterions does not fall on the charge centers of the clay particles. The result is a dipole, which on time average is zero. The direction of the dipole can have any angle between 0 and 90° with regard to the surface of the clay particle. The total dipole has a component parallel to the clay particles and one component perpendicular to the clay particles. With the applied field the parallel part disappears instantaneously because the high mobility of the counterions on the clay surface leads to charge compensation in that direction. The left over dipole is then perpendicular to the field. This dipole moment cannot disappear quickly because it requires a transfer of counterions from one side of the clay particles to the other side.

ACKNOWLEDGMENT

Parts of the experimental results of this chapter were carried out when Prof. S. Stoylov was an invited guest from the SFB 213, that was sponsored by the DFG, at the chair of Physical Chemistry I in Bayreuth. The authors would therefore like to thank Prof. Stoylov for many fruitful discussions on the electric birefringence anomaly.

REFERENCES

1. Hoffmann, H., Krämer, U., and Thurn, H., *J. Phys. Chem.*, 94, 2027, 1990.
2. Hoffmann, H., Hoffmann S., and Illner, J. C., *Prog. Colloid Polym. Sci.*, 97, 103, 1994.
3. Mueller, H. and Sackmann, B.W., *Phys. Rev.*, 56, 615, 1939; Mueller, H., *Phys. Rev.*, 55, 508, 1939; Mueller, H., *Phys. Rev.*, 55, 508, 1939; Norton, F.J., *Phys. Rev.*, 55, 668, 1939; Shah, N.J., Tompson, D.C. and Hart, C.N., *J. Phys. Chem.*, 67, 1170, 1963.
4. Laufer, M.A., *J. Am. Chem. Soc.*, 61, 2412, 1939.
5. Angel, M., Hoffmann, H., Krämer, U., and Thurn, H., *Ber. Bunsenges. Phys. Chem.*, 93, 184, 1989.
6. Krämer, U. and Hoffmann, H., *Macromolecules*, 24, 256, 1991.
7. Angel, M., Hoffmann, H., Huber, G., and Rehage, H., *Ber. Bunsenges. Phys. Chem.*, 92, 10, 1988.
8. Yamaguchi, Y. and Hoffmann H., *Col. Surf., A: Physico Chem. Eng. Aspects*, 121, 255, 67, 1997.
9. Doi, M., *J. Polym. Sci., Polym. Phys. Ed.*, 1982, 20, 1963.
10. Thurston, G. and Bowling, D.I., *J. Colloid Interface Sci.*, 30, 34, 1969.
11. Beevers, M.S., Elliot, D.A., and Williams, G., *Polym. Lett.*, 21, 13, 1980.
12. Cates, M.E., *J. Phys. II France*, 2, 1109, 1992.
13. Szabo, A., Haleem, M., and Eden, D., *J. Chem. Phys.*, 85, 7472, 1986.
14. Yamaoka, K., Tanigawa, M., and Sasai, R., *J. Chem. Phys.*, 101(2), 1625, 1994.
15. Yamaoka, K., Sasai, R., and Kohno, K., *J. Chem. Phys.*, 105(19), 8965, 1996.
16. Yamaoka, K., Fukudome, K., Matsumoto, S., and Hino, Y. J., *Colloid Interface Sci.*, 168, 349, 1994.
17. Schorr W. and Hoffmann H., *J. Phys. Chem.*, 85, 3160, 1981.
18. Schoot, P. and Cates, M.E., *J. Chem. Phys.*, 101(6), 5040, 1994.
19. Liu and Hoffmann, H., *Colloid Polym. Sci.*, 283, 24, 2004.
20. Holzheu, S. and Hoffmann, H., *J. Phys. Chem. B*, 106(17), 4412, 2002.
21. Schlagberger, X. and Netz, Roland R, *Europhysics Letters* 70, 29–135 pp. 2005.

17 Structural and Electrical Properties of Polyelectrolyte Multilayers on Colloidal Particles

Tsetska Radeva

CONTENTS

17.1 INTRODUCTION

The layer-by-layer technique, based on the alternate adsorption of anionic and cationic polyelectrolytes onto solid substrate, allows the production of a new class of nanomaterials with required properties and well-controlled thicknesses [1–5]. This technique has been extended to use synthetic or biopolymers, inorganic nanoparticles, viruses, enzymes, proteins, DNA, and other building blocks. Much of the current intense interest in the self-assembled polyelectrolyte multilayers is motivated by their vast potential applications in electronics, electrocatalysis, optical devices, sensors, selective membranes, etc. [1–5]. The polyelectrolyte multilayers have mostly been prepared on macroscopic flat surfaces, but recently this technique has also been applied to form films onto colloidal particles [6]. When dissolvable cores are used as templates, hollow polyelectrolyte shells can be produced as advanced drug delivery systems with controlled release functions [7]. The structure and electrical properties of the polyelectrolyte multilayers are crucial for all the envisaged applications.

The main driving force in the formation of the polyelectrolyte multilayers is electrostatics. As each polyelectrolyte adds to the surface, it overcompensates the surface charge, which allows the adsorption of a new layer of oppositely charged polyelectrolyte. It seems that there is no limitation on the number of layers that might be deposited in this manner [2]. The first films have been built from highly charged polyelectrolytes. Their thickness and the deposited amount of polymers have been found to increase linearly with the number of deposited layer pairs [2]. Now it is well established that the polyelectrolyte multilayers present a layered structure, with each polyelectrolyte layer interpenetrating with—three to four of the close neighboring ones [2]. This intermixing is a consequence of the intrinsic nature of the charge compensation within multilayers. The strong stability of the multilayer films is explained in terms of the formation of ion pairs between oppositely charged polymer segments. Charges on polymer segments can be balanced by those on oppositely charged chains or by small (salt) ions occluded within the film [8]. In the first case, which is termed "intrinsic" compensation, the positive charge of one polyelectrolyte is balanced by a negative charge of the other, except for the outermost layer. Small counterions of the last-adsorbed polyelectrolyte reside only at the film surface to render the system electrically neutral. Schlenoff and Dubas [9] proposed that the excess polymer surface charge, balanced by the small exchangeable ions, is spread out from the film surface through several (—two to three) layers. In the second case, denoted as "extrinsic" compensation, much of the polymer charge is balanced by salt counterions derived from the solutions used to construct the multilayers [9].

According to Schlenoff et al. [8], the film bulk does not contain small ions in the water-rinsed multilayers. Using radioanalytical methods, these authors have found that only trace amounts of small ions could be detected within a multilayer constructed via layer-by-layer deposition of two strong polyelectrolytes. However, "extrinsic" compensation within water-rinsed polyelectrolyte multilayers has been reported for films that have not a 1:1 stoichiometry of anionic and cationic groups [10–12]. The excess charge in such films must be neutralized by small counterions. An attempt to reconcile the two contradictory views about the incorporation of residual small ions into the film bulk is made in a recent study by Riegler and Essler [13]. They suggest that the ion content within the film depends on the charge density of the polyelectrolyte. If the average distance between the charges of one polyelectrolyte is less than the Bjerrum length $l_B = e^2/4\pi\varepsilon_0\varepsilon kT$ (where e is the electronic charge, ε_0 is the permittivity of vacuum, ε the dielectric constant of the solution, and kT the thermal energy), the multilayers contain small counterions that are bound to the chains of this polyelectrolyte ("Manning condensation" [14,15]). The condensed counterions neutralize the charge of the polyion chain until it decreases to a certain threshold. So, the "extrinsic" compensation might occur in the bulk of a water-rinsed film also when condensed counterions take part in the charge balance. If the charge density of the polyelectrolyte is less than the Manning's threshold value, the counterions are released during the polyelectrolyte complexation and the film remains free of small ions [13].

The problem about incorporation and mobility of small ions into the film bulk is closely connected with the mechanism of the multilayer film formation. The firm understanding of this mechanism is required not only to better control the adsorption process, but also because the presence of small ions in the film should significantly affect the electrical properties of the devices prepared from polyelectrolyte multilayers.

Irrespective of the decisive role of the electrostatics in the formation of stable multilayer films, measurements, mainly of the electrokinetic potential (ζ-potential), were carried out until now to elucidate the electrical properties of the polyelectrolyte multilayers [6,12,16]. The sign reversal of the ζ-potential after deposition of each polymer layer presents evidence of the film formation, but its values are mainly used for this purpose because of the uncertainty in the ζ-potential interpretation. Möhwald et al. [16], for instance, realized that coating of colloids with extremely different polyelectrolytes leads to almost same values of the ζ-potential.

The problem about incorporation and mobility of small ions into the film bulk is examined in a study on dielectric properties of polyelectrolyte multilayers by Durstock and Rubner [17]. They found that with increasing temperature or humidity, the low-frequency dielectric "constants" of dry multilayer films from highly charged polymers significantly grew up. Polarization of small but nonnegligible quantity of residual ions in the film bulk is supposed to explain the dielectric increments below 10^3–10^4 Hz. This explanation is similar to the one proposed by Michaels et al. [18] for the dielectric dispersion of polyelectrolyte complexes formed from the interactions of two oppositely charged polymers in a solution. (The sequential adsorption of oppositely charged polyelectrolytes onto charged substrate can be considered essentially as a controlled formation of a polyelectrolyte complex [8,9].) The frequency of the dielectric relaxation in both the investigations is assumed to depend on the size of microdomains that might exist in the film and on the mobility of the small ions in the film bulk. Apart from the increased mobility of ions with increasing temperature or humidity, an increased amount of small ions is also assumed to account for the larger increments and higher frequencies of the dielectric "constants" relaxation when films are dipped into salt solutions [17]. According to our results, however, the polarization in an electric field of small condensed counterions belonging only to the last-adsorbed polyelectrolyte is responsible for the dispersion of the electro-optical effect below 10^4 Hz [19–23]. This means that the contribution of small ions from the film bulk (if the film contains such ions) to the electro-optical effect is negligible.

From the electro-optical experiments, we can also obtain the hydrodynamic thickness of the multilayer by comparison of the rotational diffusion coefficients of naked and polymer-coated anisometric particles. The hydrodynamic thickness is most sensitive to the "loops" and "tails" in the adsorbed layer and might be twice as great as the average layer thickness. The latter is indicative of the polymer conformation and provides the basis for a more detailed understanding of the electrical properties of the adsorbed polyelectrolytes. Although certain similarity between the structure and the thickness of the multilayer films obtained on colloids and on macroscopic flat substrates is expected, this has to be experimentally verified [24]. Differences might appear because the coating procedure of colloids provides less control on the removal of the excess polyelectrolyte after deposition of each layer. Single-particle light scattering demonstrates successfully a regular film growth, but the layer thickness cannot be directly measured by this technique [6]. The thickness determination depends on assuming a refractive index of the layer, which is a function of the water content in the multilayer film. Dynamic light scattering is also not very suitable, since this technique cannot easily resolve aggregates and single particles [6]. Recently, the thickness of the polyelectrolyte multilayers on colloidal particles was determined by small-angle neutron scattering at contrast matching condition [24]. It was shown that the layer thickness remained independent of the curvature of the surface, at least down to a radius of 80 nm for the latex particles.

In this review, we present electro-optical results on the structure and the electrical polarizability of multilayers on anisometric colloidal particles [19–23]. Four topics are chosen to demonstrate the usefulness of electro-optics for *in situ* investigation of polyelectrolyte multilayers. The first is thickness determination of multilayers on anisometric colloidal particles. The electrical properties of polyelectrolyte multilayers are described in the second part, focusing on mobility of the polyelectrolyte counterions at low ionic conditions. The third topic concerns the counterions release from highly charged polyelectrolytes when they adsorb onto oppositely charged substrate. The last part presents an attempt to apply electro-optics for the calculation of the number of adsorbed polyelectrolyte chains in each top layer from the thickness and electrical polarizability data.

17.2 MULTILAYER GROWTH AS A FUNCTION OF LAYER NUMBER

17.2.1 PREPARATION OF MULTILAYER FILMS ON COLLOIDAL PARTICLES

Anionic sodium poly(4-styrene sulfonate) (NaPSS) and sodium salt of poly(acrylic acid) (NaPAA) were used for the preparation of multilayer films on ellipsoidal β-FeOOH particles. Cationic poly(allylamine hydrochloride) (PAH), poly(diallyldimethylammonium chloride)

(PDADMAC), and poly(4-vinyl pyridine) (PVP) were combined with the anionic polyelectrolytes to produce films, denoted as PSS/PAH, PSS/PDADMAC, PSS/PVP, and PAA/PAH. The films were deposited on the particle surface from pure water solutions or at $10^{-4} M$ concentration of NaCl, HCl, or NaOH when it was necessary to adjust the pH or the ionic strength of the solution. NaPSS and PDADMAC are strong polyelectrolytes with charge density not depending on the preparation conditions. The degree of ionization for the rest of the (weak) polyelectrolytes is pH-sensitive.

β-Ferric hydrous oxide particles (β-FeOOH) were prepared by acid hydrolysis of $FeCl_3$ solution [25]. The extraneous Fe^{3+} ions were removed by centrifugation in distilled water. This procedure gives ellipsoidal particles of narrow size distribution. The average dimensions a and b of the major and minor axes of the particle are determined by electron microscopy to be 285 ± 56 and 72 ± 14 nm, respectively. In acid medium, β-FeOOH particles are positively charged, with surface charge density of ca. $0.015 C$ m^2 at ionic strength 10^{-4} [26]. The concentration of the β-FeOOH suspensions was $4–8 \times 10^{-4} g$ dm^{-3} (about 10^9 particles in $1 cm^3$) in order to minimize the aggregation.

The first layer was deposited by adding NaPSS or NaPAA solution (ca. 10^{-4} mono$M$ dm^{-3}) to a suspension of β-FeOOH particles and stirring for 20 min. The concentration of the polymer was high enough to reach saturated adsorption and stabilization of the suspension against flocculation. The excess polymer was removed by centrifugation. The polymer-coated particles were redispersed by sonication in water or in a solution with the same concentration of NaCl, HCl, or NaOH as in the initial solution of polyelectrolyte. This procedure was repeated, adding the oppositely charged polyelectrolyte to the suspension of coated β-FeOOH particles [19–23].

17.2.2 HYDRODYNAMIC THICKNESS OF POLYELECTROLYTE MULTILAYERS

The existence and growth of polyelectrolyte multilayers on charged colloids takes more efforts to be proved than for the multilayer films on macroscopic flat surfaces. In this approach, a polymer solution in excess concentration of that required for saturated adsorption was added to a suspension of oppositely charged colloidal particles. The coated particles are then centrifuged and washed to ensure removal of free polyelectrolyte in solution. This procedure might provoke aggregation or partial desorption of polymer, thus spoiling the quality of the deposited films. First evidence for the successful polymer adsorption on colloidal particles was obtained from electrophoretic measurements, showing a reversal in the film charge after deposition of each new layer. The variation in the film charge with increasing layer number, however, is not a significant proof for the film formation on the particle surface. As mentioned above, the firm growth in the film thickness on colloidal particles has been verified by single-particle light scattering experiments [6]. (The latter experiments also provided evidence that no significant aggregation of the coated particles occurs as a result of centrifugation.) Information about the multilayer film growth on anisometric particles was obtained for the first time from electric light scattering measurements [19–23]. The thickness of each adsorbed layer was determined from the change in the particle rotational diffusion coefficient before and after the consecutive adsorption of both oppositely charged polyelectrolytes.

In the electro-optical experiments, suspensions of low particle concentration are used to reduce the degree of aggregation. The polyelectrolyte concentrations, on the other hand, are high enough to ensure overcompensation of the particle charge and stabilization of the suspension against aggregation. In spite of this, special efforts are made to eliminate the effect of aggregates (if they appear) in the determination of the layer thickness [19–23]. For this purpose, the time for particle disorientation (related with the rotational diffusion coefficient D_r as $\tau = 1/6D_r$) is recorded at high field strengths to ensure participation of all particles in the electro-optical effect (plateau values in Figure 17.1). According to the electro-optical theory of polydisperse systems, the contribution of aggregates (the biggest particles) is less at higher fields as compared with that of the single particles. So, the thickness of a layer adsorbed predominantly on isolated particles might be calculated at high field strengths.

Figure 17.1 shows that the time for particle disorientation after switching the applied field off depends on the field strength in a similar way for bare oxide particles and for particles coated with an increasing number of PSS layers. This means that no measurable aggregation occurs at conditions of this experiment and the increase in the particle size results mainly from the polymer adsorption onto isolated particles. Similar behavior of the τ vs. E^2 dependencies is found for all other films deposited on β-FeOOH particles [19–23].

Figure 17.2 shows a linear film growth after the deposition of four PSS/PAH [21] or five PSS/PDADMAC [20] polymer layers on the particle surface. A linear increase in the film thickness becomes evident after the deposition of the first three PSS/PVP [22] layers (Figure 17.3). This means that the effect of the substrate on the film behavior propagates over the first three to five layers. These results agree well with the findings of Ladam et al. [12], who showed in another *in situ* investigation that to achieve the linear growth regime of a PSS/PAH multilayer on flat oxide surface, one needs a precursor film of at least six polyelectrolyte layers. Our findings are in contrast, however, with those obtained for multilayers on colloidal particles. All the experiments with films on colloidal particles showed a linear increase in the film thickness from the beginning of the multilayer buildup [6,27]. We have no definite explanation for this discrepancy, but the electro-optical effect is found to increase almost in the same manner as the thickness of our multilayer films. This correlation points to an increase in the multilayer charge because of the nonlinear increase in the polymer adsorption (see the next part). In an investigation of film constructed from a polycation solution and from a binary mixture of polyanions as polyanion solution, Hübsch et al. [28] have found change from exponential to linear growth regime by adjusting the composition of polyelectrolyte mixtures. Different diffusion coefficients of both polyanions are considered to be the reason for this hybrid growth regime. The authors tried to explain the similar qualitative thickness evolution of our PSS/PDADMAC film [20] by the strong polydispersity in the size of the PDADMAC sample. However, we observed the same growth for the PSS/PAH film,

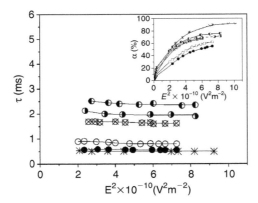

FIGURE 17.1 Dependence of relaxation time, τ, on electric field strength, E, for a suspension of β-FeOOH particles (*) and for particles coated by PSS1 (●), PSS2 (○), PSS3 (⊗), PSS4 (◓), and PSS5 (◒). Inset: the corresponding dependence of the kilohertz electro-optical effect, α, on the electric field strength. (From Radeva, Ts. et al., *J. Colloid Interface Sci.*, 266, 141, 2003. With permission.)

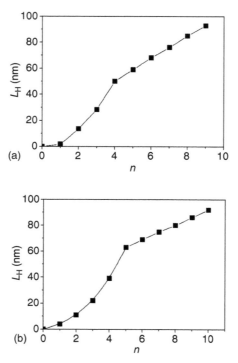

(a)

(b)

FIGURE 17.2 Hydrodynamic thickness, L_H, as a function of polyelectrolyte layer number, n. The odd layer numbers correspond to PSS deposition and the even layer numbers to (a) PAH and (b) PDAD-MAC deposition. (From Radeva, Ts. et al., *Colloids Surf. A*, 240, 27, 2004; and Radeva, Ts. et al., *J. Colloid Interface Sci.*, 266, 141, 2003. With permission.)

constructed from polymers with low polydispersity [21].

The increase in the film thickness as a function of layer number is shown in Figure 17.3 for two PSS/PVP films, prepared from polymers with strongly different molecular weights [22]. The similarity observed in the linear growth regime agrees well with other observations for multilayer systems [29]. It is in accordance with the theory for adsorption of strong polyelectrolytes on planar surfaces [30]. The thickness of the first PSS layers depends,

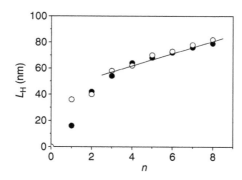

FIGURE 17.3 Hydrodynamic thickness, L_H, of multilayer films from PSS and PVP with 7×10^4 (●) and 1×10^6 (○) molecular weights. (From Radeva, Ts. et al., *J. Colloid Interface Sci.*, 279, 351, 2004. With permission.)

however, on the molecular weight of the adsorbed polymer. This indicates that some "loops" and "tails" are present in the deposited layer [31]. One possible way to explain such a "loopy" conformation of the adsorbed highly charged polymer chains is that the adsorption occurs on small particles in comparison with the length of the PSS chain. For small enough particles, the charges on their surfaces will not be sufficient to compensate for the polyelectrolyte charge. Hence, the excess charges will be expelled from the particle surface and located on polymer "loops" and "tails," protruding away into the solution [32]. This might be the case when high molecular PSS adsorbs onto β-FeOOH particles. Another reason for the "loopy" conformation of the adsorbed PSS is that some "loops" and "tails" appear when highly charged polyelectrolyte adsorbs onto a surface of low charge density [32]. This explains why the layer thickness is not so small (if compared to the diameter of a PSS chain) for the low molecular PSS as well. Since the charge density of the polymers is higher than that on the particle surface, the molecules in the next layers will adsorb in a flatter conformation. This explains the lack of dependence of the film thickness on the molecular weight of our polymers in the linear growth regime.

Hydrodynamic thickness of a bilayer in the linear growth regime is determined to be 10 nm for the PSS/PDADMAC [20], 17 nm for the PSS/PAH [21], and 8 nm for the PSS/PVP [22] multilayers. These values were obtained when the thickness was calculated at constant axial ratio $p = a/b = 4$, according to the Perrin equation for prolate ellipsoids [33]:

$$D_r = \frac{kTp^2}{4\eta v(p^4 - 1)} \left[-1 + \frac{2p^2 - 1}{2p\sqrt{p^2 - 1}} \ln \frac{p + \sqrt{p^2 - 1}}{p - \sqrt{p^2 - 1}} \right] \tag{17.1}$$

where η is the viscosity of the suspending medium and $v = 1/6 \, \pi ab^2$ is the volume of the particle. However, if one assumes a decrease in the axial ratio, $p = (a+2L_H)/(b+2L_H)$, with increasing layer number, the calculated thicknesses of a bilayer drop to 6 or 10 nm, respectively, for the first two films. Having in mind that the hydrodynamic thickness might be twice as great as the average layer thickness [34], the obtained *in situ* values seem reasonable for the thickness of a rodlike polyion layer. (It is important to note that all the steps of the experiments, including the rinsing procedure, were done with the same ionic strength solutions to avoid phenomena not related to the buildup mechanism [35].)

The formation of relatively thick films from two weak polyelectrolytes PAA and PAH, indicating that fully charged PAH chains are combined with nearly fully charged PAA chains [31], is shown in Figure 17.4. The film growth proceeds linearly with layer number after formation of about four layers, with a net increase in the thickness per layer of about 5 nm. Figure 17.4 displays, however, an irregular growth of the PAA/PAH film. The layer thickness is greater for the films ending with a PAH layer, which means that PAH is partially removed from the surface when the film is exposed to a PAA solution. The result seems unexpected, since the fully charged PAH chains are partially removed by the less charged PAA chains. But after deposition of each layer from fully charged PAH (at pH 4.5), it is rinsed with neutral water and then exposed to a PAA solution at pH 6.5. Under these conditions, the degree of PAH ionization drops to a lower value and some PAH desorption takes place. In the next adsorption step, more PAH is adsorbed than is removed in the previous one and an overall multilayer growth takes place. Irregular increase in the film thickness as a function of the polyelectrolyte layer number is also found for the PAA/PAH film, when fully charged PAA chains (at pH 7.5) are partially removed by the less charged PAH chains (at pH 6.5). The net increase in the hydrodynamic thickness per layer is about 6 nm in this case [23].

Film growth, which proceeds via a series of adsorption–desorption steps, has been reported previously for polyelectrolyte multilayers, containing one weakly charged polyelec-

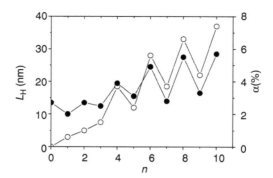

FIGURE 17.4 Hydrodynamic thickness, L_H, ($\bigcirc$) and electro-optical effect, α ($\bullet$) as a function of layer number for PAA/PAH 6.5/4.5 film. Electric field strength $E = 1.9 \times 10^4 \, \mathrm{V \, m^{-1}}$; field frequency $10^3 \, \mathrm{Hz}$. (From Radeva, Ts. et al., *J. Colloid Interface Sci.* 287, 415, 2005. With permission.)

trolyte [36–41]. The existence of a critical charge density limit [36,39] or significant mismatching of charge densities of the polyelectrolyte pair [5] have been discussed as the most important factor in determining the film growth. Both reasons cannot be applied to explain the irregular PAA/PAH film growth in the electro-optical experiment. Desorption of polyelectrolytes from a multilayer as a result of competitive binding with oppositely charged polyelectrolyte chains in a solution (polyelectrolyte complexation in a solution) is also reported by several authors [41,42]. This explanation seems more applicable to our results. The electro-optical experiments emphasize on the importance of studying each adsorption step rather than the final film formed after many deposition cycles. Such an investigation might shed light on the processes that govern the mutilayer film formation.

17.3 ELECTRICAL PROPERTIES OF POLYELECTROLYTE MULTILAYERS

The layer-by-layer deposition of the above described polyelectrolyte multilayers was first monitored by electrophoretic measurements. Figure 17.5 shows the electrophoretic mobility U_e of β-FeOOH particles, coated with PSS/PDADMAC and PSS/PAH multilayers, as a function of the layer number. Overcompensation of the particle charge is seen in both cases

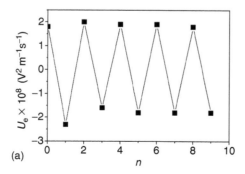
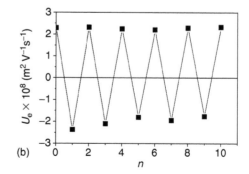

FIGURE 17.5 Electrophoretic mobility, U_e, of β-FeOOH particles covered by sequentially adsorbed layers of PSS and (a) PAH and (b) PDADMAC. (From Radeva, Ts. et al., *Colloids Surf. A*, 240, 27, 2004; Radeva, Ts., *J. Colloid Interface Sci.*, 266, 141, 2003. With permission.)

after alternate exposure to the solutions of oppositely charged polyelectrolytes. However, the electrophoretic mobility does not depend (in absolute value) on the number of the adsorbed layers and the polyelectrolyte charge density. This implies a constant level of surface charge overcompensation during preparation of the whole films. On the other hand, almost equal values of the electrokinetic charge are obtained in most previous investigations on poly-electrolyte multilayers [12,16]. One possible explanation for the U_e value being independent of the polyion charge is that condensation of counterions equalizes the "effective" charge density of the highly charged polyions used in the preparation of the multilayer films [19,43]. The "effective," not intrinsic, charge of the polymer-coated particle is determined by electrokinetics [44].

In contrast to the electrophoretic mobility, the electro-optical effect increases with increasing number of the adsorbed layer. Figure 17.6 shows the dependence of the electro-optical effect on the field frequency for the consecutive adsorbed PSS and PAH layer in a PSS/PAH multilayer [21]. We focus on three points. First, the electro-optical effect increases with the number of the adsorbed layer, showing a tendency to saturation in the linear growth regime. Second, the electro-optical effect is larger when the film ends with PAH. Third, the frequency of relaxation of the kilohertz electro-optical effect is lower when the last-adsorbed layer is from PAH.

More distinct results are presented in the next two figures. Figure 17.7 shows the increase only of the kilohertz electro-optical effect on the layer number, whereas frequency dependencies of equal (at 10^3 Hz) electro-optical effects are presented in Figure 17.8. The effects in

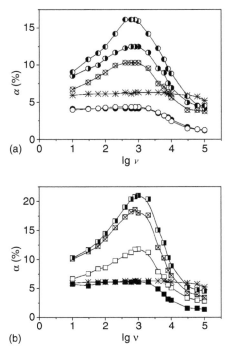

FIGURE 17.6 Frequency dependence of the electro-optical effect, α, for the suspension of β-FeOOH particles (*) and for particles with deposited layers from (a) PSS1 (●), PSS2 (○), PSS3 (⊗), PSS4 (◉), and PSS5 (◖); and (b) PAH1 (■), PAH2 (□), PAH3 (⊠), and PAH4 (◪). Electric field strength $E = 3.8 \times 10^4$ V m^{-1}. (From Radeva, Ts. et al., *Colloids Surf. A*, 240, 27, 2004. With permission.)

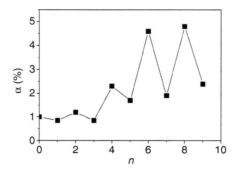

FIGURE 17.7 Electro-optical effect as a function of polyelectrolyte layer number for PSS/PAH film at $E = 1.2 \times 10^4$ V m^{-1}. (From Radeva, Ts. et al., *Colloids Surf. A*, 240, 27, 2004. With permission.)

both figures are compared at low fields, where the electro-optical effect presents a quadratic function of the electric field strength.

We notice that the electro-optical effect increases similarly to the multilayer film thickness (Figure 17.2 and Figure 17.7). This suggests that the increase in the electro-optical effect arises from the larger adsorption of the polymers on particle surface with increasing area after deposition of each next layer. The adsorbed amounts remain constant, however, in the linear growth regime, where the electro-optical effect also becomes almost independent of the layer number.

One would expect that films ending in PAH should yield larger effects than those terminated by PSS because of a higher adsorption of the longer PAH chains [38]. The higher electro-optical effect for the films ending with PAH might also be explained in terms of the larger adsorbed amount of monomoles in the PAH layers in comparison to that in the PSS layers. One rough estimation, similar to that of Riegler and Essler [13] (who have translated the layer thickness into monomoles/m^2), shows that the amounts of adsorbed PAH monomoles in all layers are two to three times higher than those in the PSS layers. For fully ionized PAH (at pH 4.5) and PSS, this means that the charge stoichiometry is not 1:1 and the PSS/PAH multilayer needs additional small counterions (Cl$^-$) for complete charge compensation [13]. Polarization in an electric field of these additional counterions might be a reasonable

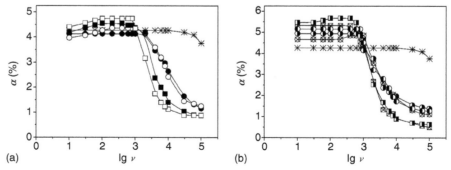

FIGURE 17.8 Frequency dependence of the electro-optical effect at low fields (equal effects at 10^3 Hz) for suspension of β-FeOOH particles (*) and for particles with deposited layers from (a) PSS1 (●), PAH1 (■), PSS2 (○), and PAH2 (□) and (b) PSS3 (⊗), PAH3 (⊠), PSS4 (◖), PAH4 (◨), and PSS5 (◓). The arrows show ν_{cr}. (From Radeva, Ts. et al., *Colloids Surf. A*, 240, 27, 2004. With permission.)

explanation for the observed larger effects when the film ends with PAH. (One has to take into account that the effect from the residual small ions into the PAH layers should be additional to the effect from polarization of the small ions belonging to the last-adsorbed polymer layer. This polymer is in excess to ensure overcompensation of the surface charge and requires small ions for neutrality.)

In contrast to the linear increase of the film thickness with layer number, however, the electro-optical effect reveals a tendency to saturation (Figure 17.7). This indicates that polarization in an electric field of the counterions belonging only to the chains of the last-adsorbed polyelectrolyte is responsible for the creation of the electro-optical effect. The electro-optical effect might be higher and of lower relaxation frequency for the films ending with PAH, since the polarization of small Cl^- counterions occurs along the contour length of the two to three times longer PAH chains than those of PSS. The contour length of the polyelectrolyte chain should play a decisive role for the effect value, if polarization of condensed counterions governs the electro-optical behavior of the adsorbed polyelectrolyte. Such an interpretation has previously been used to explain the lower relaxation frequency of the electro-optical effect from particles coated with a highly charged polyelectrolyte in comparison to the relaxation frequency of the effect from bare oxide particles [45–48].

To obtain a better estimate of the relaxation frequency, we analyze the electro-optical effects for the multilayers ending with PSS and PAH at low field strengths. As seen in Figure 17.8, a decrease in the electro-optical effect begins above 10^4 Hz for the particles without adsorbed polymer. Such a decrease starts at approximately 10^3 Hz for the particles coated with both polymers. The average distance between the charges on β-FeOOH particle surface is ca. 30 Å (at ionic strength 10^{-4}), which excludes the possibility for counterion condensation [26]. This suggests that the electro-optical effect of bare oxide particles results from polarization of the diffuse part of their electrical double layers [49–52]. The decrease in the relaxation frequency of the effect from particles coated with highly charged PSS and PAH polymers might be attributed to the polarization of condensed counterions (Na^+ or Cl^-) along the polyion chains of contour length L. The mobility of these ions is lower than the mobility of free ions because of the strong attraction to the polyion surface, which forces the condensed counterions to move along the polyion contour length [53]. An estimation according to the equation [54]

$$\nu_{cr} = \frac{4D_I}{\pi L^2} \qquad (17.2)$$

for particles coated by one PSS or PAH layer is consistent with this consideration. We used values for the PSS and PAH contour lengths 99 and 242 nm, respectively, to estimate ν_{cr}. Hayakawa and coworkers [53] obtained a value of $D_{Na^+} = 9.8 \times 10^{-7}$ cm^2 s^{-1} for the diffusion coefficient of bound Na^+ counterion in an NaPSS solution. $D_{Cl^-} = 14.7 \times 10^{-7}$ cm^2 s^{-1} is assumed to be 1.5 times higher than D_{Na^+} (like in a polyelectrolyte-free solution). The estimate gives values of ν_{cr}, ca. 13×10^3 and 3×10^3 Hz for the particles coated with PSS or PAH layer. These values are close to the experimental ones (10×10^3 and 5×10^3 Hz), shown by arrows in Figure 17.8. The good agreement between the experimentally obtained and calculated values of the relaxation frequency supports our suggestion that polarization of ions with lower mobility than that of the free ions is mainly responsible for the creation of the electro-optical effect from particles coated with highly charged polyelectrolytes. The lack of dependence of the relaxation behavior of the electro-optical effect on the layer number (or on the film thickness) demonstrates, on the other hand, that the small counterions from the film bulk do not take part in the creation of this effect (Figure 17.8b). The counterions of the polymer in the last-adsorbed layer govern the electro-optical behavior of the whole PSS/PAH multilayer.

The above proposed explanation for the dominating participation of the condensed counterions from the last-adsorbed polymer in the electro-optical effect is not at variance with presence of additional small counterions into the PAH layers of the PSS/PAH film. It only means that their mobility is negligible in comparison to that of the counterions near the surface of the last-adsorbed layer.

Similar behavior of the electro-optical effect with increasing layer number is observed for the film constructed from two strong polyelectrolytes—PSS and PDADMAC (Figure 17.9). In the initial nonlinear growth regime, the electro-optical effect increases almost in the same manner as the thickness of the PSS/PDADMAC film (Figure 17.2 and Figure 17.10). The adsorbed increments remain constant, however, in the linear growth regime, where the electro-optical effect becomes almost independent of the layer number. Similarly to the PSS/PAH film, the electro-optical effect is larger for the films ending with PDADMAC, although no deviation from 1:1 stoichiometry is expected in this case. As mentioned in Section 17.1, Schlenoff et al. [8] found that small ions did not participate in the charge balance within the PSS/PDADMAC multilayer film deposited on macroscopic flat substrate. The polymers associate completely into the bulk and only one type of counterion is present at the surface of this multilayer film. On the other hand, Caruso et al. [11] showed that a minimum of about 10% of the cationic charges of the PDADMAC in the upper region of the PSS/PDADMAC film (on latex particles) were not bound in ion-pairs with the oppositely charged sites of PSS. Some topological constraints, attributed to the more rigid nature of PDADMAC, are identified as the major cause of PDADMAC charges not taking part in ion-pair formation.

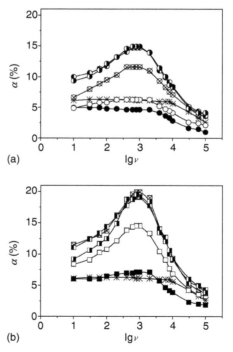

FIGURE 17.9 Frequency dependence of the electro-optical effect for suspension of β-FeOOH particles (*) and for particles with deposited layers from (a) PSS1 (•), PSS2 (○), PSS3 (⊗), PSS4 (⊙), and PSS5 (◉); and (b) PDADMAC1 (■), PDADMAC2 (□), PDADMAC3 (⊠), PDADMAC (▣), and PDADMAC5 (▣). Electric field strength $E = 3.8 \times 10^4$ V m^{-1}. (From Radeva, Ts. et al., *J. Colloid Interface Sci.*, 266, 141, 2003. With permission.)

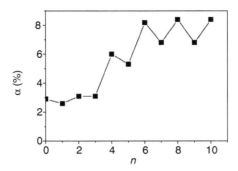

FIGURE 17.10 Electro-optical effect as a function of layer number for PSS/PDADMAC multilayer film. Electric field strength $E = 2.7 \times 10^4 \, \text{V m}^{-1}$. (From Radeva, Ts. et al., *J. Colloid Interface Sci.*, 266, 141, 2003. With permission.)

However, the results of Caruso et al. [11] are rather qualitative due to the size of the fluorescent probes used in these experiments. In the absence of salt during preparation of the multilayer film, as in our case, the results of Schlenoff et al. [8] seems more plausible.

In the last experiment, the contour length of the linear PDADMAC molecule would be three to four times greater than that of PSS. However, the used sample of PDADMAC was highly polydisperse and with branched backbones. This is the reason that we were not able to examine the dependence of ν_{cr} on the PDADMAC chain length. The strong dependence of the relaxation frequency of the electro-optical effect on the polyelectrolyte chain length is well demonstrated for the PSS/PVP films terminated by PSS with strongly different molecular weights—7×10^4 and 1×10^6 Da (Figure 17.11 and Figure 17.12). For estimation of the relaxation frequency of the electro-optical effect, we used the contour lengths 1420 and 99 nm for the high- and low-molecular-weight PSS, respectively, and the same diffusion coefficient of bound Na^+ counterion $D_{Na^+} = 9.8 \times 10^{-7} \, \text{cm}^2 \, \text{s}^{-1}$ as in the case of the PSS/PAH film. The calculation gives approximately 60 Hz and 13×10^3 Hz for the multilayers ending with high- and low-molecular-weight PSS. The good agreement between the calculated

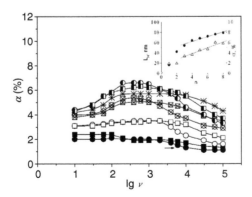

FIGURE 17.11 Frequency dependence of the electro-optical effect for β-FeOOH particles (∗) coated with PSS1 (●), PSS2 (○), PSS3 (⊗), PSS4 (⊙) and PVP1 (■), PVP2 (□), PVP3 (⊠), PVP4 (◘). Molecular weight of PSS $M_W = 7 \times 10^4$ and of PVP $M_W = 7.85 \times 10^4$. Electric field strength $E = 3.8 \times 10^4 \, \text{V m}^{-1}$. The arrow shows ν_{cr}. Inset: electro-optical effect at 3×10^3 Hz (△) and hydrodynamic thickness (●) as a function of layer number. (From Radeva, Ts. et al., *J. Colloid Interface Sci.*, 279, 351, 2004. With permission.)

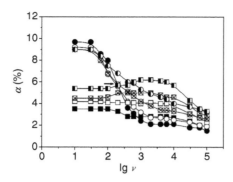

FIGURE 17.12 Frequency dependence of the electro-optical effect for β-FeOOH particles coated with PSS1 (●), PSS2 (○), PSS3 (⊗), PSS4 (◉) and PVP1 (■), PVP2 (□), PVP3 (⊠), PVP4 (▣). Molecular weight of PSS and PVP $M_W = 1 \times 10^6$. Electric field strength $E = 3.8 \times 10^4$ V m^{-1}. The arrow shows ν_{cr}. (From Radeva, Ts. et al., *J. Colloid Interface Sci.*, 279, 351, 2004. With permission.)

and the experimentally obtained values (shown by arrows in Figure 17.11 and Figure 17.12) confirms again our conclusion that polarization of condensed counterions along the chain of PSS is responsible for the creation of the electro-optical effect from multilayers ending with the highly charged polyelectrolyte.

In contrast to the highly charged polyelectrolytes, no dependence of ν_{cr} on the polyelectrolyte molecular weight is found for the films ending with weakly charged PVP (Figure 17.11 and Figure 17.12). The degree of PVP protonation in a solution at pH 3.8–3.9 (as in this experiment) is less than 10% [55]. It might be enhanced during preparation of the PSS/PVP film [56], but this additional ionization is not high enough to cause counterion condensation. An estimation of the relaxation frequency of the electro-optical effect shows that the effect results from the polarization of the diffuse part of the particle's electrical double layers when the film ends with weakly charged PVP.

Taking advantage of the results of Shiratori and Rubner [31], who have demonstrated that the composition of the PAA/PAH film surface can be tailored using pH, we prepared a PAA/PAH multilayer to probe the effect of the last-adsorbed layer on the electro-optical behavior of the whole film. According to Shiratori and Rubner [31], the film surface contains segments basically from the last-deposited polymer layer when fully charged chains of one polyelectrolyte are combined with nearly fully charged chains of the other polyelectrolyte. (PAA or PAH dominance is exemplified by contact angle measurements and absorbance determination of a cationic dye, methylene blue [31]. PAA and PAH ending multilayers are also shown to bind positively or negatively charged catalysts selectively when a pH combination is chosen that minimizes interpenetration of PAA and PAH at the multilayer surface [57].) Clear dependence of the relaxation frequency of the electro-optical effect on the contour length of the last-adsorbed fully ionized polyelectrolyte is found also for the PAA/PAH film (Figure 17.13). As in the case of PSS/PVP film, a smaller decrease in ν_{cr} is registered for the effect from PAA/PAH films ending with less charged polyions (Figure 17.14 and Figure 17.15).

17.4 COUNTERION RELEASE FROM ADSORBED POLYELECTROLYTES

The polarization of condensed counterions along the chains of the adsorbed polyelectrolyte is of particular importance when discussing the origin of the additional electro-optical effect, which appears near the range of particle rotation in suspensions containing highly charged polyelectrolyte. However, the counterion release from the polymer and the substrate surfaces

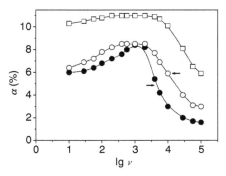

FIGURE 17.13 Relaxation frequency of electro-optical effect (the arrows) for β-FeOOH particles (□) coated by one layer of fully ionized PAH4.5 (●) and one layer of fully ionized PAA7.5 (○). (From Radeva, Ts. et al., *J. Colloid Interface Sci.*, 287, 415, 2005. With permission.)

is considered to be the true driving force for "charge-driven" adsorption, since it increases the system entropy. If the condensed counterions are not released, the driving force for adsorption should be smaller. According to the recent theory of Sens and Joanny [59], in the case of polyelectrolyte adsorption onto a substrate of low dielectric constant, the condensed counterions are not always released because of the major importance of the image-charge effect. A complete counterion release from the adsorbed polyelectrolyte occurs only for very large surface charges. The fraction of condensed counterions remains almost unchanged when highly charged polyelectrolyte adsorbs onto weakly charged substrate. Having in mind that the β-FeOOH particles are weakly charged at conditions of the electro-optical experiments, we applied the theory of Sens and Joanny to evaluate the fraction of condensed counterions on the adsorbed polyelectrolytes. The fraction of condensed counterions near the surface of the adsorbed highly charged polyion was compared with that on a polyion in salt-free solution, calculated by the theory of Manning [14,15].

In his widely cited work [14], Manning showed that the fraction of condensed counterions on a highly charged cylinder like polymer is equal to $(1-\xi^{-1})$. Here $\xi = l_B/d$ presents the charge density parameter, with d the average charge-to-charge distance along the polymer chain. For NaPSS, which has $\xi = 2.75$ ($l_B = 7.15$ in water at room temperature and $d = 2.6$ Å

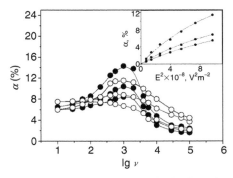

FIGURE 17.14 Frequency dependence of the electro-optical effect for the PAA/PAH 6.5/4.5 films ending with one, two, and four layers from fully ionized PAH (●) and one, two, and four layers from nearly fully charged PAA (○). Inset: electro-optical effect at 10^3 Hz as a function of field strength. (From Radeva, Ts. et al., *J. Colloid Interface Sci.*, 287, 415, 2005. With permission.)

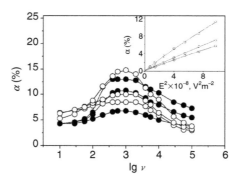

FIGURE 17.15 Frequency dependence of the electro-optical effect for the PAA/PAH 7.5/6.5 films ending with one, two, and four layers from fully ionized PAA ($\bigcirc$) and one, two, and four layers from nearly fully charged PAH ($\bullet$). Inset: electro-optical effect at 10^3 Hz as a function of field strength. (From Radeva, Ts. et al., *J. Colloid Interface Sci.*, 287, 415, 2005. With permission.)

[58]), the fraction of the condensed Na^+ counterions is 0.64. The fraction of the free Na^+ counterions of NaPSS in a solution is, therefore, $\beta = 0.36$.

According to Sens and Joanny (and contrary to the widely spread idea), most of the condensed counterions will not be released, if the Gouy–Chapman length of the double layer near the charged substrate is larger than the radius of the adsorbed polyion [59]. Equating the chemical potentials of the free and condensed counterions, Sens and Joanny obtain the fraction of free counterions β:

$$\log\left[8\xi(1-\beta)\right] = \log\left[(\kappa r)^2\right] - 2\xi\beta\log\left[\frac{rh}{L_\kappa(h)^2}\right] + 2\log\left[\frac{\kappa(h+\lambda)}{2}\right] \qquad (17.3)$$

where $\lambda = 1/(2\pi l_B\sigma)$ is the Gouy–Chapman length in a salt solution of average concentration n_0 (or Debye length κ^{-1} with $\kappa^2 = 8\pi l_B n_0$), and σ is the surface charge density. The local screening length in the Gouy–Chapman layer L_κ is small, $L_\kappa \sim \lambda/\sqrt{2}$. In Equation 17.3, r is the polyion radius and h is the distance to the surface.

The fraction of released Na^+ counterions after the adsorption of NaPSS on β-FeOOH particle surface is estimated by using the following quantities: $\sigma \sim 0.3e/(100)$ Å^2 [26], $\kappa^{-1} \sim 1000$ Å, $\lambda \sim 740$ Å, $L_\kappa \sim 520$ Å, and $r \sim 10$ Å. The distance h to the adsorbing surface is equal to the average layer thickness (~ 50Å). This estimation gives $\beta \sim 0.30$ for the fraction of free Na^+ counterions in the colloid–polyelectrolyte suspension, which closely matches $\beta \sim 0.36$ for the free Na^+ counterions in a NaPSS solution. The conclusion is that the condensed counterions are not released from the NaPSS after its adsorption on the particle surface.

The fraction of condensed counterions in adsorbed polyions is similar for highly charged PAH and PAA, since the average charge-to-charge distance for the fully ionized PAH (at pH 4.5) and PAA (at pH 7.5) is equal to that of NaPSS. For PDADMAC, the charge density parameter $\xi = 1.37$, which gives $\beta \sim 0.55$ for the free Cl^- counterions of the adsorbed polyelectrolyte. It is lower than $\beta \sim 0.70$ for the PDADMAC in a solution, indicating even some recondensation on the surface of the adsorbed polyion. When repeated, this estimation at $\sigma \sim 1e/(100)$ Å^2, the fraction of free Na^+ and Cl^- counterions increases approximately with 5%. This means that the fraction of condensed counterions remains almost unchanged in the upper adsorbed layers of the multilayer film.

Our conclusion must therefore be that the appearance of the additional effect near the range of particle rotation is due to the polarization of condensed counterions along the chains of the adsorbed highly charged polyelectrolytes [45–48]. This conclusion is also in good agreement with the theory of Cheng and Olvera de la Cruz [60], who have studied the adsorption of highly charged rodlike polyelectrolytes onto weakly oppositely charged surfaces. (These authors include lateral correlations between the adsorbed polyelectrolytes, which has not been taken into account in the Sens and Joanny consideration [59].) According to the theory of Cheng and Olvera de la Cruz, the strongly charged polyions do not release all their condensed counterions upon adsorption onto weakly charged surfaces because the counterions reduce the electrostatic repulsion energy of the adsorbed polyelectrolytes. Both presented theoretical predictions are consistent with the experimental observations described by Glinel et al. [61]. They found that the thickness increment per deposition cycle remained charge-independent in the regime where counterion condensation dominates the behavior of the polyelectrolyte multilayer. Bordi et al. [62] have also found that the highly charged NaPAA retained its condensed counterions after the adsorption on positively charged liposomes (with three times higher charge density than the β-FeOOH particles). A similar conclusion was drawn by Shin et al. [63] in an investigation of the "charge-driven" polymer adsorption onto SiO_2 surface. According to the results from our recent investigation, the condensed counterions of a highly charged NaCMC are not released during its adsorption on weakly charged β-FeOOH particle [64].

17.5   ADSORPTION OF HIGHLY CHARGED POLYELECTROLYTES

Taking advantage of the fact that the adsorbed polyions retain their condensed counterions, the number of the adsorbed chains in each layer can be calculated from the electro-optical experiments. For this purpose, the electrical polarizability of a highly charged polyelectrolyte in a solution is compared to the polarizability of a particle coated with multilayer terminated by the same polyelectrolyte. Assuming ordering of nonoverlapping polymer chains on the particle surface, this comparison should give the number of the adsorbed chains in each layer. (Such estimation should confirm the above proposed explanation for increasing of the electro-optical effect with layer number because of the increased adsorption after deposition of each next layer.)

The electrical polarizability, γ, of a highly charged cylinder like polyelectrolyte in a solution can be calculated by using Manning's equation [65,66]:

$$\gamma = \frac{e^2 L^3}{12kTd} \cdot \frac{1 - \xi^{-1}}{1 - 2(\xi - 1) \ln \kappa d} \tag{17.4}$$

For the low molecular NaPSS, with $\xi = 2.75$ and $L = 99$ nm, this estimation gives $\gamma = 6.9 \times 10^{-32}$ F m^2. The electrical polarizability of a β-FeOOH particle (coated with film terminated by PSS) is then calculated by using the equation, which presents a quadratic dependence of the electro-optical effect on the electric field strength [49]:

$$\gamma = 4kT \frac{I_0(Ka, Kb)}{A(Ka, Kb)} \frac{d\alpha}{dE^2} \tag{17.5}$$

Figure 17.16 shows the increased slope of the electro-optical effect with layer number, indicating the increase in the electrical polarizability of a particle coated with one, two, three, and four PSS layers [22]. The values of γ are presented in Table 17.1, where the calculated number of the adsorbed PSS chains is also shown.

The number of the adsorbed PSS chains in each layer can be also estimated from the thickness data (Figure 17.3). The polyions are suggested to have rodlike conformation (low ionic strength, highly charged polyion). The distance between the adsorbed polyions is

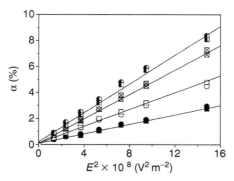

FIGURE 17.16 Electro-optical effect at 10^3 Hz as a function of the field strength for the low-molecular-weight PSS/PVP film: PSS1 (•), PSS2 (○), PSS3 (⊗), PSS4 (◐) and PVP1 (■), PVP2 (□), PVP3 (⊠), PVP4 (◪). (From Radeva, Ts. et al., *J. Colloid Interface Sci.*, 279, 351, 2004. With permission.)

suggested to be equal to the macromolecule contour length. The latter follows from the scaling theory of Dobrynin et al. [67], who state that the strong electrostatic repulsion between the adsorbed polyions forces them to get organized in a strongly correlated structure on the particle surface. The lateral correlations between the adsorbed polyions lead to nonoverlapping cells of radius that equals the half-contour length of the adsorbed macromolecule [60,67].

As shown in Table 17.1, the number of the adsorbed PSS chains, calculated from the increased particle size after deposition of each new layer, closely matches those obtained from the electrical polarizability data. Irrespective of the approximate character of both estimations (assuming additive contribution from each adsorbed polyion), they support our conclusion (see part 3) that the increase in the electro-optical effect with layer number is related to the increased adsorption of PSS in each new layer.

A comparison between the number of the adsorbed molecules, calculated from the thickness (Figure 17.4) and the electrical polarizability data (insets in Figure 17.14 and Figure 17.15) is presented in Table 17.2 for the PAA/PAH films, ending with fully charged PAA and PAH chains [23]. The number of adsorbed molecules from the polarizability data exceeds by two to three times those obtained from the thickness measurements. Shiratori and Rubner [31] have found, however, that the surface roughness of the PAH/PAA films is commensurable with the layer thickness. We may suggest, therefore, that the obtained greater number of the adsorbed PAA and PAH molecules from the electro-optical effect value is a result of the two to three times larger area of the film with rough surface.

Returning to the estimation of the adsorption from polarizability data, it should be noted that we assume unchanged optical functions in Equation 17.5 because the refractive index of

TABLE 17.1
Electrical Polarizability, γ, and Number of Adsorbed PSS Molecules in Each Layer Calculated Using Manning's Equation, n_γ, and Thickness Data, n_{L_H}, for the Low Molecular PSS/PVP Film

Adsorbed Layer	$\gamma \times 10^{31}$ (Fm^2)	n_γ	n_{L_H}
PSS1	3.0	4.3	6.6
PSS2	5.1	7.4	9.2
PSS3	7.1	10.2	10.0
PSS4	9.1	13.1	10.5

TABLE 17.2
Electrical Polarizability, γ, and Number of Adsorbed Fully Charged PAA and PAH Molecules in Each Layer Calculated from Experimental Data and Manning's Equation, n_γ, and from Thickness Data, n_{L_H}

Number of Layers	PAA			PAH		
	$\gamma \times 10^{31}$ (F m²)	n_γ	n_{L_H}	$\gamma \times 10^{31}$ (Fm²)	n_γ	n_{L_H}
1	9.9	10.4	5.0	10.7	1.0	1.1
2	11.5	12.1	6.0	14.3	1.4	1.2
3	17.8	18.0	6.5	21.4	2.1	1.3
4	18.2	19.1	7.0	23.0	2.2	1.4
5	23.8	25.0	7.5	25.0	2.4	1.4
6	29.4	30.8	8.0	–	–	–

the β-FeOOH particles is much higher ($n = 2.2$) [68] than those of the polymers ($n\sim1.5$) [2,6]. We disregard as well the contribution of the particle charge to the orientational effect. This is based on our earlier findings that the electrical properties of an adsorbed polymer in regime of surface charge overcompensation are close to the ones of a nonadsorbed polyelectrolyte [45–48]. All results presented in this review show similar electro-optical behavior for the films ending with the same polyelectrolyte. This confirms such a conclusion. On the other hand, the independence of the frequency behavior of the electrical polarizability on the layer number indicates that the small residual ions from the film bulk make no measurable contribution to the electro-optical effect. Our conclusion agrees well with findings of Xie and Granick [69], and Finkenstadt and Johnson [70], who have reported that the degree of ionization of a weak polyelectrolyte (within the film bulk) responded to the charge of the polymer included in the top layer. (This effect was shown to persist with a decay length that far exceeded the Debye length in the solution.) To explain the oscillation in the degree of polyelectrolyte ionization with layer number, the authors suggested that small ions were either localized along the polymer chains or effectively excluded from the multilayers in a way similar to their exclusion from multilayers formed from strong polyelectrolytes [8,71].

17.6 CONCLUSIONS

The layer-by-layer assembly of oppositely charged polyelectrolytes by adsorption from solution leads to the formation of ordered structures with a behavior that is expected to be dominated by the internal interfaces rather than by their volume properties. Many experimental techniques were applied to characterize the properties of the self-assembled films on the molecular scale. Experimental data collected in this review allow one to conclude that electro-optics reveals as a sensitive tool for characterizing the electrical properties of polyelectrolyte multilayers on colloidal particles. Our studies demonstrate that the number and mobility of the counterions only on the film surface govern the electro-optical behavior of the whole film. The role of small residual ions into the film bulk seems negligible. When the multilayer films are constructed from highly charged polyelectrolytes, the polarization of the condensed counterions along the chains of the last-adsorbed polymer causes the appearance of the additional low-frequency relaxation of the electro-optical effect near the range of particle rotation. This result confirms the theoretical predictions that the condensed counterions remain bound to the surface of the highly charged polyion when it adsorbs onto weakly charged substrate.

We believe that studying the electrical polarizability of polyelectrolyte multilayers on colloidal particles will gain valuable information also on the factors governing the stabilization and the electrostatic interactions in colloidal suspensions containing polyelectrolytes. The complexation between a polyelectrolyte and a charged colloidal particle of opposite sign is of interest, owing to its importance for various biological and commercial applications. The electro-optical investigations can be applied to understand more about the charge inversion and the lateral correlations between adsorbed polyions during colloid–polyelectrolyte complexation. These phenomena are of fundamental physical interest, since they may drastically alter the standard mean-field picture of the polyelectrolyte adsorption.

ACKNOWLEDGMENTS

I wish to thank the colleagues who have contributed material to this article, especially V. Milkova, K. Kamburova, and I. Petkanchin. The Bulgarian National Fund for Scientific Research (Project X-1212) is gratefully acknowledged for supporting this work.

REFERENCES

1. Decher, G., Hong, J.D., and Schmitt, J., Buildup of ultrathin multilayer films by a self-assembly process. 3. Consecutively alternating adsorption of anionic and cationic polyelectrolytes on charged surfaces, *Thin Solid Films*, 210, 831, 1992.
2. Decher, G., Fuzzy nanoassemblies: toward layered polymeric multicomposites, *Science*, 277, 1232, 1997.
3. Decher, G. et al., Layer-by-layer assembled multicomposite films, *Curr. Opin. Colloid Interface Sci.*, 3, 32, 1998.
4. Hammond, P.T., Recent explorations in electrostatic multilayer thin film assembly, *Curr. Opin. Colloid Interface Sci.*, 4, 430, 1999.
5. Bertrand, P., et al., Ultrathin polymer coatings by complexation of polyelectrolytes at interfaces: suitable materials, structure and properties, *Macromol. Rapid Commun.*, 21, 319, 2000.
6. Sukhorukov, G.B., et al., Layer-by-layer self assembly of polyelectrolytes on colloidal particles, *Colloids Surf. A*, 137, 253, 1998.
7. Donath, E., et al., Novel hollow polymer shells by colloid-templated assembly of polyelectrolytes, *Angew. Chem. Int. Ed.*, 37, 2202, 1998.
8. Schlenoff, J.B., Ly, H., and Li, M., Charge and mass balance in polyelectrolyte multilayers, *J. Am. Chem. Soc.*, 120, 7626, 1998.
9. Schlenoff, J.B. and Dubas, S.T., Mechanism of polyelectrolyte multilayer growth: charge overcompensation and distribution, *Macromolecules*, 34, 592, 2001.
10. Hoogeveen, N.G., et al., Formation and stability of multilayers from polyelectrolytes, *Langmuir*, 12, 3675, 1996.
11. Caruso, F., et al., Investigation of electrostatic interactions in polyelectrolyte multilayer films: binding of anionic fluorescent probes to layers assembled onto colloids, *Macromolecules*, 32, 2317, 1999.
12. Ladam, G., et al., *In situ* determination of the structural properties of initially deposited polyelectrolyte multilayers, *Langmuir*, 16, 1249, 2000.
13. Riegler, H. and Essler, F., Polyelectrolytes: intrinsic or extrinsic charge compensation? Quantitative charge analysis of PAH/PSS multilayers, *Langmuir*, 18, 6694, 2002.
14. Manning, G., Limiting laws and counterion condensation in polyelectrolyte solution. I. Colligative properties, *J. Chem. Phys.*, 51, 934, 1969.
15. Manning, G., Limiting laws and counterion condensation in polyelectrolyte solution. III. An analysis based on the Mayer ionic solution theory, *J. Chem. Phys.*, 51, 3249, 1969.
16. Möhwald, H., et al., From polymeric films to nanoreactors, *Macromol. Symp.*, 145, 75, 1999.
17. Durstock, M.F. and Rubner, M.F., Dielectric properties of polyelectrolyte multilayers, *Langmuir*, 17, 7865, 2001.

18. Michaels, A.S., Falkenstein, G.L., and Schneider, N.S., Dielectric properties of polyanion–polycation complexes, *J. Phys. Chem.*, 69, 1456, 1965.
19. Radeva, Ts., Milkova, V., and Petkanchin, I., Structure and electrical properties of polyelectrolyte multilayers formed on anisometric colloidal particles, *J. Colloid Interface Sci.*, 244, 24, 2001.
20. Radeva, Ts., Milkova, V., and Petkanchin, I., Electro-optics of colloids coated with multilayers from strong polyelectrolytes: surface charge relaxation, *J. Colloid Interface Sci.*, 266, 141, 2003.
21. Radeva, Ts., Milkova, V., and Petkanchin, I., Dynamics of counterions in polyelectrolyte multilayers studied by electro-optics, *Colloids Surf. A*, 240, 27, 2004.
22. Radeva, Ts., Milkova, V., and Petkanchin, I., Electrical properties of multilayers from low- and high-molecular-weight polyelectrolytes, *J. Colloid Interface Sci.*, 279, 351, 2004.
23. Radeva, Ts. and Grozeva, M., *In situ* determination of thickness and electrical properties of multilayers from weak polyelectrolytes, *J. Colloid Interface Sci.*, 287, 415, 2005.
24. Estrela-Lopis, I., et al., SANS studies of polyelectrolyte multilayers on colloidal templates, *Langmuir*, 18, 7861, 2002.
25. Streb, K.D., Elektro-optische untersuchungen zur koagulationstabilität wässirger α-Fe$_2$O$_3$ dispersionen, Ph.D. thesis, Akademie der Wissenschaften der DDR, Berlin, 1988.
26. Kanungo, S.B. and Mahapatra, D.M., Interfacial properties of two hydrous iron oxides in KNO$_3$ solution, *Colloids Surf.*, 42, 173, 1989.
27. Schwarz, B. and Schönhoff, M., Surface potential driven swelling of polyelectrolyte multilayers, *Langmuir*, 18, 2964, 2002.
28. Hübsch, E., et al., Controlling the growth regime of polyelectrolyte multilayer films: changing from exponential to linear growth by adjusting the composition of polyelectrolyte mixtures, *Langmuir*, 20, 1980, 2004.
29. Decher, G., Layered nanoarchitectures via directed assembly of anionic and cationic molecules, in *Comprehensive Supramolecular Chemistry*, Sauvage, J.-P. and Hosseini, M.W., Eds., Pergamon Press, Oxford, 1996, p. 507.
30. Van der Schee, H.A. and Lyklema, J., A lattice theory of polyelectrolyte adsorption, *J. Phys. Chem.*, 88, 6661, 1984.
31. Shiratori, S.S. and Rubner, M.F., pH-dependent thickness behavior of sequentially adsorbed layers of weak polyelectrolytes, *Macromolecules*, 33, 4213, 2000.
32. Claesson, P.M., Dahlgren, M.A.G., and Eriksson, L., Forces between polyelectrolyte-coated surfaces: relations between surface interaction and floc properties, *Colloids Surf. A*, 93, 293, 1994.
33. Perrin, F., *J. Phys. Radium*, 5, 497, 1934.
34. Varoqui, R. and Dejardin, Ph., Hydrodynamic thickness of adsorbed polymers, *J. Chem. Phys.*, 66, 4395, 1977.
35. Castelnovo, M. and Joanny, J.-F., Formation of polyelectrolyte multilayers, *Langmuir*, 16, 7524, 2000.
36. Hoogeveen, N.G., et al., Formation and stability of multilayers of polyelectrolytes, *Langmuir*, 12, 3675, 1996.
37. Lvov, Y., et al., A careful examination of the adsorption step in the alternate layer-by-layer assembly of linear polyanion and polycation, *Colloids Surf. A*, 146, 337, 1999.
38. Kolarik, L., et al., Building assemblies from high molecular weight polyelectrolytes, *Langmuir*, 15, 8265, 1999.
39. Schoeler, B., Kumaraswamy, G., and Caruso, F., Investigation of the infuence of polyelectrolyte charge density on the growth of multilayer thin films prepared by the layer-by-layer technique, *Macromolecules*, 35, 889, 2002.
40. Lukkari, J., et al., Preparation of multilayers containing conjugated thiophene-based polyelectrolytes. Layer-by-layer assembly and viscoelastic properties, *Langmuir*, 18, 8496, 2002.
41. Schoeler, B., Poptoshev, E., and Caruso, F., Growth of multilayer films of fixed and variable charge density polyelectrolytes: effect of mutual charge and secondary interactions, *Macromolecules*, 36, 5258, 2003.
42. Kovacevic, D., et al., Kinetics of formation and dissolution of weak polyelectrolyte multilayers: role of salt and free polyions, *Langmuir*, 18, 5607, 2002.

43. Graul, T.W. and Schlenoff, J.B., Capilaries modified by polyelectrolyte multilayers for electrophoretic separations, *Anal. Chem.*, 71, 4007, 1999.
44. Grosberg, A.Y., Nguyen, T.T., and Shklovskii, B.I., Colloquium: the physics of charge inversion in chemical and biological systems, *Rev. Mod. Phys.*, 74, 329, 2002.
45. Radeva, Ts., Electric light scattering of ferric oxide particles in sodium carboxymethylcellulose solutions, *J. Colloid Interface Sci.*, 174, 368, 1995.
46. Radeva, Ts. and Petkanchin, I., Electric properties of adsorbed polystyrenesulfonate I: dependence on the polyelectrolyte molecular weight, *J. Colloid Interface Sci.*, 220, 112, 1999.
47. Radeva, Ts., Electric light scattering of colloid particles in polyelectrolyte solutions, in: *Physical Chemistry of Polyelectrolytes*, Radeva, Ts., Ed., Marcel Dekker, New York, 2001, p. 305.
48. Radeva, Ts., Overcharging of ellipsoidal particles by oppositely charged polyelectrolytes, *Colloids Surf. A*, 209, 219, 2002.
49. O'Konski, C.T., Yoshioka, K., and Orttung, W.H., Determination of electric and optical parameters from saturation of electric birefringence on solutions, *J. Phys. Chem.*, 63, 1558, 1959.
50. Stoylov, S.P., *Colloid Electro-Optics*, Academic Press, London, 1991.
51. Buleva. M. and Stoimenova, M., Electro-optic investigation of oxide suspension. Mechanism of formation of the induced dipole moment, *J. Colloid Interface Sci.*, 141, 426, 1991.
52. Soimenova, M. and Radeva, Ts., Electro-optic investigation of oxide suspension. On the nature of "permanent dipole moment," *J. Colloid Interface Sci.*, 141, 433, 1991.
53. Ookubo, N., Hirai, Y., and Hayakawa, R., Anisotropic counterion polarizations and their dynamics in aqueous polyelectrolytes as studied by frequency-domain electric birefringence relaxation spectroscopy, *Macromolecules*, 22, 1359, 1989.
54. Schwarz, G., A theory of the frequency dielectric dispersion of colloid particles in electrolyte solution, *J. Phys. Chem.*, 66, 2636, 1962.
55. Pefferkorn, E. and Elaissari, A., Adsorption-desorption processes in charged polymer/colloid systems; structural relaxation of adsorbed macromolecules, *J. Colloid Interface Sci.*, 138, 187, 1990.
56. Varoqui, R., Tran, Q., and Pefferkorn, E., Polycation–polyanion complexes in the linear diblock copolymer of poly(styrene sulfonate)/poly(2-vinylpyridinium) salt, *Macromolecules*, 12, 831, 1979.
57. Wang, T.C., et al., Selective electroless nickel plating on polyelectrolyte multilayer platforms, *Langmuir*, 17, 6610, 2001.
58. Donath, E., et al., Nonlinear hairy layer theory of electrophoretic fingerprinting applied to consecutive layer by layer polyelectrolyte adsorption onto charged polystyrene latex particles, *Langmuir*, 13, 5294, 1997.
59. Sens, P. and Joanny, J.-F., Counterion release and electrostatic adsorption, *Phys. Rev. Lett.*, 84, 4862, 2000.
60. Cheng, H. and Olvera de la Cruz, M., Adsorption of rod-like polyelectrolytes onto weakly charged surfaces, *J. Chem. Phys.*, 119, 12635, 2003.
61. Glinel, K., et al., Influence of polyelectrolyte charge density on the formation of multilayers of strong polyelectrolytes at low ionic strength, *Langmuir*, 18, 1408, 2002.
62. Bordi, F. et al., Complexation of anionic polyelectrolytes with cationic liposomes: evidence of reentrant condensation and lipoplex formation, *Langmuir*, 20, 5214, 2004.
63. Shin, Y., Roberts, Y.E., and Santore, M.M., The relationship between polymer/substrate charge density and charge overcompensation by adsorbed polyelectrolyte layers, *J. Colloid Interface Sci.*, 247, 220, 2002.
64. Radeva, Ts. and Kamburova, K., Electro-optics of colloid-polyelectrolyte complexes: counterion release from adsorbed macromolecule, *J. Colloid Interface Sci.*, 293, 290, 2006.
65. Manning, G., Limiting laws and counterion condensation in polyelectrolyte solutions. V. Further development of the chemical model, *Biophys. Chem.*, 92, 65, 1978.
66. Manning, G.S. and Ray, J., Fluctuations of counterions condensed on charged polymers, *Langmuir*, 10, 962, 1994.
67. Dobrynin, A.V., Deshkovski, A., and Rubinstein, M., Adsorption of polyelectrolytes at oppositely charged surfaces, *Macromolecules*, 34, 3421, 2001.
68. Doppke, H. and Heller, W., Experimental investigation of light scattering. 9. Light scattering of flow oriented nonspherical particles, *J. Phys. Chem.*, 83, 1717, 1979.

69. Xie, A.F. and Granick, S., Local electrostatics within a polyelectrolyte multilayer with embedded weak polyelectrolyte, *Macromolecules*, 35, 1805, 2002.
70. Finkenstadt, D. and Johnson, D.D., Model of ionization response of weak polyacids in a layered polyelectrolyte self-assembly, *Langmuir*, 18, 1433, 2002.
71. Kharlampieva, E. and Sukhishvili, S., Ionization and pH stability of multilayers formed by self-assembly of weak polyelectrolytes, *Langmuir*, 19, 1235, 2003.

18 Electro-Optics of Composite Nanoparticles

M. Buleva and Stoyl P. Stoylov

CONTENTS

18.1 INTRODUCTION

Composites play essential roles in scientific fields such as structural and functional materials. According to the distribution of the components in the particles, composite particles can be classified into two different types: internally composite particles and externally composite particles [1]. Internally composite particles consist of two or more components. The homogeneous distribution of the second phase into the matrix particle could significantly affect the properties of the composite particles, and the further properties of the final product. However, such a homogeneous structure is hard to achieve through the conventional synthesis methods. Externally composite particles (coated composite particles) usually involve core particles, which are covered with a distinct coating layer. This structure of the composite particles results in changed optical, electrical, magnetic, catalytic, and other characteristics of the core particles. The possibilities of modification of these characteristics and the control of the technological processes at externally composite particles are much greater. There are many methods for covering the core particle with a layer of a different composition. Among the several techniques employed in the production of these materials in recent years, the heterocoagulation proves to be of a great importance. Various composite materials prepared via heterocoagulation are applied to obtain porous ceramics, pharmaceutical and drug delivery systems, organoinorganic composites, catalysts or catalyst carriers, adsorbents, etc. [2–8]. An understanding of the colloidal behavior of the dispersed systems of different components with regard to particle types, their chemical compositions, and morphologies, is required for design of nanostructural materials. Various experimental methods are used to investigate the structure and the properties of composites depending on the composition of disperse solution, size ratio, and the concentration ratio [9–17]. The aim of the this chapter is to illustrate the possibilities of electro-optical methods [18] in following the process of formation as well as the characteristics of composite particles (electric surface properties, size distribution, etc.). The application of electro-optics in the study of nanocomposites is of a great interest because the materials produced by hetero-coagulation of anisodiametric and spherical particles are of both theoretical [19] and practical interest [7,8]. It is expected that the high characterizing potential of electro-optical methods might find a new fertile field of application.

18.2 MATERIALS AND METHODS

18.2.1 MATERIALS

The investigation used β-FeOOH particles covered to a different degree with SiO_2 or synthetic clay Laponite.

β-FeOOH particles were prepared by acid hydrolysis of 1.8×10^{-4} M $FeCl_3$ solution containing 10^{-5} M HCl over a period of 3 weeks at room temperature [20]. The extraneous ions were removed by repeated centrifugation in distilled water until the supernatant conductivity reached the value of 10^{-4} S cm^{-1}. Highly monodispersed β-FeOOH particles were obtained by this procedure. Microphotographs show a ellipsoid like particle of major and minor axes about 270 nm and 70 nm, respectively (Figure 18.1a).

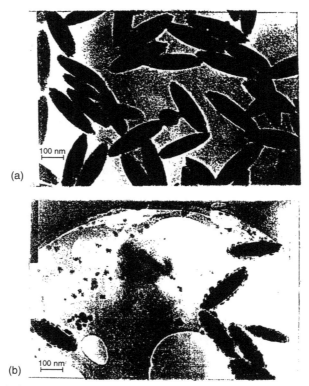

FIGURE 18.1 Transmission electron micrograph of core β-FeOOH particles (a) and composite particles (adsorbed Ludox) with R580 (b).

As in the case of all oxides and hydroxides, β-FeOOH particles in aqueous solutions adsorb mainly H^+ or OH^- acquiring positive or negative surface charge. Several authors have determined the electric double layer (EDL) of β-FeOOH [20–22]. The isoelectric point (IEP) appears at pH values approximately around 7.3. (The pH values of the IEP are between pH 6 and pH 8.) In acidic solution β-FeOOH is positively charged, whereas the surface charge is increased at lower pH values and increased ionic strength [21].

SiO_2 used was a Du Pont product Ludox HS40. The commercial product represents a 40 wt.% suspension of amorphous SiO_2 containing 0.41 w% Na_2O. At such a high particle concentration, the addition of alkaline substances and high pH values (pH = 9.7) ensures stability of the product. The Ludox particles are spherical, almost monodisperse with mean diameter about 12 nm. Figure 18.1B shows microphotograph of Ludox particles adsorbed on β-FeOOH particles.

Silica is the most investigated with respect to the EDL oxide. It represents a relatively acidic oxide: the point of zero charge (PZC) is between pH 2 and pH 3 [23–26]. At higher pH values, adsorbing hydroxyl ions negatively charges SiO_2 and the surface charge increases when the pH and ionic strength values increase. Data for the EDL of Ludox have been reported in references [23,24].

Laponite is a synthetic calcium silicate produced by Laponite Ind, similar in structure to the natural smectite clay [27]. A detailed study of the physical properties and parameters of aqueous Laponite dispersions has been carried out by Thomson and Butterworth [28]. Results obtained through a combination of methods (electron microscopy, electron diffraction, ultracentrifuge analysis, and small-angle x-ray scattering) indicate that the primary particles of this material are thin plate-like crystals of reasonably uniform size, having (plate) diameters of ca 20 nm. In dilute aqueous dispersion, the average thickness of discrete particles is 2–4 nm. It is possible to substitute magnesium ions with lithium or hydrogen ions in the crystal lattice of Laponite. As a result, the particles acquire negative surface charge that is compensated by the oppositely charged sodium ions (Na^+). This charge remains constant—it does not depend on the composition of the dispersed medium, i.e., pH values and the presence of salts. Most of the researchers accept the idea of Van Olphen [27] for positive charge on the edges at high pH values. Perkins et al. [29] have investigated the stability of Laponite sols within a broad pH range and presence of different salts and have compared it with the changes of the microelectrophoretic mobility under the same conditions. The electrophoretic mobility as a function of pH shows that the particle charge becomes more negative as the pH value increases above 6, which is explained by the adsorption of hydroxide ions of the positive edges of Laponite particles. Both Perkins et al. [29] and Thomson and Butterworth [28] mention the instability of Laponite particles in neutral and acidic media. However, the latter authors mention that the dissolution process is slowing (measured around the clock).

In our investigations, we used Laponite RD (Laporte), with mean particle size of 30 nm diameter and 1–2 nm thickness, as reported in Ref. [30].

The small Ludox and Laponite particles that were used differ both in their shape and size, as well as in their electric surface properties. As a result, under identical R ratio (see Section 18.3), it may be expected that the composites show different electric surface properties and respectively different electrophoretic and electro-optical behavior.

18.2.2 ELECTRIC LIGHT SCATTERING

The application of electric field (ac or dc) on a suspension of anisodiametric particles leads to their orientation. The orientation is due to the interaction of particle electric moments of different origins with the applied field. The electric light scattering (ELS) effect is defined in Ref. [18] as:

$$\alpha = (I_E - I_0)/I_0 \qquad (18.1)$$

where I_E, I_0 are the intensities of the scattered light when an electric field is applied and in the absence of an electric field, respectively.

At low degrees of orientation, i.e., when the energy of orientation is lower than kT, for small particles or for optically soft particles

$$\alpha = \alpha_\infty \times F(\gamma_i, \mu_{p,i}) \tag{18.2}$$

where γ_i and $\mu_{p,i}$ are the ith components of the electric polarizability and the permanent dipole moments; α_∞ is the ELS effect at very high electric fields, i.e., when all particles are completely oriented. For rotational ellipsoids, following Stoylov [18], we may present the orientation function for low degrees of orientation with the equation

$$\begin{aligned} F &= (\mu^2/kT + \triangle\gamma/kT)E^2 \\ \alpha &\sim (\mu^2/kT + \triangle\gamma/kT)E^2 \end{aligned} \tag{18.3}$$

where $\triangle\gamma = \gamma_\parallel - \gamma_\perp$ is the value of anisotropy of the electric polarizability.

The dependence of ELS effect on the field frequency gives the possibility to distinguish between different polarization mechanisms connected with permanent and induced dipole moments.

From the Equation 18.3, the initial slope $d\alpha/dE^2$ of the field dependence of ELS effect is proportional to particle polarizability at field frequencies where the contributions of "permanent" dipole moments have relaxed. The electric polarizability is connected mainly with electric double layer charges and their mobility. For oxide particles, the electric polarizability and its relaxation frequency correlate with the diffuse part of the EDL [31,32].

The transient ELS effect, following the switching off of the electric field, is connected with particles disorientation and it is given by

$$\alpha_t = \alpha_0 \exp(-6D_r t) = \alpha_0 \exp(-t/\tau) \tag{18.4}$$

where α_0 and α_t are, respectively, the steady state electro-optic effect and its value after time t, τ is the relaxation time of particles disorientation depending on particle dimensions and medium viscosity, $\tau = 1/6D_r$, where D_r is the rotational diffusion coefficient.

So, the ELS method gives the possibility to obtain electric polarizability and dimensions of the particles. The ELS measurements were made on a homemade apparatus [18].

18.2.3 MICROELECTROPHORESIS

Rank Brothers (MK II) apparatus was used for the measurement of the electrophoretic mobility, U_{ef}, in a flat quartz cell at 25°C. The electrophoretic mobility is proportional to the ζ-potential, i.e., the potential at the slipping plane [33].

18.3 PREPARATION OF THE COLLOIDAL SUSPENSIONS

In the present investigation, the composition of the composite particles is characterized by R — ratio between the bulk particle concentration of the smaller particles—Ludox or Laponite and that of β-FeOOH. The concentration of β-FeOOH particles was kept 6.5×10^{-3} g dm^{-3} (number concentration 7×10^{14} particles cm^{-3}) throughout all experiments. The concentration was selected in such a way, so as to avoid multiple optical scattering. The concentration was lower than the critical concentration of electric particle interaction by three orders of magnitude.

The dilution of the initial suspension of both components was carried out in two steps. The suspension of β-FeOOH is diluted initially to approximately 10^{-2} g dm^{-3}, adding NaCl until reaching 10^{-4} M NaCl. The pH value of the diluted suspension of β-FeOOH was approximately 5.5. The small deviations were corrected by adding HCl. The diluted suspension of Ludox might display a pH>7. In this case, HCl is added until a pH of 5.5 is reached and the ionic strength is corrected to $I = 10^{-4}$ by adding NaCl. The ratio R of the β-FeOOH and the Ludox composite was changed within the range 1–29000.

The Laponite suspensions were prepared by the dilution of the stock suspension with concentration 2.5%. This suspension was prepared just before the measurements and subjected to 60 s ultrasonic treatment to break down possible aggregates [30]. The diluted suspensions were prepared with 10^{-4} M NaCl. Correction of the pH value was not necessary, because within the entire concentration range, the pH value of Laponite suspensions used was about 5.5. The ratio R of β-FeOOH and Laponite composites was changed within the range 1–180 (approximately). The corresponding particle concentrations are calculated assuming a mean thickness 2 nm.

The examined concentration 6.5×10^{-3} g dm^{-3} of β-FeOOH and respectively that of the composite product was obtained by mixing equal volumes of the suspensions containing the two types of particles. The above mentioned dilution method ensured particles interaction at constant pH and ionic strength values for all examined disperse systems. The last condition is important in the case of oxide particles.

As a general rule, if another method not stated, the measurements with Ludox were carried out 1 h after mixing the components. In the case of Laponite, when almost all studied compositions form dispersions of unstable composites, the measurements were made at shorter intervals of time.

18.4   RESULTS AND DISCUSSION

18.4.1   INVESTIGATION OF β-FeOOH

Figure 18.2 (curve ○) shows the changes in the microelectrophoretic mobility of β-FeOOH as a function of pH values at constant ionic strength $I = 10^{-4}$. The result is in good agreement with results reported by other authors. The IEP value at pH $= 7.2$ is close to IEP in Ref. [21] and the same peculiarities are observed—a small maximum instead of a plateau within the region of high surface charges (pH < 5) and a steep decrease within the pH range 5–7.

Figure 18.3   shows the frequency response of the electro-optical effect (EOE) of a β-FeOOH suspension at several pH values. Generally, the shape of the dispersion curves coincides with the curves observed with other oxide suspensions, but the shape of the curve at low frequencies (lower than 300 Hz) is more complicated, suggesting the concurrent action of transversal and longitudinal permanent and/or slow induced dipole moments. Also, within the kHz range the observed plateau is relatively shorter. The EOE within the kHz plateau is due to the polarization of the EDL of particles and the decrease of EOE observed at higher frequencies is explained mainly by the relaxation of the charge-dependent polarizability [18].

Figure 18.4 shows the field dependencies of the EOE at a low intensity of orientating field and a frequency within the kHz plateau. The linear relationship upto $E \leq 6 \times 10^4$ V m^{-1} is observed at all investigated pH values. It corresponds to the orientation energy lower than kT (low degrees of orientation). The corresponding initial slopes are proportional to the value of anisotropy of the electric polarizability $\Delta\gamma = \gamma_{\parallel} - \gamma_{\perp} = \gamma$, with a proportionality coefficient depending on the optical properties of the particles. The determination of the absolute value of γ requires determination of the optical coefficient. As no explicit expressions for the optical coefficient are available, for particles of high refractive index, the value of γ cannot be exactly

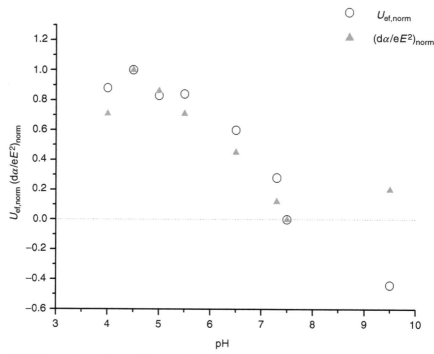

FIGURE 18.2 Electrophoretic mobility and electric polarizability of β-FeOOH at different pHs, normalized to pH 4.5.

determined. However, this value can be approximately accessed if we accept that the orientation energy and kT are approximately equal in fields at the end of the linear part. Values in the order of $10^{-14}\,\mathrm{cm}^3$ are obtained, which is lesser than the calculated value by an order of magnitude, e.g., for paligorskite [18], but near to the values reported for SiO_2 and Al_2O_3 [18,31,32]. At constant dispersity of the system, the optical coefficient is a constant and the relative changes of the initial slopes follow the relative change of the electric polarizability. As a control of the dispersity, we have used the relaxation time τ, which upto pH~6 retains its value. The electric polarizability (with accuracy up to the optic constant) as a function of pH is shown in Figure 18.2 (curve ▲). The experimental data show a qualitative correlation of the changes in the EOE to those of the electrokinetic potential. Also the relaxation frequency of polarizability (see Figure 18.3) is correlated to the value of the electrokinetic charge. The results correspond to the concept that the surface polarizability of β-FeOOH in the kHz frequency region is mainly due to the polarization of the diffuse part of the EDL. The same conclusion has been reported for other oxides [31,32]. The considerable decrease of $d\alpha/dE^2$ when approaching the IEP is accompanied by significant rise of the relaxation time τ (e.g., at pH = 7.5 at $E = 5.8 \times 10^4\,\mathrm{V\,m^{-1}}$, α_0 is only 3% and $\tau \approx 120\,\mathrm{ms}$). These changes show that an aggregation process starts when the electrostatic repulsion of particles in the dispersed system decreases. The result is in agreement with the foresight of the DVLO theory [34,35] for the stability of lyophobic colloids. Golikova et al. [17] have reported the same result for α-FeOOH.

18.4.2 INVESTIGATION OF COMPOSITE PARTICLES

When two soles with oppositely charged particles are mixed, it is expected that a hetero-coagulation process will proceed. In our specific case, when the particles have very different

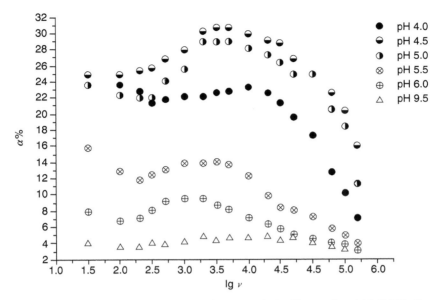

FIGURE 18.3 Frequency dependencies of the electro-optical effect α for β-FeOOH dispersions at different pHs, $E = 5.2 \times 10^4 \, \text{V m}^{-1}$.

sizes, we can consider the process as attachment of small particles onto large ones (adagulation). It is substantially to note that the small particles of Ludox and Laponite cannot be studied by ELS and microelectrophoresis because of their small sizes and low refraction index. Using the microelectrophoresis and electric light scattering, we can "see" and follow only the "core" β-FeOOH particles, as well as the hybrid particles in which β-FeOOH participates.

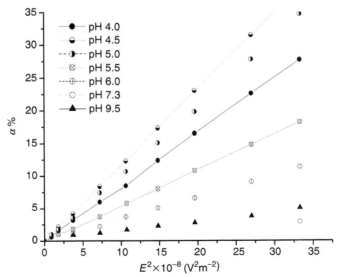

FIGURE 18.4 Field strength dependencies of the electro-optical effect α for β-FeOOH dispersions at different pHs, field frequency 3 kHz.

Thus, through the morphological changes and changes in the electric properties only of larger particles, we would follow the process of heterocoagulation, possibly the structure and shape, and at least the dimensions of aggregates within the instability region.

18.4.2.1 Investigation of β-FeOOH/Ludox Composite Particles

Figure 18.5 shows the changes of the electrophoretic mobility U_{ef} of aggregates as a function of R. The curve is rather similar to the U_{ef} of colloidal particles vs. the concentration of the specifically adsorbing ion [36,37]. An isoelectric point IEP is observed at $R\approx60$, then particles charge reversal follows and a slow increase of U_{ef} with a trend toward saturation above $R\approx600$ is seen.

The electro-optical study provided evidence that the stability of the mixed system is substantially different within the region around the IEP and for values surpassing $R\approx600$. To illustrate the electro-optical behavior of the dispersed system around the IEP, we show the results for $R\approx6$ (Figure 18.6 through Figure 18.8). The slope of the field dependencies of EOE for low degrees of orientation changes with the time (Figure 18.6). The relaxation times of hybrid particles are considerably longer than the values for β-FeOOH (Figure 18.7) and the effect at complete orientation α_∞ is different as compared with that of β-FeOOH (Figure 18.8). Both parameters (relaxation time τ and the effect at complete orientation of the particles α_∞) change with the time. It must also be noted that the shape of the curve α vs. τ (decay) is changed. It consists of two parts—abrupt part corresponding to the relaxation time of β-FeOOH and a long tail suggesting that a fraction containing much larger particles is appearing.

Within the plateau region ($R\geq600$), EOEs are constant with the observation time. The complete orientation effect α_∞ as well as the relaxation time τ of hybrid particles are also

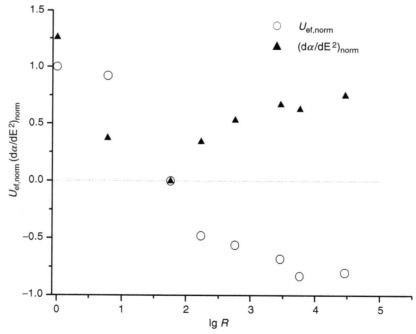

FIGURE 18.5 Electrophoretic mobility and electric polarizability of composite particles β-FeOOH Ludox with different R values (normalized to blank dispersion).

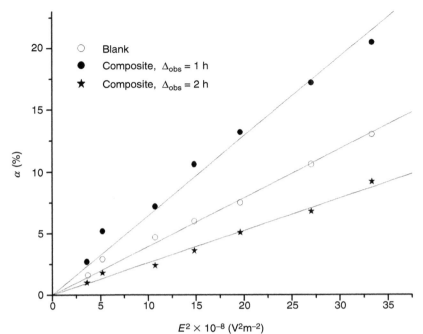

FIGURE 18.6 Field strength dependency of the electro-optical effect α for composite particles β-FeOOH and Ludox with $R6$, variable time of observation, field frequency 3 kHz.

invariable during the observation time. The observations indicate that the hybrid particles are stable in contrast of those around the IEP. Furthermore, the complete orientation effect α_∞ coincides with α_∞ of "bare" β-FeOOH (Figure 18.9) and the relaxation time τ of the hybrid particles at $R \geq 600$ is close to that of "bare" β-FeOOH (Figure 18.7). The comparison

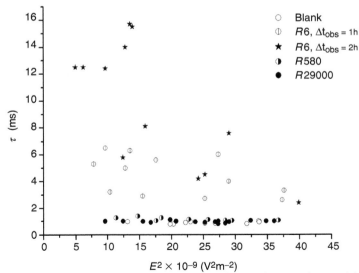

FIGURE 18.7 Field strength dependency of the relaxation time τ of composite particles β-FeOOH and Ludox with different R values.

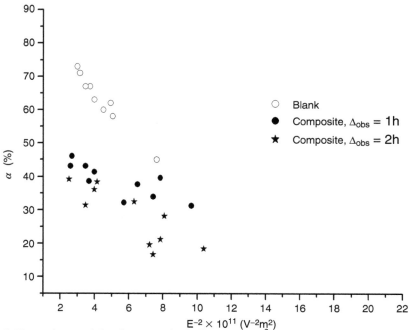

FIGURE 18.8 Dependency of the electro-optical effect α on E^{-2} for composite particles β-FeOOH and Ludox with $R6$. Field frequency 3 kHz.

between α_∞ and τ of core and hybrid particles show that at $R>600$ the hybrid particles are close in form and size to the "core" one.

When dispersions of oppositely charged particles are mixed, according to the theory developed by Derjaguin [38,39], no repulsion interactions are initiated but attraction forces appear during the overlapping of the electric double layers. The energy barrier of the interaction curve of the particles is zero and aggregation proceeds regardless of the ionic strength. Many model studies [9–17] of heterocoagulation show that the type of aggregates depends on ratio of the initial concentration of the sols, shape, and ratio of their dimensions, etc. Vincent and coworkers [14] have produced a series of publications for the interaction between small and large latex particles when the particles have opposite charge. The data indicate that at large particle number ratios (i.e., an excess of one type of particle), core-shell aggregates could be produced. However, it was also shown that, over a narrow range of particle ratio, small particles could act as a bridging flocculant for the large particles. Thus, in this particle concentration range, we may expect the formation of large fractal aggregates. This finding is supported by the research of Dumont et al. [15], who used particle-counting techniques to analyze the heteroaggregation in a similar system. These authors concluded that both the size ratio and the relative numbers of particles are important in determining the final structure of the heteroaggregates formed.

Our results show the value of R when aggregation starts is for sure higher than $R1$. At $R1$, the mixed dispersed system is stable. The relaxation time and the effect of complete orientation α_∞ do not change during the time and they are close to those of "bare" β-FeOOH (Figure 18.9). The electrophoretic mobility also does not differ essentially from that of β-FeOOH and does not change with time. Meanwhile, some changes in the low frequency dispersion of EOE and the increased surface polarizability (Figure 18.11) show that alterations might start early at this R-value in the "permanent" dipole moment and in the EDL of β-FeOOH.

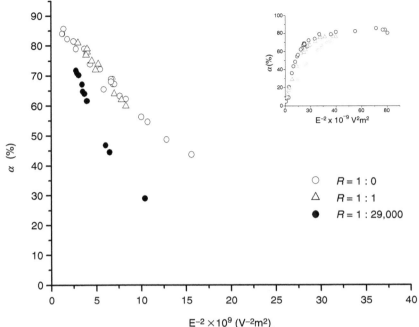

FIGURE 18.9 Dependency of the electro-optical effect α on E^{-2} for composite particles β-FeOOH and Ludox with different R values. Inset: field strength dependency of α. Field frequency 3 kHz.

Having in mind the theory of Derjaguin [38,39] and the main conclusions of Vincent et al. [14] we could describe the adagulation process in the mixed dispersions ($R>1$) in the following way.

The homogeneous sols of β-FeOOH and Ludox are stable as a result of the low ionic strength ($I = 10^{-4}$). In the mixed dispersion, as a result of adagulation "bare" β-FeOOH particles and β-FeOOH particles with Ludox particles attached are present. The flocculation would appear when two hybrid surfaces with opposite electric charges meet. It is obvious that at small R-values, as a result of the poor coverage of the core particle, many possible orientations of two "hybrid" large particles will result in a repulsion of EDL, similar to the case with homogeneous sol. With the increase of the covered surface the possibility of "effective" impacts and aggregation between large particles are also increased. If the number of small particles (expressed by R) continues to increase, such a surface coverage of large particles will be reached when "repulsion interactions" will be initiated during impact. The system will become stable, similar to a homogeneous sol. In our study such stabilization was reached surpassing $R\sim600$. A similar stabilization of AlOOH by the adhesion of Ludox is described by Iler [40], who stresses that surplus concentration is essential for the stabilization effect of small particles, rather than their high surface charge.

The considerable changes of the maximum orientation effect and of the relaxation time show that flocculation is present between the large β-FeOOH particles under the effect of small Ludox particles within the region $R = 6/173$. A smaller "step" in the change of R will probably broaden this range of flocculation.

We would like to pay attention to the reaching of IEP at values far below $R\approx600$, the number of Ludox particles geometrically required for a complete coverage of β-FeOOH. This could be explained by the complex structure of EDL of the investigated composite. The small particles have a diameter 12 nm, commensurable with the thickness of the ionic atmosphere

(k^{-1}) at the experimental conditions ($k^{-1} \approx 30$ nm at $I = 10^{-4}$). Considering this complex character, it is obvious that the assessment of the particles surface potential from the ζ-potential in this case would be not correct although it is generally accepted in the colloidal chemistry. So the ζ-potential remains the only classic characteristic of the EDL of composites [1]. This makes the characterization of the EDL by surface polarizability determined by electro-optical methods still more useful.

A surprising result is obtained within the range of instability $R = 6/173$. At $R \approx 60$ (IEP) the heterogeneous system is stable (the intensity of scattered light I_0 remains constant on time). Probably this interesting result can be explained by assuming that at a suitable distribution of small particles onto the large ones, the superposition of the electric double layers leads to $\zeta = 0$ and no electrostatic attraction between the hybrid particles exists. Chernoberezhskii et al. [41] report similar result in a homogeneous silica sol—a maximum in the stability is found at pH 2, where the electrostatic interaction was absent and only attractive forces operated between the particles. Harding [42] and Allen et al. [43] have found some peculiarities in stability of silica that are difficult to explain by the electrostatic stabilization. It might be interesting to confirm our observation at composite systems not containing silica, i.e., to manifest its more general significance.

At $R \geq 600$ stable hybrid particles are formed. Field strength dependencies of EOE of stable hybrids at low degree of orientation are shown in Figure 18.10 (the field frequency is chosen within the plateau of EOE vs. ν, assuming that only the induced dipole moment is active there). We already have noted that the complete orientation effect α_∞ of the stable hybrid particles do not depend on R (Figure 18.9). The relaxation times also do not change at $R > 600$ (Figure 18.7). The constant values of τ and α_∞ above $R \approx 600$ provide grounds to accept that the optic function in Equation 18.2 is not changed. Therefore, we can trace the changes in the surface polarizability γ of the hybrid particles by the alteration of the slope $d\alpha/dE^2$; at low orientation degrees, as we have already carried out during the study of the "bare" β-FeOOH. Data are shown in Figure 18.5 (curve ▲). It can be noted that in the commented region of R, the shape of γ vs. R is quite similar to the shape of U_{ef}, i.e., to the electrokinetically mobile charge vs. R. A correlation exists in a slight increase with a trend

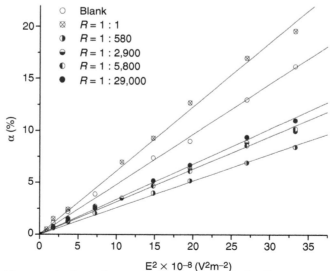

FIGURE 18.10 Field strength dependency of the electro-optical effect α for composite particles β-FeOOH and Ludox with different R values. Field frequency 3 kHz.

to saturation, as well as in the value of R, where the surface polarizability γ and the electro-kinetic potential ζ are nullified. This conformity shows that the movement of ions mainly beyond the slipping plane of EDL forms the kHz-induced dipole moment in these cases.

The frequency response curves of EOE for $R \geq 600$ are shown in Figure 18.11. It can be noted that their overall shape is not considerably changed in comparison with "bare" particles. The short kHz plateau is retained. The relaxation frequency of γ displays slight changes with R, i.e., it depends on the value of the polarizing charge as observed in the case of oxide particles [31,32]. The interesting change is noticed within the low frequency region. The minimum at about 200 Hz goes deeper at $R1$ and it disappears in the case of hybrid particles with $R \geq 600$.

This comparison of the low frequency dispersion of electro-optical effects at different R values with that of the reference blank sample provides grounds to propose the following picture for the formation of the "coverage" layer of nanoparticles. The deeper minimum at about 200 Hz, when the "first" Ludox particle is adsorbed onto β-FeOOH, corresponds to the decrease of the longitudinal dipole moment. It may be assumed that the Ludox particle is attached on the sides of β-FeOOH particle, as it is seen in Figure 18.1B. The formation of a compact layer by the smaller particles leads to the "smoothening" of the minimum at 200 Hz—the surface becomes more homogeneous and the particles more symmetric. The quite similar values of the relaxation times of the hybrid and "bare" particles provide evidence that the Ludox particles form a monolayer. This picture can be seen in the TEM micrograph (Figure 18.1B).

The results on electric polarizability and dipole moment of hybrid particles support the conclusion that the Ludox monolayer determines the surface electrical properties of the new particle. The conclusion is confirmed by the preliminary experiments on the effect of pH on the electro-optical behavior of composite particles.

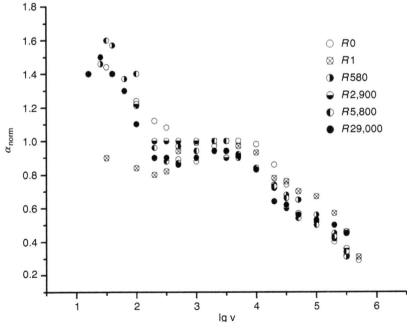

FIGURE 18.11 Frequency dependencies of the electro-optical effect α for β-FeOOH and Ludox dispersions at different R values (normalized to "plateau").

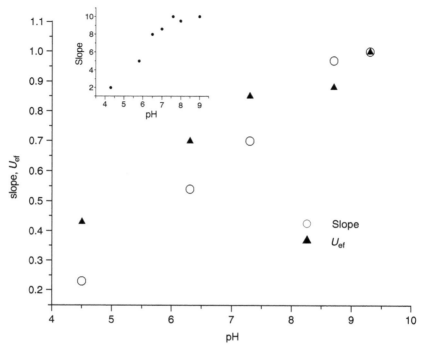

FIGURE 18.12 Electric polarizability and microelectrophoretic mobility for composite particles β-FeOOH and Ludox with R29,000 at different pHs, normalized to pH 9.5. Inset: electric polarizability vs. pH for silica particles. (From Buleva, M., Ph.D. thesis, Sofia, 1982.)

The measurements have been carried out using composite particles formed at $R = 29,000$ separated by centrifugation and again dispersed in solutions with $I = 10^{-4}$ at different pH values. The results are shown in Figure 18.12 and Figure 18.13. The influence of pH on the electric polarizability of composites and on silica particles is similar (Figure 18.12). This result is a conformation of the role of the Ludox monolayer. The correlation between the polarizability of the composite particles and the electrokinetically mobile charge confirms the above-mentioned assumption that within the kilohertz region the orientation of composites is due to the polarization of a part or of the entire diffuse layer.

The dispersion curves (Figure 18.13), however, show a completely different shape depending on ratio between the components electric charge. These differences cover both the low frequency and kilohertz region. At the same time the electrokinetic charge remains negative, i.e., predominant only in the ions at the most external part of the EDL of silica. At this stage we do not propose any explanation. We can only stress that the ELS is more informative and in greater detail reflects the structure of the EDL of the composite. It is worthwhile to use EO for the characterization of the EDL of composite particles and obviously electrokinetics is not enough as a supplementary method. The ELS is very useful when investigating similar particles with electrochemically and geometrically complex surfaces. Probably this is valid in the case of other complex surface systems, e.g., particles covered with polyelectrolyte multilayer.

18.4.2.2 Investigation of the β-FeOOH/Laponite Composites

Figure 18.14 shows the relationship between electrophoretic mobility of the composite versus the composition of the suspension. An abrupt decrease of mobility, up to the reversal of the

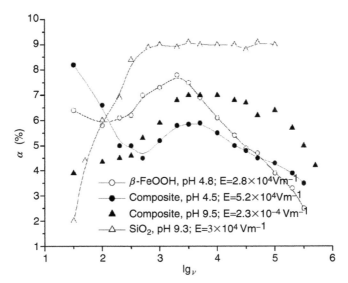

FIGURE 18.13 Frequency dependencies of the electro-optical effect α for composite particles β-FeOOH and Ludox with $R = 29,000$ at different pHs.

charge sign of particles is observed as already shown in the case of β-FeOOH and Ludox. Within the selected R range a similar clear-cut tracing of the region around IEP is absent, but it can be seen that the IEP is below and around $R = 6$, i.e., a relatively low value, as in the case of β-FeOOH and Ludox. The shape of the electrophoretic mobility curve, however, does not show a trend toward saturation, as observed in the case of β-FeOOH and Ludox.

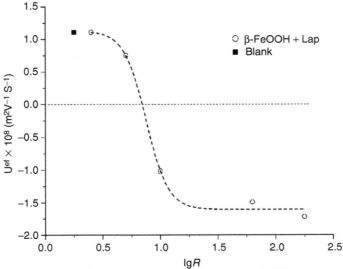

FIGURE 18.14 Electrophoretic mobility of composite particles β-FeOOH and Laponite with different R values.

The electro-optical behavior of the mixed suspensions shows the same peculiarities as observed in the case of β-FeOOH and Ludox near the IEP. The ELS characterizes all investigated compositions (with some exception at $R = 180$) as flocculating systems i.e., the formation of a stable composite by complete coverage of the β-FeOOH surface by a coating layer of small particles might not be reached.

In the case of $R = 2/5$ the relaxation times have increased up to 9–10 ms and the shape in the relaxation is even more distinctly changed (Figure 18.15). Above the IEP in the mixed suspension the flocculation process remains predominant. Only at $R = 180$ it may be accepted that stable "hybrids" are formed, but even at this composition the relaxation times are longer as compared with the value of the basic particle. The strong relationship between the relaxation times and the orientating field shows that the "hybrids" have a considerable polydispersity (Figure 18.16). In the similar manner the effects at maximum fields strength can be interpreted (Figure 18.17).

Evidently the instability of the β-FOOH and Laponite suspensions at RH1 and at higher R value makes them a less appropriate choice for investigating the structure of the composite particles, of their EDL and their relation to suspension stability. However, these composite particles present a way for studying the electrokinetic and other surface electric properties of Laponite particles, which is not possible directly for these particles alone.

A good perspective might be the study of composite particles of β-FeOOH with smaller spherical particles than Ludox. Experimental studies of such composite particles are envisaged in the near future.

18.5 CONCLUSIONS

The electro-optical studies of composite particles built from a large anisodiametric particles (e.g., several hundreds of nanometers) with much smaller spherical particles (e.g., less than

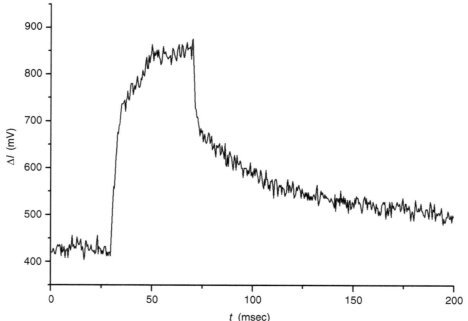

FIGURE 18.15 Transient electro-optical effect for composite particles β-FeOOH and Laponite with $R5$, $E = 5.2 \times 10^4\,\mathrm{Vm^{-1}}$, $\nu = 5\,\mathrm{kHz}$.

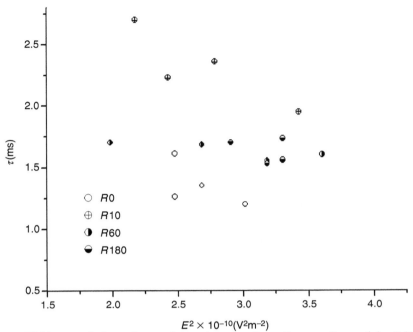

FIGURE 18.16 Field strength dependency of the relaxation time of composite particles β-FeOOH and Laponite with different R values.

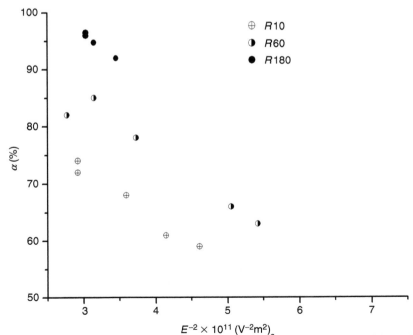

FIGURE 18.17 Dependency of the electro-optical effect α on E^{-2} for composite particles β-FeOOH and Laponite with different R values.

10–20 nm) are of great interest in at least three directions. The first direction is the deeper and wider characterization of the surface electric properties of the smaller spherical and nonspherical nanoparticles. This dramatically widens the characterization potential of electro-optical methods with respect to particles' sizes and shapes.

The second direction is the qualitatively new possibilities for an electro-optical approach to the study of the structure of the equilibrium and nonequilibrium EDL, the neutralization of surface change and their relation to particle interactions, and suspension stability.

The third direction is the utilization of the electro-optical methods for the creation of new, rapid, independent ways (electrical and optical) for the elucidation of composite particle structure. This may be done both through the determination of the particle permanent dipole moments, which is related to particle charge asymmetry and structure asymmetry and particle optical anisotropy. The determination of the composite particle asymmetry, is best demonstrated for the β-FeOOH and Ludox composite particle in this chapter for $R1$. It should be stressed that in all cases electro-optical methods give objective average information in a fraction of seconds, which is impossible with imaging methods. Furthermore, if appropriate software is used images of average composite particle size, shape, and structure might be constructed. This might be useful tool for studies of not only particle design but also its dynamics.

REFERENCES

1. Vang, H., et al, *Encyclopedia of Surface and Colloid Science*, Marcel Dekker, N.Y., 1118, 2002.
2. Osawa, C.C. and Bertran, C.A., Mullite formation from mixtures of alumina and silica sols and pH effect, *J. Braz. Chem. Soc.*, 16, 251, 2005.
3. Garino, T., Heterocoagulation as an inclusion coating technique for ceramic composite processing, *J. Am.Ceram. Soc.*, 75, 514, 1992.
4. Carlos, M.O., et al., Porous ceramics through heterocoagulation process, *Ceramica*, 51, 78, 2005.
5. Yamaguchi, K., et al., Preparation of functional core-shell particles by heterocoagulation, *Chem. Lett.*, 31, 1188, 2002.
6. Yamaguchi, K., et al., Preparation of core shell composites polymer particles by a novel heterocoagulation, based on hydrophilic interaction, *Colloid. Polym. Sci.*, 282, 366, 2004.
7. Yan-Qin, J., et al., Preparation of nanostructure materials by heterocoagulation—interaction of montmorillonite with synthetic hematite particles, *Langmuir*, 20, 9796, 2004.
8. Voorn, D.J., et al., Controlled heterocoagulation of platelets and spheres, *Langmuir*, 21, 6950, 2005.
9. Maroto, J.A. and de las Nueve, F.J., Influence of the ionic strength in the heterocoagulation process between bare and surfactant coating latexes, *Colloid Polym. Sci.*, 277, 881, 1999.
10. Glover, S., et al., Bridging flocculation studied by light scattering and settling, *Chem. Eng. J.*, 80, 3, 2000.
11. Biggs, S., et al., Aggregate structures forming via a bridging flocculation mechanism, *Chem. Eng. J.*, 80, 13, 2000.
12. Yan, Y., et al., The structure and strength of depletion force induced particle aggregation, *Chem. Eng. J.*, 80, 23, 2000.
13. Kim, A.Y. and Berg, J.C., Fractal heteroaggregation of oppositely charged colloids, *J. Colloid Interface Sci.*, 229, 607, 2000.
14. Uricanu, V., Eastman, J.R., and Vincent, B., Stability in colloidal mixtures, containing particles with a large disparity in size, *J. Colloid Interface Sci.*, 233, 1, 2001.
15. Dumont, F., Ameryckx, G., and Watillon, A., Heterocoagulation between small and large colloidal particles, Part I. Equilibrium aspects, *Colloids Surf.*, 51, 171, 1990.
16. Yates, P., et al., Heterocoagulation of particles systems: aggregation mechanisms and aggregate structure determination, presented at 6th World Congress of Chemical Engineering, Melbourne, Australia, 2001.
17. Golikova, E.V., Burdina, N.M., and Vysokovskaya, N.V., Aggregation stability of SiO_2, FeOOH, ZrO_2, CeO_2, and diamond sols and their binary mixtures. 2. The photometric study of heterocoa-

gulation of SiO_2 – FeOOH, SiO_2 – ZrO_2, SiO_2 – CeO_2, and CeO_2 – natural diamond binary systems in KCl solutions, *Colloid J.*, 64, 155, 2002 (in Russian).

18. Stoylov, S.P., *Colloid Electro-optics*, Academic Press, London, 1991.
19. Koendrink, G.H., Rotational and translational diffusion in colloidal mixtures, Ph. D. Thesis, University of Utrecht, The Netherlands, 2003.
20. Matijevic, E. and Scheiner, P., Ferric hydrous oxide sols. III. Preparation of uniform particles by hydrolysis of Fe (III)-chloride, -nitrate, and—perchlorate solutions, *J. Colloid Interface Sci.*, 63, 509, 1978.
21. Kanungo, S.B. and Mahapatra, D.M., Interfacial properties of two hydrous iron oxides in KNO_3 solution, *Colloids Surf.*, 42, 173, 1989.
22. Onoda, G.Y. and de Bruyn, P.L., Proton adsorption at the ferric oxide/aqueous solution interface. I. A kinetic study of adsorption, *Surf. Sci.*, 4, 48, 1966.
23. Allen, L.H. and Matijevic, E., Stability of colloidal silica. I. Effect of simple electrolytes; II. Ion exchange, *J. Colloid Interface Sci.*, 33, 420, 1970.
24. Laven, J. and Stein, H., The electroviscosous behavior of aqueous dispersions of amorphous silics, *J. Colloid Interface Sci.*, 238, 8, 2001.
25. Heston, W.M., Iler, R.K., and Sears, G.W., The adsorption of hydroxyl ions from aqueous solution on the surface of amorphous silica, *J. Phys. Chem.*, 64, 147, 1960
26. Tadros, Th. and Lyclema, H., Adsorption of potential-determinining ions at the silica/aqueous electrolyte interface and the role of some cations, *J. Electroanal. Chem. Interfacial Electrochem.*, 17, 267, 1969.
27. Van Olphen, H., *An Introduction to Clay Colloid Chemistry*, Wiley, N.Y., 1977.
28. Perkins, R., Brace, R., and Matievic, E., Colloid and surface properties of clay suspensions. I. Laponite CP, *J. Colloid Interface Sci.*, 48, 417, 1974.
29. Thompson, D.W. and Butterworth, J.T., The nature of Laponite and its aqueous dispersions, *J. Colloid Interface Sci.*, 151, 236, 1992.
30. Zhivkov, A. and Stoylov, S.P., Electro-optical characterization of aqueous Laponite suspensions, *Colloids Surf.*, 209, 315, 2002.
31. Buleva, M., Electric moments of colloid particles and structure of the electric double layer, Ph.D. Thesis, *Bulg. Acad. Sci.*, Sofia, 1982.
32. Buleva, M. and Stoimenova, M., Electro optic investigation of oxide suspensions. Mechanism of formation of induced dipole moment, *J. Colloid Interface Sci.*, 141, 426, 1991.
33. Hunter, R.J., *Zeta Potential in Colloid Science*, Academic Press, London, 1991.
34. Derjaguin, B.V., On the repulsive forces between charged colloid particles and on the theory of slow coagulation and stabilization of liophobic sols, *Trans. Faraday Soc.*, 36, 203, 1940.
35. Verwey, E.J.W. and Overbeek, J.Th.G., *Theory of the Stability of Lyophobic Colloids*, Elsevier, Amsterdam, 1948.
36. Wiese, G.R. and Healy, T.W., Adsorption of Al (III) at the TiO_2 – H_2O Interface, *J. Colloid Interface Sci.*, 51, 434, 1975.
37. Golub, T.P., Koopal, L.K., and Sidorova, M.P., Adsorption of cationic surfactants on silica surface: I. Adsorption isotherms and surface charge, *Colloid J.*, 66, 43, 2004 (in Russian).
38. Derjaguin, B.V., A theory of heterocoagulation, *Discuss Faraday Soc.*, 18, 85, 1954.
39. Derjaguin, B.V., Stability of colloid systems, *Uspehi Chimii.*, 48, 675, 1954 (in Russian).
40. Iler, R.K., Adsorption of colloidal silica on alumina and a colloidal alumina on silica, *J. Am. Ceram. Sci.*, 47, 194, 1964.
41. Chernoberezhskii, J.M., et al., Evaluation of the boundary layers thickness by data on particles stability and aggregation in aqueous quartz sol, in *Surface Forces in Thin Liquid Films*, Nauka, Moscow, 1979, p. 67.
42. Harding, R.D., Stability of silica dispersions, *J. Colloid Interface Sci.*, 35, 172, 1971.
43. Allen, L.H. and Matijevic, E., Stability of colloidal silica, II. Ion exchange, *J. Colloid Interface Sci.*, 33, 420, 1970, III. Effect of hydrolysable ions, ibid. 35, 66, 1971.

Part II

New Applications

C. Structured Systems

19 Electro-Optics of Colloidal Crystals

Akira Tsuchida and Tsuneo Okubo

CONTENTS

19.1 INTRODUCTION

When a suspension of monodispersed colloidal spheres is deionized exhaustively in polar solvents such as water, a crystal-like structure is formed. This structure is called as colloidal crystals, and their physical properties have been extensively studied from 1970s [1–5]. Colloidal crystals are formed by the repulsion of extended electrical double layers surrounding the spheres and the Brownian motion of the spheres. The thickness of the electrical double layers is approximated by the Debye screening length, D_l, given by the equation

$$D_l = (4\pi e^2 n / \varepsilon k_B T)^{-1/2} \tag{19.1}$$

where e is the electronic charge, n is the concentration of free-state simple ions in suspension, ε is the dielectric constant of solvent, k_B is the Boltzmann constant, and T is the absolute temperature. Note that the maximum value of D_l observed for the exhaustively deionized suspension in water is about 1 μm and longer when compared with the size of the typical colloidal particles. The importance of the electrical double layers for colloidal crystallization has been demonstrated by many researchers [1–5].

When colloidal crystals are illuminated by white light, they show very beautiful iridescence, because the interparticle distance of the crystal lattice is just in the range of optical wavelengths and each single crystal reflects light in different colors by the Bragg diffraction. As the electrical double layers are very soft and the colloidal spheres are negatively charged, the lattice spacing changes reversibly and synchronously when an alternating electric field is applied. Electro-optic effects, i.e., optical responses in an electric field, have been investigated for colloidal dispersions [6,7]. Stoimenova et al. [8–11] have studied electro-optic effects such as resonance frequencies, harmonic oscillation, and acoustic shear waves for colloidal crystals. The authors have studied the electro-optic effects of colloidal crystals by reflection spectroscopy [12–20]. Significant modulation effects are observed in the phase difference, amplitude, waveform transformation, harmonics generation, and acoustic shear waves, for example. The colloidal crystals also show electric potential storage ability as an electric energy capacitor [17]. Substantial electro-optic responses, about 100 times larger than those with white light source, are obtained by a laser light source [18].

As for the term "electro-optic effects," it is originally used for the nonlinear optical effects caused by the application of an external electric field. The electro-optic effects in the narrow sense treat the optical effects due to the Pockels and Kerr effects within higher nonlinear polarization. The nonlinear polarization is expressed by Equation 19.2.

$$P = P_0 + \varepsilon_0(\chi^{(1)} E + \chi^{(2)} EE + \chi^{(3)} EEE + \cdots) \tag{19.2}$$

where $\chi^{(1)}$ is the linear susceptibility responsible for linear absorption and refraction, $\chi^{(2)}$ is the second-order nonlinear term responsible for second harmonics generation and the Pockels effect, and $\chi^{(3)}$ is the third-order nonlinear term responsible for third harmonics generation and the Kerr effect. In general, electro-optic effects of colloidal systems are caused by the deformation of electrical double layers around the particles. The shear flow by the applied electric field induces orientation of the particles, and dipole-induced particle orientation causes the Kerr effect. In this chapter, however, the term electro-optic effects is taken in the broad sense, that is, all optical effects caused by the application of an electric field.

19.2 COLLOIDAL CRYSTALS IN AN ELECTRIC FIELD

Figure 19.1a and Figure 19.1b show typical examples of reflection and absorption spectra, respectively, measured for CS82 (diameter, 103 nm) colloidal silica suspension in water. Both reflection and absorption peak wavelengths agree excellently. The slight rise in the blue side of the absorption spectra is due to the multiple scattering of the suspension.

When incident and reflected light beams are aligned perpendicular to the measurement cell wall, the interparticle distance (D) of colloidal spheres in the crystal lattice is calculated to be

$$D = 0.6124(\lambda_p/n_s) \tag{19.3}$$

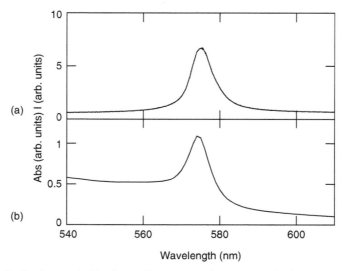

FIGURE 19.1 (a) Reflection and (b) absorption spectra for the colloidal crystals of monodispersed colloidal silica suspension ($\phi = 0.0427$).

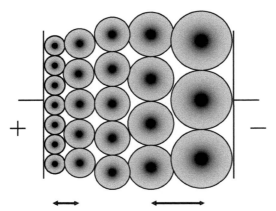

FIGURE 19.2 Schematic representation of a colloidal crystal in an electric field.

where λ_p is the peak wavelength in the reflection spectra, and n_s is the refractive index of the suspension, which is substituted by that of water for the diluted suspension. D_{obs} obtained from the Bragg peak of the reflection spectra agrees excellently with D_0 calculated from the sphere concentration of the suspension [4–5].

The application of an electric field to the colloidal crystals causes the deformation of the crystal lattice, as illustrated in Figure 19.2, due to the distortion of the electrical double layers of charged colloidal particles. In this figure, black circles and the surrounding shaded circles represent colloidal spheres and electrical double layers around the spheres, respectively. In most cases colloidal particles are charged negative, and they are attracted and repelled toward positive and from negative electrodes, respectively. This causes an increase in the particle density of colloidal spheres at the anodic side, and a decrease at the cathodic side. As a result, the interparticle distance at the anodic area decreases and that at the cathodic area increases, and then the reflection peak shifts to blue and red, respectively. Thus, when an alternating electric field is applied to the crystals, the reflection and absorption peaks should shift periodically. We have studied the electro-optic effects of colloidal crystals in detail, and many interesting effects, such as wavelength selection, harmonic generation, phase shift, resonance effect, waveform modulation, as shown in Figure 19.3, were clarified. We review the electro-optic effects of colloidal crystals in this chapter.

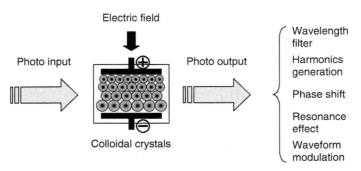

FIGURE 19.3 Electro-optic effects of colloidal crystals.

19.3 EXPERIMENTAL METHODS FOR MEASURING ELECTRO-OPTIC EFFECTS

Figure 19.4 shows electro-optic cells equipped with electrodes. These are designed and used for our electro-optic measurements [12–20]. Colloidal suspensions in the cells are sealed tightly to prevent contamination by air and other impurities. When the suspension coexisted with ion-exchange resins, the measurements are made in deionized suspension. Figure 19.4a shows a rectangular cell equipped with NESA transparent electrodes at the top and the bottom inside the cell walls. The cell has acrylic resinous sidewalls and the measurements in perpendicular (from the sidewalls) and parallel (from the top and the bottom walls) alignments against the applied electric field are available [12,18]. Figure 19.4b shows a rectangular cell equipped with two NESA transparent electrodes on a partial area at the top and the bottom inside the cell walls. It is possible to measure wave propagation outside the electrode region. The effects of cell walls on the wave propagation are also revealed by changing the cell thickness as 3, 10, and 20 mm [16,20]. Figure 19.4c presents a bridge-type cell with a pair of platinum electrodes inside. The cell is suitable for the measurements in perpendicular direction against the applied electric field. Nontransparent platinum electrodes are not suitable for the measurements across the electrode; however, the application of higher electric field is possible when they are compared with NESA electrodes. A cell shown in Figure 19.4d is similar in shape to that of Figure 19.4c, and is suitable for the measurements of wave propagation to outside areas and inside between the platinum electrodes. Many cells with different electrode separations were made, and the distributions of the modulation intensity

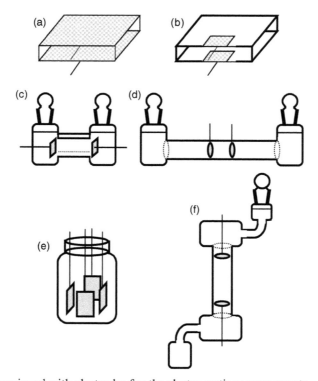

FIGURE 19.4 Cells equipped with electrodes for the electro-optic measurements.

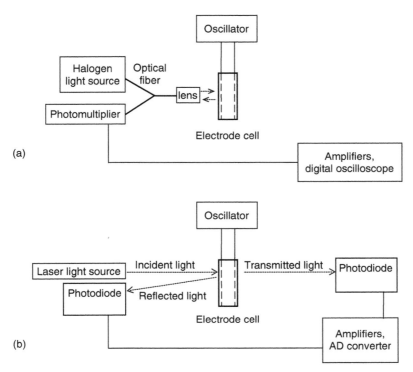

FIGURE 19.5 Schematic representations for the measurement apparatus of reflection and transmittance spectra, using (a) white light source and (b) laser light source.

were measured [17]. Figure 19.4e is a cylinder-type cell equipped with four platinum electrodes inside. Many patterns of electric field application are possible in the measurements from circumference around the cell sidewalls. The whole volume of colloidal crystals in the cell oscillates synchronously with a symmetry plane at the center of the electrodes [14,16]. Figure 19.4f is a vertical-type cell equipped with a pair of platinum electrodes at the top and the bottom areas inside the cell. This cell is appropriate for the measurements of electro-optic effects in the pseudo-microgravity. Here, the convection of solvent molecules is still active even if the charged particles are suspended freely irrespective of the earth gravity.

Figure 19.5 shows schematic representations of the instruments measuring the reflection and transmittance spectra in an electric field. The alignment (a) is used for the measurements of reflection intensity. White light from a halogen light source is focused at the inside surface of the cell wall. The reflected light intensity is detected by a photomultiplier and recorded on a digital oscilloscope. The time-resolved reflection spectra are available by substituting the photomultiplier with a multichannel photodetector. Commercially available multichannel spectroscopy apparatus records the time-resolved spectra in 20 ms time interval. The alignment b of Figure 19.5 is for the measurements of reflection and transmittance intensities with a laser light source. The matching of the Bragg wavelength of colloidal crystals with the laser wavelength enables very sensitive detection of the Bragg peak shift.

19.4 ELECTRO-OPTIC EFFECTS OF COLLOIDAL CRYSTALS

Figure 19.6 shows the time-resolved reflection spectra of colloidal crystals obtained with a sinusoidal electric field application [18]. The concentration of CS91 colloidal silica particles

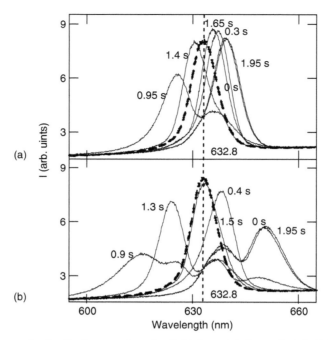

FIGURE 19.6 Time-resolved reflection spectra of colloidal crystals in an electric field of (a) $E = 5\,\text{V/cm}$ and (b) $E = 10\,\text{V/cm}$.

(diameter, 110 nm) is adjusted to be $\phi = 0.0329$ in volume fraction, in order to bring the Bragg reflection peak wavelength of the crystals in line with that of He–Ne laser at 632.8 nm. In the figure, the Bragg peak position without electric field ($E = 0$) (broken bold curve) matches well with the laser wavelength. Application of a sinusoidal electric field ($f = 0.5\,\text{Hz}$, $E = 5$, and 10 V/cm for Figure 19.5a and Figure 19.5b, respectively) causes the shift of the Bragg reflection peak to red and blue side, alternatively. The larger the applied electric field (is), the larger the swing of the peak wavelength. Reflection intensity of the laser light increases and decreases periodically. This means that the wavelength selection is attained by the application of electric field on the colloidal crystals. The Bragg peak wavelength of colloidal crystals can be changed easily by the change in the concentration of spheres. Thus, it is possible to use the colloidal crystals as wavelength-variable and tunable-filter optics.

Figure 19.7a and Figure 19.7c show the waveforms of reflection and transmittance intensities, respectively, as a function of time, measured at the same conditions as in Figure 19.6. Both waveforms are in the relation of complementary colors, and the waveform modulation by an external electric field is attained. Waveform d is the applied sinusoidal electric field and the waveform b is the reproduction from the time-resolved reflection spectra as shown in Figure 19.6b. Both Figure 19.7a and Figure 19.7b agree well, and this shows that the optimum positioning of the Bragg peak against the laser wavelength enables the prediction of modulated response waveforms. The success in prediction traces the way for the application of colloidal crystals as optical waveform modulation and phase shift devices [18].

When white light is used as the light source, prediction of the response waveform is a bit difficult. The suspension concentration, applied frequency, shape of cell, etc. sensitively affect the response waveform. However, the wide variation of waveform modulation is available by the use of white light. The solid line of Figure 19.8 shows the response waveform of the reflected light intensity when a rectangular wave of electric field with low frequency (dotted

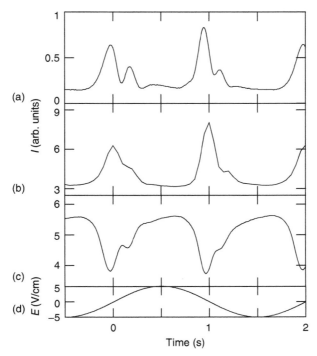

FIGURE 19.7 Waveform modulation caused by an electric field application to the colloidal crystals. (a) Change in the reflection intensity measured by a photodiode, (b) change in the reflection intensity estimated from the time-resolved reflection spectra, (c) change in the transmittance intensity, and (d) the applied sinusoidal electric field.

line, $f = 0.02\,\mathrm{Hz}$, $E = 1.8\,\mathrm{V/cm}$) is applied to the colloidal crystals of CS82 spheres ($\phi = 0.048$). It is interesting that a triangle wave with the same frequency as that of applied one is observed. The slow translational movement of colloidal spheres covered with the extended

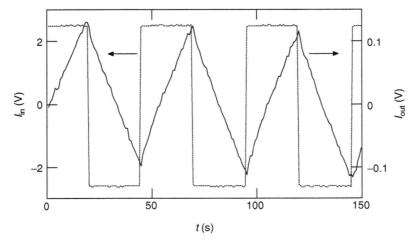

FIGURE 19.8 Reflection waveform from colloidal crystals (solid line, triangle) modulated by an applied electric field (dotted line, square).

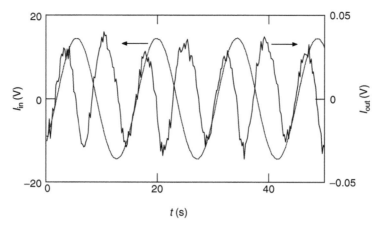

t (s)

FIGURE 19.9 Reflection waveform from colloidal crystals (solid curve, second harmonics) modulated by an applied electric field (dotted curve, sinusoidal).

electrical double layers is one of the main causes of the waveform transformation [13]. The beautiful second-harmonic generation was also observed for the same colloidal crystals as shown in Figure 19.9 by selection of the experimental parameters such as sinusoidal wave frequency and electric field strength. The solid curve of Figure 19.9 shows the response waveform of reflected light intensity when a sinusoidal wave of electric field (dotted line, $f = 0.07$ Hz, $E = 3.7$ V/cm) is applied to the colloidal crystals. The generation of the second harmonics is more substantial when a strong electric field is applied to the crystals. Harmonics of higher order than second are generated also when electric fields of low frequencies are applied. These examples demonstrate a large potential of colloidal crystals for the application of novel waveform modulation devices.

The resonance effect of colloidal crystals has been revealed for the first time by Stoimenova et al. using an electric light scattering technique [8–11]. The light scattering observations of the electrically induced changes in the colloidal crystals are made by a conventional apparatus using a photomultiplier at an observation angle of 90°. The method is sensitive to both density and shear modes. Interestingly, the response signals observed outside the electrode space were larger than those between the electrodes. Furthermore, the response signal persisted up to 1 s after the shutoff of electric field pulses. Obviously, the propagation of synchronous shear waves is observed in viscoelastic colloidal crystals. Both the amplitudes and the phases of the oscillation are varying and they are demonstrating the formation of standing waves. The response intensity of the normalized a.c. component gave a clear peak at a certain frequency, and the frequency increased with the increase in the suspension volume fraction. Although the measuring cell is asymmetric and does not enhance the precise determination of characteristic frequencies, the observed effects are so large that they permit the detection of resonance peaks.

19.6 SUMMARY

The Bragg diffraction of colloidal crystals is very sharp, because their thermal fluctuation g factors are between 0.01 and 0.1, quite similar to those of the typical metals and protein crystals. Colors of the colloidal crystals change drastically as the angles of light source and observation change slightly. Elastic modulus of colloidal crystals is very small, in the order of 10^{-2} to 10^3 Pa, compared with those of metals, from 10^{10} to 10^{12} Pa. Thus the colloidal crystal is easily distorted by weak external electric field reversibly. Many interesting applications of colloidal crystals are possible for electro-optic devices such as tunable filters, light modulator, harmonic generator, and color display, etc. We note here that the frequency response of the colloidal crystals is not high enough and less than kilohertz order. This is because the barycentric shift of colloidal particles is necessary in order to function as electro-optic devices. Contamination of ionic impurities should be avoided for the crystal formation. However, the very simple structure of colloidal crystals, consisting of only colloidal spheres and water, is suitable for the fabrication of low-cost and large-area optical devices in future.

REFERENCES

1. Vanderhoff, W., et al., *Clean Surfaces: Their Preparation and Characterization for Interfacial Studies*, Goldfinger, G., Ed., Marcel Dekker, New York, 1970.
2. Kose, A., et al., Direct observation of ordered latex suspension by metallurgical microscope, *J. Colloid Interface Sci.*, 44, 330, 1973.
3. Pieranski, P., Colloidal crystals, *Contemp. Phys.*, 24, 25, 1983.
4. Okubo, T., Extraordinary behavior in the structural properties of colloidal macroions in deionized suspension and the importance of the Debye screening length, *Acc. Chem. Res.*, 21, 281, 1988.
5. Okubo, T., Polymer colloidal crystals, *Prog. Polym. Sci.*, 18, 481, 1993.
6. Jennings, B.R., and Stoylov, S.P., Eds., *Colloid and Electro-Optics*, Institute of Physics Publishing, Bristol, 1992.
7. Stoylov, S.P., *Colloid Electro-Optics: Theory, Techniques, Applications*, Academic Press, London, 1991.
8. Stoimenova, M., and Okubo, T., Electro-optic effects in colloidal crystals of spherical silica particles, *J. Colloid Interface Sci.*, 176, 267, 1995.
9. Stoimenova, M., Dimitrov, V., and Okubo, T., Electrically induced shear waves in colloidal crystals, *J. Colloid Interface Sci.*, 184, 106, 1996.
10. Stoimenova, M., Alekov, A., and Okubo, T., Relation of electro-acoustic effects to low frequency anomalies in colloidal electro-optics, *Colloids Surf. A.*, 148, 83, 1999.
11. Stoimenova, M., and Okubo, T., Electro-optical spectroscopy of colloidal systems, in *Surfaces of Nanoparticles and Porous Materials*, Schwarz, J.A. and Contescu, C.I., Eds., Marcel Dekker, New York, 1999, p. 103.
12. Okubo, T., et al., Nonlinear electro-optics of colloidal crystals as studied by reflection spectroscopy, *J. Colloid Interface Sci.*, 199, 83, 1998.
13. Okubo, T., et al., Electro-optic properties of colloidal crystals as studied by reflection spectroscopy, *J. Colloid Interface Sci.*, 207, 130, 1998.
14. Okubo, T., et al., Electric field-induced shear waves of colloidal crystals as studied by the electro-optic reflection spectroscopy, *Colloids Surf.*, 148, 87, 1999.
15. Okubo, T., et al., Electro-optics of colloidal crystals as studied by the reflection spectroscopy, *Colloids Surf.*, 149, 431, 1999.
16. Tsuchida, A., et al., Propagation of electrooptic shear waves in colloidal crystals as studied by reflection spectroscopy, *Langmuir*, 15, 4203, 1999.
17. Tsuchida, A., Kuzawa, M., and Okubo, T., Electro-optics of colloidal crystals studied by the electric potential and reflection spectroscopy, *Colloids Surf.*, 209, 235, 2002.
18. Tsuchida, A., Shibata, K., and Okubo, T., Electro-optic effects of colloidal crystals using a laser light source, *Colloid Polym. Sci.*, 281, 1104, 2003.
19. Okubo, T., and Tsuchida, A., Spectroscopy of giant colloidal crystals, *Forma*, 17, 141, 2002.
20. Okubo, T., and Tsuchida, A., *Colloidal Crystals for Optical Components* (in Japanese), Kawaguchi, H. Ed., CMC Press, Tokyo, 2000, p. 195.

20 Electric Birefringence of Liquid Crystalline Polymers

Igor' P. Kolomiets, David Lacey, and Peter N. Lavrenko

CONTENTS

20.1 INTRODUCTION

A large amount of the current research work in the field of electro-optics is aimed at the relationship between the molecular structure and the net electrical parameters of a molecule (dipole moment, polarizability, etc.). Kerr effect in solutions of synthetic polymers has received considerable attention in the last two decades in a systematic set of studies [1,2], and for a polymer, each of the parameters is supposed to be a tensor sum of the parameters of its repeat (monomer) units. The method was widely used in molecular investigations of flexible- and rigid-chain polymers, polyelectrolytes, and biopolymers [3–6]. It was also useful in the study of low-molecular-weight compounds [7,8].

However, it appears that electro-optics could find significant application in highly ordered systems such as liquid crystalline (LC) polymers, which can form a mesophase either in a melt (a *thermotropic* mesomorphism) or in concentrated solutions (a *lyotropic* mesomorphism). Now, we briefly mention the structural peculiarities that provide the specific electro-optical properties of these polymers [3].

The thermotropic mesomorphic state in bulk may be observed for polymers composed of a combination of rigid and flexible chain fragments. The former ensures the mesogenic (and, in main, electro-optical) properties of the molecule, and the latter serves as plasticizers, which facilitate the orientationally ordered mutual arrangement of mesogenic moieties. Rigid mesogenic units can be located in both the main chain and the side chains. So, there are two basic types of LC polymers: main chain and side chain. In addition, the mesogenic units are introduced into both the main and the side chains in combined LC polymers.

Main chain LC polymers consist of repeating mesogenic (liquid crystal-like) units that form a long polymer backbone. The monomer unit must be anisotropic to enable the generation of a mesophase.

Side-chain LC polymers consist of mesogenic structural moieties appended from a polymer backbone. These polymers were synthesized in 1978 when Ringsdorf and coworkers [9] inserted a flexible spacer chain between the rigid side mesogenic unit and the polymer backbone.

The mesogenic units are usually rodlike (calamitic) but many disklike (discotic) types also exist. Fairly long spacer chains usually consist of several methylene ($-CH_2-$) units, often with ester ($-CO_2-$) or ether ($-O-$) units at the points of attachment. Given a large number of possible mesogenic units combined with different structural alternatives, it is not surprising that an enormous number of side-chain LC polymers have been prepared and evaluated.

A third class of LC polymers is called combined LC polymers. These polymers evidently combine the features of main-chain and side-chain LC polymers. The side-chain mesogenic chains can be attached to the polymer backbone.

In turn, the possibility of the formation of lyotropic polymer mesophase is usually provided by high equilibrium rigidity of their chains, favouring the orientational order in the arrangement of neighboring molecules. So many rigid-chain polymers can form a LC phase in concentrated solutions. These polymers have repeating mesogenic fragments that form a long chain. If the linking units are long and flexible, then a semiflexible polymer results; if the mesogenic fragments are directly linked all together then the polymer will be very rigid and intractable.

The degree of flexibility and the structural composition determine the mesomorphic properties of main-chain LC polymers. Main-chain LC polymers, unlike reasonably low-molecular-weight mesogens, are composed of chain units that vary in size. This distribution of chain sizes leads to wide ranges of melting to a LC phase and wide ranges over which the polymer clears to the isotropic liquid. Nematic and smectic LC phases have been found in the main-chain LC polymers. Rigid polymers of different chain sizes have difficulty packing in a layer-like manner and so usually exhibit the nematic phase. However, polymers with flexible units between the mesogenic moieties can easily arrange in a layer-like structure. The flexible spacer tends to play the same role as the terminal chains in low-molecular-weight liquid crystals, and so the longer the spacer units the greater the smectic tendency. Overall, main chain LC polymers tend to be crystalline with very high melting points.

In side-chain flexible LC polymers, the backbone has a strong tendency to adopt a random, coiled conformation. When mesogenic units are attached to the flexible polymer backbone, they will have a strong tendency to adopt an anisotropic arrangement. Clearly these two features are completely antagonistic, and where the mesogenic groups are directly attached to the backbone, then the dynamics of the backbone usually dominate the tendency for the mesogenic groups to be oriented anisotropically; accordingly, mesomorphic behaviour is not usually generated. However, if a flexible spacer moiety is employed to separate the mesogenic units from the backbone, then the two different tendencies of the mesogen (anisotropic orientation) and backbone (random arrangement) can be tolerated within the one polymer system. If this theory is correct, the backbone should influence neither the nature of generated mesophase nor their thermal stability. However, as will become apparent, the backbone does affect the mesomorphic properties of side-chain LC polymers. The spacer moiety does not totally de-couple the mesogenic unit from the backbone because studies have shown that part of the spacer group aligns with the mesogenic unit. However, enhanced decoupling is generated as the spacer moiety is lengthened. Spacer length influences the nature of the least ordered mesophase exhibited and the overall thermal stability is influenced by the polymer backbone. Mesophase with a broad temperature range tends to be favored by the use of flexible backbones (e.g., polysiloxanes); a flexible backbone gives mesogenic units more independence so that they can order more easily.

Many side-chain LC polymers have been synthesized because of the combination of a vast number of mesogenic core units available from an evaluation of structure–property relationship of the low-molecular-weight liquid crystals and many different backbone types possible (e.g., siloxanes, acrylates, methacrylates, ethylenes, epoxides). Additionally, homopolymers, Side-Chain copolymers, backbone copolymers, and side-chain/backbone copolymers considerably extend the scope of known and potential LC polymers.

Main-Chain and side-chain LC polymers are quite different in electro-optical properties. The general tendencies and conclusions on electric birefringence (EB) of solutions of rigid-chain polymers, most of which can form a mesophase, have been reviewed in papers of Tsvetkov et al. [10,11], and, therefore, they will be presented here only briefly. Corresponding references can be found in the cited papers. Most attention in the present work is paid to the recent investigations performed on the LC polymers in the last decade.

In turn, the Kerr effect, as applied to the investigation of LC polymers in a mesophase, is well reviewed by Jungnickel and Wendorff [12], and the corresponding information will not be considered here in detail as it is available in that paper. Some new results can also be found in recent papers [13–16].

So the present work deals mainly with recent molecular EB investigations on LC polymers and consists of the following parts: background, experimental, main chain polymers (thermotropic and lyotropic polymers), side-chain LC polymers (comb end-on and side-on polymers, effect of the main chain), LC hybrid dendrimers, and LC metallocomplexes.

20.2 BACKGROUND

20.2.1 DEFINITIONS AND RELATIONS

When a polymer solution or a melt is placed in an external static electric field, the polymer molecules involved are oriented and, in some instances, deformed by the action of the field on the permanent and induced dipole moments of the molecules. The polymer solution or the melt becomes birefringent, and the phenomenon is called the electro-optical or electric birefringence, or Kerr effect. The general theory of steady Kerr effect is presented in detail in the previous chapters. Therefore, we mention here only the main definitions that are used below in the interpretation of the experimental data.

In a weak field the value of EB, Δn, is proportional to the square of the field strength (Kerr law)

$$\Delta n \equiv n_{\parallel} - n_{\perp} = K_c E^2 \tag{20.1}$$

where $\Delta n = n_{\parallel} - n_{\perp}$ is the difference between the refractive indices of the specimen in the direction of, and at right angles to, the uniform electric field, E is intensity of the applied electric field (field strength, V/cm), directed at right angles to the light rays, and K_c is the Kerr constant.

The specific Kerr constant is defined by

$$K_{\rho} = \frac{\Delta n}{\rho E^2} \tag{20.2}$$

with the density of the liquid ρ. Molar Kerr constant, $_mK$, is defined (according to Debye) by

$$_mK = \frac{6n V_m}{(n^2 + 2)^2 (\varepsilon + 2)^2} \lim_{E \to 0} \left(\frac{n_{\parallel} - n_{\perp}}{E^2} \right) \tag{20.3}$$

where n is the refractive index, ε the dielectric constant, and V_m the molar volume of the sample, $V_m = M/\rho$, where M is the molecular weight of the sample with density ρ.

For a dilute two-component solution, molar Kerr constants of the solvent and solute are defined (according to Stuart) by

$$_mK_{12} = (1 - f_2)\, _mK_1 + f_2\, _mK_2 \tag{20.4}$$

where f is a mole fraction and where subscripts 1, 2, and 12 refer to solvent, solute, and solution, respectively. In the case of a polymer solution, the EB caused by the presence of the solute is characterized by the specific Kerr constant as defined by

$$K = \lim_{E \to 0, c \to 0} \frac{\Delta n_c - \Delta n_0}{cE^2} + K_0 \cdot \bar{v} \tag{20.5}$$

with EB of a solution Δn_c, EB of a solvent Δn_0, solute concentration c (g/cm³), partial specific volume of the solute in solution $\bar{v}$, and Kerr constant of the solvent K_0. The above ratio relies on the assumption that the EB of a mixture is proportional to the volume fractions of its components, without regard for the concentration dependence of intermolecular interactions. When $K \gg K_0$ and Δn_c is a linear function of E^2, instead of Equation 20.5, we have often used

$$K = \lim_{c \to 0} (\Delta n_c - \Delta n_0)/cE^2 \tag{20.6}$$

In discussion of the equilibrium EB values, we use the general theory of the Kerr effect developed for rigid dipole particles with cylindrical symmetry of optical polarizability. The main mechanism responsible for EB of solution of these particles is the rotation of their molecules as a whole, whereas the anisotropy of dielectric polarizability of the molecules does not contribute significantly to Kerr effect.

According to Langevin–Born theory, the following equation links molar Kerr constant to parameters of the polar anisotropic uniaxially symmetric molecules

$$_mK = \frac{2\pi N_A}{9} \frac{\Delta a}{45kT} \left[2\Delta a + \frac{\mu^2}{kT}(3\cos^2\beta - 1)\right] \tag{20.7}$$

where Δa is the anisotropy of optical polarizability, β the angle between the permanent dipole moment, μ, and the main axis of the molecule polarizability, kT the thermal energy.

The general theory of Kerr effect of polymers in dilute solution was developed by Stuart and Peterlin [17], Gotlib [18], Nagai and Ishikawa [19], and Tsvetkov [3]. Stuart and Peterlin examined Kuhn's random chain by assigning to each link an axially symmetric ellipsoid of polarizability, with the symmetric axis parallel to the link, and a dipole moment parallel to the link. Gotlib discussed a polymer whose one structural unit contains two skeletal atoms with a hindered rotation potential symmetric about the *trans*-disposition. According to the Stuart and Peterlin theory, both the dipole moment and polarizability in terms of the molar Kerr constant (which, aside from a trivial constant factor, is equivalent to molar Kerr constant divided by molecular weight) of Kuhn's random chain are independent of molecular weight, provided that there is no volume dipole–dipole interactions.

In accordance with Tsvetkov theory, dipole and optical properties of the kinetically rigid chains are connected with Kerr constant in coil limit, K_∞, by [3]

$$K_\infty = 2 B_1 \Delta a (\mu_0^2 \cos^2\theta) s^2/M_0$$
$$K_\infty = \lim_{M \to \infty} K \tag{20.8}$$

with optical anisotropy Δa, longitudinal component of dipole moment $\mu_0 \cos\theta = \mu_{0\parallel}$, an angle between the permanent dipole moment and the main chain contour θ, and mass of the monomer unit M_0, respectively, the number of the monomer units in Kuhn segment s, the molecular weight M. The numerical B_1 coefficient is related to the internal field by $[(\varepsilon+2)/3]^2$ according to Lorentz–Lorenz, or by $[\varepsilon(n^2+2)/(2\varepsilon+n^2)]^2$ according to Onsager. For nonpolar solvents $(\varepsilon = n^2)$, both the internal field factors are equal. For a polymer with nonpolar molecules, when the Lorentz–Lorenz internal field multiplier is used, B_1 is defined by the relation

$$B_1 = \frac{\pi N_A}{135n} \left[\frac{(n^2+2)(\varepsilon+2)}{3kT}\right]^2 \tag{20.9}$$

with N_A as Avogadro number.

The electro-optical properties of low-M polar or nonpolar liquids, as well as those of rigid-chain polymers dissolved in polar solvents, are well described with using the Onsager internal fields multiplier that leads to expression for molar Kerr constant [11]

$$_mK = \frac{2}{3} K_\rho M n \left[\frac{2\varepsilon + n^2}{\varepsilon (n^2 + 2)^2} \right]^2 \tag{20.10}$$

with K_ρ determined by Equation 20.2.

The values of optical anisotropy Δa, are available from the *Polymer Handbook* [20].

Substituting $K_{\infty, \theta=0}$, B_1, Δa, s, and M_0 into Equation 20.8 we obtain $\mu_{0\parallel}$. Equation 20.8 can also be represented in an alternative form

$$\frac{K_\infty}{(\Delta n / \Delta \tau)_\infty} = \frac{B N_A}{6kT} \frac{s\mu_{0\parallel}^2}{M_0} \tag{20.11}$$

$$(\Delta n / \Delta \tau)_\infty = \lim_{M \to \infty} \Delta n / \Delta \tau$$

with the shear optical coefficient $\Delta n / \Delta \tau$, which can be found by the flow birefringence method [3]. B is the Lorentzian internal field factor equal to $((\varepsilon + 2)/3)^2$, which coincides with the Onsager factor for a nonpolar solvent. Substituting the experimental values of K_∞, $(\Delta n / \Delta \tau)_\infty$, and s, $\mu_{0\parallel}$ can be calculated.

The values of β and θ are available from Equation 20.7 and Equation 20.8, respectively, if μ and μ_0 are determined independently.

Experimentally, the EB of the most flexible-chain polymers does not depend on M, and the K value is not usually different from that of the respective (suitable) low-molecular-weight compound that has a similar structure to the monomer (repeat) unit. This is clear because, in an external field (corrected for the medium polarization), electrical forces are applied to every polar group of the polymer chain individually but not to the macromolecule as a whole (in contrast, e.g., to an external shear flow field).

In turn, specific EB of many rigid-chain polymers was found experimentally to be dependent on the molecular weight because increase in the polymer chain length is accompanied by the changes in electric or optical anisotropy of the molecules, which are responsible for the EB effect. To explain the dependence, a suitable model should be accepted.

The theoretical interpretation of the $K(M)$ dependence was suggested by Tsvetkov [11]. In frames of Kratky–Porod' persistent chain model, the model of a chain of continuous curvature, he proposed to consider a behaviour of the macromolecules in solution under an electric field as that of macro-molecules in a frozen conformation. For such kinetically rigid chains, the EB of a polymer in solution is provided by orientation of the macromolecule as a whole, and the K value is a complicated function of x and the θ angle between the dipole moment and the optical axis of statistical segment (coinciding usually with the chain contour) as presented in Figure 20.1. This theory was very successful in explaining the unique experimental properties of the rigid-chain polymers for which not only the value but also the sign of EB are dependent on the chain length [22]. This effect is predicted by the theory at high θ (curves 7 to 10 in Figure 20.1).

The EB theory based on statistical mechanics of helical worm-like chains was developed by Yamakawa [21], who took into account the structural torsion of the persistence chain. One can use Equation 64, given in Ref. [21] for Kratky–Porod' chain with zero intrinsic curvature and cylindrically symmetric polarizabilities, to calculate the $K(M)$ function predicted for the polymer chain with contour length, L, the persistence length, a, the torsion

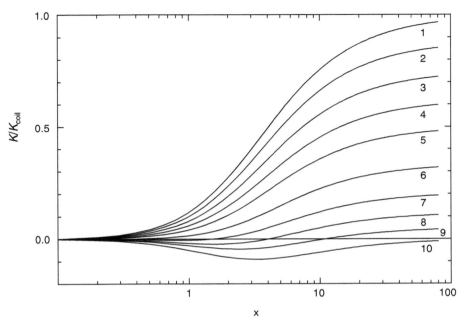

FIGURE 20.1 K/K_{coil} versus x ($x = L/a$) for persistence chains with various angles θ between dipole moment and optical axis of the monomer unit as calculated in accordance with Tsvetkov' theory (Equation 33 in Ref. [11]). $\theta = $ (1) 0°, (2) 20°, (3) 30°, (4) 38°, (5) 45°, (6) 54.7°, (7) 63°, (8) 70°, (9) 77°, and (10) 90° rad.

constant, ψ, of the transverse dipole moment, and given the θ angle. Some examples are shown in Figure 20.2a and Figure 20.2b where the ratio k/k against the reduced chain length, x ($x = L/a$), at (a) one selected value of ψ (2.8 per one segment) and various θ, and (b) one selected value of θ (90° rad) and various ψ.

Note, firstly, that both the theories lead to the similar $K(M)$ dependencies predicted for the chains with $\theta = 0$ (when the dipole moment is directed along the main axis of the macromolecule). Secondly, ~90% of the chain length effect is expected at x changing from zero to approximately 10. This is in quantitative agreement with the experimental results mentioned above.

The Gaussian limit for Kerr constant of the kinetically rigid persistence chains is given by Equation 34 in paper of Tsvetkov [11]

$$K_\infty(\theta) = K_\infty(\theta = 0)\ \cos^2\theta$$

In turn, for kinetically flexible persistence chain Yamakawa [21] obtained

$$K_\infty(\theta) = K_\infty(\theta = 0)\ (3\cos^2\theta - 1)/2$$

The Kerr constant of these chains of any chain length can be determined using Equation 33 given in Ref. [11] and Equation 64 in Ref. [21].

20.2.2 EXPERIMENTAL

The experimental methods usually used for measuring the EB rely on monitoring changes in the light beam by the null method when an elliptical compensator is used to identify the effect

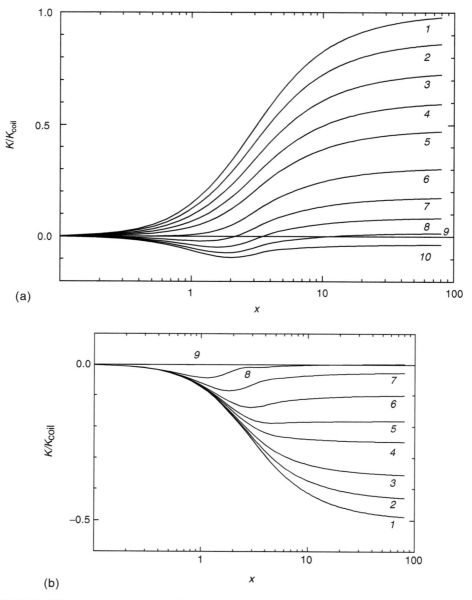

FIGURE 20.2 K/K_{coil} versus x $(x = L/a)$ for persistence chains (a) with various angles θ between dipole moment and optical axis of the monomer unit as calculated in accordance with Yamakawa' theory (Equation 64 in Ref. [21]) at the torsion ψ value of 2.8 per one segment: $\theta = (1)$ 0°, (2) 20°, (3) 30°, (4) 38°, (5) 45°, (6) 54.7°, (7) 63°, (8) 70°, (9) 77°, and (10) 90° rad and (b) the same plot at $\theta = 90°$ rad and $\psi = (1)$ 0, (2) 0.5, (3) 0.8, (4) 1.2, (5) 1.5, (6) 2, (7) 3, (8) 5, (9) at $\psi \geq 100$.

due to birefringence. There are several schemes for the EB experiment [1,22,23]. In some of them, the electrical pulses are replaced by high intensity ($\approx$ps) electromagnetic (optical) pulses from lasers. These pulses in the laser beams are much narrower and more controllable. This is important particularly in the study of small molecules in solution and when any conductive transport is not desired [25,26]. However, the EB measurements on much polymers in solution were performed in a classical scheme, and below we present a typical experimental technique that was successfully applied to investigations of polymer solutions.

A liquid under investigation is placed in a cylindrical Kerr cell whose semicylindrical titanium electrodes are fixed by molybdenum glass-soldered rods. The gap between plane-parallel electrode surfaces is $0.32 \times 7 \times 30$ mm (if other is not indicated). The cell is placed in a brass box controlled thermostatically by water through channels in the bottom of the box.

The apparatus for measuring EB has already been described in detail [27]. It is shown schematically in Figure 20.3. Electric voltage rectangular pulses formed by a G5-60 generator (Russia) and amplified by an original key scheme up to 1.2 kV are applied to a Kerr cell with a frequency of 1 Hz. The pulse duration is variable within 0.5 to 50 ms. The optical part of the setup includes an LG-78 milliwatt He–Ne laser (the wavelength $\lambda = 632.8$ nm), a polarizer (the Frank–Ritter prism), a phase modulator, the Kerr cell, a compensator (a mica plate with optical path difference $\Delta = 0.01\lambda$ mounted on a rotating limb), and an analyzer in the crossed position.

A piezoelectric converter that excited resonance vibrations of extension–compression deformation in a parallel-sided quartz plate at a frequency of 460 kHz was an active element of the phase modulator. The amplitude of vibrations of the path difference that arises in the plate due to the photoelastic effect was 0.02λ. The light beam at the outlet of the analyzer was recorded using a FEU-14 photomultiplier (Russia) and an S8-13 oscilloscope (Russia). The component of the optical signal at the modulation frequency with an amplitude proportional to the sum of path differences in the cell and compensator was selected with the aid of a V6–10 selective microvoltmeter (Russia).

The position of compensation is fixed by the zero amplitude of the signal selected. The magnitude of the effect was determined by measuring the angle of the compensator turn φ between the compensation position during pulses and the pause between them. Taking into account that the value $\Delta = 0.01\lambda$ is much smaller than the wavelength of light, the values of EB, Δn, were calculated by the formula

$$\Delta n = (\Delta/l) \; \sin 2\varphi \qquad (20.12)$$

where l is the length of light path through the field. The shape of the electric pulse and its magnitude were monitored using an oscilloscope as is illustrated in Ref. [27].

To correct the value of l that differs from the length of electrodes and to avoid systematic errors in pulse voltage determination, calibration measurements of EB of carbon tetrachloride,

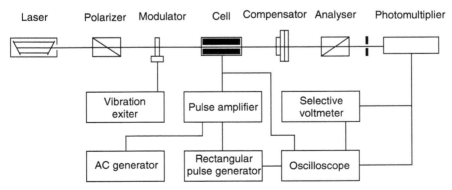

FIGURE 20.3 Apparatus for measuring the EB by the compensation method in pulsed (sinusoidal and rectangular) electric fields.

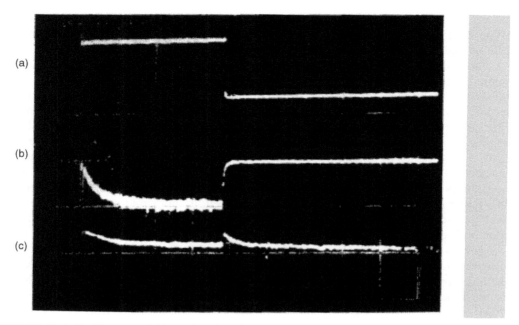

FIGURE 20.4 Oscillogram of (a) an electric pulse applied to the Kerr cell, (b) a photo-current pulse at the compensator azimuth $\varphi = 0$, and (c) the compensation position when the light intensity during the electric pulse (left) is equal to that after the pulse (right).

benzene, and toluene as well-defined liquids [23] were performed. The accuracy of Δn measurements was $\pm 2 \times 10^{-10}$.

The large EB effect can be observed on the oscilloscope screen directly, without the selective system, as illustrated in Figure 20.4. For the determination of the EB, the compensator was turned in the direction corresponding to a decrease in the amplitude of the pulsed photocurrent and acquired the position at which the light intensity during the application of the electric field is equal to that during the pause between pulses. Under these conditions of "compensation," $\Delta n = 2(\Delta/l) \sin 2\varphi$.

The EB, Δn, thus obtained was then plotted against E^2, as shown in Figure 20.5, to determine the specific Kerr constant K. The following extrapolation of the data to vanishing solute concentration is illustrated in Figure 20.6.

The experimental data were treated by the least-squares method. Experimental values are given in CGSE units, particularly, K in $cm^5 \cdot g^{-1} \cdot (300\,V)^{-2}$ and $_mK$ in $cm^5 \cdot mole^{-1} \cdot (300\,V)^{-2}$; μ is given in Debye units $(1\,D = 1 \times 10^{-18}\,g^{1/2} \cdot cm^{5/2} \cdot s^{-1})$.

20.2.3 SOLVENT EFFECT

The EB of a polymer solution is a sum of the solute and solvent effects. The solvent effect must be taken into account using relations 20.4 to 20.6. The values of reduced, K_ρ, and molar Kerr constant, $_mK$, of the common solvents are listed in Table 20.1. Usually it is preferable, when possible, to use a solvent with much lower (in comparison with the solute under investigation) K value and low permittivity, ε.

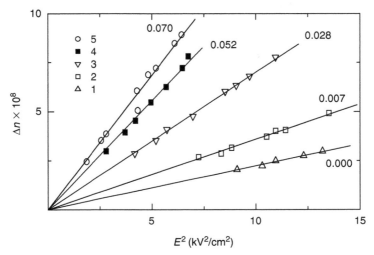

FIGURE 20.5 EB, Δn, versus square of electric field strength, E^2, as observed in dimethyl sulfoxide solution of poly(amide hydrazide) sample with molecular weight of 34 kDa. Solute concentration $c = (1)$ 0, (2) 0.007, (3) 0.028, (4) 0.052, (5) 0.070 g·dl^{-1}.

20.2.4 *M*-DEPENDENCE AND RELAXATION OF **EB**

In a polymer solution, as follows from Equation 20.7, EB is provided by dipole moment and optical anisotropy of the kinetically mobile unit (segment). A segment consists of s monomer units; its length $A = \lambda s$, mass $M_0 s$ and dipole moment $\mu_0 s$ where λ, M_0, and μ_0 are parameters

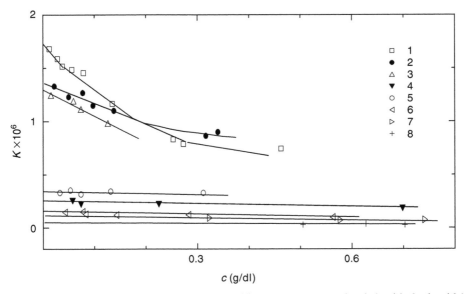

FIGURE 20.6 Concentration dependence of the specific Kerr constant of poly(amide hydrazide) samples with various molecular weights, $M = (1)$ 34, (2) 24, (3) 19, (4) 12, (5) 11, (6) 9, (7) 8.6, and (8) 5.2 kDa in dimethyl sulfoxide solution.

TABLE 20.1
Physical Properties of Common Solvents[a]

Solvent	ε	(n)	μ (D)	$10^{12} \times K_\rho$ Number-Average[b]	$10^{12} \times {}_mK$[b]
Acetone	21.5	(1.36)	2.8	90	107
Acetonitrile	37	(1.35)	3.95	353	239
Benzene	2.2	(1.50)	0	2.4	7
1-Bromonaphthalene	4.8	(1.66)	1.6	33.6	213
Carbon tetrachloride	2.23	(1.46)	0	0.3 to 0.5	—
Chlorobenzene	5.6	(1.53)	1.6	66	137
Chloroform	4.8	(1.45)	1.25	−12.7	−28.5
Cyclohexane	2.02	(1.43)	0.1	0.4	—
Cyclohexanone	15.7	(1.45)	2.8	84	184
o-Dibromobenzene	7.5	(1.61)	2.1	75	288
p-Dibromobenzene	—	—	—	—	38.1
m-Dichlorobenzene	5	(1.55)	1.37	37	74.7
o-Dichlorobenzene	11	(1.55)	2.26	170	362
p-Dichlorobenzene	2.4	(1.53)	0	10	25.8
1,2-Dichloroethane	10.4	(1.44)	1.75	—	13.4
Dimethylacetamide	40	(1.44)	3.8	300	355
N,N-Dimethylformamide	—	(1.43)	3.9	—	417
Dimethyl sulfoxide	45	(1.48)	3.9	202	144
1,4-Dioxan	2.3	(1.42)	0	0.39	1.0
Ethyl acetate	6.4	(1.37)	1.8	10.9	20.1
Ethyl methyl ketone	18	(1.38)	3.2	85	88.2
N-Methylformamide	183	(1.43)	3.86	—	210
Nitrobenzene	35	(1.56)	3.9	1960	1070
Nitromethane	36	(1.38)	3.1	55	89
Phenyl isothiocyanate	20	(1.64)	3.0	452	1230
Pyridine	12.3	(1.51)	2.2	123	150
Styrene	—	(1.55)	0.6	4 to 7	—
Tetrachloroethane	5.2	(1.45)	1.8	−11.9	−15.2
Tetrahydrofuran	7.6	(1.41)	1.6	−2.3	—
Toluene	2.38	(1.50)	0.36	3.9	—
Trifluoroacetic acid	39	(1.28)	—	36.5	—

[a] ε is dielectric permittivity, n the refractive index, K_ρ and ${}_mK$ Kerr constant as determined in neat liquid and solution, respectively; $K_\rho = \Delta n/\rho E^2$.
[b] K_ρ in cm$^5 \cdot$g$^{-1} \cdot$(300 V)$^{-2}$ and ${}_mK$ in cm$^5 \cdot$mol$^{-1} \cdot$(300 V)$^{-2}$.

Sources: From Tsvetkov, V.N. and Tsvetkov, N.V., Electric birefringence in solutions of rigid-chain polymers, *Russ. Chem. Rev.*, 62, 851, 1994; Le Fevre, R.J.W., Electro-optical constants, *Adv. Phys. Org. Chem.*, 3, 1, 1965; Shere, A.J. et al., Kerr effect and wide angle light-scattering studies of a para-aromatic polyamide in dilute solution, *Macromolecules*, 29, 2088, 1996.

of the monomeric unit. Persistence length $a = A/2$. The number of the segments in the chain is equal to $L/A = M/M_0 s$ with the chain length L. If $L < A$, then K is clearly a function of M. Figure 20.7 illustrates this dependence observed for two kinetically rigid polymers. In Gaussian limit, $L/A \rightarrow \infty$, the $K_\infty = \lim_{L/A \rightarrow \infty} K$ value is determined by Equation 20.6.

A particle is orienting in an electric field with a velocity that is limited by rotational mobility of the particle. For a macromolecule, the velocity is determined by dimensions of its mobile parts or by its own dimensions, depending on the kinetic rigidity of the chain. The EB

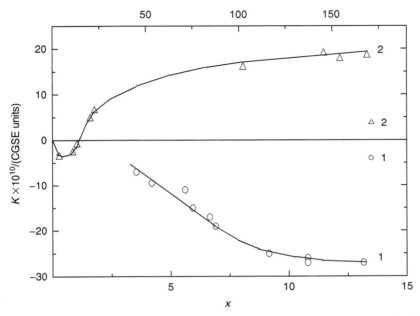

FIGURE 20.7 Specific Kerr constant versus reduced chain length for (1) cellulose carbanilates CC in dioxane solution and (2) C_{10}-aromatic polysters in dichloroacetic acid solution. Kerr effect in solutions of a thermotropically mesogenic alkylene–aromatic polyester, *Eur. Polym. J.*, 22, 543, 1986.

relaxation is, therefore, a symptom of many rigid-chain polymers. For instance, a kinetically rigid macromolecule is orienting as a whole in AC field at a given frequency, f, until its dimensions do not exceed the given value. Macromolecules with higher molecular weight, M, are not so mobile to follow the field, and a drop in the EB is observed with increasing f. The EB relaxation of macromolecules with lower M is observed at higher f. Figure 20.8 presents an impressive illustration of this classic symptom of the rigid-chain polymers.

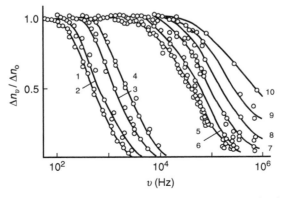

FIGURE 20.8 The relative value of EB versus frequency of AC field as obtained for cellulose carbanilate samples of various molecular weight from (1) 870 to (10) 24 kDa.

So, a time-dependent electric field yields an EB relaxation (a delay of the birefringence response to an alteration of applied electric field), which gives information on the dynamic behaviour of molecules such as molecular rotation and intramolecular motion of the polar atomic groups. Thus, the EB relaxation method has been applied to various systems as a powerful tool in the investigation of microscopic properties of the electroactive molecules.

The dispersion of EB or frequency dependence shows the presence of the relaxation phenomena in the orientational process of the macromolecules. The time of orientational relaxation τ of EB can be calculated using the Peterlin–Stuart theory of rigid dipoles orientating in an AC field by

$$\Delta n_\omega / \Delta n_0 = 1/(1 + 4\,\pi^2 v^2 \tau^2) \tag{20.12a}$$

where Δn_ω and Δn_0 is the EB in AC with frequency $v = \omega/2\pi$ and in DC $v = 0$, respectively.

In turn, the time τ of orientational dipole relaxation is determined by the rotational mobility of the macromolecule and may be related to the rotation diffusion coefficient, D_r, by

$$D_r = 1/2\tau$$

Hence, study of the EB relaxation yields the possibility the form and thermodynamic flexibility (the Kuhn segment length) of the rigid-chain polymer molecules that show the EB dispersion effect.

For kinetically flexible macromolecules τ does not usually depend on the chain length, which is the characteristic parameter of the mobile fragment of the molecule (kinetic segment). For a system of noninteracting segments, τ is a constant, and $\Delta n_\omega / \Delta n_0$ as a function of v satisfies Equation 20.12a. The dipole–dipole interactions of the kinetic segments result in an appearance of the relaxation time spectrum, and the v-dependence of $\Delta n_\omega / \Delta n_0$ becomes to be more wide than that predicted by Equation 20.12a. Polymodal τ-spectrum is observed rarely when there are several modes of the intramolecular motions.

The measurement methods for EB relaxation can be divided into two categories for the applied electric field [31]: (1) time-domain EB and (2) frequency-domain EB. In category (1), a rectangular pulse field is applied and transient raise and decay response are analyzed, while in (2) an oscillating field with an angular frequency ω is applied and DC and 2ω components of the birefringence response are detected as functions of ω. Although both categories seek the same electro-optical properties of molecules, the responses in time and frequency domains are not definitely related.

20.3 MAIN CHAIN LC POLYMERS

20.3.1 THERMOTROPIC POLYMERS

The EB effect in a polymer bulk is usually small in value and difficult to interpretate. Very few thermotropic main chain polymers were successfully studied in solution and in bulk [29–38]. Many important structure–property relationships were obtained from electro-optical investigations of aromatic polyesters in solution [39–43]. Some of them were investigated in mixed solvents [44,45]. Complicated M-dependence of the Kerr constant, as is illustrated in Figure 20.7, is one of the most remarkable properties of these polymers.

20.3.1.1 Alkylene–Aromatic Polyesters

Unique investigations of C_{10}-alkylene *para*-aromatic thermotropic polyesters (PE-10) were performed in trifluoroacetic acid solutions [22]. The structure of the PE-10 repeat unit, as shown below, consists of tricycle-ester mesogenic fragments linked together via $-(CH_2)_{10}-$ spacers.

PE-10

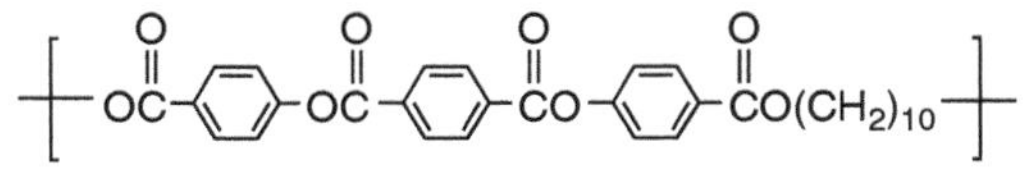

Monomer

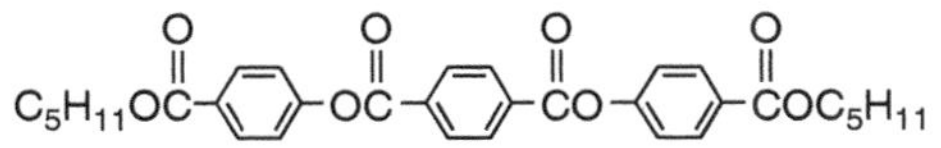

The EB of the samples and fractions of various molecular weights were evaluated and compared with properties of the monomer compound. Table 20.2 shows firstly that the Kerr constant of PE-10 is dependent on M. Over saturation ($M \to \infty$), the limit value is $K_\infty = 18 \times 10^{-10}\,cm^5 \cdot g^{-1} \cdot (300\,V)^{-2}$. Secondly, K is positive (in sign) for the high-molecular-weight fractions and negative for the low-molecular-weight ones and for the monomer compound (plot 2 in Figure 20.7). Hence, a dependence of K on M and the unique change in the K sign were detected.

This unusual behaviour was quantitatively interpreted by the theory developed for the kinetically rigid worm-like chains. The angle between the dipole moment and the main direction of the macromolecule was found to be 75° rad, and the EB effect is observed because of the large-scale orientation of the kinetically rigid polar molecules in an electrical field, in spite of a moderate equilibrium rigidity of the molecular chains (the Kuhn segment length is 50 Å [46]). This peculiarity may be considered to be specific for the structure of the macromolecules that form the thermotropic mesophase.

20.3.1.2 Aromatic Polyesters with *Para*- and *Ortho*-Phenylene Rings in Chain

Aromatic polyesters containing rigid mesogenic moieties linked via *ortho*-positions in the aromatic ring (OPPE) are also thermotropic LC compounds. The influence of mesogenic group position in the main chain (*ortho* or *para*) on the EB effect of OPPE was investigated in

TABLE 20.2
Kerr Constant of Mesogenic C_{10}-Alkylene *Para*-Aromatic Polyester, PE-10, in Trifluoroacetic Acid Solution

Polymer	M (kDa)	$10^{10} \times K$, $(cm^5 \cdot g^{-1}) \cdot (300\,V)^{-2}$	Polymer	M (kDa)	$10^{10} \times K$ $(cm^5 \cdot g^{-1}) \cdot (300\,V)^{-2}$
PE-10	68	16.5	PE-10	5	3.5
	64	16.5		2.7	−0.9
	60	17.8		1.6	−2.6
	42	15.2	PE-10, monomer	0.532	−3.2
	7.1	5.4	Solvent	—	0.365

Source: From Tsvetkov, V.N. et al., Kerr effect in solutions of a thermotropically mesogenic alkylene-aromatic polyester, *Eur. Polym. J.*, 22, 543, 1986.

dichloroacetic acid solution [47]. Kerr constant was found to be M-dependent, changing from -3.2×10^{-10} to -1.018×10^{-10} cm^5·g^{-1}·(300 V)$^{-2}$ with M increasing from 11.5 to 44.6 kDa. The effect is due to the accumulation of the longitudinal components of dipole moment. For the model compound, MC, the Kerr constant is also negative in sign ($K = -1.8 \times 10^{-10}$ cm^5·g^{-1} ·(300 V)$^{-2}$), due to the large angle (75° rad) between the dipole moment and main axis of the molecule.

OPPE

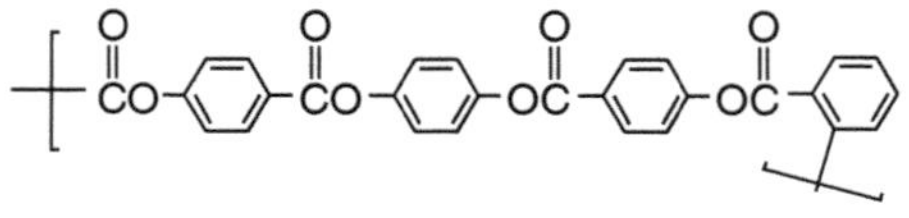

MC

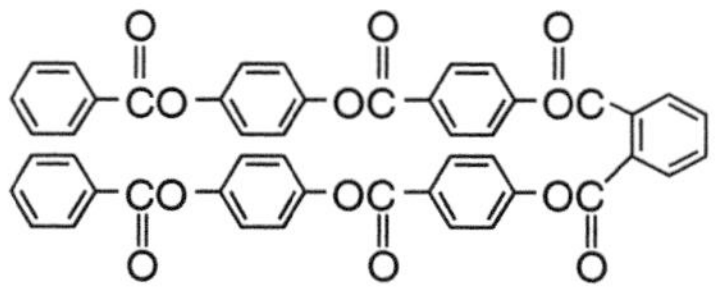

20.3.1.3   Aromatic Polyesters in Polar Solvents

Rigid-chain polymers are known to form real solutions in strongly polar solvents only. These studies provided information on the hypothesis that, in a rigid-chain polymer–polar solvent system, the effective "electro-optical" dipole is formed due to the orientation correlation between the polar bonds of the polymer chain and dipoles of the solvent molecules [11]. Such "electro-optical" dipole, with longitudinal component along the chain, provides large-scale reorientation of macromolecules under electric field.

To investigate the effect, the EB of bi-substituted aromatic polyesters (PAPE) and the model compound MC, shown below, were studied in solution in dioxane and dichloroacetic acid [45].

PAPE

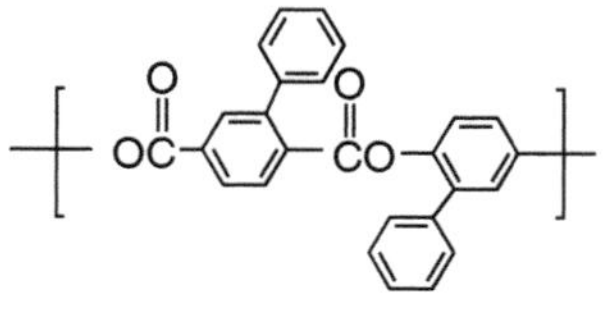

MC

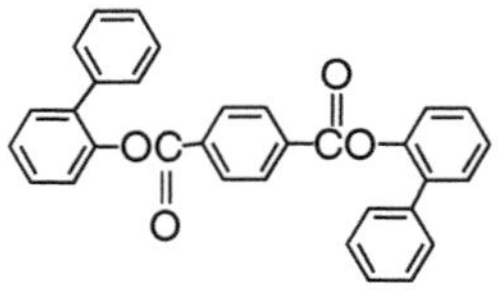

For the model compound (MC) in dioxane solution, $K = -0.18 \times 10^{-10}$ cm^5·g^{-1}·(300 V)$^{-2}$ and $\mu_0 = 1.94$ D were obtained.

The EB of PAPE in dioxane solution was found to be negative in sign and independent of M, indicating a typical flexible-chain behaviour. When a polar solvent, dichloroacetic acid, was used instead of weak-polar dioxane, great changes in EB were observed. In particular, a strong

M-dependence of EB appeared, with change in sign, from negative to positive. The EB effect was believed to be due to the large-scale molecular motions.

Methods of equilibrium and nonequilibrium EB were applied to the PAPE study by using mixed solvents (dioxane + tetrachloroethane) of various compositions, and unique results were obtained in this way [45]. A negative Kerr constant was found for PAPE in solvents with the tetrachloroethane molar fraction $f < 0.5$, and no EB dispersion was observed up to 1 MHz frequency. Hence, under these conditions, the macromolecule has no longitudinal (along the chain contour) component of the monomer unit dipole.

In contrast, at $f > 0.5$, the positive EB effect and large-scale orientation of the PAPE macromolecules in polymer solution under electric field were detected, due to the longitudinal dipole component of the monomer unit. The authors of the cited paper have found that a longitudinal component of the macromolecule "electro-optical" dipole arises when dielectric permeability of the solvent $\varepsilon > 5$. The formation of this dipole leads not only to threshold inverse in sign of the Kerr constant, but also to a change in the mechanism of the macromolecule reorientation from a low scale (in weekly polar solvents) to a large scale (in strongly polar solvents). Thus, they have discovered sharp change in character of the macromolecule reorientation under the influence of an electric field from a low scale to a large-scale due to increasing solvent polarity.

20.3.1.4 Heterocyclic Poly(ester imide)s

The effect of the electronic structure of aromatic systems containing conjugated ester groups and the dependence of this structure on donor–acceptor properties of adjacent chain elements, was studied for a series of heterocyclic poly(ester imide)s (PEI) with the structure given below [48].

The measurements of EB of PEI in dichloroacetic acid solution were carried out in a glass Kerr cell with titanium electrodes 8 cm in length along the light beam path with a gap of 0.08 cm between them. Other experimental details were the same as described in Section 20.2.2.

EB was observed in the dilute polymer solution ($c = 1.04$ g·dl^{-1}) subjected to a sinusoidal-pulsed field. EB was strongly dependent on the frequency of the field, from 1 kHz to 1 MHz, indicating the dipole orientation of the macromolecules as a whole. A large-scale rotational mechanism is provided here by significant kinetic rigidity of the chains.

Equilibrium Kerr effect is illustrated by the data given in Table 20.3. Kerr constant is positive in sign, and the absolute values of K for these polymers are rather high due to a considerable equilibrium ($A = 135$ Å) and kinetic rigidity of the macromolecules.

TABLE 20.3
Kerr Constant of Heterocyclic Poly(ester imide)s in Dichloroacetic Acid Solution

Polymer	M (kDa)	$10^{10} \times K$ $(cm^5 \cdot g^{-1}) \cdot (300\ V)^{-2}$	Polymer	M (kDa)	$10^{10} \times K$ $(cm^5 \cdot g^{-1}) \cdot (300\ V)^{-2}$
PEI(Ph)-1	17.2	65	PEI(but)-2	12	72
PEI(Ph)-4	11.5	52	PEI(but)-3	11.6	65
PEI(Ph)-7	5.1	32	PEI(but)-4	6.2	49
PEI(Ph)-8	4.6	17	PEI(but)-6	3.0	17
PEI(Ph)-9	4.6	25			
PEI(Ph)-10	1.95	4.2			
PEI(Ph)-11	1.03	2.0			
PEI(Ph)-12	0.83	0.6			

Source: From Andreeva, L.N. et al., Conformational properties of macromolecules of heterocyclic polyester imides in dilute dichloroacetic acid solutions, *J. Polym. Sci. Part B: Polym. Phys.*, 42, 12, 2004.

The dipole structure of the macromolecule is characterized by the dipole moment of the monomer unit $\mu_0 = 3.4$ D with its longitudinal component $\mu_{0\parallel} = 2.2$ D, which looks to be considerably affected by the strongly polar solvent.

20.3.1.5 Combined Main- and Side-Chain LC Polymers

In polymers containing mesogenic groups in both the backbone and the side chains, Kerr constant is the sum of positive contribution of the side chains, R, and of the negative contribution of the backbone. This was observed for the polymer with the following structure [49].

In result, Kerr constant was found to be small in value and positive in sign,

$$K = 0.046 \times 10^{-10} cm^5 \cdot g^{-1} \cdot (300\ V)^{-2}$$

as measured in dioxane solution. The EB of a polymer was related to local orientations of the chain fragments, which is due to high kinetic flexibility of the polymer chains.

20.3.2 LYOTROPIC POLYMERS

20.3.2.1 Rigid-Chain Polymers

Kerr effect offers an effective method for studying conformational, electro-optical, and electrodynamic properties of rigid-chain polymers in solution [3]. This method is particularly fruitful when applied to polymers dissolved in nonpolar solvents because, firstly, it eliminates the difficulties in choosing the "internal field" factor for a liquid, and, secondly, it permits one to obtain the true (nondistorted by the action of a strong polar solvent) electro-optical characteristics of macromolecules under study.

The equilibrium and nonequilibrium Kerr effects in solution of many rigid-chain lyotropic LC polymers were investigated. Among them, there are poly(alkyl isocyanate)s [50,51], aromatic polyamides [52], and some other polymers [53]. The results are collected in Table 20.4. Polymers with the perfect secondary structure of the macromolecules show obviously the highest EB effect. Disturbance of the structure leads to a decrease in EB as is seen, for example, for PLGA in MeOH+H_2O solution.

TABLE 20.4
Electro-Optical Properties and Dipole Moments of Lyotropic LC Polymers in Solution

Polymer (M)	Solvent[a]	$10^{10} \times K_\infty$ $(cm^5 \cdot g^{-1}) \cdot (300\,V)^{-2}$	$\mu_{0\parallel}$ (D)
Tobacco mosaic virus	Water	+ 13 600,000[b]	—
Collagen	Acetic acid	+ 430,000[b]	0.4
Poly(γ-benzyl-L-glutamate)	Dichloroethane	+ 30,000	1.6
		+ 129,600[b]	—
Poly(L-glutamic acid) (87 kDa)	MeOH	+ 146,000	—
	MeOH+H_2O (1:1)	+ 27,000	—
Poly(n-butyl isocyanate)	CTC	+ 25,000	—
		+ 28,000	—
Poly(n-octyl isocyanate)	CTC	+ 14,000	1.3
Poly(chlorohexyl isocyanate)	CTC	+ 6,400	—
Ladder polymers			
Poly(phenyl siloxane)	Benzene	−12.5	—
Poly(m-chlorophenyl siloxane)	Benzene	−15	—
Poly(phenyl-isobutyl siloxane)	Benzene	−4.9	—
Poly(phenyl-isohexyl siloxane)	Benzene	−8.5	—
Poly(amidobenzimidazole)	DMAA	+ 40 to +12,400[c]	1.5
Poly(amidohydrazide)	DMSO	+ 35,300	1.5
Aromatic polyamide[d]	THF	+ 135	—

[a]MeOH = methanol; CTC = carbon tetrachloride; DMAA = N,N-dimethyl acetamide; DMSO = dimethyl sulfoxide; THF = tetrahydrofuran.
[b]Value of $K/\bar{v}$.
[c]In line of samples with $M = 1.4$ to 45 kDa [52].
[d]~(CO-Ph(CF_3)-Ph(CF_3)-CO-NH-Ph(CF_3)-Ph(CF_3)-NH)~, sample with $M_w = 130$ kDa.

Sources: From O'Konski, C.T., *Molecular Electro-Optics*, Ed., Marcell Dekker, New York, 1976; Tsvetkov, V.N., *Rigid-Chain Polymers. Hydrodynamic and Optical Properties in solution*. Plenum Press, New York, 1989, Tsvetkov, V.N., Rjumtsev, E.I., and Shtennikova, I.N., Intramolecular orientation order and properties of polymer molecules in solutions, in *Liquid Crystalline Order in Polymers*, Blumstein, A., Ed., Academic Press, New York, 1978, p. 57; Shere, A.J. et al., Kerr effect and wide angle light-scattering studies of a para-aromatic polyamide in dilute solution, *Macromolecules*, 29, 2088, 1996; Tsvetkov, V.N. et al., Kerr effect in solution of aromatic ester with *para*- and *ortho*-phenylene cycles in chain, *Doklady Chem.*, 307, 916, 1989; Beevers, M.S., Garrington, D.C., and Williams, G., Dielectric and dynamic Kerr effect studies of poly(n-butyl isocyanate) and poly(n-octyl isocyanate) in solution, *Polymer*, 18, 540, 1977; Matsumoto, M., Watanabe, H., and Yoshioka, K., Electrical and hydrodynamic properties of polypeptides in solution, *Biopolymers*, 6, 929, 1968; 11, 1711, 1972.

TABLE 20.5
Electro-Optical Properties and Dipole Moments of Polyphosphazenes in Solution:
~ P(R)$_2$ = N ~

R	$10^{10} \times K$ (cm^5·g^{-1})·(300 V)$^{-2}$		$\mu_{0\parallel}$ (D)	
	DO[a]	EA[a]	DO[a]	EA[a]
$-$OCH$_2$ CF$_3$	—	700	—	—
$-$OCH$_2$ CF$_2$ CF$_2$ H	20	480	2.2	4.6
$-$OCH$_2$ CF$_2$ CF$_2$ CF$_2$ CF$_2$ H	—	570	—	5.2

[a]DO = dioxane; EA = ethyl acetate.
Sources: From Rjumtsev, E.I. et al., Electric and dynamic birefringence in solutions of poly-*bis*-trifluoroethoxyphosphazene, *Eur. Polym. J.*, 28, 1031, 1992; Rjumtsev, E.I. et al., Kerr effect in solutions of poly(fluoroalcoxyphosphazene)s, *Polym. Sci., Part A*, 34, 541, 1992.

20.3.2.2 Poly(phosphazene)s

High K values were obtained for a series of poly(phosphazene)s with various side groups in both dioxane and ethyl acetate solution [55,56]. As listed in Table 20.5, these values are comparable with those of typical rigid-chain polymer, but large-scale mechanism of the phenomenon is accompanied here with well-detected intramolecular movements. Unexpected K (M) dependence was found for the macromolecules in coil conformation like that of comb C$_{16}$-polymer [57].

20.3.2.3 Poly(acetylene)s

The EB properties, induced by AC fields, were studied for bi-substituted poly(acetylene)s, namely, poly(1-trimethylsilyl-1-propyne) and poly(1-trimethylgermyl-1-propyne) in dilute solution [58–60], and their unusual behaviour cannot be explained by any of the current EB theories. Table 20.6 illustrates that for a series of fractions of polymer A, whose molecular

TABLE 20.6
Electro-Optical Properties of Poly(acetylene)s in Solution:
~ C(CH$_3$) = C(R) ~

R	M (kDa)	$10^{10} \times K$ (cm^5·g^{-1})·(300 V)$^{-2}$
(A) -Si(CH$_3$)$_3$ (*cis*-): (*trans*-) = 40:60	325 to 1,040	5.9 to 12.6[a]
(B) -Ge(CH$_3$)$_3$ (*trans*- 90%)	40 to 1,650	2.7 to 15[b]
		3.5 to 20[c]
	(monomer)	0.014

[a]In toluene.
[b]In cyclohexane.
[c]In carbon tetrachloride solution.

Sources: From Rjumtsev, E.I. et al., Electrooptical properties of poly(1-(trimethylsilyl)-1-propyne) in solutions, *Polym. Sci., Part A*, 41, 765, 1999; Yevlampieva, N.P. et al., Electrooptical properties of poly[1-(trimethylgermyl)-1-propyne] in solutions, *Polym. Sci., Part A*, 44, 314, 2002.

weight is high (Gaussian M-range) and whose specific Kerr constant is dependent on M. This abnormal effect was explained by the specific dipole structure of the macromolecule resulting from specific intramolecular ordering involving the solvent molecules [58].

Different K values obtained for polymer B were explained by various stereo-regularity of the macromolecules of different fractions. Note that the K value here is two to three orders of magnitude greater than that of the monomer unit, reflecting a large-scale EB response of the macromolecule to electric field [59]. This is in contrast to the behavior of typical flexible-chain polymers for which the EB effect is usually close to that of the monomer compound.

20.3.2.4 Phenylated Poly(phenylene)s

Phenylated poly(phenylene)s were studied in solution in both polar and nonpolar solvents [61,62]. The data are summarized in Table 20.7. Over wide M-ranges, EB for nonpolar polymer molecules, dissolved in nonpolar dioxane, was found to be independent of molecular weight and no dispersion of the Kerr effect was observed within the frequency range used. In contrast, EB induced in the polymer solution in polar chloroform is several times higher, increases with increasing value of M, and is dependent on a frequency of the applied AC field. An effect of the orientation correlation between the polymer chain and dipoles of the polar solvent molecules was thus discovered. It provides the effective dipole moment per monomer unit, $\mu_0 \approx 0.8\,D$, which is responsible for the electro-optical properties of the polymers in polar solvents.

20.3.2.5 Cycle-Chain Poly(siloxane)s

EB measurements of poly(decamethyl cyclohexasiloxane), a polymer containing hexasiloxane cycles linked one to another via oxygen atoms, and containing side CH_3 groups, were performed in carbon tetrachloride solution and compared with properties found for model, nonpolymeric compound [63]. The K value of $1.9 \times 10^{-10}\ cm^5 \cdot g^{-1} \cdot (300\,V)^{-2}$ and $\mu = 12.2$ and $14.7\,D$ were obtained for the samples with degree of polymerization of 56 and 80, respectively. The stability of polar hexasiloxane cycle under the external action was considered to be responsible for the polymer's ability to form a highly organized mesophase.

20.3.2.6 Cellulose Derivatives

Cellulose derivatives present the specific family of the lyotropic polymers. Their electro-optical properties are given in Table 20.8 (degree of substitution (DS) is defined here

---

TABLE 20.7
Electro-Optical Properties of Phenylated Poly(phenylene)s in Solution:
~ mPh(Ph)$_3$-(pPh)$_n$-mPh(Ph)$_3$ ~

n	Solvent	M (kDa)	$10^{10} \times K\ (cm^5 \cdot g^{-1}) \cdot (300\,V)^{-2}$
1	Dioxane	8 to 716	0.21 ± 0.07
	Chloroform	8 to 399	1.15 to 11
2	Dioxane	14 to 200	0.37 ± 0.09

Source: From Tsvetkov, N.V. et al., Electric birefringence of a phenylated poly(phenylene) in polar and nonpolar solvents, *Vysokomolek. Soedin.*, *Ser. A*, 44, 297, 2002; Tsvetkov, N.V. et al., Hydrodynamic, conformational, and electrooptical properties of phenylated poly(phenylene) macromolecules in solutions, *Vysokomolek. Soedin.*, *Ser. A*, 46, 1695, 2004.

by number of substituents per 100 monomer units, with the maximum DS value of 300).
A separate discussion is now given for a selective number of cellulose derivatives.

20.3.2.6.1 Cellulose Carbanilate (CC)

Figure 20.8 illustrates the relaxation of EB as observed for a number of CC fractions of
various molecular weights [30]. One can see that the K value is dropping to zero with
increasing AC frequency, where the relaxation frequency is a function of M. It is a typical
property of a rigid-chain polymer.

Electro-optical properties of high-molecular-weight fractions of CC were investigated in
mixed binary solvent solutions consisting of low-conducting components of different polar-
ities, i.e., nonpolar (dioxane) and polar (ethyl acetate) solvents [64]. Equilibrium and nonequilibrium EB measurements were performed together with hydrodynamic and flow
birefringence studies. If the equilibrium rigidity of polymer molecules was found to decrease
drastically, but monotonically, with an increase in the content of the polar component, the
dependence of optical anisotropy as well as the dependence of the Kerr constant and
longitudinal component of the monomer unit dipole moment, on the molar fraction of
ethyl acetate in the mixed solvent, were also found to be nonmonotonic. This nonmonotoni-
city was interpreted to be caused by weak adsorption of ethyl acetate by the polymer, on the
basis of the concept of partial shielding of dipole moments of polymer chains by the molecules
of the polar solvent.

The behavior of CC macromolecules in one-component solvents, e.g., dioxane and ethyl
acetate, was studied independently over wide M-ranges [65]. The length of the Kuhn statistical segment, A, was estimated (250 ± 50 Å) from the measurements of the nonequilibrium
EB of low-molecular-weight fractions in a dioxane solution of CC. It was shown that
contribution of the transverse component of the dipole moment of the monomer unit in EB
is small. The values of longitudinal components of dipole moments of CC monomer unit in
dioxane and ethyl acetate solutions were found to be 0.91 and 1.3 D, respectively. The results
obtained suggest that there is a correlation between the orientations of the polar bonds of the
polymer chain and orientation of dipole moments of ethyl acetate molecules.

20.3.2.6.2 Cellulose Acetobenzoate

The EB of cellulose acetobenzoate mixed ester in dioxane solution was studied over the
M-range of 11 to 1300 kDa [66]. From a plot of the orientational relaxation times, τ versus
M and using the theory of rotational friction of kinetically rigid wormlike chains, the length
of Kuhn segment, A, was evaluated to be 300 ± 100 Å. The equilibrium value of the Kerr
constant, K, was determined from measurements using rectangular pulse fields. The sign of
the Kerr constant was negative and coincided with that of the dynamic birefringence.
Treatment of the $K(M)$ dependence, within the framework of the theory of worm-like chains,
leads one to conclude that it is the longitudinal components of dipoles of the repeating units,
$\mu_{0\parallel}$ that determines the observed EB effect, whereas the contribution of normal components
is negligible. From the $K(M)$ dependence, $A = 210 \pm 60$ Å and $\mu_{0\parallel} = 0.21$ D were determined. The data obtained illustrates the essential role of the dipole moments of the side
groups, i.e., substituents in the glycoside cycles, and in the EB phenomenon in solutions of
cellulose esters.

20.3.2.6.3 Cellulose Acetate Cinnamate

The EB of several fractions and samples of mixed acetate cinnamate cellulose esters in dioxane
solution were studied within the M-range from 7.7 to 900 kDa [67]. The experimental data are
given in Table 20.8. As follows from the study of transient Kerr effect of the cellulose esters, the
mechanism mainly responsible for the EB, as observed in solutions of these compounds at all
DSs, is the large-scale rotation of the macromolecules in the electric field, due to the presence of

TABLE 20.8
Electro-Optical Properties and Dipole Moments of Cellulose Esters in Solution

Polymer	Degree of Substitution	$10^{10} \times K_\infty$ $(cm^5 \cdot g^{-1}) \cdot (300\,V)^{-2}$	$\mu_{0\parallel}$ (D)
Benzylcellulose	220	6	0.7
Cellulose			
Acetobenzhydroxamate	285/15	−0.3	0.1
	240/60	−0.4	0.13
	170/90	−0.15	0.1
Acetobenzoate	245/55	−2.3	0.21
Acetocinnamate	200/60	−18	0.44
	140/160	−9.4	0.44
	50/250	−1.1	0.2
Acetodiphenylacetate	290/10	4.6	0.37
	150/40	4.6	0.31
	26.9/20	0.27	0.51
	115/130	3.9	0.3
	30/270	1.3	0.24
Acetomyristate	290/10	−0.7	0.5
Aceto-α-naphthylacetate	50/250	0.87	0.47
	25/275	0.18	0.2
	10/290	0.08	0.15
Acetonitrobenzoate	25/275	0.3	—
Acetophenylacetate	260/40	1.4	0.32
	150/100	1.02	0.29
	70/200	0.24	0.21
	32/246	0.21	0.26
	5/285	0.2	0.22
Acetotriphenylmethylate	145/110	87	1.53
Benzoate	216	−7.7	0.38
Butyrate	300	0.4	0.4
Carbanilate	270	−70	0.95
Diacetate	240	0.15	0.2
Diphenylacetate	290	4.8	0.32
Diphenylphosphonocarbamate	200	25	1.4
Nitrate	190	6	1
	230	−550	—
	270	−900	—
Triacetate	300	0.04	0.12
Cyanoethyl cellulose	260	250	—
Ethyl cellulose	260	25	1

Source: From Tsvetkov, V.N., *Rigid-Chain Polymers. Hydrodynamic and Optical Properties in solution*. Plenum Press, New York, 1989, p. 465; Tsvetkov, V.N., Rjumtsev, E.I., and Shtennikova, I.N., Intramolecular orientation order and properties of polymer molecules in solutions, in *Liquid Crystalline Order in Polymers*, Blumstein, A., Ed., Academic Press, New York, 1978, p. 57; Tsvetkov, V.N. and Tsvetkov, N.V., Electric birefringence in solutions of rigid-chain polymers, *Russ. Chem. Rev.*, 62, 851, 1994 (and references herein).

permanent dipole moments. The dependence of the Kerr constant on DS can be explained by the increasing the optical anisotropy of the macromolecules and the change in dipole moment of the repeat unit with increasing DS [68]. From the data summarized in Table 20.8, one can see the strong effect of the structure of the side chain on the longitudinal component of monomer dipole.

The EB of other cellulose compounds was also investigated, e.g., flax cellulose nitrate [69], cellulose cyanoethylene nitrate [70]. In particular, study of a number of fractions of flax cellulose nitrate in ethyl acetate solutions yielded $K_\infty = -150 \times 10^{-10}\,\mathrm{cm^5 \cdot g^{-1} \cdot (300\,V)^{-2}}$, $\mu_0 = 7\,\mathrm{D}$, and $\theta = 78°\,\mathrm{rad}$.

The electro-optical properties of combined cellulose derivatives [71,72] are determined by their structure and fraction composition. For example, a decrease in the fraction of nitrate substituent, from 89 to 0 wt.-%, in combined cellulose cyanoethyl nitrate was followed by a change in the Kerr constant from -93×10^{-10} to $+73 \times 10^{-10}\,\mathrm{cm^5 g^{-1}}$ $(300\,\mathrm{V})^{-2}$. On the one hand, this brilliant result is a remarkable illustration of high sensitivity of the EB effect to intramolecular ordering of the optically and electrically anisotropic fragments. On the other, it confirms the validity of the additive scheme in the determination of the dipole and optical structure of the macromolecule as accepted.

20.4 SIDE-CHAIN LC POLYMERS

We analyze now the EB data available on end-on and side-on comb polymers and effects of the main and side chains and type of attachment.

20.4.1 COMB END-ON POLYMERS

A variety of the end-on side-chain polymers are often used in molecular chemistry for new polymeric liquid crystals. Polymer backbones such as acrylic, methacrylic [73], and siloxane are often used in the design of novel polymeric liquid crystalline architectures. The structures of the spacer groups is also variable, such as methylene and oligo(ethylene oxide) [74] group moieties. Electric field effects on polymers in mesophase were studied [75], particularly, of the siloxane main chain polymers [76]. Also, detailed studies of methacrylic and acrylic LC polymers (see scheme 20.6) were performed in the isotropic phase [77,78].

Some properties of the comb polymers are provided by H-bonds arising between the side chains [79]. Here we present some new data on EB found in solution of the end-on polymers. One can see how orientation of the side groups of the macromolecule in electric field is limited by linkage function of the backbone. The higher the μ and longer the spacer chain the closer Kerr constant for a polymer and a reasonable model compound became. Hence, EB of a comb polymer is usually provided by independent rotation of the side polar residues.

20.4.1.1 Methacrylic Polymers

Experimental EB properties of some comb polymers with methacrylic main chain and mesogenic side groups are collected in Table 20.9. The values are large and the Kerr constants are negative in sign, which is typical for these polymers. For instance, $K = (-2\,\mathrm{to}\,-40) \times 10^{-10}\,\mathrm{cm^5}$ $\cdot\mathrm{g^{-1} \cdot (300\,V)^{-2}}$ for polymer C_{16} can be compared with $K = +0.21 \times 10^{-10}\,\mathrm{cm^5 \cdot g^{-1} \cdot (300\,V)^{-2}}$ as obtained for the reasonable monomer compound.

For many polymers without the mesogenic side groups, but with a Gaussian distribution of M, the Kerr constant is independent of M and is close to the value observed, in solution, of

TABLE 20.9

Electro-Optical and Dipole Moments of *End*-On Poly(R-methacrylate)s in Solution[a]

$$\sim CH_2-C(CH_3) \sim$$
$$|$$
$$CO-O-R$$

– R^b	Solvent[c]	$10^{10} \times K_\infty$ $(cm^5 \cdot g^{-1}) \cdot (300\,V)^{-2}$	$\mu_{0\parallel}$ (D)
–Ph–CO–O–C$_{16}$H$_{33}$	CTC	−0.5	0.27
–Ph–O–CO–Ph–O–C$_6$H$_{13}$	Benzene	−2	—
–Ph–CO–O–Ph–O–C$_9$H$_{19}$	CTC	−0.8 to −8	0.43
–Ph–O–CO–Ph–O–C$_{16}$H$_{33}$	CTC	−4 to −20	0.70
	Benzene	−2.2	0.28
	d	−18	0.55
	e	−40	0.82
–Ph–O–CO–Ph–O–C$_{16}$H$_{33}$ (70%)+C$_{16}$H$_{33}$ (30%)			
–Ph–O–CO–Ph–O–C$_{16}$ H$_{33}$(50%)+C$_{16}$H$_{33}$(50%)	CTC	−6	0.60
	CTC	−3.4	0.53
–Ph–CO–O–Ph–C≡N	DCAA	−4.9	0.79
–Ph–O–CO–Ph(NO$_2$)–O–C$_6$H$_{13}$	Dioxane	−7.2	0.67
–(CH$_2$)$_{10}$–CO–O–Ph–CO–O–Ph–O–C$_4$H$_9$	Dioxane	−0.9	0.4
–Ph–O–CO–Ph–O–C$_9$H$_{19}$	CTC	−3 to −7	0.80
–Ph–CO–O–Ph–O–C$_{12}$H$_{25}$	CTC	−14	0.61
–Ph–CO–O–Ph–O–CO–Ph–O–C$_{16}$H$_{23}$	Choloroform	−360	1.0

[a]Range of K is given for fractions of various M.
[b]–Ph– = –C$_6$H$_4$–
[c]CTC = carbon tetrachloride; DCAA = dichloroacetic acid.
[d]Benzene+heptane (66/34).
[e]Benzene+heptane (52/48).
Source: From Tsvetkov, V.N., *Rigid-Chain Polymers. Hydrodynamic and Optical Properties in solution*. Plenum Press, New York, 1989, p. 465; Tsvetkov, V.N. et al., Kerr effect in solutions of comb-shaped mesogenic polymers and their monomers, *Doklady Chem.*, 287, 1172, 1986.

a compound structurally similar to the side-chain. Unusually, for combs with the mesogenic side groups, even in Gaussian M-range, Kerr constant is a value strongly dependent on the length of the main chain. Figure 20.9 shows a growth in K with increasing M, which was observed for the methacrylic comb LC polymers C$_{16}$ and C$_9$ [57]. This unique phenomenon was due to the cooperative character of the interaction between kinetic units of the chain oriented in an electric field. In a recent study of EB and dielectric polarization of the C$_{16}$-polymer in dilute solution [81], it was shown that macromolecules of this polymer orientate in electric field due to induced dipole moment arising from polarization of a molecular coil. The polarization of the macromolecules was found to result from the orientation of dipole moments of the polar side groups under the action of an external electric field.

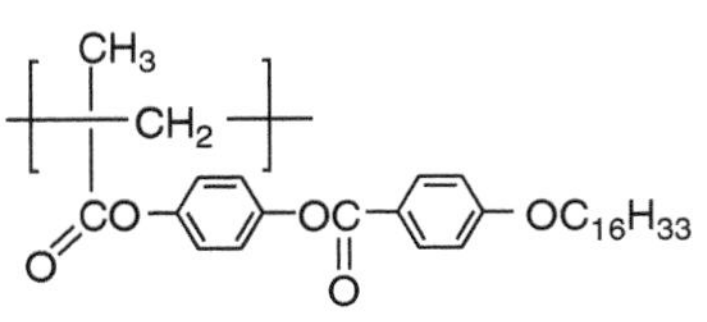

C_{16}

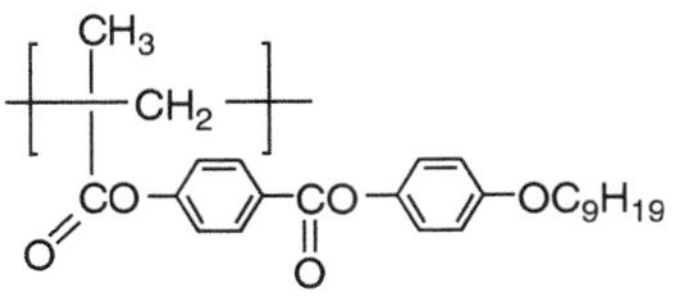

C_9

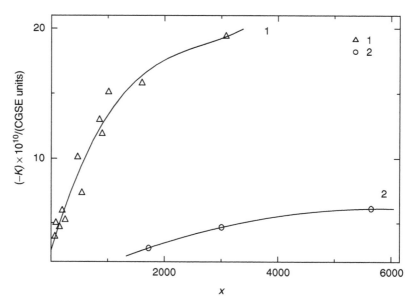

FIGURE 20.9 Kerr constant K versus x for fractions of comb polymers (1) C_{16} and (2) C_9 as obtained in carbon tetrachloride solutions (x is the reduced molecule length, $x = L/a$, L the molecule length, and a the persistence length in Kratky–Porod model).

20.4.1.2 Acrylic Polymers

Acrylic and methacrylic polymers, as illustrated in Table 20.9 and Table 20.10, are often characterized by negative K values, which are provided by an end-on attachment of the side mesogenic fragments. Because of the attachment, the optical polarizability of the repeat chain unit is greater when the side chain is orientated perpendicular to the polymer backbone, giving rise to a negative optical anisotropy and consequently a negative sign for K.

TABLE 20.10
Electro-Optical and Dipole Moments of *End*-On Poly(*R*-acrylate)s in Solution:
$\sim CH_2-CH \sim$
                $|$
         $CO-O-R$

-R[a]	Solvent[b]	$10^{10} \times K_\infty$ $(cm^5 \cdot g^{-1}) \cdot (300\,V)^{-2}$	$\mu_{0\parallel}$ (D)
$-Ph-CO-O-Ph-O-CH_3$	DCA	-4.9	0.79

[a]$-Ph- = -C_6H_4-$.
[b]DCA = dichloroacetic acid.

Source: From Tsvetkov, V.N., *Rigid-Chain Polymers. Hydrodynamic and Optical Properties in solution.* Plenum Press, New York, 1989, p. 465.

20.4.1.3 Poly(siloxane)s

Pretransition phenomena were detected via EB investigations in LC poly(siloxane)s [82].

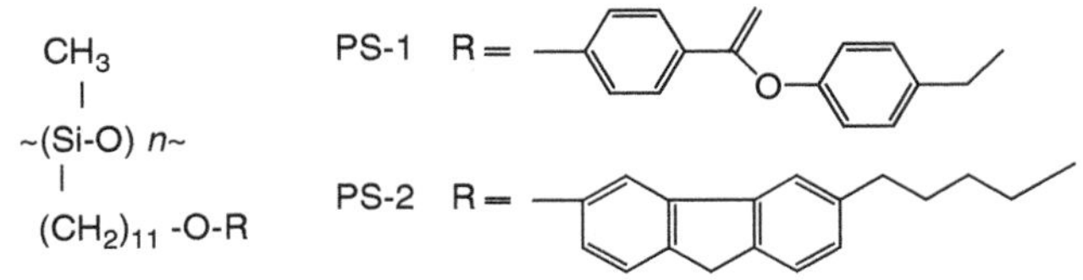

On approaching the transition temperature, the K value for both PS-1 and the model compound M-1, in the melt (1), increased several times and was positive in sign, whereas in dilute solution (2), the K value for both PS-1 and PS-2 changed sign and a sharp jump in the value of K was observed. The effects were explained by superorganization of the macromolecules with the formation of a highly ordered structure.

Outside of the pretransition range, the polymers were characterized with the K and μ values given in Table 20.11. EB of PS-1 coincides practically with that of the model compound M-1, indicating free orientation mobility of the mesogenic fragments of the macromolecules. The different K values of PS-2 and M-2 indicate steric retardations to free rotation and orientation of bulky fluorine heterocycle in the side groups.

20.4.1.4 Other End-On Polymers

High EB effect was detected (as illustrated in Table 20.12) for poly(nonyloxybenzamide styrene) [83] in benzene solution due to high optical anisotropy and dipole moment of the molecule. The authors referred the effect to ordering of the mesogenic side chains with the formation of the intramolecular LC structure similar to that observed for methacrylic polyesters.

20.4.1.5 Ionogenic Polymers

Series of the ionogenic LC copolymers of the repeat unit structure given below were investigated in solution and in the isotropic melt [84].

TABLE 20.11
Electro-Optical and Electrical Properties of Somb Poly(siloxane)s in Solution and Isotropic Melt (Given in Brackets[a])

Compound[b]	$10^{12} \times K$, $(cm^5 \cdot g^{-1}) \cdot (300V)^{-2}$	μ, (D)	θ, (deg rad)
PS-1 ($n = 100$)	$+ 34 (+ 38)$	—	—
PS-2 ($n = 40$)	$-0.8 (-1.0)$	—	—
M-1 $CH_2 = CH-(CH_2)_8-O-Ph-CO-O-Ph-C_2H_5$	$+ 31 (+ 50)$	2.8	47
M-2 $CH_2 = CH-(CH_2)_6-O-(C_{13}H_8)-C_5H_{11}$	$+ 4$	1.4	60

[a]Obtained at 15° higher the clearance temperature.
[b]–Ph– = –C_6H_4 – ; –$(C_{13}H_8)$– = heterocycle.

Source: From Rjumtsev, E.I., Yevlampieva, N.P., and Polushin, S.G., Pre-transition electrooptical properties of solutions of LC poly(siloxane)s in range of phase separation, *Russ. Optical J.*, 65, 24, 1998.

TABLE 20.12

Electro-Optical Properties of LC *End*-On Polymer in Solution:

$$\sim CH_2\text{–}CH \sim$$
$$|$$
$$R$$

–R[a]	Solvent	$10^{10} \times K_\infty$ (cm$^5 \cdot$g^{-1})·(300 V)$^{-2}$	$\mu_{0\parallel}$ (D)
–Ph–NH–CO–Ph–O–C$_9$H$_{19}$	Benzene	−140	90

[a] –Ph– = –C$_6$H$_4$–:

Source: From Rjumtsev, E.I. et al., Intramolecular LC structure in solutions of poly(nonyloxybenzamide styrene), *Vysokomolek. Soedin.*, *Part A*, 18, 439, 1976.

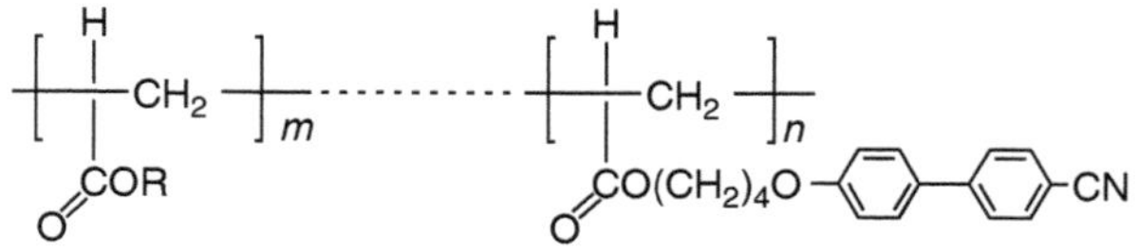

where R = H (AA) and CH$_3$ (MA), content = $m/(m+n)$.

The Kerr constant was determined using Equation 2.3. The results from this study are given in Table 20.13 where $K_0 = K_{f \to 0}$, where f is the field frequency and T^* the melt temperature. It was shown that the EB of the copolymers in solution is provided by orientation of the polar cyanobiphenyl groups, and Kerr constant is a decreasing function of the acrylic acid (AA) content.

The temperature dependence of Kerr constant and relaxation time of the orientation ordering fluctuations in the isotropic melt were investigated in the vicinity of the isotropic–nematic phase

TABLE 20.13

Electric Birefringence of the Ionogenic LC Copolymers in Tetrahydrofuran Solution at 294 K

Sample	$m/(m+n)$	$10^{10} \times K$ (cm$^5 \cdot$g^{-1})·(300 V)$^{-2}$	$10^7 \times K_0$ $(T–T^*)$, (cm$^2 \cdot$K)·(300 V)$^{-2}$
(AA)	0	4.9	1.0
	0.007	—	1.0
	0.09	5.0	1.6
	0.18	3.9	—
	0.33	3.7	—
	0.43	3.5	1.4
	0.63	—	0.8
	0.73	1.7	—
(MA)	0.10	—	1.4
	0.30	—	1.3
	0.35	3.7	—

Source: From Ryumtsev, E.I., Yevlampieva, N.P., and Kovshik, A.P., Influence of position of the alyphatic bridge fragments on polarity of LC compound molecules and their dielectric properties, *Zh. Fiz. Khim.*, 69, 934, 1995.

transition. It was shown that the relaxation time that characterizes the dipole polarization of the melt, is sharply increasing with the acrylic acid content in the copolymer. The origin of the effect may lie in the formation of the physical net of engagements in the copolymer melt as provided by H-bonds.

20.4.1.6   Low-Molecular-Weight Model Compounds

Kerr constants of simple molecules were determined in many studies. They are available, particularly from some recent works on substituted methanes gases (values of $10^{12} \times {}_mK$ from 7.5 to 363 cm^5 statvolt^{-2} mol^{-1} [86]) and others. The EB of dibromoalkanes (Br–$(CH_2)_{n-1}$ –Br, $n-1 = 3$ to 20) in cyclohexane solutions is characterized by experimental $10^{12} \times {}_mK$ values of -4.5 to 160 cm^5 statvolt^{-2} mol^{-1} [87]. A strong even–odd (zig-zag) effect was not observed experimentally with increasing the alkane chain length.

The mobility of the mesogenic units in the LC polymer molecule is usually evaluated from a comparison of the equilibrium Kerr effect of a comb LC polymer in dilute solution with that of a compound that is structurally similar to the mesogenic side chain. The properties of these compounds are grouped in Table 20.14 to Table 20.17. It is easy to see that their electro-optical properties are strongly dependent on the value and direction of dipole moment and on

---

TABLE 20.14
Electro-Optical Properties and Dipole Moments of Monomer and Model Compounds in Solution

Compound[a]	$10^{10} \times K$, $(cm^5 \cdot g^{-1}) \cdot (300\,V)^{-2}$	$\mu_{0\parallel}$ (D)
CH_3–CO–O–CH_3	2.1^b	—
CH_3–O–CO–$CH(CH_3)_2$	1.63^b	—
CH_3–CO–O–$CH(CH_3)_2$	2.45^b	—
CH_3–O–CO–C_6H_{10} –CO–O–CH_3	-2.15^b	—
CH_3–CO–O–C_6H_{10} –O–CO–CH_3	-1.85^b	—
CH_2=$C(CH_3)$–CO–O–Ph –O–CO–$Ph(NO_2)$–O–C_6H_{13}	-0.65	0.67
CH_2=$C(CH_3)$–CO–O–Ph –O–CO–Ph–O–$C_{16}H_{33}$	$+0.18$	0.52
CH_2=$C(CH_3)$–CO–O–Ph –CO–O–$C_{16}H_{33}$	$+0.15$	0.27
CH_2=$CHSi(CH_3)_3$	$+0.0144$	—
CH_2=$C(CH_3)$–CO–O–Ph –CO–O–Ph–C≡N	$+6.6$	0.79
Ph–CO–O–pPh–O–CO–pPh–O–CO–oPh– CO–O–pPh–CO– O–pPh–O–CO–Ph	-1.8	—
Ph–oPh–O–CO–pPh–CO–O–oPh–Ph	-0.18	1.94^c

[a] —Ph– = —C_6H_4 —; pPh = *para*–Ph; oPh = *ortho*–Ph.
[b] $_mK$ values (in $cm^5 \cdot 300V^{-2} \cdot mol^{-1}$)
[c] μ_0 value.

Sources: From Tsvetkov, V.N. and Tsvetkov, N.V., Electric birefringence in solutions of rigid-chain polymers, *Russ. Chem. Rev.*, 62, 851, 1994 (and references herein); Le Fevre, R.J.W., Electro-optical constants, *Adv. Phys. Org. Chem.*, 3, 1, 1965; Tsvetkov, V.N. et al., Kerr effect in solution of aromatic ester with *para*- and *ortho*-phenylene cycles in chain, *Doklady Chem.*, 307, 916, 1989; Rjumtsev, E.I. et al., Electrooptical properties of poly(1-(trimethylsilyl)-1-propyne) in solutions, *Polym. Sci., Part A*, 41, 765, 1999; Tsvetkov, V.N. et al., Kerr effect in solutions of comb-shaped mesogenic polymers and their monomers, *Doklady Chem.*, 287, 1172, 1986; Flory, P.J. et al., Optical anisotropies of aliphatic esters, *J. Phys. Chem.*, 85, 3215, 1981.

TABLE 20.15

Electro-Optical and Electrical Properties of Model Compounds with Polar Nitrile End Groups in Solution

Structural formula[a]	$10^{10} \times {}_mK$, $(cm^5 \cdot mole^{-1}) \cdot (300\,V)^{-2}$	μ, (D)	θ, (deg rad)
C_nH_{2n+1}–Ph–Ph–C≡N (n = 1 to 12)	62.2 to 73.4	4.77 to 4.9	—
C_6H_{13}–Ph–CO–O–Ph–Ph–C≡N	116 (8.6[a])	6.2	0
C_6H_{13}–Ph–CH$_2$–CO–O–Ph–Ph–C≡N	36	5.3	30
C_7H_{15}–Ph–(CH$_2$)$_4$–CO–O–Ph–Ph–C≡N	34	5.3	30
C_8H_{17}–O–Ph–CO–O–Ph–Ph–C≡N	10.7[b]	—	—
C_4H_9–(C$_6$H$_{10}$)–CO–O–Ph-Ph–C≡N	43	5.5	20
C_4H_9–(C$_6$H$_{10}$)–(CH$_2$)$_2$–CO–O–Ph–Ph–C≡N	38	5.4	30
C_4H_9–(C$_6$H$_{10}$)–(CH$_2$)$_5$–CO–O–Ph–Ph–C≡N	35	5.3	30
C_4H_9–(C$_6$H$_{10}$)–(CH$_2$)$_6$–CO–O–Ph–Ph–C≡N	32	5.2	30
C_5H_{11}–O–Ph–CO–O–Ph–C≡N	69	6.6	10
C_9H_{19}–O–Ph–CO–O–Ph–C≡N	69	6.6	10
C_7H_{15}–Ph–CO–O–Ph–C≡N	70	6.1	10
C_7H_{15}–Ph–CO–O–Ph–O–(CH$_2$)$_4$–C≡N	7	4.4	45
C_7H_{15}–O–Ph–N = N–Ph–C≡N	110	5.8	0
C_7H_{15}–Ph–N = N–Ph–O–(CH$_2$)$_4$–C≡N	1	3.7	56
C_6H_{13}–O–Ph–CH = N–Ph–C≡N	99	6.4	10
C_7H_{15}–O–Ph–Ph-C≡N	7.9[b]	5.28	—
C_8H_{17}–O–Ph–Ph–C≡N	51	5.2	10
C_6H_{13}–O–Ph–CH = CH–CO–O–Ph–C≡N	69	7.1	10
C_4H_9–O–Ph–CH = CH–CO–O–Ph–(CH$_2$)$_2$–C≡N	39	5.4	30
C_4H_9–O–Ph–CO–O–Ph–C≡N	7[b]	6.6	—

[a] –Ph– = –C$_6$H$_4$–; –(C$_6$H$_{10}$)– = cyclohexylene.
[b] Value of K in $cm^5 \cdot g^{-1} \cdot (300\,V)^{-2}$.

Sources: From Ryumtsev, E.I., Yevlampieva, N.P., and Kovshik, A.P., Influence of position of the alyphatic bridge fragments on polarity of LC compound molecules and their dielectric properties, *Zh. Fiz. Khim.*, 69, 934, 1995; Megnassan, E. and Proutiere, A., Dipole moments and Kerr constants of 4-n-alkyl-4′-cyanobiphenyl molecules (from 1CB to 12CB) measured in cyclohexane solutions, *Mol. Crystl. Liq. Crystl.*, 108, 245, 1984; Rjumtsev, E.I. et al., Value and direction of dipole moments in molecules of nematic liquid crystals containing nitrile groups, *Opt. Spektrosk.*, 41, 65, 1976 (in Russian); Laverenko, P. et al., Synthesis and molecular properties of new side chain LC polymers based on Baybis-Hillman reaction, *Mol. Crystl. Liq. Crystl.*, 338, 141, 2000.

the position of the aliphatic spacer. The data in Table 20.16 illustrates the solvent effects on the Kerr constant of the compound. Table 20.17 shows that, in accordance with Equation 2.6, the sign of Kerr constant of LC phenacyl esters [93] and hydroquinone esters [91] is reversed when the angle, θ, between the dipole moment and main axis of molecule polarizability passes its "critical" value of 54.7° rad (as calculated at negligible induced dipole moment).

In Table 20.16, the EB of 4,4′-*bis*(cyanobiphenyl) oxyalkanes in dioxane and benzene solution is compared to properties of 4,4′-(cyanobiphenyl)oxyalkanes. Nonmonotonic dependence of molar Kerr constant on the length of the flexible oxyalkane "bridge" in polymers of the first family was established and explained by the odd–even effect in intramolecular orientational polar and anisotropic order [92]. Similar, but higher in value, the odd–even EB effect was earlier observed in the isotropic phase for a nematic homologous series of 4′-n-alkyl-4-cyanobiphenyl (C_nH_{2n+1}–Ph–Ph–CN with n = 4 to 8) and 4′-n-alkoxy-4-cyanobiphenyl (C_nH_{2n+1}–O–Ph–Ph–CN with n = 3 to 9) [95,96]. The experimental data are

TABLE 20.16
Electro-Optical Properties of 4,4′-*Bis* (cyanobiphenyl)Oxyalkanes and 4,4′-(Cyanobiphenyl)Oxyalkanes in Dioxane and Benzene Solutions[a]

n	$10^{10} \times K$, (cm$^5 \cdot$g^{-1})$\cdot$(300 V)$^{-2}$ Dioxane	$10^{10} \times {}_mK$, (cm$^5 \cdot$mole^{-1})$\cdot$(300 V)$^{-2}$ Dioxane	$10^{10} \times K$, (cm$^5 \cdot$g^{-1})$\cdot$(300 V)$^{-2}$ Benzene	$10^{10} \times {}_mK$, (cm$^5 \cdot$mole^{-1})$\cdot$(300 V)$^{-2}$ Benzene
		N≡C–Ph–Ph–O–(CH$_2$)$_n$–O–Ph–Ph–C≡N		
2	−0.50	−6.3	−1.2	−14.0
3	3.26	42.4	2.6	31.1
4	1.48	19.9	1.07	13.3
5	2.66	36.8	2.23	28.6
6	3.6	51.3	3.4	45.0
7	3.82	56.1	3.53	48.1
8	4.3	65.1	4.24	59.3
9	4.6	71.3	4.5	65.0
10	5.18	82.3	4.68	69.1
12	5.9	98.2	5.6	86.7
		CH$_3$–(CH$_2$)$_n$–O–Ph–Ph–C≡N		
0	9.45	59.6	10.3	60
4	8.60	68.8	9.37	69.5
9	7.3	74	7.31	68.6

[a] –Ph– = –C$_6$H$_4$–.

Source: From Tsvetkov, N.V. et al., Electric birefringence in solutions of bis-4-cyanobiphenyloxyalkanes, *Mol. Crystl. Liq. Crystl.*, 265, 487, 1995.

given in Table 20.18 in terms of $K \times \Delta T$ where $K = \Delta n/E^2$ and $\Delta T = T - T^*$ with the second order transition temperature T^*. The results show that the even–odd effect in EB for cyanobiphenyl homologues is predominantly originates from the alternation of the expansion of the free energy per unit volume with the carbon chain length [97].

The strong odd–even oscillations of the value and sign of the Kerr constant in solution and the dielectric anisotropy, $\Delta\varepsilon$, of the nematic phase were found for dimers N≡C–Ph–Ph–O–(CH$_2$)$_n$–O–Ph–Ph– C≡N with $n = 2$ to 12 and trimers N≡C–Ph–Ph–O–(CH$_2$)$_n$–O–Ph–Ph–O–(CH$_2$)$_n$–O–Ph—Ph–C≡N with $n = 2$ to 10 [35]. These effects were explained by periodical changes in the intramolecular orientational polar and anisotropic order both in solution and in the mesophase with the variation in the length of the alkyl chain. The increase in the distance between strongly polar cyano groups leads to the weakening of correlation in the orientation of mesogenic polar fragments of dimer and trimer molecules in an electric field due to equilibrium and kinetic flexibility of the spacer chain.

20.4.1.7 Polymer Model Compounds

The properties of the LC end-on polymers under discussion are often compared with the EB of the reasonable nonmesogenic polymers. The Kerr constants of these polymers, collected in Table 20.19, are several orders lower than those of the LC polymers. As a rule, the K value of a flexible-chain polymer is independent of M excepting the oligomer M-range [99,100] and the charged state of the macromolecules [101].

TABLE 20.17

Electro-Optical and Electrical Properties of Phenacyl and Hydroquinone Esters in Solution

Structural Formula[a]	$10^{10} \times {}_m K,$ $(cm^5 \cdot mole^{-1}) \cdot (300\,V)^{-2}$	$\mu,$ (D)	θ, (deg rad)
$CH_3-Ph-CH=CH-Ph-CHO$	1.4	—	—
$CH_3-Ph-CH=CH-Ph-CH=CH-Ph-CHO$	−7.83	—	—
$C_7H_{15}-Ph-CO-O-CH_2-CO-Ph-C_6H_{13}$	2.0	2.3	49
$C_6H_{13}-Ph-CO-O-CH_2-CO-Ph-O-C_8H_{17}$	−1.5	3.6	56
$C_8H_{17}-O-Ph(F)-CO-O-CH_2-CO-Ph-C_6H_{13}$	3.2	3.3	51
$C_7H_{15}-Ph-CO-O-CH_2CO-Ph-O-C_{10}H_{21}$	−3.8	4.1	61
$C_6H_{13}-Ph-CO-O-CH_2-CO-Ph(F)-O-C_8H_{17}$	−1.3	3.6	59
$C_4H_9-Ph-CO-O-CH_2-CO-Ph-Br$	−1.3	2.6	65
$C_5H_{11}-Ph-C(CN)=CH-Ph-O-C_6H_{13}$	−5.4	3.8	64
$C_6H_{13}-O-Ph-C(CN)=CH-Ph-C_5H_{11}$	−5.4	4.0	63
$C_4H_9-(C_6H_{10})-CO-O-CH_2-CO-Ph-Br$	−1.3	2.5	67
$C_3H_7-(C_6H_{10})-Ph-CO-O-CH_2-CO-Ph(F)-O-C_8H_{17}$	−10.7	3.7	75
$C_3H_7-(C_6H_{10})-Ph-CO-CH_2-O-CO-Ph(F)-O-C_8H_{17}$	6.8	3.3	47
$C_6H_{13}-O-Ph-CO-O-Ph-O-CO-Ph-C_6H_{13}$	−0.8	3	—
$C_6H_{13}-O-Ph-CO-O-Ph-O-CO-Ph-O-C_6H_{13}$	−4.4	3.0	68
$C_6H_{13}-O-Ph-CO-O-Ph(Cl)-O-CO-Ph-O-C_6H_{13}$	−4.2	3.0	68
$C_6H_{13}-O-Ph-CO-O-Ph(CN)-O-CO-Ph-O-C_6H_{13}$	−1.0	4.2	58
$C_4H_9-O-Ph-CO-O-Ph(CN)_2-O-CO-Ph-O-C_4H_9$	−17.9	5.4	68
$C_6H_{13}-O-Ph-CO-O-Ph(NO_2)-O-CO-Ph-O-C_6H_{13}$	−7.45	4.48	63

[a] $-Ph-=-C_6H_4-$; $-(C_6H_{10})-=$ cyclohexane cycle.

Sources: From Beevers, M.S. and Goodwin, A.A., Synthesis and Kerr effect of *p*-phenylene vinylene oligomers, *J. Polym Sci., Part B: Polym. Phys.*, 34, 1133, 1996; Rjumtsev, E.I. et al., Value and direction of dipole moments in molecules of nematic liquid crystals containing nitrile groups, *Opt. Spektrosk.*, 41, 65, 1976 (in Russian); Rjumtsev, E.I. and Yevlampieva, N.P., Electrooptical properties and polar structure of LC phenacyl esters, *Mol. Crystl. Liq. Crystl.*, 251, 351, 1994; Rotinyan, T.A. et al., Molecular structure and dielectric anisotropy of some liquid crystals, *Kristallografiya*, 23, 578, 1978 (in Russian).

20.4.2 COMB SIDE-ON POLYMERS

20.4.2.1 Lateral Attachment

Macromolecules with laterally attached mesogenic side groups possess a unique ability to form thermotropic liquid crystals with a biaxial nematic phase [102] due to the hindered rotation of the side-on-fixed mesogens around their long axis. In spite of the claim that the fact of existence of this phase demands more confirmations [103], continuous attention was paid to these polymers [104–120] and the polymer-based elastomers [113]. This type of attachment leads to a strong deformation of the undisturbed coil conformation and extension of the polymer chains parallel to the phase director results.

The EB measurements of side-on polymer, LPMA, with symmetric three-ring mesogenic groups attached to the polymer backbone by its central ring via aliphatic spacer $-(CH_2)_{11}-$ were carried out in detail in a carbon tetrachloride solution and in an LC phase. These results were compared to the properties of the model compound (B) and end-on LC polymer (C). In the case of polymer (C), the mesogenic core is attached to the backbone without a spacer-group as shown in

TABLE 20.18

Electrical Birefringence of 4'-N-Alkyl-4-Cyanobiphenyl and 4'-N-Alkoxy-4-Cyanobiphenyl in Isotropic Phase[a]

N	3	4	5	6	7	8	9	Ref.
C_nH_{2n+1}–Ph–Ph–C≡N								
$10^7 \times K \cdot \Delta T$, cm$^2 \cdot$K$\cdot$(300 V)$^{-2}$	—	1.28	0.85	0.92	0.67	0.58	—	[95]
	—	—	2.24[b]	2.10[b]	1.37[b]	1.25[b]	—	[25]
C_nH_{2n+1}–O–Ph–Ph–C≡N								
$10^7 \times K \cdot \Delta T$, cm$^2 \cdot$K$\cdot$(300 V)$^{-2}$	1.48	1.14	1.1	0.83	0.77	—	—	[95]
	—	2.0	1.9	—	1.3	1.0	0.7	[96]

[a]Measured in pulsed [95] and AC field [96].
[b]$(\Delta n/\lambda E^2) \Delta T$ values in ($\times 10^{-10}$ m $\cdot$kV^{-2}).

Sources: From Coles, H.J., Laser and electric field induced birefringence studies on the cyanobiphenyl homologues, *Mol. Crystl. Liq. Crystl., Lett.*, 49, 67, 1978; Yamamoto, R. et al., The Kerr constants and relaxation times in the isotropic phase of nematic homologous series, *Phys. Lett., Ser. A.*, 69, 276, 1978; Rjumtsev, E.I. et al., Equilibrium and dynamic electrooptical properties of nematic and isotropic phases of 4'-n-alkoxy-4-cyanobiphenyl, *Russ. J. Phys. Chem.*, 69, 940, 1995.

Scheme A . The molecular parameters of the LPMA samples 1–4 of various M values are given in Table 20.20.

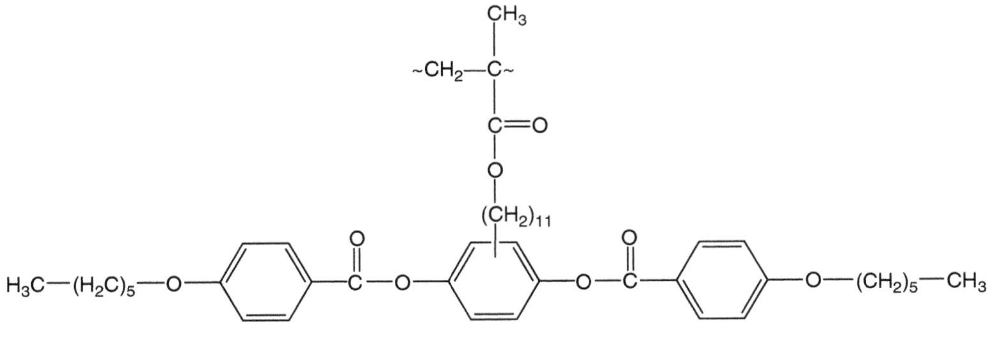

A

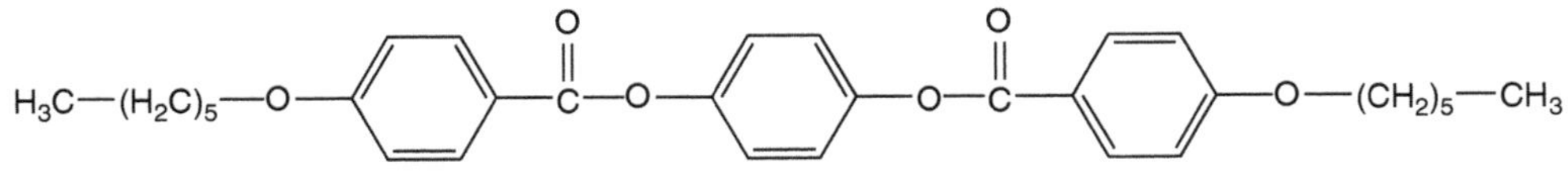

B

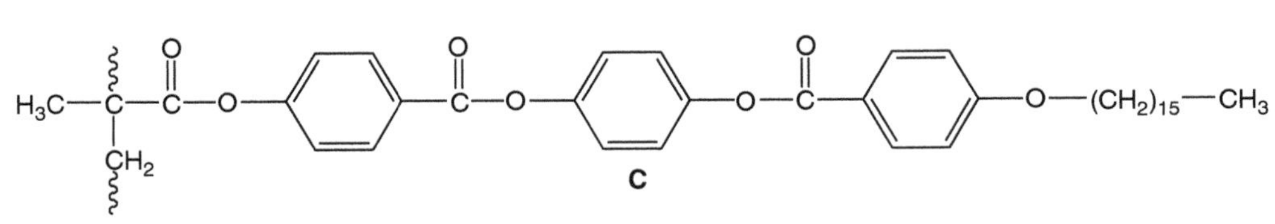

C

TABLE 20.19
Electro-Optical Properties of Some Flexible-Chain Polymers in Solution and Melt

Polymer	Solvent[a]	$10^{12} \times K \, (cm^5 \cdot g^{-1}) \cdot (300 \, V)^{-2}$
Polyethylene (low density)	(in melt)	0.57[b]
Polystyrene monomer (styrene)	—	4 to 7
Polystyrene	CTC	4.0
		0.09[b]
Poly-2,5-dichlorostyrene	Dioxane	4.0
Poly-p-chlorostyrene	Dioxane	3.5
Poly(vinyl chloride)	Dioxane	−7.7
Poly(vinylidene fluoride)	(in melt)	0.23[b]
Poly(2-vinylpyridine)	Benzene	−10
Poly(ethylene glycol)	CTC	−0.9
Poly(propylene glycol), 226 K	(in melt)	1.8[c]
Poly(dimethyl siloxane), 294 K	Benzene	1.3
	(in melt)	0.39[b]
Poly(hydromethyl-dimethyl siloxane)	Benzene	0.95
Poly(methylphenyl siloxane), 226 K, 294 K	Benzene	2.30
	(in melt)	1.6[c]
		2.24
Poly(methyl methacrylate)	Benzene	9 to 10
	Dioxane	3.2
	(in melt)	1.1[b]
Poly(buthyl methacrylate)	Dioxane	9.0
	CTC	1.3
Poly(cethyl methacrylate)	CTC	2.6
Poly(tert-butylphenyl methacrylate)	CTC	−4.0
Poly(methyl acrylate) (PA-1)	Toluene	8
Poly(butyl acrylate) (PA-4)	Toluene	9
	Decalin	3.3
Poly(decyl acrylate) (PA-10)	Decalin	3.0
PA-16	Decalin	3.0
PA-18	Decalin	3.2
PA-22	Decalin	3.1

[a]CTC = carbon tetrachloride.
[b]The value of $K/\bar{\nu}$.
[c]($\Delta n/E^2$) values in ($\times 10^{-12} \, cm^2 \cdot 300 \, V^{-2}$).

Sources: From Tsvetkov, V.N., *Rigid-Chain Polymers. Hydrodynamic and Optical Properties in solution*. Plenum Press, New York, 1989, p. 465; Beevers, M.S., Elliott, D.A., and Williams, G., Molecular motion in melt samples of poly(propylene glycol) studied using dielectric and Kerr effect relaxation techniques, *Polymer*, 21, 13, 1980; Thakur, M. and Tripathy, S., Electrooptical applications, in *Encyclopedia of Polymer Science and Technology*, Bikales, N.M., Ed., 2 ed., Wiley-Interscience, New York, 1986, 756; Beevers, M.S., Elliot, D.A., and Williams, G., Dynamic Kerr-effect and dielectric relaxation studies of a poly(methylphenyl siloxane), *Polymer*, 21, 279, 1980; Lavrenko, P.N. et al., Dynamooptical and electrooptical properties of poly(methylphenyl siloxane) in solution and bulk, *Vysokomolek. Soedin.*, 2005 (in press); Lavrenko, P.N., Kolomiets, I.P., and Vinogradova, L.V., Hydrodynamic, electrooptical and conformational properties of fullerene-containing poly(2-vinylpyridine)s in solutions, *Vysokomol. Soedin.*, 2005 (in press); Aronay, M., Le Fevre, R.J.W., and Parkins, G.M., Molecular polarisability. The molar Kerr constants and dipole moments of six polyethylene glycols as solutes in benzene, J. Chem. Soc., 7, 2890, 1960.

TABLE 20.20
Molecular Parameters[a] for LPMA Samples 1–4 as Obtained from the Velocity Sedimentation (sed) and Chromatography (GPC) Data

	4 (sed)	4 (GPC)	3 (GPC)	1 (GPC)	2 (GPC)
$M_w \times 10^{-3}$, kDa	6.8	1.19	1.257	0.68	0.49
M_w/M_n	3.8	2.73	3.22	5.77	4.51
M_z/M_w	2.1	2.28	2.02	3.52	3.01

[a]M_z, M_w, and M_n is the z-, *weight*-, and *number*-average value of molecular weight, respectively.

Source: From Lavrenko, P.N., Kolomiets, I.P., and Finkelmann, H., Optical anisotropy and conformation of macromolecules with laterally attached mesogenic side groups, *Macromolecules*, 26, 6800, 1993.

Figure 20.10 shows the EB, Δn, as a function of E^2 for various samples of LPMA of differing M values in carbon tetrachloride. They all were well approximated by linear functions to provide the Kerr constant of the solvent

$$K_0 = \Delta n/E^2 = 0.5 \times 10^{-12} \text{cm}^5 \cdot (300 \text{ V})^{-2}$$

and specific Kerr constant of LPMA as calculated by

$$K = (K_c - K_0)/c \tag{20.13}$$

where K_c is the K value for a solution with solute concentration c. The results are given in Table 20.21 together with properties of compound B and polymer C. One can see that the lateral attachment of the same mesogenic fragment via aliphatic $-(CH_2)_{11}-$ spacer (side-on polymers 1–4) leads to lower values of optical anisotropy and Kerr constant in comparison with the end-on polymer C meaning more free orientational mobility of the fragment.

It was concluded that the orientational order parameter for the mesogenic fragments in the isolated LPMA macromolecule is close to zero, meaning low ordering of the fragments with respect to the main chain contour. On average, the direction of the mesogenic group axes forms an angle (to the segment axis) with the average value close to 55° rad.

20.4.2.2 Cyanobiphenylene Polymers

To study the intramolecular axial ordering of the mesogenic side groups of an isolated macromolecule in solution, three comb LC polysiloxanes have been investigated [115]. In all polymer studied the side chain R contains an aliphatic spacer group which separates the mesogenic moiety from the backbone. A moiety contains one phenylene ring with oxy-aliphatic tail, ester linking group, and the cyanobiphenylene end group. The mesogenic moieties are attached to the aliphatic spacers, by the close to end point, through the *ortho*-phenylene ring. The main chain conformation in solution was identified to assess the hydrodynamic data using the Gaussian coil approximation. The specific Kerr constant obtained in dilute benzene solution of the polymer was compared with those of the model compound. The contribution of the backbone to the polymer electric birefringence

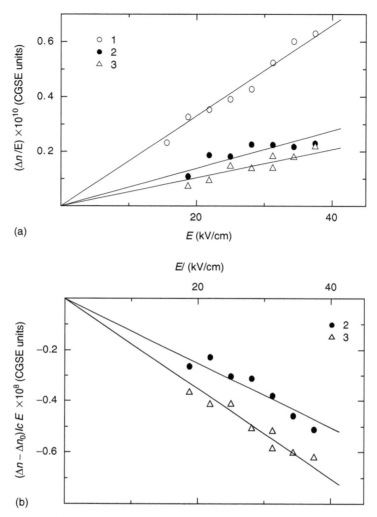

FIGURE 20.10 EB $\Delta n/E$ (a) and excess reduced EB $(\Delta n - \Delta n_0)/c\,E$ (b) versus electric field strength E in (1) solvent (carbon tetrachloride) and in solution of the LPMA samples (2) 1 and (3) IV. Solute concentration $c = $ (2) 0.75 and (3) 0.64 g·dl^{-1}.

was evaluated. The effects of temperature, solute concentration, electric field frequency, and field pulse duration were related to mobility and intramolecular ordering of the mesogenic groups in the isolated macromolecules.

The LC properties of a thermotropic comb polymer, as well as its polar structure, are significantly affected by the manner of attachment of the mesogenic fragments, e.g., cyano-biphenylene groups, to the side chains of the macromolecule [111,112,116–119]. To observe the effects of attachment, spacer length, and chemical structure of the main chain (methacrylic or siloxane backbone), Maxwell and Kerr effects in the polymer solutions were investigated, because these effects are sensitive to both the mobility and the ordering of the polar and optically anisotropic groups in the polymer chain [11,114,120,121].

TABLE 20.21
Optical, Electro-Optical, and Electrical Properties of LPMA and C Polymers and Compound B in Carbon Tetrachloride Solution

Polymer	$M_{sD}{}^a$ (kDa)	$\Delta \alpha \times 10^{25}$ (cm³)	$K \times 10^{12}$ (cm⁵·g⁻¹)·(300 V)⁻²	$\mu_{0\parallel}$ (D)	θ (deg rad)
1	1,000	−120	−40 ± 3	0.37	~ 55
4	3,000	−180	−50 ± 3	0.48	
C (in chloroform)	15 to 1,270	−4,900[b]	−360	1.0	—
Compound B	0.518	210	−29	3.0	68

[a]Obtained from the sedimentation–diffusion data.
[b]Optical anisotropy of segment.

Source: From Lavrenko, P.N., Kolomiets, I.P., and Finkelmann, H., Optical anisotropy and conformation of macromolecules with laterally attached mesogenic side groups, *Macromolecules*, 26, 6800, 1993.

Recently, this sensitivity was illustrated in a study of two novel side-chain [122] LC polymers of structure shown below [90,123].

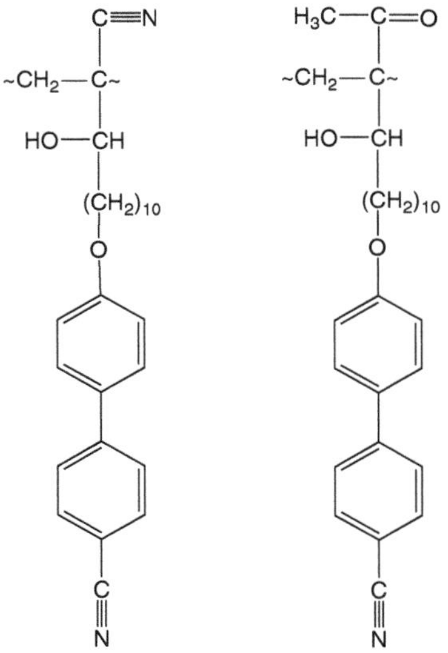

The cyanobiphenylene mesogenic moiety is attached here to the methacrylic backbone via an aliphatic spacer (an aliphatic part of the side group) by its end. Molar Kerr constant of P1 in solution was found to be very close to that of the mesogenic fragment (92%), and for P2 this parameter is lower (81%) due to the effect of the polar C=O bond in the side methyl-carboxy-substituent.

Three novel LC polymers with lateral structure of the macromolecule (side-on-fixed) polymers) as shown below were investigated [115]. A side chain of a polymer consists of an aliphatic $-(CH_2)_k-$ spacer group and a mesogenic moiety. A mesogenic moiety, in turn, consists of the one phenylene ring with oxy-aliphatic tail, ester linking group, and the cyanobiphenylene end group. Every mesogenic moiety (the mesogenic fragment with a tail) is attached laterally to the polymer backbone via a flexible methylene spacer group.

S1 $k=5$ **S2** $k=6$ **S3** $k=11$

DL-1

A low-molecular-weight thermotropic compound DL-1 was investigated as a model for the mesogenic moiety.

We consider below electro-optical properties of these LC polymers in more detail because they were found to be unusual and remarkable due to the unique combination of (1) great dipole and optical anisotropy of the mesogenic fragment and (2) type of the fragment attachment that is practically intermediate between that of side-on polymers LPMA (as presented in Scheme A, page 489) and end-on polymers P1, P2 as discussed above.

20.4.2.2.1   Electro-Optical Properties
The electric birefringence, Δn, in solution was investigated with varying frequency, pulse duration, and field strength. Figure 20.11 shows that the solvent effect, as observed in neat benzene (plot *1*), was positive (in sign) and close to the known data.

In a polymer solution, the EB was positive and larger in value than that in benzene. The $\Delta n/E$ as plotted against E for S1 and S3 in Figure 20.11a and Figure 20.11b at the pulse duration, τ, of 30 ms, is an increasing function of solute concentration (points 2–5). At every concentration, a dependence of $\Delta n/E$ on E is close to linear function, except for the most concentrated solutions.

The contribution of the solvent was then subtracted to determine the specific Kerr constant of a solute. Data are presented in Figure 20.12 in terms of the reduced excess EB $(\Delta n - \Delta n_0)/cE$, versus E as obtained under long ($\tau = 30$ ms, plots 12a and c) and short pulse duration ($\tau = 1$ ms, plot 12d). Here Δn and Δn_0 are the EB of a polymer solution and a solvent, respectively. One can see that excess EB can be approximated by a linear function of E (Kerr law) under short field pulses (plot 12d) rather than under long field pulses. Under the last condition, a liner approximation is more often available for the data obtained under a strong

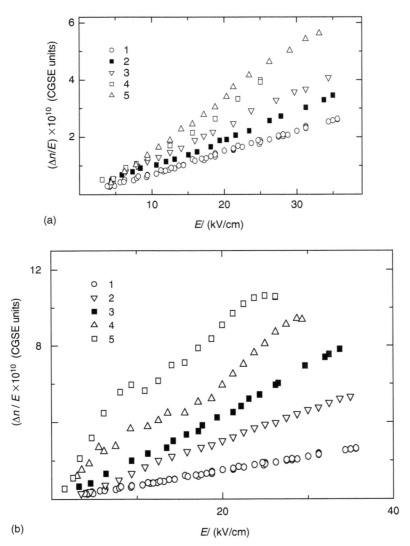

FIGURE 20.11 $\Delta n/E$ versus E for (a) S1 and (b) S3 in benzene solutions at various solute concentrations and electric pulse duration of 30 ms. Solute concentration: (a) $c = $ (1) 0, (2) 0.148, (3) 0.242, (4) 0.501, and (5) 0.562 g·dl^{-1}; and (b) $c = $ (1) 0, (2) 0.511, (3) 1.25, (4) 1.96, and (5) 2.83 g·dl^{-1}.

field, whereas various disturbances arise in a polymer solution under a weak field, all are dependent on solute concentration. Therefore, the effects of the applied electric field pulse duration, solute concentration, and the field frequency were then analyzed separately.

20.4.2.2.2 Solute Concentration Effect
Figure 20.12a and Figure 20.12c illustrate the data from the study of S1 and S3 in solution at solute concentration from 0.206 to 4.72 g·dl^{-1}. Within the point scattering, solute concentration does not obviously affect the EB value obtained under a long-pulsed field. The effect of the electric pulse duration looks here to be more significant.

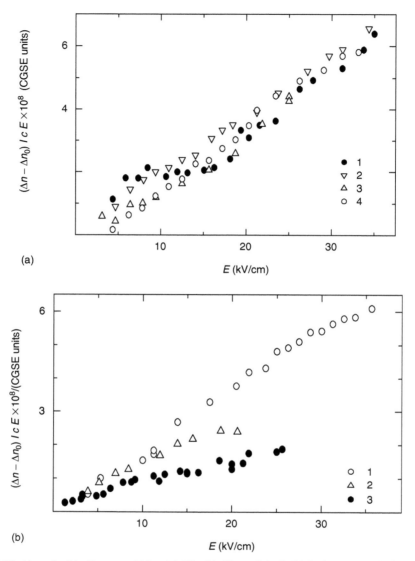

FIGURE 20.12 $(\Delta n - \Delta n_0)/c\,E$ versus E for (a) S1, (b) S2, and (c,d) S3 in benzene solution at various solute concentrations and fixed values of the electric pulse duration, τ. Solute concentrations: (a) $c = (1)$ 0.148, (2) 0.242, (3) 0.501, and (4) 0.562 g·dl^{-1}; $\tau = 30$ ms; (b) $c = (1)$ 0.502 and (2,3) 4.72 g·dl^{-1}; $\tau = (1)$ 0.6, (2) 1, and (3) 12 ms.

20.4.2.2.3 Electric Pulse Duration Effect

Figure 20.13a shows reduced excess EB $(\Delta n - \Delta n_0)/cE$ versus E as obtained for the DL-1 compound in benzene solution ($c = 0.43$ g·dl^{-1}) under the electric field pulse duration of 0.6 to 20 ms (points 1 to 4). All the plots were linear, passing through the origin in accordance with the Kerr law, $\Delta n = \text{Const·}E^2$. The differently marked points form the common function of $(\Delta n - \Delta n_0)/cE$ on E, and we conclude that the EB effect of the mesogen in solution is independent of the pulse duration. At high values of E (<30 kV·cm^{-1}), some deviation from linearity occurs and, at present, we do not know why.

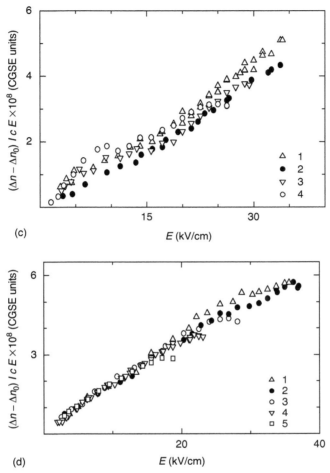

FIGURE 20.12 (continued) (c) $c = (1)$ 0.511, (2) 1.25, (3) 1.96, and (4) 2.83 g·dl^{-1}; $\tau = 30$ ms; and (d) $c = (1)$ 0.50, (2) 0.51, (3) 1.87, (4) 2.83, and (5) 4.72 g·dl^{-1}; $\tau = 1$ ms.

Another EB behaviour, as given in Figure 20.13b, was observed in a polymer solution. Within the experimental error, the $(\Delta n - \Delta n_0)/cE$ dependence on E corresponds to the Kerr law at most short pulse duration only ($\tau = 0.6$ ms for S1 and 0.005 ms for S3, respectively). Under longer pulses, the function of $(\Delta n - \Delta n_0)/cE$ on E does not obey the Kerr law. Along the pulse length, the experimental points fall down at high E; thus, specific EB is not a linear function of E, and the range of curvature shifts to smaller E (over the τ range of 0.6 to 6 ms for S1 and 0.005 to 30 ms for S3). At higher τ, the specific EB shows increasing behavior and returns to its value observed under short pulses (over the τ range of 6 to 20 ms for S1 and 30 to 50 ms for S3).

The initial slope of $(\Delta n - \Delta n_0)/cE$ on E yields specific Kerr constant defined by Equation 20.6. For all the polymers in benzene solution under the 12-ms pulses (at $c = 4.70$ g·dl^{-1}) Kerr constant was 3.0×10^{-10} cm^5·g^{-1}·(300 V)$^{-2}$. It is approximately half of the value observed

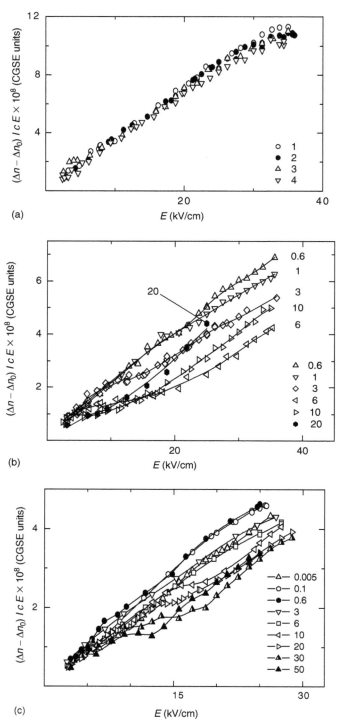

FIGURE 20.13 $(\Delta n - \Delta n_0)/c\,E$ versus E for (a) DL-1, (b) S1, and (c) S3 in benzene solutions at various pulse durations τ and fixed solute concentrations. Pulse durations: (a) $\tau = (1)$ 0.6, (2) 1.0, (3) 12, and (4) 20 ms; $c = 0.43\,\mathrm{g\cdot dl^{-1}}$; (b) $\tau = 0.6, 1, 3, 6, 10,$ and 20 ms as indicated; $c = 0.501\,\mathrm{g\cdot dl^{-1}}$; and (c) $\tau = 0.005$, 0.1, 0.6, 3, 6, 10, 20, 30, and 50 ms as indicated; $c = 1.96\,\mathrm{g\cdot dl^{-1}}$.

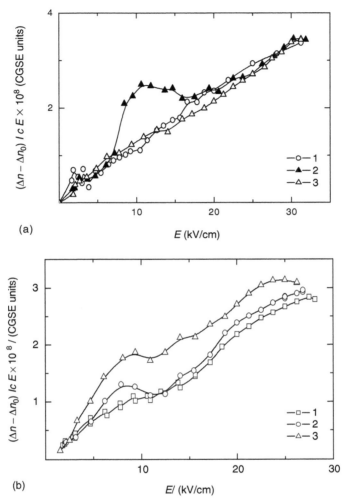

FIGURE 20.14 $(\Delta n - \Delta n_0)/c\,E$ versus E for S3 in benzene solution at various temperatures T and solute concentrations of (a) 1.87 and (b) 2.83 g·dl^{-1}. Temperatures: (a) $T = (1)$ 17, (2) 21, and (3) 27°C; $\tau = 30$ ms; and (b) $T = (1)$ 14.7, (2) 17.0, and (3) 20.0°C; $\tau = 30$ ms.

under 0.6-ms pulses. The origin of the unexpected $(\Delta n - \Delta n_0)/cE$ (E) dependence under various pulse duration remains unclear.

20.4.2.2.4 Temperature Effect

The temperature effect was qualitatively analyzed for S3 in benzene solutions at two concentrations, 1.87 and 2.83 g·dl^{-1}. Figure 20.14a and Figure 20.14b shows significant and unusual change in the EB value with increasing E as observed at different temperatures around 20°C. Initially, a rapid increase in EB with increasing E is observed, which then levels off prior to a second region, where an increase in EB with increasing E is observed. In this second region, the rate of increase of EB with E is much smaller than that found in the initial region. The effect was stable with time and repeatable for solutions of several preparations. The nature of the effect is still under question. However, the effect looks to be a phenomenon that is annihilating when experiments are carried out outside a particular temperature range, i.e., for S3 the temperature range is 15 to 25°C.

20.4.2.2.5 AC Frequency Effect

Figure 20.15 shows the EB effect as studied in a solution under alternating (AC) pulsed electric fields at different frequency, v, from 6 to 590 kHz. The effect is linear in accord with the Kerr law, at v exceeding 60 kHz. The same value of K was obtained under the high-frequency AC field and under the short-pulsed DC field (0.6 ms).

The last value was considered as characteristic of a fast process. In contrast, the K value obtained under the long-pulsed field (30 ms) is a summary result of one fast EB effect (positive in sign) and two slow (negative and positive) effects. The fast effect was attributed to the mobility of the mesogenic fragments with high proper polarizability. Under the electric field, these fragments are subjected to be oriented within the field and align parallel to it. Space orientation of the more bulky parts (domains) is usually responsible for the slow EB effect.

To confirm this observation, a detailed analysis of the output signal was carried out. Figure 20.16a and Figure 20.16b show the data obtained for polymers S1 and S3, respectively. The field strengths were 15 to 105 CGSE units for polymer S1 and 20 to 80 CGSE units for S3. For both the polymers, after a switch of a DC field voltage, a fast start jump in the EB of polymer S1 (during first time, ≤ 1 ms) was observed. The values of these jumps (from zero level) in both cases exceeded the steady EB value not higher than for 30%.

Notice the $\Delta n(t)$ values are normalized in Figure 20.16a plot to the final $\Delta n(20)$ value, whereas in the Figure 20.16b plot, they are normalized to the start of the $\Delta n(0)$ value. The higher concentration value that was available for polymer S3 (and, hence, the higher EB effect) provided us with a possibility to measure the E-dependence of the EB in front of a pulse within the first 2 μs.

The behaviour of the two polymers to the applied field at $t > 2$ ms was quite different in timescale only. In the case of the S1 polymer (short spacer group), at low field strengths (15 and 30 CGSE units), there is a further slow increase in EB during the first 2 to 3 ms, which

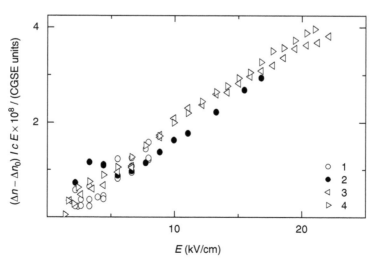

FIGURE 20.15 $(\Delta n - \Delta n_0)/c\,E$ versus E for S3 in benzene solution at various the oscillation frequencies of (1) 6, (2) 13, (3) 60, and (4) 590 kHz; $c = 2.83$ g·dl^{-1}.

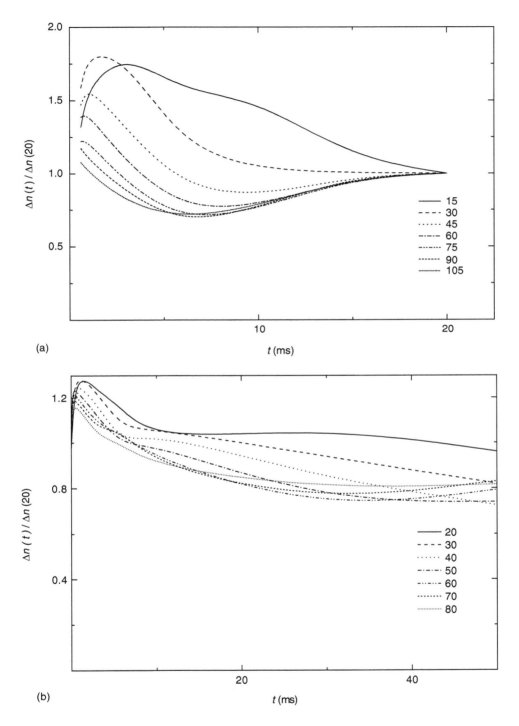

FIGURE 20.16 The time-function of reduced EB $\Delta n(t)/\Delta n$ (20) for (a) S1 and (b) S3 in benzene solution under different electric field strengths E: (a) $E = 15$ to 105 (300 V/cm) as indicated; $c = 0.501$ g·dl^{-1}; (b) $E = 20$ to 80 (300 V/cm) as indicated; $c = 1.96$ g·dl^{-1}.

then falls to a steady state at around 20 ms. At intermediate field strengths (45, 60 and 75 CGSE units), this slow rise is less prominent but the EB still falls steadily for about 10 ms, when it then rises slightly to a steady state at around 20 ms. At larger field strengths (90 and 105 GCSE units), the observations are the same as previously, but there is no obvious initial slow rise in EB.

This is in sharp contrast to that what was observed for polymer S3. In general, we see that an initial rapid rise in EB is followed by a similar rapid decrease, which then tends to level off around 10 ms. However, unlike the S1 polymer, no steady state is reached even after 45 ms. The longer transition time for approaching the steady state has to be related to longer spacer in polymer S3.

It is worth noting that the concentration of the S3 benzene solution was four times greater than that of the S1 benzene solution, and that this may have played some part in the different behaviour of the two polymers to the pulsed electric field. An explanation of this complicated behaviour of polymer molecules (aggregates?) in solution under the influence of a pulsed electric field may be the subject of further investigations. The origin of the negative EB effect is especially interesting.

20.4.2.2.6   Effect of the Side Chains

The EB effect in solution of the model DL-1 compound is illustrated by the data presented in Figure 20.13a. The value of $(\Delta n - \Delta n_0)/cE$ as a function of E yields a specific Kerr constant K of $(10.7 \pm 0.2) \times 10^{-10}$ cm$^5 \cdot$g$^{-1} \cdot$ (300 V)$^{-2}$. It does not depend on either concentration or on the pulse duration. This value is very close to 10.3×10^{-10} cm$^5 \cdot$g$^{-1} \cdot$ (300 V)$^{-2}$ known for the model compound CH$_3$–O–Ph–Ph–C≡N [115]. This means that the cyanobiphenylene group is almost solely responsible for the electro-optical Kerr effect, and that neither the aliphatic tails nor the additional benzene ring contribute appreciably to the EB effect.

Let us now compare the properties of DL-1 with other LC compounds of similar dipole structure, as given in Table 20.22. One can see that the K value for DL-1 is significantly higher that of the M7 and $_7$OCB compounds. This is mainly due to the higher optical anisotropy of

TABLE 20.22
Electro-Optical and Optical Properties of Model Compounds in Dilute Benzene Solution at 21°C

Substance[a]	$10^{10} \times K$, (cm$^5 \cdot$g^{-1})$\cdot$(300 V)$^{-2}$	$\mu_{0\parallel}$, (D)	$\Delta \alpha^{b}$, (Å^3)	Ref.
DL-1	10.7 ± 0.2	–	–	[115]
M7	8.6	6.2	19	[85]
$_7$OCB	7.9	5.28	–	[90]

[a]DL-1, C$_8$H$_{17}$ -O-Ph-CO-O-Ph-Ph-C≡N; M7, C$_6$H$_{13}$ -Ph-CO-O-Ph-Ph-C≡N; $_7$OCB, C$_7$H$_{15}$ -O-Ph-Ph-C≡N.
[b]Optical anisotropy of the molecule.

Sources: From Ryumtsev, E.I., Yevlampieva, N.P., and Kovshik, A.P., Influence of position of the alyphatic bridge fragments on polarity of LC compound molecules and their dielectric properties, *Zh. Fiz. Khim.*, 69, 934, 1995; Lavrenko, P. et al., Synthesis and molecular properties of new side-chain LC polymers based on the Baylis–Hillman reaction, *Mol. Crystl. Liq. Crystl.*, 338, 141, 2000; Lavrenko, P. et al., Hydrodynamic and electro-optical properties of novel LC polymers with laterally attached cyanobiphenylene mesogenic side groups, in *Proceedings of the Fourth International Symposium on "Molecular Order and Mobility in Polymer Systems,"* St.-Petersburg, 2002, p. 86.

the DL-1 molecule. In contrast, the DL-1 dipole moment is close to that of M7 and $_7$OCB, with the main contribution coming from the terminally-positioned CN-group.

The dipole moment μ of the S3 molecule in benzene solution was evaluated by using dipole polarization data. The permittivity increment, $\varepsilon-\varepsilon_0/c$, was determined for the S3 in benzene solution at $c = 0.34\,\text{g·dl}^{-1}$ to be $6.0 \pm 0.2\,\text{ml·g}^{-1}$. A low μ value was thus obtained, and no frequency dispersion of the effect was observed over the frequency range from 15 kHz to 1.5 MHz.

We compare the Kerr effect observed in solutions of a polymer and the model compound DL-1 now. Firstly, the intrinsic EB of compound DL-1 in solution, as found under the same electric field and solute concentration, is higher than that for a polymer, indicating the higher Kerr constant value for compound DL-1. Secondly, the same Kerr constant value was obtained for all the polymers under investigation.

This is not true for the Kerr constant value per mesogenic group. It was defined by $K_{\text{mes}} = K/x_{\text{mes}}$ (where x_{mes} is the formula weight fraction of the mesogens). The results are given in Table 20.23 where K is the experimental quantity. One can see that, within experimental error ($\sim 3\%$), the value of K_{mes} for S1 is close to that obtained for S2, both 13% smaller than that for compound DL-1. In contrast, S3 is characterized with the K_{mes} value, which is very close to that found for DL-1. This means that the $-(CH_2)_{11}-$ spacer in the S3 molecule side chain is long enough to provide sufficient free space for the polar mesogenic groups to orientate in the external electric field. The aliphatic $-(CH_2)_5-$ and $-(CH_2)_6-$ spacers in the S1 and S2 molecules, respectively, are not so long and the polymer effect (retardation to free space orientation of the side mesogenic groups through their linkage to the polymer backbone) is noticeable, even in dilute solution (in isolated state of the macromolecules). More significant effects may be expected in the LC phase.

Let us now compare the properties of two polymers, containing the same mesogenic and spacer groups but differ by their attachment of the mesogenic group to the polymer backbone, i.e., in S3 the mesogenic group is laterally attached and in P1 [13], terminally attached to the polymer backbone. We compare their EB in relation to that of DL-1. The ratio $K_{\text{mes}}^{\text{pol}}/K_{\text{mes}}^{\text{DL}}$ is 0.95 for S3 and 0.92 for P1 [90]. Thus, ignoring the difference in chemical structure of the backbone (this is reasonable since both contained a very long spacer group, $-(CH_2)_{11}O-$), we may conclude that the lateral attachment of the mesogenic side chain to the polymer backbone (S3) provides more freedom for the orientation of the mesogenic side chain in comparison with the terminal attached mesogenic side chain, polymer P1.

TABLE 20.23
Electro-Optical Properties of the Model Compound DL-1 and Polymers S1, S2, and S3 in Dilute Benzene Solution at 21°C under Electric Pulse Duration of 0.6 ms

Substance	$10^2 \times c\ (\text{g·cm}^{-3})$	$10^{10} \times K\ (\text{cm}^5\text{·g}^{-1})\text{·}(300\,\text{V})^{-2}$	$x_{\text{mes}}^{\text{a}}$	$10^{10} \times K_{\text{mes}}\ (\text{cm}^5\text{·g}^{-1})\text{·}(300\,\text{V})^{-2}$
DL-1	0.43	10.7 ± 0.2	1	10.7 ± 0.2
S1	0.50	6.4 ± 0.1	0.662	9.7 ± 0.2
S2	0.50	6.0 ± 0.2	0.648	9.2 ± 0.3
S3	0.50	6.0 ± 0.1	0.586	10.2 ± 0.2

[a]The weight fraction of the mesogenic side fragment in the repeat unit.

Source: From Lavrenko, P. et al., Hydrodynamic and electro-optical properties of novel LC polymers with laterally attached cyanobiphenylene mesogenic side groups, in *Proceedings of the Fourth International Symposium on "Molecular Order and Mobility in Polymer Systems,"* St.-Petersburg, 2002, p. 86.

20.4.2.2.7 Nonequilibrium EB

The experimental K value obviously does not depend on the solute concentration at a pulse duration of 0.6 ms only. More often, as mentioned above, the behaviour of the polymer in dilute solution, at the solute concentrations used, and in the presence of an electric field, is complicated by (a) nonlinear function of the birefringence $\Delta n/E$ on the field strength and (b) dependence of EB on the pulse duration and the solute concentration, which reflect different modes of intra- and inter-molecular movements.

Figure 20.16 illustrates very complicated but reversible changes in the impulse form as observed for polymers with (a) short and (b) long spacer groups at various values of E, from 15 to 105 (300V)·cm^{-1}. The results are presented here as related to either (a) the equilibrium EB value or (b) the start effect. One can see that an initial large jump in the EB (molecular effect) is followed by a relatively fast fall. Further slow growth in this effect is dependent on the field strength. The amplitude of the change is obviously higher for S1 and the relaxation of S3 in benzene solutions takes more than 50 ms. It is also clear that there are at least three origins of this relaxation, and two of these are associated with a positive optical birefringence and one with a negative birefringence.

This is the first time that such behaviour has been found in polymer solution [90,114,120–123]. Hence, it must be related to peculiarities of the polymers under investigation. However, relaxation of the EB effect is usually characterized by at least one to two orders of magnitude lower than their relaxation times. Even in the condensed state, rotational relaxation in the isotropic phase of the thermotropic side-chain LC polymer incorporating terminally attached cyanobiphenyl mesogenic side groups, the pretransitional T-range (which satisfactorily agrees with the de Gennes' model) is characterized with times lower than 0.5 ms [92]. The slow processes in EB are very rarely observed, e.g., for the aggregates composed of large helical poly(L-glutamic acid) molecules in aqueous methanol solvents [54] and for the bulk poly(propylene glycol) at low temperatures only [6].

For polymers under investigation, assembling of the macromolecules takes time and exceeds 20–30 ms. Such a long relaxation process can be understood if either (a) very large particles (domains) arise in the polymer solution under an influence of the electric field or (b) the effect is caused by a long-time reorganization of the macromolecular inner structure, induced by the orientation of the mesogenic moieties in an electric field via rotation of the side groups around the backbone.

There is a discrepancy between the first proposal and the fact that no aggregation was detected by hydrodynamic methods in benzene solution of the polymers in the absence of an electric field. Hence, either this is the electrically induced assembling, or these assembles exist permanently, which is not detectable by the hydrodynamic methods (<10% w/w) but is large enough to provide the EB effect.

The second proposal means that a long-time relaxation of EB reflects a change in the conformation of the macromolecular main chain under the action of the mesogenic side moieties during their orientation in an external electric field. Small differences between the start and equilibrium values of EB and independence of the relaxation behavior on the solute concentration confirms the probability of the proposal.

The origin of the negative electro-optical effect is due to the difference in polar and axial ordering of the mesogenic moieties in an electric field. For example, when the polar direction does not coincide with the main optical axis of the mesogenic moiety. In this case, the reverse of the polar moiety in an electric field is attributed to a change in the sign of the dipole (relatively to the field direction), but the initial and the final mesogenic moiety orientations are characterized by the same optical polarizability. Hence, during this reverse process,

K changes sign and magnitude. Usually it is very fast and, therefore, undetectable. The question now is why the process is so slow in the polymers under study. This will be the subject of future investigations.

In any case, the dipole–dipole interactions of the mesogenic fragments seem to be responsible for this unusual EB effects. Probably the origin of the effect is due to high dipole moment of the mesogenic cyanobiphenylene fragment, the lateral attachment of the mesogenic side chain to the polymer backbone or the flexibility of the siloxane backbone.

On average, every macromolecule consists of about 14 repeat units including 14 mesogenic groups. In comparison with the methacrylic main chain macromolecules, more compact configuration and higher form asymmetry of the macromolecules with the siloxane backbone and short spacers in the side groups are established via hydrodynamic investigations.

Electro-optical properties of the S3 polymer, with $-(CH_2)_{11}-$ spacer between the mesogenic moiety and the siloxane backbone, as studied under short-pulsed field, are close to that found for the individual model compound DL-1 ($K_{mes}^{S3}/K_{mes}^{DL}=0.95$), indicating free orientation of the mesogenic side group that is devoid of any influence from the polymer backbone, in dilute benzene solution under the influence of an external electric field. For S1 and S2 polymers with the shorter spacers, $-(CH_2)_5-$ and $-(CH_2)_6-$, respectively, the ratio K_{mes}/K_{mes}^{DL} value is noticeably lower than that for S3, reflecting the more pronounced polymer linkage effect (the linkage of the mesogenic moieties to the polymer backbone).

In turn, under a long-pulsed field, electro-optical properties of the polymers in solution are complicated with long-time relaxation caused by either the formation of large electrically induced domains within the anisotropic structure, which in turn are dependent on the field strength, or changes in the macromolecular conformation. In the latter proposal, the orientation of the mesogenic moieties in an external electric field leads not only to orientation of the macromolecule as a whole, but also to deformation of the macromolecule due to high flexibility of the siloxane chain. Thus, lateral attachment [116,124] of the mesogenic moiety, its high dipole moment, and kinetic flexibility of the $-Si-O-$ backbone can be responsible for the unusual electro-optical properties of the polymers under investigation.

20.4.2.3 Semilateral Attachment

The properties of the LC polymers with the semilateral attachment of the mesogenic fragments significantly depend on the position of the linking bond in the phenylene ring to the spacer group (see Scheme 20.13). If ortho-positioned (PMA 3), the polymer will exhibit a nematic phase that is very sensitive to an applied electric field. Polymers with a meta-positioned linkage (PMA 2) forms two smectic phases which are insensitive to an applied electric field. Polymer PMA 1, which has a para-positioned linkage, is able to form a very stable nematic and smectic phases. Additionally, polymers incorporating a terminally-positioned CN-group will posses a high and positive dielectric anisotropy.

$R = H, -OC_nH_{2n+1}, CN$ or NO_2

(a)   **PMA 1**

(b)   **PMA 2**

(*continued*)

(c)   **PMA 3**

The influence of the type of the mesogen attachment on the EB effect was studied using polymers PMA 1 to PMA 3 [120].

The EB data are given in Table 20.24. The negative Kerr constant obtained for PMA 1 is explained by direct end-on attachment of the mesogenic fragments to the backbone. As a result, the polar axis of the mesogenic fragment is oriented in a direction perpendicular to the macromolecule contour (direction of maximum polarizability). Lateral (PMA 3) and semi-lateral (PMA 2) attachment leads to more free orientation of the polar fragments in the electric field, the direction of the polar axis is approaching that of polarizability, and positive (in sign) Kerr effect is observed. Note that the K value for PMA 2 is close to that of PMA 3, meaning the effect of attachment (passing from lateral to semilateral type) on fragment mobility is equivalent to the effect of twofold shorter spacer length.

20.4.3 EFFECT OF MAIN CHAIN

20.4.3.1 Methacrylic Chain

The EB of methacrylic main chain is close to the effect from *n*-alkanes. Table 20.25 with experimental molar Kerr constants, as converted from data of Stuart [126], shows that EB is

TABLE 20.24
Electro-Optical Properties of Cyanophenyl-Containing Poly(R-methacrylate)s in Solution

Polymer	Solvent[a]	$10^{10} \times K$, $(cm^5 \cdot g^{-1}) \cdot (300\,V)^{-2}$
PMA 1	DCAA	−4.9
PMA 2	Benzene	+ 3.0
PMA 3	Benzene	+ 2.8

[a]DCAA = dichloroacetic acid.

Source: From Lavrenko, P. et al., Optical anisotropy of comb-like macromolecules of lateral and semilateral structure, *Macromol. Chem. Phys.*, 199, 207, 1998.

TABLE 20.25
Molar Kerr Constants of *n*-Alkanes

n	$10^{12} \times {}_m K^a$	*n*	$10^{12} \times {}_m K^a$	*n*	$10^{12} \times {}_m K^a$
4	10.089	7	23.931	11	45.297
5	15.264	8	27.918	14	56.556
6	19.206	9	32.895	15	72.585

^{a}In $cm^5 \cdot mol^{-1} \cdot (300\ V)^{-2}$.

Source: From Stuart, H.A., *Die Struktur des Freiem Moleküls*, Springer-Verlag, Berlin, 1952, 460.

positive in sign and low in value. Therefore, the contribution of the methacrylic main chain can often be ignored.

20.4.3.2 Siloxane Chain

The contribution of the siloxane backbone to the Kerr effect of comb-shaped polymers in solution is specified by the fact that a siloxane polymer chain is characterized by a significant difference in valence angles of atoms along the chain and an alternating direction of bond dipole moments. In poly(dimethyl siloxane) (PDMS), for example, the valence angle at the oxygen atom is more than 30° higher than that found at the silicon atom [127]. As a result, the plane *trans*-configuration of the chain possesses its own curvature, and the PDMS chain in this configuration would close on itself within 11 units. X-ray analysis of a sample of PDMS crystal (performed at 183 K) showed that the chain is in the extended helix conformation. Inturn, dielectric investigations of the dimethyl siloxane oligomers and the polymers in the homology line led to the conclusion that the helix conformation of the chain may also be retained in dilute solution.

In a side-chain LC poly(siloxane)s, the side groups strongly influence their liquid crystallinity due to high flexibility and low anisotropy of the polymer backbone. Experimental data from the studies of Tsvetkov [3] illustrated this effect on Kerr and Maxwell constants, which characterize the induced birefringence in solution of the poly(siloxane)s. For example, the optical anisotropy of the PDMS per repeat unit $(0.96 \times 10^{-25}\ cm^3)$ and poly(methylphenyl siloxane) $(-13.5 \times 10^{-25}\ cm^3)$, both determined by the flow birefringence method, differ from one another with more than a factor 10 and have opposite signs. Hence, the siloxane backbone contributes a relatively low anisotropy to macromolecule polarizability. This fact agrees well with the data on optical anisotropy of the ladder poly(siloxane)s with the various side groups [3,128].

Low optical anisotropy of the siloxane chain is confirmed by the data on the Kerr constant obtained for poly(siloxane)s in the bulk. Thus, replacement of 25% of the methylene side groups in the PDMS chain by the phenylene rings leads to threefold growth in the Kerr constant [129]. However, if optical anisotropy must fall to give a negative value with increasing substitution of the methylene groups by the phenylene rings, then the Kerr constant must increase in keeping with the positive sign for Kerr constant. This observation shows that the role of dipole moments of the Si–O bonds in the polymer backbone ($\mu_{Si-O} = 1$ D), which can be oriented by an electric field, might be essential in affecting the value and sign of the Kerr constant of the polymer.

To evaluate the contribution of the polymer backbone to the overall molecular parameters of these types of polymers, electro-optical and hydrodynamic properties of the linear poly(methylhydro-dimethyl siloxane) (PMS), poly(methylphenyl siloxane) (PMPS), and poly(dimethyl siloxane) (PDMS) were investigated in dilute solution and in melt [129,130].

20.4.3.2.1 *Poly(methylhydro-dimethyl siloxane)*
The structure of PMS polymer used in this study [130] is given below.

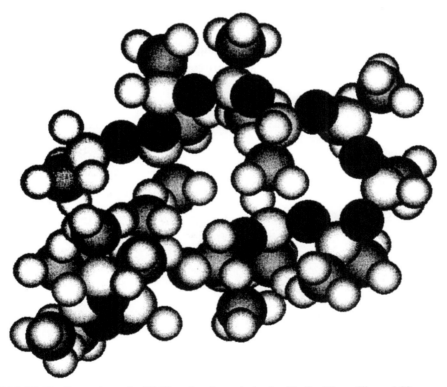

The molecular weight of the PMS sample, $M = 950 \pm 50$, is low. Nevertheless, in isolated state, the PMS molecule is significantly coiled, as is seen in Figure 20.17, where one possible configuration of the PMS polymer is represented.

The characteristics of PMS were compared with the properties of low-molecular-weight PDMS and PMPS. The average ratio of phenylene and methylene substituents in the PMPS chain is clear from the structure (where x is the degree of polymerization).

FIGURE 20.17 Conformation of a PMS molecule optimized with the HyperChem 6.03 program. The minimum volume of the occupied cell is $15.33 \times 12.66 \times 9.29$ Å.

PDMS 7 $(H_3C)_3SiO$ —$\left[\begin{array}{c} CH_3 \\ | \\ SiO \\ | \\ CH_3 \end{array}\right]_7$— $Si(CH_3)_3$

PMPS H_3C —$\left[\begin{array}{ccc} CH_3 & CH_3 & CH_3 \\ | & | & | \\ SiO & SiO & SiO \\ & | & \\ & CH_3 & \end{array}\right]_{(x-1)/3}$— $Si(CH_3)_3$

The content of aromatic units in the PMS product did not exceed 1–2% as found from the spectroscopy data. (this test is clearly very important in investigations of such compounds as PMS and PDMS with low proper polarity and anisotropy because the EB effect is very sensitive to the presence of the optically anisotropic moieties).

The dependence optical birefringence Δn on the electric field strength is shown in Figure 20.18 as the ratio $\Delta n\,(E)/E$ plotted against E. In these scales, quadratic dependence $\Delta n(E)$ is a linear function that provides a reliable extrapolation of the experimental data to vanishing field ($E = 0$). They were obtained for neat benzene (plot 1) and two benzene solutions of PMS, of various concentrations at 394 K. One can see, firstly, that all points on the plots lie on straight lines that go through the origin. This is a clear evidence of the quadratic function. Secondly, the effect observed in solution is evidently lower than that in the solvent, and its value is decreasing with increasing solute concentration.

The relation $\Delta n/E$ was calculated by using

$$\Delta n/E = (\Delta n_c - \Delta n_0)/c\,\bar{v}\,E + K_0\,E \qquad (20.14)$$

where Δn_c is EB of the solution, Δn_0 is EB of the solvent taken as $\mathcal{K}_0 E^2$, $c\bar{v}$ is the volume fraction of a polymer in solution, and $\mathcal{K}_0$ is Kerr constant for neat benzene. This equation is

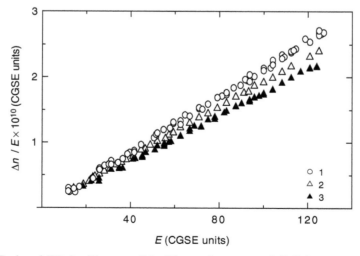

FIGURE 20.18 Reduced EB $\Delta n_c/E$ versus E in (1) neat benzene and (2,3) benzene solution of PMS. Solute concentration $c = $ (2) 15.8, (3) 29.0 g·dl^{-1}.

deduced from the fact that EB in a binary mixture is proportional to volume part of a component, provided that the concentration dependence of intermolecular interactions are ignored.

Figure 20.19 shows the $\Delta n/E$ value as a function of E for the two polymer solutions in benzene (plot 3). Points 3 represent the data obtained at the different solute concentrations. Within experiment limits, the dependence is an indistinguishable, meaning that the EB of the polymer is independent of solute concentration. This confirms the validity of using the formula above.

A set of points 3 was approximated by the least mean square fit of the data, and the slope of this curve yielded specific Kerr constant of PMS in benzene solution

$$K = \Delta n \, \bar{v}/E^2 = (0.91 \pm 0.02) \times 10^{-12} \text{cm}^5 \cdot g^{-1} \cdot (300 \text{ V})^{-2}$$

The properties of PMS are given in Table 20.26.

Molar Kerr constant of the solute per monomer unit is then given by

$$_mK/x = 6n \, M_0 \, K/(n^2 + 2)^2 \, (\varepsilon + 2)^2 \tag{20.15}$$

where n and ε are the refractive index and the dielectric constant of the medium, and M_0 the molecular weight of the solute per repeat unit. The difference between the external and internal electric fields is taken into account by the use of the Lorentz–Lorenz relation valid for a nonpolar liquid. For PMS in benzene solution ($n_D = 1.5010$, $\varepsilon = 2.3525$) at 294 K it leads to

$$_mK/x = (1.6 \pm 0.1) \times 10^{-12} \text{ cm}^5 \cdot \text{mole}^{-1} \cdot (300 \text{ V})^{-2}.$$

The Kerr constant for PMS is at least three-fold higher than the Kerr constant for PDMS ($M = 1500$) in bulk and does not exceed $0.3 \times 10^{-12} \text{ cm}^5 \cdot \text{mol}^{-1} \cdot (300 \text{ V})^{-2}$ at 295 K [129]. Segment optical anisotropy $(\alpha_1 - \alpha_2)$, for PDMS in benzene solution is $4.7 \times 10^{-25} \text{ cm}^3$. Taking

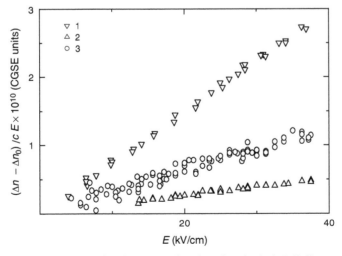

FIGURE 20.19 Reduced specific EB $(\Delta n - \Delta n_0)/c \, E$ (plot 3) and ratio $\Delta n/\rho E$ (1,2) versus E for (1) PMPS and (2) PDMS in bulk [131] and for (3) PMS in benzene solution. Solute concentration $c = 15.8$ and $29.0 \text{ g} \cdot \text{dl}^{-1}$ (the points are not marked out).

TABLE 20.26
Electro-Optical, Optical, and Hydrodynamic Properties of PMS in Solution

Solvent	$\mathcal{K} \times 10^{12}$, $(\text{cm}^5 \cdot \text{g}^{-1}) \cdot (300\,\text{V})^{-2}$	R_η (Å)	R_H (Å)	$A_0 \times 10^{10}$ $(\text{erg} \cdot \text{K}^{-1} \cdot \text{mole}^{-1/3})$	$(dn/dc)_{546}$ $(\text{cm}^3 \cdot \text{g}^{-1})$
Benzene	0.91 ± 0.02	6.6	6.7	2.9 ± 0.1	$-(0.10 \pm 0.01)$
Chloroform	—	7.8	7.4	3.1 ± 0.1	$-(0.04 \pm 0.01)$

Source: From Kolomiets, I.P., Lacey, D., and Lavrenko, P.N., Electrooptical and hydrodynamic properties of a low-molecular-mass copolymethylhydrodimethylsiloxane in solutions, *Polym. Sci., Ser. A.*, 47, 319, 2005.

into account that approximately half of the chain units in PMS are identical to the PDMS units then the interpretation of the PMS behaviour can be deduced by using the literature data on the optical anisotropy and thermodynamic flexibility of the PDMS macromolecule.

In discussion of the equilibrium EB value and dipole structure, the general theory of Kerr effect developed for rigid dipole particles with axial symmetry of optical polarizability was employed. It was concluded that the main mechanism responsible for EB in solution of these particles is the rotation of their molecules as a whole, and the anisotropy of dielectric polarizability of the molecules does not contribute significantly to the Kerr effect. The dipole and optical properties of such kinetically rigid particles were connected with the Kerr constant by Equation 20.8. Substitution of $n = 1.50$, $\varepsilon = 2.3$, and $T = 294\,\text{K}$ to Equation 2.9 yields $B_1 = 2.107 \times 10^{50}$ CGSE units.

The finite length of the PMS molecule was then taken into account. For a persistent chain with length L, the function of K on the reduced chain length, $x \equiv 2L/A$ (A the Kuhn segment length) was obtained by averaging of the squared dipole moment of the molecule on all the chain conformations. The function is represented in Figure 20.1 for various angles θ between optical axis and dipole moment of the molecule. For $\theta = 0$ at $x = 3$ (because $L/A = 1.5$ for PMS as determined below) we have

$$K_{\infty,\theta=0} \approx K_{\exp}/0.4 = 2.28 \times 10^{-12} \text{ cm}^5 \cdot \text{g}^{-1} \cdot (300\,\text{V})^{-2}$$

Substitution of $K_{\infty,\theta=0} = 2.28 \times 10^{-12}$, $B_1 = 2.107 \times 10^{50}$, $\Delta a = (1/5)\,4.7 \times 10^{-25}$ (all in CGSE units), $s = 5$, and $M_0 = 67$ into Equation 2.8 yields $\mu_{0\parallel} = 0.39$ D. These results were compared with Kerr effect of PDMS.

20.4.3.2.2 *Poly(dimethyl siloxane) and Poly(methylphenyl siloxane)*

Figure 20.19 shows the excess optical birefringence as a function of the electric field strength obtained in benzene solution of PMPS (plot 1) and in melt of low-molecular-weight PDMS (3) as compared with that of PMS (2) [129]. Linear approximation yields the Kerr constant of PDMS 7 in bulk melt

$$K \equiv \Delta n/E^2 = (0.388 \pm 0.003) \times 10^{-12} \text{cm}^2 \cdot (300\,\text{V})^{-2}.$$

The kerr constant of PDMS in bulk is evidently close to that in dilute solution, with both resulting in twice the effect of PMS. The origins of the effect may be found in the difference in the molecular weight or in the configuration of the skeletal chain, namely, in higher part of the *gouch*-sequences, which provide the higher longitudinal component of dipole moment.

Thus, the dipole moment per PMS monomer unit was found to be $\mu_0 \approx 0.4\,\mathrm{D}$ that is close to $\mu = 0.29\,\mathrm{D}$ known for PDMS. This result is in accordance with the conception that the *gouch*-configurations of the siloxane chain units ~Si–O~ are predominant [127]. This must be true because of the presence of nonzero longitudinal (along the chain) component of dipole moment $\mu_{0\parallel}$ is conditioned by this configuration only.

The value obtained, 0.4 D, exceeds $\mu_{0\parallel} = 0.2\,\mathrm{D}$ found earlier for ladder poly(methylbutene siloxane) (PMBS) [128]. This is in qualitative agreement with conformation of the PMBS chain where dipole moments of the transversal Si–O–Si bonds diminish the longitudinal component of the dipole moment of the PMBS monomer unit. Unfortunately, quantitative comparison of the dipole structures of PMS and PMBS is complicated with uncertain defects in the ladder structure of the PMBS chains.

The Kerr constant of PMPS in the melt was obtained from linear approximation of points 1 in Figure 20.17 to be

$$K_\rho = \Delta n / \rho E^2 = (2.24 \pm 0.09) \times 10^{-12}\,\mathrm{cm^5 g^{-1}}(300\,\mathrm{V})^{-2}.$$

The values of EB of PMPS in benzene and decalin solution are close to one other and to that of PMPS in melt,

$$(K_{\mathrm{sp}})_{\mathrm{benz}} = (2.30 \pm 0.02) \times 10^{-12}\mathrm{cm^5 \cdot g^{-1} \cdot (300\,V)^{-2}}$$
$$(K_{\mathrm{sp}})_{\mathrm{dek}} = (2.23 \pm 0.02) \times 10^{-12}\mathrm{cm^5 \cdot g^{-1} \cdot (300\,V)^{-2}}.$$

Hence, Kerr constant of PMPS is independent of aggregate state of the polymer.

Kerr constant of PDMS is obviously three to six times lower than that of PMS and PMPS. The contribution of the siloxane backbone to Kerr effect of a siloxane LC side-chain polymer is positive in sign and low in value. However, this value would be significant and, therefore, should be taken into account when investigating substances with low Kerr constants.

20.5 LC HYBRID DENDRIMERS

20.5.1 CARBOSILANE-BASED END-ON DENDRIMERS

Carbosilane denritic matrices of different generations, with numbers from 1 to 5, were modified with cyanobiphenylene mesogenic fragments bonded as the terminal groups to the matrix via flexible aliphatic spacers $-(\mathrm{CH_2})_{10}-$ [132–134]. As a result, symmetric compounds, whose structure is shown in Figure 20.20, were obtained with amount of the mesogenic arms from 8 to 128. The EB method was used to evaluate orientation mobility of the mesogenic fragments in these end-on hybrid LC dendrimers because contribution of matrix is low, $0.17 \times 10^{-10}\,\mathrm{cm^5 \cdot g^{-1} \cdot (300\,V)^{-2}}$. The K value per one fragment is close to that of comparable-low-molecular weight compound, $(4 \pm 2) \times 10^{-10}\,\mathrm{cm^5 \cdot g^{-1} \cdot (300\,V)^{-2}}$. Analysis of the data collected in Table 20.27 leads to conclusion that the K value is independent of the generation number, and, hence, EB of the end-on hybrid dendrimers in solution is caused by short-range orientation of the mesogenic end fragments in an electric field, as it takes place in the typical comb end-on LC polymers with long spacers. The behaviour of the side-on hybrid LC dendrimers [136] is expected to be similar to that of the side-on combs.

The C_{60} fullerene effects on the polarity and electro-optical properties of hybrid LC malonate-based dendrimers and hemi-dendrimers as investigated in dilute benzene solution [137–139] in comparison with fullerene-free analogs are not considered here as described in detail in Chapter 22 of this book.

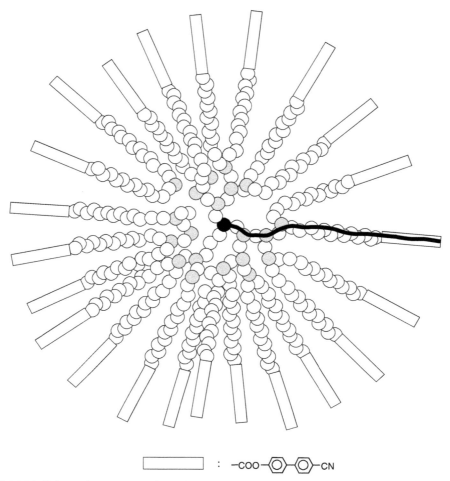

FIGURE 20.20 Schematic representation of hybrid dendrimer with mesogenic cyanobiphenyl terminal fragments. One arm of the star-like molecule is selected.

TABLE 20.27
Kerr Constant of Carbosilane-Based Hybrid LC *End*-On Dendrimers as Obtained at 21°C in Toluene and Chloroform Dilute Solution

Sample (Generation)	$K \times 10^{10}$ (cm$^5 \cdot$g^{-1})$\cdot$(300 V)$^{-2}$	
	Toluene	Chloroform
R 8 (1)	2.63	–
R 16 (2)	2.68	4.11
R 32 (3)	2.68	–
R 64 (4)	–	4.32
R 128 (5)	–	4.21

Source: From Lezov, A.V. et al., Dualism in hydrodynamic behavior of LC carbosilane dendrimers in dilute solutions, *Doklady Chem.*, 362, 638, 1998.

20.5.2 Cylindrical Dendrimers

First- and second-generation cylindrical dendrimers based on α-aspargic acid with widely varying molecular weights were studied by the methods of equilibrium and nonequilibrium EB [140]. It was shown that macromolecules of the second generation undergo reorientation under the electric field primarily according to the large-scale mechanism, whereas the reorientation of the first-generation dendrimers proceeds by both large- and small-scale mechanisms. It was discovered that the value of molecular weight strongly affects the equilibrium Kerr constant of the cylindrical dendrimer of the second generation. As a result, the Kuhn segment length of its macromolecules was found to be $A = 90$ Å. The combined dipole and conformational analyses of the molecular structure of the dendrimers was performed. It was demonstrated that the cylindrical dendrimers of low generations are distinguished by a rather low equilibrium rigidity of the macromolecules along with an appreciable kinetic rigidity.

20.6 LC METALLO-COMPLEXES

Lanthanide LC complexes with a variable content belong to the novel photoactive supramolecular and multifunctional materials for advanced photonic devices. The EB method was successfully used in the determination of space structure of these compounds. In particular, the metal–mesogenic complexes of (1) Pr^{3+}, (2) Dy^{3+}, (3) Yb^{3+} with 3-fluoroacetylcamphor were studied in carbon tetrachloride solution [141], yielding the μ and $_mK$ values given in Table 20.28. The significant magnitudes of these parameters were obtained in spite of high symmetry of the individual complex geometry. This is mainly due to dimerized form of the complexes, for which, via graphic analysis of the EB data in terms of $_mK$ versus μ^2, the conformations of the mobile particles were studied.

With the same electro-optical techniques under a pulsed electric field, LC metal–mesogenic Dy- and Ho-complexes with Schiff bases and nitrate anions were investigated,

TABLE 20.28
Electro-Optical Properties and Dipole Moments of Metallomesogenic Compounds

Compound	$10^{12} \times K$, CGSE unit[a]	$10^{12} \times {}_mK$, CGSE unit[a]	μ (D)
Pr^{3+} complex (1)	–	25.90	6.39
Dy^{3+} complex (2)	–	21.92	5.75
Yb^{3+} complex (3)	–	20.65	5.50
$Dy(LH)_3 (NO_3)_3$	9	–	5.50
$Dy(LH)_3 (CF_3 SO_3)_3$	–	–	7.59
$Dy(LH)_3 (C_{12} H_{25} -O-SO_3)_3$	4	–	5.36
$Ho(LH)_3 (NO_3)_3$	5	–	6.57
$Ho(LH)_3 (CF(CF_2)_5 CH_2 SO_4)_3$	4	–	9.73
LH: $C_{12} H_{25} -O-C_6 H_3 (OH)-CH = N-C_{16} H_{33}$	16	–	1.73

[a] K and $_mK$ are given in $cm^5 \cdot g^{-1} \cdot (300\,V)^{-2}$ and in $cm^5 \cdot mole^{-1} \cdot (300\,V)^{-2}$, respectively.

Sources: From Abdullin, I.R. et al., Electrooptical properties and structure of 3-trifluoroacetylcamphorates of lantanoides, *Rep. Russ. Acad. Sci., Chem.*, 2, 486, 1990; Yevlampieva, N. et al., Metallomesogenic Dy- and Ho-complexes: polarizability, polarity and structure, presented at *Eighth International Symposium on Metallomesogenes*, Namur, Belgium, May 28–31, 2003, p. 52.

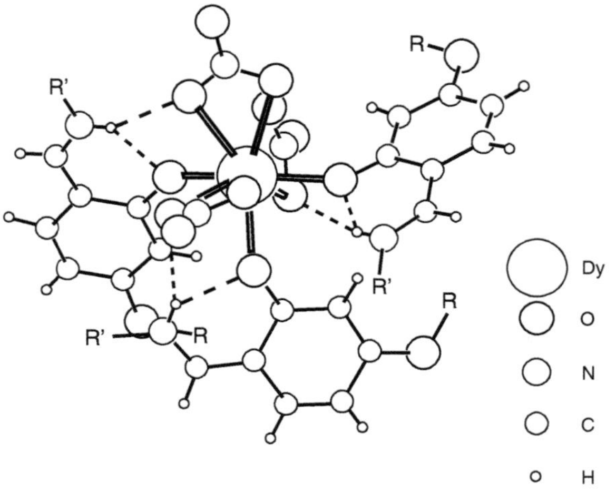

FIGURE 20.21 The spatial structure of complex of Dy with Schiff bases and nitrate counterions; R and R' are the end alkylene substituents.

on the one hand, and Schiff bases and perfluoroalkylsulfate anions, on the other [142]. These are interesting materials due to their paramagnetic properties, anomalous high magnetic anisotropy, and highly chromatic intense luminescence. Polarizability, polarity, and structure of the particles were also studied in carbon tetrachloride solution. The specific Kerr constants and the permanent dipole moments obtained for these compounds in benzenc solutions are given in Table 20.28, along with the values found for the model (ligand) compound LH.

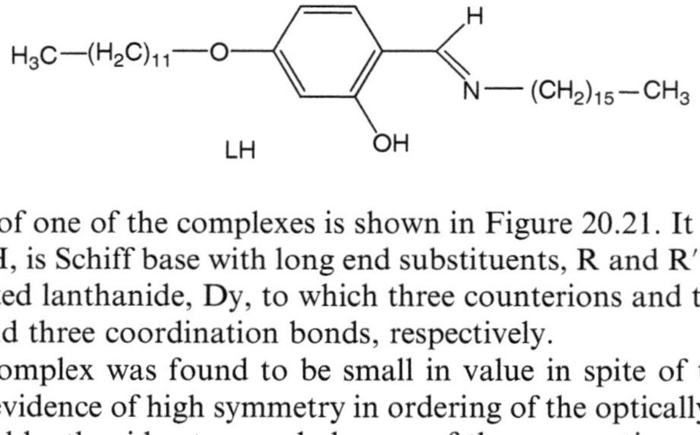

The space structure of one of the complexes is shown in Figure 20.21. It is $[Dy(LH)_3 (NO_3)_3]$ where the ligand, LH, is Schiff base with long end substituents, R and R'. The central atom is a ninefold-coordinated lanthanide, Dy, to which three counterions and three organic ligands are bound via six and three coordination bonds, respectively.

The EB of the complex was found to be small in value in spite of the very high dipole moment. This is an evidence of high symmetry in ordering of the optically anisotropic ligands relative to the central lanthanide atom and absence of the aggregation effects [143].

20.7 CONCLUSIONS

"In this review we tried to show that the electro-optical method is a powerful tool of molecular physics. Last two decades saw a successful developments in experimental techniques and new theories in the field of optical and dielectric properties.

The method is particularly powerful when combined with other classic methods of molecular hydrodynamics and optics. As a result, due in large part to the works of Tsvetkov et al. [3], researches have identified and characterized the fundamental physical properties of polymers.

Significant EB results for new LC compounds with a well-defined structure were obtained. Much attention was paid to hybrid LC dendrimers, star-like polymers, fullerene derivatives, and metallo-mesogenic complexes. Because of the form geometry, the EB effect is provided by polarizability (including hyperpolarizability) rather than by orientation of the particle as a whole under the influence of an electric field. Interestingly, hybridization of such symmetric materials of these compounds allows them to be included into a strongly ordered mesophase. On the one hand, this confirms the highly ordered form of the materials. On the other hand, their incorporation into a mesophase presents a unique way of finely modifying the LC phase, which can be controlled by the EB method.

A new interpretation of the EB effect observed in solutions of comb LC end-on polymers with crystal-like ordering of the mesogenic side groups was developed. For these, the main role of the induced dipole moment of the macromolecule was established. Advanced application of the EB technique was done in studies of novel comb polymers with the dendronized side groups.

Being very sensitive to space orientation of polar and optically anisotropic groups, the EB method yielded some important conclusions on the nature and physical mechanisms of mesomorphism of the comb side-on polymers which possess a unique possibility to form the biaxial nematic phase. Based on these polymers, very advanced elastomers have been manufactured. Future success might be expected in the EB investigations of combined LC polymers and comb polymers with complicated architectures as provided by various attachments of mesogenic units, intermediate between lateral, semi-lateral and end-on LC polymers.

ACKNOWLEDGMENT

I.P.K. and P.N.L. are grateful for the partial financial support from the Education Ministry of Russian Federation, grant No. AO147.

REFERENCES

1. O'Konski, C.T., Electrooptics, in *Encyclopedia of Polymer Science and Technology*, Bikales, N.M., Ed., 1st ed., Wiley-Interscience, New York, 1968.
2. O'Konski, C.T., *Molecular Electro-Optics*, ed., Marcell Dekker, New York, 1976.
3. Tsvetkov, V.N., *Rigid-Chain Polymers. Hydrodynamic and Optical Properties in Solution*, Plenum Press, New York, 1989.
4. Wijmenga, S.S., van der Touw, F., and Mandel, M., Scaling relations for aqueous polyelectrolyte salt solutions. 4. Electric birefringence decay as a function of molar mass and concentration, *Macromolecules*, 19, 1760, 1986.
5. Jennings, B.R., Electro-optical methods for characterisation of polymers and biopolymers, *Pure Appl. Chem.*, 54, 395, 1982.
6. Beevers, M.S., Elliott, D.A., and Williams, G., Molecular motion in melt samples of poly(propylene glycol) studied using dielectric and Kerr effect relaxation techniques, *Polymer*, 21, 13, 1980.
7. Beevers, M.S. and Williams, G., Electro-optical Kerr-effect in solutions of benzylidene aniline and its derivatives, *J. Chem. Soc. Faraday Trans.*, Part II, 72, 2171, 1976.
8. Beevers, M.S. and Goodwin, A.A., Synthesis and Kerr effect of *p*-phenylene vinylene oligomers, *J. Polym Sci.*, *Part B: Polym. Phys.*, 34, 1133, 1996.
9. Finkelmann, H. et al., LC polymers with biphenyl-moieties as mesogenic group, *Makromol. Chem.*, 179, 2541, 1978.
10. Tsvetkov, V.N., Rjumtsev, E.I., and Shtennikova, I.N., Intramolecular orientation order and properties of polymer molecules in solutions, in *Liquid Crystalline Order in Polymers*, Blumstein, A., Ed., Academic Press, New York, 1978, p. 57.

11. Tsvetkov, V.N. and Tsvetkov, N.V., Electric birefringence in solutions of rigid-chain polymers, *Russ. Chem. Rev.*, 62, 851, 1994 (and references herein).

12. Jungnickel, B.-J. and Wendorff, J.H., The Kerr effect as applied to the investigation of polymers, in *Chemistry and Physics of Macromolecules*, Fischer, E.W. et al., Eds., VCH Verlag, Weinheim, 1991, p. 349.

13. Jakli, A. and Saupe, A., Electro-optic effects in smectic A phase, *Mol. Crystl. Liq. Cryst.*, 222, 101, 1992.

14. Li, J.-F., et al., Electrooptic investigations of enantiometric mixtures of the antiferroelectric liquid crystal TFMHPOBC, *Liq. Crystl.*, 23, 255, 1997.

15. Kim, J.-H., Petschek, R.G., and Rosenblatt, C., Electro-optic response of surface-induced nematic order above the nematic-isotropic phase transition temperature, *Phys. Rev., E*, 60, 5600, 1999.

16. Xu, M. and Yang, D.-K., Electrooptical properties of dual-frequency cholesteric liquid crystal reflective display and drive scheme, *Jpn. J. Appl. Phys.*, 38, 6827, 1999.

17. Stuart, H.A. and Peterlin, A., Optishe Anisotropie und Form von Fadenmolekülen. II. Künstliche Doppelbrechung, *J. Polym. Sci.*, 5, 551, 1950.

18. Gotlib, Y.Y., Theory of birefringence in polymers, Ph.D. Dissertation, Gertsen Leningrad Pedagogical Institute, 1956; *Trans. Gertsen Leningrad State Pedagogical Institute*, 141, 1958.

19. Nagai, K. and Ishikawa, T., Internal rotation and Kerr effect in polymer molecules, *J. Chem. Phys.*, 43, 4508, 1965.

20. Tsvetkov, V.N., Andreeva, L.N., and Tsvetkov, N.V., Anisotropy of segments and monomer units of polymer molecules, in *Polymer Handbook*, Grulke, E., Brandrup, J., and Immergut, E.H., Eds., 4th ed., Part VII, Wiley, New York, 1998.

21. Yamakawa, H., Shimada, J., and Nagasaka, K., Statistical mechanics of helical wormlike chains. X. Dipole moments, electric birefringence, and electric dichroism, *J. Chem. Phys.*, 71, 3573, 1979.

22. Tsvetkov, V.N. et al., Kerr effect in solutions of a thermotropically mesogenic alkylene–aromatic polyester, *Eur. Polym. J.*, 22, 543, 1986.

23. Le Fevre, R.J.W., Electro-optical constants, *Adv. Phys. Org. Chem.*, 3, 1, 1965.

24. Thakur, M. and Tripathy, S., Electrooptical applications, in *Encyclopedia of Polymer Science and Technology*, Bikales, N.M., Ed., 2nd ed., Wiley-Interscience, New York, 1986, p. 756.

25. Coles, H.J., Laser and electric field induced birefringence studies on the cyanobiphenyl homologues, *Mol. Crystl. Liq. Crystl. Lett.*, 49, 67, 1978.

26. Idrissi, A. et al., Temperature dependence of the reorientational dynamics and low-frequency response of aqueous urea solutions investigated by femtosecond optical Kerr-effect spectroscopy and molecular-dynamics simulation, *Phys. Chem. Chem. Phys.*, 5, 4666, 2003.

27. Tsvetkov, V.N. et al., Application of modulation of elliptic light polarization for study of electric birefringence of polymer solutions in pulsed fields, *Vysokomol. Soedin., Ser. A*, 25, 1327, 1983.

28. Tsvetkov, V.N., Kolomiets, I.P., and Lezov, A.V., Electric birefringence of solutions of aromatic poly(amide hydrazide) in dimethyl sulphoxide, *Eur. Polym. J.*, 18, 373, 1982.

29. Shere, A.J. et al., Kerr effect and wide angle light-scattering studies of a *para*-aromatic polyamide in dilute solution, *Macromolecules*, 29, 2088, 1996.

30. Lavrenko, P.N. et al., Electric birefringence, conformation and rigidity of the molecules of cellulose esters in solutions, *J. Polym. Sci., Part C, Polym Symp.*, 44, 217, 1974.

31. Dejardin, J.-L., *Dynamic Kerr effect*. World Scientific, Singapore, 1995.

32. Tsvetkov, V.N., Andreeva, L.N., and Filippov, A.P., Electrooptics of thermotropic mesogenic polymers in solution and in bulk, *Mol. Crystl. Liq. Crystl.*, 153, 217, 1987; *Eur. Polym. J.*, 24, 379, 565, 1988.

33. Tsvetkov, N.V., Tsvetkov, V.N., and Skorokhodov, S.S., Electrooptical properties of mesogenic polymer of combined structure in dilute solutions and nematic phase, *Vysokomol. Soedin., Ser. A*, 38, 1032, 1996.

34. Tsvetkov, V.N. et al., Influence of length of flexible chain fragments on electroopical properties of mesophase formed by chain molecules, *Vysokomol. Soedin., Ser. A*, 37, 1255, 1995.

35. Tsvetkov, V.N. and Tsvetkov, N.V., Electrooptical properties of mesogenic chain molecules in solutions and in nematic state, *Macromol. Symposia*, 113, 27, 1997.

36. Tsvetkov, N.V. et al., Electrooptical and dielectric properties of comb-like mesogenic polymers in solutions and nematic melts, *Polym. Sci., Part A*, 42, 714, 2000.

37. Tsvetkov, N.V. et al., Electrooptical properties of linear aromatic polyesters containing *bis*-(4,4′)- or *bis*-(4,3′)-biphenyl fragments in main chain in solutions and nematic melts, *Polym. Sci., Part A*, 42, 721, 2000.

38. Tsvetkov, N.V. et al., Electrooptical properties of main-chain mesogenic polymers in nematic state and dilute solutions, *Mol. Crystl. Liq. Crystl.*, 373, 227, 2002.

39. Tsvetkov, N.V., Pogodina, N.V., and Zuev, V.V., Optical and electrooptical properties of aromatic polyesters with the main chain mesogenic groups in dilute solutions, *Vysokomolek. Soedin., Part A*, 38, 1133, 1996.

40. Tsvetkov, N.V. et al., Optical and electrooptical properties of combined mesogenic polymers in dilute solutions and mesophase, *Vysokomolek. Soedin., Part A*, 38, 1831, 1996.

41. Tsvetkov, N.V. et al., The influence of length of flexible fragments on electrooptical characteristics of the chain mesogenic molecules in solutions, *Vysokomolek. Soedin., Part A*, 37, 1265, 1995.

42. Bushin, S.V. et al., Hydrodynamical, optical and electrooptical properties of alkylenaromatical polyester in solutions, *Polym. Sci., Part A*, 44, 431, 2002.

43. Bushin, S.V. et al., Hydrodynamical, optical, and electrooptical properties of aromatic polyester with azobiphenylene fragments in chain, *Vysokomolek. Soedin., Part A*, 45, 54, 2003.

44. Tsvetkov, N.V. et al., Electrooptical properties of bisubstituted aromatic polyester in mixed solvents, *Vysokomolek. Soedin., Part A*, 40, 1577, 1998.

45. Tsvetkov, N.V. et al., Electro-optics of aromatic polyether in mixed solvents and effective "electrooptical" dipole of macromolecules, *Doklady Chem.*, 361, 507, 1998.

46. Tsvetkov, V.N., Andreeva, L.N., and Lavrenko, P.N., Diffusion and flow birefringence of alkylene-aromatic thermotropic polyesters, *Eur. Polym. J.*, 20, 817, 1984.

47. Tsvetkov, V.N. et al., Kerr effect in solution of aromatic ester with *para*- and *ortho*-phenylene cycles in chain, *Doklady Chem.*, 307, 916, 1989.

48. Andreeva, L.N. et al., Conformational properties of macromolecules of heterocyclic polyester imides in dilute dichloroacetic acid solutions, *J. Polym. Sci. Part B: Polym. Phys.*, 42, 12, 2004.

49. Tsvetkov, V.N. et al., Molecular characteristics of polymers with mesogenic groups in the backbone and side chains in dilute solutions, *Vysokomolek. Soedin., Part A*, 36, 983, 1994.

50. Tsvetkov, V.N. et al., Electric birefringence in polybutylisocyanate solutions, *Eur. Polym. J.*, 26, 163, 1990.

51. Beevers, M.S., Garrington, D.C., and Williams, G., Dielectric and dynamic Kerr effect studies of poly(*n*-butyl isocyanate) and poly(*n*-octyl isocyanate) in solution, *Polymer*, 18, 540, 1977.

52. Tsvetkov, V.N. et al., Electric birefringence in solutions of rigid-chain aromatic polyamides, *Eur. Polym. J.*, 26, 575, 1990.

53. Tsvetkov, V.N. et al., Dynamo-optical and electro-optical properties of solutions of diphenyl-substituted poly-*p*-phenylene terephthalate in polar and nonpolar solvents, *Eur. Polym. J.*, 27, 319, 1991.

54. Matsumoto, M., Watanabe, H., and Yoshioka, K., Electrical and hydrodynamic properties of polypeptides in solution, *Biopolymers*, 6, 929, 1968; 11, 1711, 1972.

55. Rjumtsev, E.I. et al., Electric and dynamic birefringence in solutions of poly-*bis*-trifluoroethoxyphosphazene, *Eur. Polym. J.*, 28, 1031, 1992.

56. Rjumtsev, E.I. et al., Kerr effect in solutions of poly(fluoroalcoxyphosphazene)s, *Polym. Sci., Part A*, 34, 541, 1992.

57. Tsvetkov, V.N. et al., Intramolecular liquid-crystal order in polymers with chain side groups, *Eur. Polym. J.*, 9, 481, 1973.

58. Rjumtsev, E.I. et al., Electrooptical properties of poly(1-(trimethylsilyl)-1-propyne) in solutions, *Polym. Sci., Part A*, 41, 765, 1999.

59. Yevlampieva, N.P. et al., Electrooptical properties of poly[1-(trimethylgermyl)-1-propyne] in solutions, *Polym. Sci., Part A*, 44, 314, 2002.

60. Yevlampieva, N.P., Yakimanskii, A.V., and Rjumtsev, E.I., Electrooptical method for determination of average length of stereo-blocks in structure of poly(1-trimethylsilyl-1-propine)s and poly(1-trimethylgermyl-1-propine)s. *Polym. Sci., Part B*, 48, 51, 2006.

61. Tsvetkov, N.V. et al., Electric birefringence of a phenylated poly(phenylene) in polar and nonpolar solvents, *Vysokomolek. Soedin., Ser. A*, 44, 297, 2002.
62. Tsvetkov, N.V. et al., Hydrodynamic, conformational, and electrooptical properties of phenylated poly(phenylene) macromolecules in solutions, *Vysokomolek. Soedin., Ser. A*, 46, 1695, 2004.
63. Yevlampieva, N.P., Makarova, N.P., and Rjumtsev, E.I., Molecular properties of mesomorphic poly(decamethyl-cyclohexasiloxane) in solution, *Mol. Crystl. Liq. Crystl.*, 352, 133, 2000.
64. Tsvetkov, N.V. and Didenko, S.A., Conformational, optical, and electric properties of solutions of cellulose carbanilate in mixed solvents, *Vysokomolek. Soedin., Ser. A*, 35, 1640, 1993.
65. Tsvetkov, N.V., Didenko, S.A., and Tsvetkov, V.N., Optical, dynamic, and electric properties of solutions of cellulose carbanilate in dioxane and ethyl acetate, *Vysokomolek. Soedin., Ser. A*, 35, 1625, 1993.
66. Tsvetkov, N.V. et al., Kerr effect of cellulose acetobenzoate solutions in dioxane, *Polym. Sci.*, 34, 545, 1992.
67. Tsvetkov, N.V. et al., Equilibrium and nonequilibrium Kerr effect in dioxane solutions of cellulose acetate cinnamate, *Polym. Sci.*, 34, 697, 1992.
68. Tsvetkov, N.V. et al., Influence of degree of substitution on optical and electric properties of some aromatic cellulose esters in dioxane, *Doklady Chem.*, 326, 678, 1992.
69. Kolomiets, I.P. et al., Dynamic and electrical birefringence in solutions of flax cellulose nitrates, *Vysokomolek. Soedin., Part A*, 29, 2494, 1987.
70. Tsvetkov, N.V., Kutsenko, L.I., and Didenko, S.A., Flow and electric field birefringence of cellulose cyanoethyl nitrate in cyclohexanone, *Vysokomolek. Soedin., Part A*, 37, 1300, 1995.
71. Tsvetkov, N.V. et al., Optical and electrooptical properties of cellulose esters with alyphatic side substituents, *Vysokomolek. Soedin., Part A*, 37, 1306, 1995.
72. Tsvetkov, N.V., Didenko, S.A., and Tsvetkov, V.N., Optical and electrooptical properties of combined cellulose esters, *Doklady Chem.*, 337, 483, 1994.
73. Tsvetkov, N.V. et al., Electrooptical and dielectric properties of comb-like poly(methacrylate)s of various dipole architecture of mesogenic groups, *Vysokomolek. Soedin., Part A*, 42, 1108, 2000.
74. Duran, R. and Gramain, P., Synthesis and tacticity characterization of a novel series of liquid crystalline side chain polymers with oligo(ethylene oxide) spacers, *Makomol. Chem.*, 188, 2001, 1987.
75. Goozner, R.E. and Finkelmann, H., Electric field effects on a LC side chain polymer with a negative dielectric anisotropy, *Makromol. Chem.*, 186, 2407, 1985.
76. Takahashi, K. et al., Electro-optical effects in new polymeric liquid crystals, *Mol. Crystl. Liq. Crystl.*, 8, 33, 1991.
77. Eich, M. et al., Pretransitional phenomena in the isotropic melt of a mesogenic side chain polymer, *Polymer*, 25, 1271, 1984.
78. Rjumtsev, E.I. et al., Rotation relaxation in a side chain LC polymer in the vicinity of the nematic-isotropic transition temperature, *Liq. Crystl.*, 21, 777, 1996.
79. Tsvetkov, N.V. et al., Optical and electrooptical properties of comb-like polymer with intramolecular H-bonds, *Polym. Sci., Part A*, 39, 1281, 1997.
80. Tsvetkov, V.N. et al., Kerr effect in solutions of comb-shaped mesogenic polymers and their monomers, *Doklady Chem.*, 287, 1172, 1986.
81. Lezov, A.V. et al., Nature of electro-optical effect in solutions of a comb-shaped polymer with mesogenic side groups, *Polym. Sci., Ser. A.*, 45, 656, 2003.
82. Rjumtsev, E.I., Yevlampieva, N.P., and Polushin, S.G., Pre-transition electrooptical properties of solutions of LC poly(siloxane)s in range of phase separation, *Russ. Optical J.*, 65, 24, 1998.
83. Rjumtsev, E.I. et al., Intramolecular LC structure in solutions of poly(nonyloxybenzamide styrene), *Vysokomolek. Soedin., Part A*, 18, 439, 1976.
84. Polushin, S.G. et al., Electrooptical and hydrodynamic properties of ionogenic LC copolymers in solutions and melts, *Vysokomol. Soedin, Ser. A*, 43, 817, 2001.
85. Ryumtsev, E.I., Yevlampieva, N.P., and Kovshik, A.P., Influence of position of the alyphatic bridge fragments on polarity of LC compound molecules and their dielectric properties, *Zh. Fiz. Khim.*, 69, 934, 1995.
86. Burnham, A.K., Buxton, L.W., and Flygare, W.H., Kerr constants, depolarization ratios, and hyperpolarizabilities of substituted methanes, *J. Chem. Phys.*, 67, 4990, 1977.

87. Khanarian, G. and Tonelli, A.E., A Kerr effect and dielectric study of alfa, omega-dibromoalk-anes, *J. Chem. Phys.*, 75, 5031, 1981.
88. Flory, P.J. et al., Optical anisotropies of aliphatic esters, *J. Phys. Chem.*, 85, 3215, 1981.
89. Megnassan, E. and Proutiere, A., Dipole moments and Kerr constants of 4-*n*-alkyl-4′-cyanobiphe-nyl molecules (from 1CB to 12CB) measured in cyclohexane solutions, *Mol. Crystl. Liq. Crystl.*, 108, 245, 1984.
90. Lavrenko, P. et al., Synthesis and molecular properties of new side-chain LC polymers based on the Baylis–Hillman reaction, *Mol. Crystl. Liq. Crystl.*, 338, 141, 2000.
91. Rjumtsev, E.I. et al., Value and direction of dipole moments in molecules of nematic liquid crystals containing nitrile groups, *Opt. Spektrosk.*, 41, 65, 1976 (in Russian).
92. Tsvetkov, N.V. et al., Electric birefringence in solutions of *bis*-4-cyanobiphenyloxyalkanes, *Mol. Crystl. Liq. Crystl.*, 265, 487, 1995.
93. Rjumtsev, E.I. and Yevlampieva, N.P., Electrooptical properties and polar structure of LC phenacyl esters, *Mol. Crystl. Liq. Crystl.*, 251, 351, 1994.
94. Rotinyan, T.A. et al., Molecular structure and dielectric anisotropy of some liquid crystals, *Kristallografiya*, 23, 578, 1978 (in Russian).
95. Yamamoto, R. et al., The Kerr constants and relaxation times in the isotropic phase of nematic homologous series, *Phys. Lett., Ser. A.*, 69, 276, 1978.
96. Rjumtsev, E.I. et al., Equilibrium and dynamic electrooptical properties of nematic and isotropic phases of 4′-*n*-alkoxy-4-cyanobiphenyl, *Russ. J. Phys. Chem.*, 69, 940, 1995.
97. Chen, F.L. and Jamieson, A.M., Odd-even effect in miscibility of main-chain liquid crystal polymers with low molar mass nematogens, *Liq. Crystl.*, 15, 171, 1993.
98. Aroney, M., Le Fevre, R.J.W., and Parkins, G.M., Molecular polarisability. The molar Kerr constants and dipole moments of six polyethylene glycols as solutes in benzene, *J. Chem. Soc.*, 7, 2890, 1960.
99. Champion, J.V., Meeten, G.H., and Southwell, G.W., Electro-optical Kerr birefringence of polystyrenes in dilute solutions, *Polymer*, 17, 651, 1976.
100. Kelly, K.M., Patterson, G.D., and Tonelli, A.E., Kerr effect studies of the poly(oxyethylenes), *Macromolecules*, 10, 859, 1977.
101. Tricot, M. and Houssier, C., Electrooptical studies of sodium poly(styrenesulfonate). 1. Electric polarizability and orientation function from electric birefringence measurements, *Macromolecules*, 15, 854, 1982.
102. Hessel, F., Herr, R.-P., and Finkelmann, H., Synthesis and characterization of biaxial nematic side chain polymers with laterally attached mesogenic groups, *Makromol. Chem.*, 188, 1597, 1987.
103. Praefcke, K., Can thermotropic biaxial nematics be made real? *Mol. Crystl. Liq. Crystl.*, 364, 15, 2001.
104. Zhou, Q.F., Li, H.M., and Feng, X.D., Synthesis of LC polyacrylates with laterally substituted mesogens. *Macromolecules*, 20, 233, 1987; 22, 491, 1989.
105. Kapitza, H. and Zentel, R., Combined LC polymers with chiral phases, 2. Lateral substituents, *Makromol. Chem.*, 189, 1793, 1988.
106. Hardouin, F. et al., Evidence for a jacketed nematic polymer, *J. Phys. II*, 1, 511, 1991.
107. Ringsdorf, H. et al., Molecular engineering of LC polymers, in *Chemistry and Physics of Macro-molecules*, Fischer, E.W. et al., Eds., VCH, Weinheim, 1991, p. 21.
108. Petzl, G., Thermodynamic behavior and physical properties of thermotropic liquid crystals, in *Liquid crystals*, Stegemeyer, H., Ed., Springer, New York, 1994, p. 51.
109. Zentel, R., LC polymers, in *Liquid crystals*, Stegemeyer, H., Ed., Springer, New York, 1994, p. 103.
110. Pugh, C. et al., Induction of smectic layering in nematic liquid crystals using immiscible compon-ents. 2. Laterally attached side-chain LC poly(norbornene)s and their low-molar-mass analogues with hydrocarbon/oligodimethylsiloxane substituents, *Macromolecules*, 31, 5188, 1998.
111. Terrien, I. et al., Side-on fixed LC polymers: New stationary phases for high performance liquid chromatography, *Mol. Crystl. Liq. Crystl.*, 331, 431, 1999.
112. Kim, G.-H., Pugh, C., and Cheng, S.Z.D., A new route to stabilize the smectic C phase in a series of laterally attached side-chain LC polynorbornenes with a one-carbon spacer, *Macromolecules*, 33, 8983, 2000.
113. Sigel, R. et al., Near-critical behavior of the dynamic stress-optical coefficient in LC elastomers, *Macromolecules*, 26, 4226, 1993.

114. Lavrenko, P.N., Kolomiets, I.P., and Finkelmann, H., Optical anisotropy and conformation of macromolecules with laterally attached mesogenic side groups, *Macromolecules*, 26, 6800, 1993.

115. Lavrenko, P. et al., Hydrodynamic and electro-optical properties of novel LC polymers with laterally attached cyanobiphenylene mesogenic side groups, in *Proceedings of the Fourth International Symposium on "Molecular Order and Mobility in Polymer Systems,"* St.-Petersburg, 2002, p. 86.

116. Hessel, F. and Finkelmann, H., A new class of liquid crystal side chain polymers. Mesogenic groups laterally attached to the polymer backbone, *Polym. Bull.*, 14, 375, 1985.

117. Oulyadi, H. et al., ^{13}C NMR investigation of molecular order in LC polysiloxanes, *Macromolecules*, 23, 1965, 1990.

118. Oulyadi, H. et al., ^{13}C NMR investigation of local dynamics in nonoriented mesomorphic polysiloxanes, *Macromolecules*, 24, 2800, 1991.

119. Rousseau, D. et al., Conformation in solution of side-chain LC polymers as a function of the mesogen-graft amount, *Polymer*, 44, 2049, 2003.

120. Lavrenko, P. et al., Optical anisotropy of comb-like macromolecules of lateral and semilateral structure, *Macromol. Chem. Phys.*, 199, 207, 1998.

121. Lavrenko, P. et al., Temperature influences ordering of mesogens in comb-like LC poly(methylsiloxane) macromolecules with side groups consisting of "phenyl benzoate" moieties in solution, *Macromol. Chem. Phys.*, 198, 3581, 1997.

122. Richtering, W.H. et al., Solution behavior of two LC polymers of different architectures, *Colloid Polym. Sci.*, 267, 568, 1989.

123. Lacey, D. et al., Mobility of mesogenic side-chains in LC polymers based on the Baylis–Hillman reaction, *Macromol. Chem. Phys.*, 200, 1222, 1999.

124. Lavrenko, P., Yevlampieva, N., and Finkelmann, H., Biaxiality phenomenon in molecular properties of comb LC polymer with lateral structure: first evidence, *Mol. Crystl. Liq. Crystl.*, 299, 259, 1997.

125. Takenaka, S., Molecular structure and liquid crystalline properties: synthesis and thermal properties of laterally attached LC polymers, in *Proceedings of the International Conference on LC polymers*, Beijing, 1994, p. 133.

126. Stuart, H.A., *Die Struktur des Freiem Moleküls*, Springer-Verlag, Berlin, 460, 1952.

127. Flory, P.J., *Statistical Mechanics of Chain Molecules*, 2nd ed., Hanser Publishers, München, 1989.

128. Tsvetkov, V., Molekulare Konformation, hydrodynamische, dynamooptische und elektrooptische Eigenschaften von Polymeren mit Leiterstruktur in Loesungen, *Makromol. Chemie*, 160, 1, 1972.

129. Beevers, M.S., Elliot, D.A., and Williams, G., Dynamic Kerr-effect and dielectric relaxation studies of a poly(methylphenyl siloxane), *Polymer*, 21, 279, 1980.

130. Kolomiets, I.P., Lacey, D., and Lavrenko, P.N., Electrooptical and hydrodynamic properties of a low-molecular-mass copolymethylhydrodimethylsiloxane in solutions, *Polym. Sci., Ser. A.*, 47, 319, 2005.

131. Lavrenko, P.N. et al., Dynamooptical and electrooptical properties of poly(methylphenyl siloxane) in solution and bulk, *Vysokomolek. Soedin.*, 2006 (in press).

132. Lezov, A.V. et al., Dualism in hydrodynamic behavior of LC carbosilane dendrimers in dilute solutions, *Doklady Chem.*, 362, 638, 1998.

133. Rjumtsev, E.I., Yevlampieva, N.P., and Lezov, A.V., Kerr effect of carbosilane dendrimers with terminal mesogenic groups, *Liq. Crystl.*, 25, 475, 1998.

134. Lezov, A.V. et al., Electrooptical properties of carbosilane dendrimers with terminal mesogenic groups in solution, *Russ. J. Phys. Chem.*, 77, 1050, 2003.

135. Lavrenko, P.N., Apparent polymer homology of dendrimers of various generation, *Vysokomolek. Soedin., Ser. B*, 43, 1440, 2001.

136. Pastro, L. et al., End-on and side-on nematic LC dendrimers, *Macromolecules*, 37, 9386, 2004.

137. Yevlampieva, N.P. et al., Electrooptical behavior of fullerene-containing LC dendrimers in solutions, *Chem. Phys. Lett.*, 382, 32, 2003.

138. Deschanaux, R. et al., Novel fullerene-containing liquid crystalline dendrimers and hemi-dendrimers: synthesis and characterization, *Fullerenes, Nanotubes and Carbon Nanostructures*, 12, 193, 2004.

139. Lavrenko, P. et al., Hydrodynamic investigations of the C_{60} fullerene effects on LC malonate-based dendrimers and hemidendrimers in solution, *Prog. Colloid Polym. Sci.*, 127, 61, 2004.

140. Tsvetkov, N.V. et al., Large-scale reorientation of cylindrical dendrimers in electric fields, *Vysokomolek. Soedin.*, *Ser. A*, 45, 253, 2003.

141. Abdullin, I.R. et al., Electrooptical properties and structure of 3-trifluoroacetylcamphorates of lantanoides, *Rep. Russ. Acad. Sci., Chem.*, 2, 486, 1990.

142. Yevlampieva, N. et al., Metallomesogenic Dy- and Ho-complexes: polarizability, polarity and structure, presented at *Eighth International Symposium on Metallomesogenes*, Namur, Belgium, May 28–31, 2003, p. 52.

143. Lavrenko, P.N. et al., Hydrodynamic size of complexes of Holmium and Dysprosium with Schiff bases and various counterions in solutions, *Russ. J. Phys. Chem.*, 79, 408, 2005.

144. Lavrenko, P.N., Kolomiets, I.P., and Vinogradova, L.V., Hydrodynamic, electrooptical and conformational properties of fullerene-containing poly(2-vinylpyridine)s in solutions, *Vysokomol. Soedin.*, 2006 (in press).

21 Frequency-Domain Electro-Optic Response of Chiral Smectic Liquid Crystals

Yasuyuki Kimura and Reinosuke Hayakawa

CONTENTS

21.1   INTRODUCTION

Liquid crystal is an intermediate state of matter between a liquid and a solid. Their properties are also intermediate between those of a liquid and a solid; they can flow (viscous) and resist deformation (elastic). Most molecules that show liquid crystalline state have anisotropic shapes such as rods, disks, and ellipsoids. Therefore, their physical properties such as elastic constant, viscosity, permittivity, and refractive index are anisotropic. Especially, the anisotropy of refractive index and permittivity makes it possible to utilize liquid crystals for flat panel displays for TVs, personal computers, digital watches, calculators, and cellular phones.

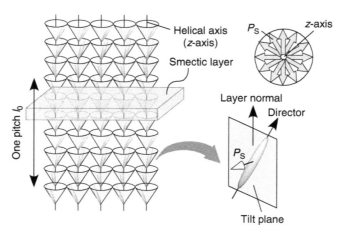

FIGURE 21.1 Schematics of molecular arrangement in the SmC* phase of a ferroelectric liquid crystals (FLCs). FLC molecules form layered structure and tilt from the layer normal homogenously within each layer. A spontaneous polarization appears in the direction perpendicular to the tilt plane defined by layer normal and director. The direction of tilt proceeds in helical fashion and its period (pitch) is l_0.

In 1975, R.B. Meyer and his collaborators designed and realized ferroelectricity in liquid crystals based on simple symmetrical consideration [1]. Since then, ferroelectric liquid crystals (FLCs) have attracted much attention of researchers for their interesting physical properties and their technological potential for display applications [2,3], for example, the surface-stabilized FLC (SSFLC) device [4]. In SSFLC device, a high-speed switching whose characteristic time is as short as microseconds (about 10 ms in a conventional nematic LC device) can be realized by utilizing the reversal of spontaneous polarization. The ferroelectricity appears in the chiral smectic C (SmC*) phase consisting of anisotropic molecules with the component of the dipole moment perpendicular to their longitudinal axis. In the SmC* phase, the constituent chiral molecules tilt from the direction normal to the smectic layers and the tilt direction proceeds in the helical order along the layer normal as illustrated in Figure 21.1. In each smectic layer, a spontaneous polarization P_S appears perpendicular to the tilt plane, which is defined by layer normal and director due to the loss of mirror symmetry about the tilt plane. But the averaged macroscopic polarization over one helical pitch disappears due to the helical arrangement of local spontaneous polarization as shown in Figure 21.1.

There are two thermal fluctuations of the director in the SmC* phase in the low-frequency region (under several MHz) as schematically shown in Figure 21.2. One of them is Goldstone (phase) mode, which is the fluctuation of the azimuthal angle ϕ of the molecular tilt and the other is soft (amplitude) mode, which is the fluctuation of the magnitude of the tilt angle θ. These two modes are well studied by the dielectric relaxation spectroscopy both theoretically and experimentally [3]. As the increment of Goldstone mode is usually larger and its relaxation frequency is rather low (several tens of Hz to several kHz) in the SmC* phase, the contribution of this mode is large in dielectric spectrum. But it vanishes in the Smectic A (SmA) phase as the helical structure disappears. The temperature dependence of the permittivity is also studied theoretically by using the phenomenological theory of Landau-type for the SmA–SmC* phase transition [2,3].

Recently, the dielectric spectroscopy has been extended to the nonlinear regime and applied to the studies on polymers with large permittivity [5,6] or ferroelectricity [7]. In these studies, it is found that the nonlinear spectra offer more useful information on the dynamics of polymers and their phase transition than the one obtained by linear dielectric measurement alone. For example, the negative sign of the third-order nonlinear dielectric increment for

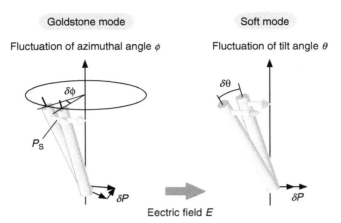

FIGURE 21.2 Schematics of typical low-frequency relaxation modes of directors in the SmC* phase: Goldstone mode and soft mode.

polar polymers suggests that the origin of this nonlinearity is mainly due to the saturation of permanent dipole moments under an electric field. In this case, the increment of nonlinear permittivity gives information on the value and the number density of effective dipoles, which cooperatively responds to the field [5,6].

In the case of chiral smectic liquid crystals including FLCs and antiferroelectric liquid crystals (AFLCs), we have applied nonlinear dielectric spectroscopy to these substances and succeeded in obtaining detailed and new information on their properties, dynamics, and phase transitions [8–12].

It is observed that the linear electro-optical response in the frequency-domain for liquid crystals, which is the response to a sinusoidal applied electric field, is known to offer equivalent information to the one obtained by the dielectric relaxation spectroscopy. But there are some advantages in the electro-optical measurement over the dielectric measurement to be mentioned below:

(1) The magnitude of electro-optic response depends not only on an applied electric field but also on the optical anisotropy of material. As the liquid crystals have large optical anisotropy in general, it is expected that this greatly enhances the sensitivity of measurement.
(2) The influence of conductivity and its temporal fluctuation is less serious compared to dielectric measurement at low frequencies.
(3) The measurement configuration can be selected rather flexibly. In the case of dielectric measurement, the area of the electrodes imposes restrictions on the measurement configurations.

Despite these advantages of electro-optic measurement over dielectric measurement, most studies are performed in time-domain, where a pulse or step electric field is applied, and the measurements in the frequency-domain for chiral smectic liquid crystals are also poor [3,13–16].

In this chapter, we extend electro-optic measurement in frequency-domain to the nonlinear regime and apply to FLCs in the SmC* phase and AFLCs in the antiferroelectric smectic C_A^* (SmC$_A^*$) phase. We also discuss the electro-optic frequency spectrum theoretically by using a simple torque balance equation of chiral smectic liquid crystals concerning the azimuthal angle ϕ, which was previously used in the analysis of the nonlinear dielectric relaxation spectra of FLCs [9]. Further, we compare the theoretical results with those predicted from a general phenomenological theory of nonlinear response. In the SmC_A^* phase, the nonlinear spectra offer new information on the dynamics of AFLCs and such information cannot be obtained from linear electro-optic spectrum.

21.2 THEORETICAL CALCULATION OF THE NONLINEAR ELECTRO-OPTIC RESPONSE OF THE GOLDSTONE MODE IN THE SMC* PHASE

21.2.1 APPROXIMATION OF THE INTENSITY OF THE TRANSMITTED LIGHT THROUGH A LIQUID CRYSTAL CELL

We regard a homogeneously aligned (smectic layers ⊥ cell surfaces) FLC cell as the series of the homogenously tilted uniaxial crystals whose optical axis coincides with the local director n of molecules. The direction of n is defined by the tilt angle θ against the layer normal direction, which is set to z-axis and the azimuthal angle $\phi(z)$ around the z-axis as is illustrated in Figure 21.3. The direction of the incident and outgoing light is taken as y-axis. The intensity of the transmitted light I through an FLC cell between crossed polarizers is obtained by averaging the local intensity $I(z)$ given as [17]

$$I(z) = \frac{I_{in}}{2} \sin^2\left(\frac{\pi \Delta n(z) d}{\lambda}\right) \sin^2\left(2(\psi + \alpha(z))\right). \tag{21.1}$$

over one helical pitch. In Equation 21.1, I_{in} is the intensity of the incident light, λ is the wavelength of the incident light, d is the thickness of the sample, ψ is the angle between the direction of one polarizer and z-axis. The local birefringence $\Delta n(z)$ and the local apparent tilt angle $\alpha(z)$ are respectively given as [17]

$$\Delta n(z) = \frac{n_\parallel n_\perp}{\sqrt{n_\perp^2 + (n_\parallel^2 - n_\perp^2)\sin^2\theta \sin^2\phi(z)}} - n_\perp, \tag{21.2}$$

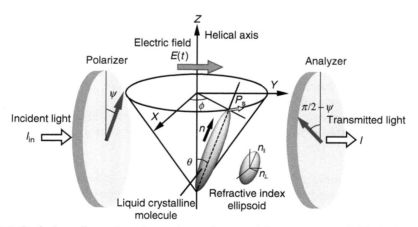

FIGURE 21.3 Optical configuration of experimental setup. A homogenous cell filled with a ferroelectric liquid crystal is sandwiched between a pair of crossed polarizers. Smectic layer is parallel to a cell surface and helical axis is set to Z-axis. The direction of one of the polarizers is set at the angle Ψ relative to Z-axis. The direction of incident light and that of an electric field $E(t)$ is set to Y-axis. The refractive index of liquid crystal parallel to director is $n_\parallel$ and that perpendicular to director is $n_\perp$ (uniaxial approximation).

$$\alpha(z) = \tan^{-1}(\tan\theta\cos\phi(z)), \tag{21.3}$$

where $n_\parallel$ and $n_\perp$ are the refractive indexes parallel and perpendicular to the molecular long axis, respectively (see Figure 21.3). According to Equation 21.1, one can notice that there are two contributions to the electro-optic response of FLCs. One is due to the change in the averaged apparent tilt angle and the other is due to the change in the birefringence. Both originate from the deformation of the helical structure by applying an electric field.

When the tilt angle θ is assumed to be small enough, which is a likely assumption, the electro-optical response to a sinusoidal electric field can be written from the lower-order terms with respect to θ as

$$
\begin{aligned}
\frac{\langle I\rangle}{I_{in}} = &\left[\frac{1}{4}\sin^2\Theta - \frac{\pi n_\parallel d}{8\lambda}\left(\frac{n_\parallel^2}{n_\perp^2}-1\right)\sin^2\theta\sin(2\Theta)\langle\sin^2\phi\rangle + \cdots\right]\\
&+ \sin(4\psi)\left[\tan\theta\sin^2\Theta\langle\cos\phi\rangle + \cdots\right] + \cos(4\psi)\left[-\frac{1}{4}\sin^2\Theta + 2\tan^2\theta\sin^2\Theta\langle\cos^2\phi\rangle\right.\\
&+ \left.\frac{\pi n_\parallel d}{8\lambda}\left(\frac{n_\parallel^2}{n_\perp^2}-1\right)\sin^2\theta\sin(2\Theta)\langle\sin^2\phi\rangle + \cdots\right],
\end{aligned}\tag{21.4}
$$

where Θ is defined as $\Theta \equiv \Delta n d/\lambda$ in terms of the birefringence $\Delta n = n_\parallel - n_\perp$ and $\langle\cdots\rangle$ means the spatial averaged value over one helical pitch. The terms in the first square brackets are the components independent of the direction of polarizers. The terms in the second square brackets will disappear at $\psi = 0$, $\pi/4$, $\cdots$ and those in the third brackets will disappear at $\psi = \pi/8$, $3\pi/8$, $\cdots$. We limit the following discussion to the results for $\psi = \pi/8$ and $\psi = 0$. The approximated presentation for the averaged intensity $<I>$ is respectively given as

$$
\begin{aligned}
\frac{\langle I\rangle}{I_{in}} = &\frac{1}{4}\sin^2\Theta + \tan\theta\sin^2\Theta\langle\cos\phi\rangle - \frac{\pi n_\parallel d}{8\lambda}\left(\frac{n_\parallel^2}{n_\perp^2}-1\right)\sin^2\theta\sin(2\Theta)\langle\sin^2\phi\rangle\\
&+ \cdots\ (\psi = \pi/8),
\end{aligned}
$$

and

$$\frac{\langle I\rangle}{I_{in}} = 2\tan^2\theta\sin^2\Theta\langle\cos^2\phi\rangle + \cdots\ (\psi = 0). \tag{21.6}$$

21.2.2 DYNAMIC RESPONSE OF GOLDSTONE MODE UNDER WEAK ELECTRIC FIELD IN THE SMC* PHASE

Time and spatial dependence of the azimuthal angle $\phi(z,t)$ under an electric field can be calculated by a simple torque balance equation for the elastic deformation discussed by Handschy et al. [18]. In this model, the deformation of the smectic layers is neglected and the bookshelf geometry (the layer normal is parallel to the glass plates) is assumed. In addition, we consider not only the ferroelectric interaction between spontaneous polarization P_S and electric field E applied parallel to y-direction but also the dielectric interaction. The torque balance equation we utilize is given as

$$K\frac{\partial^2\phi(z,t)}{\partial z^2} - \gamma\frac{\partial\phi(z,t)}{\partial t} = P_S E \sin\phi(z,t) + \frac{1}{2}\Delta\varepsilon_a E^2\sin 2\phi(z,t), \tag{21.7}$$

where K is the torsional elastic constant, γ is the rotational viscosity, and $\Delta\varepsilon_a$ is the dielectric anisotropy. In the following discussion, the tilt angle θ is assumed to be constant under weak electric field, which is a good approximation except at the temperature near the SmA–SmC* phase transition. The values of K, γ, and $\Delta\varepsilon_a$ are also assumed to be constant even under an electric field.

In the absence of an applied field ($E=0$), the solution of Equation 21.7 under the boundary conditions $\phi(z=0,t)=0$ and $\phi(z=l_0,t)=2\pi$ is given by the uniform change of $\phi(z, t)$ along z-axis:

$$\phi_0(z) = q_0 z, \tag{21.8}$$

where q_0 is the wave number of helical structure defined as $q_0 = 2\pi/l_0$ (l_0: helical pitch).

If the applied field E is weak, the solution of Equation 21.7 can be written in the perturbed form to the field-free solution of Equation 21.8. We further assume that the helical pitch is l_0 even under the weak applied field and the boundary condition for $\phi_0(z)$ is also applicable.

The solution $\phi(z,t)$ is given as a sum of the terms proportional to powers of E and $\phi_0(z)$ as

$$\phi(z, t) = q_0 z + \sum_{k=1}^{\infty} \phi_k(z, t), \tag{21.9}$$

where the term $\phi_k(z,t)$ is proportional to E^k. By inserting the solution obtained from Equation 21.9 into Equation 21.7, the latter can be decomposed into a set of infinite numbers of partial differential equations as

$$K\frac{\partial^2 \phi_1}{\partial z^2} - \gamma\frac{\partial \phi_1}{\partial t} = P_S E \sin q_0 z, \tag{21.10}$$

$$K\frac{\partial^2 \phi_2}{\partial z^2} - \gamma\frac{\partial \phi_2}{\partial t} = P_S E \phi_1 \cos q_0 z + \frac{1}{2}\Delta\varepsilon_a E^2 \sin(2q_0 z), \tag{21.11}$$

$$K\frac{\partial^2 \phi_3}{\partial z^2} - \gamma\frac{\partial \phi_3}{\partial t} = P_S E\left(-\frac{1}{2}\phi_1^2 \sin q_0 z + \phi_2 \cos q_0 z\right) + \Delta\varepsilon_a E^2 \phi_1 \cos(2q_0 z). \tag{21.12}$$

By solving a set of Equation 21.10 through Equation 21.12, we have

$$\phi_1(z, t) = C_{11} \sin q_0 z, \tag{21.13}$$

$$\phi_2(z, t) = C_{22} \sin(2q_0 z), \tag{21.14}$$

$$\phi_3(z, t) = C_{31} \sin q_0 z + C_{33} \sin(3q_0 z), \tag{21.15}$$

where

$$C_{11} = -\frac{P_S}{\gamma} \int_{-\infty}^{t} dt_1 E(t_1) e^{-(t-t_1)/\tau}, \tag{21.16}$$

$$C_{22} = \frac{P_S^2}{2\gamma^2} \int_{-\infty}^{t} dt_2 \int_{-\infty}^{t_2} dt_1 E(t_1)E(t_2)e^{-(4t-t_1-3t_2)/\tau} - \frac{\Delta\varepsilon_a}{2\gamma} \int_{-\infty}^{t} dt_1 E(t_1)^2 e^{-4(t-t_1)/\tau}, \qquad (21.17)$$

$$\begin{aligned}
C_{31} = \frac{P_S^3}{8\gamma^3} &\left[3 \int_{-\infty}^{t} dt_3 \int_{-\infty}^{t_3} dt_2 \int_{-\infty}^{t_3} dt_1 E(t_1)E(t_2)E(t_3)e^{-(t-t_1-t_2+t_3)/\tau} \right. \\
&\left. - 2 \int_{-\infty}^{t} dt_3 \int_{-\infty}^{t_3} dt_2 \int_{-\infty}^{t_2} dt_1 E(t_1)E(t_2)E(t_3)e^{-(t-t_1-3t_2+3t_3)/\tau} \right] \\
&+ \frac{P_S \Delta\varepsilon_a}{4\gamma^2} \left[-2 \int_{-\infty}^{t} dt_2 \int_{-\infty}^{t_2} dt_1 E(t_1)E(t_2)^2 e^{-(t-t_1)/\tau} \right. \\
&\left. + \int_{-\infty}^{t} dt_2 \int_{-\infty}^{t_2} dt_1 E(t_1)^2 E(t_2)e^{-(t-4t_1+3t_2)/\tau} \right]
\end{aligned} \qquad (21.18)$$

$$\begin{aligned}
C_{33} = -\frac{P_S^3}{8\gamma^3} &\left[\int_{-\infty}^{t} dt_3 \int_{-\infty}^{t_3} dt_2 \int_{-\infty}^{t_3} dt_1 E(t_1)E(t_2)E(t_3)e^{-(9t-t_1-t_2-7t_3)/\tau} \right. \\
&\left. - 2 \int_{-\infty}^{t} dt_3 \int_{-\infty}^{t_3} dt_2 \int_{-\infty}^{t_2} dt_1 E(t_1)E(t_2)E(t_3)e^{-(9t-t_1-3t_2-5t_3)/\tau} \right] \\
&+ \frac{P_S \Delta\varepsilon_a}{4\gamma^2} \left[2 \int_{-\infty}^{t} dt_2 \int_{-\infty}^{t_2} dt_1 E(t_1)^2 E(t_2)e^{-(9t-t_1-8t_2)/\tau} \right. \\
&\left. + \int_{-\infty}^{t} dt_2 \int_{-\infty}^{t_2} dt_1 E(t_1)^2 E(t_2)e^{-(9t-4t_1-5t_2)/\tau} \right]
\end{aligned} \qquad (21.19)$$

and τ is the relaxation time of linear response defined as $\tau = \gamma/Kq_0^2$.

The averaged values $\langle\cos\phi\rangle$ and $\langle\sin^2\phi\rangle$ over one helical pitch can be calculated in the expanded form with respect to the powers of E by Equation 21.9, Equation 21.13 through Equation 21.15 as

$$\langle\cos\phi\rangle = -\frac{C_{11}}{2} + \left(-\frac{C_{11}C_{22}}{4} - \frac{C_{31}}{2} + \frac{C_{11}^3}{16} \right) + \cdots \qquad (21.20)$$

and

$$\langle\sin^2\phi\rangle = 1 - \langle\cos^2\phi\rangle = \frac{1}{2} + \left(-\frac{C_{11}^2}{4} + \frac{C_{22}}{2} \right) + \cdots \qquad (21.21)$$

21.2.3 LINEAR AND NONLINEAR ELECTRO-OPTIC RESPONSE OF THE GOLDSTONE MODE IN THE SMC* PHASE

By replacing $\langle\cos\phi\rangle$, $\langle\sin^2\phi\rangle$, and $\langle\cos^2\phi\rangle$ in Equation 21.5 and Equation 21.6 with Equation 21.20 and Equation 21.21, the intensity of the transmitted light I is given as a sum of the terms, respectively, proportional to the powers of an applied field $E(t)$,

$$I = I_{dc} + \sum_{k=1}^{\infty} I_k, \qquad (21.22)$$

where I_k is the term proportional to $E(t)^k$. For $\psi = \pi/8$, I_1, I_2, and I_3 are respectively given as

$$\frac{I_1(t)}{I_{\text{in}}} = \frac{1}{2} \tan\theta \sin^2\Theta \left(\frac{P_E}{\gamma}\right) \int_{-\infty}^{t} dt_1 E(t_1) e^{-(t-t_1)/\tau}, \tag{21.23}$$

$$\frac{I_2(t)}{I_{\text{in}}} = \frac{\pi n_\parallel d}{32\lambda} \left(\frac{n_\parallel^2}{n_\perp^2} - 1\right) \sin^2\theta \sin(2\Theta) \left[\left(\frac{P_S}{\gamma}\right)^2 \left\{\int_{-\infty}^{t} dt_2 \int_{-\infty}^{t} dt_1 E(t_1) E(t_2) e^{-(2t-t_1-t_2)/\tau}\right. \right.$$
$$\left. \left. - \int_{-\infty}^{t} dt_2 \int_{-\infty}^{t_2} dt_1 E(t_1) E(t_2) e^{-(4t-t_1-3t_2)/\tau}\right\} + \frac{\Delta\varepsilon_a}{\gamma} \int_{-\infty}^{t} dt_1 E(t_1)^2 e^{-4(t-t_1)/\tau}\right] \tag{21.24}$$

$$\frac{I_3(t)}{I_{\text{in}}} = -\frac{1}{16} \tan\theta \sin^2\Theta \left[\left(\frac{P_S}{\gamma}\right)^3 \left\{3 \int_{-\infty}^{t} dt_3 \int_{-\infty}^{t_3} dt_2 \int_{-\infty}^{t_3} dt_1 E(t_1) E(t_2) E(t_3) e^{-(t-t_1-t_2+t_3)/\tau}\right.\right.$$
$$+ \int_{-\infty}^{t} dt_3 \int_{-\infty}^{t} dt_2 \int_{-\infty}^{t} dt_1 E(t_1) E(t_2) E(t_3) e^{-(3t-t_1-t_2-t_3)/\tau}$$
$$- 2 \int_{-\infty}^{t} dt_3 \int_{-\infty}^{t} dt_2 \int_{-\infty}^{t_2} dt_1 E(t_1) E(t_2) E(t_3) e^{-(5t-t_1-3t_2-t_3)/\tau}$$
$$\left. - 2 \int_{-\infty}^{t} dt_3 \int_{-\infty}^{t_3} dt_2 \int_{-\infty}^{t_2} dt_1 E(t_1) E(t_2) E(t_3) e^{-(t-t_1-3t_2+3t_3)/\tau}\right\}$$
$$+ \frac{2\Delta\varepsilon_a P_S}{\gamma^2} \left\{\int_{-\infty}^{t} dt_2 \int_{-\infty}^{t} dt_1 E(t_1)^2 E(t_2) e^{-(5t-4t_1-t_2)/\tau}\right.$$
$$- 2 \int_{-\infty}^{t} dt_2 \int_{-\infty}^{t_2} dt_1 E(t_1) E(t_2)^2 e^{-(t-t_1)/\tau}$$
$$\left.\left. + \int_{-\infty}^{t} dt_2 \int_{-\infty}^{t_2} dt_1 E(t_1)^2 E(t_2) e^{-(t-4t_1+3t_2)/\tau}\right\}\right]. \tag{21.25}$$

The linear response I_1 is given by a convolution integral of an applied electric field and an exponentially decaying function of time with the characteristic time τ. Such a representation agrees with that of general linear response for a relaxational system [19]. The nonlinear responses I_2 and I_3 are given by a sum of convolution integrals composed of the product of the electric field applied at multitime points and exponentially decaying functions of time. The terms in I_2 and I_3 related to the ferroelectric response are respectively composed of dual and triple integral about two and three time variables.

According to the phenomenological theory of nonlinear response [20], a nonlinear response $Y(t)$ to a stimulus $X(t)$ can be generally written as a sum of convolution integrals of $X(t)$ at multiple-time points and nonlinear after-effect functions $\alpha_n (t - t_1, t - t_2, \cdots, t - t_n)$, which characterizes the nonlinear response as

$$Y(t) = \sum_{n=1}^{\infty} \int_{-\infty}^{t} dt_1 \int_{-\infty}^{t} dt_2 \cdots \int_{-\infty}^{t} dt_n \alpha_n (t - t_1, t - t_2, \cdots, t - t_n) X(t_1) X(t_2) \cdots X(t_n). \tag{21.26}$$

where $\alpha_n (t - t_1, t - t_2, \cdots, t - t_n)$ has symmetric property with respect to the permutation of the time variables. From Equation 21.23 through Equation 21.25, we can evaluate the after-effect functions $\alpha_n (t_1, t_2, \cdots, t_n)$ for the linear and nonlinear electro-optic response. In the following parts, the discussion for FLCs is limited to the case of $\Delta\varepsilon_a = 0$.

In the case of linear response, the after-effect function $\alpha_1(t_1)$ is obtained from the response $Y_1(t)$, which is the first term of $n = 1$ in Equation 21.26, to the impulse input $X(t) = e_1 \delta (t - \tau_1)$ as $Y_1 (t) = e_1 \alpha_1 (t - \tau_1)$. On the other hand, by substituting $E(t)$ in Equation 21.23 with the impulse input $E(t) = e_1 \delta (t - \tau_1)$, the output $Y_1(t)$ is calculated as

$$Y_1(t) = \frac{P_S}{2\gamma} \tan\theta \sin^2\Theta e_1 e^{-(t-\tau_1)/\tau}. \tag{21.27}$$

Therefore, the linear after-effect function $\alpha_1 (t_1)$ is given as

$$\alpha_1(t_1) = \frac{P_S}{2\gamma} \tan\theta \sin^2\Theta e^{-t_1/\tau}. \tag{21.28}$$

We can extend this procedure to the nonlinear response by utilizing the electric field given as the sum of the impulse input at multitime points $\tau_1, \tau_2, \tau_3, \cdots$:

$$E(t) = \sum_{n=1}^{m} e_n \delta(t - \tau_n). \tag{21.29}$$

The term proportional to $e_1 e_2$ in the second-order response in Equation 21.26 to an electric field given as a sum of two impulses (up to $n = 2$ in Equation 21.29) is $2\alpha_2 (t - \tau_1, t - \tau_2) e_1 e_2$. Therefore, the second-order after-effect function $\alpha_2 (t_1, t_2)$ is calculated as

$$\alpha_2(t_1, t_2) = \frac{\pi n_\parallel d}{64\lambda} \left(\frac{n_\parallel^2}{n_\perp^2} - 1\right) \sin^2\theta \sin(2\Theta) \left(\frac{P_S}{\gamma}\right)^2 \left\{2e^{-(t_1+t_2)/\tau} - e^{-(t_1+3t_2)/\tau}\right\}(t_1 > t_2). \tag{21.30}$$

In a similar way, the term proportional to $e_1 e_2 e_3$ in the third-order response in Equation 21.26 to the electric field given as a sum of two impulses (up to $n = 3$ in Equation 21.29) is $6\alpha_3 (t - \tau_1, t - \tau_2, t - \tau_3) e_1 e_2 e_3$. Therefore, the third-order after-effect function $\alpha_3 (t_1, t_2, t_3)$ can be calculated as

$$\alpha_3(t_1, t_2, t_3) = \frac{1}{48} \tan\theta \sin^2\Theta \left(\frac{P_S}{\gamma}\right)^3 \left\{3e^{-(t_1+t_2-t_3)/\tau} + -3e^{-(t_1+t_2-t_3)/\tau}\right.$$

$$\left. -e^{-(t_1+3t_2+t_3)/\tau} - 2e^{-(t_1+t_2+3t_3)/\tau} - e^{-(t_1+3t_2+t_3)/\tau}\right\} \quad (t_1 > t_2 > t_3) \tag{21.31}$$

The dependence of the calculated after-effect functions $\alpha_1, \alpha_2, \alpha_3$ on time for $t_1 = t_2 = t_3 = t$ are shown in Figure 21.4. The second-order function $\alpha_2 (t, t)$ slowly decreases with time at short timescale but more rapidly at large timescales. The third-order after-effect $\alpha_3 (t, t, t)$ shows maximum at short timescale, but after that, it decreases with time and approaches to the linear one at large t.

In the case of $\psi = 0$, the terms proportional to odd powers of E, I_1, and I_3 will disappear in its electro-optic response due to the symmetry and I_2 is given as

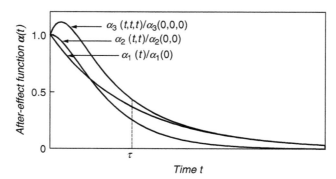

FIGURE 21.4 Temporal evolution of the normalized after-effect functions $\alpha_1(t)/\alpha_1(0)$, $\alpha_2(t,t)/\alpha_2(0,0)$, and $\alpha_3(t,t,t)/\alpha_3(0,0,0)$.

$$\frac{I_2(t)}{I_{in}} = \frac{1}{2}\tan^2\theta\sin^2\Theta\left[\left(\frac{P_S}{\gamma}\right)^2\left\{\int_{-\infty}^{t}dt_2\int_{-\infty}^{t}dt_1 E(t_1)E(t_2)e^{-(2t-t_1-t_2)/\tau}\right.\right.$$

$$\left.\left. -\int_{-\infty}^{t}dt_2\int_{-\infty}^{t_2}dt_1 E(t_1)E(t_2)e^{-(4t-t_1-3t_2)/\tau}\right\} + \frac{\Delta\varepsilon_a}{\gamma}\int_{-\infty}^{t}dt_1 E(t_1)^2 e^{-4(t-t_1)/\tau}\right].$$

$$(21.32)$$

The corresponding after-effect function $\alpha_2(t_1, t_2)$ is the same one of Equation 21.30 except the numerical factor.

21.2.4 Calculation of Linear and Nonlinear Electro-Optic Relaxation Spectra

From Equation 21.23 through Equation 21.25 and Equation 21.32, we can calculate the electro-optic response under an arbitrary electric field $E(t)$. When we apply a sinusoidal electric field $E(t) = E_0\cos\omega t$ of the angular frequency ω, the intensity of the transmitted light I becomes sum of the fundamental and the higher harmonic components of ω and their amplitudes are given as the power series of E_0,

$$\frac{I(t)}{I_{in}} = \frac{I_{dc}}{I_{in}} + \sum_{m=1}^{\infty}\sum_{n=m}^{\infty}a_{nm}^*\left(\frac{E_0}{2}\right)^n\exp(im\omega t) + \text{c.c.}$$

$$(21.33)$$

where "c.c." represents the complex conjugate of the preceding term. When E_0 is small, the contribution of the term proportional to E_0^m, a_{mm}^*, becomes dominant term in the respective mth-harmonic component.

The linear and nonlinear electro-optic spectra a_{11}^*, a_{22}^*, and a_{33}^* for $\psi = \pi/8$ are respectively obtained from Equation 21.23 through Equation 21.25 as

$$a_{11}^* = \frac{P_S}{2Kq_0^2}\tan\theta\sin^2\Theta\frac{1}{1+i\omega\tau},$$

$$(21.34)$$

$$a_{22}^* = \frac{\pi n_\parallel d}{64\lambda}\left(\frac{n_\parallel^2}{n_\perp^2}-1\right)\sin^2\theta\sin(2\Theta)\left\{\frac{P_S^2}{(Kq_0^2)^2}\frac{3+i\omega\tau}{(1+i\omega\tau)^2(2+i\omega\tau)} + \frac{\Delta\varepsilon_a}{Kq_0^2}\frac{1}{2+i\omega\tau}\right\}, \quad (21.35)$$

$$a_{33}^* = \frac{1}{8} \tan\theta \sin^2\Theta \left\{ -\frac{P_S^3}{(Kq_0^2)^3} \frac{3 + 5i\omega\tau - \omega^2\tau^2}{(1 + i\omega\tau)^3(2 + i\omega\tau)(1 + 3i\omega\tau)} \right.$$

$$\left. + \frac{3P_S\Delta\varepsilon_a}{(Kq_0^2)^2} \frac{1}{(1 + i\omega\tau)(2 + i\omega\tau)(1 + 3i\omega\tau)} \right\}. \tag{21.36}$$

For $\psi = 0$, we can calculate as

$$a_{11}^* = 0, \tag{21.37}$$

$$a_{22}^* = \frac{1}{4} \tan^2\theta \sin^2\Theta \left\{ \frac{P_S^2}{(Kq_0^2)^2} \frac{3 + i\omega\tau}{(1 + i\omega\tau)^2(2 + i\omega\tau)} + \frac{\Delta\varepsilon_a}{Kq_0^2} \frac{1}{2 + i\omega\tau} \right\}, \tag{21.38}$$

$$a_{33}^* = 0. \tag{21.39}$$

The linear spectrum a_{11}^* of Equation 21.34 shows a single relaxation of Debye-type as shown in Figure 21.5a, which is equivalent to the linear dielectric relaxation ε_1^* of the Goldstone mode [3]. The profiles of the nonlinear spectra a_{22}^* and a_{33}^* given by Equation 21.35 and Equation 21.36 are respectively shown in Figure 21.5b and Figure 21.5c. In Figure 21.5b and Figure 21.5c, we only consider the ferroelectric interaction and set $\Delta\varepsilon_a = 0$. The real part of a_{22}^*, a_{22}', takes a minimum value at $\omega\tau = 1.50$ and approaches to a dc value after changing its sign at $\omega\tau = 0.88$. Its imaginary part a_{22}'' shows a similar profile to that of a_{11}'', but it takes a maximum value at $\omega\tau = 0.54$ in contrast to that a_{11}'' takes the maximum at $\omega\tau = 1.0$. The profile of a_{22}^* seems to be more complicate than that of a_{11}^*, but it can be illustrated simpler in the Cole–Cole representation (imaginary part versus real part) as a spiral over two quadrants around the origin as shown in Figure 21.6. The real part of a_{33}^*, a_{33}', takes a maximum value at $\omega\tau = 0.71$ and approaches the dc value after changing its sign at $\omega\tau = 0.39$. The imaginary part of a_{33}^*, a_{33}'', takes the two extremums in opposite sign, respectively at $\omega\tau = 1.66$ and 0.26, and changes its sign at $\omega\tau = 1.19$. The Cole–Cole representation of a_{33}^* becomes a spiral over three quadrants around the origin as shown in Figure 21.6. From the symmetry of the response, the odd harmonic components have their origin in the change of the average optical axis $\alpha(z)$ and the even harmonic ones originate from the change in the anisotropy of the refractive index $\Delta n(z)$. But there also appear the mixing terms of these two contributions in the higher harmonic components.

In the case of the nonlinear dielectric response of Goldstone mode in FLCs [9], the response directly relates the spatial average of $P_S \cos\phi$ over one helical pitch and takes the same form as the leading term of the linear and third-order nonlinear spectra in the electro-optic response. On the contrary, the even-order harmonic components disappear in the dielectric spectra due to the symmetry of dielectric response to an electric field.

21.2.5 COMPARISON WITH THE PHENOMENOLOGICAL NONLINEAR RESPONSE THEORY

In this section, we discuss the calculated nonlinear spectra in the framework of a general phenomenological theory of nonlinear response [20]. According to the theory, the linear and nonlinear complex frequency spectra $a_n^*(\omega_1, \omega_2, \cdots, \omega_n)$ are generally given by the multitime Fourier transform of the after-effect function $\alpha_n(t - t_1, t - t_2, \cdots, t - t_n)$ as

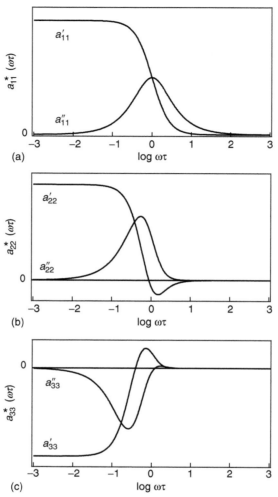

FIGURE 21.5 Frequency dependence of the linear $\alpha_{11}^*(\omega\tau)$, the second-order $a_{22}^*(\omega\tau)$, and the third-order nonlinear spectra $a_{33}^*(\omega\tau)$, where $a_{mm}^* = a_{mm}' - ia_{mm}''$ and we set $\Delta\varepsilon_a = 0$.

$$\alpha_n^*(\omega_1, \omega_2, \cdots, \omega_n) = \int_0^\infty dt_1 \int_0^\infty dt_2 \cdots \int_0^\infty dt_n \alpha_n(t_1, t_2, \cdots, t_n) \exp\left(-i\sum_{k=1}^n \omega_k t_k\right). \quad (21.40)$$

When we notice the symmetric property of $\alpha_n(t - t_1, \ t - t_1, \ \cdots, \ t - t_n)$ on time variables, Equation 21.40 can be rewritten as

$$\alpha_n^*(\omega_1, \omega_2, \cdots, \omega_n) = \sum_{\omega_1, \omega_2, \cdots, \omega_n} \int_0^\infty dt_1 \int_0^\infty dt_2 \cdots \int_0^\infty dt_n \alpha_n(t_1, t_2, \cdots, t_n) \exp\left(-i\sum_{k=1}^n \omega_k t_k\right),$$
$$(21.41)$$

where the summation is taken over possible permutations of $\omega_1, \omega_2, \cdots, \omega_n$. From Equation 21.28, Equation 21.30, and Equation 21.31, the linear $\alpha_1^*(\omega_1)$, second-order $\alpha_2^*(\omega_1, \omega_2)$, and

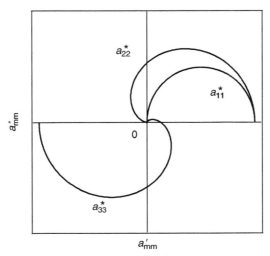

FIGURE 21.6 Cole–Cole representation of the linear a_{11}^*, the second-order a_{22}^*, and the third-order nonlinear spectra a_{33}^* ($\Delta\varepsilon_a = 0$).

third-order nonlinear spectra α_3^* (ω_1, ω_2, ω_3) for $\Psi = \pi/8$ in the case of $\Delta\varepsilon_a = 0$ are respectively, calculated as

$$\alpha_1^*(\omega_1) = \frac{P_S\tau}{2\gamma}\tan\theta\sin^2\Theta\frac{1}{1+i\omega_1\tau}, \tag{21.42}$$

$$\alpha_2^*(\omega_1,\omega_2) = \frac{\pi n_\parallel d}{64\lambda}\left(\frac{n_\parallel^2}{n_\perp^2}-1\right)\sin^2\theta\sin(2\Theta)\left(\frac{P_S\tau}{\gamma}\right)^2$$
$$\times\sum_{\omega_1,\omega_2}\left[\frac{1}{(1+i\omega_1\tau)(1+i\omega_2\tau)}-\frac{1}{(1+i\omega_1\tau)\{4+i(\omega_1+\omega_2)\tau\}}\right], \tag{21.43}$$

$$\alpha_3^*(\omega_1,\omega_2,\omega_3) = -\frac{1}{96}\tan\theta\sin^2\Theta\left(\frac{P_S\tau}{\gamma}\right)^3\sum_{\omega_1,\omega_2,\omega_3}\left[\frac{1}{(1+i\omega_1\tau)(1+i\omega_2\tau)(1+i\omega_3\tau)}\right.$$
$$+\frac{3}{(1+i\omega_1\tau)(1+i\omega_2\tau)\{1+i(\omega_1+\omega_2+\omega_3)\tau\}}$$
$$-\frac{2}{(1+i\omega_1\tau)(1+i\omega_3\tau)\{4+i(\omega_1+\omega_2)\tau\}}$$
$$\left.-\frac{2}{(1+i\omega_1\tau)\{4+i(\omega_1+\omega_2)\tau\}\{1+i(\omega_1+\omega_2+\omega_3)\tau\}}\right]. \tag{21.44}$$

The profile of the second-order nonlinear spectrum α_2^* (ω_1, ω_2) for ω_1, $\omega_2 > 0$ is shown in Figure 21.7. In order to obtain the whole profile of α_2^* (ω_1, ω_2) experimentally, we have to use the sum of sinusoidal electric fields with two different frequencies as is performed for the measurement of electric birefringence (Kerr effect) [21]. As we usually apply a sinusoidal electric field with a single frequency and measure fundamental and higher harmonic

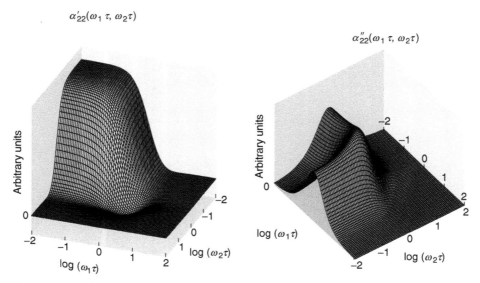

FIGURE 21.7 Dependence of the calculated real $\alpha_{22}'(\omega_1, \omega_2)$ and imaginary part $\alpha_{22}''(\omega_1, \omega_2)$ of the second-order nonlinear electro-optic spectrum $a_{22}^*(\omega_1, \omega_2)$ on frequencies.

components of the response in frequency-domain nonlinear electro-optic measurement, we set $\omega_1 = \omega_2 = \omega$ in Equation 21.43 and this reproduces Equation 21.35. On the other hand, when we set $\omega_1 = -\omega_2 = \omega$, we can obtain the second-order nonlinear spectrum which contributes to dc component of the response, and we find its imaginary part has vanished. In a similar way, if we set $\omega_1 = \omega_2 = \omega_3 = \omega$ in Equation 21.44, it reproduces Equation 21.36.

In the case of linear relaxation, the after-effect function is an exponentially decay function. Though the after-effect function for nonlinear relaxation cannot generally be discussed, it is natural to extend the linear after-effect function to nonlinear relaxation as

$$\alpha_n(t_1, t_2, \cdots, t_n) = \Delta\alpha_n \prod_{k=1}^{n} \frac{1}{\tau_k} \exp\left(-\frac{t_k}{\tau_k}\right), \tag{21.45}$$

where $\Delta\alpha_n$ is the increment of the nth order nonlinear spectrum and τ_k ($k = 1, 2, \cdots, n$) is the relaxation times. The Fourier transformed nonlinear relaxation spectrum is given by inserting Equation 21.45 to Equation 21.40 as

$$\alpha_n^*(\omega_1, \omega_2, \cdots, \omega_n) = \frac{\Delta\alpha_n}{\displaystyle\prod_{k=1}^{n}(1 + i\omega_k\tau_k)}. \tag{21.46}$$

When we use a field with a single frequency ω and the all relaxation times set to a single relaxation time $\tau_1 = \tau_2 = \cdots = \tau_n$, Equation 21.46 is reduced to more simple form as

$$\alpha_n^*(\omega, \omega, \cdots, \omega) = \frac{\Delta\alpha_n}{(1 + i\omega\tau_n)^n}. \tag{21.47}$$

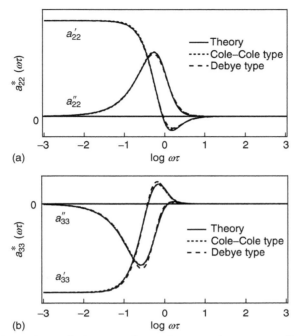

FIGURE 21.8 Comparison of the calculated nonlinear electro-optic spectrum $a_{22}^*(\omega\tau)$ and $a_{33}^*(\omega\tau)$, with Equation 21.47 and Equation 21.48.

This formula or its modified form is often utilized for the analysis of the experimentally obtained nonlinear dielectric relaxation spectra [5,6,8]. It is important to compare this rather simple phenomenological result to Equation 21.35 and Equation 21.36. From the experimental point of view, the slightly modified form of Equation 21.47 is utilized for the analysis as

$$\alpha_n^*(\omega, \omega \cdots \omega) = \frac{\Delta\alpha_n}{\left\{1 + (i\omega\tau_n)^{\beta_n}\right\}^n}, \qquad (21.48)$$

where β_n represents the broadness of the distribution of relaxation time. This is usually called "Cole–Cole parameter" in the linear relaxation and takes a real number between 0 and 1. In Figure 21.8, the best-fitted curves of Equation 21.48 to a_{22}^* and a_{33}^* are respectively shown as broken lines. The obtained best-fitted parameters are $(\tau_2, \beta_2) = (1.10\tau_1, 1.00)$ and $(\tau_3, \beta_3) = (1.53\tau_1, 1.00)$ by keeping β to 1.0 and $(\tau_2, \beta_2) = (1.11\ \tau_1, 1.02)$ and $(\tau_3, \beta_3) = (1.50\ \tau_1, 0.97)$. As the agreement between the best-fitted curves and the theoretical ones is good, we can use the phenomenological spectrum of Equation 21.48 as an approximated formulation to the nonlinear electro-optic spectrum. But the physical meaning of the shift of the relaxation time among τ_1, τ_2, and τ_3 cannot be explained by the phenomenological theory.

21.3 EXPERIMENT

A sinusoidal electric field generated from a synthesizer equipped with a FFT analyzer (Advantest R9211) is applied to a homogeneously aligned liquid crystal cell (smectic layers are perpendicular to the cell surfaces). The transmitted light intensity I through a cell was

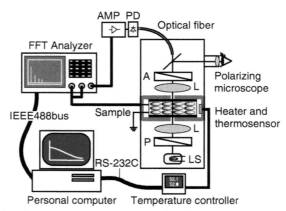

FIGURE 21.9 Schematic of a measurement system for frequency-domain nonlinear electro-optic spectrum. LS: halogen lump, L: lens, P: polarizer, A: analyzer, PD: photodiode, and AMP: amplifier.

detected and converted into a voltage signal by a high-speed PIN photodiode module (Hamamatsu S3887) attached to a polarizing microscope (Olympus BHSP) as illustrated in Figure 21.9. The optical response I is decomposed into the frequency components by the FFT analyzer as

$$I = I_{dc} + \mathrm{Re}\left[\sum_{m=1}^{\infty} I_m^* \exp(im\omega t)\right], \tag{21.49}$$

where I_m^* is the Fourier-decomposed complex amplitude of the mth harmonic component. The direction of analyzer and a cell was set to make an angle of $\Psi = \pi/8$ or 0 with the layer normal direction. The direction of $\Psi = 0$ was chosen as the direction where the transmitted light intensity under crossed polarizers takes a minimum value around the rubbing direction in the SmA phase.

The FLC used in this study was CS1017 provided by Chisso Corporation, which is a commercial mixture with a short helical pitch of 1 μm and shows SmC* phase at room temperature. The sample was sandwiched between two glass plates with ITO electrodes. The area of the electrode is 1 cm². The surfaces of the plates were coated with polyimide and rubbed unidirectionally to attain a homogenous alignment of molecules. The polyethylene terephthalate (PET) film was used as spacers and the thickness of the cell was 25 μm. It is determined by the capacitance of an empty cell. The sample FLC was heated up to the isotropic phase and introduced to the cell by a capillary. After the sample was slowly cooled to SmA phase, we checked the alignment within the cell under a microscope and chose a well-aligned region where the light pass through. The cell was set up in a holder whose temperature was controlled by a temperature controller (CHINO DJ-11) within $\pm 0.1°C$.

21.4 EXPERIMENTAL RESULTS FOR NONLINEAR ELECTRO-OPTIC SPECTRA OF FLCS

As is discussed in Equation 21.33 the fundamental and higher harmonic components of ω, I_m^* ($m = 1, 2, \ldots$) is given as a sum of terms proportional to the powers of an applied electric field E_0^n ($n \geq m$) as

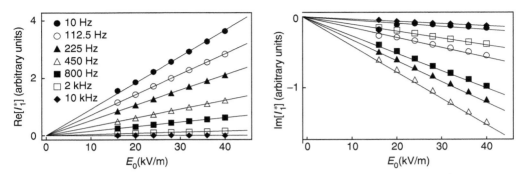

FIGURE 21.10 Dependence of real and imaginary part of the fundamental component I_1^* of the electro-optic response of CS1017 on applied electric field E_0 (54.0°C). The best-fitted lines are drawn as solid ones.

$$I_m^*(\omega) = \sum_{n=m}^{\infty} a_{nm}^* \left(\frac{E_0}{2}\right)^n. \qquad (21.50)$$

When E_0 is small, the dominant term in I_m^* is one proportional to E_0^m. The dependence of the complex amplitude of the fundamental I_1^*, the second-order I_2^*, and the third-order component I_3^* for $\Psi = \pi/8$ on the applied field E_0 are respectively shown in Figure 21.10, Figure 21.11, and Figure 21.12. It is found that I_1^*, I_2^*, and I_3^* are proportional to E_0, E_0^2, and E_0^3 as discussed above. The best-fitted lines are respectively drawn as solid ones in Figure 21.10, Figure 21.11, and Figure 21.12. We can obtain the linear and nonlinear electro-optic spectra a_{mm}^* from the slopes of those best-fitted lines as

$$a_m^*(\omega) \equiv a_{mm}^*(\omega) = \lim_{E_0 \to 0} \frac{I_m^*(\omega)}{E_0^m} \cdot 2^{m-1} \equiv a_{mm}' - i a_{mm}''. \qquad (21.51)$$

The linear spectrum a_1^* drawn in Figure 21.13 shows a nearly single relaxation profile of Debye-type and is well fitted to a modified form of Debye relaxation (Cole–Cole spectrum) with a parameter β_1 characterizing distribution of relaxation time given as

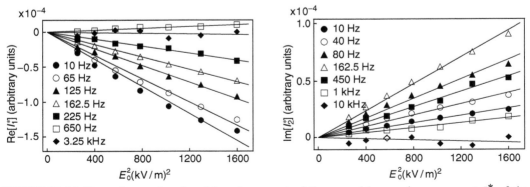

FIGURE 21.11 Dependence of real and imaginary part of the second harmonic component I_2^* of the electro-optic response of CS1017 on applied electric field E_0^2 (54.0°C). The best-fitted lines are drawn as solid ones.

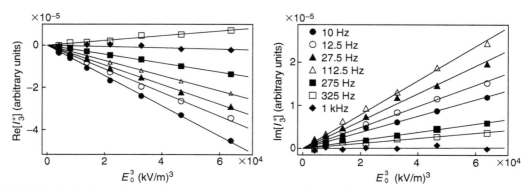

FIGURE 21.12 Dependence of real and imaginary part of the third harmonic component I_3^* of the electro-optic response of CS1017 on applied electric field E_0^3 (54.0°C). The best-fitted lines are drawn as solid ones.

$$a_1^*(\omega) = \frac{\Delta a_1}{1 + (i\omega\tau_1)^{\beta_1}}, \qquad (21.52)$$

where Δa_1 and τ_1 are the increment and the relaxation time of the linear spectrum. The best-fitted curve with $\tau_1 = 0.58$ ms and $\beta_1 = 0.88$ is drawn as a solid line in Figure 21.13. As the contribution of the soft mode to the electro-optic response seems to be negligibly small at this temperature (54.0°C), a_{11}' approaches to zero at high frequencies.

In the same way, the nonlinear spectra a_2^* and a_3^* are obtained from Figure 21.11 and Figure 21.12 with Equation 21.51 are as shown in Figure 21.14 and Figure 21.15. We can analyze the obtained spectra of a_2^* and a_3^* by the modified theoretical spectra of Equation 21.35 and Equation 21.36 in which $i\omega\tau$ is replaced by $(i\omega\tau_n)^{\beta n}$ and $\Delta\varepsilon_a$ is set to zero as

$$a_2^* = \frac{\Delta a_2\left\{3 + (i\omega\tau_2)^{\beta_2}\right\}}{\left\{1 + (i\omega\tau_2)^{\beta_2}\right\}^2\left\{2 + (i\omega\tau_2)^{\beta_2}\right\}}, \qquad (21.53)$$

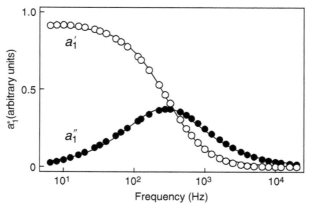

FIGURE 21.13 Frequency spectrum of the linear electro-optic response a_1^* ($\equiv a_1' - ia_1''$) at 54.0°C for CS1017 calculated from Figure 21.10 utilizing Equation 21.51. The best-fitted line of Equation 21.52 is drawn as a solid line.

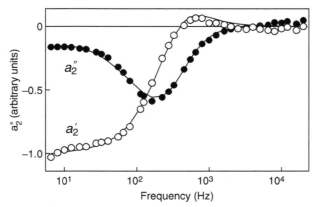

FIGURE 21.14 Frequency spectrum of the second-order nonlinear electro-optic response a_2^* ($\equiv a_2' - ia_2''$) at 54.0°C for CS1017 calculated from Figure 21.11 utilizing Equation 21.51. The best-fitted line of Equation 21.53 is drawn as a solid line.

$$a_3^* = \frac{\Delta a_3 \left\{ 3 + 5(i\omega\tau_3)^{\beta_3} + (i\omega\tau_3)^{2\beta_3} \right\}}{\left\{ 1 + (i\omega\tau_3)^{\beta_3} \right\}^3 \left\{ 2 + (i\omega\tau_3)^{\beta_3} \right\} \left\{ 1 + 3(i\omega\tau_3)^{\beta_3} \right\}}, \tag{21.54}$$

where Δa_i and $\tau_i (i = 2,3)$ are the increment and the relaxation time of the second- and third-order nonlinear spectrum. The nonlinear spectra a_2^* and a_3^* are well fitted by Equation 21.53 and Equation 21.54 with $\tau_2 = 0.42$ms, $\beta_2 = 0.90$, $\tau_3 = 0.49$, and $\beta_3 = 0.89$. The best-fitted curves are drawn as solid lines in Figure 21.14 and Figure 21.15. We find that the sign of Δa_1 is positive and that of Δa_3 is negative, which makes good agreement with those expected from Equation 21.34 and Equation 21.36. From Equation 21.35, the sign of Δa_2 depends on that of $\sin\Theta$, which depends on cell thickness, wavelength of light, and the value of birefringence. The best-fitted values of τ_is and β_is ($i = 1, 2, 3$) are not strongly depend on i, which confirms the theoretical prediction that the shape of the nonlinear spectra is determined by the relaxation time τ of the linear spectrum.

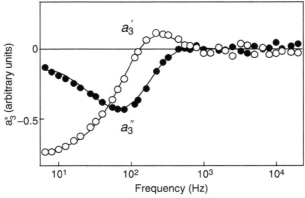

FIGURE 21.15 Frequency spectrum of the third-order nonlinear electro-optic response a_3^* ($\equiv a_3' - ia_3''$) at 54.0°C for CS1017 calculated from Figure 21.12 with utilizing Equation 21.51. The best-fitted line of Equation 21.54 is drawn as a solid line.

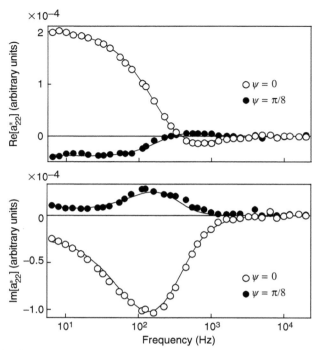

FIGURE 21.16 Angle Ψ dependence of frequency spectrum of the second-order nonlinear electro-optic response a_2^* ($\equiv a_2' - ia_2''$) at 54.0°C for CS1017. The best-fitted line of Equation 21.53 is drawn as a solid line.

The frequency spectrum of a_2^* measured at $\Psi = 0$ is shown in Figure 21.16 with that at $\Psi = \pi/8$ for comparison. It is found that the relaxation frequencies of both spectra are the same and the signs of Δa_2 are opposite. From Equation 21.38, the sign of Δa_2 at $\Psi = 0$ has to be positive and this makes good agreement with experimental results. According to Equation 21.37 and Equation 21.39, a_1^* and a_3^* will be zero. Though the magnitude of Δa_1 and Δa_3 at $\Psi = 0$ is much smaller than those at $\Psi = \pi/8$, we can clearly observe their profiles. This is partly due to the failure of uniaxial approximation to its optical property in the SmC* phase and the existence of deformation of layers in a sample.

21.5 NONLINEAR ELECTRO-OPTIC RESPONSE OF THE GOLDSTONE MODE IN THE SmC$_A^*$ PHASE

In 1989, Chandani et al. found that 4-(1-methylheptyloxycarbonylphenyl) 4'-octyloxybiphenyl-4-carboxylate (called MHPOBC in short form) show the tri-stable electro-optical switching which is originated from antiferroelectricity in a phase designated as SmC$_A^*$ phase [22]. In the SmC$_A^*$ phase, molecules form layered structure and tilt from the layer normal direction is the same as in the SmC* phase. But the direction of tilt in neighboring layers is opposite (anticlinic structure) and the unit cell is made up of two layers as schematically shown in Figure 21.17. Due to this anticlinic structure, spontaneous polarizations in the neighboring layers point to the opposite direction and they are locally cancelled. In addition to such

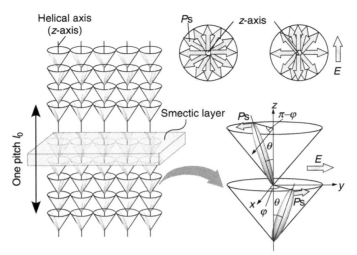

FIGURE 21.17 Schematics of molecular arrangement in the SmC$_A$* phase of antiferroelectric liquid crystals (AFLCs). AFLC molecules form a layered structure and tilt from the layer normal homogenously within each layer. A spontaneous polarization and direction of tilt in the neighboring layers are nearly opposite. But their directions modulate in helical fashion and form a double twisted helix with a period (pitch) l_0. When a weak electric field is applied the molecules, they will align as long axis of permittivity ellipsoid being parallel to the field without break of antiferroelectric order.

antiferroelectric local order of molecules, the direction of tilt is modulated and forms helicoidal structure due to the chirality of the molecules.

As the AFLCs show such double helical structure, the fluctuation with doubled wavenumber (corresponding the half pitch) of helical structure becomes the most fundamental one in AFLCs, which is usually called antiferroelectric Goldstone mode. Therefore, the linear electro-optic response of this mode is not expected to be observed in the SmC$_A^*$ phase. The dominant response to a small electric field is originated from the coupling between an electric field and the dielectric anisotropy $\Delta\varepsilon_a$. The corresponding torque balance equation can be written as

$$K\frac{\partial^2\phi(z,t)}{\partial z^2} - \gamma\frac{\partial\phi(z,t)}{\partial t} = \frac{1}{2}\Delta\varepsilon_a E^2 \sin 2\phi(z,t). \tag{21.55}$$

The linear and nonlinear spectrum calculated from Equation 21.55 is given by setting $P_S = 0$ in the Equation 21.34 through Equation 21.39. We find the linear a_1^* and the third-order nonlinear spectrum a_3^* show no relaxation behavior and the leading term of the second-order nonlinear spectrum a_2^* is calculated for $\Psi = \pi/8$ as

$$a_2^* = \frac{\pi n_\parallel d}{128\lambda}\left(\frac{n_\parallel^2}{n_\perp^2} - 1\right)\sin^2\theta\sin(2\Theta)\frac{\Delta\varepsilon_a}{Kq_0^2}\frac{1}{1+2i\omega\tau_a}, \tag{21.56}$$

and for $\Psi = 0$ as

$$a_2^* = \frac{1}{8}\tan^2\theta\sin^2\Theta\frac{\Delta\varepsilon_a}{Kq_0^2}\frac{1}{1+2i\omega\tau_a}, \tag{21.57}$$

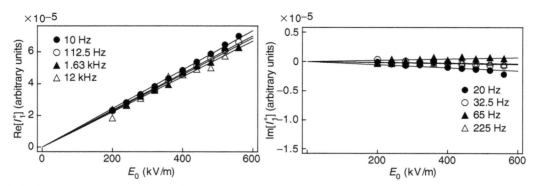

FIGURE 21.18 Dependence of real and imaginary part of the fundamental component I_1^* of the electro-optic response of (R)-MHPOBC on applied electric field E_0 (100°C) in the SmC$_A^*$ phase. The best-fitted lines are drawn as solid ones.

where τ_a given by $\tau_a = \gamma/(4Kq_0^2)$ is the relaxation time of the Goldstone mode in the SmC$_A^*$ phase. The spectrum of a_2^* is a single relaxation of Debye-type, which corresponds to the antiferroelectric Goldstone mode with wavenumber of $2q_0$.

The experimental setup we utilized is the same one as is shown in Figure 21.9. An AFLC we used is (R)-MHPOBC and it shows the SmC$_A^*$ in a wide temperature range and its phase sequence is [22]

$$\text{Iso} - 148°C - \text{SmA} - 122 - \text{SmC}_\alpha^* - 120.9 - \text{SmC}^* - 119.2 - \text{SmC}_\gamma^* - 118.5 - \text{SmC}_A^* - 63 - \text{SmI}_A^*.$$

The dependence of the complex amplitude of the fundamental I_1^* and the second-order I_2^* component on the applied electric field E_0 for $\Psi = \pi/8$ at 100°C are shown in Figure 21.18 and Figure 21.19. It is found that I_1^* and I_2^* are respectively proportional to E_0 and E_0^2 as in the SmC* phase. The best-fitted lines are drawn as solid ones in Figure 21.18 and Figure 21.19. We can obtain the linear a_1^* and the second-order nonlinear electro-optic spectrum a_2^* from the slopes of those best-fitted lines as shown in Figure 21.20 and Figure 21.21. The linear spectrum a_1^* is almost independent of frequency except at high frequencies. This is partly due to the contribution of the soft mode or the molecular rotational relaxation around its short

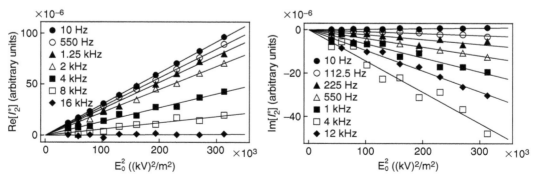

FIGURE 21.19 Dependence of real and imaginary part of the second harmonic component I_2^* of the electro-optic response of (R)-MHPOBC on applied electric field E_0^2 (100°C) in the SmC$_A^*$ phase. The best-fitted lines are drawn as solid ones.

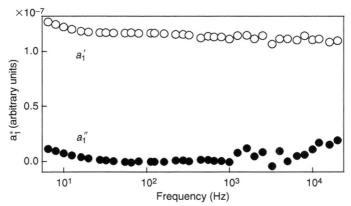

FIGURE 21.20 Frequency spectrum of the linear electro-optic response a_1^* ($\equiv a_1' - ia_1''$) at $100°C$ for (R)-MHPOBC calculated from Figure 21.18 utilizing Equation 21.51.

axis, those are also reported in linear dielectric measurements [23,24]. The second-order spectrum a_2^* shows a single relaxation of Debye-type, which corresponds to the antiferro-electric Goldstone mode. We can analyze a_2^* by Cole–Cole spectrum as

$$a_2^* = \frac{\Delta a_2}{1 + 2(i\omega\tau_a)^{\beta_a}}. \qquad (21.58)$$

The best-fitted curve of the Cole–Cole function Equation 21.58 with $\tau_a = 21$ μs and $\beta_a = 0.98$ is drawn a solid line in Figure 21.21. The positive sign of the increment of a_2^* predicted by Equation 21.56 is confirmed. As the unit cell in the SmC_A^* phase is two smectic layers, there are two kinds of Goldstone mode corresponding to the fluctuations of azimuthal angle in the neighboring layers which are respectively in-phase (acoustic mode) or out-of-phase (optical mode). The detected fluctuation of director concerns the acoustic-like Goldstone mode and this mode couples to the dielectric anisotropy, which can be observed only in the nonlinear spectrum.

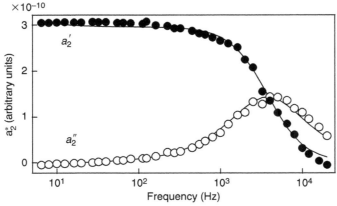

FIGURE 21.21 Frequency spectrum of the second-order nonlinear electro-optic response a_2^* ($\equiv a_2' - ia_2''$) at $100°C$ for (R)-MHPOBC calculated from Figure 21.19 utilizing Equation 21.51. The best-fitted line of Equation 21.58 is drawn as a solid line.

21.6   CONCLUSIONS

The nonlinear electro-optical spectroscopy in the frequency-domain is applied to the Goldstone mode of a ferroelectric liquid crystal in the SmC* phase and that of an antiferroelectric liquid crystal in the SmC$_A^*$ phase. We have also calculated the nonlinear electro-optic response in both cases theoretically by using a perturbation solution of a torque balance equation. It is found that the nonlinear spectra obtained experimentally make good agreements with theoretical ones. We have discussed the availability of this method to research of dynamics of chiral polar smectic liquid crystals. This method is found to be useful to investigate the various types of electrical interaction in the liquid crystals and various mesophases.

ACKNOWLEDGMENT

One of the authors (Y.K.) thanks for the financial support by a Grant-in-aid for Scientific Research from Japan Society for the Promotion of Science and from Ministry of Education, Culture, Sports, Science, and Technology of Japan.

REFERENCES

1. Meyer, R.B., Liébert, L., Strzelecki, L., and Keller, P., Ferroelectric liquid crystals, *J. Phys.* (France), 36, L69, 1975.
2. Lagerwall, S.T., *Ferroelectric and Antiferroelectric Liquid Crystals*, Wiley-VCH, Weinheim, 1999.
3. Musevic, I., Blinc, R., and Zeks, B., *The Physics of Ferroelectric and Antiferroelectric Liquid Crystals*, World Scientific, Singapore, 2000.
4. Clark, N.A. and Lagerwall, S.T., Submicrosecond bistable electro-optic switching in liquid crystals, *Appl. Phys. Lett.*, 36, 899, 1980.
5. Furukawa, T., Tada, M., Nakajima, K., and Seo, I., Nonlinear dielectric relaxations in a vinylidenecyanide vinylacetate copolymer, *Jpn. J. Appl. Phys.*, 27, 200, 1988.
6. Furukawa, T. and Matsumoto, K., Nonlinear dielectric relaxation spectra of vinylacetate, *Jpn. J. Appl. Phys.*, 31, 840, 1992.
7. Furukawa, T., Nakajima, K., Koizumi, T., and Date, M., Measurements of nonlinear dielectricity in ferroelectric polymers, *Jpn. J. Appl. Phys.*, 26, 1039, 1987.
8. Kimura, Y. and Hayakawa, R., Experimental study of nonlinear dielectric relaxation spectra of ferroelectric liquid crystals in the smectic C* phase, *Jpn. J. Appl. Phys.*, 32, 4571, 1993.
9. Kimura, Y., Hara, S., and Hayakawa, R., Nonlinear dielectric relaxation spectroscopy of ferroelectric liquid crystals, *Phys. Rev. E*, 62, R5907, 2000.
10. Kimura, Y., Hayakawa, R., Okabe, N., and Suzuki, Y., Nonlinear dielectric relaxation spectroscopy of the antiferroelectric liquid crystal 4-(1-trifluoromethyl-pheptyloxycarbonyl) phenyl 4′-octyloxybiphenyl-4-carboxylate, *Phys. Rev. E*, 53, 6080, 1996.
11. Kimura, Y., Isono, H., and Hayakawa, R., Critical behavior of nonlinear permittivity in the smectic-A phase of chiral liquid crystals, *Phys. Rev. E*, 64, 060701(R), 2001.
12. Kimura, Y., Isono, H., and Hayakawa, R., Nonlinear dielectric response of antiferroelectric liquid crystals in the smectic C$_\alpha^*$ phase, *Eur. J. Phys. E*, 9, 3, 2002.
13. Glogarova, M., Dielectric and electro-optic properties of chiral smectics with a dipolar order, in *Modern Topics in Liquid Crystals*, Buka, A. Eds., World Scientific, Singapore, 1993, p. 271.
14. Skarabot, M., Blinc, R., and Musevic, I., Lifshitz point in the phase diagram of a ferroelectric liquid crystal in an external magnetic field, *Phys. Rev. E*, 61, 3961, 2000.
15. Skarabot, M., Blinc, R., Heppke, G., and Musevic, I., Electro-optic response of an antiferroelectric liquid crystal in submicron cells, *Liq. Crystl.*, 28, 607, 2001.
16. Kimura, Y., Kobayashi, N., and Hayakawa, R., Electrooptical response of ferroelectric liquid crystal in thick free-standing films, *Phys. Rev. E*, 64, 011705, 2001.

17. Blinov, L.M. and Chigrinov, V.G., *Electrooptic Effects in Liquid Crystal Materials*, Springer-Verlag, New York, 1994.
18. Handschy, M.A., Clark, N.A., and Lagerwall, S.T., Field-induced first-order orientation transitions in ferroelectric liquid crystals, *Phys. Rev. Lett.*, 51, 471, 1983.
19. Kubo, R., Toda, M., and Hashitsume, N., *Statistical Physics II*, Springer-Verlag, Berlin, 1985.
20. Nakada, O., Theory of non-linear responses, *J. Phys. Soc. Jpn.*, 15, 2280, 1960.
21. Hosokawa, K., Shimomura, T., Frusawa, H., Kimura, Y., Ito, K., and Hayakawa, R., Two-dimensional spectroscopy of electric birefringence relaxation in frequency domain: measurement method for second-order nonlinear after-effect function, *J. Chem. Phys.*, 110, 4101, 1999.
22. Chandani, A.D.L., Gorecha, E., Ouchi, Y., Takezoe, H., and Fukuda, A., Antiferroelectric chiral smectic phases responsible for the tristable switching in MHPOBC, *Jpn. J. Appl. Phys.*, 28, L1265, 1989.
23. Hiraoka, K., Takezoe, H., and Fukuda, A., *Ferroelectrics*, 147, 13, 1993.
24. Merino, S., de la Fuente, M.R., González, Y., Pérez Jubindo, M.A., Ros, B., and Puértolas, J.A., *Phys. Rev. E*, 54, 5169, 1996.

22 Electro-Optic Behavior of Fullerene-Containing Compounds

Natalia P. Yevlampieva

CONTENTS

22.1 INTRODUCTION

Fullerenes or cage carbon particles (molecules) are in the focus of scientific interest since their discovery in 1985 [1]. Each fullerenes' family member (C_{32}, C_{44}, C_{50}, C_{58}, C_{60}, C_{70}, C_{76}, C_{78}, and so on) represents a closed carbon network (cage) whose surface is formed by the fused hexagons and pentagons (Figure 22.1). Their three-dimensional shape of high symmetry is unrivaled among other natural molecular objects as well as their unique π-conjugated system. Accumulation of the experimental data on physical properties of fullerenes and fullerenes derivatives during the last decade has clarified very good perspectives of the application of these compounds for different purposes, for instance, as the photosensitive highly switching materials in photonics [2], as the basic compounds for the construction of novel elements for the devices in optoelectronics and telecommunications [3], and as the bioactive compounds in medicine [4].

The structural specificity of fullerenes elevates them above other allotropic forms of carbon—diamond and graphite—and separates them into a new class of spherically aromatic multiatom carbon molecules. In contrast with the nanosized diamond, graphite, or carbon soot,

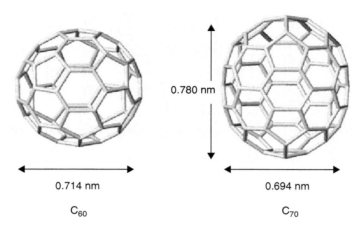

FIGURE 22.1 Schematic representation of fullerenes C_{60} and C_{70} with indications of their dimensions. (From Kroto, H.W. et al., *Nature*, 318, 162, 1985; Shinar, J., Vardeny, Z.V., and Kafafi, Z.H., Eds., *Optical and Electronic Properties of Fullerenes and Fullerene-Based Materials*, Marcel Dekker, New York, 1999, p.392; Hirsch, A., Brettreich, M., and Wudl, F., *Fullerenes: Chemistry and Reactions*, Wiley, New York, 2004, p. 435.)

which make up the dispersions only in organic solvents, fullerenes are limitedly soluble in a wide range of solvents. This is also a reason to consider them as the molecular form of carbon. The solubility of cage carbon particles in organic solvents provides a possibility to study their properties by various methods, including optical and electro-optical Kerr effect, in solutions.

Fullerenes C_{60} and C_{70} became available in bulk quantities earlier than the other family members [5]. Because of their real accessibility these fullerenes are the most commonly used particles for scientific investigations and practical applications. The present chapter is mainly devoted to properties of C_{60} and C_{60} derivatives with the covalent type of linkage in the solution.

One of the modern trends in the field of fullerenes' studies is connected with chemical modification of fullerenes for different applications, as the modification, more often than not, leads to improvement of physical properties of fullerenes useful for the practice [6]. First of all, the investigations of chemically modified fullerenes or fullerene-containing compounds have a priority status in current research in modern materials science. Besides the interest in fullerenes and fullerene-containing compounds, it is very important to know their solution behavior, that is closely connected with technologies of purification, separation of different fullerenes or isomeric fullerene adducts from each other, thin film formation from the solution, and other technologies, based on the solubility of substances. As it will be shown further in the chapter, a study of electro-optical properties of fullerene-containing compounds may be very useful for the understanding of its solution behavior.

This contribution demonstrates the capability of electro-optics for the decision of practical and fundamental, physical and chemical tasks in the study of fullerene-containing compounds as of the class of novel objects whose properties are not yet systematically known. Section 22.2 of this chapter is devoted to a brief review of properties of fullerenes themselves detected by optical Kerr effect and electro-optical Kerr effect in solutions, Section 22.3 describes electro-optical behavior of C_{60} adducts (low-molecular weight C_{60} derivatives), and Section 22.4 presents analysis of the behavior of polymeric C_{60} derivatives in electric field.

22.2 PROPERTIES OF FULLERENES C_{60} AND C_{70}

22.2.1 Optical Kerr Effect Data

The substances traditionally used as the nonlinear optic materials often consist of aromatic compounds or of the compounds with a high degree of conjugation in their chemical structure. Fullerenes, as the molecular objects with the highly conjugated π-electron system, have been, first of all, tested as the compounds potentially intended for the nonlinear optics [7–9].

Optical Kerr effect caused in solution by a strong electric field of light beam from a powerful laser is the method now widely used to estimate the nonlinear response of the compounds for photonic applications [10]. The investigations of optical Kerr effect in the solutions of C_{60} and C_{70} by means of nondegenerative four-wave mixing technique have shown that fullerenes possess ultrafast (sub-picosecond time regime) optical response time, but its nonlinear optic properties are not outstanding [7–9,11]. As a rule, nonlinear optic properties of the substance are characterized by a value of the second order hyperpolarizability (third order polarizability) γ or by the cubic susceptibility value $\chi^{(3)}$, which are connected with each other as $\gamma = \chi^{(3)}/(NL)$, where N is the molecule concentration in solution, and $L = [(n^2+2)/3]^4$ is the local field correction factor in the solvent with the refractive index n.

Table 22.1 presents some data on γ values of C_{60} and C_{70}, received experimentally in Refs. [7–9] in comparison with the substances normally used as the nonlinear optic materials [11]. The result of experimental γ determination has been confirmed by the quantum-chemical calculations of γ value for C_{60} and C_{70} in vacuum [12–14]. Calculated γ values were not in contradiction to the experimental data (see Table 22.1). Quantum-chemical analysis explained

TABLE 22.1
Second-Order Hyperpolarizability γ of Fullerenes C_{60}/C_{70} and Some Standard Nonlinear Optic Substances

S.No	Object	Method, Conditions	γ, ESU	Reference
1	C_{60}	FWM[a], benzene solution, femtosecond time regime	3.7×10^{-35}	7
2	C_{70}	SGH[b], toluene solution, sub-picosecond time regime	1.3×10^{-33}	8
3	C_{60}	*ab initio* quantum-chem. calculation	5.5×10^{-35}	12
4	C_{70}	*ab initio* quantum-chem. calculation	7.54×10^{-35}	12
5	C_{60}	AM-1 quantum-chem. calculation	2.7×10^{-35}	14
6	Si (film)	SGH[b]	$\sim 10^{-32}$	11
7	CS_2	FWM[a]	1.5×10^{-36}	7
8	Anthracene	AM-1 quantum-chem. calculation	1.7×10^{-35}	14
9	Naphthacene	AM-1 quantum-chem. calculation	4.5×10^{-35}	14
10	Liquid crystals	SGH, FWM	$\sim 10^{-27}$–10^{-32}	11

Notes: Polarization P of the substance in electric field is determined as $P = \alpha E + \beta E^2 + \gamma E^3 + \ldots$, where α, β, γ are the polarizabilities of the first, second, and third order, respectively. The mean γ value is determined as $\gamma = 1/5 \sum_{i,j} \gamma_{iijj}$, i,j, $\epsilon\{x, y, z\}$.
Data on α and β values for C_{60} and C_{70} one can find in [7,12,14].
[a]FWM—nondegenerative four-wave mixing technique; 2 dye lasers pumped with an excimer laser at $3080 \times 10^8 \, \text{cm}^{-1}$ were used in [7].
[b]SGH—electric field-induced second harmonic generation.

the relatively low hyperpolarizability value of fullerenes by the spherical aromaticity of these molecules (in contrast with planar aromatics, for instance, object 8 in Table 22.1) following from the high symmetry of their structure [14].

It should be mentioned that in spite of cage carbon particles' relatively low hyperpolarizability characteristics, they can be classified as the compounds with good nonlinear optic properties, because C_{60} and C_{70} possess excellent nonlinear laser light absorbtion characteristics [3,15]. This property allows the use of fullerene-containing materials in manufacturing of effective optical limiters and optical modulators for the photonic devices [2,3]. A discovery of fullerenes property to absorb light more strongly in the optically excited state than in the ground sate (reverse saturable absorption) was one of the unexpected features of spherically aromatic compounds. The most important modern applications of fullerenes such as the optical limiters and active medias in laser technologies were based mainly on their saturable absorption in the near infrared and visible light scale and on their ultrafast reaction on the external fields [2,3].

Our study of C_{60} by means of an ordinary electro-optical Kerr effect in solution [16] caused by the low power laser light source (He–Ne laser, 1–1.5 mW power) at the relatively small rectangular pulsed electric field strength (less than 10^4 V cm^{-1}) has shown that this method may help to understand fullerenes behavior in solutions. Other details of experimental equipment can be found in Refs. [17,18].

22.2.2 ELECTRO-OPTICAL BEHAVIOR OF C_{60} IN *N*-METHYLPYRROLIDONE AND *N*-METHYLMORPHOLINE

For fullerene C_{60} solvents, heterocyclic solvents *N*-methylpyrrolidone (MP) and *N*-methylmorpholine (MM) are good enough, as these solvent's ability to dissolve this fullerene is comparable with that of toluene and benzene, traditionally used for extracting C_{60} from the fullerene soot [19].

From the structural point of view, nonpolar and optically isotropic C_{60} molecules (due to a high symmetry its permanent dipole moment and optical anizotropy of polarizability Δa have a close to zero values according to quantum-chemical data [14]) with low optical hyperpolarizabily γ value could not give any significant input to the electro-optical effect of the solvent in our conditions (see the end of Section 22.2.1), but the experimental result has shown the difference of the effect value in pure solvents and in MP and MM solutions of C_{60}. The surprise became clear when attention was given to the dependence of electro-optical effect value on the time of storage of these solutions and on the initial concentration of solute. Figure 22.2 through Figure 22.5 show the results of monitoring of C_{60} solution properties in MP and MM during several weeks by the methods of electro-optical Kerr effect, viscometry, translation diffusion, and photoluminescence. All these methods detected that the properties of C_{60} solutions in MP and MM have slowly changed with the time.

On the whole, the performed study [17,18] has confirmed that C_{60} stimulates a slow developing aggregation process in organic solvent media, which was demonstrated earlier by Chu's group, who investigated C_{60} solution behavior in benzene by the light scattering technique [20]. It was possible to destroy the aggregates of C_{60} in benzene by simple shaking of the cell with the solution [20], but C_{60} aggregates in MP and MM were proved to be significantly stable, which we were able to conclude when the experiments were repeated after the shaking or the dissolving of solutions [17]. It may also be noticed that C_{60} solutions in MP and MM were colored in yellow-brown and continued to be transparent for the light beam during the month and more. A precipitation of the aggregates was not observed during the period of experiment described in Refs. [17,18].

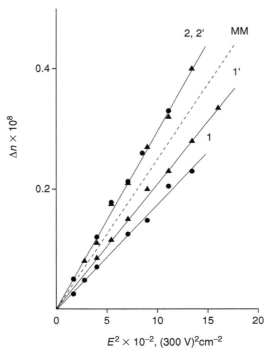

FIGURE 22.2 Dependence of the optical birefringence value Δn versus the squared electric field strength E^2 for C_{60} solution in methylmorpholine (MM) at the solute concentration $c = 7.2 \times 10^{-4}$ (1, 2) and 3.2×10^{-4} g cm^{-3} (1′, 2′), time of the storage of solution 5 days (1, 1′) and 21 days (2, 2′).

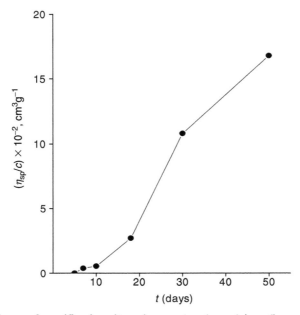

FIGURE 22.3 Dependence of specific viscosity values $\eta_{sp}/c = (\eta - \eta_o)/\eta_o c$ (here η and η_o are viscosity values of the solution and the solvent, respectively) [16] with the time t for C_{60} solution in methylpyrrolidone (MP) at the solute concentration $c = 3.3 \times 10^{-4}$ g cm^{-3}.

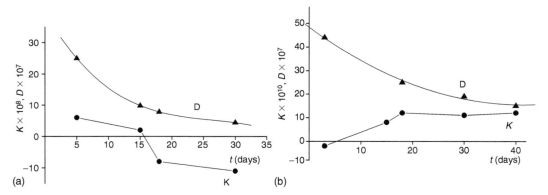

FIGURE 22.4 Change with the time of the translation diffusion coefficient $D \sim 1/R$ (where R is the hydrodynamic radius) [16] and Kerr constant value K ($K_{E \to 0} = (\Delta n - \Delta n_o)/E^2 c$) (in CGS units) for C_{60} solution in MP at the solute concentration $c = 3.3 \times 10^{-4}$ and $c = 2.18 \times 10^{-4}$ g cm^{-3} (diffusion) (a), and C_{60} solution in MM at the solute concentration $c = 3.0 \times 10^{-4}$ g cm^{-3} (b).

The above-mentioned type of electro-optical behavior of C_{60} in solutions is connected with the following reasons. Fullerenes due to electron deficiency, determining nature of their chemical activity, are strong π-electron acceptors. Acceptor type activity is realized in fullerenes' ability to form charge transfer complexes with any electron donors from the closest neighbors [6]. MP and MM as representatives of the secondary amines can be electron donors for C_{60}. With the passage of time an appearance of a strong photoluminescence of C_{60} solutions in MP and MM with the shift of the maximum of the spectra (see Figure 22.5) is the evidence that electronic structure of fullerene molecules change. The observed change in photoluminescence spectra, presented in Figure 22.5 is typical of C_{60} charge transfer complex formation [6]. It is widely known that the compounds with complex type of chemical linkage are able to aggregate in solution. The polarity and optical anizotropy of aggregates were the main reasons of the electro-optical effect, which have been observed in solutions of C_{60} in MP and MM [17,18].

22.3   ELECTRO-OPTICAL AND POLAR PROPERTIES OF THE LOW-MOLECULAR WEIGHT C_{60} DERIVATIVES

As a consequence of the above-described properties of C_{60} in solutions, it might be assumed that behavior of fullerene saturated by electrons, which will no longer be so active at the donor–acceptor level in the liquid media, will differ from C_{60} behavior. Our study of some C_{60} derivatives has confirmed this suggestion [21–24]. We did not observe any aggregation phenomena with time for the low- and high-molecular weight C_{60} derivatives with the covalently bonded fullerene in organic solvents. It should be noticed here that in general the solubility of C_{60} derivatives is much better than that of the initial fullerene.

The saturation of fullerenes may be achieved by their chemical modification. Modification of cage carbon molecules can only be performed by means of different addition reactions because of substitution reactions, inherent for the planar aromatics with hydrogen periphery, cannot be realized for fullerenes [6]. Fullerene C_{60} contains 30 C = C bonds of equal activity before the start of chemical reaction, but this situation changes when the first covalent attachment of the addend to the fullerene cage will be done. Under the interaction of the addend with the fullerene cage, C_{60} reconstructs itself through the alteration of the carbon-carbon bond length [25]. As a result, the chemical activity of the adduct will not be equal to

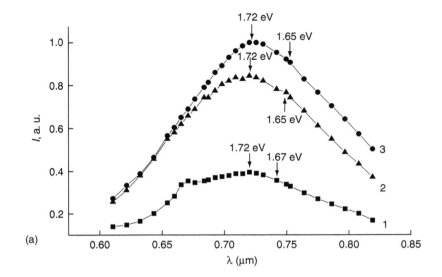

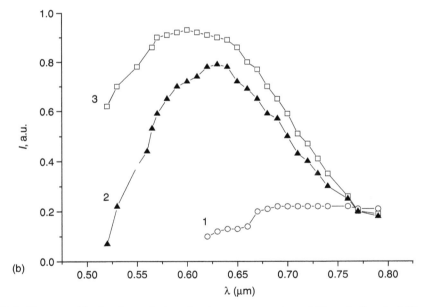

FIGURE 22.5 Change with the time of photoluminescence spectra of C_{60} solution in MP at concentration 0.1×10^{-2} g cm^{-3} (a) and 0.01×10^{-2} g cm^{-3}(b). The spectra have been measured after 2 (1), 15 (2), and 38 days after the preparation of solution (3).

that of the initial fullerene. In compliance with the rule of regioselectivite attachment to C_{60} [6], not more than six addends of the same chemical structure can be grafted to fullerene's sphere unless a special effort is undertaken to increase number of addends. The degree of donor–acceptor activity of fullerene derivative in solution directly connected with the degree of saturation of C_{60} in this compound. The stable mono-adducts of C_{60} were synthesized by the reaction of cycloaddition (Figure 22.6) [26].

22.3.1 MONO-ADDUCTS OF METHANOFULLERENES AND FULLEROPYRROLIDINES

The samples of methanofullerenes (MF) and fulleropyrrolidines (FP), whose structures are presented in Figure 22.6, have been investigated by electro-optical Kerr effect in the rectangular pulsed electric field in chloroform and toluene solutions; the permanent dipole moments of the samples have been received experimentally by Gutgenheim–Smith method [27] in toluene and calculated by means of density functional theory (DFT), which is ab initio quantum-chemical method. For details please see Ref. [21]. The synthesis of such compounds has been undertaken with the purpose to enforce the fullerene's bioactivity through the attachment of additional bioactive functional groups.

Molecular dipole moment values μ have been calculated from the following equation:

$$\mu^2/M = 27kT[(\varepsilon-\varepsilon_o)/c - (n^2-n_o^2)/c][4\pi N_A(\varepsilon_o^2+2)^2] \tag{22.1}$$

Here k is the Boltsmann constant, T is the absolute temperature, N_A is the Avogadro number, $(\varepsilon-\varepsilon_o)/c$ and $(n^2-n_o^2)/c$ are the increments of dielectric permittivity ε and the squared refractive index n, respectively, c is the solute concentration, M is the molecular weight of the solute. A subscript "o" corresponds to the solvent.

In Figure 22.7 are shown the concentration dependences of $(\varepsilon-\varepsilon_o)$ for MF and FP samples. The slopes of these dependencies have been used for the determination of the increments $(\varepsilon-\varepsilon_o)/c$. Values of $(n^2-n_o^2)/c$ and experimental dipole moments μ_{exp} of C_{60} derivatives (Equation 22.1) are presented in Table 22.2 together with the result of theoretical calculation of μ_{cal} for these compounds. The correlation between μ_{exp} and μ_{cal} is not bad taking into account that the structure of addends allows isomerization, when the theoretical values μ_{cal} in Table 22.2 correspond to their *trans*-configurations as for the most probable [21].

Electro-optical properties of MF and FP have been studied under the following conditions. A compensatory technique with the photoelectric registration of optical birefringence of solution under the treatment of the rectangular pulsed electric flield was applied. Impulse duration of 2 ms and voltage range 100–700 V were used in this series of measurements. A thin mica plate compensator with its own optical phase difference $0.01 \times 2\pi$ was used. A glass cell with the titanium electrodes of 2 cm in light path length and with the gap between electrodes of 0.05 cm was employed. He–Ne laser (1.5 mW power) with the wavelength of 632.8 nm was used as light source. Specific Kerr constant K (see Equation 22.2) as the characteristic of the equilibrium electro-optical properties each sample was experimentally determined.

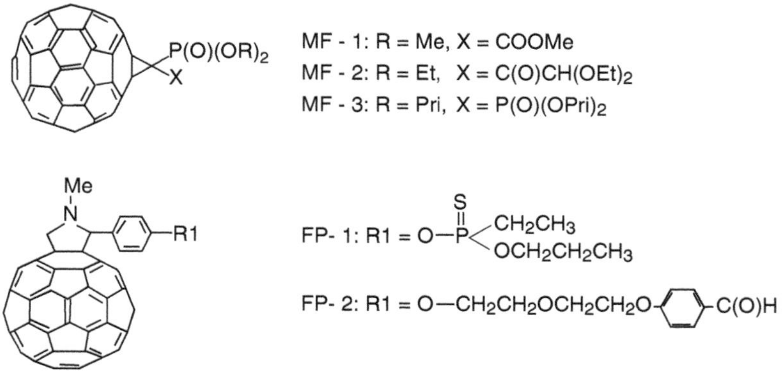

FIGURE 22.6 Chemical structure of mono-adducts of methanofullerenes (MF) and fullereopyrrolidines (FP). Here Me is methyl, Et is ethyl, and Pri corresponds to propyl-iso.

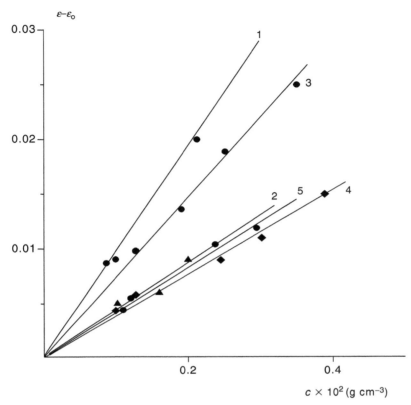

FIGURE 22.7 Concentration dependencies of the difference $(\varepsilon-\varepsilon_o)$ between the dielectric permittivity values of solution and solvent for the C_{60}-mono-adducts MF-1 (1), MF-3 (2), MF-2 (3), FP-1 (4), FP-2 (5) in toluene.

$$K = \lim_{\substack{c\to 0 \\ E\to 0}} (\Delta n - \Delta n_o)/E^2 c \qquad (22.2)$$

Here $(\Delta n - \Delta n_o)$ is the difference between the optical birefringence value of the solution with the solute concentration c and of the solvent, correspondingly, E is the electric field strength. Figure 22.8a and Figure 22.8b demonstrate both stages of extrapolation according to Equation 22.2.

Taking the theoretical values of dipole moments μ_{cal} and values of anizotropy of optical polarizability Δa for samples MF-1–3 and FP-1, 2, it was not difficult to theoretically estimate K values. Equation 22.3 was used for this purpose. Equation 22.3 follows from the theory [28] that is true for the specific Kerr constant of the spheroid in form particles with which fullerene derivatives under consideration are similar.

$$K = \frac{\pi N_A (n_o^2 + 2)^2 (\varepsilon_0 + 2)^2}{1215 k T n_o} [\frac{2\Delta a^2}{M} + \frac{\mu^2 \Delta a}{M k T}(3\cos^2\beta - 1)] \qquad (22.3)$$

Here β is the angle between the principal optical axis and the permanent dipole moment of the spheroid particle; the other symbols are same as explained above (see Equation 22.1).

K_{cal} were calculated according to Equation 22.3 with the proviso $\beta = 0$, that follows from quantum-chemical analysis of K_{cal} MF and FP compounds structure. The comparison of

TABLE 22.2

Molecular Mass M, Increment of the Squared Refractive Index Value $(n^2 - n_o^2)/c$, Experimental and Calculated Permanent Dipole Moment μ, Experimental and Calculated Values of Specific Kerr Constant of MF and FP Samples (Figure 22.6) in Toluene and Chloroform

Sample	M^a	$(n^2 - n_o^2)/c$, cm³ g⁻¹	$\Delta a \times 10^{24}$, cm³	$\mu_{(exp)}/\mu_{(calc)}$, D	$K^b \times 10^{10}$, cm⁵ g⁻¹ (300 V)⁻², Toluene (experimental)	$K \times 10^{10}$, cm⁵ g⁻¹ (300 V)⁻², Toluene (calculated)	$K \times 10^{10}$, cm⁵ g⁻¹ (300 V)⁻², Chloroform (experimental)	$K \times 10^{10}$, cm⁵ g⁻¹ (300 V)⁻², Chloroform (calculated)
MF-1	900	0.502	14.61	7.8/5.6	6.3 ± 0.3	4.3	12 ± 1	10.0
MF-2	1001	0.479	16.57	5.5/5.6	3.2 ± 0.3	2.2	6.8 ± 0.5	5.1
MF-3	1063	0.541	21.62	7.3/8.6	5.3 ± 0.4	4.8	10.9 ± 0.5	11.1
FP-1	1020	0.568	17.0	5.3/5.9	–	2.1	4.8 ± 0.4	5.9
FP-2	1061	0.563	17.0	5.4/ -	–	2.1	5.4 ± 0.4	5.8

a Molecular mass corresponds to the structural formula of the samples.

b Kerr constant for FP-1 and FP-2 samples in toluene has not been determined because of the difference between the electro-optical effect value in the solution and in the solvent was very small at the solute concentration suitable for the measurements. FP toluene solutions have been more coloured in comparison with MF.

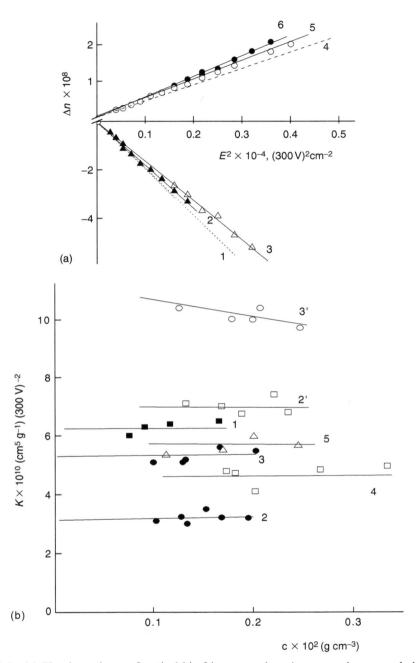

FIGURE 22.8 (a) The dependence of optical birefringence values Δn versus the squared electric field strength E^2 for sample FP-1 at the solute concentrations 0 (1), 0.202 (2), 0.334×10^{-2} g cm^{-3} (3) in chloroform, and for sample MF-3 in toluene at the solute concentrations 0 (4), 0.101 (5), 0.203×10^{-2} g cm^{-3} (6). (b) Concentration dependencies of specific Kerr constant value K for mono-adducts MF -1 (1), MF-2 (2), MF-3 (3) in toluene and MF-2 (2′), MF-3 (3′), FP-1 (4), FP-2 (5) in chloroform.

experimental K values and K_{cal} presented in Table 22.2 appears to have a good correlation between them.

Data on the polar and electro-optical properties of methanofullerenes show that sample MF-1 with the addend of the most asymmetric structure has the largest dipole moment value (Table 22.2); its specific Kerr constant has the largest value among the similar structure samples MF-2, 3. A substitution of the ethyl-group in MF-2 by the more asymmetric isopropyl-group in MF-3 leads to increasing of the adduct's polarity and elevates its K value. All experimental facts point toward the conclusion that the behavior of MF and FP in electric field has the orientational nature typical of the polar spheroid in form particles.

Thus, the results of the study of electro-optical behavior and the determination of the polarity of fullerene mono-adducts, synthesized by the cycloaddition methods, made it possible to conclude, that the transition from C_{60} to C_{60} derivative leads to transformation of nonpolar and optically isotropic particle (C_{60}) to another particle (adduct), anisotropic in properties and possessing a permanent dipole moment. Without doubt, this transformation is based on the destruction of the initial symmetry and electronic structure of C_{60} molecule under the attachment of the addend. Good correlation between the experimental and theoretical characteristics of the polarity and electro-optical properties of MF and FP should be considered as a confirmation of the fact that mono-adducts of C_{60} behave themselves in solution in electric field similar to the classical solid nanosized particles without any aggregation phenomena. This is a very important result for the further analysis of fullerene derivatives' properties.

22.3.2   Multi-Adducts of C_{60}—Buthylamine-Fullerene

Reaction of C_{60} with the primary amines is the easiest way to modify the fullerene because this type of reactions does not need any special conditions, sometimes they are capable to undergo without the catalysts at room temperature [6]. Photophysical properties of C_{60}-amino-adducts are better than those of fullerene [29]. Due to the listed reasons, the synthesis of amine-derivatives of C_{60} (C_{70}) and their properties investigation are the subject of special interest of modern materials science. A product of C_{60} reaction with any primary amine, as a rule, represents a mixture of adducts with different number of addends and separation of these adducts from each other is practically intractable [6].

Electro-optical Kerr effect has been used for the analysis of the content of buthylamine-derivative of C_{60}—sample (BA-C_{60})—(Figure 22.9) synthesized by the injection of fullerene into liquid buthylamine [22]. At the same experimental conditions that were described earlier in Section 2.1 a specific Kerr constant value $K = (1.4 \pm 0.2) \times 10^{-10}\,cm^5\,g^{-1}\,(300\,V)^{-2}$ and specific dielectric polarization $\mu^2/M = (1.6 \pm 0.4) \times 10^{-38}$(CGS units) in toluene were determined for BA-C_{60} sample. The experimental values $(\varepsilon - \varepsilon_0)/c = 2.19 \pm 0.06$ and $(n^2 - n_o^2)/c = 0.323 \pm 0.005$ have been received for BA-C_{60} and have been used then for the calculation of μ^2/M from Equation 22.1. Figure 22.10a and Figure 22.10b demonstrate some experimental dependencies.

Semiempirical quantum-chemical method PM-3 [30] was applied for the optimization of molecular geometry of C_{60} buthylamine-adducts with the varied number of addends from 1 to

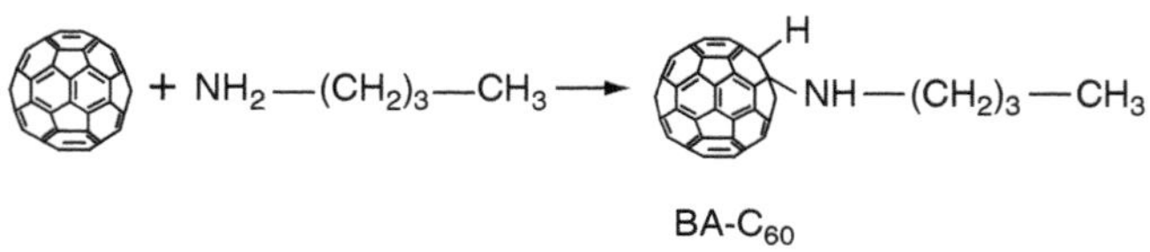

BA-C_{60}

FIGURE 22.9 Scheme of the synthesis of buthylamine-fullerene. The mono-adduct is shown.

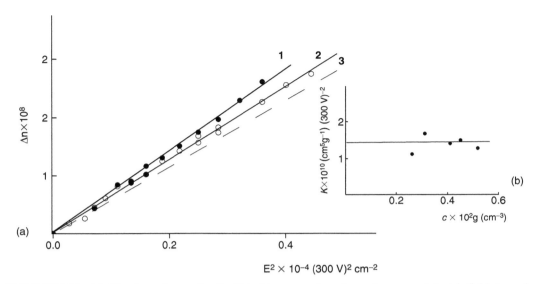

FIGURE 22.10 (a) The dependence of optical birefringence values Δn vs. the squared electric field strength E^2 for sample BA-C_{60} in toluene at the solute concentration 0.518 (1), 0.311×10^{-2} g cm^{-3} (2) and pristine toluene (3). (b) Concentration dependence of specific Kerr constant value K for the sample BA-C_{60} in toluene.

6 (Figure 22.11). Such molecular characteristics as the dipole moment μ or μ^2/M value, anizotropy of optical polarizability Δa, and molar Kerr constant value K_M were calculated for the fully optimized model adducts of C_{60} with different number of buthylamine-arms m,

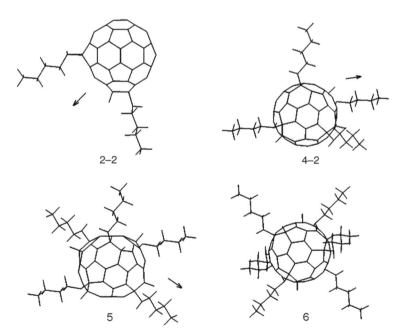

FIGURE 22.11 The optimized BA-C_{60} molecules with the addend number $m = 2, 4, 5, 6$ (semiempirical quantum-chemical method PM-3, MOPAC program). The designations of the adducts are the same as in Table 22.3 and Table 22.4. Dipole moments of the polar adducts are shown by the darts.

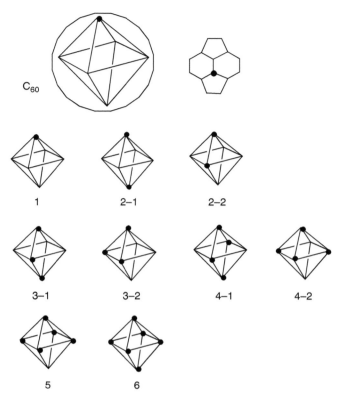

FIGURE 22.12 Scheme of the regioselectivity of addends' attachment to fullerene C_{60} according to [6]. The designation of adducts and isomers corresponds to the model molecules with the different number of addends presented in Table 22.3 and Table 22.4.

attached to fullerene sphere in accordance with the regioselective rule (Figure 22.12). Results are presented in Table 22.3 and Table 22.4. The following equations (Equation 22.4) were used for the calculation of K_M:

TABLE 22.3
Energy of Formation ΔH_f Calculated by Semiempirical Quantum-Chemical Method PM-3 for Model BA-C_{60} Molecules with Different Addend Number m (Figure 22.12)

m	ΔH_f, kcal mol^{-1}
0	811.7
1	763.8
2–1	716.1
2–2	715.6
3–1	667.7
3–2	667.1
4–1	619.7
4–2	618.9
5	570.5
6	521.8

TABLE 22.4

Molecular Mass M, Dipole Moment μ, and Anisotropy of Optical Polarizability Δa, Calculated by Quantum-Chemical Method PM-3, θ_1, θ_2 and K_M Values Calculated According to Equation 22.4, Specific Dielectric Polarization Values μ^2/M for the Model Molecules BA-C_{60} with Different Addend Number m

m	M	μ^a, D	$\Delta a^b \times 10^{24}$ cm^3	$\theta_1 \times 10^{34}$	$\theta_2 \times 10^{34}$	$K_M \times 10^{10}$ cm^5 (300 V)$^{-2}$	$K_M^{av} \times 10^{10}$ cm^5 (300 V)$^{-2}$	$(\mu^2/M) \times 10^{38}$ (CGS units)
1	793	2.44	10.06	1.108	13.69	55.96	55.96	0.75
2–1	866	0.0008	20.50	4.6	<0.001	17.40	46.67	~10^{-7}
2–2		3.37	10.36	1.175	13.01	53.99		1.31
3–1	939	2.32	18.15	3.60	0.67	16.15	20.35	0.57
3–2		4.00	1.63	0.029	7.019	26.66		1.70
4–1	1012	0.001	21.43	5.027	~0.0001	19.01	56.93	~10^{-7}
4–2		3.12	10.74	1.26	16.30	66.41		0.96
5	1085	2.12	11.06	1.34	12.44	52.11	52.11	0.39
6	1158	0.003	2.37	0.0615	~0.0001	2.33	2.33	~10^{-7}

Note: Designations 1, 2–1, 2–2, 3–1, 3–2, 4–1 and so on correspond to the scheme of regioselectivity of C_{60}, presented in Figure 22.12.

$^a \mu = (\Sigma \mu_i^2)^{1/2}$, $i = 1, 2, 3$.

$^b \Delta a = \{[(a_1-a_2)^2 + (a_2-a_3)^2 + (a_3-a_1)^2]/2\}^{1/2}$.

$^c K_M^{av}$ is the averaged K_M value.

$$K_M = 2\pi N_A (\theta_1 + \theta_2)$$

$$\theta_1 = (45kT)^{-1}[(a_1 - a_2)^2 + (a_2 - a_3)^2 + (a_3 - a_1)^2]$$

$$\theta_2 = (45k^2T^2)^{-1}[(\mu_1^2 - \mu_2^2)(a_1 - a_2) + (\mu_2^2 - \mu_3^2)(a_2 - a_3) + (\mu_3^2 - \mu_1^2)(a_3 - a_1)] \quad (22.4)$$

Here θ_1 and θ_2 are the "deformation" and "orientation" modes of the electro-optical effect in solution of anisotropic particles, μ_i are the projections of the dipole moment on the Cartesian coordinates, axis 1 of which is the primary axis of the optical and dielectric polarizability of the particle, a_i are the main components of optical polarizability tensor.

The linkage between the specific Kerr constant K and molar Kerr constant K_M is expressed by the following equation:

$$K = [(n_0^2 + 2)^2(\varepsilon_0 + 2)^2 K_M]/M \, 6n_0 \quad (22.5)$$

Data of Table 22.3 show that polar and nonpolar isomeric adducts, for instance, 2–1 (nonpolar), 2–2 (polar) or 3–1 (nonpolar), 3–2 (polar) and so on, have equal probability to be realized due to the equality of their energy of formation ΔH.

The calculated values of θ_1 and θ_2, presented in Table 22.4, indicate that for polar model isomers their electro-optical behavior will have a single-valid orientation nature. But the nonpolar isomers will also contribute any nonzero input to the common effect in solution of the mixed isomers due to the existence of a small deformational polarizability of these particles.

A comparison of the experimental value K and μ^2/M values of BA-C_{60} with the model adducts has shown that they are close to the calculated characteristics of most polar model adducts, but at the same time experimental characteristics cannot be correlated with any concrete model compound presented in Table 22.4. Thus, the experiment has demonstrated that sample BA-C_{60} must necessarily contain polar isomers. The specific dielectric polarization μ^2/M of the mixture of polar and nonpolar particles is not a simple physical value for analysis, but electro-optical properties give a good opportunity for analyzing the inhomogeneity of studied BA-C_{60} sample.

Keeping in mind the property of K_M to be an additive value in multicomponent solution and using Equation 22.6, we calculated K_M of mixture of adducts with a presumption that every isomeric adduct from Table 22.4 is present in the mixture in equal molar fraction amount.

$$K_M = \sum_{m=0}^{m=6} w(m) K_M^{av}(m) \qquad (22.6)$$

Here $w(m)$ is the molar fraction of model adduct, $K_M^{av}(m)$ is the mean value of molar Kerr constant for isomeric adducts (polar and nonpolar) with the same m.

Calculated according to Equation 22.6 $K_M = 34 \times 10^{-10}$ cm^5 (300 V)$^{-2}$ correlates with $K_M = 38 \times 10^{-10}$ or 37×10^{-10} cm^5 (300 V)$^{-2}$, received according to Equation 22.5 from the experimental specific Kerr constant $K = (1.4 \pm 0.2) \times 10^{-10}$ cm^5 g^{-1} (300 V)$^{-2}$ of BA-C_{60} at $M = 939$ or 963 g mol^{-1}, which are the number-averaged and weight-averaged M values of all possible isomers with m from 1 to 6. More details can be found in Ref. [22].

Thus, the analysis of electro-optical behavior of buthylamine-fullerene derivative in toluene solution has made it possible to conclude that a product of C_{60} reaction with the primary amine is the mixture of multi-adducts, which contains all possible regioisomers with the number of addends from 1 to 6 in equally probable quantities.

The equilibrium conditions of C_{60} reaction with the primary amine, which were realized for the synthesis of BA-C_{60}, could be shifted to the yield of adducts with any definite m value by the change of reaction conditions, as was shown in Ref. [22].

22.4 ELECTRO-OPTICAL BEHAVIOR OF FULLERENE-CONTAINING POLYMERS

The synthesis of polymeric C_{60} derivatives has been undertaken for different purposes. For instance, polymer may be used as fullerene-containing matrix for the manufacturing of film-forming material [31]. Other trends were connected with the use of high symmetry of C_{60} for the design of star-like polymers [6,32]. In spite of the difference in purposes of C_{60} modification with polymers there is one common phenomenon: a strong enough influence of covalently bonded fullerene on the parent polymer properties that cannot be useful in all the cases. A study of electro-optical behavior of fullerene-containing polymers was able to contribute to understanding of this phenomenon [23–24].

22.4.1 STAR-LIKE POLY-VINYLPYRROLIDONES WITH C_{60} AS THE CORE

At the beginning of this chapter a bioactivity of C_{60} has been pointed out, but this property of C_{60} is difficult to utilize in practice because cage carbon particles are practically insoluble in aqueous media [19]. With the aim to produce novel bioactive water-soluble fullerene containing objects, the water-soluble polymer widely used in medicine and pharmacology—polyvinylpyrrolidone (PVP)—was selected to modify C_{60}. Two star-like PVP-C_{60} samples were synthesized according to scheme presented in Figure 22.13 [23,33]. As one can see, these samples are the amino-derivatives of C_{60} because the attachment of PVP-branches to fullerene was realized through the amino-end group of polymer chains. From the conclusions on

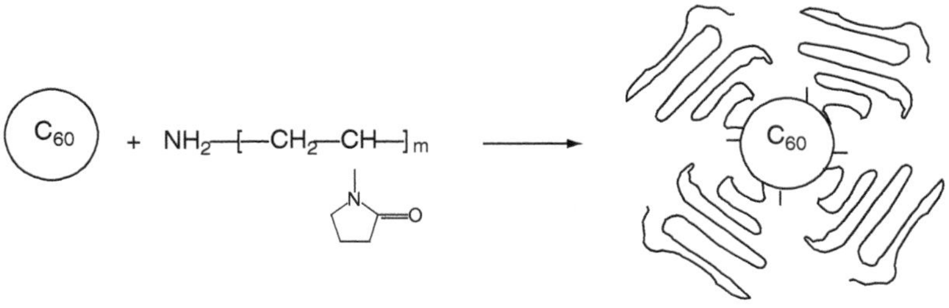

FIGURE 22.13 Scheme of the synthesis of the star-like polymeric derivatives PVP-C_{60}. The degree of polymerization of the PVP-branches was equal to 40 for sample PVP-C_{60}-1 and to 70 for sample PVP-C_{60}-2.

structure of amino-adducts of C_{60} made in the previous part, one may be sure that PVP-C_{60} samples have strong structural inhomogenity connected with the variation of the branch numbers of these polymeric stars.

PVP-C_{60} samples have been investigated by electro-optical Kerr-effect and hydrodynamic methods in chloroform solution in comparison with the linear fullerene free PVP samples of the same averaged molecular weight M. The main results of hydrodynamic study are presented in Table 22.5. Data of this table show that macromolecules PVP-C_{60} could be considered as more compact (see R_h values in Table 22.5) and as having a higher molecular density M/V if their molecular characteristics are compared with the same one of PVP samples.

Figure 22.14 demonstrates a great difference in electro-optical properties of PVP and PVP-C_{60} samples. Fullerene free PVP samples have negative sign of specific Kerr-constant K, whose value doesnot depend on the molecular weight of the sample and on the concentration of the solute. Such behavior is typical of flexible chain polymers [16]. Kerr constants of PVP-C_{60} have a positive sign and K have different values for the samples PVP-C_{60}-1 and PVP-C_{60}-2 (see Table 22.6).

With the goal to explain the transformation of electro-optical behavior of fullerene-containing PVP, a special synthesis of the model star-like compound A-C_{60}, analogous to the core of the star-like PVP-C_{60}, has been undertaken under the conditions similar to that used for the synthesis of polymeric star-like PVP-C_{60}. A chemical structure of the arm of A-C_{60} corresponds to $-NH-(CH_2)_2-S-(CH_2)_3-CH_3$. The electro-optical properties of sample

TABLE 22.5
Hydrodynamic Characteristics of the Star-Like PVP-C_{60}–1, 2 Samples and Their Fullerene Free Analogues PVP-1, 2 of Linear Structure in Chloroform Solutions at 21°C

Sample	$M^a \times 10^{-3} g$ mol^{-1}	$[\eta] cm^3 g^{-1}$	$R_\eta^b \times 10^8 cm$	R_{star}/R_{linear}	$V^c \times 10^{20} cm^3$	$(M/V) \times 10^{-23} g$ $mol^{-1} cm^{-3}$
PVP-1	10.0	12.0	26.7		8.0	1.3
PVP-C_{60}-1	9.7	7.5	22.6	0.847	4.8	2.0
PVP-2	38.0	21	50.2		53.0	0.7
PVP-C_{60}-2	35.7	15	44.0	0.877	35.7	1.0

[a]Molecular mass M of the samples was determined by translation diffusion and sedimentation methods [16].
[b]The hydrodynamic radius R_η was calculated as $(3[\eta] M/10\pi N_A)^{1/3}$, where $[\eta]$ is the intrinsic viscosity value.
[c]$V = 4 \pi R_\eta^3/3$ is the hydrodynamic volume, M/V is the molecular density.

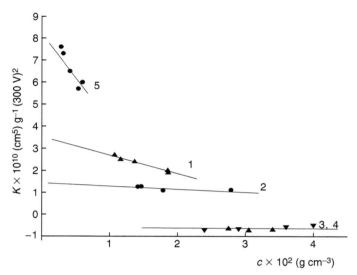

FIGURE 22.14 Concentration dependence of specific Kerr constant value K for star-like samples PVP-C_{60}-1 (1), PVP-C_{60}-2 (2), their linear structure analogues PVP-1 (3), PVP-2 (4), and model compound A-C_{60} (5) in chloroform.

A-C_{60} have been studied in chloroform solution also. Experimental specific Kerr constant value for the model compound A-C_{60} was equal to $K = +8 \times 10^{-10}\,cm^5\,g^{-1}\,(300\,V)^{-2}$, which value together with the results presented above in Section 22.3.1 and Section 2.3.2, was a sufficient evidence that the star-like PVP-C_{60} samples contain polar and optically anisotropic particle—fullerene adduct in the center of macromolecular coil, which may be responsible for a new electro-optical properties of fullerene-containing PVP if compared with the parent PVP.

Addressing again to the additivity of the molar Kerr constant K_M and presuming that the inputs of core and polymeric branches of PVP-C_{60} to the electro-optical effect in solution of the star-like fullerene derivatives under investigation are independent, it is easy to show that fullerene weight fraction in the content of PVP-C_{60} samples (see the remarks to Table 22.5) is sufficient to explain the Kerr-constant sign alteration in PVP-C_{60} solution. The transformed Equation 22.6 has been used for the quantitative estimations in this case.

$$K_M = [6n_o/(n_o^2 + 2)^2(\varepsilon_o + 2)^2]\,(K_1 M_1 w_1 + K_2 M_2 w_2) \tag{22.7}$$

TABLE 22.6
Specific Dielectric Polarization μ^2/M and Specific Kerr Constant K of the Linear PVP-1, 2, Star-Like PVP-C_{60}-1, 2 and Model Sample A-C_{60}

Sample	$(\mu^2/M) \times 10^{38}$ (CGS units), Chloroform-CCl$_4$ (1:1)	$K \times 10^{10}$, cm^5 g^{-1} (300V)$^{-2}$, Chloroform
PVP- 1	2.2	−0.8
PVP- 2	2.5	−0.8
PVP-C_{60}-1	5.2	+3.5
PVP-C_{60}-2	3.6	+1.5
A-C_{60}	5.9	+8.0

Here the subscriber index 1 is related to the model sample A-C_{60}, the subscriber index 2 is related to polymer, w_1 and w_2 are the molar fractions of fullerene and PVP in PVP-C_{60} sample, respectively.

So, using $w_1 = 0.0885$, $w_2 = 0.9115$ (that correspond to the weight fraction of fullerene 1.5% for PVP-C_{60}-1), $K_1 = +8 \times 10^{-10} \, cm^5 \, g^{-1}$ (300 V)$^{-2}$, $M_1 = 720$ (as for C_{60}), $K_2 = -0.8 \times 10^{-10} \, cm^5 \, g^{-1}$ (300 V)$^{-2}$, and $M_2 = 111$, because of a monomer unit of the flexible chain polymer determines its electro-optical properties in electric field, we can calculate for PVP-C_{60}-1 in chloroform solution $K_M = 0.0118 \times (8 \times 10^{-10} \times 720 \times 0.0885 \, ^{-} \, 0.8. \times 10^{-10} \times 111 \times 0.9115.) = 0.01118 \times 10^{-10} \times (509.76 - 80.94) = +4.794 \times 10^{-10}$. This K_M value will give the specific Kerr constant value for PVP-C_{60}-1 $K = 3.86 \times 10^{-10} \, cm^5 \, g^{-1}$ (300 V)$^{-2}$, which can be received from Equation 22.5 if we presume that for PVP-C_{60}-1 as for PVP the same monomer unit mainly determines its electro-optical properties. The latter value is close to experimental K value presented in Table 22.6.

Experimental determination of the specific dielectric polarization value μ^2/M for the samples under investigation also has shown that fullerene incorporation to the PVP by means of C_{60} reaction with the end amino-group of the polymer leads to significant increasing of this characteristic for PVP-C_{60} in comparison with PVP (see Table 22.6). In relation to dielectric properties we could not exclude the influence of fullerene on the additional polarization of polymer due to difference in response time of the fullerene core and polymeric branches on the electric field treatment.

As a conclusion it may be noted that the presence of covalently bonded fullerene is the reason of the observed change of electro-optical and dielectric properties of the star-like polymers, the deviations of their properties from that of the parent PVP will depend on fullerene weight fraction in PVP-C_{60} samples. Data on PVP-C_{60}-1 characteristics clearly confirm the latter: the deviations from the parent PVP properties have been more significant for this sample in which content fullerene weight fraction was larger than in PVP-C_{60}-2.

22.4.2 FULLERENE-CONTAINING DENDRIMERS

The previous Section 22.4.1 illustrated that the presence of fullerene increases the absolute values of dielectric polarizability and electro-optical Kerr constants of the flexible chain polymer when incorporated through a covalent linkage at center of polymer coil. Here are presented data on electro-optical properties of the strongly polar dendritic macromolecules with different fullerene position in their structure (Figure 22.15). These fullerene-containing compounds are liquid crystals with high photoconductivity characteristics potentially designed as the materials for application in solar batteries [24,34]. Data of Table 22.7 allow us to conclude that the influence of fullerene on the polarity and electro-optical Kerr constant K of these samples is not as drastic as in the previous case of PVP, but still visible. These data show that fullerene position in the structure of macromolecule and the type of fullerene linkage (here the cycloaddition reaction was used which does not lead to strongly polar fullerene derivative) to the basic compound may be important for the final result of polymer modification with C_{60}.

22.5 CONCLUSION

Optical and electro-optical Kerr effect methods contributed significantly to the general study of fullerenes and their derivatives. The results on electro-optical behavior of C_{60} adducts and C_{60} containing polymers, received by electro-optical Kerr effect technique in solutions, have demonstrated that this method is especially fruitful for the study of the molecular properties of fullerene derivatives when fullerene state inside these compounds may be considered as

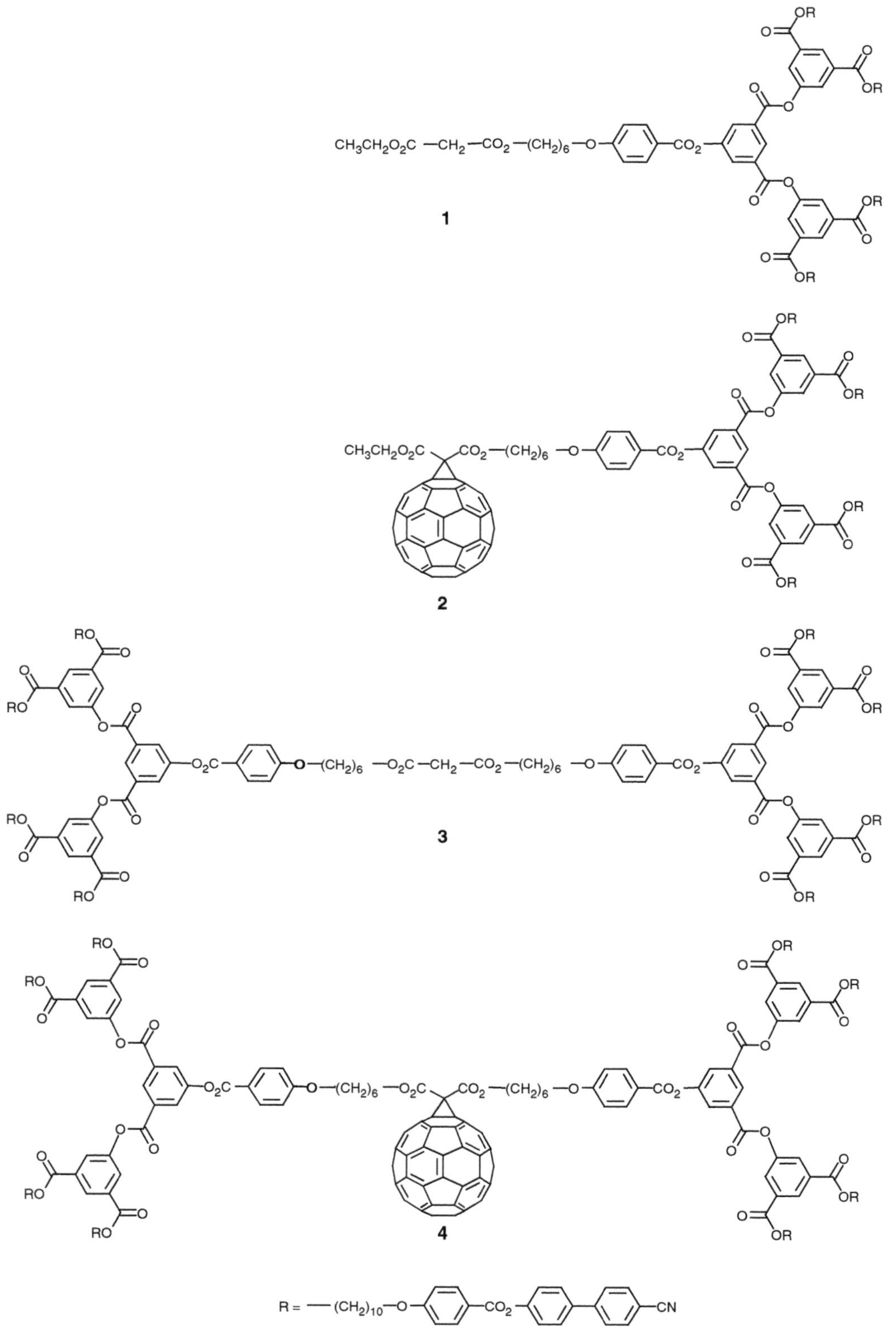

FIGURE 22.15 Chemical structure of fullerene-containing liquid crystalline dendrimers and its fullerene free analogues. The designations of the samples correspond to the same in Table 22.7.

TABLE 22.7
Dielectric Pirmittivity Increment $(\varepsilon - \varepsilon_o)/c$, Squared Refractive Index Increment $(n^2 - n_o^2)/c$, dipole moment value μ, specific dielectric polarization μ^2/M and Specific Kerr Constants K of Dendritic Samples 1, 3, Its Fullerene Containing Analogues 2, 4 in Benzene Solutions

Sample	M^a	$(\varepsilon - \varepsilon_o)/c$	$(n^2 - n_o^2)/c$	μ, D	$(\mu^2/M) \times 10^{38}$, CGS units	$K \times 10^{10}$ cm^5 g^{-1} × (300 V)$^{-2}$
1	2579	9.53	0.10	13.9	7.5	4.9
2	3299	11.38	0.16	17.1	8.9	6.3
3	5158	9.02	0.13	19.1	7.0	5.5
4	5878	7.61	0.15	18.6	5.9	7.9

^{a}Molecular mass M calculated according to structural formula.

equivalent to solid polar particle. Such a state of C_{60} cannot be realized for every fullerene derivative, because it depends on the type of fullerene chemical linkage, number of addends, and symmetry of their distribution over fullerene surface, an existence of a donor-acceptor interaction between fullerene and addends. For instance, a highly symmetrical 6-arm C_{60} adduct may be nonpolar and very weakly optically anizotropic (please, see the last line in Table 22.4). A study of homogeneous 6-arm star-like C_{60} derivatives without structural defects may be a very interesting task for electro-optics.

Electro-optical study has shown that the concrete fullerene's state in the fullerene-containing compound should be considered for the analysis of its electro-optical properties. From this fact the common conclusions follow that the type of addends—carbon cage surface interaction—finely determines physical properties of fullerene derivative, and that the derivative's physical properties in every case may be significantly different in relation to the initial fullerene and to the initial addend.

In the present chapter the electro-optical properties of the compounds only with covalent type of C_{60} linkage have been discussed. This selection states on the fact that chemical structure of these compounds is well defined and can always be confirmed by the spectral methods. But many fullerene derivatives with another type of linkage between C_{60} and C_{70}, and a base compound are known [6]. Among them are the complexes of C_{60} and C_{70} with the polymeric or low molecular mass donors of electrons. The increase of the second-order hyperpolarizability of these derivatives nearing some orders of magnitude with respect to C_{60} has been established by optical Kerr effect [35]. A study of electro-optical behavior of the complex-type polymer-C_{60} and C_{70} derivatives, for instance, could also bring interesting results.

REFERENCES

1. Kroto, H.W. et al., A new form of elemental carbon—buckminsterfullerene—C_{60}, *Nature*, 318, 162, 1985.
2. Schon, J.H. et al., A superconducting field-effect switch, *Science*, 288, 656, 2000.
3. Shinar, J., Vardeny, Z.V., and Kafafi, Z.H., Eds., *Optical and Electronic Properties of Fullerenes and Fullerene-Based Materials*, Marcel Dekker, New York, 1999, p. 392.
4. Bosi, S., Da Ros, T., and Prato, M., Fullerene derivatives: an attractive tool for biomedical applications, *Eur. J. Med. Chem.*, 38, 913, 2003.
5. Murayama, H. et al., Fullerene production in tons and more: from science to industry. *Fullerenes Nanotubes Carbon Nanostruct.*, 12, 1, 2004.

6. Hirsch, A., Brettreich, M., and Wudl, F., *Fullerenes: Chemistry and Reactions*, Wiley, New York, 2004, p. 435.

7. Geng, L. and Wright, J.C., Measurement of the resonant third-order nonlinear susceptibility of C_{60} by nondegenerate four-wave mixing, *Chem. Phys. Lett.*, 249, 105, 1996.

8. Weng, Y. and Cheng, L.T., Nonlinear optical properties of fullerenes and charge-transfer complexes of fullerenes, *J. Phys. Chem.*, 96, 1530, 1992.

9. Kaizar, F. et al., Third order nonlinear properties of fullerenes, *Proc. SPIE*, 2284, 58, 1994.

10. Ahmanov, S.A. and Nikitin, S. Yu., *Physical Optics*, Moscow Univ. Press., Moscow, 1998, p. 656.

11. Kuyzyk, M.G. and Dirk, C.W., Eds., *Characterization Techniques and Tabulations for Organic Nonlinear Optical Materials*. Marcell Dekker, New York, 1998.

12. Jonsson, D. et al., Electric and magnetic properties of fullerenes, *J. Chem. Phys.*, 109, 572, 1998.

13. Luo, Yi., On the scaling law of the second static hyperpolarizabilities of fullerenes, *Chem. Phys. Lett.*, 289, 350, 1998.

14. Lin, Y.-T. and Lee, Sh.-L., Structural effects on the low cubic hyperpolarizability of C_{60}: a scaling of conjugation in three-dimensional curvature of π-conjugation systems, *Int. J. Quantum Chem.*, 75, 457, 1999.

15. Kost, A. et al., Optical limiting with C_{60} in poly(methyl methacrylate), *Opt. Lett.*, 18, 334, 1993.

16. Tsvetkov, V.N., *Rigid-Chain Polymer Molecules*, Plenum, Consultants Bureau, New York, 1989.

17. Yevlampieva, N.P. et al., Aggregation of fullerene C_{60} in *N*-methylpyrrolidone, *Colloids Surf. A*, 209, 167, 2002.

18. Yevlampieva, N.P. et al., Aggregation of fullerene C_{60} in *N*-methylpyrrolidone and *N*-methylmorpholine, *Russ. J. Phys. Chem.*, 74, 1314, 2000.

19. Ruoff, R.S. et al., Solubility of C_{60} in a variety of solvents, *J. Phys. Chem.*, 97, 3379, 1993.

20. Ying, Q., Marecek, J., and Chu, B., Slow aggregation of buckminsterfullerene (C_{60}) in benzene solution, *Chem. Phys. Lett.*, 219, 214, 1994.

21. Yevlampieva, N.P. et al., The polarity characteristics and structure of phosphorylated methanofullerenes and fulleropyrrolidines, *Russ. J. Phys. Chem.*, 79, 1314, 2005.

22. Yevlampieva, N.P. et al., The synthesis, polarity and electrooptical properties of amine-derivative of fullerene C_{60}, *Russ. J. Gen. Chem.*, 75, 795, 2005.

23. Yevlampieva, N.P. et al., Star-like fullerene containing poly(vinylpyrrolydone) derivatives: chloroform solution properties. *Fullerenes Nanotubes Carbon Nanostruct.*, 12, 353, 2004.

24. Yevlampieva, N.P. et al., Electrooptical properties of liquid-crystalline fullerene containing dendrimers in solutions, *Chem. Phys. Lett.*, 382, 32, 2003.

25. Zverev, V.V. and Nuretdinov, I.A., Analysis of the structure of methanofullerene $C_{61}H_2$ by quantum-chemical methods, *Russ. J. Phys. Chem.*, 76, 1228, 2002.

26. Nurettdinov, I.A. et al., The synthesis of phosphorilated methanofullerenes, *Russ. Chem. Bull.*, 12, 2083, 2000.

27. Oehme, F., *Dielectrische Messmetoden zur quatitativen Analyse und fur chemische Structurbestimmungen*. Chemie, Weinheim, 1962.

28. O'Konsky, Ch.T., *Molecular Electro-optics*, Marcel Dekker, New York, 1978.

29. Janaki, J. et al., Thermal stability of fullerene-amine adduct, *Thermochimica Acta*, 356, 109, 2000.

30. Schmidt, M.W. et al., General atomic and molecular electronic structure systems, *J. Comput. Chem.*, 14, 1347, 1993.

31. Chen, Yu. et al., Polymeric modification and functionalization of [60]fullerene, *Eur. Polym. J.*, 34, 137, 1998.

32. Mathis, C., Audouin, F., and Nuffer, R., Controling the number of arms of polymer stars with a fullerene C_{60} core. *Fullerenes Nanotubes Carbon Nanostruct.*, 12, 341, 2004.

33. Yevlampieva, N.P. et al., Electrooptical and molecular properties of star-shaped fullerene-containing derivatives of poly(*N*-vinylpyrrolidone) in solutions, *Polym. Sci., Ser. A*, 46, 822, 2004.

34. Dardel, B., Guillon, D., and Deschenaux, R., Fullerene containing liquid crystalline dendrimers, *J. Mater. Chem.*, 11, 2814, 2001.

35. Koudoumas, E. et al., Large enhancement of nonlinear optical response of reduced fullerene derivatives, *Chem. Eur. J.*, 9, 1529, 2003.

23 Drying Dissipative Structures of Colloidal Dispersions

Tsuneo Okubo

CONTENTS

23.1 INTRODUCTION

Most structural patterns in nature and experiments in the laboratory form via self-organization, accompanied with the dissipation of free energy and in the nonequilibrium state. Among several factors in the free energy dissipation, evaporation and convection induced by the earth's gravity are very important. Several papers on the pattern formation in the course of drying of the monodispersed colloidal suspensions have been reported so far [1–16]. Most of the papers have studied the liquid-like suspensions in the particle distribution. Electrostatic interparticle interactions have been pointed out as one of the important factors for the dissipative structures. Hydrophobic and hydrophilic interactions are also demonstrated to be important for the drying process [14–16]. Gelbart and coworkers [4,5,7] examined the mechanism of solvent dewetting in annular ring structures formed by drying a diluted metal colloid on a substrate. Shimomura and Sawadaishi have studied intensively the dissipative patterns in the processes of film formation by drying the polymer solutions [17].

In previous papers from our laboratory [18,19], drying dissipative patterns on a cover glass have been observed for colloidal crystal suspensions of colloidal silica and monodispersed polystyrene spheres, which are hydrophilic and hydrophobic in their surfaces. The colloidal crystal is undoubtedly one of the most simple and convenient systems for the study of dissipative structures on the laboratory scale [20–24]. For example, accurate structural information on the processes of dissipative pattern formation is available for colloidal crystal suspensions by use of reflection spectroscopy in real time. Quite similar macroscopic and microscopic dissipative structural patterns formed between the colloidal silica and polystyrene spheres. The broad ring patterns of the hill accumulated with the spheres and the spoke-like and ring-like cracks were formed in the macroscopic scale. From these observations, the existence of the small circle convection cells proposed by Terada et al. [25–27] was supported. The primitive patterns of valleys were formed already in the concentrated suspensions before dryness and they grow toward fine cracks in the course of solidification. Branch-like fractal patterns of the sphere association were observed in the microscopic scale. Capillary forces between spheres at the air–liquid interface and the different rates of convection flows of water and spheres at the drying front were important for these pattern formations. Drying dissipative structures have also been studied for a series of colloidal silica spheres ranging from 29 nm to 1 mm in diameter [28].

Macroscopic and microscopic structural patterns were studied in the course of drying the suspension of Chinese black ink on a cover glass and in a dish [29]. The clear broad ring and spoke-like patterns of the rims accumulated with particles were formed especially in the central region of the pattern. The convection of water and colloidal particles in the different rates under gravity and the translational and rotational Brownian movement of the particles were important for the macroscopic pattern formation. Microscopic patterns were influenced strongly by the translational Brownian diffusion of the particles and the electrostatic and between the particles and the substrate in the course of solidification of the particles. For Chinese ink direct observation of the convection flow was made mainly in a glass dish. The drying patterns of the aqueous suspensions of monodispersed bentonite particles were investigated in detail [30].

The drying dissipative structures have been studied for the linear-type macrocations, poly (allylamine hydrochloride) (PAL) [31]. Macroscopic broad ring patterns, in which the polymers accumulate densely on the outside edge, were formed. Furthermore, beautiful string-like fractal patterns were observed in the microscopic scale.

Drying experiments were made for n-dodecyltrimethylammonium chloride (DTAC), which is one of the typical cationic detergent molecules [32]. Broad ring patterns of the hill accumulated with detergent molecules were formed around the outside edges of the film in the macroscopic scale. Star-like, blanch-like, arc-like, and small block-like microstructures were also observed. The convection of water and detergents at different rates under gravity and the translational and rotational Brownian movement of the latter were important for macroscopic pattern formation. Microscopic patterns were determined by the translational Brownian diffusion of the detergent molecules and the electrostatic and the hydrophobic interactions between detergents and between the detergents and substrate in the course of the solidification. Furthermore, drying structural patterns were studied for a series of anionic detergent molecules, sodium n-alkyl sulfates (n-alkyl=n-hexyl, n-octyl, n-decyl, n-dodecyl, n-hexadecyl, and n-octadecyl) [33].

From these studies on drying dissipative structures, macroscopic broad ring patterns for various solutions and suspensions were surprisingly found to be similar to each other. Interestingly, microscopic patterns such as branch-like, string-like, arc-like, and small block-like ones were reflected in the shape, size, and flexibility of the solute molecules. In this chapter, dissipative structures of many kinds of suspensions and solutions including colloidal dispersions are discussed and reviewed as systematically as possible.

It should be mentioned here that study on the drying dissipative structures of the colloidal crystal suspensions in an electric field is highly exciting since the dried film should show the characteristic electro-optical properties. We have already reported the preliminary results of the drying dissipative structures in an electric field in the tenth international conference on the electro-optics (ELOPTO-2003) in New Orleans. In near future we would like to investigate the electro-optics of the dried films, which are prepared by the drying dissipative structural technique.

23.2 EXPERIMENTAL

Colloidal silica spheres, CS22p, CS45, CS82, CS161, CS301, and CS1001 were kindly donated from Catalyst and Chemicals Company (Tokyo). Monodisperse polystyrene spheres, D1W52 were manufactured by Dow Chemical (Indianapolis, IN). The diameter (d_o), standard deviation (δ) from the mean diameter, and the polydispersity index (δ/d_o) were compiled in Table 23.1. These colloidal samples were first purified several times using an ultrafiltration cell (Model 202, Diaflo-XM300 membrane, Amicon Company), and then deionized by coexistence with a mixed bed of cation- and anion-exchange resins for more than five years before use. The colloidal samples released an amount of alkali from their surfaces. It takes a long time before complete deionization is achieved, as the deionization proceeds between the solid and the liquid phases one after another, i.e., first between colloidal and water and then between water and the resins. The values of d_o and δ were determined by the courtesy of Nippon Synthetic Rubber Company (Tokyo) in part and by using electron microscopy by the authors. The surface densities of strongly acidic charges of these spheres (Table 23.1) were determined by the conductometric titration with a Horiba Model DS-14 conductivity meter (Kyoto).

Monodisperse bentonite particles, code 2p7f, were obtained by the fractional repeated centrifugation technique. The details of the sample preparation were described in the previous papers [34,35]. The diagonal diameter of 2p7f, measured using electron microscopy, was 324 nm. These colloidal samples were deionized with a mixed bed of cation- and anion-exchange resins for more than two years. The deionized suspension thus obtained was light blue and transparent, which clearly supports the formation of the liquid-like distribution of the particles by the extended electrical double layers. It should be mentioned here that the crystal-like distribution of the particles was not formed since the single crystals were not observed for the suspension.

Two kinds of stock suspensions of Chinese black ink, Fueki FV3612 (Fueki Paste Company, Tokyo, Japan) and Kaimei SE-1702 (Kaimei Company, Urawa, Japan), were purchased. These colloidal particles were highly polydispersed and their mean diameters were 33.0 ± 6.4 and 32.8 ± 9.2 nm, respectively, which were determined from transmission electron microscope measurements (Hitachi type H8100, Tokyo, Japan). The stock suspensions were deionized with a mixed bed of ion-exchange resins for more than six months before preparing the sample suspensions.

PAL was a gift from Nittobo (Tokyo). The degree of polymerization was about 1000. The macrocation sample was purified further by the dialysis using a Visking tube against water, which was purified by a Milli-Q plus (Millipore Corporation, Bedford, MA).

TABLE 23.1
Characteristics of Colloidal Silica and Polystyrene Spheres Used

Sphere	d_o (nm)	δ (nm)	δ/d_o	Charge Density ($\mu C/cm^2$)
CS22p	29	6	0.21	0.48
CS45	56.3	7.6	0.13	0.30
CS82	103	13.2	0.13	0.38
CS161	183	18.6	0.10	0.47
CS301	311	22.4	0.072	0.40
CS1001	1090	45	0.041	—
D1W52	88	6.2	0.07	1.14

The DTAC for ion pair chromatography was purchased from Tokyo Kasei Industries (Tokyo, Japan) and was used without further purification. Sodium *n*-alkyl sulfates (*n*-alkyl=*n*-hexyl (S6S), *n*-octyl (S8S), *n*-decyl (S10S), *n*-dodecyl (S12S), *n*-hexadecyl (S16S), and *n*-octadecyl (S18S)) were purchased from Aldrich Chemical Company (Milwaukee, WI) and were used without further purification.

Potassium chloride (99.9%), calcium chloride dihydrate (99.9%), and lanthanum chloride heptahydrate (99.9%) were obtained from Wako Pure Chemical Industries (Tokyo, Japan). These simple electrolytes were used without further purification. Ethanol (99.5%), used for rapid scanning liquid chromatography, was also obtained from Wako Pure Chemical Industries.

The water used for the sample preparation was purified by a Milli-Q reagent grade system (Milli-RO5 plus and Milli-Q plus, Millipore Corporation, Bedford, MA).

The aqueous suspension or solution of 0.05 to 0.2-ml volume was dropped carefully and gently on a micro cover glass (30 × 30 mm, thickness no. 1, 0.12 to 0.17 mm, Matsunami Glass Company, Kishiwada, Osaka) in a glass dish (60 mm in diameter, 15 mm in depth, Petri Company, Tokyo). The cover glass was used without further rinsing in most of the cases. The extrapolated value of the contact angle for pure water was $31 \pm 0.2°$ from the drop profile of a small amount of water (0.2, 0.4, 0.6, and 0.8 μl) on the cover glass. A pipette (1 ml, disposable serological pipette, Corning Lab. Sci. Co.) was used for the dropping. The macroscopic and microscopic observations were made for the film formed after the solution or suspension was dried up completely on a cover glass in a room air-conditioned at 25°C and 65% humidity.

To observe the convection flow in the suspension of Chinese black ink directly, 0.5 to 3.0 ml of the suspensions was set in a dish (30 mm in diameter and 15 mm in height) on a Shamal hotplate (HHP-401, Iuchi Seiei Dou Company, Tokyo, Japan). The suspension temperature was raised and an equilibrium temperature was reached within 10 min at 36, 48, 58, 72, and 80°C when the temperatures of the plate surface were set at 50, 75, 100, 125, and 150°C, respectively.

Macroscopic dissipative structures were observed with a digital HD microscope (type VH-7000, Keyence Company, Osaka) and a Canon EOS 10 camera with macrolens (EF 50 mm, aperture, $f = 2.5$) and a life-size converter EF. Microscopic structures were observed with a laser 3D profile microscope (type VK-8500, Keyence) and a Metallurgical microscope (Axio-vert 25CA, Carl-Zeiss, Jena GmbH). The observations of the microscopic patterns were also made with an atomic force microscope (type SPA400, Seiko Instruments, Tokyo) and with a transmission electron microscope (Hitachi, H8100).

23.3   RESULTS AND DISCUSSION

23.3.1   Macroscopic Drying Patterns of Colloidal Crystal Dispersions

Figure 23.1 shows the typical pattern formed in drying the colloidal crystal of CS82 spheres at the sphere concentration of 0.0333 in volume fraction [18]. The spoke-like cracks were formed from the outside edge toward the center. Clearly the cracks were formed in the process of shrinking of the wetted films [36]. There were quite a few spheres in the central region, whereas a broad ring-like region was occupied with the spheres. The spoke-like cracks and broad ring patterns were also observed for the deionized aqueous suspensions (colloidal crystal) of monodisperse polystyrene spheres [19].

The angles between the adjacent spoke-like cracks (θ) increased as sphere concentration increased. The θ-values correlate deeply with the convection flow of water and spheres as described below, and further with the rigidity of the dried film separating from the cover glass. Interestingly, the cracks were introduced in the final stage of drying the colloidal

FIGURE 23.1 Patterns formed in the drying process of CS82 spheres on a cover glass at 25°C (in water, 0.1 ml, $\phi = 0.0333$, 1-mm bar length).

suspensions along the outer edges first, where the dryness proceeds in advance. Then the cracks turned and developed straight toward the center of the film. It is impressive that the new cracks developed successively keeping the same angle with the adjacent old cracks. The θ-value should increase as the rigidity of the wetted film increases.

A main cause for the broad ring formation is due to the convection flow of the solvent and the colloidal spheres. Especially important is the flow of the colloidal spheres from the center toward the outside edges in the lower layer of the liquid drop, which was observed directly from the movement of the very rare aggregates of the spheres [18]. The flow is enhanced by the evaporation of water at the liquid surface, resulting in the lowering of the suspension temperature in the upper region. When the spheres reach the edges of the drying frontier at the outside region of the liquid, a part of the spheres will turn upward and go back to the center. However, the movement of most of the spheres may stop at the frontier region because of the disappearance of water. This process must be followed by the broad ring-like accumulation of the spheres near the round edges. It should be noted here that the importance of the convection flow of colloidal spheres in the ring formation has often been reported in the process of film formation [9,36]. Figure 23.2 shows the side view of the film. The left-hand side of the figure shows the center of the film. The thickness of the dried film of the silica spheres is shown as a function of the distance from the center, r in Figure 23.3 [18]. The experiments were made directly by taking a close-up CCD picture of the film section, which was stripped off from the cover glass and the section of the cracks appeared often. Clearly, the thick ring pattern appeared at the outside edge.

At low sphere concentrations, several fine circles and islands of sphere regions distributed around the outer regions, though the graph showing this was omitted in this chapter. The area (S) of the dried film was very small compared with that of the initial state of the drop when the sphere concentration is lower than the critical sphere concentrations of 0.02 (in volume

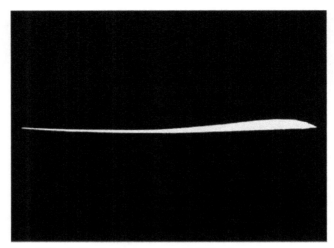

FIGURE 23.2 Side view of film in the drying process of CS82 spheres on a cover glass at 25°C (in water, 0.1 ml, $\phi = 0.0333$, 1-mm bar length).

fraction) for the silica and polystyrene spheres. The S-values agreed when the sample liquid with sphere concentrations higher than 0.02 (in volume fraction) was set on a cover glass. The influence of salt on the macroscopic patterns was studied (Table 23.2). The shape of the outer edge changed greatly from smooth to rough with the addition of sodium chloride for CS82 and D1W52 spheres. The reason for this observation is not clear yet. A slight increase in the air-suspension interface tension by the addition of sodium chloride will be one of the main reasons. The time for completion of the drying observed with the naked eye, T, increased significantly with the addition of NaCl. It should be noted that the film at time T is still wet and not dried completely. Interestingly, the area decreased in the presence of sodium chloride. This is explained by the increase of surface tension with sodium chloride. The θ-values

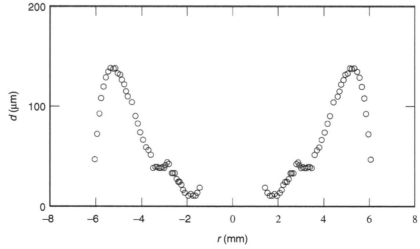

FIGURE 23.3 Thickness of the film formed for CS82 spheres as a function of the distance from the center at 25°C (in water, 0.2 ml).

TABLE 23.2
Influence of Salt on T, S, θ, and d_{max}. $\phi = 0.0221$, 0.2 ml

[NaCl]	T	S	θ	d_{max}
(M)	(min)	(mm^2)	(degree)	(μm)
0	216	183	2.3	63
1×10^{-5}	267	142	4.3	93
1×10^{-4}	282	106	9.0	127
1×10^{-3}	258	122	9.1	100

increased greatly with the addition of sodium chloride. This is due to the decrease in the nearest-neighbor intersphere distance with sodium chloride through thinning of the electrical double layers around colloidal spheres followed by an increase in the elastic modulus of the film. It should be mentioned further that the maximum value in the film's thickness, d_{max}, increased with the addition of sodium chloride. The reason for this observation is not clear yet.

Figure 23.4 shows the typical patterns formed in drying a series of suspensions of colloidal silica spheres, ranging from 29 to 1090 nm in diameter and at the concentration of 0.0333 in volume fraction [Ref. 28]. The broad rings and the spoke-like cracks were observed. Clearly, for small spheres, film is transparent but so many cracks are observed. When size increases, film became white and the intercrack angle increased. The number of the cracks decreased sharply as

(a)   (b)

(c)   (d)

FIGURE 23.4 Patterns formed for colloidal silica spheres at 25°C. (a) CS22p, (b) CS82, (c) CS161, and (d) CS1001 (in water, $\phi = 0.0333$, 0.1 ml, 5-mm bar length).

sphere size increases at the same sphere concentration in volume fraction. Increase in size will result in the increase in the elastic modulus of the film and then in the decrease in the crack number.

23.3.2 MICROSCOPIC DRYING PATTERNS OF COLLOIDAL CRYSTAL DISPERSIONS

The most extended microscale patterns of polystyrene and silica spheres are shown in Figure 23.5a and Figure 23.5b, respectively. Clearly, the stratified structures form from macro- to microregions. The structures appearing in the figure are clearly fractal patterns. Figure 23.5a and Figure 23.5b look to be earthworm-like and branch-like, respectively. White parts are occupied with the colloidal spheres and the black regions are vacant on a cover glass. These microscopic patterns must be made by the movement of spheres to associate with each other, especially in the final stage of drying processes. The atomic force microscope observation clarified that a large number of colloidal spheres associate on a cover glass making the fractal pattern, though the picture showing this was omitted in this article. The driving force for the association will be the van der Waals force and further hydrophobic intersphere attraction. The fractal dimensions of the patterns for D1W52 spheres, estimated by counting the number of squares covering colloidal spheres [38], were 1.45 irrespective of the sphere concentration. These values are similar to those of colloidal crystals of silica spheres, between 1.3 and 1.9 [18], and slightly larger than the fractal dimension of Koch curve, 1.26.

Interestingly, the whole single crystals were observed during the drying processes. The size of single crystals increased in the course of drying, which clearly shows that the contamination of the suspensions with air, especially with carbon dioxide, progresses with time. It is well known that the crystal size becomes largest at the salt concentration of crystal melting

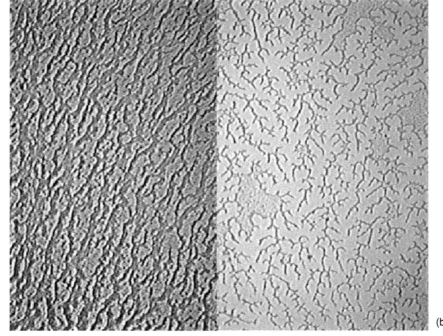

(a)   (b)

FIGURE 23.5 Patterns formed in the drying process of D1W52 spheres (a) and CS82 spheres (b) on a cover glass at 25°C (in water, 0.1 ml, ϕ = (a) 8.85×10^{-4} and (b) 0.00133, 10-μm bar length).

[21,23]. As time proceeded, the crystal size began to decrease after reaching the maximum value. The single crystals were not recognized in the wet-drying state. However, the peak appeared even for the wet film and for the completely dried film in the reflection spectroscopy, as will be described below.

The reflection spectra were taken during the drying of colloidal crystal of CS82 spheres at the center of the pattern [18]. The peak wavelength at the beginning stage of drying shifted to the longer wavelengths, and then decreased with time, especially drastically in the final stage. These observations show that the sphere concentration decreases slightly at the initial stage and then increases. Just before completion of drying at the center, the reflection spectra become broad. However, the reflection peaks remained even at the state of dryness. These observations clearly show that almost all the spheres at the central region move outside by convection flow. It is interesting to note that the dried film still shows the reflection peaks indicating the existence of the crystal structures in the film.

The branch-like fractal patterns were observed for a series of colloidal silica spheres ranging from 29 to 311 nm, which is the only limitation of sphere size and concentration. Generally speaking, these fractal patterns appeared often on the area, where there are not enough spheres to cover it. The fractal dimensions of the patterns were 1.2 ± 0.1 and were insensitive to sphere size and sphere concentration.

23.3.3 DRYING DISSIPATIVE STRUCTURES OF ANISOTROPIC SHAPED COLLOIDAL SUSPENSIONS OF MONODISPERSE BENTONITE PARTICLES

The drying patterns formed in the bentonite suspensions of 2p7f particles at the concentrations ranging from 4.6×10^{-4} to 4.6 mg/ml were studied [30]. The broad ring patterns were observed and their widths increased sharply with the particle concentration. A main cause for the broad ring formation is again due to the different rates of convection flow of water and the colloidal particles, where the rate of the latter will be slower than that of the former under gravity. Especially, flow of the particles from the center toward the outside edges in the lower layer of the liquid drop, which was observed on a digital HD microscope directly from the movement of the very rarely occurred aggregates of the particles, is important [29]. We should note here that there appeared a hill in the center in addition to the broad ring, especially at the high particle concentrations. These hills in the central area have not been observed for the suspensions of any kind of spherical particles hitherto. The rotational movement must be highly restricted for the anisotropic-shaped particles, such as plate-like ones here, and the sliding movement will be major especially in the area close to the substrate plane. This restricted Brownian movement must be correlated deeply to the appearance of the hill in the center. However, the mechanism of the coexistence of the two patterns is not yet clear because the visual observation of the convection cells was not easy experimentally.

The surfaces of the broad ring of the extended patterns look very rough, which may support the random accumulation of the plate-like particles. However, much more extended microscopic patterns of the broad ring parts demonstrate that the particles are stacked rather regularly forming the short hills oriented circularly. The extended patterns of the central area were very beautiful fractal ones. Most extended patterns showed the short and long hills distributed radially and in the fractal features. It should be mentioned here that no spoke-like and circle cracks were observed for 2p7f particles.

The thickness of the film was observed directly using a 3D profile microscope. Surprisingly, the d–r profiles support the existence of the broad ring patterns and do not support existence of the round hills in the center except the suspension of the highest particle concentration. However, the flat and thick regions were observed clearly for the bentonite particles. It should be recalled that the central region of the film of the spherical particles was extremely thin always [18,19,33]. The areas (S) covering the particles in the dried film were 9.7, 19, 58, 59, 54, and $60\,mm^2$ at particle concentrations (w) of 4.6×10^{-4}, 0.046, 0.092, 0.46, 0.92, and $4.6\,mg/ml$, respectively. S remained very small when w is smaller than $0.05\,mg/ml$, but increased sharply with further increase in the particle concentration. This tendency corresponds to the fact that most of the particles move to the broad ring regions and dried up, but parts of the particles distribute among the whole film area including the central area. The main reason for this observation will be the fact that in the translational mode the plate-like particles are difficult to move compared with the spheres, which move easily by the rotational diffusion on a substrate.

Figure 23.6 shows the typical examples of the pattern formation processes of the bentonite suspension. The drying frontier moves from right to left (as shown in the figures) with time. The border between the liquid and the solid regions is the frontier zone of drying. Soon after setting the suspensions, vague wrinkle patterns appeared. It is clear that the patterns, which are composed of the accumulated particles, had already been formed in the suspension phase, though they were not so fine and clear. As time elapsed, the drying frontier moved to the left side, and the vague patterns that were formed in the liquid phase became clearer and finer in the course of drying. These observations suggest strongly that the patterns grow and are already fixed in the suspension phase. Clearly, the drying patterns kept the same structures basically as those formed in the liquid phase.

Broad ring patterns in the deionized suspension shifted toward the single round hill pattern by the addition of sodium chloride as is clear from Figure 23.7. The broad ring disappeared completely when the concentration of NaCl is $0.001\,M$. The suspension became whitey turbid when the salt concentration is higher than $1\times10^{-4}\,M$. Association of the

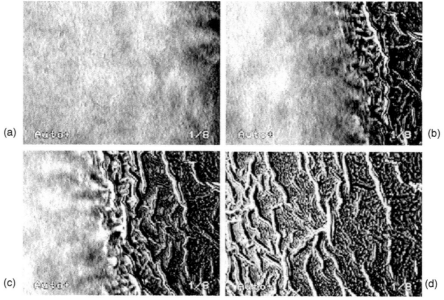

FIGURE 23.6 Patterns formed in the drying process of 2p7f particles on a cover glass at 25°C. (a) 46 min 30 s after dropping the suspension, (b) 48 min, (c) 48 min 45 s, (d) 49 min 50 s (in water, $w = 4.6\,mg/ml$, 0.05 ml, 10-μm bar length).

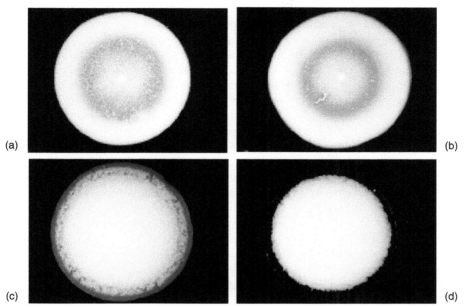

FIGURE 23.7 Patterns formed in the drying process of 2p7f particles on a cover glass at 25°C. [NaCl] = (a) 0 M, (b) 1.0×10^{-5} M, (c) 1.0×10^{-4} M, (d) 1.0×10^{-3} M (in water, w = 4.6 mg/ml, 0.05 ml, 2-mm bar length).

plate-like particles is highly plausible. Substantial decrease in the translational movement of the associated particles is one of the main reasons for the very interesting change in the patterns. The area of the patterns first increased slightly, but turned to decrease as the concentration of NaCl increased.

Influence of the inclination of the cover glass upon the patterns was also examined. Location of the pattern center shifted to the lower position as the inclination angle increased. Furthermore, the shape of the outside broad ring patterns changed from circle to the egg-shaped. Distortion of the outside patterns of the bentonites was not so significant compared with that of spheres. This tendency is consistent with the idea that the translational movement of the anisotropic-shaped particles should be much lower than the spherical particles.

23.3.4 DRYING DISSIPATIVE STRUCTURES OF CHINESE BLACK INK

Patterns formed in the drying suspensions of Chinese ink on an unrinsed glass have been studied. The broad rings are observed clearly in the pictures at the outer edges irrespective of the concentration. The width of the broad ring regions increased when the ink concentration increased. The width, W increased significantly when the particle concentration increased. W increased linearly as a function of logarithms of the particle concentration. It should be mentioned here that the macroscopic broad ring patterns were observed for all the suspensions or solutions examined in the laboratory: colloidal silica [18], polystyrene spheres [19], ionic detergent of DTAC [32], plate-like bentonite [30], rodlike palygorskite, PAL [31], sodium chloride, and other simple electrolytes, for example.

The spoke-like pattern was clearly observed at a particle concentration at 1×10^{-5}, 0.01, and 0.1 g/ml. It should be recalled that the spoke-like patterns of cracks were observed for colloidal crystal suspensions of silica [18] and polystyrene spheres [19].

However, the patterns for Chinese black ink were not cracks but ridge lines. Chinese ink contains the glue component and the film must be strong enough not to make cracks. The origin for the spoke-like patterns, whether cracks or ridge lines, is the many circle convection cells formed in the liquid phase in the course of drying of the suspension [18,19,25,31,32].

The pattern area, S increased first as the particle concentration increased and attained a certain critical value. The dependency of the S-values suggests that the interfacial tension of Chinese ink on a cover glass decreases as the concentration of the ink particle increases and that the Chinese ink is slightly surface-active. The drying times, T increased slightly as the particle concentration increased, especially at low-suspension temperatures; however, at elevated temperatures above 45°C, T was quite insensitive to the particle concentration. This suggests that at high temperature the water activity (vapor pressure) is high and is not influenced very much by the coexisting colloidal particles.

At the special experimental condition of $w = 0.001$ g/ml at 25°C, the convection flow of the Kaimei particles was visible with the naked eye. At 1 min after setting, Chinese ink particles meet together and the associated particles distribute at random. However, after 36 min, the particles began to orient themselves as a result of the convection flow. Surprisingly, after 136 min, the spoke-like lines both in the liquid and in the solid phases just coincided with each other. This observation suggests strongly that the spoke lines were already formed in the suspension phase by the convection flow. The drying frontier at the border between gray (center part) and black regions (edges) moved inside with time. The clear-cut spoke lines appeared especially around the outside region of the liquid area.

A further study on the convection flow in the course of drying was made using a large amount of the ink suspension in a dish on a hot plate keeping the suspension temperatures between 25 and 80°C. The Cyclic convection cells around the outside edges of the suspension, the grain-like convection patterns, and the smoke-like convection structures were observed after setting the sample suspensions at $w = 1 \times 10^{-5}$, 5×10^{-5}, and 1×10^{-4} g/ml on a hot plate for 21, 15, and 21 min, respectively. Interestingly, the patterns of the dried films were spoke-like lines, fine circles, and the surface patterns of the Japanese earthenware called "Shigaraki Yaki," respectively. Clearly, the convection flow of the particles and further the difference in the flow rates between the particles and the solvents play an important role for the pattern formation.

23.3.5 DRYING DISSIPATIVE STRUCTURES OF IONIC AND NONIONIC DETERGENTS

Cationic surfactant of DTAC [32] and a series of the anionic ones of sodium n-alkyl sulfates (SnS) [33] are studied. At low concentrations of DTAC solutions, the pattern area shrank in the center and the broad ring regions distributed around the outer edges. At the highest concentration of 0.01 M, a broad ring pattern occupied with a large amount of detergent molecules was observed at the outer edge.

Figure 23.8 shows close-up details at 2×10^{-5} M. Figure 23.8a through Figure 23.8d are the extended patterns observed from the edge to the center. Clearly, the patterns are cross-like irrespective of the location. However, the patterns become large and complex and the blocks are also observed.

The microscopic patterns of DTAC molecules coexisting KCl, CaCl$_2$, and LaCl$_3$ were also studied. When the concentrations of the detergent and salt are comparable to each other, tree branch and arc-like patterns were observed, which are entirely different from those of the

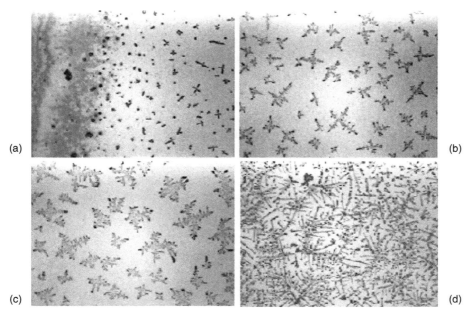

FIGURE 23.8 Patterns formed in the drying process of DTAC detergents on a cover glass at 25°C. The patterns from (a)–(d) correspond to those from the edge border to the center (in water, 0.1 ml, 2×10^{-5} M, 20-µm bar length).

pure detergent or salt solutions. The patterns are also similar to the microscopic structures of PAL [31]. These results suggest that the micelle structure of DTAC molecules is in the form of a long rod in the solution just before the solidification.

For all the solutions of S8S examined, the pattern area also shrank to the center and the broad ring regions distributed around the outer edges. This shrinking phenomenon looks to correlate intimately to the critical micelle concentration (cmc) of the detergent solution. When the solute concentration is lower than cmc, the initial solution area should shrink without solidification until the solution is concentrated in the course of drying and its concentration reaches the cmc. For the higher concentrations of the detergents than the cmc, the shrinking of the solution area stops. In other words, the solution area shrinks when the contact angle between the solution and the substrate is large and the shrinking stops when the angle is small enough at the concentrations of detergent higher than cmc. The shrinking of the observed solution area strongly supports the cmc value of S8S being higher than 0.01 M. The reference value of cmc for S8S has been reported to be ca. 0.1 M [39], which agrees with the predicted value from this work. It should be noted here that the pattern area of the sodium poly-α-L-glutamate and hydrochloride of poly-L-lysine shrank when their concentrations were lower than the critical concentration, m^*, where these polymers form the structured conformation at the air–water interface, and their surface tensions start to decrease sharply as polymer concentration increases [40,41].

The pattern area of S12S also shrank toward the center and the broad ring regions formed in the outer edges, which are similar to the features of S8S. It should be noted here that the dried pattern area increased as S12S concentration increased and the area at 0.01 M was not so far from the initial liquid area. These observations support that the cmc value of S12S is not far from 0.01 M. The reference value of cmc for S12S is 0.007 M [39]. Fine multirings were observed. The fractal patterns composed of crosses and branches, string-like rings, and spokes were observed. These micropatterns are also quite similar to those of S8S, though the string-like patterns were observed much more often compared with S8S.

The pattern area of S16S shrank toward the center when the detergent concentrations are between 1×10^{-4} and 1×10^{-3} M. However, no shrinking observed at 0.01 M, which suggests strongly that the cmc of S16S is between 1×10^{-4} and 1×10^{-3} M. The reference value of cmc for S16S solution is ca. 5×10^{-4} M [39], which agrees excellently with the prediction. Thus, it is firmly concluded that the shrinking of the pattern area always occurs when the surface tension and then the contact angle between solution and substrate are high. The broad ring regions were observed for all the solute concentrations of the examined solutions, and the macroscopic patterns are quite similar to those of S8S and S12S.

The microscopic patterns of S16S, composed of small blocks and strings, were observed, though the patterns apt to change from blocks to strings as the detergent concentration increases.

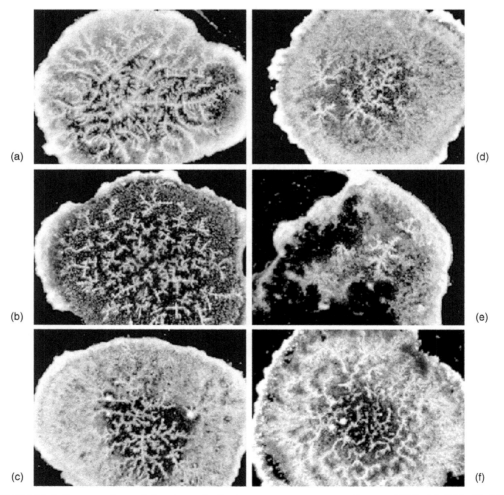

FIGURE 23.9 Patterns formed for the anionic detergents on a cover glass at 25°C. (a) S6S, (b) S8S, (c) S10S, (d) S12S, (e) S16S, (f) S18S (in water, 0.1 ml, $w = 1 \times 10^{-7}$ M, 200-μm bar length).

Figure 23.9 shows the close-up details of the microscopic patterns of a series of anionic detergents from S6S to S18S at 1×10^{-7} M. Surprisingly, quite similar patterns of branch-like strings were observed in each case. This observation supports that the shape of detergent molecules from S6S to S18S is essentially rodlike and then the similar patterns of branch-like strings were formed.

23.4 SUMMARY

Macroscopic and microscopic dissipative structural patterns form in the course of drying a series of the colloidal dispersions including colloidal crystals of monodispersed colloidal silica spheres and polystyrene spheres. Drying patterns of nonspherical particles of bentonite have been observed on a cover glass. The broad ring patterns of the hill accumulated with the particles are formed around the outside edges in the macroscopic scale. The microscopic fractal patterns are formed. The pattern area and the time for the dryness have been discussed as a function of detergent concentration and number of carbons of the detergents. The convection flow of water accompanied with the particles occurs at the drying frontier between solution and substrate in the course of dryness, and interactions among the particles and substrate are important for the macroscopic pattern formation. Microscopic patterns are determined mainly by the translational Brownian movement of the particles, and the electrostatic and hydrophobic interactions between particles and between the particle and the substrate in the course of solidification. Drying dissipative patterns of cationic and anionic detergents have also been reviewed.

ACKNOWLEDGMENTS

I thank the Ministry of Education, Science, Sports and Culture for grants-in-aid for Scientific Research on Priority Area (A) (11167241) and for Scientific Research (B) (11450367). Drs. Komatsu and Nishida are acknowledged deeply for their kind help in providing a series of colloidal silica spheres. The author appreciates the coworkers cited in our papers.

REFERENCES

1. Vanderhoff, J.W., Bradford, E.B., and Carrington, W.K., The transport of water through latex films, *J. Polym. Sci., Polym. Symp.*, 41, 155, 1973.
2. Nicolis, G., and Prigogine, I., *Self-Organization in Nonequilibrium Systems*, Wiley Interscience, New York, 1977.
3. Cross, M.C., and Hohenberg, P.C., Pattern formation outside of equilibrium, *Rev. Mod. Phys.*, 65, 851, 1993.
4. Ohara, P.C., Heath, J.R., and Gelbart, W.M., Self-assembly of submicrometer rings of particles from solutions of nanoparticles, *Angew. Chem., Int. Ed. Engl.*, 36, 1078, 1997.
5. Ohara, P.C., Heath, J.R., and Gelbart, W.M., Interplay between hole instability and nanoparticle array formation in ultrathin liquid films, *Langmuir*, 14, 3418, 1998.
6. Uno, K., Hayashi, K., Hayashi, T., Ito, K., and Kitano, H., Particle adsorption in evaporating droplets of polymer latex dispersions on hydrophilic and hydrophobic surfaces, *Colloid Polym. Sci.*, 276, 810, 1998.
7. Gelbart, W.M., Sear, R.P., Heath, J.R., and Chang, S., Array formation in nanocolloids, *Faraday Discuss.*, 112, 299, 1999.
8. van Duffel, B., Schoonheydt, R.A., Grim, C.P.M., and De Schryver, F.C., Multilayered clay films: atomic force microscopy study and modeling, *Langmuir*, 15, 7520, 1999.
9. Maenosono, S., Dushkin, C.D., Saita, S., and Yamaguchi, Y., Growth of a semiconductor nanoparticle ring during the drying of a suspension droplet, *Langmuir*, 15, 957, 1999.

10. Brock, S.L., Sanabria, M., Suib, S.L., Urban, V., Thiyagarajan, P., and Potter, D.I., Particle size control and self-assembly processes in novel colloids of nanocrystalline manganese oxide, *J. Phys. Chem.*, 103, 7416, 1999.

11. Nikoobakht, B., Wang, Z.L., and El-Sayed, M.A., Self-assembly of gold nanorods, *J. Phys. Chem.*, 104, 8635, 2000.

12. Ge, G., and Brus, L., Evidence for spinodal phase separation in two-dimensional nanocrystal self-assembly, *J. Phys. Chem.*, 104, 9573, 2000.

13. Chen, K.M., Jiang, X., Kimerling, L.C., and Hammond, P.T., Selective self-organization of colloids on patterned polyelectrolyte templates, *Langmuir*, 16, 7825, 2000.

14. Lin, X.M., Jaenger, H.M., Sorensen, C.M., and Klabunde, K.J., Formation of long-range-ordered nanocrystal superlattices on silicon nitride substrates, *J. Phys. Chem.*, 105, 3353, 2001.

15. Kokkoli, E., and Zukoski, C.F., Surface pattern recognition by a colloidal particle, *Langmuir*, 17, 369, 2001.

16. Ung, T., Liz-Marzan, L.M., and Mulvaney, P., Optical properties of thin films of $AuSiO_2$ particles, *J. Phys. Chem.*, B, 105, 3441, 2001.

17. Shimomura, M., and Sawadaishi, T., Bottom-up strategy of materials fabrication: a new trend in nanotechnology of soft materials, *Curr. Opin. Colloid Interface Sci.*, 6, 11, 2001.

18. Okubo, T., Okuda, S., and Kimura, H., Dissipative structures formed in the course of drying the colloidal crystals of silica spheres on a cover glass, *Colloid Polym. Sci.*, 280, 454, 2002.

19. Okubo, T., Kimura, K., and Kimura, H., Dissipative structures formed in the course of drying the colloidal crystals of monodispersed polystyrene spheres on a cover glass, *Colloid Polym. Sci.*, 280, 1001, 2002.

20. Kose, A., Ozaki, M., Takano, K., Kobayashi, Y., and Hachisu, S., Direct observation of ordered latex suspension by metallurgical microscope, *J. Colloid Interface Sci.*, 44, 330, 1973.

21. Okubo, T., Extraordinary behavior in structural and dynamic properties of colloidal and flexible macroions in deionized solution. Important role of Debye-screening length, *Acc. Chem. Res.*, 21, 281, 1988.

22. Okubo, T., Polymer colloidal crystals, *Prog. Polym. Sci.*, 18, 481, 1993.

23. Okubo, T., Recent advances in the kinetic and dynamic properties of colloidal crystals, *Curr. Top. Colloid Interface Sci.*, 1, 169, 1997.

24. Okubo, T., Crystalline colloids, in *Encyclopedia of Surface and Colloid Science*, Hubbard, A., Ed., Marcel Dekker, New York, 2002, p. 1300.

25. Terada, T., Yamamoto, R., and Watanabe, T., Experimental studies on colloid nature of Chinese ink, Part I, *Sci. Paper Inst. Phys. Chem. Res. Jpn.*, 27, 173, 1934; *Proc. Imper. Acad. Tokyo*, 10, 10, 1934.

26. Terada, T., Yamamoto, R., and Watanabe, T., Experimental studies on colloid nature of Chinese ink, Part II, *Sci. Paper Inst. Phys. Chem. Res. Jpn.*, 27, 75, 1934.

27. Terada, T., and Yamamoto, R., Cataphoresis of Chinese ink in water containing deuterium oxide, *Proc. Imper. Acad. Tokyo*, 11, 214, 1935.

28. Okubo, T., Yamada, T., Kimura, K., and Tsuchida, A., Drying dissipative structures of the deionized aqueous suspensions of colloidal silica spheres ranging from 29 nm to 1 μm in diameter, *Colloid Polym. Sci.*, 283, 1007, 2005.

29. Okubo, T., Kimura, H., Kimura, T., Hayakawa, F., Shibata, T., and Kimura, K., Drying dissipative structures of Chinese black ink on a cover glass and in a dish, *Colloid Polym. Sci.*, 283, 1, 2005.

30. Yamaguchi, T., Kimura, K., Tsuchida, A., Okubo, T., and Matsumoto, M., Drying dissipative structures of the aqueous suspensions of monodisperse bentonite particles, *Colloid Polym. Sci.*, 283, 1123, 2005.

31. Okubo, T., Kanayama, S., Ogawa, H., Hibino, M., and Kimura, K., Dissipative structures formed in the course of drying an aqueous solution of poly(allylamine hydrochloride) on a cover glass, *Colloid Polym. Sci.*, 282, 230, 2004.

32. Okubo, T., Kanayama, S., and Kimura, K., Dissipative structures formed in the course of drying the aqueous solution of n-dodecyltrimethylammonium chloride on a cover glass, *Colloid Polym. Sci.*, 282, 486, 2004.

33. Kimura, K., Kanayama, S., Tsuchida, A., and Okubo, T., Drying dissipative structures of the aqueous solution of sodium *n*-alkyl sulfates on a cover glass, *Colloid Polym. Sci.*, 283, 898, 2005.

34. Matsumoto, M., Transient electric birefringence of colloidal particles immersed in shear flow, Part II: The initial response under the action of a rectangular electric pulse and the behavior at a low alternating electric field, *Biophys. Chem.*, 58, 173, 1996.

35. Matsumoto, M., Transient electric birefringence of colloidal particles immersed in shear flow, Part III: a preliminary report on determination of anisotropy of electric polarizability, *Colloids Surf., A*, 148, 75, 1999.

36. Colina, H., and Roux, S., Experimental model of cracking induced by drying shrinkage, *Eur. Phys. J., E*, 1, 189, 2000.

37. Latterini, L., Blossey, R., Hofkens, J., and Vanoppen, P., Ring formation in evaporating porphyrin derivative solutions, *Langmuir*, 15, 3582, 1999.

38. Falconer, K.J., *The Geometry of Fractal Sets*, Cambridge University Press, New York, 1985.

39. Mittal, K.L., and Fendler, E.J., Eds., *Solution Behavior of Surfactants. Theoretical and Applied Aspects*, Vol. 1, Plenum Press, New York, 1982.

40. Okubo, T., Onoshima, D., Kimura, K., and Tsuchida, A., Drying dissipative structures of biological polyelectrolyte solutions, publication in preparation.

41. Okubo, T., and Kobayashi, K., Surface tension of biological polyelectrolyte solutions, *J. Colloid Interface Sci.*, 205, 433, 1998.

Index

Milton Keynes UK
Ingram Content Group UK Ltd.
UKHW050307111024
449327UK00043B/2075